Construction Technology for Builders

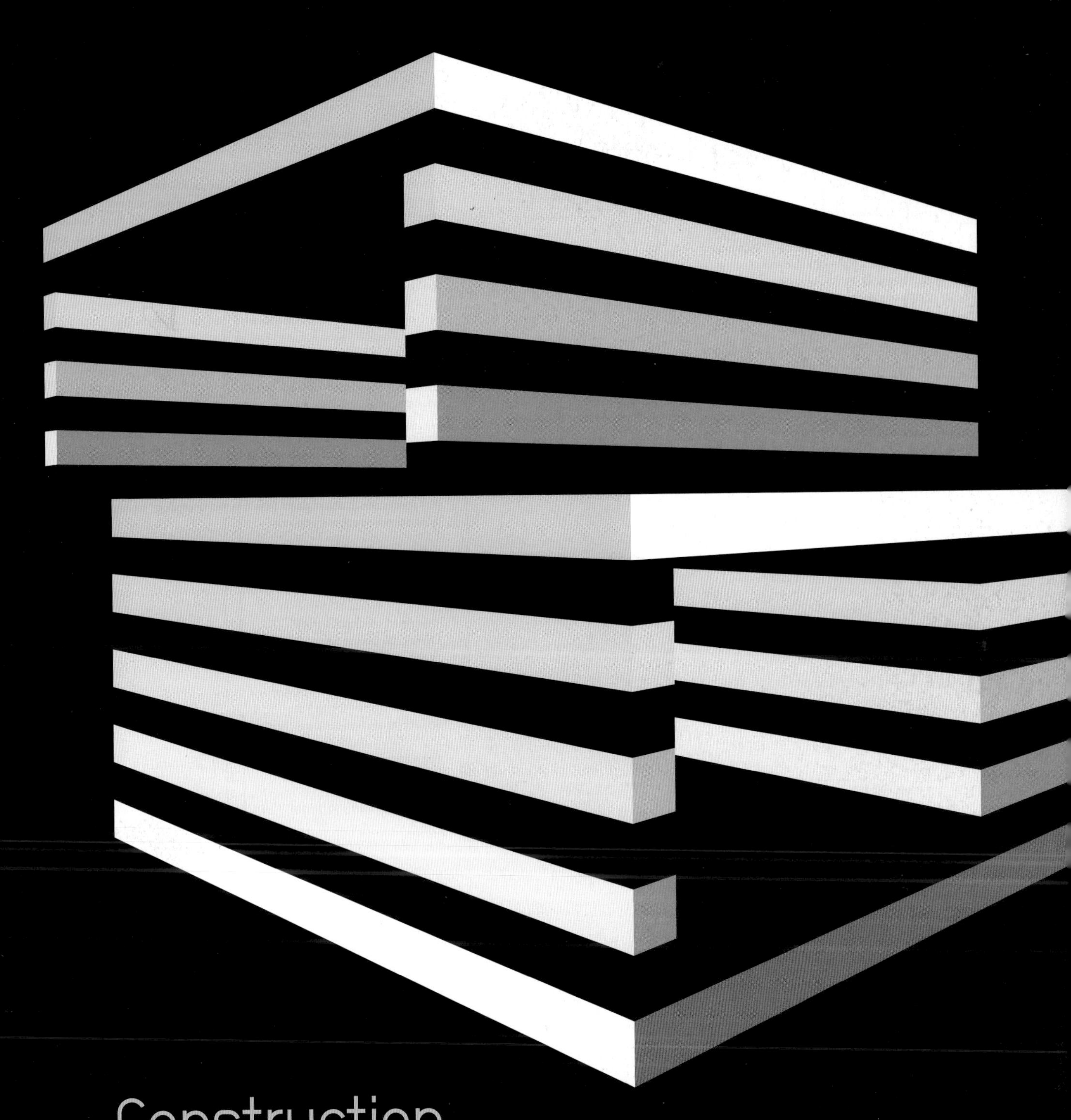

Construction Technology for Builders

1E

Glenn P. Costin

Construction Technology for Builders
1st Edition
Glenn P. Costin

Head of content management: Dorothy Chiu
Senior content manager: Sophie Kaliniecki
Content developer: Samantha Brancatisano
Project editor: Raymond Williams
Text designer: Nikita Bansal
Cover designer: Chris Starr (MakeWork)
Editor: Greg Alford
Proofreader: Sylvia Marson
Permissions/Photo researcher: Liz McShane
Typeset by Cenveo Publisher Services

For product information and technology assistance,
in Australia call 1300 790 853;
in New Zealand call 0800 449 725

For permission to use material from this text or product, please email
aust.permissions@cengage.com

National Library of Australia Cataloguing-in-Publication Data
ISBN: 9780170416047
A catalogue record for this book is available from the National Library of Australia

Cengage Learning Australia
Level 7, 80 Dorcas Street
South Melbourne, Victoria Australia 3205

Cengage Learning New Zealand
Unit 4B Rosedale Office Park
331 Rosedale Road, Albany, North Shore 0632, NZ

For learning solutions, visit cengage.com.au

Printed in China by 1010 Printing International Limited.
1 2 3 4 5 6 7 24 23 22 21 20

BRIEF CONTENTS

CONTENTS

Guide to the text

As you read this text you will find a number of features in every chapter to enhance your study of Construction Technology for Builders and help you understand how the theory is applied in the real world.

CHAPTER-OPENING FEATURES

A list of **Elements** gives you a clear sense of what topics each chapter will cover. It will indicate what you should be able to do after reading the chapter within that part.

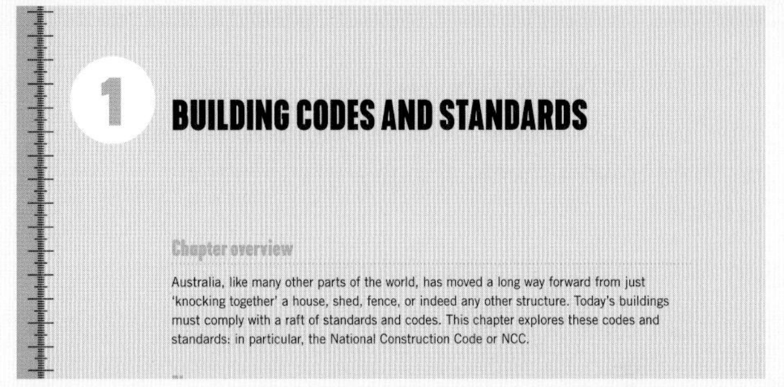

1
BUILDING CODES AND STANDARDS
Chapter overview
Australia, like many other parts of the world, has moved a long way forward from just 'knocking together' a house, shed, fence, or indeed any other structure. Today's buildings must comply with a raft of standards and codes. This chapter explores these codes and standards: in particular, the National Construction Code or NCC.

FEATURES WITHIN CHAPTERS

Learning tasks encourage you to practically apply the knowledge and skills that you have just read about.

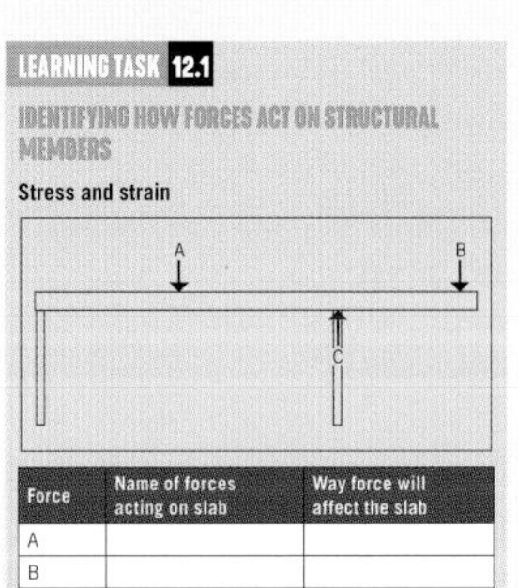

LEARNING TASK 12.1
IDENTIFYING HOW FORCES ACT ON STRUCTURAL MEMBERS
Stress and strain
A
B
C
Force
Name of forces acting on slab
Way force will affect the slab
A
B
C
Using the information shown on the suspended floor slab shown above, identify the name of the 'effect' of the forces acting on the slab at A, B and C. Describe the way those forces will affect the suspended slab member.

Dead loads

FIGURE 12.12 Examples of loads on a typical two-storey construction

Loads 1: Live and dead loads

The purpose of this section is to identify and define the loads common to most domestic structures. The section begins with ***dead loads*** and ***live loads*** before addressing ***wind loads***. A range of other common loads important to structural design are then discussed followed by the identification of ***load paths***. The section closes with a range of basic load calculations based upon Newton's laws and the concept of statics and equilibrium discussed earlier.

You should note from the outset that when dealing with loads on structures you should refer to the following standards:

- AS/NZS 1170:2002 Structural design actions and its parts:
  - General principles (1170.0:2002)
  - Permanent, imposed and other actions (1170.1:2002)
  - Wind actions (1170.2:2011)
  - Snow and ice actions (1170.3:2003)
  - Earthquake actions in Australia (AS only

Case studies provide step by step instructions on how to perform specific tasks/processes.

CASE STUDY

Multiple classifications

Consider a multi-storey Class 7a carpark with a Class 5 office included in the 2nd storey. The office floor area is only 9% of the total floor area for that level. In this case the whole storey may be considered as Class 7a. If that was the sole alternate use element in the whole structure, then the whole building is Class 7a.

However, in the next level of car parking above, there is a larger office. In this case it takes up 14% of the total floor area. This level of the carpark must then be duel classified as Class 5/7a. Likewise, the building as a whole (assuming no other alternative use sections of the building exist) will be Class 5/7a.

The exception to this ruling is the Class 4 element: a Class 4 dwelling is a Class 4, irrespective of its percentage of the storey's floor area.

While a building is being designed it may be unclear who the tenant or tenants will be, or how they may ultimately use the facility. This is why classifying a building is a form of risk management. You identify the

behalf of the appropriate building authority (local council or independent company). However, having an understanding of these sections, and building classifications generally, can help you discuss the merits and cost implications of a particular class with the building certifier to the benefit of all parties.

LEARNING TASK 1.1

NATURE OF BUILDINGS AND THEIR CLASSIFICATION

1 An old industrial building has been used to store a number of differing types of wholesale goods on pallets and raking systems. You have been approached to get council approvals for a redevelopment that will allow most of the building to be used as office space, with a small conference room also included.
   - What is the current classification?
   - What will the classification be should the proposed redevelopment go ahead?
2 You are on acreage in a country area and have a large family home on that land. There is some discussion that you would like to build a large, 100 m^2 pergola-type structure, with internal roof

END-OF-CHAPTER FEATURES

Chapter summaries highlight the important concepts covered in each chapter as well as link back to the key competencies.

SUMMARY

The aim of this chapter was to take the reader through a carefully guided exploration of both Volumes One and Two of the NCC. In particular, how they pertain to all 10 classes of buildings as defined by this code, and how the code interrelates with the Australian standards. Reading the NCC and the relevant standards requires experience to ensure that you have covered all the inflections and nuances that are within them. The 2019 edition of the NCC is far superior to the previous renditions in ensuring greater cohesiveness between Volumes One and Two, making them more user friendly. However, to be fully assured that you have accurately interpreted the requirements will still require some time. The examples offered in this chapter give you a brief taste of some of the simpler elements of their structure and what you might expect when ensuring compliance. Don't be afraid to explore much deeper into other areas of your projects – be assured, this is critical knowledge for you in your role as either builder or supervisor.

The **references and further reading sections** provide you with a list of each chapter's references as well as links to important text and web-based resources.

REFERENCES AND FURTHER READING

Australian Building Codes Board (ABCB), *Evidence of Suitability Handbook 2018*, ABCB.
Australian standards, **https://infostore.saiglobal.com**
National Construction Code, 2019, **https://ncc.abcb.gov.au/ncc-online/NCC**

Guide to the online resources

FOR THE INSTRUCTOR

Cengage is pleased to provide you with a selection of resources that will help you prepare your lectures and assessments. These teaching tools are accessible via **http://login.cengage.com**.

SOLUTIONS MANUAL

The Solutions Manual includes solutions to end-of-chapter worksheets and answers to in text activities.

WORKSHEETS

All chapter worksheets are available as writeable pdfs for your students.

COMPETENCY MAPPING GRID

The downloadable competency mapping grid demonstrates how the text aligns to the Certificate IV in Building and Construction (Building).

There are additional resources for this text so contact your Cengage Learning Consultant for more information.

FOR THE STUDENT

Visit **http://www.cengagebrain.com** and search for this book to access the bonus study tools available on the Construction Technology for Builders companion website.

The website contains resources for each chapter, including:

- Chapter worksheets

AUSTRALIAN STANDARD UNITS OF MEASURE

The structural principles of construction derive from the school of engineering. As such, these principles require a firm grasp of common mathematical processes and notation. In particular, those of length, area, volume and mass, as well as force and pressure.

The basic units

These are the units with which you will mostly be familiar, those of length, mass and time. These all have lower case letters as their symbols:

- Length: **m** for metre
- Mass: **g** for gram
- Time: **s** for seconds

Larger and smaller values are denoted by prefixes to these basic units; For values less than a million, the prefixes are also in lower case

For values of a million or greater, the prefixes are in capitals, i.e.:

- mega (M) one million (10^6) × the basic unit
- kilo (k) one thousand or (10^3) × the basic unit
- milli (m) one thousandth (10^{-3}) × times the basic unit (note the '–3')

One millimetre (mm) is therefore one thousandth part of a metre.

One kilogram (kg) is 1000 grams.

Scientific and engineering notation

Scientific notation was developed so that very large, or very small, numbers could be expressed in a manner that was easier to understand at a glance than one with lots of zeros in front or behind.

For example

In the case of 1 000 there are three zeros after the '1'. This number can be expressed as 1×10^3 which simply means $1 \times 10 \times 10 \times 10$ which equals 1 000.

Note the correlation – the equation has three '10s' in multiply mode, the resultant number has three zeros and so we write 10^3.

In the case of 1 000 000 there are 6 zeros. So, we can write it as 1×10^6. I.e. $1 \times 10 \times 10 \times 10 \times 10 \times 10 \times 10$ which equals one million.

The little numbers '3' and '6' attached to the 10 in the examples above (i.e. 10^3 and 10^6) are known as *indices* or *index* numbers.

But the number doesn't have to be all zeros. We can apply this approach to any number.

For example

In a number like 64 875 352.0 you would take the first number '6' and count back the number of digits that follow after it until you reach the decimal point. In this case there are seven (7) so our number could look like:

6.4875352×10^7

Only this hasn't really helped us as we still have all the numbers to look at. So, we reduce those numbers down to required or desired *level of accuracy*. In our example, we will use the first four (4) numbers of the original.

So, our long number now looks like this:

6.488×10^7

Note the last '8'. This has come about through what is known as '*rounding*'. If the number after the last figure you want is 5, or greater, then you round up. If less than 5 then you round down (i.e. leave the last number alone).

- So: 6.487**5** becomes 6.48**8**

Alternatively, if we had a number like 4.2572, then:

- 4.257**2** becomes 4.257

This applies no matter how many numbers follow after the one you wish to stop at.

- i.e.: 4.257**2**7983 still becomes 4.257 the numbers after the '2' are ignored
- and 6.487**5**7983 still becomes 6.48**8** the numbers after the '5' are ignored

The number 6.488×10^7 is an expression of '**scientific notation**'. In '**engineering notation**', however, you will find that only indices divisible by 3 are used. That is:

1×10^3; 1×10^6, 1×10^9

and so on.

So, our number would look like this:

64.88×10^6

Working from our original number of 64 875 352.0, in engineering notation its best to work from the decimal point outwards. Because in engineering the indices must be divisible by 3, we count in groups of 3.

64 **875 352**.0 $\Rightarrow 64.88 \times 10^6$

Two groups of three, so the decimal point has moved six places, hence our index number is '6'.

Note: In using scientific or engineering notation for very small numbers the same basic rules apply.

For example

0.**001** $\Rightarrow 1.0 \times 10^{-3}$

The decimal point has moved one group of three to the right or 'backwards' so index becomes negative 3 or '–3'.

NO matter how small the number the same approach of counting backwards in groups of three can be used.

0.000064875352 $\Rightarrow$ 0.**000 064** 875 352 $\Rightarrow 64.88 \times 10^{-6}$

Note: In engineering notation you can have up to three numbers to the left of the decimal point in your final expression; i.e.:

523.34×10^6

Units of force

The definition of force is:

Force = mass × acceleration

or

$F = ma$

Where acceleration is measured in metres per second squared or m/s^2.

Where the acceleration is due to gravity, that acceleration is $9.8\ m/s^2$.

This figure is frequently rounded up to $10\ m/s^2$ for simplicity, plus it adds a bit of 'head room' in our calculations to allow for minor errors.

Hence, the units involved in our equation of force are:

Force = kg × m/s^2 or kg.m/s^2 (where the '.' means '×' or multiply)

This is rather a messy looking suffix to put after a number, so it has been given a name after an historical figure associated with it. In this case, Sir Isaac Newton.

So our 'unit of force' derived from this combination is known as a '**newton**' or **N**.

Where:

$1 \text{ N} = 1 \text{ kg} \times 1 \text{ m/s}^2$

Since gravity is 10 m/s^2 (when rounded off) the force of 1 kg under the gravitational pull of the earth is:

$1 \text{ kg} \times 10 \text{ m/s}^2 = 10 \text{ N}$

or

$0.1 \text{ kg} \times 10 \text{ m/s}^2 = 1 \text{ N}$

Construction projects are generally much too heavy to use units as small as a newton (N). Instead we generally use kilonewtons (kN), i.e.:

1 kN = 1 000 N

Mass vs weight

Mass and weight are two different concepts which are frequently misused.

- Weight is a force. Mass is not.

Mass is measured and stated in kg but is not a force in itself. Mass, when combined with acceleration gives force. This force we call weight. A stationary 10-kg object in deep space will still have a mass of 10 kg. It will have no 'weight', however, so long as no acceleration is acting upon it.

Therefore, if you have a *load* (a mass) in kilograms (kg) you must multiply it by the gravitational pull of Earth to get the *force* being applied. That is, multiply it by 9.8 or 10 to convert it into newtons of force.

A useful conversion is:

$10 \text{ kg/m}^2 = 0.1 \text{ kN/m}^2$ (i.e divide kg/m^2 by 100 to get kN/m^2)

Units of pressure

Pressure is a force applied evenly over a defined area. For example, a concrete footing applies pressure to the foundation material, a brick column puts pressure on a footing and so on.

The units for calculating pressure are newtons of force per square metre of area. This is another combination where a relevant historical person's name is use. In this case, French mathematician Blaise Pascal. So, we refer to this combination as a Pascal (Pa). That is:

$1 \text{ Pa} = 1 \text{ N/m}^2$

and

$1 \text{ Pa.m}^2 = 1 \text{ N}$

As this could be equated to, say, 100 g of sand spread over 1.0 metre square of concrete, this unit of measure is too small for construction purposes. So instead, the units of **kPa** (1 000 Pascals) or **MPa** (1 000 000 Pascals) are used. Most students of construction will already be familiar with MPa as the statement of concrete's compressive strength – e.g. 20 MPa or 25 MPa concrete for paths and basic footings.

INTRODUCTION

Welcome to the first edition of *Construction Technology for Builders*. A long time in the making, this book is designed specifically for advanced students of building, particularly those seeking to gain a builder's licence, understand the building process and structural principles more fully, or seeking to gain a greater grasp of construction management.

There is wealth of material in this text that requires you to engage with various standards and codes. It is important that you have ready access to these throughout your studies. The most significant is the National Construction Code, a multi-volume document of which the first two are frequently referenced in various chapters. Accessing these, and the Australian standards – of which a significant number will need to be sourced – is outlined below.

In addition to the main chapters, each of which aligns with a specific competency unit in the Certificate IV in Building and Construction, there is a glossary of terms and a number of appendices at the end of the book. The glossary you will find invaluable when some of the more technical terms are necessarily used. The appendices provide additional learning or explanatory material, or information such as plans and specification documents that are referenced in the examples and exercises.

A book of this type is a huge undertaking that relies upon a broad range of skilled professionals to create. It is never the work of one person. Yet it remains, arguably, a work in progress, with a further volume yet to be compiled. Feedback on this first edition will be welcomed so that over time it may be improved and eventually cover all the most commonly required areas of study. But that is another story in the making…

Accessing the National Construction Code (NCC)

The NCC is a free to download suite of documents. New editions are compiled every three years with the current edition being 2019. This edition will therefore remain current until May 2022.

Access is via the internet, go to: https://ncc.abcb.gov.au

You will need to create a free login account before you can download the documents.

Once this is done, please download and save to your computer or portable drive the following documents:

- NCC 2019 Volume One
- NCC 2019 Volume Two

These documents do not have an expiry date so you may open them through Adobe or Foxit pdf Readers whenever you require. Due to their size and complexity, they are best viewed electronically rather than in print form.

Accessing Australian standards

Accessing the many Australian standards referenced in the chapters is also via the internet. The complexity is that all Australian standards must be purchased individually – and they are expensive. There are, however, two avenues by which you may gain free access.

Your TAFE or university library

Most TAFE campus or university libraries will have access to the Australian standards database. Your enrolment provides you with free access to any standard you might require.

State and national libraries

For those using this book for study through a private registered training organisation (RTO) or simply for their own benefit (such as an owner builder) there is still a way to access the standards for free.

The Australian federal government has ensured that, due to the excessive cost, the standards have been made available for free through the National Library of Australia and the State Library of each state or territory. Membership is free in each case, and may be completed online. Once a member, you may access the Australian standards database and again download any standard required.

Note 1: This access is paid for by the library or organisation in question. Due to the cost, access is limited, in some cases to only a few (3 or 4) concurrent users at any one time. This means you should access and download the standards you want onto your computer and then log off as early as you can, otherwise others cannot get into the system.

Note 2: Australian standards are best viewed through Foxit pdf Reader instead of Adobe. You will have less issues with reading and exploring them with that application. Foxit pdf Reader is freeware that comes as part of the standard software package on many Windows PCs from major companies such as HP, Acer and ASUS. It may also be downloaded free from the developer's website: https://www.foxitsoftware.com/pdf-reader/

National Association of Steel-framed Housing (NASH) standards

These standards are referred to only in passing, so you do not need to access them in any specific manner. However, if you require them you will most likely have to pay, or find a library that holds them in hard or single access electronic copy. Some TAFEs and universities that deliver construction-based courses do hold them, so it is best to ask before having to purchase.

Australian standard units of measure

The following pages outline the standard units of measure used in Australia and New Zealand (and many other countries). You should refer to this when dealing with any of the chapters that involve calculations, particularly Chapters 12 and 13 that deal with structural principles.

ABOUT THE AUTHOR

Glenn Costin PhD, BEd, BA, Cert IV Building and Construction, Cert IV TAE, Certificate of Trade, Carpentry & Joinery, is currently a Senior Lecturer within Deakin University's School of Architecture and Built Environment. Prior to this Glenn was at Riverina Institute of TAFE for 24 years where his duties included delivery of all levels of Construction trade, post trade and pre-apprenticeship courses. In addition Glenn taught across a range of fields and courses including computing, environment, social inclusion and arts. Glenn was also heavily involved in Worldskills Australia as the chief judge and national and regional designer of carpentry for a decade. Outside of education, Glenn designs heritage renovations, additions and new homes from both a sustainability and energy efficiency or zero carbon perspective. In addition, Glenn has worked internationally in a range of countries and travelled extensively.

ACKNOWLEDGEMENTS

Dr. Igor Martek PhD, University of Melbourne; Master of International Relations, Australian National University; MBA, AGSM University of NSW; Bachelor of Architecture (Hons.) University of Melbourne School of Architecture and Built Environment, Deakin University.
CH 2 Legal requirements for building and construction projects
CH 4 Work health and safety

Sharon Rumble is a CPA qualified accountant who has been teaching with TAFE NSW for the last 10 years and is currently teaching small business finances for a range of TAFE NSW qualifications including Certificate IV Building and Construction, Certificate IV New Small Business and Certificate IV Leadership and Management.
CH 9 Small business finances

Josef Fritzer Bachelor of Building, and Bachelor of Housing, University of Western Sydney, is qualified in assessment planning, designing and validation and has taught in the vocational sector
Chapter learning tasks
Worksheets

Glen Rodgers Bachelor of Architecture (Honours), Bachelor of Arts (Anthropology), Cert IV Training and Assessment, Cert IV NatHERS assessment, is a sessional tutor in the Deakin University School of Engineering and Built Environment and a registered architect with a broad range of experience in ecologically sustainable design, community development, contract administration, and client liaison.
CH 11 Simple building sketches and drawings

UNIT CONVERSION TABLE

Pressure	Multiply by...	Equals...
kPa	0.145	psi (lbs/in^2)
Psi (lbs/in^2)	6.895	kPa
kPa	4	Inches WG
Inches WG	0.25	kPa
kPa	10	mb
mb	0.1	kPa
Inches Hg	13.6 x 0.25	kPa
Heat energy & power	**Multiply by...**	**Equals...**
MJ	947.8	BTU
BTU	0.001055	MJ
kWh	3.6	MJ
MJ	0.2778	kWh
kWh	3412	BTU
BTU	0.2931	kWh
MJ/m^3	26.76	BTU/cu.ft
BTU/cu.ft	37.37	MJ/m^3
Volume	**Multiply by...**	**Equals...**
m^3	35.32	cu.ft
cu.ft	0.128	m^3
m^3	1000	L
Imp. Gallon	4.546	L
L	0.22	Imp. Gallon
US Gallon	3.785	L
L	0.2642	US Gallon
Imp. Gallon	0.8326	US Gallon
US Gallon	1.201	Imp. Gallon
Area	**Multiply by...**	**Equals...**
mm^2	0.01	cm^2
m^2	10.764	ft^2
ft^2	0.0929	m^2
Length	**Multiply by...**	**Equals...**
m	3.281	ft
ft	0.3048	m

Abbreviations			
kPa	(kilopascals)	m^3	(cubic metres)
psi	(pounds per square inch)	cu.ft	(cubic feet)
Inches WG	(inches water gauge – also "WG)	m^2	(square metres)
Inches Hg	(inches mercury – also "Hg)	mm^2	(square millimetres)
mb	(millibars)	ft^2	(square feet)
MJ	(megajoules)	L	(litres)
kWh	(kilowatt hour)		
BTU	(British thermal units)		

LIST OF FIGURES

COLOUR PALETTE FOR TECHNICAL DRAWINGS

Colour name	Colour	Material
Light Chrome Yellow		Cut end of sawn timber
Chrome Yellow		Timber (rough sawn), Timber stud
Cadmium Orange		Granite, Natural stones
Yellow Ochre		Fill sand, Brass, Particle board, Highly moisture resistant particle board (Particle board HMR), Timber boards
Burnt Sienna		Timber – Dressed All Round (DAR), Plywood
Vermilion Red		Copper pipe
Indian Red		Silicone sealant
Light Red		Brickwork
Cadmium Red		Roof tiles
Crimson Lake		Wall and floor tiles
Very Light Mauve		Plaster, Closed cell foam
Mauve		Marble, Fibrous plasters
Very Light Violet Cake		Fibreglass
Violet Cake		Plastic
Cerulean Blue		Insulation
Cobalt Blue		Glass, Water, Liquids
Paynes Grey		Hard plaster, Plaster board
Prussian Blue		Metal, Steel, Galvanised iron, Lead flashing
Lime Green		Fibrous cement sheets
Terra Verte		Cement render, Mortar
Olive Green		Concrete block
Emerald Green		Terrazzo and artificial stones
Hookers Green Light		Grass
Hookers Green Deep		Concrete
Raw Umber		Fill
Sepia		Earth
Vandyke Brown		Rock, Cut stone and masonry, Hardboard
Very Light Raw Umber		Medium Density Fibreboard (MDF), Veneered MDF
Very Light Van Dyke Brown		Timber mouldings
Light Shaded Grey		Aluminium
Neutral Tint		Bituminous products, Chrome plate, Alcore
Shaded Grey		Tungsten, Tool steel, High-speed steel
Black		Polyurethane, Rubber, Carpet
White		PVC pipe, Electrical wire, Vapour barrier, Waterproof membrane

PART 1

CODES AND STANDARDS

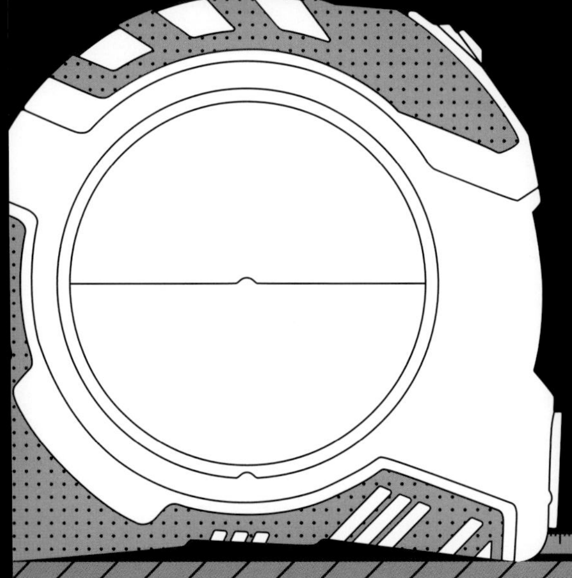

1 BUILDING CODES AND STANDARDS

Chapter overview

Australia, like many other parts of the world, has moved a long way forward from just 'knocking together' a house, shed, fence, or indeed any other structure. Today's buildings must comply with a raft of standards and codes. This chapter explores these codes and standards: in particular, the National Construction Code or NCC.

Elements

This chapter provides knowledge and skill development materials on how to:

1. access and interpret relevant codes and standards
2. classify buildings
3. analyse and apply a range of solutions to a construction problem for compliance with the NCC
4. apply fire protection requirements.

To gain the most from this chapter, you must have access to the National Construction Code, Volumes One and Two. These codes are available for free download, after free registration from the following internet site: https://ncc.abcb.gov.au/ncc-online/NCC.

Download the following documents:

- NCC 2019, Volume One (Building Code of Australia Class 2 to Class 9 Buildings)
- NCC 2019, Volume Two (Building Code of Australia Class 1 and Class 10 Buildings)

Introduction

This chapter introduces Volumes One and Two of the National Construction Code (NCC), the legal guide by which buildings constructed in Australia are governed. This is a large and complex document which will take you time to become confident using. The chapter begins with a brief history and overview of the NCC and Australian standards as a whole. The next section then discusses the different classifications of buildings, it being these classes that determine which part of the code applies to any given structure. The following two sections provide guidance on accessing and interpreting the various clauses of the code, particularly how to identify the standards relevant to your project. These sections also explore the two pathways to meeting the NCC's requirements as a whole; i.e. through the 'deemed-to-satisfy' provisions or by way of a 'Performance Solution'. The final section of the chapter will look at fire protection measures as outlined in both volumes.

History and purpose of the National Construction Code

Before looking too deeply into the National Construction Code (NCC), it is worth gaining a basic understanding of its history and purpose. Australia's search for a national set of standards and codes began in 1965, but it was not until 1988 that the first Building Code of Australia (BCA) volume was published; and it wasn't until the mid 1990s that all states and territories actually became signatories to the code. That is, Australia's construction codes are actually a very recent regulatory innovation. Up to this point, the BCA was what is called a 'prescriptive' code, meaning that it defined the what, when and how of all building works and, in so doing, set minimum building standards.

In 1996, the BCA96 was introduced; this reflected an important shift in the aim of the document as it was now 'performance'-based rather than prescriptive; thereby allowing alternative approaches and encouraging innovation. Again, there was a delay until 1998 before all states and territories adopted this new code. In 2003, annual amendment cycles were introduced and so from 2004 the BCA became BCA 2004, BCA 2005 and so on.

In 2011, the regulations governing the BCA and the Plumbing Code of Australia (PCA) were consolidated, giving rise to the National Construction Code or NCC. As of 2016, the annual amendment cycle was changed to a three-year cycle, reflecting both the acceptance and the stability of the current codes.

The NCC consists of three main volumes, and two secondary texts as follows:

Primary volumes:

- NCC, Volume One – Building Code of Australia Class 2 to Class 9 Buildings
- NCC, Volume Two – Building Code of Australia Class 1 and Class 10 Buildings
- NCC, Volume Three – Plumbing Code of Australia (All building classifications).

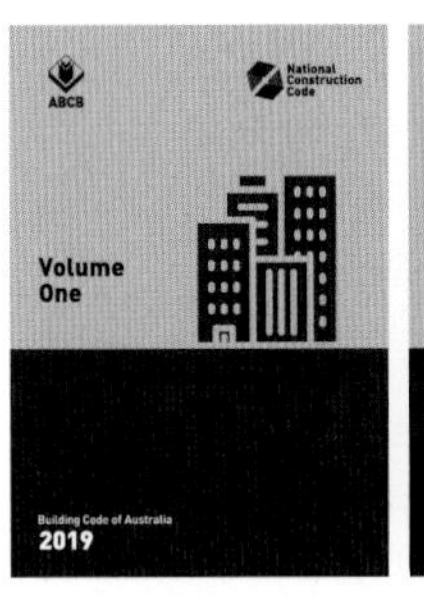

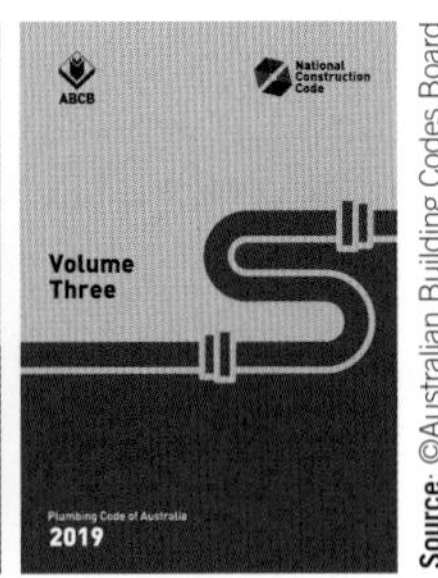

Source: ©Australian Building Codes Board

Secondary texts:

- Guide to NCC Volume One (provides clarification, illustrations and examples)
- Consolidated Performance Requirements (provides guidance on the above volumes).

Source: ©Australian Building Codes Board

Note: Only Volumes One and Two are discussed in this chapter.

The purpose of the NCC is to provide minimum standards by which buildings and associated structures (fences, pools and the like) may be constructed, with consideration to:

- occupant health and safety
- amenity and accessibility
- bushfire survivability
- sustainability and energy efficiency
- structural integrity
- climate and geographical location
- innovation.

As a 'national' construction code, the above is applicable to all Australian states and territories. It is also the avenue by which relevant Australian standards are given authority; i.e. unless an Australian standard is called up by the NCC it has guidance value only to the construction industry.

To ensure that the NCC's purpose is based upon sound factual data, its elements are constantly tested for workability, practicality and effectiveness – being restrictive only so much as to be in the public's and industry's best interests. This is exampled by the NCC being a 'performance'-based document, with two pathways of compliance: deemed-to-satisfy (DTS) provisions and Performance Solutions.

Deemed-to-satisfy provisions

Deemed-to-satisfy (DTS) provisions are known and modelled (within the NCC) ways of creating a particular part of a structure. For example, the size, shape and appropriate reinforcement of a concrete slab will be offered for a particular soil type, given load, width of structure and strength of concrete. Such an example will be fully documented within the code. Alternatively, the code may specify an Australian standard (such as AS 1684 Residential timber-framed construction) as the means by which a particular Performance Requirement of the code may be met.

Performance Solutions

Performance Solutions, previously known as 'alternative' solutions, are those that have been specifically developed by the builder, designer, or product or material supplier. Such being the case, they are not documented within the NCC. Rather, they directly respond to the Performance Requirements of the NCC and are shown to do so by one or more of the assessment methods offered. These assessment methods are described within the general requirements section at the front of each of the NCC volumes. There are four methods offered, any combination of which may be used to determine compliance:

- Evidence of suitability – i.e. evidence, as per the general requirements, is supplied
- Verification methods – tests, calculations, inspections or the like as deemed appropriate within the NCC or by an appropriate authority (as defined by the NCC)
- Comparison with the DTS provisions – the proposed solution is compared with existing deemed-to-satisfy examples offered within the NCC
- Expert judgement – a qualified and experienced person judges that a particular approach complies with the Performance Requirements.

Over time, solutions alternative to those already documented within the NCC (i.e. the DTS provisions) may be fed into the NCC volumes and become, of themselves, DTS provisions.

The main point you should understand from this section is that you have a choice: you can follow the examples provided in the NCC, *or* you can develop one of your own – *provided* that you can show that it satisfies the Performance Requirements for that particular element of the structure.

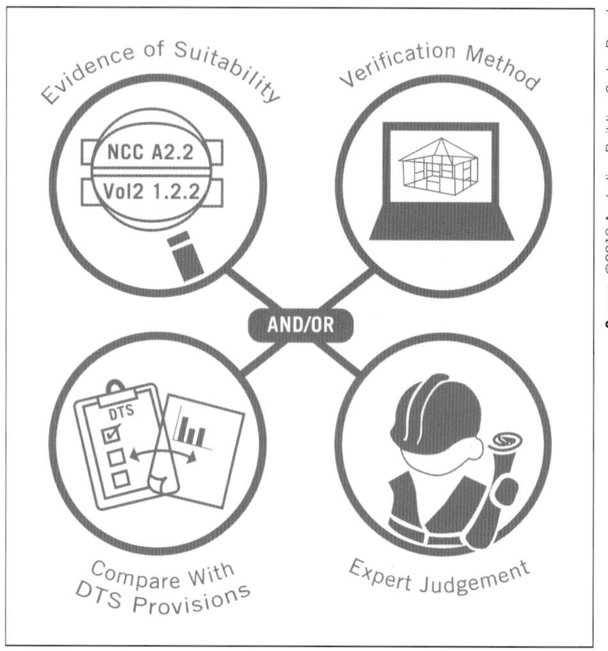

Source: ©2019 Australian Building Codes Board

Types of standards

Standards are globally recognised as the means by which we structure and define both the quality and the approach to much of our daily lives. Originally, standards were industry and commerce focused; today they are integral to almost all aspects of our society, including product safety and reliability, legal systems and workplace safety. In the construction industry there is an Australian standard to cover every aspect of the structure, including the materials and equipment used to create that structure – right down to the sand in the mortar and the brush used to apply the paint.

There are basically three types of standards that you may encounter: international, regional and national. Common international standards will have the prefix ISO (International Organization for Standardization) or IEC (International Electrotechnical Commission). Regional standards are those developed and adopted by an economically aligned 'zone', such as the European Union or EU. Australia and New Zealand often adopt the one standard and these will have the prefix AS/NZS.

Australian standards

National standards are the ones you will most commonly come across in your daily work. Australian standards are developed by Standards Australia, the government and internationally recognised peak body for standards in this country. This organisation develops and revises standards based upon public input, international comparison and through access to some 9000 volunteer technical committee members. The resultant standards are currently distributed through a separate, now privately owned, organisation called Standards Australia International Global – better known as SAI Global. Baring Private Equity Asia, the current owner of SAI Global, has committed to continuing the online distribution and sale of Australian standards until 2023.

Australian standards are prefixed with simply AS. You can identify genuine Australian standards by the logo of Standards Australia and the 'wordmark' Australian Standard®. Joint Australian and New Zealand standards will have both the Australian and New Zealand logos as well as the trade mark Australian/New Zealand Standard™.

Individual standards are identified by name, number, part number (if applicable) and year of approval as per the examples below:

- AS/NZS 4505:2012 Garage doors and other large access doors
- AS 1684.2 – 2010 Residential timber-framed construction. Part 2: Non-cyclonic areas

Other prefixes you may come across include HB (Handbook), and SA TS (Standards Australia Technical Specification), such as:

- HB 195-2002 The Australian Earth Building Handbook
- SA TS 101:2015 Design of post-installed and cast-in fastenings for use in concrete

Standards Australia are not the only source of standards referenced within the NCC. The National Association of Steel-Framed Housing (NASH) is an important example. NASH develop and market all the referenced steel framing standards listed within the NCC.

Source: © Standards Australia Limited Reproduced by Cengage under licence by Standards Australia.

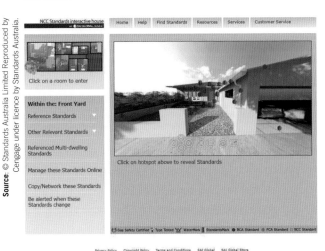

FIGURE 1.1 SAI Global's interactive NCC House, showing the Australian standards involved in various building elements

Aside from being the access point for standards, SAI Global also host a very useful web-based interactive by which you can explore how those standards are applied to the domestic house by the NCC (https://bca.saiglobal.com/ProductsServicesPage.asp?path=house). It is strongly recommended that you take the opportunity to use it as way of raising your awareness of the number of standards that can apply to any one area of a house.

Legal authority of the NCC and Australian standards

No code or regulation has legal effect unless legislation has been passed enforcing it. Each Australian state and territory has passed an Act of Parliament giving legal power to the NCC.

The NCC, in turn, gives legal effect to those Australian standards applicable to the Performance Requirements.

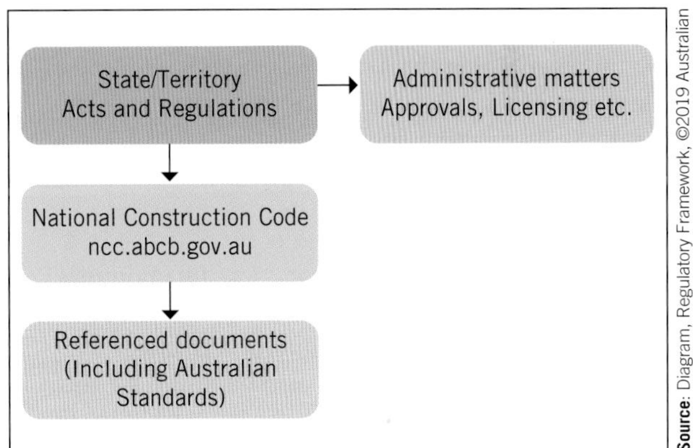

Source: Diagram, Regulatory Framework, ©2019 Australian Building Codes Board, https://ncc.abcb.gov.au

FIGURE 1.2 Regulatory path leading to the adoption of a standard (Australian or otherwise)

The administration and compliance supervision of the NCC is therefore the role of the individual states and territories, not the federal government or the Australian Building Codes Board (ABCB) who develop it. In the main, compliance and administration of the NCC falls upon local government authorities (councils) or private building certifiers.

Note that it is no longer a requirement that building approvals and inspections be conducted by, or through, the local council. Private certifiers are commonly available in most regions of Australia and are competitive in price and quality of service.

The NCC classes of buildings

The elements for this unit of competency are ordered so that identifying a building's classification follows after accessing and interpreting codes. For comprehension and logical flow, you should become familiar with the NCC's classes of buildings before proceeding further.

Determining the nature of a building

The NCC works through the systematic application of Performance Requirements to specifically identified classes of buildings and other structures. The determination of a classification depends much upon the nature of the structure. By 'nature', we mean the form and function, or purpose, of the structure, a building, or a designated part of a building. In many cases this may seem straightforward – and is; in other cases, the nature of a structure is less clear cut.

In general, the nature of a building falls into one of four categories, it is either a:

- private dwelling
- public building
- commercial building
- industrial building.

However, there are structures that don't fit any of these categories – such as fences, farm buildings and sheds, swimming pools, laboratories, private hospitals, car parks, bushfire shelters and many more.

In determining the nature of a building for the purposes of the NCC, therefore, you must consider carefully its function and use.

Function vs use

The function of a structure differs from its use in a few simple but important factors. The function of a farm shed, for example, could be to provide protection against sun, rain, dust and wind for a tractor. However, the same shed may also be 'used' for the mechanical servicing of the tractor by mechanics or other farmhands employed on the property and set up accordingly; i.e. it is a workshop. Used as shelter only, the shed fits a very different category to that of a workshop. Likewise, a medical practitioner's 'rooms' serve as a place for his or her practice. In general, this would include the physical treatment of only minor medical complaints. If, however, the rooms are used to carry out treatments that could render patients unconscious or non-ambulatory – i.e. the patient must stay resident at the rooms for some period after treatment – then its use is considered to be very different by the NCC. Function and form may be effectively the same in each instance; the use, however, is very different and so the NCC classification changes.

You must also consider all possible uses. This is because you may not be aware at the outset who the tenant may be, or exactly what they may use the building for. The farm shed offered earlier is the classic example. In these cases, you take the scenario that imposes the most onerous requirements – i.e. that the shed will most probably be used for maintenance, not just storage.

Having determined the nature of a building, you are then in a position to identify its NCC classification. This, in turn, will determine factors such as access and egress requirements, light and ventilation, fire control measures and a raft of other important design and construction factors.

Identifying a building's classification

An important element of deciding a building's classification is that, in doing so, you are identifying the risks associated with its intended use and purpose. Getting it right means you will not over- or under-specify the building – either of which has cost implications. Additionally, correctly identifying a building's classification ensures that the purposes of the NCC outlined earlier are being met; i.e. health and safety, amenity, energy efficacy, sustainability and structural integrity.

The NCC provides for 10 classes of buildings, along with a number of sub-classifications. Each classification provides for the often highly divergent natures of the buildings, and hence the design elements and materials that must be incorporated into them. The NCC outlines these classifications in Part A6 of both Volumes One and Two. These NCC definitions tend to be brief but are quite specific. For the purposes of introduction, a more simplified set of descriptions is initially offered in Table 1.1 below. These will then be discussed in more detail.

NCC building classes: some expanded commentary

Table 1.1 gives you a rough guide to the classes of buildings defined by the NCC. Further commentary is required, however, for a full understanding of each. We will begin with defining the concept of one 'storey' of a building, before following on with the expanded descriptions of each building class.

Storey: NCC definition

The NCC definition of one 'storey' of a building is interesting and provides for some peculiar interpretation in one or two instances. You will find it in both NCC Volumes One and Two, within Schedule 3, Definitions (Vol. One, p. 518; Vol. Two, p. 663). It reads as follows:

Storey means a space within a building which is situated between one floor level and the floor level next above, or if there is no floor above, the ceiling or roof above, but not —

(a) a space that contains only —

(i) a lift shaft, stairway or meter room; or

(ii) a bathroom, shower room, laundry, water closet or other sanitary compartment; or

(iii) accommodation intended for not more than 3 vehicles; or

(iv) a combination of the above.

(b) a mezzanine.

TABLE 1.1 NCC building classifications

Classification		Description
Class 1	Class 1a	Single dwellings such as a detached house, or an attached row of houses, units, town houses or the like, separated by a fire-resisting wall continuous from ground to roof apex.
	Class 1b	Boarding or guest houses/hostels with max floor area of less than 300 m^2. Generally, less than 12 occupants, but can include 4 or more single dwellings on the one allotment for use as short-term holiday accommodation.
Class 2		Apartment buildings. Usually multi-residential structures with each apartment considered a sole-occupancy unit or SOU (see below). Often multi-storey, but may be single storey when the units share same subfloor or roof space not separated by a fire-resisting wall.
Class 3		Residential buildings that don't fit Class 1 or 2. Larger guest houses/hostels, dormitory-style accommodation, detention centres, workers quarters, or care facilities not considered to be Class 9a.
Class 4		A dwelling in an otherwise non-residential building of Class 5 through to 9. There can only be one Class 4 dwelling in a building. A caretaker's residence in an office complex is a common example.
Class 5		Office buildings used commercially or professionally by lawyers, accountants, doctors, government bodies and the like. Excludes buildings Classed 6, 7, 8 or 9.
Class 6		Shops, restaurants, cafés and the like. Retail or service outlets to the general public. Includes shopping centres, public laundries, bars and funeral parlours.
Class 7	Class 7a	Buildings serving as carparks except private garages of 3 cars or less.
	Class 7b	Storage or wholesale display buildings such as warehouses.
Class 8		Laboratories, factories or workshops used for trade, sale or gain such as production, repair, maintenance, altering, packing or cleaning.
Class 9 Public Buildings	Class 9a	Public buildings including health-care facilities (e.g. hospitals) and clinics in which patients may become unconscious or otherwise unable to move without assistance. Can include a laboratory without the need for multi-classification (laboratories are normally considered Class 8).
	Class 9b	Assembly buildings such as theatres, churches, night clubs, schools and preschools, sports facilities, gyms, train and bus stations.
	Class 9c	Residential aged care facilities in which 24-hour care services are provided.
Class 10	Class 10a	Non-habitable buildings or structures such as sheds, carports and private garages. There can only be one storey or level of garage space within any single building.
	Class 10b	Non-habitable buildings or structures such as fences, antenna masts, retaining walls, swimming pools and the like.
	Class 10c	Bushfire shelter when associated with, but not attached to, a Class 1a building.
SOU		A Sole-Occupancy Unit (SOU) is not a class of building. It is a part of a building otherwise classed 1b, 2, 3, 5, 6 or 9 that is intended for the exclusive use of its owner(s), tenant or lessee. There may be multiple SOUs within the one building.
Storey		*See definition in text.*

This suggests that a private garage for three cars, coupled with a laundry, toilet and bathroom, located as the first level of a domestic house on a level block of land does not constitute a 'storey'; i.e. such a house is in fact a single-storey building. This only becomes an issue if you are working in a state where the low-rise definition of licenced work applies. And only in instances such as a Class 1b guest house that is sited above a private garage for three or less cars – this is because a two-storey guest house is a Type B construction (see Chapter 13, 'Low-rise construction: a definition') and therefore outside your scope of works. You should check with your local building authority for their interpretation before accepting to contract such works.

Class 1a

The most common form in this class is the typical Australian suburban home (Figure 1.3). However, Class 1a also includes homes that are attached to each other. Buildings of this type include terrace, row, or town houses, as well as units. What defines a cluster of attached dwellings as Class 1a is that each dwelling is separated by a vertical fire-resisting wall that is continuous from the ground through to the roof (see Figure 1.3). This includes any basements or subfloor garages. Class 1a buildings may be more than one storey, indeed there is no limit defined within the NCC; however, it is uncommon to see them exceed three storeys. Likewise, there is no limit to the number of dwellings or homes in any single development; i.e. a set of row or

Source: Glenn Costin

FIGURE 1.3 Class 1a: suburban home

town houses could run the full length of a street. The key limitation to both Class 1a and Class 1b buildings is that they cannot be constructed above, or below, any other class of building except a Class 10a private garage. Nor can they be stacked one upon the other.

Class 1a defining characteristics:

- a residential dwelling
- attached dwellings may be attached horizontally only
- attached dwellings must be separated by a fire-resisting vertical wall running from ground to roof
- cannot be constructed above or below any other class of building except a Class 10a private garage.

Class 1b

Boarding and guest houses or hostels (Figure 1.4) with a maximum floor area of less than 300 m^2 is the common definition for this class. In general, we are looking at less than 12 occupants overall, though this is not a definitive limit, and that occupants are transitory (short term). In a boarding house (or lodge), occupants may be resident for extended periods of time, but they are not regarded as tenants in that they do not hold a leasing agreement; as such, the buildings still fit this classification. A bed and breakfast also fits this class on the basis of short-term accommodation for which there is typically no lease agreement.

Source: Peter Redmond

FIGURE 1.4 Class 1b: boarding or guest houses and hostels

This class also includes four or more single dwellings on the one allotment for use as short-term holiday accommodation; again, on the basis of there being typically no lease agreement. Common examples include individual cabins in caravan parks, resorts, and on farm-stay type properties. Note that up to three single dwellings on the one allotment retain the Class 1a classification.

As with Class 1a, Class 1b buildings cannot be constructed above, or below, any other class of building except a Class 10a private garage. Nor can they be stacked one upon the other.

Class 1b defining characteristics:

- a residential dwelling
- occupants are typically transient (generally short term) without a lease
- structured such that not more than 12 people would ordinarily occupy it
- floor area less than 300 m^2
- cannot be constructed above or below any other class of building except a Class 10a private garage.

Class 2

These are apartment buildings: multi-residential structures with each apartment considered a sole-occupancy unit (SOU). They can be, and often are, multi-storey. However, a single-storey cluster of units that are situated above a common subfloor or basement, or have a common ceiling space, or are built above a common Class 7a carpark, all fit the Class 2 category. The latter example is a multi-classification building.

Class 2 defining characteristics:

- a residential dwelling
- apartments are defined by the NCC as sole-occupancy units or SOUs
- no limit to height of building, number or size of units
- may share a common subfloor or ceiling space
- can be constructed above or below another class of structure except a Class 1.

Class 3

These are larger guest houses or hostels where long term and/or transient people may reside; length of stay is therefore not important. They exceed the floor area limitations of a Class 1b and include dormitory-style accommodation, detention centres and workers quarters. This class also includes care-type facilities not considered to be Class 9a, such as shared accommodation for people with a disability, the elderly (but not infirm), children and refugees.

Class 3 defining characteristics:

- a residential dwelling
- occupants may be long term or transient without a lease
- floor area larger than 300 m^2
- can be constructed above or below another class of structure except a Class 1.

Class 4

A Class 4 building is not really a building at all, but rather an integrated residential element of an otherwise non-residential structure – structures classified Class 5 through to 9. A Class 4 'building' is effectively an SOU. However, there can only be one Class 4 dwelling in any given building – that includes the whole building, not just a single storey of that building. A caretaker's residence in an office complex is a common example.

Class 4 defining characteristics:

- a residential part of non-residential building
- is a sole-occupancy unit (SOU)
- not limited by size
- only one Class 4 element in any one building.

Class 5

These are office buildings used commercially or professionally by lawyers, accountants, doctors, government bodies and the like. There is generally more than one office in the building, making each office an SOU. They are *not* shops, however (which are Class 6). They are also not doctor's surgeries where patients are likely to undergo medical treatment that may leave them unconscious or non-ambulatory (unable to move without assistance); in such cases the building is Class 9a.

Class 5 defining characteristics:

- a non-residential building
- is an office, or is made up of offices, used for commercial or professional purposes
- not defined by size
- can be constructed above, below or within another class of non-residential building.

Class 6

Shops, restaurants, cafés and the like all fit this classification. These are buildings that serve as retail or service outlets to the general public. Where there are multiple shops, each 'shop' is an SOU by definition of the NCC. Class 6 buildings include shopping centres, public laundries, bars and funeral parlours. They are not defined by size, only by purpose; so they can be incredibly small (Figure 1.5), independently large, or form part of a multi-class building.

Class 6 defining characteristics:

- a non-residential building
- a retail or service outlet, or is made up of SOUs that have retail or service purposes
- not defined by size
- can be constructed above below or within another class of non-residential building.

Class 7a

These are buildings that serve as carparks. A garage associated with a building classed other than Class 1 (a or b) that holds more than three vehicles is classified as a 7a carpark: a typical example being a garage space under a Class 2 building that holds more than three vehicles.

Source: Glenn Costin

FIGURE 1.5 Class 6: shops, restaurants and cafés

Class 7a carparks can also be separate single- or multi-storey buildings. A farm shed where more than three tractors (or other farming vehicles) are parked also fits this category.

Class 7a defining characteristics:

- a non-residential building, and not associated with a Class 1 building
- designed for the express purpose of storing vehicles
- defined by the minimum number of vehicles stored (more than three)
- can be constructed independently, or above, below, or within another class of non-residential building.

Class 7b

Generally, a Class 7b building will be some form of warehouse or storage facility. However, this classification also includes buildings that not only store, but also display goods or produce for wholesale purposes. Class 7b buildings are not 'shops'; i.e. the owner/occupiers/tenants do not engage in retail to the general public, but rather offer bulk supply to those who do. Class 7b buildings can include multiple SOUs provided they are occupied for wholesale transactions.

Class 7b defining characteristics:

- a non-residential building
- a building used for storage, and/or the display of produce or goods sold wholesale
- not defined by size
- can be constructed independently, or above, below, or within another class of non-residential building.

Class 8

These are factories or workshops in which some form of production, repair, maintenance, packing, assembling or other handicraft or processes takes place. In all cases, such activities are for the purposes of trade, sale or gain. This classification includes mechanics workshops, abattoirs, canneries and other food processing factories. Farm sheds where tractors or other farming equipment are stored, but also where mechanics are employed to service them, would fit this category. It also includes laboratories unless they are constructed as part of a Class 9a building.

Class 8 defining characteristics:

- a non-residential building
- factories, laboratories or workshops, the outputs of which are for trade, sale or gain
- not defined by size
- can be constructed independently, or above, below, or within another class of non-residential building.

Class 9a

These are public buildings and are generally some form of health-care facility, such as hospitals and day surgery clinics in which patients may become unconscious or otherwise unable to move without assistance. Laboratories that form an integral part of the health-care facility are also Class 9a. This allows the whole building to go under the one classification despite laboratories normally being regarded as Class 8.

Class 9a defining characteristics:

- a public building
- a health-care facility
- not defined by size
- can be constructed independently, or above, below, or within another class of non-residential building.

Class 9b

Class 9b are also public buildings. In this case, they are some form of assembly building such as theatres, churches, night clubs, schools, preschools and childcare centres, sports facilities, gyms, and train and bus stations. The reason for people gathering is immaterial, it is the fact that they may gather within the space that is important, as it raises questions of access and egress (capacity to get out) as well as fire control. They can be built within buildings of another classification: a cinema within a shopping complex or a chapel within a hospital being common examples.

Class 9b defining characteristics:

- a public building
- a place of gathering or assembly
- not defined by size
- can be constructed independently, or above, below, or within another class of non-residential building.

Class 9c

As the third class of public buildings, Class 9c covers aged care facilities. This is a form of residential accommodation for the elderly who may require personal care services such as bathing, nursing, dressing and eating. In addition, they must be provided with staff assistance 24 hours a day for purposes of emergency evacuation.

Class 9c defining characteristics:

- a public building
- a residential aged care building
- not defined by size
- can be constructed independently, or above, below, or within another class of non-residential building.

Class 10a

All Class 10 buildings are deemed non-habitable buildings or structures. Class 10a includes sheds, carports and private garages. Farm sheds fall into this category when they are not deemed to be either Class 7a or 8 (see above). Note that a garage is only deemed a 'private garage' when it is:

- associated with a Class 1 building – attached to it or free standing – in which case there is no limit to the number of vehicles it may house
- a single storey of a building with a maximum capacity of three vehicles. Only one such garage storey is permitted within any given building
- a separate single-storey garage associated with another building of any class that is limited to the parking of three vehicles. There may be multiple such separate garages associated with any one building (limited only by local council regulation).

Class 10a defining characteristics:

- a non-habitable building
- sheds, carports and private garages
- not limited by size, except when deemed a private garage not associated with a class 1 building – in which case limited to three-vehicle storage
- can be a separate structure, or when a private garage as a single storey within another building.

Class 10b

These are non-habitable buildings or structures such as fences, antenna masts, retaining walls, swimming pools, wading and paddling pools, spas and the like. They may be attached to or associated with any other class of building.

Class 10b defining characteristics:

- a non-habitable structure
- typically fences, retaining walls, antenna masts and swimming pools
- not limited by size
- can be attached to or associated with buildings of another class.

Class 10c

This is a very specific class, which contains only private bushfire shelters. A private bushfire shelter is defined as a structure associated with, but not attached to, or part of, a Class 1a building. Though deemed non-habitable, a private bushfire shelter is designed to be occupied as a last resort refuge from the immediate threat to life by an approaching bushfire.

Class 10c defining characteristics:

- a private bushfire shelter associated with a Class 1a building.

Buildings of multiple classifications

Buildings may be designed for multiple uses. In such cases the building will generally also need to hold multiple classifications. This means that a building can be classified as a Class 6/4 building, for example, or

Source: Newspix/Stuart McEvoy

FIGURE 1.6 Class 10c: bushfire shelters

even a Class 5/6/7 building. Class 1 buildings being the exception: Class 1 buildings cannot be built above or below any other dwelling, or class of building, excepting a Class 10a private garage.

Common examples of mixed-use buildings include Class 2 apartments constructed over Class 7a carparks or Class 6 shops; or a combination of all three. By definition, Class 4 parts of a building are always located within a building of a different classification.

Every part of a building must be classified separately. However, the percentage of a single storey's floor area enclosed by a particular activity determines if it needs to be classified separately or not. When that area is less than 10%, then it may be considered as auxiliary to the main activities of that floor, level or storey of the building. See the Case study 'Multiple classifications'.

CASE STUDY

Multiple classifications

Consider a multi-storey Class 7a carpark with a Class 5 office included in the 2nd storey. The office floor area is only 9% of the total floor area for that level. In this case the whole storey may be considered as Class 7a. If that was the sole alternate use element in the whole structure, then the whole building is Class 7a.

However, in the next level of car parking above, there is a larger office. In this case it takes up 14% of the total floor area. This level of the carpark must then be duel classified as Class 5/7a. Likewise, the building as a whole (assuming no other alternative use sections of the building exist) will be Class 5/7a.

The exception to this ruling is the Class 4 element: a Class 4 dwelling is a Class 4, irrespective of its percentage of the storey's floor area.

While a building is being designed it may be unclear who the tenant or tenants will be, or how they may ultimately use the facility. This is why classifying a building is a form of risk management. You identify the function and possible uses of a building, particularly multi-use buildings, then classify accordingly. The building is then designed such that each separately classified part of the structure meets the requirements of its classification. In this manner the building is able to be used safely for any or all of the identified purposes.

It should be noted at this point that in carrying out new works within, or additional to, an existing building, use and function must again be considered. Where there is a change of purpose, there is a change of classification. This is something that is often not considered when converting an existing private garage to a self-contained unit, bedroom with ensuite, or a dedicated private workshop to one for commercial gain.

A final note on NCC classifications

Not all buildings fit the above classification system easily; aircraft hangers and aircraft boarding concourses being some complex examples, farm sheds being more rudimentary examples. The NCC attempts to account for these less easily defined structures through:

- NCC, Volume One, Section G – Ancillary Provisions and
- NCC, Volume One, Section H – Special Use Buildings

A careful reading of these sections may help in defining a particular structure if you are in doubt. Ultimately, however, the defining of a building's class is the responsibility of a building certifier acting on behalf of the appropriate building authority (local council or independent company). However, having an understanding of these sections, and building classifications generally, can help you discuss the merits and cost implications of a particular class with the building certifier to the benefit of all parties.

LEARNING TASK 1.1

NATURE OF BUILDINGS AND THEIR CLASSIFICATION

1. An old industrial building has been used to store a number of differing types of wholesale goods on pallets and raking systems. You have been approached to get council approvals for a redevelopment that will allow most of the building to be used as office space, with a small conference room also included.
   - What is the current classification?
   - What will the classification be should the proposed redevelopment go ahead?
2. You are on acreage in a country area and have a large family home on that land. There is some discussion that you would like to build a large, 100 m² pergola-type structure, with internal roof

>>

>>

heights of around 3.4 m, that you could use to hold family functions, weddings and other meetings. What classification will the council place on this type of structure?

3 A building you have recently inspected is approximately 3000 m² and has a steel fabrication workshop at the rear of the building, approximately 2500 m²; has large, open rooms acting as display and marketing rooms; and worker amenities and lunch rooms with some offices in the 500 m² area at the front of the building over open under-croft parking.
   - Determine the classification/s that might apply to this type of building. Provide reasons for your classifications.

Accessing and interpreting codes and standards

At the beginning of this chapter you were directed to the internet site of the Australian Building Codes Board (ABCB) from which you could download the NCC volumes. This is the simplest and most logical means of accessing the code. It's free, and it's fast. Unfortunately, this does mean that to read the code you must always have a digital device of some kind on hand; in general, this means a laptop due to the need for a decent screen size and faster searching. Hard copies are not available directly from the ABCB. The Master Builders Association of Victoria (MBAV) also produces hard copies for purchase, for those who prefer this format, although these can be expensive and need to be updated every three years.

Australian standards are not so freely available. Aside from a handful of standards available through agencies such as the MBAV, in general Australian standards are only available through SAI Global.

Fortunately, most Registered Training Organisations (RTOs) offering the course you are undertaking will provide you with free online access via their library or online portal. Likewise, the Australian National Library, and all state libraries also currently offer free access to the SAI Global database of standards once you become a member of that library (which is free). In either case, you may download the standards for free and store them for a limited period on your computer or digital storage device. Unfortunately, you can only view them for five days, after which the Digital Rights Management (DRM) software renders the file unusable.

These documents are large and complex. In the next section we will look at how to identify the general and specific provisions that are relevant to any individual project.

NCC suite:

- Digital download only – https://www.abcb.gov.au/ncc-online/NCC
- Hard copy – https://www.mbansw.asn.au/store

Australian standards:

- Hard copy or digital download – https://infostore.saiglobal.com
- Digital download – your RTO library or web portal, Australian national and state libraries

Note: Digital copies (PDF) of Australian standards are limited to five days, after which you will need to download them again.

Identifying the Performance Requirements

Once you have identified the correct classification of the building you are working on, the next step is to identify the relevant standards that the building will need to comply with.

The only mandatory requirements of the NCC are the Performance Requirements (PR), and the supporting General Requirements (GR). All other elements of the NCC are advisory only. It is therefore critical that you can locate, identify and understand the PRs as they pertain to the particular part of the structure you are considering.

General Requirements (GR)

The General Requirements sections of the NCC are where the classification of buildings, described above, and the application of the NCC generally is outlined. The General Requirements (sometimes referred to as 'General Provisions') are outlined in Section 1 of NCC, Volume Two, and Section A of NCC, Volume One.

A reading of the General Requirements will aid you in developing or choosing an acceptable design. It will also aid in interpreting the Performance Requirements and how you might verify your approach to the requirements.

Performance Requirements (PR)

As mentioned earlier, the NCC is performance-based, rather than prescriptive. This means that it does not lay out a set of 'rules' or instructions describing a single way of doing things that must be followed. Instead, it lays out standards that must be met, while allowing flexibility and encouraging innovation in the way these might be achieved. This encourages innovation, since builders are given the freedom to do things in different ways, so long as they achieve a level of quality and safety that is acceptable according to the standards. They are therefore restrictive only where necessary to ensure the best interests of the public and industry.

The Performance Requirements are found under Section 2, Performance Provisions. Performance Requirements specify all those aspects by which a given building element will be evaluated for compliance with the code. A Performance Requirement *does not* specify 'how' you are to go about achieving that goal. Instead you are given one of two options:

- prove that you can achieve the requirement by your own approach
- use one of the deemed-to-satisfy (DTS) approaches in the latter sections relevant to that element.

Performance Requirements are outlined in the NCC, under Section 2 of Volume Two. In Volume One you need to check the subheadings for 'Performance Requirements' under each Section B to J.

Alternatively, you can download the 'Consolidated Performance Requirements' from the NCC website. This document holds all the PRs from both BCA-focused volumes, and the Plumbing Code of Australia (PCA, being Volume Three of the NCC).

To quickly find the Performance Requirement for any given element of your building its best to have some idea of the structure and terminology of the NCC volumes. As an example, the main headings for Section 2, Volume Two are given below. In addition, the subparts for one of these are also exampled. Though the two NCC volumes differ somewhat in their respective layouts, these headings will allow you to quickly access the PRs in either volume for the more common elements of your building. Logic will serve you in most instances; however, only time and experience will really allow you to find areas quickly – unless, of course, you are using a digital copy – in which case you can use the search function of the PDF file.

- 2.1 Structure
- 2.2 Damp and weatherproofing
- 2.3 Fire safety
- 2.4 Health and amenity
  Performance Requirements
  - P2.4.1 Wet areas
  - P2.4.2 Room heights
  - P2.4.3 Facilities
  - P2.4.4 Light
  - P2.4.5 Ventilation
  - P2.4.6 Sound insulation
  - P2.4.7 Condensation and water vapour management

  Verification Methods
  - V2.4.2 Room or space height
  - V2.4.4 Verification of suitable natural light
  - V2.4.5 Verification of suitable indoor air quality
  - V2.4.6 Sound insulation
  - V2.4.7 Verification of condensation management
- 2.5 Safe movement and access
- 2.6 Energy efficiency
- 2.7 Ancillary provisions and additional construction requirements

CASE STUDY

The bedroom window

You intend to build a bedroom in a Class 1a home and you want to know the size of the timber window required. The room will be close to the property boundary fence.

Lighting is part of Health and Amenity so you look up that section of the Performance Requirement in the NCC (Volume Two, 2019, p. 58) as this is the *volume* that deals with Class 1 & 10 buildings. In so doing, you find the following:

P2.4.4 LIGHTING

> a) A *habitable room* must be provided with *windows*, where appropriate to the function or use of that part of the building, so that natural light, when available, provides an *average daylight factor* of not less than 2%.

The first thing you notice is that some of the terms are in blue and italicised. If working with a digital copy, you can click on them and you will be taken to the appropriate definition page. This definition page is part of the General Requirements and is there to aid in you getting the correct interpretation. If you are working with a hard copy, then you would go to the definitions in the General Requirements and search alphabetically.

First, you check 'habitable room' to see if this applies to a bedroom: the definition is found in Schedule 3, Definitions on page 509, and is given in part below:

> **Habitable room** means a room used for normal domestic activities, and –
> Includes a bedroom, living room, lounge ... [continues with multiple other listings]

The definition clearly includes bedrooms. So now you know that P2.4.4 (a) applies to your situation, but it does not tell you how to achieve it. This is where you start looking at the two options outlined at the beginning of this section; i.e. will a DTS solution work in your instance, or do you want – or do you have to – come up with something else?

Determining relevant deemed-to-satisfy provisions

As we have said, the NCC is not 'prescriptive'. As a result, there are two pathways of compliance. The first is to follow one of the 'deemed-to-satisfy' solutions that are offered by the NCC itself. The second is 'Performance Solutions' which will be covered in the following section.

Deemed-to-satisfy (DTS) solutions

DTS provisions are known and approved ways of creating a particular part of a structure in order to meet the necessary standards, which are modelled in the NCC. The deemed-to-satisfy (DTS) provisions provided in the two NCC volumes are what might be called your 'go to' solutions. They are prescriptive: they tell exactly what you must do to satisfy any given PR. In the majority of the structures you will build, these will be the simplest and easiest means by which you can be assured of satisfying the Performance Requirements for any given element of your building.

For example, the NCC will provide a fully documented example of the size, shape and appropriate reinforcement of a concrete slab for a particular soil type, given load, width of structure and strength of concrete. Alternatively, the code may specify an Australian Standard (such as AS 1684 Residential timber-framed construction) as the means by which a particular Performance Requirement of the code may be met.

DTS solutions themselves hold two compliance pathways:

- **acceptable construction practices**: those practices held by the NCC to be the most common forms of construction practice in Australia
- **acceptable construction manuals**: these are the documents referenced as such in the DTS solutions provided in the two NCC volumes; e.g. an Australian standard.

DTS solutions are those outlined in Section 3 Acceptable Construction of Volume Two of the NCC. Again, in Volume One you must check the relevant subheadings referring to 'deemed-to-satisfy provisions' in each of the sections.

CASE STUDY

The bedroom window

Working through the bedroom window example offered above, we established that the relevant Performance Requirement is P2.4.4 (a), and therefore the window must be constructed in a way that ensures that 'natural light, when available, provides an *average daylight factor* of not less than 2%'.

The easiest option is to go the DTS provisions in Section 3. A quick hunt through the index brings you again to Part 3.8 on Health and Amenity, and subsection 3.8.4 on 'light'.

Reading the section, you will find that **3.8.4.2 Natural Light** (pp. 283–6) is extensive and highly specific, with very precise detail on all the parameters you need to achieve the PR. The first page of it looks like this:

Health and amenity

PART 3.8.4 LIGHT

Appropriate *Performance Requirements*:

Where an alternative light system is proposed as a *performance solution* to that described in **part 3.8.4**, that proposal must comply with–

(a) *Performance Requirement* **P2.4.4**; and

(b) the relevant *Performance Requirement* determined in accordance with **1.0.7**

Acceptable construction practice

3.8.4.1 Application

Compliance with this acceptable construction practice satisfies *Performance Requiremnt* **P2.4.4** for light.

3.8.4.2 Natural light

Natural light must be provided to all *habitable rooms*, in accordance with the following.

(a) Natural light must be provided by–

  (i) *windows*, excluding *roof lights* that–

    (A) have an aggregate light transmitting area measured exclusive of framing members, glazing bars or other obstructions of not less than 10% of the floor area of the room; and

    (B) are open to that sky or face a court or other space open to the sky or an open verandah, carport or the like; or

  (ii) *roof lights* that–

    (A) have an aggergate light transmitting area measured exclusive of framing members, glazing bars or other obstructions of not less than 3% of the *floor area* of the room; and

    (B) area open to the sky; or

  (iii) a proportional combination of *windows* and *roof lights required* by (i) and (ii)

In following the information provided you will satisfy the PR P2.4.4.

However, with further reading of the six pages of DTS description you note that clause **3.8.4.2 Natural Light (b)** (p. 285) states the following:

(b) A window required to provide natural light that faces a boundary of an adjoining allotment must not be less than a horizontal distance of 900 mm from that boundary.

Unfortunately, your window is going to include a window seat which will push the window itself out to only 750 mm from the boundary. This means that the DTS provision cannot be applied 'directly' to your situation.

A DTS solution must be followed to the letter in order to comply with the NCC. However, as the case study above shows, deemed-to-satisfy solutions will not work in every situation. This is where the 'non-prescriptive' approach to the development of the NCC comes to the fore. You are, by regulation, able to come up with another approach that will satisfy the Performance Requirement. We will discuss how to achieve this in section 3 of this chapter.

Accessing and interpreting the relevant Australian standards

The various paths to accessing the Australian standards have been outlined above, and in the introductory passages to this book (p. 2). As stated in these sections, it is best if you use one of the many free digital access paths and download the documents as you need them, or simply view them online. Remember, once you have finished your course you can still access these standards via membership with your state library.

Likewise, identifying which standard you need to access is fairly easy. They are listed as relevant in each of the NCC volumes in direct relation to both the DTS provisions and, on occasion, in the Performance Requirements

CASE STUDY

The bedroom window

In the case of the bedroom window discussed previously, in addition to knowing the required size of the window, you will also need to know how to make the window, or how others must make the window. This too is covered in the NCC, and to find the relevant information we must look at Part 3.6 Glazing. In particular, Part 3.6.0.

This section outlines what are described as ***Acceptable construction manuals***. Part (a) of this clause is given below:

3.6.0

(a) *Performance Requirements P2.1.1* and *P2.2.2* are satisfied for glazing and *windows* if designed and constructed in accordance with AS 2047 for the following glazed assemblies in an *external wall*:

  (i) *Windows* excluding those listed in (b)

  (ii) Sliding and swinging glazed doors with a frame, including French and bi-fold doors with a frame

  (iii) Adjustable louvres

  (iv) *Window* walls with one-piece framing

From this statement you know that either you, or your window manufacturer, must work to the Australian Standard AS 2047 to understand the requirements for how the window must be constructed – this is the ***Acceptable construction manual***.

Note: the NCC will not state the year of the standard, only the number; it is your responsibility to check that you are consulting the most up-to-date version of the standard. Note that there have been two amendments to AS 2047, one in 2016, and the other in 2017. The first file on the list is simply entitled AS 2047:2014; it does not mention the addition of the two amendments.

Structure of an Australian standard

Australian standards can vary quite markedly in their structure or format, depending upon the subject matter. However, in the main they will be set out as follows:

- ***Front page*** (p. 1): the name, number and publishing history of the standard are shown here. Under the standard's number on this page (top right corner) there may be a statement in brackets akin to:

(Incorporating Amendment Nos 1 and 2).

Always check after downloading or opening online if the document you are viewing actually includes all of the amendments. In those standards with amendments included, you can identify the amended clauses, parts or sections, by the amendment number (A1, A2, etc.) coupled with a vertical dash showing down the left-hand margin.

Language in the NCC and Australian standards

There are some very specific words or terms that are associated with the structure of an Australian standard and which you must understand the meaning of. While there are many, the following will suffice to 'get you in', so to speak, allowing you to find your own way with time and experience.

- **Normative**: Within standards, normative means required practice, rules or something that must be followed. Many standards list associated 'normative' standards in *Section 1 – Scope and General*. This means these standards must be followed when called up in specific sections of the standard you are currently reading. A 'normative' appendix likewise is something that must be followed, much in the same manner as a DTS provision in the NCC.
- **Informative**: Meaning for information only. It is advice or additional information aimed to help clarify or example some aspect of the standard. Such information does not have to be strictly followed; you may do something else if it is equal or better.
- **Performance**: This word, and the one above, have much the same meaning when used in the NCC. A 'performance' standard, or section of a standard, is one that offers clear aims and objectives that must be achieved to be acceptable – but does not prescribe the solution; i.e. like the NCC it allows you to come up with your own innovations. Performance standards can sometimes give reference to other standards that may give a definitive solution, acting much like a DTS provision does for the NCC. On occasions this ***Normative*** standard is the only allowable solution – this will be clearly stated in both standards. **Note**: Normative standards do not have to be prescriptive (see below), they too can be written as performance standards.

- **Prescriptive:** This type of standard, or section of a standard, is one that outlines exactly what must be done to be acceptable and how to do it. You have no options.
- **Must, Shall, Should:** These words are very carefully chosen by those authoring a standard. ***Must*** *is never used* in a standard, but is in legislation, specifications and throughout the NCC. ***Must*** is deemed to equate with ***Shall*** which *is used* in standards and means just that – you are required to do something or else you are deemed to be outside the legislation or performance aims of the standard. ***Shall*** is a ***Normative*** and ***Prescriptive*** term. ***Should***, on the other hand, is an ***Informative*** word – it fronts a suggestion or recommendation given in advisory notes or informative appendices. Such suggestions or recommendations are not a requirement; i.e. you do not have to follow them.

CASE STUDY

The bedroom window

Looking back at our bedroom window example, clause 3.9.2.6, Protection of openable windows – bedrooms (NCC 2019, Vol. 2, p. 327) is prescriptive in that it states clearly that windows '***must***' be protected, and then clearly describes how and in what manner. This clause is complete in itself. Part 3.6 Glazing (of the same volume), however, makes multiple references to AS 2047 as one means of satisfying the performance requirements. Below, we will look at just one clause of this standard.

2.3.1.2 WINDOW RATINGS, AS 2047, P. 13

The following shall apply:

(a) Window assemblies shall be rated in accordance with Table 2.1 and by achieving the test results detailed in clauses 2.3.1.3 to 2.3.1.7 to the appropriate test in AS/NZS 4420.1.

(b) The supporting members shall have sufficient thickness of materials and strength to ensure the performance of the window-operating hardware.

Windows and doors with 25% or more of the width of a single panel or pane within 1200 mm of the building edge shall be classified as corner windows and doors.

NOTE: Appendix C nominates window ratings or design wind pressures for each window and door assembly.

Clause 3.6.0 of Volume Two of the NCC, on the other hand, seems prescriptive in stating that windows '… designed and constructed in accordance with AS 2047 …' satisfy Performance Requirements P2.1.1 and P2.2.2. However, this simply provides one path of compliance, it does not preclude other options that you come up with yourself. As such, it remains a performance-based clause.

The standard AS 2047, though, is normative. This is because if it is to be used to satisfy the Performance Requirements it must followed exactly. Yet within the standard there are both informative and normative clauses. Take for example, Appendix C, shown below:

APPENDIX C

NOMINATION OF WINDOW RATINGS AND DESIGN WIND PRESURE

(Informative)

Nomination of window ratings or design wind pressures for each window and door assembly should be as follows:

(a) For housing, the purchaser should nominate—
  (i) the window rating:
  (ii) the window exposure classifications; and
  (iii) whether the window is a corner window when ordering the window assemblies.

(b) For other residential buildings, the purchaser should nominate the design wind pressures when ordering the windows assemblies.

(c) For commercial buildings, the purchaser should nominate the design wind pressures for the window asssemblies when ordering the windows.

(d) The manufacturer of the window assemblies should verify the window assemblies meet the window rating or design wind pressures as provided by the purchaser.

This is an ***informative*** appendix in both title and wording. Note the use of the word ***should*** rather than ***shall*** in all instances. ***Shall*** is never used in any informative part of a standard.

Our example window has been specified as being of timber. AS 2047, Section 2, Performance, provides a clause for generic timber windows (2.2.2, p. 12). This clause is added for the express purpose of enabling small joineries to construct windows without resorting to expensive testing in each individual case. The clause states:

> **2.2.2 Generic timber windows**
> Generic timber windows shall comply with the requirements of Appendix F.

Appendix F begins at page 48, and continues with highly descriptive detail to page 64. Part of the first page is given below:

APPENDIX F

GENERIC TIMBER WINDOWS

(Normative)

F1 GENERAL

The generic timber designs have been developed to provide a compliance path for small timber joineries that cannot justify the expense of proprietary testing.

The generic designs, referenced as an acceptable alternative, have been tested by CSIRO as being capable of achieving an ultimate load of 1000 pa for windows, an ultimate load of 700 pa for sliding doors, and a water resistance capacity of 150 pa. The results are based on engineering calculations and laboratory testing, and

>>

>> assume that the profiles nominated are assembled in accordance with accepted trade practices, but cannot exempt a manufacturer's responsibilty to produce products that are fit for the purpose for which they are intended.

It may be seen from the title that this is a normative appendix. If you choose to follow it, then you must follow it exactly. You cannot deviate or bring in alternatives to any element. But note – you don't have to follow it at all if you don't want to: you can build a timber window and then test and prove that it satisfies AS 2047 yourself. The appendix is simply provided so you don't have to go to such extremes for one window of a generic (typical or common) nature.

LEARNING TASK 1.2 ACCESS AND INTERPRETING THE NCC AND STANDARDS INFORMATION

Access the NCC 2019, Vol. One & Vol. Two, and using the terminology used in the NCC, provide answers to the following questions.

1 Where in NCC 2019, Vol. Two, can the 'Referenced documents' be found?

2 a What does the term 'mezzanine' mean in building terms?
  b Where in the NCC, Vol. Two, did you find this information?

3 From the NCC 2019, Vol. Two, determine which clause/s relate to:
  a the Performance Requirement/s that relate to room heights
  b the Verification Method/s that are available in determining room height requirements
  c the deemed-to-satisfy options: specify the following minimum room heights nominated in the NCC 2019 and compare these to those provided in the NCC 2016:
    (i) kitchen
    (ii) internal stairway
    (iii) bathroom
    (iv) basement used occasionally as a games room
    (v) stairway leading from one level to another.

4 You have a luxury bathroom with a vanity with two basins, an 1800 x 700 mm spa bath and large shower that can fit two people. An exhaust fan and heater are also available for use in the room. Look at the areas named below and provide the NCC reasoning as to what this room is considered to be in terms of NCC definitions:
  a a wet area
  b a sanitary compartment.

5 Where in the NCC 2019, Vol. Two, can information about fire-resistance levels (FRLs) be found?

6 Where can information and the requirements for Fire-resisting construction be found in the NCC?

Review design solutions for effectiveness and compliance

When analysing a solution – or more typically, a range of solutions – to any element of a construction project, there are a number of factors you need to take into account. One obvious task is to identify a range of alternatives in the first instance: this will include both DTS and performance-based solutions. Another task is to determine the assessment methods and associated documentation by which a solution is deemed to satisfy the NCC Performance Requirements. What we will look at first, however, is the broader criteria by which you will ensure your chosen construction methods comply with the NCC.

Performance Requirements: the compliance criteria

As stated previously, to comply with the NCC, you must satisfy the Performance Requirements. Compliance with the PRs is outlined in Part A2 of Volume One and likewise in Part A2.0 of Volume Two. As discussed earlier in this chapter, there are two paths open to you; and these are similarly described in both cases. The extract below is taken from Volume Two (p. 164); and assessing compliance through these two paths will be discussed later in this section. For now, it suffices to know that in both cases they are assessed by that concept in both volumes known as Assessment Methods.

A2.2 Performance Solution

(1) A Performance Solution is achieved by demonstrating—
  (a) compliance with all relevant Performance Requirements; or
  (b) the solution is at least equivalent to the Deemed-to-Satisfy Provisions.

(2) A *Performance Solution* must be shown to comply with the relevant *Performance Requirements* through one or a combination of the following *Assessment Methods*:
  (a) Evidence of suitability in accordance with Part A5 that shows the use of a material, product, plumbing and drainage product, form of construction or design meets the relevant Performance Requirements.

(b) A Verification Methods including the following:

(i) The Verification Methods provided in the NCC.

(ii) Other Verification Methods accepted by the appropriate authority that show compliance with the relevant Performance Requirements.

(c) Expert Judgement

(d) Comparison with the Deemed-to-Satisfy Provisions

From the previous sections you should have a reasonable understanding of Performance Requirements and the shift of the NCC volumes from a prescriptive approach to being performance based. In undertaking this shift, the Performance Requirements had to be written in one of two ways: qualitatively or quantitatively.

Quantitative Performance Requirements

The word 'quantitative' means you are dealing with quantities: numbers, measures, weights and the like. You are dealing in the measurable. Very few PRs will be stated in this manner, and most of these will still be identifiably 'non-prescriptive' despite any numbers that may be in the text. For example:

P2.2.1 Rainwater management

a) *Surface water*, resulting from a storm having an *average recurrence interval* of 20 years and which is collected or concentrated by a building or *sitework*, must be disposed of in a way that avoids the likelihood of damage or nuisance to any *other property*.

As may be seen, this is still not prescriptive in that it does not state how you are to achieve the goal. It is quantitative only in that it describes a volume of water derived from a specific 20-year interval. It cannot give the volume as it will change with location. However, by looking up the definition of 'average recurrence interval' you find it invokes rainfall intensity which, in turn, leads you to **Tables 3.5.3.1a to h** (NCC, Volume Two, pp. 200–202) from which you can derive the necessary quantity of rain expected from a 1-in-20 year storm over a five-minute period – in Albury that's about 12 mm (139 mm ÷ [60 min ÷ 5 min]). Taking, as an example, a surface area of 300 m^2, that means 3.6 m^3 or 3600 litres – in five minutes.

The NCC avoids using quantity values in PRs because it is hard to ensure such values cover all the various building types and design options; i.e. quantitative language can tend to be more prescriptive, and therefore restrictive of innovation, despite the intended aim of the PR.

On the other hand, there are times when it is the best, or only, way to specify clearly the required outcome, as shown in the above example.

Qualitative Performance Requirements

These form the majority of PRs in all the NCC volumes. Qualitative means that it is the quality, not the quantity, of something that is being considered. PRs defined in this manner are seeking compliance with a concept rather than a defined value. Sticking with the NCC, Volume Two, the Performance Requirement P2.4.2 Room heights is a typical qualitative statement:

P2.4.2 Room heights

A room or space must be of a height that does not unduly interfere with its intended function.

The requirement doesn't specify a particular height requirement for any room, since this may vary widely depending upon the nature of the building. The appropriate height is therefore determined by whether or not it 'unduly interferes with its intended function' – which, you will note, is a more subjective criterion.

Negotiating qualitative language

Qualitative language can at times seem vague in its expression. There are two phrases in particular which you need to understand that you will come upon frequently, particularly in the NCC, Volume One. These are 'To the degree necessary' and 'Appropriate to'.

- **To the degree necessary**: as the phrase suggests, you only need to carry through the specified Performance Requirement as necessity dictates. At times, this may mean you don't have to do anything at all. It is totally dependent upon context and circumstance.
- **Appropriate to**: this phrase will generally be followed by a list of criteria which must be considered when developing or assessing a Performance Solution.

Both these phrases are exampled in Part F2 Sanitary and other facilities, page 243 of the NCC, Volume One:

FP2.1 Personal hygiene facilities

Suitable sanitary facilities for personal hygiene must be provided in a convenient location within or associated with a building, to the degree necessary, appropriate to –

(a) The function or use of the building; and

(b) The number and gender of the occupants; and

(c) The disability or other particular needs of the occupants

The use of the phrases '***to the degree necessary***' and '***appropriate to***' clearly help to make evident what is actually required to achieve this PR. Sanitary facilities, be they toilets, showers, hand basins or the like, must be provided to the '***degree necessary***'. Which may mean that on your particular building (e.g. a Class 7a carpark with space for only four vehicles) none are needed at all. '***Appropriate to***' lists the function and use along with other factors, all of which may be used in supporting this assessment for your carpark project.

PART 1

1

CASE STUDY

The bedroom window

The Performance Requirement for this building element and project has already been identified and discussed earlier in this chapter, but is repeated here for convenience. The Performance Requirement in this case is a qualitative requirement:

P2.4.4 LIGHTING

a) A *habitable room* must be provided with *windows*, where appropriate to the function or use of that part of the building, so that natural light, when available, provides an *average daylight factor* of not less than 2%.

You will notice firstly the phrase 'where appropriate' has been used, but without an apparent listing. However, it is followed by the words 'function' and 'use', which serve the same purpose; that is, you must ensure that the level of natural light available to the room is adequate for how the room is intended to be used.

Your next question is, of course, how to determine what is 'appropriate', and by what criteria this is determined. The first thing to check is if there are any associated Verification Methods that might offer some insights. In this NCC Volume Two example, a Verification Method for Part 2.4.4(a) is promoted on the following page; i.e. V2.4.4 as found on pages 60 & 61.

Moving to the NCC, Volume One, and looking at this same issue (provision of natural light) you will note that the relevant clause from Part F4 Light and Ventilation is very similar:

FP4.1

Sufficient openings must be provided and distributed in a building, appropriate to the function or use of that part of the building so that natural light, when available, provides an average daylight factor of not less than 2%.

As with the NCC, Volume Two, the Verification Methods closely follow the Performance Requirements – in this case beginning on page p. 268. Again, there is nothing in this section that covers provision of natural light. However, at this point the difference between Volumes One and Two is marked. The DTS provisions for a given section of Volume One immediately follow the Performance Requirement and Verification clauses. As with the Class 1a bedroom window example above, it is to these that you would need to turn for guidance on what constitutes an 'appropriate' level of light or illuminance.

Determining appropriate solutions: DTS options

So far, we have used a bedroom window to example the determining of a solution for a chosen building element. We have also noted that on occasions the DTS solutions may not be applicable due to some issue of your particular context – in the window example it was the distance to the boundary. We will look more closely at what are called Performance Solutions to that problem in the next section. In this section we will look at those areas of the DTS provisions that provide multiple options.

CASE STUDY

The bedroom window

In the discussion above, we noted that AS 2047 is the acceptable construction manual for the construction of our bedroom window. However, it should be noted that there are alternatives within the Glazing DTS provisions themselves that you could look at: Clause 3.6.0 (b) offers AS 1288 as an alternative manual, but unfortunately the clause excludes the type of window being used in the example. So, in this case it will require being satisfied by the Performance Solution path.

In other elements there are sometimes some genuine alternatives portrayed. Framing, for example, is described in timber or steel; likewise, the subfloor may be approached in a number of different ways that are all recognised as valid by the NCC. Timber framing, through the recognition of AS 1684 as the acceptable construction manual, acknowledges multiple paths by which bracing or hold down might be achieved, aside from the framing itself. But these are rarities. Generally, most DTS provisions will be case specific; i.e. it will give you one example that reflects your particular building element, after which, if you cannot comply exactly, you must develop your own Performance Solution. Your task, as the builder, is therefore to 'collect' a range of DTS solutions such that you have one to cover all the various elements of your particular construction. You also need to evaluate these for appropriateness to your company, work team skill base, client requirements and context – including access to material and product supply.

In practice, all of the above becomes part of the specifications associated with the building plans (see Chapter 5). The example below is drawn from a typical Class 2–9 specification, in this case the first subclause of the work section covering door and window seals.

1.1 STANDARDS

Seals general

- Quality management for manufacture: To ISO 9001.
- Acoustic applications: To AS 1191 or AS/NZS ISO 717.1.
- The NCC cites ISO 717-1:1996 and AS/NZS 1276.1 for testing of construction required to have a certain R_w rating.

>>

>>

- Fire door assemblies: To AS 1530.4 and AS 1905.1.
- Smoke door assemblies: To ***NCC, Vol. One Spec C3.4***, AS 1530.7 and AS 3959 for silicon flame retardant PVC and TPE weather seals with a Flammability Index not more than 5 to AS 1530.2 providing BAL 40.
- Combined fire and smoke door assemblies: To ***NCC, Vol. One Spec C3.4***, AS 1530.4, AS 1905.1, AS 1530.7 and AS 3959 for weather seals providing BAL 40.
- Weather and energy saving seals for proprietary windows and door assemblies: To AS/NZS 4420.1 clause 5 and clause 6, and AS 2047.
- Door bottom and perimeter seals for glazed external doors: To AS 2047.
- Threshold plates: To AS 1428.1 Threshold plates.

Note in the above the references to an NCC specification (bold italics). Specification C3.4 (*not* Sub Part C3.4) is a DTS provision found on pages 150–1 of the NCC, Volume One. In addition, the specifier has brought in a number of relevant Australian standards such as AS 3959 Construction of buildings in bushfire-prone areas. Even the quality of manufacture of the seals is governed by an international standard – ISO 9001; this limits procurement (purchase and supply) of products to those of an acceptable quality. In turn, this limits the likelihood of product failure and hence expensive call backs.

Performance Solutions

Where DTS solutions are not suitable, or you choose not to use them, the second pathway to compliance is to come up with a Performance Solution. Performance Solutions, previously known as 'alternative' solutions, are those that are not documented within the NCC, but have been developed by the builder, designer, or product or material supplier, and have been shown to meet the Performance Requirements of the NCC.

Performance Solutions are not just used when the DTS solution isn't available: it may be that you, your client, the designer and/or the engineer have developed a new approach that is more cost effective, or simply better, stronger or quicker.

The development of a Performance Solution is a four-stage process that requires collaboration on the part of a project's key stakeholders. The four stages are:

- Step 1: preparation of a performance-based design brief (PBDB)
- Step 2: Analysis, modelling and/or testing
- Step 3: Collation and evaluation of results by stakeholders (the initial assessment report)
- Step 4: Preparation of the final report.

The exact format of the documents will depend greatly upon your company's standardised systems; however, the content of each of these phases of the process should reflect the steps outlined below.

CASE STUDY

The bedroom window

In the bedroom window example, the issue was the reduced distance from window to boundary. In this case you have to explore alternatives simply because there is no DTS provision that matches your needs.

Your options could include:

- a larger window based upon the amount of light lost by being 150 mm closer to the fence than the DTS approach is configured to
- inclusion of a roof light (skylight/window), that also increases the volume of light proportionately
- reflective boundary panelling
- argument that the fencing is below window sill height and that the other property can't be built close to the fence at that section due to an easement of some kind
- argument that the window faces a road, path or some other area that can't be built upon.

These are just suggestions, and they would need to be ratified by the appropriate building authority or building certifier. However, you have options to explore – you are not faced immediately with a dead end. You are allowed to be innovative – provided that the *purpose* of the PR is met.

As with the DTS solutions, you would still need to evaluate these for appropriateness to your company's work practices. You then need to document your proposed approach so that it can be considered for approval by the relevant building authority.

Step 1: The performance-based design brief (PBDB)

The preparation of a PBDB is a collaborative affair between key stakeholders of the project. Its purpose is to document the agreed parameters and desired outcomes of the proposal. A design framed around this brief has a greater chance of approval as the considerations of all parties have already been incorporated. The PBDB should therefore outline:

- the design objectives
- key stakeholders – client, builder, architect/designer, relevant building authority, engineers
- agencies or authorities as relevant (water, environment, health, fire, etc.), and neighbours and the like

- justification for the need for the Performance Solution
- types of evidence required
- the relevant Performance Requirement(s)
- the closest corresponding DTS provision(s)
- agreed criteria for acceptance by stakeholders.

Determining the relevant Performance Requirements can, on occasions, be fairly straightforward. In most cases, however, the task needs much wide-ranging consideration. The NCC has a process, outlined in each of the volumes – Parts A2 of both Volume One and Volume Two – that must be followed. In summary, this requires that you identify:

- the Performance Requirements from the sections or parts to which the proposed Performance Solution applies
- any other Performance Requirements that might affect, or be affected by, the development or end use of the proposed Performance Solution.

In some instances, the Performance Requirements are met through a combination of the proposed Performance Solution and a DTS solution. In such cases the above is expanded to include the relevant DTS provisions, sections and parts.

Step 2: Analysis, modelling and/or testing

Having developed a design based upon this PBDB, it then needs to be tested, analysed and/or modelled to ensure that it satisfies the Performance Requirements. The Assessment Methods will be discussed in the next section.

It is generally only after developing a design that an appropriate assessment method can be settled upon, though sometimes this can be envisaged earlier and included in the design brief.

Step 3: Initial assessment report

Once the suggested Performance Solution has been tested or analysed, an initial assessment report is then prepared and submitted for review by the various stakeholders. This may require amendments to the design based upon feedback. Once stakeholder approval has been gained, a final report can then be submitted to the relevant building authority.

Step 4: Final report

The final report should show clear evidence that the Performance Solution complies with or exceeds the NCC Performance Requirement(s) identified in the original performance-based design brief.

How the documentation will ultimately look will depend upon your company policy and formats. It may also vary between appropriate building authorities. It will definitely vary with the perceived consequences of a solutions failure.

The typical documents included in the final report are:

- The original performance-based design brief, including:
  - the Performance Solution sought
  - key stakeholders
  - relevant Performance Requirements and DTS provisions
- Evidence documenting (as relevant):
  - the intended function of the room
  - a plan showing the location building and the issues involved
  - a plan of the building detailing the element being considered
- Outline of methods of assessment used, which may include one or a combination of:
  - evidence of suitability
  - Verification Methods employed
  - expert judgement where used
  - comparison with DTS provisions
- Evidence related to the evidence, certification and credentials of any third parties who contributed to the report or testing, including:
  - a registered testing authority's report
  - certificates of conformity and current certificates of accreditation
  - certification from appropriately qualified persons
  - certification from national, international or co-national accreditation bodies
  - an evaluation of the results
- The conclusions:
  - confirmation of key stakeholder approval of design
  - design limitations and conditions.

Ratification of the Performance Solution

Generally, the Performance Solution documentation outlined above will be developed in the design stage of a project, and submitted to the relevant building authority as part of the building application process. However, there will be times when a DTS solution, approved in the application, cannot be complied with due to unforeseen circumstances. In these cases, you must repeat the same procedure to develop and document an alternative Performance Solution that is then ratified by the relevant building authority.

Likewise, there may be cases where you intended to use a DTS solution, but discover during the building project that the DTS provision has not been strictly adhered to after a particular building element has been constructed. In this case, the relevant building authority must review the work both through inspection and through documentation of the as built structure/element. In so doing, a case shall be made that either:

- the 'as built' element is demonstrated to be equal to, or better than, the DTS solution
- that remedial works will need to be conducted.

Either of these scenarios will normally involve an application for a variation to the existing permit.

Performance Solutions: methods of assessing compliance

The previous section outlined the path by which a performance-based solution is developed and presented for approval. We will now look at the ways Performance Solutions are assessed to determine whether or not they comply with the Performance Requirements of the NCC.

The NCC briefly outlines four avenues from which evidence might be obtained so that the proposal can be assessed for compliance. These include:

- evidence of suitability
- verification methods
- expert judgement
- comparison with DTS provisions.

We will now look at each of these in turn.

Evidence of suitability

The Australian Building Codes Board (ABCB), aside from producing the NCC, produces a range of non-mandatory (informative only) handbooks, some of which are directly accessible from their website. One such handbook is dedicated to outlining what constitutes evidence of suitability. Titled *Evidence of Suitability Handbook 2018*, it provides a framework designed to reflect a hierarchy of rigour (Figure 1.7).

The levels of rigour (quality or strictness of evidence) are arranged such that the highest or best is at top, graduating downwards in quality to the bottom; i.e. the lowest quality of evidence. The general approach that is promoted by the ABCB in this document is shown in Table 1.2.

TABLE 1.2 Testing a Performance Solution

Performance Solution context	Level of rigour most likely to ensure approval
New, innovative, or previously untested or tried components Performance Solutions where consequences of failure are high	High
Products with extensive history of sound performance in construction; i.e. low probability of failure	Medium
Components or Performance Solutions where the consequences from a failure are low	Low

The options given in Table 1.2 are not mutually exclusive; i.e. a CodeMark Australia certificate of conformity may be coupled with an expert's report or a product technical statement. Alternatively, a certificate from a certification body could be submitted with reports from both an accredited testing laboratory and that of an expert.

The clauses in Volume One (A2.2) and Volume Two (1.2.2) require that suitability is shown at three levels:

- materials, construction and design
- calculation methods
- documentary evidence.

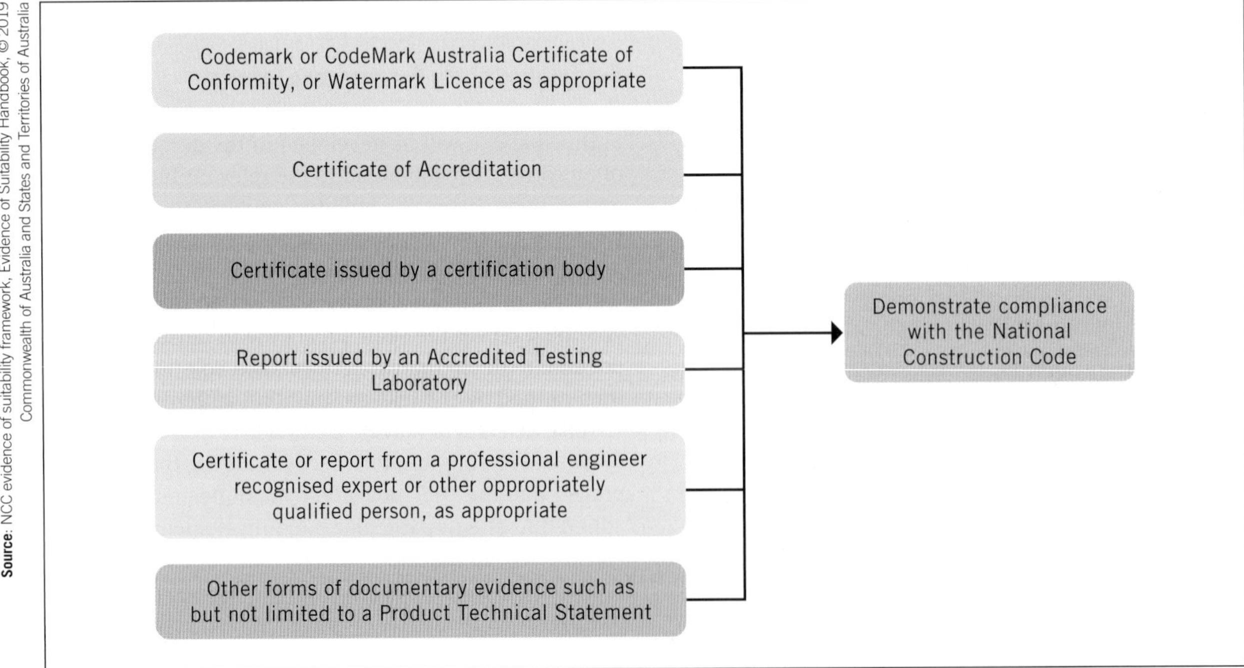

Source: NCC evidence of suitability framework, Evidence of Suitability Handbook, © 2019 Commonwealth of Australia and States and Territories of Australia

FIGURE 1.7 Hierarchy of rigour

This means that you must be able to demonstrate that your evidence:

- is of the appropriate rigour via the appropriate quality of documentation
- covers the materials and construction and design approach
- uses an ABCB protocol compliant calculation method.

This latter point is covered by what is known as Verification Methods.

Verification Methods

The second method of proving that a Performance Solution meets the Performance Requirements of the NCC is via a '**Verification Method**' which is a '...test, inspection, calculation or other method that determines whether a *Performance Solution* complies with the relevant *Performance Requirements*' (NCC, Vol. Two, p. 519).

Verification Methods are to be found in each of the NCC volumes, as follows. In Volume One they typically follow immediately after the Performance Requirements, which may be listed collectively for the whole Section, or divided up by Part.

In previous editions of the NCC, not all sections included Verification Methods. However, this has been rectified in the main with the 2019 publication. When a Verification Method is not offered you must rely upon the remaining strategies of comparison to the DTS provisions, expert judgement and evidence of suitability.

What a Verification Method actually looks like is highly variable. They can be extensive formulas, wordy descriptions, or written descriptions coupled with diagrams. For example, NCC, Volume One and NCC, Volume Two describe similar testing procedures in their Verification Methods for wire barriers: under Section D Access and Egress (DV1, p. 123) in Volume One, and Part 2.5 Safe Movement and Access (V2.5.1, p. 63) in Volume Two. In both cases the description spans three pages. The diagram of the testing equipment, though not overly complex, is not easily interpreted without reading the associated text and is shown below in Figure 1.8.

Even this relatively simple procedure – testing the possible deflection in a wire balustrade – demonstrates that being equipped for establishing any given Performance Requirement by means of a Verification Method is not for everyone.

The ABCB produce two tables (one for each NCC volume) that provide a link between the NCC Performance Requirements, the NCC Verification Methods and the NCC DTS provisions. These tables are available on the document 'Developing a Solution Using the NCC' which can be accessed at: https://www.abcb.gov.au/Resources/Publications/Education-Training/Developing-a-Solution-using-the-NCC.

This is the easiest way to quickly identify when and if there is a Verification Method applicable to the particular section of the NCC for which you are required to produce a Performance Solution. In addition, there are a range of 'Supporting Documents',

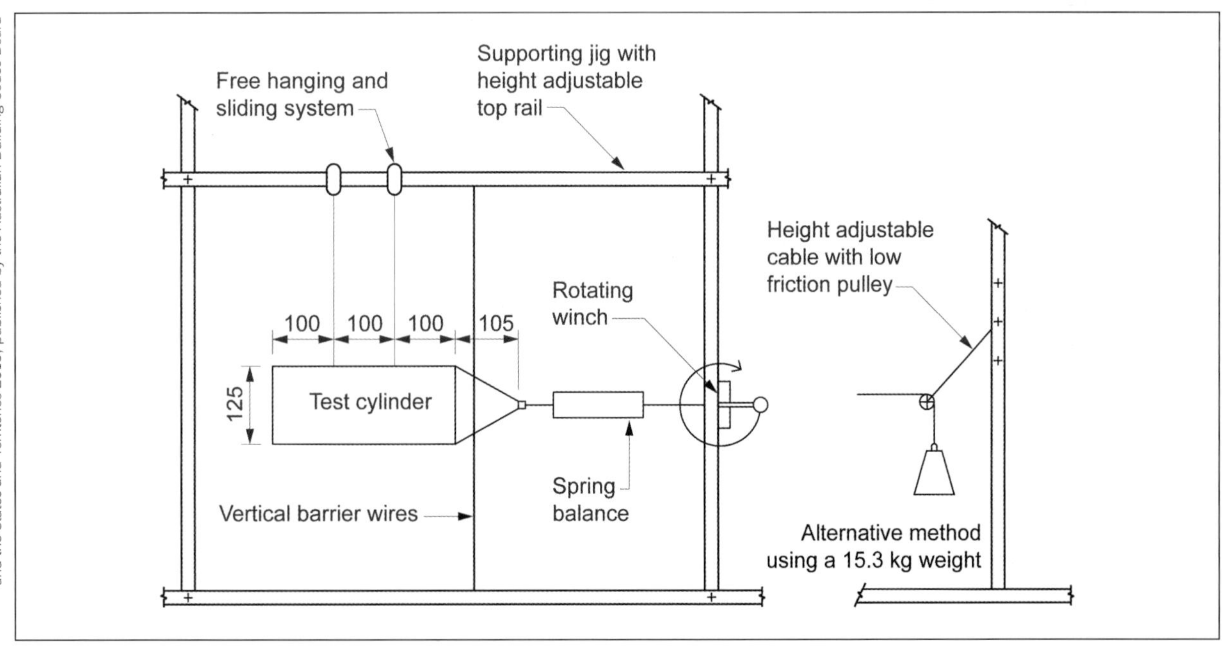

FIGURE 1.8 Apparatus for testing wire balustrades

again produced by the ABCB, that expand upon specific Verification Methods (not all) and how best to employ them as part of a Performance Solution application. These too may be found by searching the ABCB website.

As shown above, Verification Methods can be quite complex in nature, providing strict test procedures, tables, engineering formulas and the like. This is why, when embarking upon the development of a performance-based design brief you must ensure you have access to the right professional advice. It is also why expert judgement is almost always part of the final report of any successful Performance Solution.

Expert judgement

Of the four paths by which a Performance Solution may be reached, this is the most vague in its definition within the NCC, since the definition of an 'expert' is quite subjective. It is also often the most important. According to the NCC definitions, expert judgement is derived from someone who holds '*... the qualifications and experience to determine ...*' if a Performance Solution, or a part of such a solution, complies with the Performance Requirements. Ultimately, this often means that it falls upon the relevant building authority to accept, or not, the background of any given individual as being sufficient to make the determination. They, in making this judgement, will be looking not just at an individual's history and credentials, but also for peer recognition; i.e. if those who also work in that field hold that person to be an expert. Given this subjectivity in even determining who is an expert, it is not often that expert judgement will suffice as the only evidence in matters where the consequences of failure are high.

On the other hand, when the consequences of failure are high, it is expert opinion that will be critical to counterbalance what may be over-enthusiasm for a formula – as most experienced builders will attest, the engineering may say one thing, whereas practice frequently suggests another. It is therefore a combination of sound, experience-based advice and quality engineering that provides the most reliable solution.

Despite the above, expert judgement is frequently called for when a Performance Solution cannot be quantifiably benchmarked; i.e. it is not something that can be easily tested through computer modelling or engineering calculation. It is effectively a subjective opinion based upon the personal history of an individual, or preferably a team of individuals, coupled generally with an extensive review of literature and construction history, regionally, nationally and internationally.

Expert judgement is best used in an application for a Performance Solution in the following manner:

- in initially developing the proposed solution and the performance-based design brief
- when identifying appropriate materials or products as part of the proposal
- as guidance in the selection and application of Verification Methods
- when testing through calculation or computer-based modelling is not possible
- in comparing the proposed Performance Solution to the DTS provisions
- in relating the proposed solution to the history of construction practice
- when determining the consequences of failure of the solution to structure, occupants, environment, context (other structures and people), and related infrastructure
- in discussions with the appropriate building authority
- in preparing the final report.

Used in this manner, expert judgement becomes invaluable.

Comparison with DTS provisions

This is by far the simplest route by which a Performance Solution proposal may be demonstrated to comply with the Performance Requirements. A DTS, by definition (as well as name), has already been shown to comply with the requirements. Expert judgement (outlined above) is generally the means by which this comparison is made. On occasions, when deviation from the DTS is minor, that expert judgement will be made simply from within the appropriate building authority. On other occasions it may require more extensive input to demonstrate the correlation.

A DTS comparison is undertaken by effectively subjecting both the DTS provision and the Performance Solution to the same level of analysis, thereby testing them both using the same methods. This effectively provides a benchmark, or set of data, to which the proposed solution may be compared. On occasions this can still be an elaborate process using computer-based modelling systems. As with the previously listed mechanisms used to determine compliance of a Performance Solution, the level, depth, extent and cost of analysis will tend to reflect the consequences should the solution fail in practice. That is, extreme risks to structure, occupants, environment, context (other structures and people) and related infrastructure will require a significantly higher level of analysis. On the other hand, our bedroom window example, with the only risk being a slight reduction in available light to a room, means a fairly low level of proof would be required.

CASE STUDY

The bedroom window

The parameters for this example have been given so far as simply:

- Class 1a detached house
- a bedroom
- a window seat arrangement
- distance from widow to the boundary of 750 mm

To this we shall add:

- dimensions of the bedroom as being 4500 mm × 5000 mm, with the long side (5.0 m) running along the external wall
- ceiling height shall be taken as 2700 mm
- proposed window is 1500 mm × 3200 mm; sill height 1000 mm above finished floor level
- that the house has been constructed on stumps and bearers. The finished floor level of this room to ground level below the window is 600 mm
- that the boundary adjoins a wide public easement running parallel and sloping down away from the house to creek frontage
- the fence on this boundary is 2000 mm high.

Assuming the performance-based design brief has already been completed, a DTS comparison for the bedroom window would include the following:

- The applicable Performance Requirement, in this case:
  - P2.4.4 Lighting (NCC, Vol. Two, p. 58)
  - P2.4.5 Ventilation (NCC, Vol. Two, pp. 58–9)
  - P2.3.1 Spread of fire (NCC, Vol. Two, p. 54).
- The applicable DTS provisions, in this case:
  - 3.8.4 Light (NCC, Vol. Two, p. 283)
  - 3.8.5 Ventilation requirements (NCC, Vol. Two, pp. 287)
  - 3.7.2 Fire separation of external walls (NCC, Vol. Two, p. 233).
- Calculations of floor area and resultant window and ventilation areas based upon DTS provision clause 3.8.4.2 & 3.8.5.2 that require 10% of the floor area as light, and 5% of the same area as ventilation, in this case:
  0.1×4.5 m $\times 5.0$ m $= 2.25$ m^2 unobstructed glazed area
  0.5×2.25 m$^2 = 1.125$ m^2 ventilation area (i.e. ½ window area).

From the above, it can be seen, firstly, that a window of the required size will easily fit into the available wall area of the bedroom in normal circumstances. Likewise, that the proposed window of nominal size 1.5 m × 3.2 m already offers twice the required light, even when allowance is made for the frame. Taking the window seat arrangement into account, meaning that the glazed area would not begin until approximately 1.0 m off the floor, this still leaves more than adequate space for creating a window of the necessary size.

What can also be seen, however, is that there is more than one Performance Requirement that will need to be met, and others that may need to be considered. P2.3.1, for example, needs consideration because the window effectively pushes the external wall at this point closer than 900 mm to the boundary. Normally this would mean the window would have to be non-openable with a fire-resistance level (FRL) of not less than -/60/- (meaning it must be able to resist flames and gases from entering the building for 60 minutes. See 'Fire protection', p. 26 below). However, the first part of the relevant DTS provision reads as follows:

> **3.7.2.2 External walls of Class 1 Buildings**
> An external wall of a Class 1 building, and any openings in that wall, must comply with 3.7.2.4 if the wall is less than –
> (a) 900 mm from an allotment boundary other than the boundary adjoining a road alignment or other public space

As the wall in question faces a public space (public creek easement) this clause needs no further consideration.

The proposed size of the window can now be approached, and the arguments laid out. Included in your arguments will be:

- The window sill will be approximately 1600 mm from the external ground line.
- The fence height, at 2000 mm means that available light is only being restricted by approximately 400 mm of the lower portion of the window.
- That the size of the window will provide more than twice the required light under the relevant DTS.
- The reduced distance to the boundary is only one-quarter or 25% of the DTS stated distance.
- That being bounded by a public easement that cannot be built over means that no future reductions in light are possible (excepting trees or bushes – which are not considered in the provisions).

This, in itself would probably suffice, though some level of expert judgement would strengthen the argument.

While the above may (dependent upon approval by the appropriate building authority) cover the size of the window with regards to the distance to boundary, the actual design of the window would need to be discussed from other aspects. These would include the window seat arrangement and hence the probability of human impact on the glass – clause 3.6.4 Human impact safety requirements; and possibly clause 3.9.2.6 Protection of openable windows (just to dismiss it).

The proposed size of the window would also have to be considered from an energy efficiency perspective, taking into account clause 3.12.2.1 External glazing, and clause 3.12.2.2 Shading. This would be necessary to ensure that the proposed window size would not need to be reduced in order to comply with the NCC Glazing calculator (see Chapter 14 on energy efficiency).

Documenting the solution

How the documentation will ultimately look will depend upon your company policy and formats. It may also vary from between appropriate building authorities. It will definitely vary with the perceived consequences of a solutions failure. The content will reflect that which has been outlined in the sections above, particularly Performance Solutions (p. 20). The key points regarding the final report are reiterated here for convenience.

A final report must show clear evidence that the Performance Solution matches or exceeds the Performance Requirements. Hence it will include:

- The performance-based design brief covering:
  - the Performance Solution sought
  - key stakeholders
  - relevant Performance Requirements and DTS provisions
- Evidence documenting (as relevant):
  - the intended function of the room (a bedroom in this case)
  - a plan showing the location building and the issues involved
  - a plan of the building detailing the element being considered
  - an outline of the assessment methods, analysis, modelling or testing
  - a registered testing authority's report
  - certificates of conformity and current certificates of accreditation
  - certification from appropriately qualified persons
  - certification from national, international or co-national accreditation bodies
  - an evaluation of the results
  - evidence of suitability
  - Verification Methods employed
  - expert judgement where used
  - comparison with DTS provisions
- The conclusions
  - confirmation of key stakeholder approval of design
  - design limitations and conditions.

The final report for the bedroom example would include, therefore, all of the details discussed in the previous section of the chapter. Each of the Performance Requirements would be described in full. Of the clauses, only those immediately relevant would be copied out fully, the others would be mentioned by their clause or part numbers only. You would, however, need to include the window part of the energy audit (see Chapter 14) that has been conducted on the plans, to satisfy the appropriate building authority that the window size would not need to be reduced due to excessive solar gain.

LEARNING TASK 1.3

ASSESSING COMPLIANCE AND PROPOSING DESIGN SOLUTIONS

1. Stairs: you have a set of plans that show the information below. Using the NCC, Vol. Two, Table 3.9.1.1, ensure there is enough space to have a straight flight of stairs fit into a space shown as:
   - a 2850 mm ground floor level (GFL) to first floor level (FL1)
   - b 3.42 m long total going distance is available at GFL
   - c 1.0 m hall width (between walls) is indicated where stair is to be located.
2. Show your method to calculate the required space for the proposed stairs, so as to comply with any NCC or Australian standards requirements.
3. Will there be enough space to fit this straight flight of stairs into the space provided?
4. What two (2) solutions can you offer the client, should the stairs not fit in the space currently shown on the plans?

Fire protection

Fire in a building presents a significant risk to life and property – including neighbouring buildings, the environment and infrastructure. For this reason, fire prevention, containment and suppression systems form a significant part of the NCC. Irrespective of which volume is being considered, the two primary requirements of the NCC are:

1. To facilitate safety of the building's occupants (those inside the building at the time of the fire)
2. Minimise the spread of fire within the building, and harm to life and property external to the building.

To this end, NCC, Volume One provides a number of Performance Requirements (additional to the Governing Requirements of Section A) and DTS provisions spread among its various sections – particularly Sections B, C, D, E and G, and to a lesser extent H. Volume Two holds two Performance Requirements in Part 2.3 and four more in Part 2.7. These cover:

- protection from spread of fire
- fire detection and early warning
- heating appliances
- construction in bushfire areas
- private bushfire shelters
- construction in alpine areas.

In addition to the above, the NCC describes fire-resistance levels or FRLs for specific elements of a building (see 'Fire-resistance level', p. 28). The role of designers and builders is to determine these required levels of fire resistance and ensure that the various elements identified for both passive and active fire safety measures are in place. In combination, these measures constitute what is known as a fire safety system. Such a system is designed to:

- warn people of an emergency; and/or
- provide for safe evacuation; and/or
- restrict the spread of fire; and/or
- extinguish a fire.

We'll begin this section by looking at what constitutes these measures, how to identify them and how to apply them.

Passive and active fire safety measures

When thinking of fire control in buildings, most people will envisage fire extinguishers, ceiling mounted sprinklers, fire hose reels, smoke detectors and the like. These elements are actually only one part of the equation when dealing with fires in a building – that part referred to as active fire safety measures. They are important, as they help to detect a fire in the first instance, while also being our first resource in attempting to supress a fire before it gains dangerous proportions.

Just as important, and in many ways more so, are the unseen elements of a building that act to stop the spread of fire: to contain it to within the area in which it began. These elements are known as passive fire safety measures.

Passive fire safety measures

Passive fire safety measures mean those that do not act, react, or initiate any action; i.e. they are static. They form part of the building and/or the design of the building, and do not require electrical or mechanical intervention to perform their function. This form of measure includes:

- elements of the building fabric
- compartmentation – dividing large buildings such that one part can be isolated from another
- open spaces between buildings and potential sources of fire.

The elements of a building's fabric included in a fire safety system will typically have one or more of the following capacities:

- to maintain their stability and loadbearing roles for a nominated period of time
- are non-combustible and can resist the passage of flames and hot gases or fumes
- provide some level of insulation such that the temperature on the non-fire side is not raised beyond specified limits.

These reflect the three aspects of the fire-resistance levels (FRLs) to be discussed in greater depth later in this section. The most common building elements to be considered include:

- floors
- walls
- roofs
- stairwells
- service shafts
- doors and windows.

Compartmentation

Compartmentation is a factor of design. It is the considered use of appropriate elements of the building fabric, such as those listed above. The purpose of compartmentation is to separate one or more parts of a building from each other, or the remainder of the building, by barriers to fire, heat and fumes. Any linking doorways or other openings between '***fire compartments***' are equally protected by self-closing doors or similar barriers also capable of resisting the spread of fire.

The NCC requires different levels of compartmentation for different building classes and types of construction. The elements used to make the separation will therefore have an FRL rating applicable to the Performance Requirements for that building class.

Exit travel distances

Though not listed under Section C – Fire resistance, the exit travel distances defined in D1.4 (Section D – Access and Egress, p. 129) are derived from fire studies, and form a significant part of passive fire safety design. They are therefore a critical part of occupant safety and would be reviewed in any assessment of a building's design, pre- or post-construction.

Exit travel distances are based on a historical building code from the 1970s, with recent studies supporting these existing measures by determining how far able-bodied people, both young and elderly, are able to walk in a smoke-filled corridor when visibility is down to 25 m.

The requirement for exit travel distance differs for the various classes of buildings. In addition, it differs within each class depending upon other aspects of a building's design, such as the number of available exits from any given location, what form those exits take, and to where they lead.

At this point you should also be aware of what the term 'exit' means as it pertains to the NCC. An exit, as defined by the NCC, does not necessarily open directly to the outside of the building. To be an exit it need only provide egress – a passage out – to a road or open space. This means an exit can be any one, or a combination of, the following:

- an internal or external stairway
- a ramp
- a fire-isolated passageway
- a doorway opening to a road or open space.

As you are already aware, the two NCC volumes are structured very differently: this holds for how they handle the issue of fire protection. Volume One consolidates the ***passive*** fire safety measures listed above in Section C Fire resistance. Volume Two, on the other hand, spreads both ***passive*** and ***active*** fire safety measures through Part 3.7 Fire safety: the passive clauses being described in Part 3.7.1 Fire properties for materials and construction and, in a more specific sense, in Parts 3.7.2 Fire separation of external walls, through to 3.7.5 Smoke alarms and evacuation lighting. Part 3.10.5 then covers Construction in bushfire-prone areas. The bulk of Volume One's ***active*** fire safety measures are consolidated in Section E Services and equipment and are discussed below.

Active fire safety measures

Active fire safety measures are those that help to supress a fire before it gains dangerous proportions. Active fire safety measures tend to fall into one of two groups: those that are automatically activated, and those that are activated manually. Measures in the automatically activated group include:

- smoke detectors and alarms
- sprinkler systems
- emergency lighting
- air pressurisation systems
- smoke extraction vents
- magnetic door closers.

Those that are activated manually include:

- fire reels and hydrants
- break-glass fire alarms
- fire extinguishers and blankets.

Some classes and types of buildings will also require a Fire Control Centre. This is a dedicated area – or room for buildings over 50 m in height (called a Fire Control Room, with its own special requirements of construction) – for the express purpose of directing fire-fighting operations or other emergency procedures.

In the NCC, Volume One, Section E Services and equipment, the Performance Requirements are set out for each type of active fire safety measure. Included in this section are the DTS provisions, and in some cases Specifications. This is an extensive section occupying page 170 through to page 226 and includes emergency lighting and photoluminescent exit signs. An example of the detail of this clause is offered below; note that it includes reference to Australian standards, as is typical with this document more generally.

E1.3 Fire hydrants

(a) A fire hydrant system must be provided to serve a building –

(i) having a total *floor area* greater than 500 m^2, and

(ii) where a *fire brigade* is –

(A) no more than 50 km from the building as measured along roads; and

(B) equipped with the equipment capable of utilising a fire hydrant.

(b) The fire hydrant system –

(i) must be installed in accordance with AS 2419.1, except –

(A) a Class 8 *electricity network substation* need not comply with clause 4.2 of AS 2419.1 if –

(aa) it cannot be connected to a town main supply; and

(bb) one hour water storage is provided for firefighting ...

The NCC, Volume Two takes a very different approach given that the focus of this volume is on residential buildings. As stated earlier, the Performance Requirements can be found in Part 2.3. Take note in accessing these requirements as many specify critical factors that will need consideration if a Performance Solution is being developed. The DTS provisions are then found in Part 3.7.5 Smoke alarms and Part 3.10.7 Boilers, pressure vessels, heating appliances, chimneys and flues.

Determining required level of fire resistance

The concepts of the ***fire resistance***, ***level of fire resistance*** and ***fire-resistance levels (FRLs)***, can get rather daunting for those faced with the NCC for the first time. Each volume tackles these concepts similarly in terms of definition but, except for FRLs, differ in terms of how they are addressed in each volume. As FRLs are common to both volumes, and they frequent much of our later discussions, we shall begin with these. We'll then look at two other concepts as they pertain to each volume individually.

Fire-resistance level – FRL

The fire-resistance level of a building element is a three-part numerical code such as 60/60/60. It is used to express the capacities of an element to withstand the effects of fire for a specified period of time. An example may be 60/90/40; with each number being a grading in minutes for one of the three criteria listed below, and expressed in the order they are listed:

a) structural adequacy
b) integrity
c) insulation.

For example, an FRL of 60/90/40 means that the building element (wall, floor, door or the like) shall:

- remain structurally adequate – meaning be able to maintain stability and loadbearing capacity, as determined by AS 1530.4, for 60 minutes
- remain sufficiently whole (maintain its integrity) to be able to resist the passage of flames and gases for 90 minutes

- keep the temperature on the surface opposite to that exposed to fire below specific limits specified in AS 1530.4 for 40 minutes. For example, insulate one room from another, or the storey above from the heat from a fire in the storey below, or from one side of a door to the other.

An FRL rating does not have to have gradings in all three parts of the rating. For example, a window may have a rating of –/60/–. This means the window is not expected to hold any loads, nor is it expected to keep the heat in or out (i.e. it has no insulation factor). But it is expected to keep flame and smoke at bay for 60 minutes.

Both volumes will nominate elements of a building to have specific FRLs in a variety of clauses. In both volumes they are formally introduced (aside from in the definitions) in Clause A5.4 Fire-resistance of building elements (Vol. One, p. 25; Vol. Two, p. 27) with direction to Schedule 5. They then appear frequently in diagrams dealing with anything from boundary distances to ceilings. For an appreciation of the FRLs deemed to be achieved by a range of building elements, look to Table 1 of Schedule 5 on pages 685–8.

Fire resistance: NCC, Volume One

Fire resistance is more fully addressed in Volume One under a section of that name: Section C – Fire resistance. This section focuses upon three key areas:

- the structural stability of a building's various elements
- the reduction of the spread of fire both within a building and to other buildings
- the performance of materials and assemblies exposed to fire.

The section holds nine Performance Requirements that expand upon these areas, many of which will hold a number of 'critical factors'. Critical factors being a list of details generally following the phrase 'appropriate to' in the Performance Requirement; i.e. you, and/or the designer, must establish which critical factors are required (are appropriate) for the building under consideration, and to what level they should be applied. An example of critical factors in the first Performance Requirement is shown below, which incidentally defines fire resistance as a concept for building.

CP1 Structural stability during a fire

A building must have elements which will, to the degree necessary, maintain structural stability during a fire appropriate to–

a) the function or use of the building; and
b) the *fire load;* and
c) the potential *fire intensity*; and
d) the *fire hazard*; and
e) the height of the building; and
f) its proximity to *other property*; and
g) any active *fire safety systems* installed in the building; and
h) the size of any *fire compartment*; and
i) *fire brigade* intervention; and
j) other elements they support; and
k) the *evacuation time.*

As stated above, Performance Requirement CP1 effectively defines this volume's approach to the concept of fire resistance. It also suggests that buildings differ in the level of fire resistance required by way of the phrase you have dealt with in other sections of this chapter; i.e.: '… ***to the degree necessary*** …'. As may be seen from the wording, CP1 applies to all classes of buildings described in Volume One; i.e. Classes 2 through to 9. This is not the case with all of the Performance Requirements. This is where determining the required ***level of fire resistance*** for the building becomes important.

Level of fire resistance

The level of fire resistance required for each building is not a number, code or singular DTS provision. Rather, it is an interpretation of the various sections of the NCC, and the appropriate application of each clause as it applies in any given instance. You must read each of the nine Performance Requirements carefully, therefore, to find which clauses apply in the specific case of your building class and type, as well as the particular elements and materials from which it is made up. For example:

CP3 Spread of fire and smoke in health and residential care buildings

A building must be protected from the spread of fire and smoke to allow sufficient time for the orderly evacuation of the building in an emergency.

Application:

CP3 only applies to –

(a) a *patient care area* of a Class 9a *Health-care building;* and
(b) a Class 9c building

Another example is:

CP5 Behaviour of concrete external walls in a fire

A concrete *external wall* that could collapse as a complete panel (e.g. tilt-up and pre-cast concrete) must be designed so that in the event of fire within the building the likelihood of outward collapse is avoided.

Limitation:

CP5 does not apply to a building having more than two *storeys* above ground level

After the nine Performance Requirements there are four very specific Verification Methods: CV1 through to CV4. The first three are applicable only to CP2, and only apply to clause (a)(iii) in each case.

The fourth applies to all of the Performance Requirements (CP1 to 9) when a building is designed to Schedule 7 (p. 699 and effective 1 May 2020). These Verification Methods deal with avoiding the spread of fire between buildings on the same allotment (CV1), and those on adjoining allotments (CV2).

Fire safety: NCC, Volume Two

Volume Two deals with resistance to fire somewhat differently to Volume One. The bulk of the measures are dealt with under the heading 'Fire safety', as opposed to 'resistance', which in itself alludes to a shift in focus.

Fire resistance vs fire safety

As the explanatory information in Part 2.3 Fire safety of Volume Two outlines, the objective of this part is to:

(a) safeguard the occupants from illness or injury by alerting them of a fire in the building so they may safely evacuate; and –

(b) avoid the spread of fire

As can be seen from the above, the concept of fire resistance is similar to that of Volume One, but differs in focus. Due to its need to deal with multiple building classes and construction types, the focus in Volume One tends to be upon the building and its structural capacities. Volume Two, on the other hand, focuses more upon the occupants, and the structure's capacity to protect those occupants – even in the advent of a bushfire. This is to be understood, even when a specific Performance Requirement does not mention the occupants at all, for example:

P2.3.1 Spread of fire

(a) A Class 1 building must be protected from the spread of fire from –

(i) Another building other than an associated Class 10 Building; and

(ii) The allotment boundary, other than a boundary adjoining a road or public space

So, when a Class 1 building is on fire, the aim is to alert the occupants so that they can make good their escape and/or for them to extinguish the fire. When the fire threat is from without, the aim is for the dwelling to act as protection.

This does not suggest that Volume One is not concerned about the occupants, as that is certainly not the case. Volume One seeks to protect the occupants by ensuring the structure of a building can withstand the effects of a fire, stop the spread of fire and fumes within, and provide mechanisms for quelling the fire. With regards to these last point two points, note that there are no requirements within Volume Two regarding extinguishing a fire, nor stopping its spread *within* a building – only between buildings.

Level of fire resistance

With regards to fire resistance, Volume Two – in the main – only splits its requirements between Classes 1 and 10a; two distinctly different forms of building. Aside from the location of smoke alarms and emergency lighting, there is no variation to the requirements due to size or height between a Class 1a or Class 1b dwelling. This is where the two NCC volumes take a significant departure from each other. Rather than varying levels of fire resistance based upon classes and types of buildings, Volume Two effectively assumes the one class, and overlays a single level of minimum requirement: variation then comes with proximity to boundaries and other structures.

However, variation also comes when a dwelling is deemed to be in a **designated bushfire prone area**. In this case, the level of fire resistance is based upon the calculated **bushfire attack level (BAL)**. The BAL for a given location can produce a significant variation to a dwelling's level of required fire resistance.

The issues associated with Class 1 or 10a buildings constructed in designated bushfire prone areas is dealt with in Performance Requirement P2.7.5 and Part 3.10.5 Bushire areas. Part 3.10.5 begins simply enough by introducing two standards that form the Construction Manuals: AS 3959 and NASH Standard – Steel Framed Construction in Bushfire Areas. Despite NSW having a sizable variation to this clause, the underpinning requirements remain clear. There are six BALs, each based upon the risk of ember attack, radiant heat and direct flame contact. This is expressed in kilowatts/m^2 and is in turn developed from aspects of the allotment and surrounds defined within AS 3959 such as:

- a location's fire danger index (FDI) – taken from state and territory overlays
- the classification and proximity of nearby vegetation
- the slope of ground under the vegetation.

The six BALs are: Low, 12.5, 19, 29, 40 and FZ. The determination of a BAL then guides you to that section of AS 3959 that defines the required building materials, elements and practices that must be adhered to in order to comply with the NCC. Once again, this is where the concept of FRLs come into effect significantly, driving much of what the buildings elements must reflect in terms of resistance to fire, and thereby shaping the building design's overall response to fire safety.

Checking existing structures for compliance

Checking an existing structure for compliance is not a matter of reverse application of the NCC, but starting once more with the basics derived from the previous sections above. This means following the full procedure – from determining the class of building, through to determining the level of fire resistance and, from this, the fire safety measures that should be in place. Having determined the parameters of your evaluation, you can now check that each of the building's elements are as required.

The usual mode for carrying out this procedure is by means of a compliance sheet, notes and photographs. These are then compiled as a final report with recommendations for any areas that are deemed to be insufficient. Such studies are best conducted by registered fire engineers, who in the worst cases may actually condemn a building, or restrict access until such time as remedial measures are put in place.

LEARNING TASK 1.4

RESIDENTIAL FIRE REQUIREMENTS

1 You are looking to build a granny flat close to an existing building. What is the required minimum distance from the existing building?

2 Specify the NCC, Vol. Two, clauses where you found your answer to question 1:
   a Performance Requirement clause
   b Verification Methods clause
   c DTS provisions clause

3 You have been asked to provide a solution that will allow you to build the granny flat around 1.6 m away from an existing wall, with a small window adjacent, that provides light to a bathroom window in the existing house. **Note:** you are *not* in a bushfire zone.

   What advice will you give your client, with supporting information that allows you to do this, while ensuring compliance with the NCC, Vol. Two?

SUMMARY

The aim of this chapter was to take the reader through a carefully guided exploration of both Volumes One and Two of the NCC. In particular, how they pertain to all 10 classes of buildings as defined by this code, and how the code interrelates with the Australian standards. Reading the NCC and the relevant standards requires experience to ensure that you have covered all the inflections and nuances that are within them. The 2019 edition of the NCC is far superior to the previous renditions in ensuring greater cohesiveness between Volumes One and Two, making them more user friendly. However, to be fully assured that you have accurately interpreted the requirements will still require some time. The examples offered in this chapter give you a brief taste of some of the simpler elements of their structure and what you might expect when ensuring compliance. Don't be afraid to explore much deeper into other areas of your projects – be assured, this is critical knowledge for you in your role as either builder or supervisor.

REFERENCES AND FURTHER READING

Australian Building Codes Board (ABCB), *Evidence of Suitability Handbook 2018*, ABCB.

Australian standards, **https://infostore.saiglobal.com**

National Construction Code, 2019, **https://ncc.abcb.gov.au/ncc-online/NCC**

2

LEGAL REQUIREMENTS FOR BUILDING AND CONSTRUCTION PROJECTS

Chapter overview

Some of the legal matters that impact builders and the building industry include licensing and registration, workplace safety, contracts, insurance, payroll, industrial relations and dispute resolutions. This chapter provides an overview of these legal areas as they impact on the building profession.

Elements

This chapter provides knowledge and skill development materials on how to:

1. apply the laws relating to builder licensing or registration
2. apply WHS legislation and provisions on site
3. apply the codes, Acts, regulations and standards relevant to construction
4. comply with insurance and regulatory requirements for housing construction
5. apply legislation to financial transactions
6. meet building contract obligations
7. apply industrial relations policies and obligations relevant to housing construction
8. apply dispute resolution processes.

The Australian standards that relate to this chapter are:

- AS 1720 Timber structures
- AS 3600 Concrete structures
- AS 4100 Steel structures.

Introduction

In the past, you learned a trade, and then went ahead and worked in it. These days, government reaches into just about every corner of the workplace with rules and regulations, setting out what can and can't be done, and setting out procedures for just about all that we do. You can't just start working in a trade, and you can't just do things as you see fit. You have to follow the rules; and there are many of them.

The good thing is that these laws protect people. Clients are protected from dodgy builders, employees are protected from exploitation, and everybody is protected by work health and safety (WHS) standards. The difficulty is that you have to know the law and your legal obligations. Good people can get into legal trouble by failing to follow the rules, even when their intentions were honourable.

This chapter sets out to familiarise you with the range of laws that impact the building and construction industry in Australia. It will help you to be aware of the obligations and standards you personally, or your company, need to comply with.

Builder licensing or registration

People who engage a professional to do something for them, whether it is a dentist, pizza maker or builder, take comfort knowing that the person is qualified to do the work they do. How do people know they can trust these professionals? Because government demands that those offering their skills and services are properly trained and licensed.

Identifying licensing and registration legislation

There are two types of legal compliance that you need to know about when working in a trade or profession. First, you personally may need to fulfil certain requirements. These include having a recognised educational qualification (a **licence**), along with appropriate **registration**. Doctors, accountants, lawyers, as well as tradespeople, generally need to show that they have a relevant qualification, as well as registration with the authority that sets and oversees the professional standards of its members. With regards to the construction industry, the qualification must often be matched with a requisite amount of broad-based experience. Indeed, in Victoria, qualifications are not required at all; experience is the key factor for the domestic sector, qualifications being deemed just one way of demonstrating a partial level of competence. But it remains that without the appropriate registration, it would be illegal to work in that profession, just as it is illegal to drive a car without a licence. The question, of course, is which licence do you need for your given field of work?

Establishing the appropriate licence class and qualifications

In Australia, each state and territory has its own set of legal requirements regarding the qualifications and licence needed to work in any particular trade. Being qualified in one state, therefore, doesn't necessarily qualify you to work in other states. The requirements to practice as a construction contractor or building supervisor are set out in state/territory Acts, which may be amended from time to time.

A useful website for viewing all Australian laws is Lawlex: https://lawlex.com.au/Legislation/Browse

From this website you can select the state or territory you are interested in. You can look for the main Act, known as the Principal Act, which sets out the law in detail. Then check for amendments which lists selective changes to particular parts of the Principal Act that have been made since the Act was first passed.

Embedded in each Act will be further useful links. The key Acts for each state related to building are shown in Table 2.1.

These Acts are rather longwinded and at times it is difficult to decipher the exact requirement for a licence for a particular trade or skill area. It is frequently

TABLE 2.1 State building Acts

State	Legislation	Website
Victoria	*Building Act 1993*, No. 126 (VIC)	https://lawlex.com.au/Legislation/CDHP/1860
New South Wales	*Home Building Act 1989*, No. 147 (NSW)	https://lawlex.com.au/Legislation/CDHP/4001
Queensland	*Building Act 1975*, No. 11 (QLD)	https://lawlex.com.au/Legislation/CDHP/13233
South Australia	*Building Work Contractors Act 1995*, No. 87 (SA)	https://lawlex.com.au/Legislation/CDHP/28418
Western Australia	*Home Building Contracts Act 1991*, No. 61 (WA)	https://lawlex.com.au/Legislation/CDHP/15262
Tasmania	*Building Act 2016*, No. 25 (TAS)	https://lawlex.com.au/Legislation/CDHP/157127
Australian Capital Territory	*Building Act 2004*, No. 11 (ACT)	https://lawlex.com.au/Legislation/CDHP/76340
Northern Territory	*Building Act 1993*, No. 29 (NT)	https://lawlex.com.au/Legislation/CDHP/26825

best to go directly to the government body that is responsible for issuing licences to get the clear facts. Organisations such as NSW Fair Trading, the Victorian Building Authority, or Tasmania's Consumer, Building and Occupational Services are typical examples of these state authorities. They issue the licences, they know which one(s) you need and, most importantly, the paths to obtaining them.

Classifications for builders, supervisors and managers

As stated above, each state and territory has its own requirements for obtaining a relevant licence with the following commonalities:

- application
- application fee
- ASIC certificate confirming your business registration
- nominated qualified supervisor consent declaration form
- any other requirements.

Specific licensing requirements for each state/territory are listed on government websites; however, a comprehensive digest can be found on the Skills Certified Australia website (see the list of websites at the end of the chapter). In some states and territories there are multiple 'building' licences to contend with and the term can be confusing. In Victoria, for example, there is the singular domestic builders' licence (unrestricted), but then a multitude of 'restricted' builder's licences covering very specific fields within the industry, such as waterproofing, fence building, kitchen installation and so on.

Apart from the building contractor licence, there are also supervisor and tradespersons qualifications and licences for specific areas of work, particularly those areas of high risk to the individual doing the work and/or those around them. Examples include:

- high-risk work licence
- heavy vehicle licence
- electrical licence
- plumbing licence
- gas-fitting licence
- ground water licence
- pesticide licence
- waste disposal licence
- asbestos removal licence.

There are also licences or short-term permits allowing work that engages with:

- heritage properties
- public lands, roads and the like
- public infrastructure (telecommunications, water, sewerage and the like)
- waterways and other areas of high risk to the environment and/or wildlife.

Licensing your business

The second area of compliance relates to setting up a business. Of course, when you first start work, it is likely that you will do so as an employee of somebody else's business, but someday you may start a business of your own. Should you wish to start your own business, your accountant will help you with advice. There are a number of things you will need to consider.

Businesses, on the one hand, can bring in more money than just a salary, but they can also bring in less. They are risky, and businesses do go bankrupt. A key step is to develop a business plan. Every business will have financial obligations – bills to pay – whether it is making money or not. These include rent, insurance, equipment leases, or even salary obligations to employees. Revenue – money coming in – may not be so steady. If the jobs coming in are not enough to pay the bills coming in, then the business has a cash-flow problem, and without enough money in the bank to cover lean times, the business will fail. A business plan helps to confirm that expected revenues will be enough to cover expected costs.

Once the decision to go ahead with setting up a business is made, it will be necessary to get the business registered. The important thing to understand here is that running a business comes with many legal obligations that you must fulfil. Being unaware of your obligations is no excuse. If you are found to be in breach, you could be fined or, worse still, jailed. Running a business means you have to know the relevant laws and follow them.

The rest of this section will talk you through the main legal points you need to know when operating a business.

Legal structures of businesses

There are many types of legal structures that businesses can take on. The four most common are sole trader, partnership, private company and trust. Each has its own level of complexity to register and administer, and each has its own set of advantages and disadvantages.

Sole trader

The **sole trader** is the simplest form of business. It consists of one person – the self-employed individual. It is relatively easy to set up with no formal processes. The life of the business is limited to the life of the person who set it up. The advantage is that by setting up as a sole trader the individual is effectively employing themselves – they no longer need to be hired. A disadvantage is that it is not easy to borrow money to develop the business, since only the assets owned by the sole trader can be used as security. A further downside is that should the sole trader be unable to pay debts, their own personal property – such as the family home – can be seized. When a business goes bad, an employee may lose their job, but a sole trader may lose all they own.

Partnership

A **partnership** is where two or more people get together to run a business. As with the sole trader, partnerships are relatively easy to set up. Also, as with the sole trader, a partnership is not itself a separate legal entity. Thus, each partner is personally liable for all the debts incurred by the partnership. That is, if one partner runs off with all the money and leaves debts unpaid, the other partner has to make good on all those debts, and again like the sole trader, could lose all they own personally. This is why proper checks and balances are needed to make sure everything is above board. The advantage of a partnership is that two or more people can bring more money into the business, as well as a bigger range of skills to run the business. Usually, in a partnership, each partner will have an area of expertise: design, engineering, construction management, marketing, etc. Individual responsibilities and rules of operation will be set out in a partnership agreement.

Private company

Private companies are quite different from sole traders and partnerships in that they are viewed in law as an individual entity. That is, companies can own assets, make investments, earn money, pay taxes, sue and be sued, and even break the law. They are therefore difficult to set up and must pass many stringent requirements before they are allowed to operate. Moreover, they must also disclose their dealings to the public and are subject to stricter accounting, auditing and reporting obligations.

The key advantage of companies is that investors can buy shares in the company knowing that the maximum amount of money they stand to lose is the amount they have invested, and no more. Investors need not work in the company, though of course key shareholders are often company employees. Shares can be bought and sold at any time, and if the company needs more cash, it can issue more shares. In this way, companies can finance growth effectively. One of the greatest disadvantages of a company, however, is that it pays double tax. If you are a sole trader, you earn money from the business and pay tax once, as an individual. If you own a company, the company will pay you a salary. First the company pays tax on its profits, but then you also will have to pay tax on the salary your company paid to you. Remember, though you own the company, the company is effectively an individual at law, and employs you.

Trust

A trust is where a person – the trustee – is required to hold and manage assets for the benefit of others. Consider a child whose rich parents have died. The child inherits the wealth but is in no position to make decisions on how to manage that wealth. In this case a trustee may be appointed – and indeed the government offers this service through the various state trustees offices. On the other hand, a family business may be run as a trust. Here the senior family members act as stewards, looking after the interests of the family's junior members until the day they inherit the business.

Registering your business

No matter the business structure, the business must be registered for a number of things. The key registrations are set out below.

ABN

The first step to registering a business is to get an Australian Business Number (ABN). This number is unique for every business and identifies it for all business, tax and legal purposes. It is the number that will appear on all formal communications and transactions – sales and purchases. Have a look at the next invoice you receive, and it will have the ABN of the company that sent you the invoice.

Business name

Next the business will need to register its name. Like the ABN, the business name is unique for the business, similar to a 'username' on a computer or website. Registration is through the Australian Security and Investment Commission (ASIC). A fee is paid to claim a name, with a smaller annual fee due to retain use of the name.

GST

The Goods and Services Tax (GST) is the government's way of taxing people for the things they buy. It is currently set at 10%, though has varied, and indeed does vary in different countries. Effectively, if you buy something that costs $100, you will have to pay $110, with the company pocketing the $100 as revenue, while collecting the extra $10 for the government. Currently, if your business is very small, you need not collect GST from customers. However, if you expect your revenues to exceed $75,000, you need to register for GST and then to charge GST to your customers. At the end of the financial year you will need to total up the GST you have collected from customers, then subtract the total GST you have paid to suppliers, passing on the difference to the tax office. Sometimes businesses get caught out at tax time because the amount they have to pay can be quite high.

PAYG

A building contractor, for example, may have employees. Employees also pay tax, but in Australia some tax – usually the pro-rata amount predicted that the employee will owe over the year – is withheld by the employer each time salaries are paid, and forwarded instead to the tax office. Employers pay tax on behalf of their employees. This system is called Pay As You Go (PAYG). Employers need to register in order to participate in PAYG.

Other registration requirements

Payroll tax is collected by each state and territory on the wages a company pays. Fringe benefits tax is collected on benefits an employee receives, such as company car or low interest loans. Fuel tax credits are also issued on plant and equipment used. For all of these, a company needs to be registered with the relevant authority. Again, your accountant will help.

LEARNING TASK 2.1

RESEARCHING RELEVANT LICENSING AND REGISTRATION LEGISLATION

For the state or territory you live in, research what licensing and registration is required to operate as a:

1 builder
2 company.

WHS legislation and provisions on site

Worldwide, the construction industry is one of the most dangerous sectors, with a relatively high incidence of severe workplace injuries and death. On the other hand, Australia has one of world's best records of work safety in construction. Partly, this is due to the high safety standards set in Australia. Partly also, it may be due to the heavy punishments that are given to companies and managers when those work safety standards are breached.

Identifying WHS legislation and regulations previsions

The term occupational health and safety (OHS) has recently been replaced by the term **workplace health and safety (WHS)**. For all practical purposes the terms can be used interchangeably and this is often the case on construction sites. The key point to remember is that construction sites are dangerous places, and those visiting or working on construction sites have an increased exposure to physical harm.

While it may seem like common sense that an individual is responsible for their own safety, at law, site supervisors have a '**duty of care**' and will be called to account for any injuries that occur on their watch. Consequently, supervisors will restrict access to sites to only those people who absolutely must be there, and ensure that those working on site fully comply with all the safety standards required. This will include appropriate warnings of danger, use of protective equipment and proper safety training.

The *Workplace Health and Safety Act*, and codes of practice, can be found at: https://www.business.gov.au/risk-management/health-and-safety/work-health-and-safety

Business owners, contractors, subcontractors, workers, manufacturers and suppliers, as well as visitors, all have a responsibility to themselves and others to comply with safe practices and procedures while on site, and while construction, maintenance, renovations, commissioning or demolition is underway.

Regulations and codes applicable to on-site construction

Each state and territory has its own legislation regarding WHS and OHS, but they all require:

- safe premises
- safe working environments
- safe facilities, machinery and materials
- safe systems of work
- appropriate information, instruction, training and supervision of personnel.

Under the Act, employers have a 'duty of care,' and are required to provide a safe workplace for all who enter – not only employees. Under the Act, businesses undertaking construction work are those which:

- design and build a structure
- commission construction work
- perform the role of principal contractor
- manage and control the workplace where construction work is carried out
- carry out high-risk construction work.

Site safety signage requirements

Construction sites are dangerous places, but the nature of the dangers may vary from site to site. Consequently, as you approach a construction site you will see a large board displaying all the safety issues to be encountered on that particular site. If you have ever been to a construction site you will be familiar with this sort of board and its many warnings. Inside, too, specific warning signs will be positioned as needed.

The signage will let you know of dangerous activities, equipment, and materials and chemicals that may be encountered. Imagine a fire breaking out. When the firemen arrive, just by looking at the board they will know what sort of chemicals may be involved in the fire, and instantly know what flame retardant to use in putting out the fire.

The site itself should be fenced off, not only to deter trespassers, but to keep people away from the danger. At the site entrance there should be a statement to the effect of 'authorised access only,' or, 'all visitors report to the site office'. You should also see the name of the principal contractor on a development site, along with their phone number. This information makes it possible to contact the key responsible person, should there be an emergency – even after hours.

The types of signs vary. Basically, they are designed to be understood visually and instantly, much like pedestrian lights at a street intersection – green man for go; red man for stop. Some examples of signage include:

- warning signs – e.g. 'high voltage'
- safety signs – e.g. 'danger do not enter'
- hazard signs – e.g. 'asbestos removal'
- dangerous material signs – e.g. 'flammable liquids'

- prohibition signs – e.g. 'no smoking'
- restriction signs – e.g. '10 k/hr zone'
- mandatory signs – e.g. 'protective helmet'
- emergency signs – e.g. 'emergency exit'.

CASE STUDY

Construction site tragedy

Carlton beer was named after the Carlton and United Breweries, established in 1864 in the former North Melbourne Brewery, situated on the corner of Swanston Street and Victoria Parade, Melbourne. The multi-storey brick construction was for many generations an icon of the city. It consisted of an imposing freestanding brick warehouse, protected along the Swanston Street frontage by a high brick wall.

Despite efforts to preserve the building and its surrounds as a heritage site, permits were granted for its demolition, and in 2013 Grocon tore down the brewery that had been standing for a century and a half.

Late one Friday afternoon, as people walked down Swanston Street towards the city, the perimeter wall collapsed, killing three people.

An investigation was held and Grocon, along with Aussies Signs, who had been contracted by Grocon to secure the site, were found guilty of negligence, and each fined $250,000.

As it turns out, the main brewery tower had been protecting the brick wall from strong westerly winds. Once this building was demolished, the wall became exposed, and a strong wind gust that fateful Friday afternoon brought down the wall that had been standing for well over 100 years.

What do you think? Could Grocon have foreseen these unfortunate events?

LEARNING TASK 2.2 RESEARCHING MODEL CODES OF PRACTICE FOR CERTAIN CONSTRUCTION-RELATED TASKS

Visit the Safe Work Australia website: https://www.safeworkaustralia.gov.au/resources_publications/model-codes-of-practice

Choose an area of work that is relevant or of interest to you, such as demolition, manual lifting, excavation, falls from heights, etc. Select and download the relevant code of practice from the Safe Work Australia website. Summarise into a table the recommended safe work practices for that work task of construction-related activity.

The following headings could be used:

Task or work activity	Potential danger or harm doing the task/work	Page/s that detail this danger in the code of practice (CoP)	Suggested method from the CoP that could be used to lower or remove the danger from the task /work

Legal requirements relevant to construction

We have looked at legislation regarding licensing and registration, as well as workplace health and safety. But there are also lots of regulations covering the actual building of buildings, the materials used in their construction and standards of construction. We will review these here.

Researching legal requirements applicable to a particular project

As the above and Chapter 1 of this book have demonstrated, there are a range of state and federal laws and associated regulations governing the construction industry. The reason government gets involved is because it also has a duty of care to ensure that people are delivered buildings that are safe, healthy, structurally sound, won't easily burn, or be compromised by strong winds. In addition, buildings must comply with strict minimum energy requirements.

The National Construction Code

The Council of Australian Governments (COAG) set up the National Construction Code (NCC), initially released in 2011, to establish minimum performance standards that any building must meet. These standards are reviewed on a three-year cycle, with minor amendments released on 1 May annually. For more comprehensive information on the NCC, see Chapter 1.

Australian standards

The NCC frequently 'calls up' (lists and gives power to) key standards by which specific elements must be constructed, or materials used or arrayed. The bulk (but not all) of these are Australian standards. Standards Australia is the peak body that oversees the development of these standards. It benchmarks its requirements to both national and, where possible or relevant, international expectations. Other standards are also produced within Australia for the Australian building sector, and may have equal force or legitimacy, but without the Australian standards banner; e.g. the NASH standard for steel framing.

From a legal perspective, it is important to know which of these standards is relevant to each aspect of a building project. The key point here is that a standard (Australian standard or otherwise) only has force in the construction industry if it is called up in the NCC. If it is not called up, then it may be used to *inform* some aspect of your project, but any other standard or clause within the NCC will have greater legal force and so must be followed first.

Other legislation impacting building construction

As mentioned already, you may think that working in the building trade, you need only to know about building, but the government now more than ever has views about how we should behave. With this in mind, be aware that there are a number of laws that lay out government expectations.

A list of all the legislation can be found on the Federal Register of Legislation webpage: https://www.legislation.gov.au/

How we compete in business, how we manage information we receive from others and how we deal with people are all matters the government requires us to comply with.

Some of the relevant legislation on these matters can be found at the websites in Table 2.2.

Carrying out construction processes to meet legal requirements

Once you have won a building project you will need to lay out the steps involved in working through the project. You will need to set out the actual activities that must be undertaken, and you will need to consider what other professionals will be needed to do the work. This section looks at these considerations.

Preparing the construction process

Before you even begin on a project, there are preliminary things that must be sorted out.

- Is the project a new construction or a renovation?
  - If new, check the land use, zoning and planning regulations for the site.
  - If not new, check for planning and building approvals.
- Establish what the client's budget is.
  - Check that the client can afford what they are after – that their expectations are reasonable.
  - Clarify the risks and uncertainties in costs – contingency monies are needed to cover any unforeseen extra costs, or budget overruns.
- Ensure that the client has all the necessary approvals and surveys in place.
  - If these have not been carried out, it may be up to the builder to arrange for these.
- Make sure the client gets a geotechnical report.
  - A geotechnical report provides a pathway to site soil classification for the purposes of footing design. In the domestic sector this responds to AS 2870 and such classes as A, S, M, D, etc. (see 'The footings' in Chapter 13). In the commercial sector there is no such classification and the engineer must design the footing based on the loads and forces involved. The report therefore has legal ramifications if not conducted, or poorly conducted, as does the engineer's designs based upon that report.
- Only now are you ready to commence work.

Planning permit

Even before a design is completed, a project will need go through council planning approval. Whether or not a planning permit is required is dependent upon a range of factors, such as the proposed use of the land or building, demolition, heritage overlays and general zoning such as residential or commercial. The planning permit ensures that the proposed building fits within the categories of building limited by the zone. For example, you cannot just build a hospital in the middle of a suburban area. Similarly, you may be not able to build a residential house in an area set aside for factories. In such cases, the planning permit must be obtained before the building permit, as it confirms to the permit holder what exactly can be built on the site.

Building permit

After the planning permit has been approved, the design process may be completed. This includes all engineering and final construction drawings. This is an expensive phase, up to 10% of the build price, hence the need to get planning approval first. Only very

TABLE 2.2 Other legislation affecting building construction

Act	Website
Competition and Consumer Act 2010	https://www.legislation.gov.au/Details/C2017C00369
Privacy Act 1988	https://www.legislation.gov.au/Details/C2014C00076
Australian Human Rights Commission Act 1986	https://www.legislation.gov.au/Details/C2018C00050
Age Discrimination Act 2004	https://www.legislation.gov.au/Details/C2018C00022
Disability Discrimination Act 1992	https://www.legislation.gov.au/Details/C2016C00763
Racial Discrimination Act 1975	https://www.legislation.gov.au/Details/C2014C00014
Sex Discrimination Act 1984	https://www.legislation.gov.au/Details/C2014C00002

minor structures, such as small garden sheds and the like, may be constructed without a **building permit**. To build without a permit is a serious offence and the building may be torn down – even if it would have complied with a permit.

The building permit does a number of things. First, it confirms that the proposed work will conform to all the regulations, codes and standards: in so doing it validates all the plans and drawings that form the basis for the proposed construction. Second, it specifies when an occupancy permit is to be granted, and when all inspections including the final inspection are to be undertaken. It will nominate the builder (owner-builder or otherwise), the architect/designer and, at times, the engineer(s) involved. This is a very important legal step, as should something go wrong during the build or after completion, there is a legal trail of 'fault' to be pursued. That is, this is one of the many paths towards ensuring that the building is not only designed correctly, but also built correctly. In addition, it confirms that the builder undertaking the works is qualified, registered and has the necessary insurance covers.

Other permits, approvals and certificates

As the building progresses, further checks that things are being done properly are needed. The way government makes these checks is to require the builder to lodge further paperwork. These may include development applications, statements of environmental effects, reports on alterations to existing buildings, applications for hoarding and scaffolding.

For smaller jobs where a building permit was not needed, the builder may still have to obtain certificates. This may be for demolition work, construction of a small, secondary building (like a shed), internal alterations (like a shop fit-out), street awnings, business signs and the like.

Carrying out the construction process

With the permits obtained, the actual process of construction can begin. Planning the construction of a building is complex and requires a high level of experience and skill. The main issues from a legal perspective are:

- identify site
- secure financing
- get surveys
- obtain preliminary approvals
- get a certified design – usually from an architect
- obtain permits
- find and appoint contractor
- establish building timeline
- identify the appropriate contract.

Many of the above are self-explanatory with regards the legalities. Obtaining permits, for example, has been covered previously and it is clear that to operate without them in place will carry a legal burden in the way of fines or have other negative implications such as wasted time. Other factors, however, may seem simple, but can be extremely problematic if not carried through correctly. Getting the land surveyed and the correct lot or site identified seems simple. So simple that many builders do it themselves. The problem comes when they find that lot number pegs have been moved around by playful children, who do not understand the implications of their actions (or not so playful adults who do …). Building on the wrong block of land means that the finished building technically belongs to the owner of that land, not you or your client.

Establishing the timeline for the project is another critical factor, whereby an error of judgement can lead to significant cost implications for your client, which they will want to push back on you through your contract.

And then there is the contract itself. This critical document outlines the entire legal framework for the project, or at the very least points to that legal framework. The choice of contract and the wording within it is thus a vital part of the preliminary work – for more on this side of the construction industry, see Chapter 3.

Consultants and experts

Though the builder will know much, they may not be expert in all areas relevant to the project being built. For this reason, they may engage expert consultants to help out. There are many kinds of experts that the builder may call on, including:

- surveyor
- engineer
- architect
- accountant
- finance broker
- solicitor – expert in property law
- quantity surveyor
- project manager.

For certain situations, engaging an expert will be required by law. For example, if the house being built is standard, along with the site conditions, then the builder need only follow the NCC. However, if the soil is exceptionally unstable, then it will be mandatory to call in a specialist engineer to design the footings.

LEARNING TASK 2.3

IS THERE GENDER DISCRIMINATION IN THE BUILDING INDUSTRY?

The building industry remains an area of work dominated by men, with a smaller proportion of women working in the field than in other industries, such as education, health or tourism.

One view is that the building industry is among the most discriminatory, with a work culture that positively dissuades women from joining. Another view is that men are naturally drawn to working with 'things' while women are naturally drawn to working with 'people'.

>>

>>

Read the information provided in the links and provide a response to the questions posed.

1 What do you think explains the lack of women in the building sector?
   - https://www.uts.edu.au/news/tech-design/why-are-so-few-girls-choosing-careers-construction
2 What is your view of governments' attempts to introduce more women into the building and construction industry?
   - https://www.abc.net.au/news/2018-10-17/women-in-construction-goverment-can-do-more/10382932
   - https://www.women.nsw.gov.au/strategies-and-resources/women-and-work
   - http://www.tglaw.com.au/construction/2018/06/22/nsws-ten-point-plan-for-construction/

Insurance and regulatory requirements

Businesses are required to have insurance. This is another area of the law in which those in the building industry must comply. Buildings are expensive propositions, and if something goes wrong, the costs will be huge. Injuries may also happen, and once again, people need to be protected against the costs of getting them back on their feet.

Think of car insurance. There are basically three levels of car insurance. Comprehensive car insurance protects your car as well as the other car. If there is an accident, the insurance covers the costs of all repairs. Third party property just protects the other car. Here, you are protected from the costs of repairing the other car – which might be a Ferrari – when the costs may be more than you can pay. At worst, you might be down your own car, but at least you won't go bankrupt. Both these insurances are voluntary. Third party personal, however, protects any people involved in an accident. This insurance is compulsory, and is paid at the time of yearly registration. Here, the government makes sure that anybody injured in a car accident, no matter who is at fault, will have their medical costs covered. Compulsory insurance is how government seeks to protect people.

It is the same in the building industry. The government does not force you to take on any and every type of insurance, but when there is a possibility that your actions may adversely affect others, you can be sure the government requires you to have a policy to cover that.

Insurance cover

You need to protect the people you work with, both the public and employees. Professional indemnity, public liability and workers compensation insurances protect against unfortunate accidents, while superannuation contributes to supporting people in retirement. These are discussed below.

Professional indemnity

As a builder, you will enter into contracts that promise to deliver a building to a certain standard and within a certain time period. What happens if you don't? Well, you might be sued. How are the dollar amounts in such court cases calculated? Imagine an apartment building is delayed one month over the contract deadline. Because the building is completed late, the owner will lose one month's rent for the whole building. He would be entitled to sue the builder for that amount – though in all likelihood the penalty for being late would have been written into the building contract.

In this case, the builder's **professional indemnity** insurance would cover the late penalty. Professional indemnity insurance protects the builder from any claims related to the quality of work, delays or even advice given that proves unsound resulting in financial loss.

Such insurance may seem like a 'get out of jail free card' that allows builders to be less careful than they might otherwise be. However, it is important to note that, just like ordinary insurance, premiums can go up with claims, while if you prove to be consistently unprofessional, you may lose your registration altogether.

Public liability

Mistakes can be honest mistakes, but they can still be classed as negligence – doing (or not doing) something that leads to damage when you should have known better. Negligence is where something goes wrong when it could have been reasonably avoided. **Public liability** insurance protects against being found negligent. There are four classes of negligent acts:

- negligence leading to injury or death
- negligence leading to emotional or psychological distress
- negligence leading to misinforming others, causing them to make mistakes
- negligence leading to property damage – such as fire.

Workers compensation insurance

Each state and territory has laws that award monetary compensation to employees who are injured, disabled or killed on the job. For example, providing for the cost of rehabilitation and life-long lost income where a person is severely incapacitated. Benefits may also be provided to dependents of such injured workers – to the wife and child of a man killed at work who was the primary breadwinner.

The idea of workers compensation is to provide speedy financial relief for people in distress, and save them from having to seek financial compensation through lengthy, expensive and uncertain court proceedings. In certain states, workers compensation laws also protect fellow employees from liability. Again, this is to streamline resolution – looking to find just closure for victims, rather than apportioning blame on individuals. Of course, it needs to be said that while individuals may be protected from liability, the company itself may have to answer, if evidence of negligence is found leading to the injury at hand.

Safe Work Australia Act (2008) https://www.legislation.gov.au/Details/C2009A00084

Superannuation guarantee levy

Though not exactly insurance, superannuation is another means of protecting workers. **Superannuation** is a system of forced saving that aims to leave people with sufficient (or near-sufficient) funds for their retirement. Companies are obliged to pay superannuation for all eligible employees to the minimum amount of 9.25% of gross income.

Only certain categories of people are exempt from receiving superannuation. These are people:

- paid less than $450 per month
- aged over 70 years or over
- aged under 18 years, working less than 30 hours per week
- domestic or private workers (housemaids), working less than 30 hours per week.

Contract law and its application

When contracting as a professional to perform work on behalf of others, such as constructing a building, the law states you need to be properly insured. The range of insurances you will need depend on the structure of your business and the type of work you carry out. Minimum requirements are set out in the Acts. These will be different for each state and territory, as listed in Table 2.1 earlier.

The previous section covered insurance matters related to people in the construction industry – the supervisors who may need insurance because of the mistakes they make, and the employees who may need insurance because of the risks they face at work. But we still need insurance for the actual building project itself, and this is discussed here.

Builders warranty insurance

Known in some states as domestic building insurance, this is a compulsory cover that must be taken out by all builders undertaking construction of homes or other domestic building work. A copy of this insurance must be provided to the client before any deposits may be taken on a project. The purpose of the insurance is to cover the client in event that you are unable to complete a project because:

- you become insolvent
- disappear
- die.

The cover is for structural defects or incompletions up to $300,000 for six years; and non-structural defects for up to two years. Note that it is only for the above circumstances; if you are still active in the construction industry then it is you who must carry out the works, or pay for others to do so.

Homeowner insurance

Pretty much everybody that owns a home will have it insured. Certainly, if your house is mortgaged it will be a condition of the bank loan that the house is insured.

However, that insurance is unlikely to cover any additional works that are carried out on the building – the building works themselves or the resultant completed renovations. For this reason, a builder about to work on a home renovation is required to advise the homeowner to:

- notify their home insurance provider of the works
- clarify whether home contents are covered for damage and theft over the construction period
- provide their lender with a certificate of insurance for the additional works
- increase the value of their insurance policy to take account of the building works.

Construction works insurance

Apart from carrying insurance for themselves and their company, a builder must insure each building project. Construction works insurance covers any loss or damage suffered during the building works in progress. Such insurance will be a condition of the loan used to finance the building. It also makes good business sense to protect against any calamity – rare as they might be.

There are, however, no standard policies, and builders do need to take care to consider what inclusions and exclusions they want. On the one hand, you want to be covered for all reasonable risks, but on the other hand, each inclusion may add significantly to the overall policy cost. Construction works insurance does not come cheap.

Some of the standard inclusions to consider in a construction insurance policy are:

- theft, malicious damage
- storms, flood, water damage
- fire and earthquake
- subsidence and landslip
- ancillary and temporary works – such as scaffolding, sheds, formwork
- plant and machinery – such as mixers, hoes, diesel engines, saws, tools.

Additional inclusions include:

- owner-supplied materials –if the owner provides items used in the works
- materials in transit –if specialist materials are brought in
- off-site fabrication –if parts are manufactured or assembled off site
- escalation costs –if the contract is subject to severe inflationary cost increases
- mitigation expenses –if costs are incurred in preventing threats to the works
- removal of debris –if removal of damaged items is required
- professional fees –if additional services are required to repair damaged items.

LEARNING TASK 2.4

WHAT DOES IT COST TO HAVE ONE EMPLOYEE? A THOUGHT EXPERIMENT

Let's say you start work for a company, and they agree to pay you $1000 per week. Well, what does it cost the company to have you as an employee? The answer is: more than $1000 – quite a bit more. But how much more? Can you figure it out?

For a start, there is an extra $92.50 in superannuation. There are insurance costs to protect you, and also to cover any mistakes you make. On top of that there is the rent for the space you work in, utilities, and the cost of the equipment you use. What about when you go on holidays, or are sick, or in lean times when there is no work around. You collect your salary as before, but without generating revenue. And redundancies are expensive too.

By some estimations the cost to keep someone in a job is 50% over the salary they pocket.

Financial transactions

Government needs money to operate. When that government is a democracy, such as we have in Australia, it means not only that it's up to us to choose who governs us, but also how much of our money politicians get to spend and what they get to spend it on. In other words, our tax money buys a service – good governance. In theory at least, if we don't like the service, we can change our government, just as we might change restaurants.

The upshot is that government then puts the burden back on citizens to report how much money they make, so that tax can be claimed from everybody. When running a business, the list of things that need to be reported to the Australian Tax Office (ATO) is quite extensive. The key reporting items for a small business, such as that of a registered builder, are discussed below.

Payroll systems

The average employee has tax deducted from their salary by their employer. This is so ordinary people won't be stuck with a huge bill from the ATO at the end of the financial year. But everybody still needs to fill in a 'tax return' every year, and there are certain deductible expenses that an individual may claim, which can result in a tax refund.

There is a similar requirement for businesses, with important differences. First, tax is not necessarily deducted along the way, so businesses need to be careful to set aside sufficient funds to fulfil any tax obligations that will likely arise at the end of financial year. Second, running a business incurs a whole range of expenses, and it is revenue minus expenses, or profit, that is taxed. Consequently, businesses try to report as many expenses as possible, so that profit is minimised (at least on paper). However, these expenses have to be provable, and managing the receipts – or evidence of expenses – is a significant administrative task. Third, the ATO also requires businesses to manage and report on the tax-related matters of employees, contractors and other entities with whom financial transactions have occurred. This too requires significant administration.

Pay As You Go (PAYG)

Businesses are required to deduct tax from employees and to send the withheld tax on to the ATO, either weekly or monthly, according to the size of the organisation. These regular payments to the ATO are called PAYG instalments, and offset the final tax burden the employee will be obliged to pay at year's end. Consequently, the amount withheld is calculated based on the projected tax the employee will owe over the financial year.

PAYG is withheld for:

- wages and salaries, commissions, bonuses and allowances paid to employees
- social security payments
- compensation and accident payments
- termination payments
- unused leave payments
- suppliers who do not quote an ABN, at the maximum tax rate (49%).

Where non-cash benefits are paid to employees (deemed a substitute to taxable income), then businesses will have to register for and remit fringe benefits tax (FBT).

Items liable for fringe benefits tax

- car owned or leased by employer, but used by an employee
- housing paid for or subsidised by employer, but used by an employee

- loans provided by employer at lower than market rate for an employee
- airline transport paid for by employer, for an employee's private use.

The link for the Australian government's fringe benefits tax webpage can be found at: https://www.ato.gov.au/General/Fringe-benefits-tax-(FBT)/In-detail/FBT---a-guide-for-employers/

Payroll tax

Apart from being required to withhold tax on behalf of employees, businesses are themselves taxed when the total payment to employees exceeds a certain amount. This is argued to help very small businesses, which do not pay that tax. For example, in NSW, a business paying over $750,000 annually in salaries has to pay 5.45% tax on salaries paid.

Payments liable for payroll tax

- directors fees
- bonuses and commissions
- contractor and consultant wages
- apprentice and trainee wages
- third-party payments
- fringe benefits
- allowances
- salary sacrifice
- shares and options
- superannuation
- termination payments

Payments not liable for payroll tax

- workers compensation
- parental paid leave
- jury duty payments
- reimbursements

GST systems

Society needs services that no individual can pay for by themselves, such as roads, schools, hospitals, police, army, etc. So, to pay for these things, government collects tax from its citizens. In some countries the tax rate is higher, and in some countries the tax rate is lower. This rate should be a reflection of the extent of services offered – meaning in high-taxed countries, such as Australia, we get free or subsidised medical care, education, retirement pensions and so on, while in other countries with lower tax, such as the USA, you have to pay for these benefits out of your own pocket. In Australia, it costs about $110,000 a year to keep a person in jail, which is paid for by the taxpayer. In some countries, prisoners actually have to pay for themselves – they don't get to eat if their family refuse to bring in or pay for meals.

The Goods and Services Tax (GST) was introduced through legislation in June 1999. The tax is an idea borrowed from the UK and aims to relieve poorer people of their tax burden. Rather than taxing people's earnings at a high rate, earnings are taxed at a lower rate, with the shortfall made up by taxing luxury items that people buy. In other words, you get to keep more of the money you earn, but when you go out to buy things – unless those things are classified as an essential items such as milk and bread – they will cost you more.

Buildings and building products are classified as non-essential items, and are therefore subject to GST. The current GST tax rate is 10%.

The key point here is that if you own a business, and your business has an annual turnover of over $75,000, you will be required to collect GST on the goods and services you sell, and to pass that money on to the government. You effectively become the government's tax collector. Businesses can register for GST at the time they register their ABN, though they can also do so subsequently.

The link for registering for GST can be found at: https://www.ato.gov.au/Business/GST/Registering-for-GST/

As a business, you are required to add a 10% levy on all (non-essential) items you sell, and to pass that on to the government. You are also required to report on your GST collection, including:

- providing accounts of taxable sales
- issuing tax invoices for goods sold, as well as collecting tax invoices for goods purchased
- lodging a business activity statement (BAS) at the end of each financial year, along with net collected GST payment submissions to the Australian Tax Office.

LEARNING TASK 2.5

CALCULATING HOW MUCH TAX YOU SHOULD WITHHOLD FOR AN EMPLOYEE

You run a business, and you employ someone who you pay $4000 gross per month.

They are a resident, this is their only job, they have no outstanding student loans (HECS, etc.), they claim the full Medicare levy, and they have a partner and two dependent children.

Visit the Australian government PAYG calculator website, below, and calculate the PAYG monthly withholding tax you are to collect for forwarding to the ATO on behalf of your employee.

- https://www.ato.gov.au/Calculators-and-tools/Host/?anchor=TWC&anchor=TWC/questions#TWC/questions

Building contract obligations

Contracts for the construction industry are covered extensively in Chapter 3; however, it is relevant to explore the topic here to some degree from the legal perspective. Contracts are essential in business, and in fact unavoidable. Technically, a contract is formed whenever one party offers to do something or provide something for a price (called a consideration), and the other party agrees. A contract does not need to be written, it can be verbal, or even a reciprocated action. Buying an ice-cream – where you hand over the money to the shop-owner and walk away with your mango splice – is a contractual exchange.

But when large sums of money are involved, and where the stakes are high, like in a building project, the contract has to be in writing. If problems or disagreements arise, then the contract becomes the referee to the disagreement, laying out the rules for resolving the situation. Thus, a contract specifies exactly who does what, where, when, for whom, to what standard and for what amount of money. Just as important as saying what will be done, is making clear what won't be done. That is, contracts must specify the scope of the works. And it should also make explicit the remedies (and penalties) should either party fail to honour their contractual obligations.

Selecting the correct form of contract

In Australia, for residential works, it is standard practice to use Housing Industry Association (HIA) contracts. There are two types of contracts:

- For residential building works costing between $5000 to $20,000 (essentially renovations, refurbishments or small extensions)
- For residential building works costing over $20,000.

Clearly the second type is more extensive in its provisions, given the greater costs involved.

Types of contracts

Contracts may be standard or customised. Standard contracts, complying with the idiosyncratic legal differences that occur from state to state, may be downloaded from the HIA website, though you will have to pay for them: https://hia.com.au/contracts/

An instructive YouTube video on accessing HIA contracts can be found at: https://www.youtube.com/watch?v=S22Nj1_c44s

Builders are restricted to taking a deposit from clients of not more than 10% of the estimated costs, and must provide the client with a copy of the Consumer Building Guide for the state/territory in which the works are being done. For further information on contracts in the construction industry, see Chapter 3.

Though standard contracts are commonly used, it is also possible to write a customised contract. Why would you write a customised contract? The worst thing that a contract can be is unclear. Standard contracts are familiar – they are tried and tested. In particular, they have been tested in the courts and are shown to be robust, precise and comprehensive. However, if the works being conducted are unique in some way, requiring extra or specialised works not otherwise covered in a standard contract, then customisation is warranted. Usually, customisation takes the form of an amendment or addition to a standard contract, covering the extra scope of works.

It is important to note that contracts cannot be written to lessen the legal protections afforded clients. A contract seeking to undermine statutory consumer rights would be illegal.

Fee payment method

An important question for both the client and the building contractor is the fee. How much is the contractor to be paid by the client? This is not an easy question to answer as there are many ways to calculate the fee, each with risks and advantages to both parties. What needs to be avoided is a fee structure that turns out to be punishing to one side or the other. If the builder finds he is not making enough money to cover his own costs, he may start taking short-cuts that will compromise the final quality and performance of the building. Worse still, he may go bankrupt, which would be terrible for the client as well, since the building would be left unfinished.

There are three common ways in which fees are structured. These are hourly rate, fixed sum and percentage of the total building cost. These fee structures can be combined, or used for different stages of the project.

Hourly rate

As the name suggests, hourly rates are a fixed dollar amount, per hour. This is safe for the builder, but when the duration of work that needs to be undertaken is unknown, this can be risky for the client. Obviously, the builder has a disincentive for efficiency, and rather an incentive to string out the work for as long as possible in order to inflate his fee.

Consequently, hourly rates tend to be used in preliminary consultations, much like when visiting a professional advisor. They tend also to be used when the work is limited in scope and duration – such as when producing sketch drawings, or when engaging a tradesperson for an installation.

The rates themselves are usually dictated by market forces.

Percentage of total building cost (cost plus)

It should be noted from the outset that cost-plus contracts are not promoted as best practice for domestic (housing) projects in any state or territory. Some states, such as Victoria, go further and prohibit cost-plus contracts for domestic projects under $1 million, with other states using $500,000 as their threshold.

When agreeing on the fee to be paid to a contractor for delivering a complete building, the cost-plus contract nominates a percentage over and above the cost of the building itself. For small projects, such as a house, this can be around 15%; for larger projects, such as a high-rise apartment building, the percentage will be much less – perhaps around 6%. (This is because the time and effort for even small projects can be relatively substantial.)

While the percentage was once mandated for all professionals by industry associations, based on the value of the project, fixed percentages was seen as an anti-competitive barrier and have since been lifted.

The idea behind this fee structure is that the final cost of the building, or even the scope of the project, may not be known with certainty from the outset. If it turns out that more work is required, then the builder gets paid more for that extra work. This provides surety for the builder, but it generates great risk for the client as there is no ceiling upon the final price to be paid; hence, the state limitations for domestic works where homeowners may find themselves considerably out of pocket.

Fixed price (lump sum)

However, there will come a point when the preliminary work has been completed and a final design brief and scope of works established. At that point there will also be a fixed budget set for the project. With a final cost for the project known, the fee to be paid to the builder is also established. Consequently, it is also common practice at this point to convert the cost-plus fee to a fixed fee. This gives added certainty to the client, and also now provides an incentive for the builder to stick to the budget. Even if the building costs increase from here on, the builder will not be receiving any increases in payment for his services.

Contractual obligations

Undertaking the contracted works is a two-stage process. First, all the necessary documentation related to the building project needs to be put in place. Then, the project itself has to be managed.

Obtaining compliant documentation

Before the building works can begin, a number of documents must be produced or collected.

Land title certificate

The land title identifies the owner of the land it describes. From the title, a land surveyor can set out pegs on the actual site, confirming the exact location of the title boundaries. By law, buildings must be contained within a minimum set distance from all boundaries, easements and encumbrances.

It happens more frequently than you might expect that buildings are built in the wrong place. Even if a building is mistakenly built only a few centimetres over a boundary line, the trespassing portion will have to be removed.

Building and planning permits

A building permit will have to be obtained for the proposed building, before construction begins, confirming that the planned building conforms to all the relevant building regulations.

Similarly, a planning permit may also be needed when a building is to be erected in a place that has been specially zoned. This permit confirms that the proposed building usage type is allowed for that zone.

The NSW planning permit process is set out here: https://www.planning.nsw.gov.au/about-us/nsw-planning-portal

The Victorian planning permit process is set out here: https://www.planning.vic.gov.au/permits-and-applications/do-i-need-a-permit

Building plans (construction drawings)

As most people realise, the standard way to relate the designer's vision for the proposed building to the builder is through working drawings. The drawings are meant to show exactly what the builder is to build. They are also used by the building surveyor in granting the building permit, the quantity surveyor in estimating the bill of quantities, and the various trades people in setting out the electrical, plumbing and fixtures.

The plans are also used to cost the building.

Specifications

The specifications complement the building drawings. They are written documents that cover all the requirements of the building that cannot be explained through the working drawings. It is the drawings and specifications, together, that form the contract – the promised deliverables.

Importantly, if there is a discrepancy between the drawings and the specifications, it is the specifications that overrule. Consequently, while the working drawings are the more accessible and readily interpretable, the specifications document should not be ignored as it will detail all the points that cannot be stated in the drawings.

Engineering reports

The greatest cost uncertainty in a building are usually the footings. A site may be stable, or unstable, or even extremely unstable. Building on problem sites can add significantly to the cost of a building. More to the point, the builder needs to know just how stable the site is, and for this the expertise of a soil engineer is needed.

Delivering the contracted works

Obviously, the main contractual obligation is to construct the building works, as specified. However, the steps along the way that deliver the finished building works need to comply with certain procedural requirements.

Progress schedule

As the project progresses, the builder will expect to be paid for each stage completed. That the work associated with each stage is indeed complete must be verified, usually by an independent inspector, such as the architect.

Statutory warranties

Like other expensive products, buildings are covered by warranties: six years for major defects and two years for minor defects. If something goes wrong during the warranty period, the builder has to make good at their own expense.

Completion date

The completion date signifies when the building is handed over from the builder to the client. Just like with the sale of a car, the transfer of ownership implies a transfer of responsibility. Post completion date, warranties come into effect and insurance responsibilities move to the owner.

Prime cost items

While it might be obvious that the builder is responsible for pouring the concrete slab, it may not be clear who is paying for the bedroom curtains. However, it should be clear. All the building's fittings and appliances should be listed, along with who – the builder or the client – is responsible for purchasing and installing them.

Additions and variations

As careful, thorough and even experienced as you might be, it is almost certain that changes to the original design will be required as the building progresses. This may be because specified materials prove unavailable, or because some deficiencies in design are uncovered, or even because the client simply has changed their mind on a feature. All such changes must be agreed, along with price adjustments, documented and signed by all involved.

Site access

Site access must be controlled, not only for security, but for safety. It is the responsibility of the lead contractor to manage the site, providing safety training, ensuring warning signage is in place, enforcing use of protective wear and diligently overseeing WHS standards. Even the client visiting the site must comply with the lead contractor's directives.

Termination of contract

Though not desirable, there may be extreme circumstances warranting a termination of contract – either instigated by the client or the builder. Usually such an event is inconvenient and expensive for both sides. Thus, while unlikely to occur, the way termination is to be instigated and managed must be spelled out clearly in the contract.

Compensation for loss or damage

When something goes wrong, whether due to the actions of the builder or client, resulting in a financial loss to the other side, the manner by which compensation is to be calculated and paid out must also be clear in the contract.

Conditions of the contract

Once the building has been completed, there needs to be a handover, in which the client verifies that all the works have been completed, as required.

For insurance and other purposes, this may be a specified date, though there will also be a warranties period initiated from that date in which any deficiencies that manifest will have to be corrected by the builder. Warranty periods will have different durations for different components of the building – heating, ventilation and air-conditioning (HVAC), concrete slab, fittings. There must also be a process by which control of the building is passed from the builder to the owner. This step is known as commissioning.

A third-party inspector will also be called in to verify that the finished building is constructed as per the building permit and is indeed fit for occupancy. In the final hand-over meeting, keys, certificates, warranties and other documents are given to the owner.

Connection and commissioning

Plant and equipment should all be operating. For example, the electrical authority will have conducted a final testing and inspection of the installed electrical system.

Licences, certificates and registrations

Certificate of occupancy and any other licences necessary for the owner to take occupancy are handed over.

Defect liability period

A defects liability period will come into effect, as established in the contract.

As constructed drawings

A set of drawings of the building, as completed, are handed over.

LEARNING TASK 2.6

REVIEWING THE CONSUMER BUILDING GUIDE IN YOUR STATE/TERRITORY

As a builder, you are obliged to inform your client of their rights. This is covered in the consumer building guides applicable to each state and territory.

Download a copy of the NSW consumer building guide from: https://www.fairtrading.nsw.gov.au/__data/assets/pdf_file/0019/382123/Consumer_Building_Guide.pdf

>>

>>

Or download a copy of the Victoria consumer building guide from:

https://www.consumer.vic.gov.au/housing/building-and-renovating/plan-and-manage-your-building-project/domestic-building-consumer-guide

Read through the guide and respond to the following:

1 Make a list of the contractor obligations (in your state) to the client; i.e. list the rights and protections your client has, as well as their obligations, according to this guide.
2 How would you explain all of these obligations (from the list you have made) in plain English to a client from a non-English-speaking background?

Industrial relations policy

Australia's form of government is classified as a socialist democracy. As such, Australia has very strong protections in place for employees, tending to see workers as the vulnerable party in the employer–employee relationship. Consequently, government has put in place a range of measures that businesses must comply with in order to uphold the rights of workers. As an employer, you will need to know about your legal obligations to employees.

Identifying industrial relations policies and obligations

There are effectively two types of people who will work for you. They are:

- independent contractors
- employees.

Employees and contractors

Employees work directly under the supervision of the employer, are given tasks to complete and are paid regularly on an ongoing regular basis. Contractors, or subcontractors, on the other hand, run their own businesses, have an ABN and determine their own work schedules, the work they will perform and how they will perform it.

In the construction industry, because work is intermittent, often lasting only as long as the project, there is a tendency to see a greater ratio of contractors to employees than in other industries. In fact, because the obligations of a business to an employee are much stricter than to a contractor, there is a tendency to prefer to engage people on a contractual basis, wherever possible.

Once a person is employed, it is not that easy to fire them. Moreover, employers must provide employees with a work contract, insurance cover, a legal minimum wage, manage their tax and superannuation, grant them holiday and sick leave, as well as other leave, and provide them with a safe work environment with all the associated equipment they need.

Ironically, the effort of government to provide such protections to employees is a reason why businesses are actually reluctant to take on fixed staff, unless absolutely necessary. A way around this is to employ people on a daily or weekly basis, or even to just hire contractors for everything. Even so, the government stipulates that when a person works exclusively for one business, they are not in fact an independent contractor, but have become an employee, and must be afforded the rights and protections given an employee.

Unions

Not mentioned so far are the unions. Historically, construction was an area that was largely low-skilled, dependent on manual labour and was piecemeal – irregular and uncertain. Under such conditions the greatest bargaining power lay with the companies who, given the large pool of willing workers, could easily exploit them with low wages and poor conditions.

This environment gave rise to unions. The idea of unions is that while an individual cannot really bargain alone for better working conditions, a collective of all workers can – they become a force companies in need of labour cannot ignore. Unions demanded and got better working conditions for their members.

If you are ever in Melbourne's CBD, you can see evidence of the power of unions at the top end of Russell Street, where it meets Victoria Parade. On the north-east corner is the Trades Hall, where former Prime Minister Bob Hawke began his career; on the south-west corner you will see a monument with the numbers 8-8-8 on it. This signifies that the day is divided into eight hours of work, eight hours of rest and eight hours of sleep. The eight-hour work day was a right won by unions, and we now take it for granted that employers cannot demand we work for as long as they tell us to. Of course, they once could.

Workplace agreements

Also, as we have seen, government too has set in place protections for workers. The combination of laws and union demands create the standards that a business must offer its workers. Those standards, expressed as policies and procedures, should be put into force as a 'workplace agreement'. While an employment contract is between the private individual worker and their company, workplace agreements lay out the conditions for all workers and their company.

An important corollary is that companies should only engage with other companies which treat their employees ethically and according to law. (Which is why there is public outrage sometimes when we hear of companies buying products or services from overseas firms who mistreat, exploit, or underpay their workers.)

Applying relevant awards

A contract is an agreement made between two parties. An employment contract is where one party agrees to perform specified work for a set renumeration paid by the other party. I may offer you work for $10 per hour and you may agree. The difficulty, however, is that this particular arrangement is illegal. As with so many things, the government intervenes and demands that any employment contract must meet certain minimum standards; such as minimum pay per hour. These minimum standards are called awards.

Thus, in employing anybody you need to be aware of your obligations, just as a prospective employee needs to be aware of their rights.

National Employment Standards (NES)

The minimum standards for employment are set out in the National Employment Standards (NES). These may be found under the *Fair Work Act*, set out here: https://www.legislation.gov.au/Details/C2017C00323

The NES covers issues such as:

- maximum weekly hours of work
- requests for flexible working arrangements
- annual leave
- long service leave
- public holidays
- parental leave and entitlements
- personal care and compassionate leave
- community service leave
- notice of termination and redundancy.

Workplace agreements

Workplace agreements are discussed above. The other consideration in regards to workplace policy relate to equal opportunity legislation in both recruitment and in professional development. A business should have an equal employment opportunity policy that leads to a statement of principles and objectives. The business then implements that policy by educating employees so that a welcoming work culture is created, while managing grievances fairly.

Equal employment opportunity

Questions of equality, equity and diversity are ideological issues. They are controversial and have a polarising effect on society. For example, the term 'equality' sounds moral, and few would oppose it, but it has different meanings to different people. 'Equality of opportunity' means everybody has an equal chance to get ahead, but given different levels of ability, not everybody will – this is a rightist view. 'Equality of outcome' means that various identifiable groups in society must be equally represented, and when some groups don't succeed, this must be corrected – this is a leftist view.

The issue of women in the workplace – equal employment opportunity – is an example of ideology moving into the realm of business practices. Should companies take on a set quota of women, as a principle, or should they continue to hire the best people they can find? As stated, the answer to this is controversial and polarising.

Even so, it is clear that politicians, both Liberal and Labor, are finding themselves leaning increasingly left on the issue, and this is reflected in ongoing measures placed into law to increase the participation rates of those groups in society that are seen as less empowered than 'white, heterosexual males'. Certainly, construction stands out as a male-dominated industry, and consequently as a target for progressive change.

The *Equal Employment Opportunities Act* can be downloaded at: https://www.legislation.gov.au/Details/C2004C00712/026f40ef-ad72-4b70-a23a-16476215f494

Recruitment process

As part of a business' policies and procedures, there should be a transparent and fair recruitment process. The process needs to be documented such that claims against unfair employment practices can be defended.

Some of the key steps involved in a recruitment process include:

- clarify the precise needs of the position being advertised
- develop the position description
- advertise the position
- shortlist candidates – matching applications against position requirements
- prepare for the interview – ensure the interview questions and process is similar for all, and that a number of interviewees make up the interview panel
- selection – should be based on the best qualified candidate for the job, as defined
- notify candidates – both those that were unsuccessful as well as the one chosen.

Discrimination and harassment

An employer cannot discriminate. For example, you cannot ask an employment candidate about their ethnic background, religious beliefs, sexual preferences, their age, whether they are married, or whether they plan to have children. You cannot sexually harass or bully people either.

There are a whole range of laws protecting individuals against discrimination (see Table 2.3).

Training agreements

Construction sites are dangerous places, incurring above-average levels of serious injuries. It is therefore important that staff are well trained in their role, as well as in matters of WHS.

TABLE 2.3 Laws against discrimination

Act	Website
Australian Human Rights Commission Act 1986	https://www.legislation.gov.au/Details/C2017C00143
Age Discrimination Act 2004	https://www.legislation.gov.au/Details/C2018C00322
Disability Discrimination Act 1992	https://www.legislation.gov.au/Details/C2018C00125
Racial Discrimination Act 1975	https://www.legislation.gov.au/Details/C2016C00089
Sex Discrimination Act 1984	https://www.legislation.gov.au/Details/C2017C00383
Fair Work Act 2009	https://www.legislation.gov.au/Details/C2017C00323

Work skills related training

All workers must have an appropriate level of skill for the work they do. Much construction work is trades related, and as such apprenticeships are characteristic of the industry. On the one hand, apprentices can become loyal employees after their apprenticeships are complete, but on the other, they require an investment in time and effort to be properly trained. Gone are the days when apprentices could be exploited as cheap labour.

Apprenticeships generally require both on the job and off the job (educational) training. Employers taking on apprentices should document exactly what they are offering an apprentice in terms of training, and what the apprentice's work obligations to the employer are in return.

Safety training

When someone is seriously injured at work, the employer will be investigated to audit whether they have honoured their duty of care. It is easy to overlook your WHS obligations, but should something go wrong a negligent employer will find themselves facing the full force of the law, with a hefty fine, compensation or even jail time.

Remember, even if the injured worker was technically at fault in their own injury, the employer will be investigated to confirm that they did all they legally should to avert the injury. Here are some guidelines on what steps are needed to minimise safety risks:

- identify all risks, both in terms of severity and likelihood of happening
- remove all possible risks
- for remaining risks, further limit the severity and likelihood using substitute materials (e.g. less dangerous solvents), and mechanical controls (e.g. barriers and locks)
- minimise risks again with administrative controls (warning signs, protective equipment)
- review, revise and improve on a regular basis.

Making legal information available to employees

Every business must have its own clear policies and procedures. These should be followed by all, including the managers. After all, a business should not be the extension of any particular individual, running on whims and moods, like a dictatorship.

Guidelines on business industrial relations policies can be found for NSW at: http://www.industrialrelations.nsw.gov.au/

The Employsure fair work in Australia guide can be found at: https://employsure.com.au/guides/fair-work-australia/

Workplace policies should cover the following issues:

- code of conduct
- recruitment procedures
- health and safety policy
- drug and alcohol policy
- non-smoking policy

CASE STUDY

Workplace injury

In a recent publicised case, a worker who had refused to wear a safety glove ended up slicing off his own hand. The employer's defence was that they had provided the glove to the worker, but the worker had habitually and consistently taken the glove off, despite warnings.

The problem had gotten so bad the employer fired the worker for not wearing the glove, but the worker was reinstated on the grounds of unfair dismissal – the employer had fired the worker without going through the proper procedures of issuing written warnings.

It was after the worker was reinstated that he lost his hand.

Even so, the employer was found liable for the worker's injury. In the judgment handed down against the employer, the judge noted that it was the responsibility of the employer to provide appropriate and sufficient training to the injured worker on wearing the glove, which the employer had not done.

What do you think of this outcome?

- anti-discrimination and harassment policy
- grievance handling policy
- discipline and termination policy
- internet and email policy
- mobile phone policy
- use of social media.

LEARNING TASK 2.7 FAIR WORK STATEMENT (FWS)

Every employee must be given a Fair Work (Information) Statement (FWS) by their employer:

- https://www.fairwork.gov.au/employee-entitlements/national-employment-standards/fair-work-information-statement
- https://www.fairwork.gov.au/ArticleDocuments/723/Fair-work-information-statement.pdf.aspx

Have a look at the statement and its requirements. As an employer, how would you go about giving your employees their FWS? What information would it contain?

Dispute resolution

Disputes and the pathways to avoid and/or solve them are covered extensively in in Chapter 7; however, some cursory discussion is warranted here from the legal angle. While nobody wants disputes, it should also be recognised that they are inevitable and a natural part of life and doing business. It is important that businesses have a process for dealing with disputes that is as constructive as possible, and that leaves all parties to a dispute satisfied that justice has been done. Disputes can arise between people in a firm, or they may arise with people dealing with the firm.

Organisational dispute resolution processes

Businesses should have a formal dispute resolution process. Since disputes occur between people and emotions get involved, it is also necessary that those handling disputes have good people skills.

Did you know that there are basically only three causes of disputes? When looking to intervene to resolve a dispute, one should begin by deciding where the cause of the dispute lies. The first cause is a difference of view over what the facts are. In settling this sort of dispute, one needs to find agreement on where to find the facts. Once we all agree on this we can go and check on the facts. Disagreements over the facts of an issue are the easiest to resolve, particularly when everything said and done is documented. Keeping records of all decisions made will avoid the biggest source of disputes.

The second cause of disputes has to do with relationships. A dispute may arise when someone feels offended, slandered or insulted by another. Here, we need to uncover people's hurt feelings and the cause for the hurt. Resolving these sorts of disputes is a matter of finding a way to reconcile the parties involved.

The third cause of disputes is the hardest to resolve. Here, one side simply wants power and control over the other side. These sorts of disputes are the most vicious. You will probably want to resort to legal measures or some outside help to give you the leverage that offsets the aggressors' power-play.

Workplace disputes

Here is a simple five-step process to follow in resolving workplace disputes:

- identify the source of the conflict
- look beyond the incident to see if there is a deeper, longer-standing issue
- ask all sides for solutions
- discuss these solutions to find one which everybody can accept
- find an agreement on what should be done going forward.

Unfair dismissal

Once you employ somebody as a permanent employee, you cannot just fire them at will. You need both a good reason, and to follow a fair process. The fair process required usually demands that a number of warnings be given – verbal, written and second written – which stipulates the employee's failures, as well as specific remedial measures they must take to get back on track. If an employee complains of **unfair dismissal** to the Fair Work, Ombudsman and the employer cannot show documentation of a fair dismissal process, the worker will be reinstated and the employer may have to pay back-pay.

To make a claim of unfair dismissal, the employee must have been working with the company for more than six months. This is generally called a probation period. For smaller businesses (with less than 15 employees) this is relaxed to 12 months.

Customer complaints

Complaints from customers may be of a personal nature, just as may occur with staff; however, they are more likely to be systemic – having something to do with problems in the way your business is operating. For example, if you get lots of calls that the roofs of the houses you have built are leaking, then there is likely something wrong with the way you are building roofs. The point here is not that complaints are an annoyance, but rather that they are important feedback on your business, without which you could finally land in serious trouble.

Client complaints

Here is a straightforward process for managing client complaints:

- receive complaint
- document complaint in a form
- send complaint to relevant employee responsible
- investigate cause of complaint
- decide on remedial measures
- take corrective action on the problem
- add notes of corrective action on complaint form
- management should correct business practices to ensure no repeat of the issue.

Documenting disputes and outcomes

People make mistakes, and hopefully they learn from them. Sociologists and evolutionists argue that this is the primary reason animals and humans have memory. Companies make mistakes too, but can they learn from their mistakes – do businesses have memories? Well, in smaller businesses the owners are basically the company. What the business knows is what the owner knows – they are one and the same.

However, as companies get larger, they need a formal knowledge management system. In such cases, information has to be stored, processed and analysed so that lessons can be extracted to improve company performance. A complaints register can also be used as a means for learning what the company is doing well, and not so well. Understanding what a business is doing wrong is essential in preventing the business from making fatal mistakes. It is also helpful in uncovering new ways to be even more competitive.

LEARNING TASK 2.8 LEARNING FROM COMPLAINTS

Geelong lies about 70 km south-west of Melbourne. It is now one of the fastest growing areas in Australia, with immense opportunities for local domestic builders. You are thinking of expanding your own operations into the region.

However, most of the new developments are being built on what was swamp-land. Soil in the area is highly reactive clay. A review of new home buyer chat rooms reveals customers are unsatisfied with their new homes. In particular, dozens of houses are developing large cracks in their slabs.

On further investigation you find that waffle slabs are the most common system being used, and that engineers are saying that while these technically conform to the codes for reactive soils, they are in fact under-designed.

How would you go about competing in this market? Going in with a similar product and risk complaints, or try to persuade cash-strapped families to pay more for a stronger slab? Which approach will get you the business?

SUMMARY

This chapter covered the legal side of working on building and construction projects. While you may be skilled in your trade, a great builder, or a successful businessman, this chapter will have made clear to you that you also need to know and follow the law.

There are many laws and legal requirements that you must comply with when working in the construction field. These include licensing and registration, WHS, building permits, codes and Acts, contracts, insurance, obligations to employees, payroll, GST, discrimination, as well as handling disputes.

Hopefully, as you read through that list, they all now mean something to you, and you understand why they are important.

REFERENCES AND FURTHER READING

Building Code of Australia (BCA) can be purchased from the HIA: **https://hia.com.au/Shop/bca**

All Australian laws can be found at the federal register of legislation: **https://www.legislation.gov.au/Home**

Also see Lawlex at SAI Global: **https://lawlex.com.au/**

NSW planning permit process: **https://www.planning.nsw.gov.au/about-us/nsw-planning-portal**

Skills Certified Australia: **https://www.skillscertified.com.au/**

Victorian planning permit process: **https://www.planning.vic.gov.au/permits-and-applications/do-i-need-a-permit**

Workplace Health and Safety Act, and codes of practice: **https://www.business.gov.au/risk-management/health-and-safety/work-health-and-safety**

3 CONSTRUCTION CONTRACTS

Chapter overview

Virtually all construction work undertaken in Australia is required by law to be bound by a contract. Therefore, no matter what your role is within the industry, you will be engaged in contracted works of some kind. What this means, however, is slightly more complex than a simple piece of paper signed by two people. What is on the paper, how that paper becomes, or fails to become, a legally binding agreement, is often highly complex. That it doesn't even need to be written is frequently not understood. And how easy it is to breach a contract, unfortunately, can sometimes be all too simple. This chapter sheds light on construction contracts and other contractual issues through the following elements.

Elements

This chapter provides knowledge and skill development materials on how to:

1 identify and analyse the essential elements, sections and clauses of a business contract
2 select an appropriate contract for the works to being undertaken
3 prepare a contract.

Introduction

For some time now, contracts have been a legal requirement for most levels of building works in Australia. Depending upon the state or territory in which you reside, there will be some allowance for minor works to be conducted without a contract. In NSW, for example, this will be for works with a total cost (including materials, labour and GST) of less than $5000. In Queensland the figure is a mere $3300, while in Tasmania it is set much higher at $20,000. In all cases, it is required by the various state or territory legislation that contracts be in writing prior to the commencement of work. Yet despite this stipulation, it is possible to find yourself with contractual obligations even though no signed document exists between you and someone else.

It is the purpose of this chapter to guide you through the Australian building and construction contract landscape. In so doing, we will be looking at:

- what a contract actually is, how and of what it is composed, where you can get one from
- the rights and the responsibilities on each side of a contract
- the types of contracts and how to select the right one
- what you need to put into different contracts
- the factors around the termination of a contract
- what the final contract should look like.

We can take these pretty much in order, so we shall begin with the essentials.

Business contracts: the essentials

A **contract** is an agreement, a type of promise, between two or more 'parties'. The 'parties' being individuals, companies, organisations or groups that are in a legal position to make such agreements. This correctly suggests that not everyone, or every group, is actually able to make a legally binding contract. What 'form' a contract takes will differ depending upon the purpose of the contract, and as far as building and construction goes, the costs involved.

Types of building and construction contracts

Building and construction contracts in Australia tend to fall into one of two groupings: domestic contracts – for dwellings, garages, fences and the like (i.e. Class 1 and Class 10 structures); and commercial contracts (i.e. most other classes of buildings). Contracts for the domestic housing sector are regulated by the relevant state or territory Act, such as:

- Victoria: *Domestic Building Contracts Act 1995 (2018)*
- Western Australia: *Home Building Contracts Act 1991 (2018)*
- NSW: *Home Building Act 1989 (2018)*.

(Links for these and those of other states may be found at the end of this chapter and/or in the introduction to the book.)

These Acts dictate only the domestic sector, however; but in so doing, they limit the type and form that contracts may take when used in this area. We will discuss the specifics of domestic contracts later in the chapter; for the moment we will expand upon the various types of contracts in a more general sense.

Common types of contracts

In the building and construction industry, contracts can take a variety of forms. We use them for large and small domestic works, commercial jobs, even minor renovations and fencing. The most common contract is known as a **fixed price**, **fixed sum**, or **lump sum** contract. There are, however, several others that fall into one or the other of the classifications below:

- lump sum with rise and fall
- **cost plus**
  - cost plus a fee
  - cost plus a percentage
- guaranteed maximum price
- bespoke or custom contracts.

Excepting the 'bespoke' or custom contracts, none of the above may be used in the domestic construction industry except for in special, set circumstances. Use outside of these set circumstances can lead to heavy penalties under Australian law. We will discuss the sources of law surrounding contracts in Australia shortly; for the moment we quickly look at each type of contract because though they mostly cannot be used in the domestic sector, they are most certainly available to the commercial sector.

Fixed price contracts

Known also as a lump sum contract, this is the most common type of contract you will come across in any sector of the building and construction industry. Almost all domestic houses will be built on this basis. This is because state and territory regulations governing this sector require that, with very limited exceptions, all domestic building contracts must be undertaken to a fixed price.

Despite this, many home buyers find that their 'fixed' price can increase upon final payment. The features that allow this to occur legally within the scope of the contract are:

- variations
- prime cost items
- provisional sums
- taxes and duties.

These items will be discussed later in the chapter, for now it suffices that you be aware that 'fixed' price is not absolute: it can be changed (up or down) based upon certain provisions within the document.

Rise and fall contracts

A rise and fall contract is one that is designed to reduce the risk burden held by the builder. A fixed price contract, aside from the provisions mentioned above, cannot be raised simply because the cost of materials, labour, or transport went up. The idea behind rise and fall contracts is that should any of these or other nominated factors go up, or down, then by using a formula explicitly described in the contract the total price shall rise or fall accordingly. This means that the client may have to pay more than the original quoted price, or can gain; i.e. pay less, should prices fall.

Such contracts reduce a builder's need to include high margins in their quotes just to ensure they are not caught by an upsurge in prices. Again, this may mean that the initial quoted price, and the final price, is lower under this form of contract than it may have been with a fixed price contract.

On the other hand, it means all risk for price fluctuations falls to the client. The client may gain, or they may lose, the 'gamble' is theirs, not the builder's.

State and territory regulations and domestic building Acts vary marketedly with regards to what contracts are acceptable under their jurisdiction. Western Australia, for example, explicitly prohibits rise and fall contracts in the domestic sector. NSW and Victoria, on the other hand, make no mention of this type of contract, and hence there is no prohibition.

Cost plus

Cost-plus contracts are a deceptively simple contract framed around the actual costs incurred by the builder. There are two forms of this contract: cost plus an agreed fee, and cost plus a percentage.

- **Cost plus a fee:** In this case, the contract stipulates that the builder will receive a set amount, the fee, on top of the actual cost incurred and documented.
- **Cost plus a percentage:** This type of contract states that the builder will receive an agreed percentage of the actual costs additional to those costs.

As with rise and fall contracts, some states and territories disallow this form of contract for domestic housing; or, if allowed, only in special circumstances. The reason for this restriction is because contracts of this nature have no 'ceiling' to the final cost. This means that clients can find themselves paying many thousands more than they had originally envisaged or planned for. In addition, banks will seldom, if ever, agree to a domestic mortgage framed around a cost-plus building contract.

You should note also that in some states and territories, cost-plus contracts are prohibited for domestic building work under a certain project value. Victoria, for example, expressly excludes cost-plus contracts for any project under $1 million. It is important to check with the relevant authority of the state in which a project is being undertaken as to the validity of this form of contract.

Guaranteed maximum price

This is in many ways similar to the cost-plus contract. **Guaranteed maximum price (GMP) contracts** allow the builder to be paid for their actual expenditure plus a fee. However, in GMP contracts there is a ceiling – the maximum price. The benefit to the client is that any cost overruns (additional expenses) are carried by the builder, while any savings or cost underruns means the price is lowered; i.e. the client pays less overall.

While these contracts are gaining favour in some quarters, there are some significant concerns from all sides. From the builder's perspective, the contract offers no incentive for increased profits by coming in on time or at reduced costs, unless the contract provides for profit sharing. Even so, this assumes a profit; in cases where there are cost overruns, the burden is on the builder to pay. From the client's and financier's (bank's) perspective, the works may be compromised as the builder, sensing or suffering cost overruns, starts to cut corners and lower quality. Additional complexity can occur with variations to the contracted works. The GMP only applies to the originally contracted works, so the GMP contract must then be changed accordingly – in some cases this means constant new versions of the contract rather than simpler variations to the other types mentioned above.

Custom contracts

Fixed price, cost plus, and rise and fall contracts are fairly common to the industry depending upon the state or territory you are in. This being the case, you can generally access a standardised contract suitable to your context from a number of sources such as the Housing Industry Association (HIA) or the Master Builders Australia (MBA) – known in each state as Master Builders Association. In addition, many public authorities – such as NSW Fair Trading – and local government bodies also distribute standard contracts. These contracts will generally suffice for most domestic and light commercial works. For larger or more complex projects, however, it is frequently necessary to adapt or customise the contracts, even if using Standards Australia versions (see below).

The generally accepted reason for a customised or 'bespoke' contract is the need to offset risk, or at least to make the risks more equitably shared among the parties concerned. This means that most changes in the contracts will align with areas such as:

- extensions of time
- handling of variations
- payment schedules
- penalties associated with delays.

Significant legal advice is required when customising a contract as even seemingly minor changes to wording can have significant implications to the interests of either party. In the case of domestic contracts, the relevant state or territory Act dictates the content of contracts to this sector and so must be adhered to even when creating a custom contract.

Standard contracts

A research report published by Melbourne University in 2014 (Sharkey et al. 2014) suggests that over two-thirds of all building and construction contracts use standard forms developed by industry organisations or Standards Australia. Australian standard contracts (the AS 4000 series) were shown to dominate overall, particularly in the commercial sector, being used in nearly half of all the projects they surveyed. It was also found that in most commercial cases (80%), it is the client or 'principal' (or their legal representative) who determines the type of contract and its source.

As stated above, there are many sources of standard contracts open to builders and clients. Many are free, others are available for a fee from organisations such as the MBA or HIA. Australian standard contracts are available through SAI Global (the access point for Australian standards generally); however, a fee must be paid for other than reference copies.

Irrespective of where a standard contract originates from (MBA, Standards Australia, etc.) they frequently have some form of alphanumerical code: AS 4300 and AS 4000 being common Australian standard contracts, while a predominant NSW government contract is GC21, The MBA jointly publish their contracts with the Australian Institute of Architects under the title of Australian Building Industry Contracts or the ABIC suite with codes like ABIC MW-2008. On the other hand, many of the free domestic contracts from public authorities such as NSW Fair Trading or Consumer Affairs Victoria will simply have a title outlining its target usage; e.g. NSW Home Building Contract for Work over $20,000.

Key terms of contracts

The 'terms' of a contract are the key components that outline the rights and obligations of each party. These terms are usually defined within the document by the 'clauses' of the contract. Indeed, in some state domestic regulations, the word 'term' has been removed and replaced by 'clause' to reduce confusion. Some terms are **expressly** stated, while others may be ***implied***. We need to look quickly at each of these types of contractual terms individually before looking at terms more generally.

Express terms

Good contracts are those in which key terms are 'express' – the legal word for *clearly defined and stated* – and implied terms are minimised. This is because implied terms are by their very nature ill-defined and overly open to interpretation. While express terms can be verbal, in the construction and housing industry they will always be written and voluntarily included in a contract. Though they vary, each state and territory has some form of building contract Act. This regulates what terms must be express in a domestic building contract.

Implied terms

Despite the above, there will always be some level of implied terms surrounding a building contract. These include the legislation that frames the contract, some commonly accepted terminology, and in the very action of developing a written contract in the first instance.

Section 8 of the Victorian *Domestic Building Contracts Act 1995*, for example, entitled Implied warranties concerning all domestic building work, sets out conditions for materials, workmanship, compliance with the Building Act and the like. Likewise, the Western Australian *Home Building Contracts Act 1991* frames the implied conditions for defect liability and building permits.

Similar conditions form part of the implied terms for contracts in all states and territories. They are implied because, though such issues may not be 'express' within the terms of the contract, the 'Act' states that they are terms of the contract nonetheless.

Common contractual terms

Using the example of domestic contracts (Consumer Affairs Victoria, *Building contract for new homes*, p. 18; see References at the end of this chapter), you will notice that a contract typically starts with a range of preliminary material prior to the contract proper being reached. In the Victorian example, the terms of the contract do not begin until Part B, some 16 pages in. The NSW offering is much simpler, with the terms starting at page 6.

The terms of a contract typically begin or end with a clause called Definitions; in the Victorian example it is ***Clause 1***. The purpose of this clause is to render the terms of the contract as clearly (express) as possible and thereby reduce the implied; even seemingly simple concepts such as a 'Business Day', or 'Defect', are therefore clearly defined.

Clause 2 of this contract seeks to further reduce implied terms by offering guidance on how the various elements of the contract are to be interpreted. A reading of Clause 2.8 is an example:

2 Interpretation

Clause function: This clause provides guidance about how the contract is to be interpreted.

....

2.8 A reference to 'include' or 'including' or 'for example' in a list does not mean that items not listed are not included.

Clause 3 then goes on to outline how the two parties involved in the contract will communicate with each other formally or '...for the purposes of this contract...'. This drills right down to the time an email is *deemed* to have been received based upon the day and time it was sent.

It is only at ***Clause 4*** that the contract details actually begin, though each of the previous three clauses are considered part of the contract 'terms'.

There can be many terms to a contract, and they may vary dramatically from one contract to another. In the Victorian contract for example, you will find some 43 clauses, two schedules, and a further six clauses in Part C that expand upon legislation relevant to some key clauses of Part B. These all form part of the 'Terms' of the contract.

It is important that you familiarise yourself with the terms of both example contracts so that you get a feel for how clauses are phrased, and the manner in which contract documents are formatted. All terms of a contract are important, otherwise they would not be there, so it is difficult to state that some are 'key', without suggesting that therefore some are not. However, it is clear that some are more important than others to one party or the other. The General Conditions is one of these more important terms of a contract that need careful study by both parties.

Conditions to contracts

The General Conditions of a contract hold a raft of information covering the specific responsibilities of both parties. A condition in a contract outlines:

- particular circumstances that when done, or not done, may trigger some other action; or
- they stipulate the circumstances under which a particular action can be taken or should be taken.

Breaching a condition can lead to the contract becoming null and void, one party being in the position to sue the other, or the final price being altered. Conditions are therefore fundamental to the application of a contract.

Common domestic conditions

Domestic contract conditions will include requirements of the builder to establish the state of the site foundations, the location of the site, surveys and the like. There will also be a range of subclauses that require either party to perform a particular action, such as the owner to supply a copy of the title, within a set period of time. The Victorian example looks like this:

12.4 The *Builder* may give the *Building Owner* a notice that within 10 *Business Days* after receiving the notice, the *Building Owner* must give the *Builder* satisfactory evidence of the:

a) *Building Owner*'s title to the *Building Site*

b) dimensions of the *Building Site*.

...

12.6 If the *Building Owner* fails to provide satisfactory evidence following a request under Clause 12.4, the *Builder* may suspend the *Work* under Clause 35, or take action to terminate this contract under Clause 37.1.

There may be a number of other conditions that have similar consequences – i.e. the termination of the contract – scattered among the subclauses. It is important for both parties to ensure that they are aware of each of these and act accordingly.

Other common domestic contract conditions are those surrounding variations to the final contracted price. As stated previously, rise and fall contracts and cost-plus contracts have very limited application in the domestic sector. However, there are express conditions by which a contract price can be altered while remaining within regulations imposed by state or territory Acts governing domestic contracts. Contracts will generally offer a warning to this effect, such as the Victorian example below (Building contract for new homes, Victoria, p. 27):

Warning: Changes to the Contract Price

The price of this contract may be altered as a result of:

- the actual cost of *Prime Cost Items* and *Work* for which *Provisional Sums* have been specified being more or less than the estimates set out in the contract (see Clause 24 and Clause 25)
- *Variations*, including those required by council or a registered building surveyor (see Clause 23)
- interest on overdue payments (see Clause 27.2).

Ensure that you fully understand how the clauses dealing with these matters may affect the *Contract Price*.

Note that the NSW domestic housing contract ($20,000) expressly includes increases to taxes as do those of some other states. Most state and territory Acts, however, include this provision (changes in taxes) as an allowable change to the final contracted price. This then becomes an *implied term* of the contract (see p. 56).

The phrases *prime cost items* and *provisional sums* are discussed further in Chapter 5; for now, you simply need to understand that both these represent costings that are not fixed, and so can change as the job progresses. The concept of 'variations', however, is best addressed now.

Variations

Variations to contracted work are common in our industry. As the work unfolds, the client, the builder or even the architect may find cause to make changes to the proposed structure. Sometimes it is simply a case of material or product availability. A contract is set up to allow for such eventualities by way of a 'variations' clause. This will be a very distinct and well-defined clause outlining the process by which changes to the contracted works may occur. The main point is that a variation must be costed, and the implications to time lines and other effects to the final structure be fully understood by the client. It will also be required that any such change be fully documented and signed prior to the works being undertaken.

The key point of a variation clause is that it allows changes to the contract itself, which in turn may change the final price.

Commercial conditions

There are a significant number of other conditions that may be applied to both domestic and commercial contracts. From a commercial perspective, the full listing of appropriate conditions can be found in the Standard Australian Contract document AS 4000 General conditions of contract. This standard is part of the 'AS4000 series' suite of documents offering standard contracts for all elements of major and minor construction works – though mainly works for which a ***superintendent**** is required. In the main they are not dissimilar to the domestic sector, differing mostly in detail and phrasing.

One example of a shift in phrasing and detail is ***Latent conditions*** – Clause 25 of AS 4000. Latent conditions are conditions of the site and surrounds (except weather) that differ from those that could reasonably have been anticipated. Examples can include large boulders, old unmapped infrastructure (pipe, lines, etc.), mineral springs, footing incursions from neighbouring properties, cultural finds of significance and the like.

Latent conditions are an important inclusion as they may legitimately raise the price of a contract through an unavoidable variation. It is therefore important that you and your client are conscious of this condition within the contract: for the builder, because subcontractors may need to be called up to fix this condition; for the client, as they will ultimately wear the cost.

Sticking with the foundations and excavation, it should also be noted that any finds of value – minerals, fossils, coins, treasure or the like – are, according to a minor condition common to commercial contracts, the property of the client or 'principal' – See AS 4000 subclause 24.3.

* Note: a *superintendent* differs from a supervisor in being engaged by the client or principal and has two roles: act to ensure contractual obligations are followed through; and to act as an independent certifier. So, on the one hand they can suspend work on behalf of the principal, and on the other certify delay costs to the contractor. Needless to say, superintendents are hired on the basis of scrupulous honesty.

Warranties

The concept of warranty is addressed in two distinct ways in a contract. There is the concept you will be familiar with whereby the works are 'warranted' against defects for a stated period of time; and there is the perhaps less familiar concept of a promise or guarantee by the builder that the works undertaken will be done in a specified manner. Both are discussed here, dealing with the latter first.

As outlined above, the conditions of a contract may be viewed as the essential terms of the agreement. Breach of a condition can lead to termination of the contract. Warranties, on the other hand, may be viewed as a promised manner of conduct – it is not that they are unessential, it's just that breaching them will not generally lead to termination. Instead they will most likely lead to claims for damages; e.g. for some level of restitution, lowering of price or rectification.

Many warranties will be expressly stated as a clause of the contract, others will be implied by various state and territory Acts and/or regulations. The latter are known as statutory warranties.

Statutory warranties

Statutory warranties are those implied by the Acts and regulations dealing with building activities in each state or territory. In the domestic sector these will generally require the work to be undertaken:

- skilfully and carefully
- with quality new materials
- lawfully
- to the plans and specifications provided
- in a timely manner, and if a dwelling, it will be suitable for habitation
- such that provisional sums and prime cost items will be calculated with care and skill
- whereby the builder will provide everything necessary to complete the project
- whereby the owner must pay for the works so conducted.

On occasions, statutory warranties may be called up by the General Conditions of a contract to give them more clarity; Clause 3 part (a) of the NSW Home building contract (p. 7), being an example. The word 'warrant' will also appear in various parts of the contract; however, it will generally do so in regard to matters mentioned previously under statutory warranties such as prime cost items or preliminary sums.

In the commercial sector, the same basic principles apply. However, with regards to design and construction contracts, the contract will also warrant that you:

- examine the preliminary designs provided by the principal and that they are suitable for the stated project requirements
- will carry out the design obligations to comply with the project requirements
- will complete the works so that they will be fit for their stated purpose in compliance with the requirements of the contract.

Warranty periods

In the domestic sector, the builder is required to warrant their work for a period of time dictated by the relevant state or territory Acts and regulations. In most cases this is six years from completion, except for the ACT and SA where it is five years. That is, the statutory warranties expressed above are covered for this period of time. Builders warranty insurance (see Chapter 2, p. 42) is a government-backed scheme designed to cover homeowners over this period for instances when builders become insolvent, die or otherwise cease to operate.

Other contractual terms

While it's not within the scope of this chapter to cover each and every term in a contract, there are just a few additional terms or clauses of which you should be specifically aware.

Dispute resolution clauses

Disputes and misunderstandings are common in the construction industry, mainly due to misinterpretation rather than any attempt by one party to cheat the other. Understanding this to be the case, governments (state, territory and federal) have successfully reduced the number of court cases by ensuring there are defined – express – clauses in contracts dictating the path by which disputes should be handled. Typically, this will be:

- document the issues for each other to study in writing
- talk to each other honestly in a timely fashion
- continue with your obligations under the contract
- only if the above fails, then lodge a complaint with the building complaints authority of your state or territory.

Commercial contracts generally skip the last point and have their own plan for mediation; then, failing that, arbitration. These paths will be discussed in Chapter 7 Resolve business disputes.

For an example clause of this type, see Clause 34 of the Victorian Building Contract (pp. 55–6).

Exclusion clauses

These seek to limit the damages that one party may obtain from another for breaching some part of the contract. Generally, these clauses are prohibited in any Australian domestic building contract, but they are sometimes found in commercial contracts. Courts do not favour these clauses as they are often wordy and deliberately hard to decipher. Indeed, courts tend to treat these clauses in such a way that they work against the party it was designed to benefit. This being the case, both you and the client should be wary of entering into contracts where exclusion clauses have been inserted.

That stated, claims by either party that they did not fully understand a clause or condition in a contract, or did not 'read that bit', is a not a valid means of denying a liability or obligation, or pursuing claims against the other. Limitation of liability clauses will find their way into contracts, particularly those of larger commercial works. In addition, they are invariably in contracts associated with your project, such as the delivery of materials. A transport company's limited liability clause could mean, for example, that the monetary value of materials or components for your job, damaged in transit, may not be fully recoverable.

Conditional clauses

It is common for contracts to include conditions under which the contract may be terminated by either party. In reviewing any of the contract examples offered in this chapter, you will find clauses to this effect – such as Clause 25 or 26 of the NSW Home building contract (pp. 22 & 23). You will find several other conditions leading to termination within the document. These are known as **conditions subsequent**; i.e. something that happens after the contracted works have begun.

However, there are also **precedent conditions**. Some contracts are set up so that something has to happen before the contract fully comes into affect. That is, something must be in place, or some action must occur first: only after this has happened will work begin. If the action does not occur within a specified timeframe, then the contract can be cancelled. An example of this is Clause 17 of the NSW Home building contract (p. 18) whereby the contract may be terminated if council approval is not gained within 60 days of sign off.

Entire agreement clauses

Entire agreement clauses are a common term in larger construction contracts. The purpose of such a clause is to consolidate the promise being made between the two parties. That is, to make it clear and succinct, without adding or excluding anything. You will find, therefore, that the language in such clauses tends not to be exclusionary, but rather of a summary and positive nature. Another way to look at this type of clause (it has other names such as *integration clause* or *merger clause* and even *boilerplate clause*) is as a declaration by both parties that the contract represents the entirety of the agreement being made. The wording of this clause can be quite brief and remarkably straightforward; an example offered by Lachlan McKnight, CEO of LegalVision, is as follows:

> This agreement is the entire agreement between the parties in relation to the subject matter and replaces all previous representations or proposals not contained in this agreement

Source: https://legalvision.com.au/entire-agreement-clause/, accessed 14/7/2019

This clause comes into its own when previous dealings have been conducted in which alternative works or services have been discussed. These prior dealings are made redundant with the inclusion of this clause.

Note: Such a clause does not exclude implied terms, such as those conferred on any contract through a state or territory Act or regulation.

Unfair contractual terms

The Australian Consumer Law (ACL) is a national law covering, and administered by, all states and territories. This law allows courts to limit or make void any term or condition in a contract seen by that court to be unfair. It takes into account the relative strength of each party in the bargaining process leading up to the creation of the contract. It also includes what is referred to in law as unconscionable conduct in business transactions. This is conduct by one party towards the other that is clearly unfair and, what is more, the party doing it knows this. That is, the conduct is immoral or deliberately aimed to deceive without quite being fraudulent. Such clauses will never pass in court, particularly if those being deceived are in a significantly less powerful position than those who would profit by the clause.

The intent to enter a legal relationship

For a contract to be formulated between two parties there must first be intent. With a written contract the intention often tends to be taken for granted – no two parties would normally go to the effort of formulating a legally binding contract without there being an intention for it to be so. The courts, however, do not hold this to be an absolute. In addition, some minor building works can be lawfully undertaken without a written contract – under $5000 in most states and territories, though it varies. In such cases the 'contract' may to be verbally agreed on, or simply implied.

The question of intent in contracts therefore remains. It is important that this intent is clearly identified and understood to exist by both parties. Under Australian contract law, intention is no longer presumed, even in commercial agreements (all building works, minor domestic or otherwise, are considered legally to be commercial agreements). Under such law, intent must be proved objectively *and* not disproved subjectively. This reasonably new approach to contract law means that even when a written contract exists (the objective), it may be shown that legally the contract is not binding – or more correctly, no contract actually exists despite the paperwork – due to some words, deeds or actions on behalf of one or both parties (the subjective).

The simplest example of the above would be a full marriage ceremony conducted by a priest or other appropriate person on a stage with actors playing the roles of bride and groom. The event may be concluded, and documents even signed, but legally neither the bride nor the groom could demand of the other that a legally binding marriage had actually occurred. This is because clearly the subjective (personal) intent of at least one participant was otherwise; i.e. in their mind it was a theatrical play.

Rights, responsibilities and liabilities of contracted parties

Once the intent to enter a legal relationship is confirmed by both parties, you need to be sure of what each party's rights and responsibilities or obligations will be in a chosen contract. Most importantly, you must also identify any liabilities that might arise should an obligation not be met. The aim of any sound contract is to provide a fair balance between the rights and responsibilities of the client/principal and those of the builder/contractor. These are reflected in the terms and particularly the General Conditions of a contract. Good contracts will typically have notes associated with each clause to aid parties to understand their rights and obligations.

An example of a joint obligation clause and supplementary notes is Clause 32.1 of the Victorian building contract for new homes:

32 Completion notifications – *Opinion Procedure*

Clause function: This clause states the process the parties will follow when giving an opinion about Completion of Work.

32.1 The *Building Owner* must give a notice to the *Builder* within 10 *Business Days* of receiving the *Builder's* opinion that *Completion of Work* has been achieved, that the *Building Owner*:

a) accepts the opinion, or

b) does not accept the opinion, giving detailed reasons.

....

Clause 32 note

This clause provides for the *Builder* to notify the *Building Owner* that *Completion of Work* has been reached.

The *Builder* should be satisfied that all *Work* is completed before providing the *Completion of Work* notice.

The *Building Owner* should be satisfied that the *Work* is complete and is performed in accordance with the contract before accepting the *Builder's* opinion that *Completion of Work* has been achieved. The *Building Owner* may consider engaging a competent professional to assist them to establish *Completion of Work* has been achieved.

Other contract clauses to which particular attention should be made are those surrounding rectification, defects, progress payments and liquidated damages. Each of these carry liabilities for those parties who fail in their obligations. Of these we need to look closer at two: progress payments default fees and liquidated damages.

Progress payments default fees

Progress payments are monies owed by the principal or client to the builder/contractor for works completed to a nominated stage. While important in any type of construction, large or small, these payments are of particular importance to smaller home building contractors who will not have access to large cash reserves. Without payments being on time, a builder's ability to carry out other works may be compromised. Most likely there will also be bank interest to be paid.

Progress payments are therefore often 'incentivised' in contracts developed by some state government bodies and some domestic builder's organisations such as the HIA or MBA; i.e. the client is encouraged (given an incentive) to make payments on time. This is done by way of default interest on the amount owing – generally based upon bank standard variable rates plus not more that 5% (i.e. something close to 10% per annum). For an example of this see Condition 20 (p. 2) of the QBCC General Conditions of New home construction contract.

From the builder's perspective, default fees for late payments are an important and legitimate survival tactic. You need to let your client know that these terms are in the contract and that they are fair. You should also be aware that you cannot put down excessive interest rates. Doing so will only lead to court (should it be disputed), or building dispute bodies, finding against you on the basis that the figure is a 'penalty rate' not a legitimate recovery of costs.

Liquidated damages

Liquidated damages refers to set costs that may be incurred by your client if you fail to deliver the works on time. Like the overdue payments clause in penalty rates (above) these figures must be legitimate and not punitive (i.e. acting as a penalty above the cost actually accrued). Most construction contracts will have a liquidated damages clause and so you should be aware of it, and the rates agreed to, prior to signing. For an example see Clause 24 (p. 11) of the QBCC New home construction contract.

A common liquidated damages clause in a contract is the cost of accommodation; for example rent, that your client may have to pay while waiting to enter their new home. On larger projects, it may be loss of business. The fit-out renovation of a bank, for example, may be on a very tight schedule – perhaps just a weekend – after which the bank expects to be trading again first thing on the Monday morning. It is not unknown in these instances for the rate to be set per hour, not day or week.

Some domestic contracts will not have a liquidated damages clause, but simply a 'loss or damage' clause. This is not quite the same thing, in that it will not provide a schedule of rates per day/week, but an estimate of costs the client expects might be incurred, and for which you, the builder, might be liable. Though not a particularly helpful clause – in that you don't know precisely what or if you must pay – it does at least alert you to the possible costs. For an example of this see Clause 15 (p. 42) of the Victorian contract for new homes.

Whichever sector you intend to conduct your works in, be sure to check this clause and negotiate appropriate rates. At the same time, be aware of the project end dates and be confident that you can complete construction within the times you specified.

Common contractual procedures

Many phrases and words specific to construction contracts have already been discussed above. There are more to follow, so at the end of this book you will find a glossary with a range of words or phrases common to the industry. What has been only briefly discussed, however, are the procedures that arise from, lead to, or are otherwise woven around these words and phrases; as they do likewise from the terms, clauses and conditions of a contract.

Each term, clause or condition of a contract, as well as some key phrases, gives rise to a procedure. These procedures will often be conducted or initiated by means of some agreed form that has already been adopted or developed as part of the contract. Using the QBCC New home construction contract as an example, the contract 'package' comes in three parts:

- Consumers Building Guide
- General Conditions
- Schedules and Forms.

The General Conditions in this contract state the when and by whom of any procedure that must take place. The schedules and forms are the means by which these procedures will then be conducted. From this you will begin to understand that contractual life means a bit more than the odd phone call: there is a way of conducting your activities, and this way is dictated by the contract. Sticking with the QBCC contract, this may be exampled by the simple procedure surrounding the formal start of the building works:

17 Commencement and performance of the Works

17.1 The Contractor must commence work under this Contract at the Site on or before the Starting Date.

17.2 Within 10 business days after the date on which work under this Contract commences on Site, the Contractor must give a written notice to the Owner (such as QBCC Form 1 - Commencement Notice) stating:

(a) the date on which work under this Contract commenced on Site; and

(b) the Date for Practical Completion.

...

The procedure is therefore laid out, you must start on or before the agreed date, and then formally advise your client of this start date using *QBCC Form 1* or something like it. The QBCC package has seven different forms that cover such activities as: calls for extension of time, progress claims, disputes variations, defects and completion. Each couples with its relevant clause outlining the procedure of which it is a part. The Victorian version of this domestic contract is similarly structured and holds some 18 individual forms which again respond to specific procedures laid out in the contract.

The NSW Home building contract, on the other hand, provides no forms as such. Instead, each clause lays out the specifics of the procedure and what must be contained in any given communication. The 'procedure' of Clause 13 Variations (p. 15) of this document says:

Procedure for variations

Before commencing work on a variation, the contractor must provide to the owner a notice in writing containing a description of the work and the price (including GST). If not otherwise specified the price will be taken to include the contractor's margin for overheads, supervision and profit. **The notice must then be signed and dated by both parties to constitute acceptance.**

...

This particular example also demonstrates the power of a single word as it applies to a construction contract. The word 'variation' is quite specific and important in construction as you have already learnt (see p. 58). It is important because it has the capacity to change the contract price – up or down. Therefore, the moment the word *variation* is mentioned, it invokes a procedure: a procedure similar to Clause 13 shown just above. This is the case with many of the words and phrases offered in the glossary. It is worth your time exploring the use and implications of each by a careful study of contracts applicable to your own state or territory.

Breaching a contract

Breaching a contract refers to an obligation in the contract not being fulfilled by one of the parties. Three things may happen at this point:

- the offending party will be advised of the breach and given an opportunity to rectify it
- failing rectification or refusal to rectify, the appropriate contractual liability will be applied
- the contract might be terminated.

A breach may be said to have occurred when a ***condition*** of a contract has not been complied with, or when an implied statutory warranty or duty has not been adhered to. This may be caused by either party, the builder or client, and the implications will depend upon the severity, time of occurrence and which particular clause is involved. Breaches of a statutory warranty will be dealt with in accordance with the applicable state or territory regulation or Act. The damages available to the innocent party may include but are not limited to:

- the cost of someone else fixing the problem, including labour and materials
- any loss of profit or other losses suffered as a consequence of the breach
- costs associated with relocation of furniture, etc. and rental expenses.

Breaches of a condition will be dealt with according to the terms of the contract. Clause 37.1 of the Victorian building contract for new homes (p. 57) typifies an approach to what are described as substantial breaches:

37.1 If the *Building Owner* or the *Builder* is in substantial breach of this contract, the other party may give a notice clearly identifying the breach and stating that party's intention to terminate this contract if the breach is not remedied within five *Business Days*. If the breach is not remedied within that time, the party giving the notice may terminate this contract.

Examples of what constitutes a substantial breach in the case of either the builder or the client are also given in this clause. One of each are shown below:

Building Owner

a) failing to provide satisfactory evidence of title to the *Building Site* after being required to do so under Clause 12.4

Builder

a) failing to perform or progress the *Work* or the *Approval Work* in accordance with this contract

It is important to note that not all breaches are 'substantial' and not all substantial breaches need lead to termination of the contract. In all cases, when you suspect a breach has occurred, either by yourself or by the client, both parties are encouraged by the regulations to talk it through first. Most people are trying to be honest and are trying to do the right thing, but mistakes happen and circumstances change. Talking it through can save a lot of time and money.

Legislative requirements

This is a chapter in itself, as the legislation and ensuing regulations dealing with construction vary extensively between each state and territory. In addition, these regulations vary between domestic and commercial activities. It is therefore not within the scope of this chapter to cover all elements for each state; however, a general overview is possible, coupled with guidance on where you should look for your particular region's requirements.

The main purpose of any regulatory framework pertaining to construction is to:

- provide or call up standards for the construction and maintenance of buildings
- provide protection to homeowners for whom building works are being undertaken

- cover homeowners against financial loss through compulsory builder-funded insurance
- regulate building practices through licensing or permit systems
- reduce disputes and provide mechanisms for their resolution.

As a general rule, the regulations governing a particular type of contract in any given state or territory will be plainly identified. Frequently, this will be at the specific point in the contract to which it is relevant. On other occasions there will be a particular clause or term of the contract that calls up the regulations as an overarching framework. An example of the latter from AS 4000 (p. 12) is offered below:

11 Legislative requirements

11.1 Compliance

The *Contractor* shall satisfy all *legislative requirement* except those in *Item* 19 (a) or directed by the *Superintendent* to be satisfied by or on behalf of the *Principal*.

The *Contractor*, upon finding that a *legislative requirement* is at variance with the *Contract*, shall promptly give the *Superintendent* written notice thereof.

...

There are a few things to note in the above example. First, any key word or phrase for which there is a definition is in italics. Second, there is a provision here to revoke the need to satisfy every legislative requirement; and that those parts of the legislation that will be revoked may be found in something called 'Item 19 (a)'.

Because you can access this contract via the paths laid out at the beginning of this chapter, it is recommended that you review the definitions provided. Likewise, you will then be able to find 'Item 19 (a)'; being an item of Part A, an annexure of the contract, where exclusions may be entered.

The fact that not all legislation need be followed by a commercial contract is important to know. You should also be aware that in excluding a particular piece of legislation you should have sound reasons, and be sure that it does not give rise to an argument of unfair or inappropriate contract conditions.

LEARNING TASK 3.1

UNDERSTANDING RESIDENTIAL/COMMERCIAL CONTRACTS

Obtain a copy of the Office of Fair Trading (OFT) residential building contract in your state.

After reading through it and any other guidance notes provided by your state OFT, develop and document a 'guidance checklist' (or flowchart) on the systematic process that should be followed by the builder to ensure that both client and builder clearly understand what their obligations are in a domestic residential building contract.

Selecting a contract

Given the various types of contracts outlined earlier in this chapter (see 'Types of building and constructions contracts', p. 55) and the state and territory variations governing their use, selecting the right contract can be daunting. For domestic builders, often the best avenue is by selecting a standard contract from either a builder's organisation such as the MBA or HIA, or from a state government authority such as the NSW Department of Fair Trading. On commercial works, standard contracts remain your best option but may need to be sourced from SAI Global.

Even so, choosing the right contract may depend upon the type of construction you are undertaking. Getting it right will save time, and possibly money, for both parties.

Contracts and types of construction

Types of construction, or more correctly, building classifications, are discussed in Chapter 1 Building codes and standards. As you saw in Table 1.1 there are 10 classes of buildings and a number of subclasses. Unfortunately, the current array of standard contracts available have not been developed to be identified directly with any specific class or even cluster of classes. At least not explicitly. This means you must choose informatively.

Your path to choosing an appropriate contract document generally flows like this:

- determine the type of construction you are undertaking
- identify the state or territory in which the works shall be undertaken
- identify a standard contract that matches the above criteria.

Sounds simple enough; let's look at each in turn.

Determine the type of construction

Is the work you are about to undertake domestic or commercial? This is the initial division you will need to make. If its commercial then you have a very different set of contracts to choose from. Within the commercial sector there is still a variety of different building classes and types of structure, so these too will have implications for your contract choice. If your company is both designing and constructing then this affects your choice too. Another division is if it is a government or non-government project.

If it is a domestic structure, is it an isolated domestic house or works associated with same; e.g. swimming pools, garages, fences and the like, or is it a row of town houses? As stated above, just determining that it's a Class 1a building is not enough, contracts haven't been written from that perspective. If it is not an independent house, but still a dwelling (Classes 1 and 2), then there is a high probability that it will need to be considered commercial from the perspective of contract selection, depending upon which state or territory you are in.

Identify the state or territory in which the works shall be undertaken

For those working in any of Australia's major cities this will seem like a rather self-evident fact and not worthy of comment. For those living on state borders, however, it is an issue easily missed for those new to the industry. Building licences are state bound and it often takes some young builders a little time to recognise they are limited to within a state boundary when they have lived their whole lives crossing from one state to the other every day. More experienced builders moving into a border area can also fall foul of the changes in regulations from one side of a river or line to another.

The message is therefore quite simple: make sure the contract you are selecting applies to the location of the site upon which the work will be undertaken. If it's on the Queensland side then a NSW domestic contract should not be used; if NSW, then the Victorian one is not appropriate and so on for each border district. At the very least, keeping alert can save you having to write the contract out again.

Identify a standard contract that matches the above criteria

Having identified the state or territory, and the type of structure, you can now explore the range of available contracts. For domestic, single dwelling and associated projects, most standard contracts from state government bodies such as NSW Fair Trading will be appropriate. The cost of the works needs to be considered as each state usually has two contracts: one framed for work up to a maximum of 'x' (often $20,000); the other for works over that figure – the upper limit is not usually stated, but a figure of $500,000 is commonly mentioned in state and territory Acts as the point at which other contracts types may be considered.

If it is a dwelling, but not a single isolated home, but rather rows of town houses or other multi-residential Class 1a, Class 1b, or Class 2 buildings, then these contracts are not really suitable. The QBCC residential contract does include duplexes, but this is not common to other states or territories. In such cases, you will need to consider more commercially-oriented contracts.

Within the commercial sector the government or non-government division will define your contract choices. If the latter, most likely a Standards Australia contract will suit – one of the AS 4000 suite, chosen on the basis of being construct only, or design and construct (AS 4300). If it is a government contract, then the choice will probably be made for you under the terms of the tender documents you quoted on: in NSW, and on some Queensland projects, with a value of $1 million or greater then this will probably be a GC21, or MW21 for lower priced works.

For those who are members of a builder's association such as the HIA or MBA, your path is much easier. In any of the aforementioned cases, these organisations have a range of appropriate standardised contracts to which they may guide you. The MBA has the well-respected ABIC suite framed around project type and pricing. In addition, these organisations can provide well-informed legal advice for works where they do not have a standard contract, or how a standard contract can be amended to suit.

If you are not a member of such organisations and intend to undertake commercial work, then you should always seek legal advice from agencies that understand and have experience within the construction sector in your state or territory.

Documents that make up a contract

As has been demonstrated previously in this chapter, there is more than one document that makes up a contract. In some standard contracts the 'contract' comes as a package; i.e. it holds all the necessary forms and parts that constitute the legal aspects – such as forms and annexures. However, in construction this seldom comes close to being the whole of a contract.

In construction, what makes up a complete contract will vary markedly depending on the type of work being undertaken. There is no longer an easy division between domestic or commercial, but rather a statement of the scope and nature of the work. In each case the missing element from a contract 'package' will therefore be the description of these works. As you will learn in Chapter 5 Plans and specifications, this description can be as simple as a sketch plan for a piece of minor fencing, or a full building information management (BIM) suite of electronic files in a variety of formats; the content of which may be in constant flux as the 'as built' data is factored in.

That stated, there are some generalisations that may be made. For domestic housing, the general requirements are:

- the contract itself – generally a standard industry or government body contract (See, 'Types of building and construction contracts' p. 55.)
- procedural forms associated with this contract (See 'Common contractual procedures' above, p. 62.)
- any ancillary documents that the standard contract calls up, such as a consumer building guide, and (except Tasmania) some form of home warranty insurance
- a full set of plans and specifications giving all the required details of the home being built. This will include:
  - engineering details
  - material, window and door schedules

- energy audit
- bushfire (BAL) audit
- site and floor plans, elevations, sections, details, footing diagrams, etc.
- council approvals, such as planning and construction permits.

Once we move away from the domestic sector, things vary greatly. The construction of a school fence is a commercial project. It may be very long and cost over $100,000 to complete, but it may not need very much by way of diagrams to adequately describe it. Alternatively, a small community skate park for the local council, also a commercial venture and of the same value ($100,000), will require a vast array of diagrams and highly detailed specifications.

Again, however, some generalisations can still be made for this sector. These include:

- public liability insurance (see Chapter 2 Legal requirements for building and construction projects)
- plans and specifications (see Chapter 5 Plans and specifications)
- bill of quantities (see Chapter 10 Site surveys and set out procedures)
- tender documentation – if applicable (see Chapter 8 Tender documentation)
- development approval documents (see Chapter 5 Plans and specifications)
- project time and milestone schedules (see Chapter 5 Plans and specifications)
- labour (human resource) schedules (see Chapter 10 Site surveys and set out procedures)
- equipment and site services schedules (see Chapter 5 Plans and specifications).

The chapters listed give expanded commentary on each of these common requirements of commercial contracts. Through reading these you will also gain greater insights into the applicability of each item to specific project types. Learning to identify the scope of the work you are undertaking, and the type of structures involved, will aid you in determining the exact needs of any individual project. If you are dealing with a tender, or a government project, you will be aided significantly by the documents to which you would be responding.

In all cases, it is wise to seek legal advice either from a builder's organisation such as the HIA or MBA, or if not a member, from a legal agency experienced in the construction sector.

Contracts: offering and accepting

Having reached this point in the chapter, you will have a fair idea that the average construction contract is a bit more than a piece of paper with a couple of signatures on it. You will also be aware that there are numerous regulations and Acts in place dictating both contract content and execution. These same laws require that there must be three things in place before a contract can even be deemed to exist. These are:

- an offer
- an acceptance
- a consideration.

We will look at each of these in turn.

The offer

An **offer** is some form of promise made by one party to another. The 'promise' is that they will do, or not do, something in return for which the other party will do, or not do, something in return. In the construction industry this is typically some form of construction works in return for a specified amount of money.

The offer itself is not a contract, it is just the offer. A tender or a quote is an offer. These document the works that will be undertaken in a specific context over a specified timeframe in return for which a specified cost is stipulated.

You should note that there is a difference between an *offer* and an *'invitation to deal'*. An offer is a set amount of action from one party for a set amount of return from the other. An invitation to deal, if accepted, is only an invite to become involved in further negotiations. No contracted works arise from such an agreement.

The acceptance

The **acceptance** is a clear stated agreement by one party to the offer made by the other. Once the acceptance is made, an agreement of terms has been struck. In construction this may be by way of the client signing the contract. However, in smaller domestic works it may also be verbal, between the builder and a subcontractor. In verbal cases, if there is a dispute and an ensuing court case, then the evidence of agreement may be a quoted price and simply the beginnings of the specified works.

When any written contract is involved, the offer and the acceptance of that offer must match each other. That is, whatever was on the written offer – amounts, timelines, materials, etc. – must be replicated exactly in the contract documents upon which the acceptance is made.

The consideration

The **consideration** is the return given by one party as payment for that promised by the other. In construction this generally means the contract price for the works described. For a contract to be valid if challenged in court, proper consideration must be shown to exist in the agreement. The word 'consideration' is taken to mean that the party doing the work has considered the proposed return as appropriate for the labour, time, materials or other costs expended.

In verbal agreements, the concept of appropriate consideration often comes up in disputes. For example, a client may take a builder's offer of doing a pergola for '15' as being $1500 and accept. The courts would state the $1500 was clearly not proper consideration for the project delivered and favour the builder's argument that it represented $15,000 – ignoring for the moment that the builder should have had a written contract in place for works in this price range.

Capacity to form an agreement

Contracts can only exist if both parties have the legally identified capacity to make such an agreement. Without this legally identified capacity, a contract is generally not binding. To be considered capable, a party must be able to understand and interpret their obligations and liabilities as held within the contract documentation. Legally identified lack of capacity stems from well-known issues arising from vulnerability and hence exploitation. In some cases, however, it also applies in purely legalistic terms with regards to bankrupts, companies and even the Crown.

As with many other factors surrounding contracts, the rules or laws vary between states and territories. However, the basic principles are much the same. Areas where capacity to make a contract are known to be challenged are:

- minors
- mental disorders
- intoxication
- companies
- bankrupts
- the Crown.

The latter two can still make contractual arrangements; however, they are limited and dependent upon a range of legalities. In certain instances, minors too can be party to a contract, and in some instances be held to it. With regards to construction contracts, however, it is highly unlikely the courts will favour a builder in attempting to enforce it.

You may be surprised to see companies in the list above. In the main, companies are treated as individuals and so have the same capacity to form an agreement. The question arises as to who has the right to sign on behalf of a company. Small builders will generally only have the one director or, if two, it may be a partnership. In such cases, the individual, or both partners, would sign. If dealing with a much larger organisation, however, then knowing who has the authority to sign becomes of significant importance. In such cases legal advice may be required, plus evidence provided by the company as to appropriate authority, the use of a company seal or other indicators of capacity.

Factors of consent

Consent to a contract is more than a single signature at the bottom of a page. Depending upon the type of contract, and the terms and conditions included in the contract, there will be a variety of areas needing evidence of acknowledgement from either or both parties.

In the main these will be areas where liabilities, enforceable under the contract, may be incurred. These have been discussed before and include:

- interest upon progress payments that are late
- variation clauses
- conditions leading to termination of the contract
- liquidated damages.

In each case, either a signature or initials will be required as evidence that the terms have been sighted and understood.

Other factors around consent have been mostly covered in the previous section on 'capacity'. Some unlikely issues that may still arise are those of:

- misrepresentation
- undue influence
- duress
- unconscionable conduct.

Misrepresentation

Misrepresentation is when either an individual, or an element within the contract, represents as factual but is not. For example, that an individual can sign on behalf of a party or is that party; or that a part of the contract will be undertaken when it will not be. This is an area that borders upon, or may be considered directly as, fraud; in which case the courts would hold that no consent had been given.

Undue influence

Though minors and the elderly are particularly susceptible to this issue, it can apply to either party. **Undue influence** is when one party takes advantage of their authority, presumed authority, or position of power or wealth to gain consent to a contract, when without that influence the contract would not have existed. For example, some contractors have in the past misused their presumed superior knowledge of structure or materials to convince clients to have works undertaken that were not necessary. Conversely, politically or economically well-positioned 'clients' have encouraged builders to undertake low-priced works on the promise of future works that are never forthcoming. Once again, this form of consent is questionable within the courts.

Duress

Being forced to sign a contract under threat of violence or damages is known as **duress**. Aside from being a criminal offence, the courts will agree that appropriate consent was not given.

Unconscionable conduct

Unconscionable conduct is said to occur when one party takes advantage of another in a manner that is deceitful, underhanded, ruthless or unethical; i.e. without a sense of conscience. In such cases, the disadvantaged party has been unable to protect their own interests through some special advantage the other has over them.

Though uncommon, each of the above factors have been brought before the courts in the past, and no doubt will continue to be presented in the future. Your task, both as an honest builder and as a member of your community, is to prevent these issues from arising upon any project you are involved with.

LEARNING TASK 3.2

CHANGES TO CONTRACT PRICE AND EXTENSIONS OF TIME

You have been asked to price and suggest an acceptable contract for the construction of a \$1.35 million home, on a 1000 m^2 block of land, with extensive landscaping also included in the works, in a capital city of Australia.

Explain your reasoning for the contractual arrangements that you will use to ensure that there is little chance of a dispute arising due to the type of contract being used.

Contract preparation

Frequently, domestic builders feel that the preparation of the contract is a task undertaken by themselves. And that the finished contract is then presented to the client solely for their signature of acceptance. Commercial builders know a different story: depending upon the scale and scope of a project, the preparation of the contract is an exercise shared between the designers, subcontractors, site supervisors and, most importantly, the client. The reality is that it should always be a shared process and the contract, prior to signing, be a jointly approved construct – even when it is a standard housing contract from a state government body or industry association.

If the contract agreement process is a shared one you radically reduce the likelihood of unnecessary termination. But often this is not enough: jointly produced contracts have failed in the past despite good intentions. This is generally due to a lack of experience in such documents by one or both parties. To prevent this you need expert advice, which forms the first part of our discussion in this section.

Expert advice

Builders often make the error of thinking that because the client has come to them, then they – the builder – are the expert. You are, to a degree; but most certainly not on all things building, and definitely not on all things contractual. The biggest thing you have to learn is knowing when you don't know. Once you have that bit of wisdom, then it's the, not always simple, matter of finding someone trustworthy who does.

Areas you may need assistance with are also broader than you think, even on domestic contracts. And where you will find that assistance from can be surprising to some. What follows is a brief outline of just some of the sources you may need to turn to, ensure your project will run smoothly based upon the contract you will be signing.

Tradespeople and subcontractors

Frequently, these are the first people to go to for establishing a price and estimating timelines. Depending upon their experience base, they can also provide you with a wealth of knowledge regarding the applicability of materials, fixing processes and the like. This will help you in interpreting the plans and specifications of projects and therefore frame certain elements of the contract.

Architects and designers

These are the people who put the design together originally. Any questions you have concerning the plans and specifications you would put directly to them. Some large projects may have to be undertaken in stages; these are the people who can help with developing the scope of each stage, and how this might be framed in the contract.

Site supervisors and superintendents

As with tradespeople and contractors, these people have a wealth of experience, and they can inform timelines and therefore contractual obligations framed by progress. They have a very good idea of what can and can't be done within a particular timeframe based upon a given labour force and in a given context; i.e. location, season, political/social/cultural environment.

Quantifiers, estimators and accountants

These are people skilled in dealing with the numbers surrounding your project. They will be of invaluable assistance in the development or interpretation of rise and fall formulas that will need to be applied, where such clauses exist.

Councils, town planners and building inspectors

These are particularly important sources of information regarding permits and requirements concerning site access (road closures, footpath closures, school crossing implications and the like), noise pollution, obligations to neighbouring properties and the like. Also, for stages of inspection and expected standards.

Workplace health and safety (WHS) authorities

These people are not your enemy. They are out to make your life and the lives of those involved in your project safe and injury free. They are a very important source for knowing your obligations and in establishing, in conjunction with all of those mentioned above, safe processes, site access, material handling and so forth.

All of which have implications for the development of your contract timelines. WHS compliance adds a huge cost to a project, which must be passed on to the client. Joint discussions with the designers can also help in developing alternative approaches to the project design that may limit risks.

Lawyers and builders' organisations

The 'go to' for contract selection from day one. On large or small projects these people and organisations have a knowledge base that can be critical to your success or failure, if only in how they aid you in framing the contract initially.

Manufacturers and suppliers

Once again, important for timelines, but also for site access requirements, availability and any specialised skills and equipment that may be required for use or installation. All will influence your contracted timelines and the staging of progress payments.

Government

If it is a government project then logically you will need to discuss the project and then contract with their representatives. These people will generally have dealt with many contracts previously so you can gain significant insights from them. State government agencies or organisations such as the NSW Fair Trading or QBCC are also great sources of information around the applicability or application of standard contracts that they can supply.

Other agencies or organisations

Aside from the contract design and conditions, most of the information you will need includes ensuring you have the correct permits, issues concerning timelines, project staging, and implications for other stakeholders associated directly and indirectly with the project's progress and outcomes. To this end, when engaging with any of the above, question them as to who else might be worth talking to. Organisations and companies involved with service infrastructure such as power, water, sewerage and telecommunications, for example, may all need to be engaged with at some point – and this is better done before a contract is compiled, than after.

Terminating a contract

In previous discussions of the various clauses and conditions of contracts the concept of contract termination has been mentioned on many occasions. Most of this previous discussion has focused upon breach of contract. However, this is only one of the ways by which a contract may be terminated, or more correctly **discharged**. Others include:

- satisfactory completion (known as performance or discharge by performance)
- agreement
  - acceptance of partial performance
  - substituted contract (rescission or novation)
- frustration
- termination for convenience
- breach
  - actual
  - anticipatory, repudiation
  - discharge by accord and satisfaction.

Discharge by performance

This means that the terms of the contract have been fulfilled precisely and to the satisfaction of both parties. All works have been carried to completion, and all monies owed have been paid in full.

Agreement

Projects, or the parties involved in the contract, may reach a point at which one or the other cannot, or choose not, to continue. That is, the contract has been ***partially performed***. Depending upon which party initiates the discharge, and for what reasons, then some compensation may be due. For example, the client may no longer be in a position to pay for the whole project and needs to end the contract: in this case the builder – assuming they agree to the discharge – would generally be compensated. Alternatively, it may be the builder who needs to withdraw from the agreement with the client, in which case it would be the client who would be compensated.

Rescission or **novation** occurs when a new contract replaces the old – rescission being the discharging of the original contract – with the purpose of substituting it with a new contract, the novation.

Frustration

Frustration of a contract means that, without the fault of either party, it has become impossible to perform the contracted works. This means it is not about one party being difficult, or that the project slows down too much, or that it becomes too expensive – it is that something unexpected has occurred after the contracted works began. Ultimately, this too tends to lead to a discharge by agreement, though there are occasions when it ends up in court due to the non-initiating party disagreeing with the other party's perception that things had reached such a point.

Termination for convenience

This is a clause that may be inserted into a contract under that title. It is generally only found in large contracts of extensive duration. This clause allows the client or principal to **terminate** a contract at any time regardless of the performance level of the contractor or building team. Despite the action being completely driven by the principal, it is sometimes viewed as a form of discharge by agreement as the builder will be compensated as per the already agreed upon details in the contract.

Breach

This has been discussed previously, but in brief it means that one party or the other has failed in their obligations as stipulated in the contract. Breaches do not always end in termination of a contract, but it is one of the ways. A breach can be ***actual***, or ***anticipatory***: actual, in that one party has actually not done as obligated; anticipatory, in that one party has somehow indicated that they will not carry out a future obligation.

The latter of these two is known as ***repudiation***. It is remarkably easy to accidentally repudiate a contract, so take this as a warning. If you say to a client that you are not going to do something, and insist on this stance – then, if it's in the contract, you have effectively repudiated that contract. You are in anticipatory breach, and so possibly out of a project, and in the wrong. Conversely, if your client gets angry with you and tells you to get off the site, then it is they who have repudiated the contract.

A breach can also lead to what is known as ***discharge by accord and satisfaction***. This may occur after a breach when one party agrees to release the other from the contract, provided they perform another action. That is, like novation, the original contract is agreed as discharged, and a new contract is put in its place. The difference on this occasion is that the change was initiated by a breach, and the offending party is, to a degree, put in a position whereby acceptance is the preferred option.

Rise and fall amounts

Rise and fall contracts were covered at the start of this chapter. As outlined, this form of contract is designed to reduce the risk carried by the builder; if the cost of materials, labour, or transport goes up, or down, then the contract total changes accordingly.

Calculating the actual rise and fall amounts can only be done while determining each scheduled progress payment. However, the establishment of the formula for this calculation must be done as part of the initial contract. The formulas can be quite complex depending upon what it is that will be included within the rise and fall clause. If it is just one or two fluctuating materials, then the approach is simplified, such as stating the price at the time of contractual agreement, and then adding or subtracting the difference at the time of progress payment calculation.

On large-scale projects, where labour, materials, and transport are all to be considered, then a set formula must be developed. An example may be found in the current edition of the National Cost Adjustment Provision (e.g. NCAP2 – where '2' is the edition). Other formulas may be derived through the use of Australian Bureau of Statistics inflation and price indexes such as the consumer price index (CPI), Producer Price Indexes (PPIs), International Trade Price Indexes (ITPIs) and the Wage Price Index (WPI).

Given the complexity of these calculations, advice should be sought for larger projects from experienced quantifiers and accountants in the preparation of this section of your contract.

Progress payments

Progress payments were first discussed earlier in this chapter. From this you will be aware that any project greater than $20,000 is likely to include progress payments. These will be scheduled in the contract at specific stages of completion. In preparing the contract you will need to identify the stages and gain agreement from all parties that these timings and amounts are acceptable. In considering all parties, you should include subcontractors, suppliers and workers and your capacity to pay them – any failure in this capacity will stall the project and leave you open to breach.

An example of a common domestic housing contract progress payment schedule can be found at Clause 12, p. 14 of the NSW Fair Trading Home Building Contract. Filled out with a typical breakdown for this type of structure, with a contract price of $275,000, it would look as follows:

TABLE 3.1 Example of domestic progress payment schedule

1	Base stage 20% (Less 5% deposit)		
	(less deposit: $ 13,750.00)		$ 41,250.00
2	Framing stage 20%		$ 55,000.00
3	Lockup stage 25%		$ 68,750.00
4	Fixout stage 20%		$ 55,000.00
5	Practical completion 15%		$ 41,250.00
6			$
7			$
8			$
9			$
10			$
(If space is insufficient, attach a sheet referring to this schedule)		Total	$261,250.00

You will note that the final total is short by $13,750, which matches the deposit of 5%. Combined they will match the total contracted agreed price.

When you review the contract this is drawn from (NSW Fair Trading, p. 14), you should also notice a place at the bottom of the page for the initials of both yourself and your client. This the required contractual evidence that both parties agree to the stages set, and the amounts to be paid at each stage.

In larger commercial projects the same principle applies, only there will be more stages and most probably more money involved at each one.

Extension of time (EOT)

We have so far only briefly mentioned extension of time (EOT) under the heading 'Custom contracts'. This may have suggested that EOTs are a clause of these contracts only. This is not the case – in fact, avenues for EOT clauses are in almost all standard contracts. Extension of time refers to a need to push back the planned handover or practical completion of the project to a later date.

Practical completion is usually set at a specified number of days, weeks or months from the actual start date. Generally, there will be some compensation that will need to be paid to the client or principal for any days or weeks (sometimes hours) past this point. What is known and understood by all involved in construction projects is that unforeseen events, often external to the project, can cause delays. The EOT clause allows you, as the contractor, to ask for delay based upon these events.

In the domestic sector, the relevant state or territory housing Act describes the circumstances for which a builder may claim an EOT in that jurisdiction. Such claims must be made in writing to the client or owner who must respond in kind. If the client rejects your request, then the matter is dealt as a dispute. Clause 7 of the NSW Fair Trading Home Building Contract offers a good example of how this functions in the domestic sector.

In the commercial sector, it is the superintendent that evaluates the EOT claim. In reviewing Clause 34 of the Australian standard contract AS 4300, you will note first the statement *qualifying cause of delay*. This is defined in Clause 1 Interpretation and construction of contract as:

a) any act, default or omission of the Superintendent, the Principal or its
b) consultants, agents or other contractors (not being employed by the Contractor); or

other than

i) a breach or omission by the Contractor;
ii) industrial conditions or inclement weather occurring after the date for practical completion; and
iii) stated in Item 23

In reading Clause 34 you will find that 'Item 23' allows the principal to list any other factors for which they will disallow an EOT request.

In preparing the contract you must therefore make yourself aware of these clauses and, particularly, to any provisions such as 'Item 23' that may impact your timelines, or capacity to extend those timelines should need arise. These provisions are negotiable, but only during the creation of the contract. Upon signing, you have agreed to the conditions and must abide by them.

The final contract

As outlined earlier in this chapter, a construction contract is much more than just the signed papers; it includes plans, specifications, schedules, warranties and much more. You need to gather these together, analyse them to ensure they are factual, current and project relevant. It is best to create a checklist of what is required for the specific type and context of the project to hand. You can have standardised checklists for each type of project, but with a change of context, some items may not be required, or others may need to be sourced. This is where the expert advice comes in for both parties.

Once you are satisfied with the contract as a package, you must meet again with your client, and any professional support you and they require, and check with them again for confirmation of agreement on the terms and conditions set. This will include progress payment schedules, the scope of work under contract, timelines and so on until all aspects are covered. Only then do you begin the joint sign-off process. Remember, many elements or clauses of a contract must be initialled to confirm the agreement of those parts, the progress payment schedule being just one example.

If you are operating within a large organisation, there will be specific processes that you will need to follow to which only your company will be privy. This means that the contract may have to go through a range of approval stages. These stages will most likely include accounts and procurement departments, as well as your construction department and, of course, head office. If your company uses advanced Building Information Modelling (BIM) systems, then this forms part of the contract, much as a standard set of plans would. The difference being that with BIM, the contract would need to address how this system is integrated into the project, and how the principal (your client) is informed by, and can inform, that system. For more information on BIM, see Chapter 5 Plans and specifications.

Once everything is signed off, copies must be retained by both parties. Large corporations may have requirements that copies be held by relevant departments as required. Small builders will sometimes request that their solicitors or lawyers securely retain the originals, keeping only copies for their own office and site supervisors.

LEARNING TASK 3.3

PREPARE DOCUMENTS TO BE INCLUDED WITH CONTRACTS

Obtain a plan set (or similar) that has been approved by a local council. Create a simple project schedule (bar chart or similar) that provides guidance on what target dates will be used as the basis for requesting progress payments from the other party (client) to the contract. Confirm that this will be the document that will be used to:

1 provide the client with advice on anticipated 'target' dates for completed work
2 provide the basis (documentation and reasoning) for any changes to the planned construction schedule requiring 'an extension of time'.

Nominate and clarify the anticipated progress payment schedule, including anticipated dates for completed work from your construction schedule.

SUMMARY

Although the point at which a project's value triggers a contract varies from state to state, the reality is that a contract is required for virtually all construction work undertaken in Australia. As a builder or contractor, it is therefore a given that you will be engaged in contracted works of some kind. As has been seen, what this means can be quite complex and it is imperative that you know how to interpret a contract's content, as what is on that paper becomes a legally binding agreement. And if this means you find yourself out of pocket at a project's end, it is generally due to a failure to understand just what it is you signed up for. This chapter also showed how this lack of understanding can lead to very basic or very significant breaches of a contract. While acknowledging that contracts do need to be changed at times, what has been made clear in this chapter is when, how and why a contract may be changed appropriately; likewise, when a contract may be terminated altogether through the actions of one party or another; or as desired by both.

There are two elements that have not been discussed within the chapter: the *administration* and *finalisation* of the contract. You will find both areas covered extensively in Chapter 6 On-site supervision.

Having worked through this chapter you should now be able to identify a range of contract types, know from where to source them and when to apply them. Most importantly, you should now have a clear understanding of the main contents of a contract document, particularly the clauses on dispute resolution. Knowing these clauses is the best means of avoiding significant conflict during a project, and so advancing a project to a successful conclusion.

REFERENCES AND FURTHER READING

Australian standard contracts, AS 4000, AS 4300 series © Standards Australia Limited Reproduced by Cengage under licence by Standards Australia.

Building Contract for New Homes © State of Victoria (Consumer Affairs Victoria) 2016. Licensed under Creative Commons Attribution 4.0 International licence

Sharkey, J. et.al. (2014), *Standard Forms of Contract in the Australian Construction Industry: Research Report*, University of Melbourne, Melbourne. Available at: **http://law.unimelb.edu.au/__data/assets/pdf_file/0007/1686265/Research-Report-Standard-forms-of-contract-in-the-Australian-construction-industry.pdf**, accessed 13 March 2018.

Useful weblinks

http://www.constructionlawmadeeasy.com/Index

https://www.fairtrading.nsw.gov.au/

https://lawhandbook.sa.gov.au/ch23s11s02.php

https://www.qbcc.qld.gov.au/contracts/new-home-construction-contract

WORK HEALTH AND SAFETY

4

Chapter overview

Work health and safety (WHS) is especially important in the construction sector, as it is one of the most potentially dangerous industries to work in. Construction sites can be chaotic, if not properly controlled, with heavy machinery in use, building components being moved about, dangerous materials and chemicals in use, along with lots of contractors and tradespeople busy at their jobs.

It is a fact that most severe injuries and deaths in the workplace occur on construction sites. But it also a fact that, worldwide, Australia has one of the best construction safety records. This safety record is the result of a strong WHS culture, which took decades to develop. How then did we get so good at being safe? Partly it has to do with strict safety laws that are rigorously enforced; and partly it has to do with companies maintaining effective WHS strategies.

This chapter discusses what is involved in putting together an effective WHS strategy.

Elements

This chapter provides knowledge and skill development materials on how to:

1. determine areas of potential risk in the building and construction workplace
2. inspect and report on areas of specific risk
3. advise on implementation of control measures at the building and construction workplace
4. establish and review communications and educational programs.

The Australian standards that relate to WHS safety signs are:

- AS 1216 Class labels for dangerous goods
- AS 1318 Use of colour for marking physical hazards and the identification of certain equipment in industry
- AS 1319 Safety signs for the occupational environment.

Introduction

Safety is everybody's concern. Certainly, it is common sense that we should be careful when using a power saw or when working on a scaffold. But accidents typically happen when a number of things go wrong at the same time, and when bad luck steps in.

About a year ago in Melbourne, as a project was nearing completion, a construction crane was being dismantled. As an established safety practice, a perimeter was set around the crane and workers stood away at a safe distance. There had been a problem with the crane cabin door which needed a spanner to fix, and the spanner had been left in the cabin. As the crane cabin came down, the cabin tipped, the door opened and the spanner fell out. The safe perimeter had been established precisely as a precaution against this kind of event. However, at the very moment the spanner came down, one of the workers had taken off his helmet, and as luck would have it, the spanner hit a cross beam and ricocheted off to the side and hit the worker in the head. He was hospitalised, but thankfully made a full recovery.

When nothing ever seems to go wrong, people can become complacent and prone to think that certain safety measures may be overkill. Establishing a large perimeter around the crane in the above story may have seemed so to some workers. But it is this sensitivity and a persistent adherence to safe practices – even when it may seem unnecessary – that has kept Australians relatively safe on construction sites.

Safety on site is thus the result of strict discipline, vigilance and adherence to a set, proven safety strategy. In this chapter you will learn how to develop such a safety strategy.

Areas of potential risk

Construction sites present risks and dangers that are not found elsewhere. Think of the heavy equipment and materials being moved around; think of the power tools being used and of people working at heights. If something were to go wrong, the injuries could be serious, or even fatal. It is therefore important that any and all risks are identified, and then prioritised according to their chance of happening and the level of injury they could cause. Evaluating construction site hazards should be done in line with company policy and according to the law. Then, once all that could go wrong is clearly identified, it is necessary to put in place measures and practices that reduce the risk of injuries to the absolute minimum.

Identifying risks

Construction involves a whole range of activities, each with its own set of risks. The site supervisor needs to be aware of what tasks are being undertaken on site at any one time, and ensure all work is being done in compliance with WHS regulations. Examples of potentially dangerous construction activities include demolition work, excavation, working at heights, working with loud machinery, working in confined spaces, as well as working with hazardous materials, such as asbestos removal.

Hazards

A **hazard** is something that can hurt you. It can be something you are doing, like welding, where you could damage your eyes if you don't wear the correct protective gear. It can also be the place you are in, such as the construction site itself, where an object may fall on top of you. On-site hazards include toxic chemicals as well as machinery. Procedural hazards involve how these are dealt with in the workplace. Workers need to know how to properly lift heavy objects, what protective clothing to wear (helmet, boots, ear-muffs, high-vis vests, etc.) and how to use dangerous equipment safely.

Hazards may have an immediate or long-term effect. A cut or fall wll have an instantaneous consequence, but hearing loss from constant exposure to loud noise, or lung damage from long exposure to caustic chemicals will not be immediately noticeable, though the damage may ultimately be as debilitating. Hazards, too, need not only be physical. Bullying and workplace harassment can create psychological trauma, while the construction industry itself is notorious for its stressful long hours, uncertain employment conditions and poor work–life balance.

All these hazards must be identified. They also need to be assessed for the impact they would have, some being more dangerous than others. As it turns out, construction sites don't have the largest number of WHS incidents, but they do have the largest number of workplace deaths. Government office workers record the largest number of single days off for illness, but construction site injuries require much longer recovery periods. Construction-related injuries are not the most frequent, but they are the most serious.

Risks

The list of potential hazards can be pretty endless: fire, eletrocution, unsafe working platforms or scaffolds, overloaded forklifts, collapsing cranes or falling objects, and unguarded machinery or cutting tools. Each carries with it a potential level of injury; some able to inflict greater injury than others. A hammer dropped on your boot is perhaps less of a problem than you falling from a scaffold. But each hazard carries with it a likelihood (or probability) of eventuating. You'd expect dropped hammers to be a more likely occurance than people falling. The term **risk** refers to this level of expected likelihood of a WHS event occuring. Usually, high-frequency events are less serious than the rarer mishaps; or so you'd hope. (Thankfully, raging fires or nuclear power plant meltdowns come few and far between.)

Risk, however, is not only determined by the nature of the hazard itself. It is also affected by how often the task is carried out, how many people are involved and how well they are trained and prepared. Consequently, the danger presented by each and every hazard is modified by that hazard's risk of occurence. Simply, on-site WHS is challenged as the number of hazards rises, and as the risk of their occurence increases.

Risk control

With hazards and their risks identified, a system needs to be put in place to manage WHS. This is called **risk control**. To be effective, risk control should follow a pre-prepared, systematic process. Construction sites must have a 'WHS management system' (WHSMS), and these systems are ideally based on standards set by the International Standards Organization (ISO).

The goal of a WHSMS is to pursue continuous improvement, where knowledge and experience gained from past hazard and risk events are used to update current safety strategies. The process can be visualised as a cycle of *planning, doing, monitoring, reviewing, learning* and *replanning*.

Managing WHS risks

Managing WHS on a construction site requires getting the right people involved, and ensuring that potentially dangerous work is undertaken safely.

The people

There are a range of personnel that have responsibilities as well as a legal obligation to ensure work is being carried out in accordance with WHS legislation. These are the workers and contractors, not just the site supervisor or managers. Each must manage those aspects of WHS that fall within the scope of their work duties. A construction project will also have a WHS committee, with managers and workers (usually represented by union officials) tasked with carrying out the WHSMS.

The work

Examples of construction work that can be dangerous have already been discussed. Indeed, the range of potentially harmful events is as wide-ranging as the variety of projects being built. Key risky activities include operating machines or tools, handling hazardous materials or chemicals, working with electrical or other high-energy components, working at heights or within confined spaces, as well as being exposed to pyschological or emotional stresses that come directly from the demanding work being done, or as a result of operating within a toxic work culture of bullying or unreasonable working conditions.

Each set of risk-carrying activities will be associated with a particular set of construction occupations. Concreters, electricians, bricklayers, plumbers, engineers, excavators, welders, roof tilers, even painters, and just about anybody working on site, will be exposed to some risks.

Evaluating hazards and potential risk areas

Worldwide, the construction industry is one of the least safe industries to work in. Construction accidents account for more deaths than occur in any other sector of the economy. Australia, however, has one of the world's best records for construction safety. Safety standards in this country are world's best practice.

How did we become so safe? It is because Australia has some of the strictest regulations regarding WHS, and they are implemented, monitored and managed by vigilant regulatory organisations. Simply put, should something go wrong and people get injured, punishment for breaches of WHS in Australia are swift and severe. There are strong incentives, legal and financial, as well as moral, for making sure that accidents don't happen.

WHS Act

Incentives that keep the construction industry relatively safe begin with the law. The *Work Health and Safety Act* was introduced in 2011. Previously, each state had its own set of legislation covering WHS (formerly OH&S). The WHS Act was introduced to unify all the different state laws around the country into one set of national standards. Each of the states and territories has done so, with the exception of Victoria and Western Australia, which are expected to consolidate their individual laws under the Act in due course.

Additional sources of WHS information

Apart from the Act, associated regulations and accompanying codes of practice, there are also a number of other resources construction managers may refer to in order to strengthen on-site safety. These include:

- workplace policies and processes
- construction industry bodies, associations and groups
- unions
- safety specialists
- international construction safety standards
- related industry safety standards
- insurance agents and actuaries.

Duty of care

The WHS Act sets out the legal obligations of a firm and its representatives for ensuring a safe work environment. It also sets out punishments for breaches of the Act. What happens when a serious accident occurs? Of course, there is the trajedy of injury itself; the trauma to the injured as well as to the victim's family and dependants. Nobody wants to see that happen. But the threat of punishment for breaching the Act is arguably the stronger motivator. Managers who fail in their duty of care may end up in jail, the project itself may be delayed with penalty costs incurred, and then there are the fines imposed on the construction company, along with damaging publicity and loss of reputation and business.

All stakeholders in a construction project will have responsibilities and associated duties of care. Making a workplace safe takes time, money and energy. But knowing you might go to jail or end up bankrupt is no small incentive in keeping you focused at all times on preventing accidents.

Approaches to remediation

If you are going to make your worksite safe, you first need to know what the hazards and risks are. Only then will you be in a position to control and mitigate the dangers. Audits and surveys can help identify potential threats, while safety plans can prepare you to deal with issues should problems arise.

WHS strategy

Safety begins with a strategy that ensures risks are identified and managed. Usually, such strategies come in the form of an appropriate document, handbook or risk assessment template. And, usually, it will comprise five steps:

- identify the threats
- measure the potential impact of the threats (risk)
- identify options for eliminating (or reducing) the threats
- evaluate the options (according to cost and effectiveness)
- select the optimal threat-mitigation actions and implement them.

Though sound in principle, this five-step process will need to be carried out, and properly. To do that, a further five questions need to be answered:

- Who is going to be responsible for managing the WHS strategy assessment?
- What is the scope of the assessment?
- How will the assessment be recorded and documented?
- Who is going to make the judgement calls about threat levels, options available and final actions to be taken?
- And how often (and by whom) will follow-up WHS reviews be undertaken?

Certainly, it is unlikely that strategies will need to be created from scratch, as established organisations will have procedures in place. However, there will always be room for improvement. WHS officers should look to learn from internal experiences, as well as to keep abreast of ongoing safety developments within industry.

WHS management plan

The law requires that all high-risk construction work be covered by a WHS management plan. The plan will include specific processes for collecting, evaluating and reviewing work risks. It will identify all hazards and specify control measures. It will then also lay out the precise procedures that must be followed when carrying out any high-risk activity.

Supervisors and workers will have access to this management plan. Moreover, building inspectors and other authorised personnel have the right to request the plan to assess whether the WHS management plan is satisfactory and being complied with.

Supplementing the management plan, safe work procedure documents provide precise instructions on how to safely undertake work activities that have been assessed as involving potential risks.

Safety audits and surveys

The effectiveness of a WHS management plan needs to be verified; not just to confirm that it complies with legislation, but also to ensure that it delivers a safe work environment. Audits are performed by experienced employees or WHS specialists. They may also be undertaken by external inspectors testing for breaches in the law. Auditors compare stipulated work processes and procedures with those they actually observe taking place.

Equipment, too, can be used in the audit. If construction noises are high and hearing loss is considered a possible hazard, then sound-level meters could be used to establish the extent of the risk. Other instruments that might come in handy include light meters, temperature-measuring devices, humidity meters, gas meters or even radiation meters.

Of course, it is the workers doing the job every day that have the best grasp of what actually happens on site. In order to find out what is really going on, it makes sense to ask them. This can be done through surveys. Questions can be asked verbally, and responses recorded, or workers can be given a questionnaire to fill in. The sorts of things asked include:

- behaviour – *'What do you actually do when you do your job?'*
- knowledge – *'What sorts of hazards and risks do you see around you?'*
- opinions – *'What do you think should be done to make things safer?'*

Job safety analysis (JSA) and safe work method statements (SWMS)

Two effective methods for identifying and mitigating the risks associated with any work activity are the job safety analysis (JSA) and safe work method statements (SWMS). Depending upon the state or territory (or commercial enterpise), one or the other (or on occasions both) will be deployed. These clearly identify the risks associated with a particular job, and provide instruction on how to carry out the job safely.

The difference between the two methods is marginal; however, the JSA generally focuses upon risk identification, while the SWMS is generally considered the better tool for determining the safest path to undertaking the work. Whichever system is used, they should always be completed in consultation with those who will actually do the task; in particular, with those who are experienced in doing the task or similar tasks. That is, you should wherever possible seek advice backed by extensive experience.

Both documents should include:

- a checklist of job-related hazards and risks, and paths for their elimination or reduction
- instructions on how to go about the job
- the names of personnel permitted to carry out the job (training and certificate determined)
- the names of supervisors able to provide authorisations and approvals
- worksheets to record job activities
- steps to be taken in the event of an incident.

LEARNING TASK 4.1

MANAGING WHS WITHIN EACH OF THE CONSTRUCTION TRADES

Visit the WorkSafe website:

- https://www.worksafe.vic.gov.au/industries

Under 'Industries', select the 'Construction' link. It will show information on preventing injuries in construction work. Scroll down to 'Information for construction trades'. There you'll find a list of trades: bricklaying, carpentry, concreting, etc.

Pick a trade. Try to imagine all the things that could go wrong in that trade, and then try and think of what could be done to reduce the hazards and risk.

Then open up the link and see what it says. Did you anticipate all that could go wrong? Do you think the WorkSafe site recommendations would be effective? What more could be done to make the trade even safer?

Go through this exercise for a few trades, until you get a feel for how the construction industry attempts to manage WHS.

Inspecting and reporting on areas of specific risk

Construction can be dangerous, and injuries that do happen on a construction site can be serious. In order to minimise the risk of injury, certain precautions are needed. First, all risks must be identified, and this should be done for every trade involved. Second, experts in WHS, both workers within the company as well as outside consultants, should be called in to advise on how best to manage the identified risks. Finally, policies, work procedures and inspection reports need to be written up that inform everybody involved on what best practice and the law requires.

Inspecting the workplace

The task of identifying and managing risks is a continuous one. Because the workplace is ever-changing, so too new risks can be expected to emerge all the time. Consequently, it is helpful to think of on-site risk mitigation as a cyclical process; as each stage of the process is completed, the next stage needs to be carried out in an endless, repeating loop (see Figure 4.1).

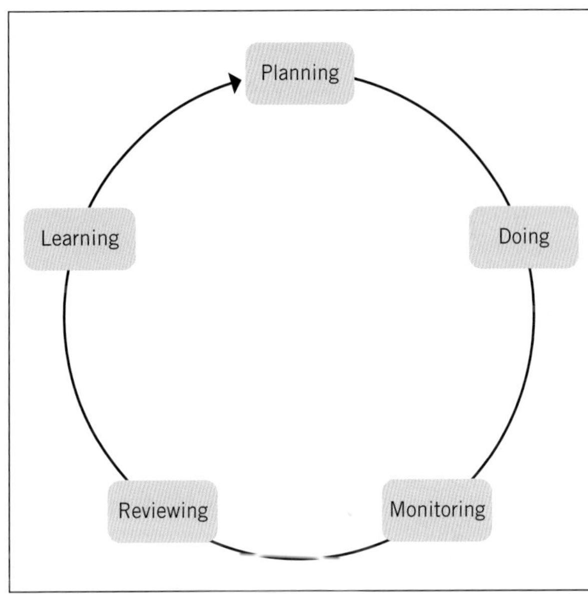

FIGURE 4.1 The structured risk management process cycle

It is also important to remember that companies are required by law to have a WHSMS. In the event a serious WHS incident occurs, an inspector will attend to report on the incident and review the WHSMS. Companies that do not have a documented WHSMS that can be shown to the inspector, or who have breached their own standards as set out in the WHSMS, can expect to be penalised by the law.

Principles of risk management

The following is a set of risk management principles that every company should follow. Risk management should:

- be part of every business decision and business activity
- address every area of uncertainty
- be made according to the best available information
- be tailored to the specific needs of the company and its personnel
- generate added value to the company
- be designed with input from company personnel and experts

- be monitored regularly for compliance and effectiveness
- be regularly updated in response to changing needs and new insights.

WHS risk assessment

WHS is concerned with making the workplace safe. There are two concepts that need to be understood: 'hazard' and 'risk'. The hazard is the thing that potentially poses a danger, while the risk is the likelihood of the hazard being a danger. The two concepts go hand in hand. In fact we can think of the level of danger that exists in the workplace as being the product of 'hazard impact' × 'likelihood of the impact occuring'.

To help you understand, consider this example. What is the danger of an asteroid hitting the Earth and destroying all life? Well, an asteroid hitting the Earth may be a serious event, potentially wiping out most life forms, as happened millions of years ago with the dinosaurs. But the chance of that happening next Tuesday, or even in your lifetime, remains almost zero. That is, the risk is incredibly small. For that reason, even though the impact would be catastrophic, the danger is also consequentially small. (But, looking out into the future far enough, the odds are there that a comet or asteroid will one day hit us again.)

To help quantify workplace dangers it is useful to fill out a risk matrix chart. The chart shows the danger and its level of impact on one axis, and the risk of the danger occuring across the other axis. Mapping the two together indicates the magnitude of the danger. In other words, small dangers that are almost certain to happen may be considered as large a safety risk as big dangers that will almost never happen. As can be seen in Figure 4.2, severe risks that are certain to happen pose the greatest danger and must be dealt with as an immediate priority.

Assessing hazards for personnel

We can think of hazards as activities or processes that can go wrong and injure people. But we can also think of the people themselves and the hazards that they are likely to encounter in the course of their particular line of work. Anybody might get bitten by a snake, but a snake handler, despite their care and expertise in handling snakes, are still more likely than others to get bitten. Thus, when assessing WHS dangers we should also identify:

- the range of trades that will be active on site
- the jobs each trade will be undertaking
- the dangers associated with each job.

Seeking expert advice

By now you will have got the idea that workplace safety is important and that you need to do all you can to ensure the workplace is safe. But, of course, even with lots of experience, you will not be able to anticipate all the hazards, and know all the best ways to control them. For that reason, construction managers need to get WHS experts involved. Just as a project manager needs to know only enough about electrical work to know whether the electricians on site are doing a good job, so the project manager also needs to know just enough about WHS to make sure the experts charged with that responsibility are carrying out their duties properly.

Larger companies typically employ full-time WHS officers who are expert in the field. The role of the WHS specialist is to ensure a company is compliant in all aspects of workplace safety. Union representatives, as well as the trades people themselves will also be valuable sources of advice. Sources of information that can indicate the state of a company's WHS are depicted in Figure 4.3.

Inspection report completed according to best practice and statutory obligations

Australian law tends to emphasise preventative measures. That is, it will prosecute people for failing to follow a prescribed directive even where no bad outcome results. Think of seat belts in cars, or bicycle helmets. In Australia these must be worn, and failure

RISK HAZARD MATRIX			*Consequence*				
			Severity of outcome				
			Minimal	*Minor*	*Significant*	*Major*	*Severe*
Likelihood	Chance of risk occurring	*Certain*	5	10	15	20	25
		Likely	4	8	12	16	20
		Moderate	3	6	9	12	15
		Unlikely	2	4	6	8	10
		Rare	1	2	3	4	5

Hazard impact (risk assessment) matrix

FIGURE 4.2 Hazard impact (risk assessment) matrix

FIGURE 4.3 Sources of safety management information

to do so is an offense, even when no injury occurs. This intervention in personal behaviour is why Australia is sometimes labelled a 'nanny state,' but it is also why Australia has such a relatively good safety record.

Thus, in Australia, the law requires that construction workplaces have all the appropriate WHS documents, policies and records in place. You not only will face investigation and possible prosecution if someone you are responsible for suffers a severe work-related injury, but you could also be prosecuted for simply failing to comply with WorkSafe standards or keeping accurate, up-to-date records.

For this reason it is important to keep safety inspection reports while managing on-site WHS.

Inspection reports

Inspections should be carried out regularly. WorkSafe inspectors are empowered to visit work sites at any time to check for compliance. For this reason, companies should conduct their own internal inspections at regular intervals so as to ensure that when the external inspectors arrive, no problems will be found.

Workplace walk-throughs

A '**walk-through**' is the term used to describe the physical inspection of a worksite. It is the best way to discover what is actually going on with regards to a company's WHS policy, emergency response capabilities and record keeping. It is also a practical way to find out what workers really think about the company's WHSMS. During a walk-through, notes should be taken on what is being done right and what needs to be improved.

Inspection results

The inspection report generated by a walk-through or other inspection should be used to improve the current WHSMS. The report should comment on:

- compliance with WHS legislation, and where adjustment is needed
- hazards that need attention and risks that need to be reduced
- strengths and vulnerabilities of current operating procedures
- documented evidence (notes, photos, sketches) of the worksite showing both compliance and areas needing attention
- directions and specific notices for rectification of any WHS lapses or weaknesses.

Managing on-site WHS

Identifying WHS hazards and risks is only one part of the story. Despite doing everything to limit the possibility of something going wrong and people getting hurt, accidents do happen. The other part of the story is responding quickly and effectively to the accident when it occurs.

Emergency plans

What would you do if a fire broke out at work? What if someone collapsed to the floor? It is all very well to have a WHSMS, but the real test is whether in a time of emergency people spring into action and deal with the situation in the best and most effective manner. In other words, does your company have an emergency plan?

Equipment, procedures and training should all be in place. There should be fire extinguishers, first-aid kits and defibrillators on hand; and there should be people around who are trained to used them. There should also be nominated emergency wardens deputised to take charge in an emergency.

Induction and training

Obviously, if people are going to respond effectively in an emergency, they need to know what to do. This invovles a safety induction when they begin work, as well as ongoing training appropriate to their specific job duties and safety responsibilities. Some areas of training relevant to WHS include:

- identification of hazards
- safety signage reading
- safe handling and storage of dangerous materials
- control of access to authorised personnel
- use of safety clothing and safety equipment
- use of fire-fighting equipment

- safe operation of machinery
- traffic control
- working at heights
- working in confined spaces
- working with dangerous substances.

Licensing

Some jobs are so dangerous that training is not enough. To prove that you can do these jobs safely you will need a licence. Examples of work that is deemed high-risk – and where a licence is needed – include: forklift driver, crane operator, scaffolder, rigger and boiler operator.

Personal protective equipment (PPE)

When working in a high-risk environment, you will need to wear safety gear or **personal protective equipment (PPE)**. What safety gear you need will depend on the job you are doing and the place you are in. Basic gear includes: hard hat, protective goggles, ear protection, gloves and steel-capped boots. But you may also need a safety harness when working on a roof or at heights. You may need fire-resistant clothing when working at a steel mill, you may need a life-jacket when working on an ocean rig, or you may need to wear breathing apparatus when sanding or spray-painting surfaces. Sunburn is something else to watch out for.

Workplace environment

Tasks need to be made safe and the people doing them need to be made safe. However, the general workplace also needs to be made as safe as possible. This includes:

- fencing off the construction area
- securely storing hazardous materials
- keeping the site clear of debris, and providing sufficient bins
- marking out traffic corridors and storage zones on factory floors
- regularly sevicing and cleaning washroom and other amenities
- providing proper hazard warnings and signage.

LEARNING TASK 4.2 CREATING A RISK ANALYSIS WORKSHEET

The following video describes how to create a risk analysis worksheet, using a step-by-step approach. Take a look at the video:
https://www.youtube.com/watch?v=olMKwMzEcyU

Now apply the detailed explanations given in the video to a construction activity, such as working at heights, which is often required in the construction industry.

You are required to ensure subcontractors on your site have been properly informed and trained as to safety issues and risks involved in installing roof trusses on a two-storey residential building.

Access the Safe Work Australia link below, which provides information about hazards and risks involved with working safely at heights and then respond to the questions given.
https://www.safeworkaustralia.gov.au/heights

1. Create a risk analysis worksheet or similar list that you might use to compare with the identified hazards and risks the subcontractor has provided you with as part of his SWMS.
   - You can compare your list of identified hazards and risks with those your classmates have prepared. The person with the most complete list should have been the PCBU (supervising person responsible for the worksite).
2. Was your list the most complete one, or did you just skip over important dangers and hazards that in this job might result in injuries to the workers carrying it out?
3. If your list was the most detailed one identifying the hazards and risks for the installation of roof trusses, check it to see if any other yet to be identified hazards might be added to it from others in your work group.

LEARNING TASK 4.3 CONSTRUCTION OCCUPATION WHS HAZARDS AND RISKS

Every construction project will have its own set of dangers, hazards and risks.

- Step 1 – Find a partner and think of a specific construction project. Now list all the trades that might be needed to complete the project. (Some examples might be: carpenter, demolisher, concreter, excavator, boiler maker, crane driver, glazier, site foreman, among many others.) Individually, rank each of those trades in order of how dangerous you think their work is. Compare your ranking with that of your partner. Discuss any differences and see if you can come to an agreement on which trades have the most dangerous jobs.
- Step 2 – For the trades with the greatest danger, what specific tasks make their trade so dangerous? What do you think they do to keep themselves safe? What more could be done to make their work even safer?

LEARNING TASK 4.4

CONDUCTING A WORKPLACE SAFETY INSPECTION

Take a look at the video on tips for conducting work place safety inspections, by WorkSafe Victoria: https://www.youtube.com/watch?v=Ruj0IRezVXE

What are the 'top tips' for conducting a safety inspection offered in the video?

Think about how you would go about conducting a safety inspection at a place you have worked in. Write down a list of the steps and activities you would perform. Show your work safety inspection plan to a partner and ask them to comment.

LEARNING TASK 4.5

GETTING A HIGH-RISK WORK (HRW) LICENCE TO DO DANGEROUS JOBS

Go to the Safe Work Australia website for licences: https://www.safeworkaustralia.gov.au/licences

Read and understand what you need to do to be licensed to work in an HRW area. The usual procedure for getting a licence is to gain the appropriate accredited qualification through a registered training organisation (RTO), and then to be assessed.

Pick a high-risk job related to construction (such as fork-lift driver, asbestos remover, etc.). Then find out what course you would need to complete in order to get a licence for that HRW.

You can find courses at: https://training.gov.au/

How long would it take you to get the qualification?

Implementation of control measures

Once we know the hazards, their risks, and the trades and people involved, we need to go about eliminating those hazards. We need to follow a procedure if we are to minimise dangers effectively. First, we need to review the inspection reports and make recommendations. Second, we need to make sure we remain compliant with statutory requirements. Third, we need to then implement the agreed upon control measures. Finally, we need to constantly monitor the work environment to check that the measures continue in place and that they are effective.

The inspection report and recommendations

WHS management is a cyclical process. We look for potential dangers, do what we can to eliminate them, see how effective our actions have been, then look again for new dangers and even better ways to manage the old ones. Figure 4.4 shows the cyclical nature of a workplace WHSMS.

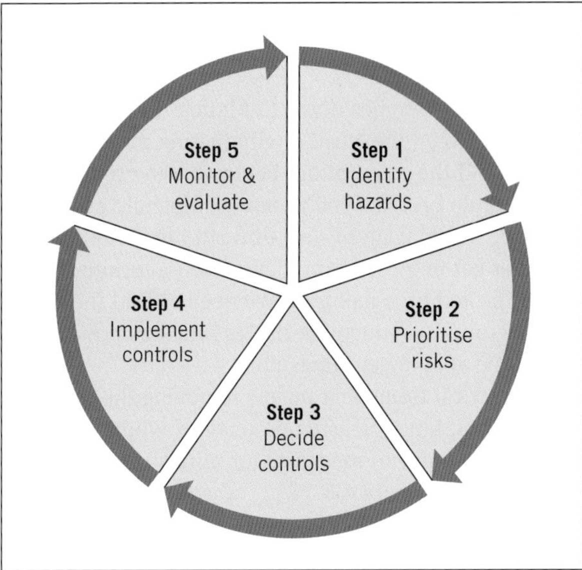

FIGURE 4.4 WHSMS cycle of risk assessment and management

Scheduling planning actions

Any workplace will have a least some potential hazards, and not all of them will be as dangerous as each other. Thus, in mitigating these dangers, it is imporant to prioritise the safety measures that are to be taken. Those dangers that can cause serious harm to people are the ones that must be eliminated immediately. Having identified the hazards, and their risks, you will be able to produce a 'risk hazard matrix,' as discussed earlier, and with that, generate a quantitative value for each danger. These numbers will help you prioritise the order in which dangers are to be dealt with:

- immediately
- in the short term (within weeks)
- in the long term (within months).

Hazard control hierarchy

In Australia, young children up to 7 years of age must sit in booster seats when riding in a car; and they must sit in the back. This is to give them extra protection in case of a car accident. The fine for failing to do this runs to hundreds of dollars. This requirement is an example of hazard control. However, if we really wanted to prevent all injuries from car accidents with certainty, the answer would be to ban children from riding in cars altogether. Of course, this is not practical, and so we have to accept some risk. In fact, children under the age of 7 years are allowed to ride in the front seat if the back seats are taken up by other younger children. And, when riding in a taxi, children are exempt from having to sit in a booster seat. The trick in hazard control is to find the right balance between reducing risk while still being able to get things done.

The things we need to consider when finding this right balance include:

- How severe is the danger?

- How likely is it to happen?
- How easy is it to control?
- Who would it affect?

In the end it comes down to finding the right balance between the 'cost' of eliminating the risk entirely, and the 'cost' of not being able to do what you wanted to do because you're now completely avoiding that risk. Think again of how difficult life would be if we never got in a car again; never used a forklift, hoist, saw, or picked up a nail gun. We need to find the right middle ground – reduce the risk as much as possible, but still be able to get things done.

In practical terms that means removing the dangers that you can, but increasing protections when you cannot. The range of measures for mitigating dangers is summarised in Figure 4.5.

Consultation on statutory requirements

Workplace safety concerns everybody on site. They are the ones exposed to any dangers, and they are the ones with first-hand knowledge of the dangers. Therefore, these people should be consulted when making the workplace safer.

People able to contribute to WHS

The list of people who could contribute to WHS include:

- workers
- managers, site supervisors
- self-employed personnel and contractors
- manufacturers and designers
- suppliers and installers
- WHS representatives and committee members
- clients and visitors.

Workers

Perhaps it is the workers who are most exposed to workplace dangers, since they are the ones on site more often. According to the WHS Act, they have a duty of care to do what they can to preserve their own health and safety, as well as that of others. They must also follow any WHS instructions given by supervisors and cooperate

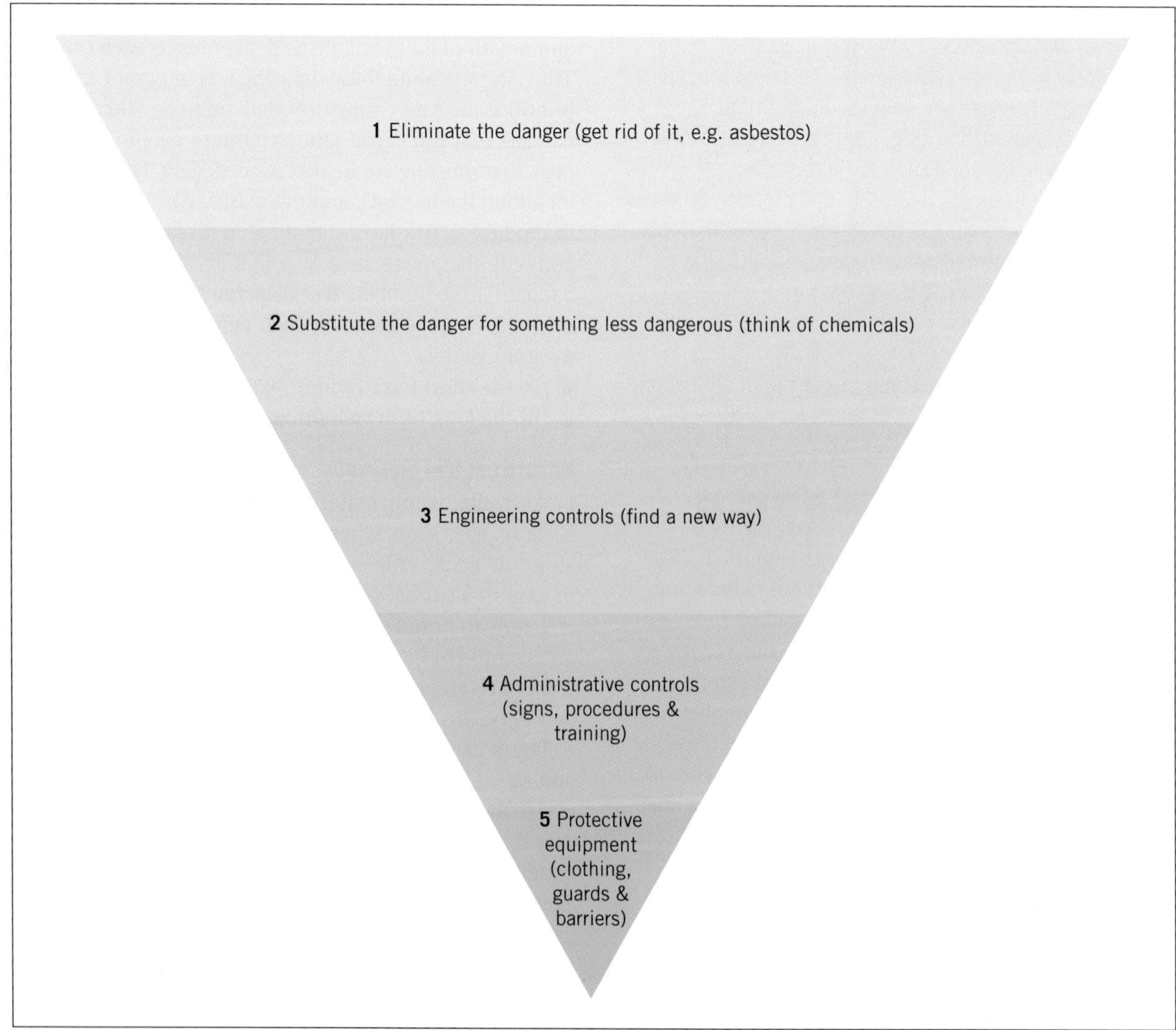
1 Eliminate the danger (get rid of it, e.g. asbestos)
2 Substitute the danger for something less dangerous (think of chemicals)
3 Engineering controls (find a new way)
4 Administrative controls (signs, procedures & training)
5 Protective equipment (clothing, guards & barriers)

FIGURE 4.5 Hierarchy of measures for controlling hazards

with company WHS policies and procedures. Any WHS incidents, no matter how minor, must be reported.

Managers

Persons in control of a workplace have a legal responsibility to make the site safe. A WHSMS must be in place and enforced diligently. Materials handling and equipment usage, as well as work practices, must all comply with legal requirements.

Contractors and businesses

Businesses working with one another must ensure that they all comply with WHS practices. An outside contractor, for example, stepping on site, must ensure they are following the set WHS procedures for that workplace.

Manufacturers and installers

Section 25 of the WHS Act specifies that all plant, substances and goods brought into and used at a worksite must be tested, analysed and certified as being safe and able to perform as specified.

WHS personnel

WHS representatives have additional duties above that of other employees or contractors. They speak for all workers and therefore must be particularly vigilant as to WHS. They need to monitor safety on site, investigate complaints and breaches, and proactively look for issues that require attention.

WHS committees have a role in facilitating cooperation between the unions and management. They assist in developing standards and improved work practices, ensure compliance with regulations, and assist in investigations of WHS breaches, or when incidents arise.

Implementing control measures

Once information has been gathered on hazards and their risks, and a WHSMS developed, the plan has to be implemented. And just as with gathering information, implementations should be done in consultation with all the relevant personnel: workers, managers, contractors and the WHS committee members. Key questions to consider at this stage are:

- How are workers going to be involved?
- How will information, expectations and responsibilities be disseminated?
- How will workers' views be heard and considered?
- How will various options be narrowed down to an agreed course of action?
- How will new information and risks be gathered, and incorporated into the plan?

It is important to remember that something like WHS only works when everybody is involved, and when everybody involved 'buys-in' and believes in the WHSMS. For that reason, even though WHS is largely a management responsibility, it requires careful consultation and cooperation. The human element in communicating WHS expectations and fostering everyone's participation should not be underestimated. The five-step process for implementing the risk control plan is shown in Figure 4.6.

Monitoring the effectiveness of control measures

With the WHS plan implemented, the next consideration is monitoring and reviewing. It is not enough to have a plan in place; the plan itself should be monitored to ensure it is working as expected – picking up all the hazards and dealing with them effectively. Despite best efforts in the planning and implementation stage, there will always be things that were missed, things that could be done better, and new risks that come up as workplace conditions and circumstances evolve.

Monitoring of control measures

Three aspects of monitoring and reviewing the risk control measures are scope, frequency and method.

Scope

Scope here refers to the effectiveness with which the plan, along with various people's roles, obligations and duties with respect to that plan, have been communicated and acted upon. In other words, did everybody get the message and are they all acting on the WHS plan as needed to make it work? This needs to be checked. Where people still don't know what to do, they should be informed; where they are not acting on their duties, they should be trained.

Frequency

Checking on the plan is an ongoing process; and the question here is, how often should checks be made?

- How often should evaluations be conducted?
- How often should workers and other stakeholders be consulted for WHS input?
- How often should WHS information be updated and disseminated?

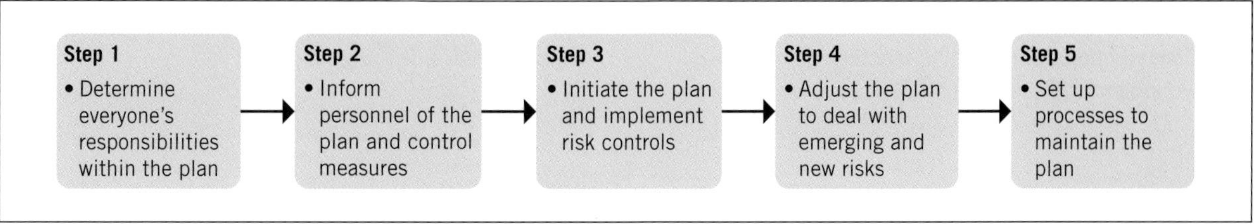

Step 1
• Determine everyone's responsibilities within the plan
Step 2
• Inform personnel of the plan and control measures
Step 3
• Initiate the plan and implement risk controls
Step 4
• Adjust the plan to deal with emerging and new risks
Step 5
• Set up processes to maintain the plan

FIGURE 4.6 Five-step risk control plan implementation process

Method

Then, of course, the above steps of checking on the WHSMS plan and on gathering and disseminating information in turn also need to be evaluated. In other words, the system for checking on the system also needs to be reviewed from time to time.

Continuous improvement

When all is said and done, the WHSMS should be considered as a continuous cycle of finding hazards, evaluating their risks, prioritising these dangers, finding remedial measures, implementing these measures, and then checking to see it is all working as expected. There is never an end-point, and this cycle remains in perpetual change as new dangers appear and new methods of dealing with them are developed. And all the while, everybody on site must continue to play an active and engaged role at every stage of the process.

CASE STUDY

The next disaster is the one you don't see coming

If you go to YouTube and look at all those 'car-crash' videos, it's fun trying to anticipate how the accident you know you're about to see will happen. Usually, it's sudden, unexpected, and leaves the victims with little chance to escape. The army saying goes, 'the shot that'll kill you is from a sniper you'll never see'.

So, how do you prepare for the disaster you don't see coming?

When the Tohoku Earthquake struck Japan in 2011, it resulted in a 'level 7' meltdown of three nuclear reactors at the Fukushima Daiichi Nuclear Power Plant complex. These plants had been built to the world's highest quake-resistant standards. Yet they failed. As a result of the massive radiation spill, whole cities had to be evacuated and abandoned. Eight years on, radiation levels remain high and people still haven't been able to return to their homes.

In hindsight, it is easy to say that the power plant wasn't built strong enough, and that it should have been built to withstand the quake and the tsunami that followed. This is akin to critics saying that the twin towers of the World Trades Center, in New York, that collapsed in 2001 in the attack of 9/11, should have been built strong enough to withstand the impact of high-speed passenger aircraft.

The problem is balancing the probability of a disaster happening against the cost to prevent it. Who could have guessed planes would be flown into those buildings? And who would have guessed the Fukushima plants would have been hit by one of the strongest earthquakes ever recorded in history?

The Tohoku Earthquake was magnitude 9.1. (The scale is logarithmic, with each level being 10 times the previous level.) The tsunami waves reached over 40 metres high. More than 400,000 buildings were destroyed, with another 700,000 severely damaged. Sixteen thousand people drowned; 230,000 people remain displaced. The cost to the Japanese economy is estimated at US$240 billion; the costliest disaster in history.

If that doesn't impress you, think about this. The earthquake was so strong that it moved the whole island of Honshu, Japan, 2.4 metres to the east, and shifted the Earth's rotational axis by 10 cm.

So, what's the answer? Avoid the risk – and forget about nuclear power plants altogether? That's what some countries have done. Or, build the next nuclear plant even stronger, gambling that it will be strong enough to withstand the next, even bigger, unexpected monster quake?

LEARNING TASK 4.6 DOCUMENTING AND MONITORING THE EFFECTIVENESS OF CONTROL MEASURES

Identify a situation at your workplace where an injury or near miss has occurred.

Provide a 'written summary' of what happened, similar in style to the points covered in the 'case study' above.

Note: This is the **inspection report defining the problem**.

Having completed your 'case study' summary, provide a strategy to:

a **Consult others** potentially impacted/affected or working in the same situation at your workplace, seeking their input as to how to eliminate or lower the risk/s associated with your case study.

b Document how you would go about **implementing** any proposed control measures based on the consultation process you have undertaken.

c Document the approach you will be taking to **monitor and control** the effectiveness of your control methods – be sure to 'nominate' realistic timeframes and nominate the people responsible for making sure those timeframes are being met.

Communication and educational programs

Once the WHSMS has been worked out and we know what everybody should do, the final task is to communicate the policy and duties to everyone. This requires consultation with appropriate personnel about developing a communication strategy, then delivering the education and training to the right people, followed by a review of the effectiveness of the communications and education program.

Key elements of a strong WHS communication strategy include:

- clear mission and purpose
- precise message to convey
- clearly identified people to communicate to
- the right set of means to communicate with
- the right set of people to undertake the WHS communication.

Strategies for communicating WHS policy and practice

It's one thing to have a clear WHS strategy, with all the hazards identified and all the safety measures worked out, but it's another getting everybody to follow through and do what's needed. If the plan is going to be implemented, people need to know about it. This may seem a small matter, but in many ways this final step is the hardest. There may be a huge number of people involved in any one project, and many of them on site for just a short period of time. In the ebb and flow of people, how do you keep track of how well they're all informed about company WHS policy?

But more than that, even if everybody gets the memo, reads the signs and does the training, how committed are they to doing what they should? Are the workers following through and taking WHS seriously? People don't buy into a company policy because they are ordered to – they have to believe in it – and for that to happen the communication has to be ongoing and tailored to the deeply felt safety concerns of the workers themselves.

Communication strategy

A good WHS communication strategy needs to inform all the right people about all the right things. It also needs to be done in an easily comprehensible manner that gets people to behave as expected. Simply, a WHS communication strategy needs to be clear, to the point, with precise outcomes. The communication process should:

- provide workers with information that keeps them safe
- enable workers to be involved in the decision process and to raise concerns
- incorporate feedback and updates, and distribute that as well.

In fact, it a legal requirement that workers must be consulted when management introduces new work procedures.

Communication plan

A good communication plan identifies:

- the target audience that is to be informed
- what they need to know
- how they will be informed
- the outcomes of the communication (what people are to do as a result)
- and, how those outcomes will be measured (how do we know people are complying?).

Thus, in developing a communication, plan there are a number of things to consider:

- Who will you need to inform – not just employees, but other stakeholders as well?
- What are the current ways in which information is disseminated?
- What other options for communication are available?
- How effective are those alternatives?
- What is it we specifically want to achieve through the messages we send?
- What are the costs; who will be responsible; what are the timetables for delivery?
- How will we collect communication coming back the other way (feedback)?
- How will we know we are getting through to people?

Communication strategies and educational programs

WHS legal requirements and industry best practice require that the WHSMS be disseminated to employees and other relevant personnel through effective communication and training. The aim is to inform workers of WHS codes of practice, and ensure those practices are carried out in the workplace.

Communication strategies

There are various ways in which WHS communication can be provided. Chiefly, verbal instructions, written information and signs.

Verbal

Informing personnel of on-site hazards, risks and dangers, can be done quickly and effectively through face-to-face exchanges. Talking is often preferred by both the giver and receiver of information; it is perhaps the most natural way of communicating. Importantly, when speaking to someone, the other party has the opportunity to question and confirm their understanding of the situation. Also, too, the person giving the message has the opportunity to confirm that their audience does indeed understand what is being said. Talking is a two-way exchange, and that is what makes it effective. Some occasions where verbal communication might be used include:

- on-the-job instructions – such as explaining how a machine should be used
- pre-start meetings – highlighting new developments, changes or dangers
- 'toolbox talks' – focusing on a particular safety issue that has come up
- one-off sessions – debriefing an event, mishap or incident that has come up
- WHS meetings – disseminating specific information or discussing concerns.

Written

Though verbal communication may be the most effective in one-on-one communication, written communication acts as a record that can be referred to time and again. It is therefore particularly suited for recording detailed instructions, or for conveying information to many people in many places. Written documents also serve as tangible evidence that WHS responsibilities are being undertaken as required by law. Written information allows readers to:

- read at their convenience
- read over and over
- keep a record.

Written documents come in a variety of forms:

- written reports – incidents, observations, memos
- WHS bulletins – WHS news items, notices and updates.

They should also meet certain standards:

- be clear, concise and relevant
- use language appropriate to the audience
- use correct spelling and grammar.

Signage

There are three Australian standards that cover the use of safety signs:

- AS 1319 – Safety signs for the occupational environment
- AS 1216 – Class labels for dangerous goods
- AS 1318 – Use of colour for marking physical hazards and the identification of certain equipment in industry.

Safety signs can be word-only signs ('DANGER'), or they can be an image (lightning symbol to indicate high electrical voltage). There are seven categories of signs:

- prohibition (must not do) – e.g. 'no smoking'
- restrictions – e.g. 'local traffic only'
- mandatory – e.g. 'safety helmets must be worn'
- hazard warning – e.g. 'hot surfaces'
- (danger hazards) – e.g. 'asbestos removal in progress'
- fire – indicating location of alarms, pumps and other facilities
- emergency information – indicating where emergency exits and safety equipment are located.

Education programs

The above forms of communication convey specific, concise information on targeted WHS issues. They do not provide the wider, general awareness and competencies needed to make employees safe in the workplace. For that, the workers must be educated. Consequentially, a key requirement of a WHS-compliant site is that all its workers receive appropriate WHS training.

There will likely be a WHS component in the training an apprentice receives as part of their trades qualification. However, once on the job, an employee can expect further site-specific WHS training:

- WHS induction
- implementation of new risk-management processes
- implementation of new work processes
- operation of new equipment and tools
- materials handling and storage
- regular WHS refresher courses.

Reviewing the effectiveness of communication and educational programs

When all this is done, the final step is to monitor and review the WHS strategy and educational program, and confirm it is working as expected – and where not, to make it right. There are a number of ways feedback can be gathered:

- data reports – injuries, near misses, frequencies
- meeting reports – comments and recommendations from worker and committee meetings
- stakeholders – gather feedback from all people with a vested WHS interest
- audits – risk assessment and workplace inspections.

Consultation process

Once the reviews have been conducted, the WHSMS needs to be modified to take on the changes that are needed. The process is an ongoing cycle of 'do–review–update–repeat' (see Figure 4.7). WHS never ends; mangers and workers alike need to be vigilant and adaptive at all times. The workplace safety game is an ever-changing one, and WHS-savvy employees always keep their eye on the ball.

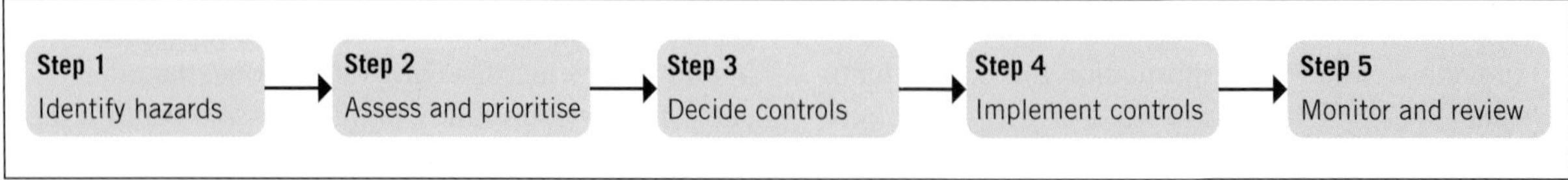

FIGURE 4.7 Steps in the ongoing repeating loop of WHS risk control, with consultation at every stage

LEARNING TASK 4.7 COMMUNICATION AND EDUCATIONAL PROGRAMS

View the following video:

- https://www.safeworkaustralia.gov.au/media-centre/construction-worker-takes-safety-moment
  and/or
- access induction training content from: https://www.safeworkaustralia.gov.au/doc/induction-construction-work
  and/or
- access your state/territory WHS authority documentation or code of practice relating to construction WHS safety communication/education strategies.

You are a supervisor at your building site, acting as a team leader to four others.

1. Develop/document a simple (maximum) two-page strategy (method) for communicating WHS issues and your organisational/site requirements to your work team.
2. Document a simple strategy that you might use into the future to ensure that you have evidence of providing ongoing WHS training to your work team as part of your organisational WHS training requirements.
3. Explain what you might do (how you would collect evidence) to see if your team training activities are effective.

SUMMARY

This chapter covered workplace health and safety (WHS) – formerly occupational health and safety (OH&S). Construction can be dangerous, and accidents can be serious or even life-threatening. Consequently, the message in this chapter is that everybody has a responsibility to ensure workplace safety, both for themselves and for others. To make sure that happens, Australia has enacted comprehensive laws mandating that every workplace complies with set safety standards, and that all companies put into operation a WHS management system (WHSMS).

Developing a WHSMS involves a number of steps. First, all workplace dangers must be identified, and the severity of those dangers, along with the likelihood of them transpiring, must be assessed. Second, a priority list for danger mitigation must be developed, along with options for dealing with each and every danger. Third, once the best approach for dealing with the dangers is decided on, they must be acted upon. Fourth, the right people need to be brought in to act. Finally, danger, hazard and risk mitigation effectiveness must be monitored and reviewed to confirm that the workplace is indeed safe.

At every stage, impacted workers and other stakeholders all need to be consulted for their views and insights. At the same time, they also need to be trained, educated and informed so that they can go about their work as safely as possible. At every stage, management and workers alike must remain vigilant for any emerging threats, and be ready to respond. Only in this way can we expect to maintain Australia's good workplace safety record.

REFERENCES AND FURTHER READING

Australian Standard®, *Safety Signs for the occupational environment*, AS 1319–1994:
https://www.saiglobal.com/PDFTemp/Previews/OSH/as/as1000/1300/1319.pdf

Safe Work Australia publications can be obtained from:
https://www.safeworkaustralia.gov.au/resources_publications/all

Safe Work Australia national standards:
https://www.safeworkaustralia.gov.au/resources_publications/national-standards

Safe Work Australia virtual seminar series:
https://www.safeworkaustralia.gov.au/vss

WHS / OH&S Acts, regulations and codes of practice:
https://www.business.gov.au/risk-management/health-and-safety/whs-oh-and-s-acts-regulations-and-codes-of-practice

PLANS AND SPECIFICATIONS

5

Chapter overview

Architectural and construction drawings are arguably a language unto themselves: if you are not fluent, you are like a traveller in a foreign land, totally dependent upon an interpreter and never quite sure that what you are being told is fact, fiction, or just what the interpreter thinks you would like to hear. Being able to fluently interpret plans and specifications is therefore at the core of what it is to 'be' a builder.

This chapter expands upon the key elements of reading and interpreting plans and specifications. Its content covers both low-rise residential and commercial projects, allowing you to confidently engage with clients, architects and other trades people as well as being able to interpret documents for the purposes of quantity and costing estimates.

Elements

This chapter provides knowledge and skill development materials on how to:

1. identify types of drawings and their purposes
2. apply commonly used symbols and abbreviations
3. locate and identify key features on a site plan
4. identify and locate key features on drawings
5. correctly read and interpret specifications
6. identify structural and non-structural aspects to the specification.

In addition, the Australian standards of relevance to this chapter are:

- Australian standards set AS 1100, particularly AS 1100.301–2008 (including Amdt 1–2011) Technical drawing – Architectural drawing

  Includes:
  - AS 1100.101–1992 (R2014) Technical drawing – General principles
  - AS 1100.101–1992/Amdt 1–1994 Technical drawing – General principles
  - AS 1100.201–1992 (R2014) Technical drawing – Mechanical engineering drawing
  - AS 1100.201–1992/Amdt 1–1994 Technical drawing – Mechanical engineering drawing
  - AS 1100.301–2008 (includes amdt 1–2011) Technical drawing – Architectural drawing

- AS 1100.301–2008/Amdt 1–2011 Technical drawing – Architectural drawing
- AS 1100.401–1984 (R2014) Technical drawing – Engineering survey and engineering survey design drawing
- AS 1100.401–1984/Amdt 1–1984 Technical drawing – Engineering survey and engineering survey design drawing
- AS/NZS 1100.501–2002 (R2014) Technical drawing – Structural engineering drawing

• Handbook HB 50 – 2004.

Introduction

Plans and specifications are effectively a language, a language full of its own quirks and odd phrases that seem to make sense to those who know it, and not much to those who don't. And like a language it has many dialects, and undergoes change over time: or in this case, types of drawings and forms of documents; all varying depending upon when and where they were drawn, their context, or the purpose of their use. This chapter explores these various types of construction industry drawings, how to read them and how to extract and correlate the necessary information from those other documents that often accompany them.

The first section offers an overview of the various types of drawings, their purpose, the key aspects to be found within them, and the Australian standards applicable. The section also addresses the concept of 'scale' as it applies to construction drawings. The second section covers common abbreviations and symbols used in architectural drafting and specifications. The third and fourth sections look more closely at the reading of various plans and how to find relevant information within them. The fifth and sixth sections deal with the interpretation of various aspects of the specifications documents.

Types of drawings and their purposes

Your own experience tells you that there are many different ways to depict something through drawings. The purpose of the drawing – that is, what is it that we actually want to convey to the viewer – determines which type of drawing to use. Because of these varying purposes, in construction we actually use a broad array of drawing types, from simple sketches through to highly detailed construction drawings developed with the aid of computers.

The purpose of this section is, therefore, to introduce you to the various drawing types we use, their purposes, advantages and limitations. The section also covers the basics of how these drawings interrelate and, to a lesser degree, how they are developed or drawn.

Types of drawings: their purposes and advantages

Orthographic drawing

For the purposes of architecture and construction there are basically two categories of drawings that will be presented to you in two-dimensional form – that is to say, on a flat sheet or as one of the many common computer documents such as a PDF or JPG image on a computer. These two categories are known as pictorial and orthographic drawings. As it is important to know the difference, we shall cover them first.

These two categories of drawings are distinctly different and have very different purposes due to their differing target 'audiences', and the expectations of these different groups. As it is orthographic drawings that you, as the builder or trades person, must become the most familiar, it is with these we will begin.

Orthographic drawings offer the viewer a flat plane depiction of a subject (Figure 5.1). These drawings are useful for describing true lengths and proportions, although what they show is not how we actually 'see' an object with our eyes. It is orthographic drawings that give us the typical set of 'plans' commonly used to construct domestic and commercial structures.

Figure 5.1 compares how we might see a three-dimensional figure with the eye (on the left) to an orthographic drawing of the same object (on the right). Note that when we look at this object, the middle section would block out a portion of the upper and

Source: Glenn Costin

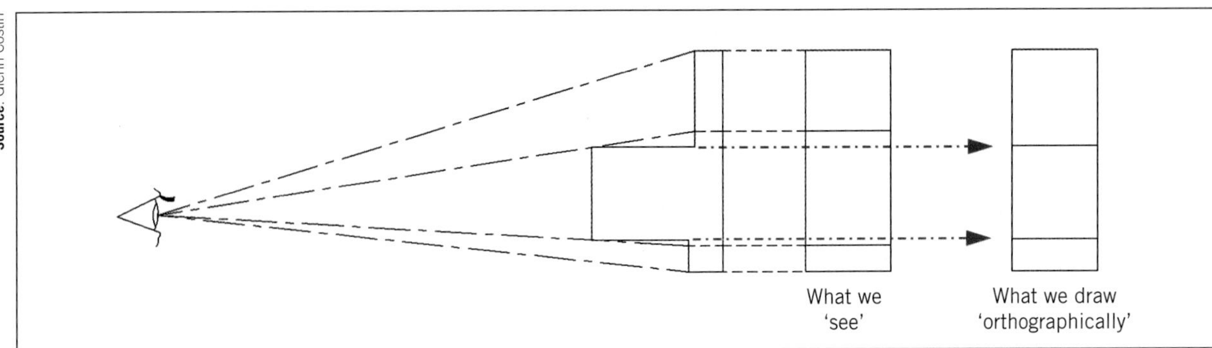

FIGURE 5.1 An orthographic drawing

lower sections because they are stepped back. This means our mind perceives these sections as being shorter than they really are. Orthographic drawing overcomes this issue by showing the actual sizes of each surface in relation to each other. It allows us to put accurate dimensions to each part of the drawn object and show exactly where those dimensions begin and end.

However, orthographic drawings offer only one side, face or 'plane' of a structure at a time (i.e. top, side, front and the like). The front and side views being known as 'elevations' (see Figure 5.4). This means that while we get the real relationship between the various parts of an object, the object doesn't actually look very real. Its flat on the page and, for those not used to reading this form of drawing, the object is difficult to understand as a three-dimensional (3D) object. This is where pictorial drawings come in.

Pictorial drawings

Pictorial drawings attempt to overcome the flat plane issues of orthographic depiction by representing the missing third dimension: they attempt to provide a sense of depth. There are various forms of pictorial drawings used in construction, the most common being isometric, oblique, and perspective representations. Pictorial drawings offer three faces within the one view, as if the viewer were standing slightly to one side and above (or below) the object.

In isometric drawings all dimensions retain their full-length relationship to each other, much as orthographic drawings do. However, to achieve the '3D' effect in isometric drawings, an object's side surfaces are represented as being at 30 degrees to the horizontal base line (see Figure 5.2).

Oblique drawings achieve a slightly different 3D effect by drawing the front face of the object much as it would appear orthographically, but then drawing one side angling back at 45°. Unlike the isometric or orthographic drawings, however, the side dimensions are generally measured up the angle at only half their actual length (Figure 5.2). Though this shortening of the receding dimensions is not always used, it tends to offer a more realistic representation of the object. The principle of using half-length dimensions is sometimes referred to as 'cabinet' oblique, as against 'cavalier' oblique when full lengths are used. In either case, the viewer is provided with a more easily visualised image of the subject.

Source: Glenn Costin

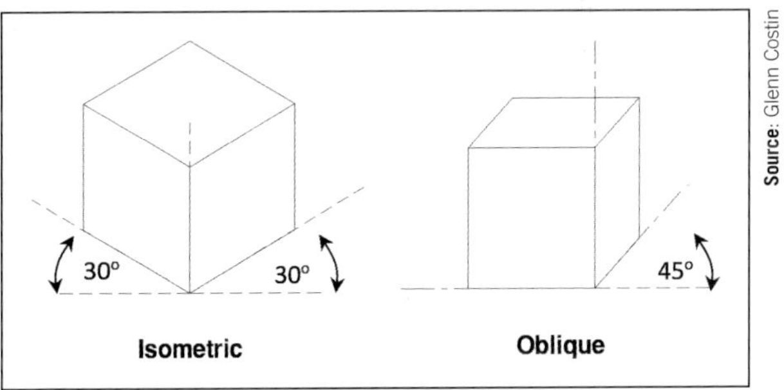

FIGURE 5.2 Isometric and oblique drawings

Perspective drawing is a more advanced form of pictorial drawing that provides a very close approximation to how an observer really sees an object. There are various forms of perspective drawing. In its simplest form, the viewer sees things vanish back to one point only. In its most realistic form, things vanish back to three points. Generally, drawings are worked to two vanishing points on the 'horizon', as shown in Figure 5.3. Perspective drawing is commonly used by architects to present their ideas or concepts to clients and/or the public for comment and feedback (see Figure 5.7).

When drawing pictorially we draw in proportion: that is, the various dimensions bear a direct relationship to each other in length. Naturally, we can seldom draw a construction project, or even a small element or detail of that project, full size; i.e. where 1 metre (1 m) of building is 1 metre on the paper – it just won't fit. To make it fit the page, and to ensure the relationship of length to height and width remains true, we need to reduce the sizes, and do so equally. We do this by dividing all the lengths by a chosen number, be it 2, or 5, or 10, or 100, or indeed any number we wish. So long as we always use the same number for all the measurements on any particular view the proportions will be accurate. The number we choose

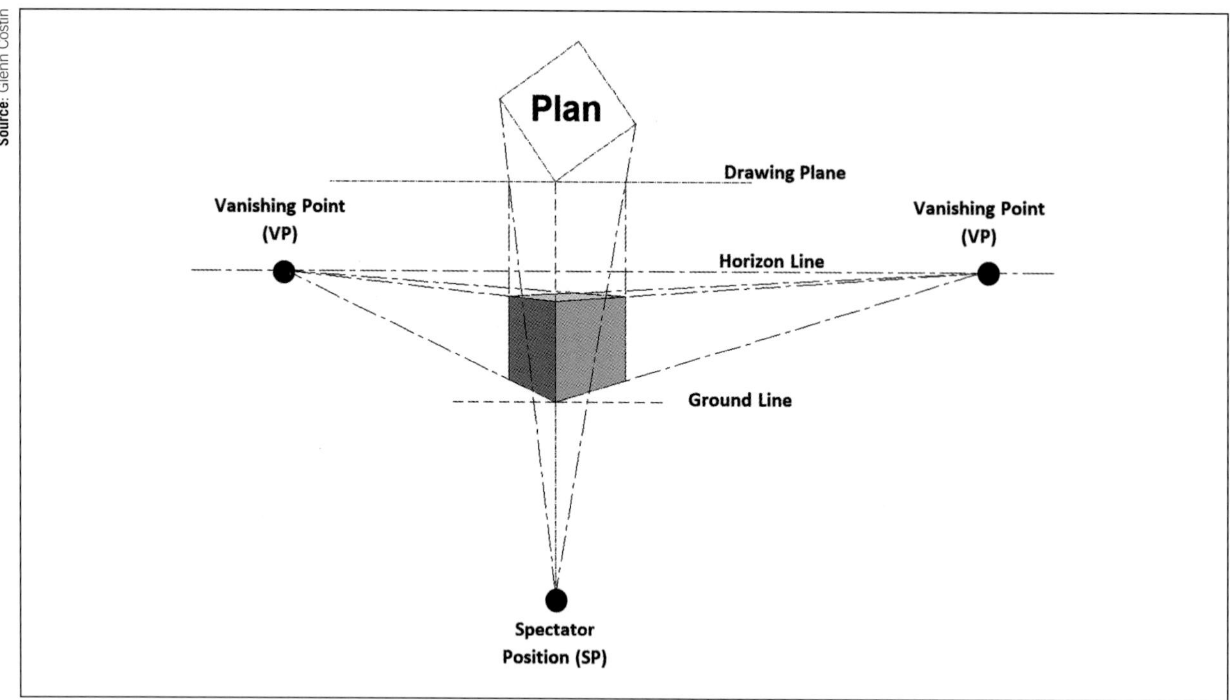

FIGURE 5.3 Two-point perspective

to divide by depends upon the size of the drawing we want to produce, or the size of the paper we are drawing or printing on. This is known as scaling, or drawing to scale.

The way scale is expressed is also important. If we need to divide everything by 2 (half size), for example, we write this as 1:2 (spoken, 'one *is to* two'). If we need to divide things by 100, then we write it as 1:100 (spoken, 'one *is to* a hundred', or 'one *is to* one hundred'). This latter example means that what we have drawn is one-hundredth the size of the original object. Note that the larger the number you are dividing by, the smaller the size of the drawing you produce. That is, the larger the number, the *smaller* the scale. For example, 1:500 is five times *smaller* than 1:100.

Given that you will seldom be drawing, but rather reading and deriving information and sizes from drawings already produced, this concept is important to you in its reverse. For example, if you have a plan given to you that is drawn at 1:100, then you know that you can measure any length on that drawing and multiply it by 100 to know what size it is in reality. For example, if you measure a length of 25 mm on a plan drawn at 1:100, then multiply 25 × 100 and you get its true length – in this case 2500 mm or 2.5 m.

Pictorial drawings, be they isometric, perspective or oblique, are seldom considered, or required to be, to scale. The concept, however, is the same: take all the dimensions for that particular view and divide them by your chosen number – your 'scale' – before plotting them on the drawing.

Orthographic drawings, on the other hand, are almost always drawn to scale. In the construction industry there are a number of common scales used, indeed required to be used, based upon the Australian standards set AS 1100 and applicable council requirements. It is also a requirement of this standard that the scale used on a drawing be shown clearly on each page, or beneath each drawing on the page if multiple scales are used. Figures 5.12 and 5.16 offer examples of how these might appear on the plan.

Sketches

Sketches are drawings you will encounter during the design phase of a project. Typically sketch plans illustrate the general layout, form and aesthetics of a residence, building and/or site. Being able to sketch is a skill that most builders should develop to at least a basic level that allows them to present concepts and ideas to clients, particularly those seeking to change something during the construction process (known as a 'variation').

Initial sketches

Initial sketches are drawings used to help the designer and client develop a construction proposal. They are frequently done on site or in front of the client. Such sketches are seldom drawn to scale, providing only such information as will help a client 'see' where the designer is going with their ideas. They can also help a client understand the implications of their own ideas for a structure (Figure 5.5).

Sketch plans and preliminary drawings

Also known as 'roughs', sketch plans are generally done to scale. This offers true proportions for the client, council, or other key stakeholders to consider. These may be created by 'pencil and stick' (drawn by hand)

Source: Glenn Costin

FIGURE 5.4 Typical construction drawing elevation

Source: Glenn Costin

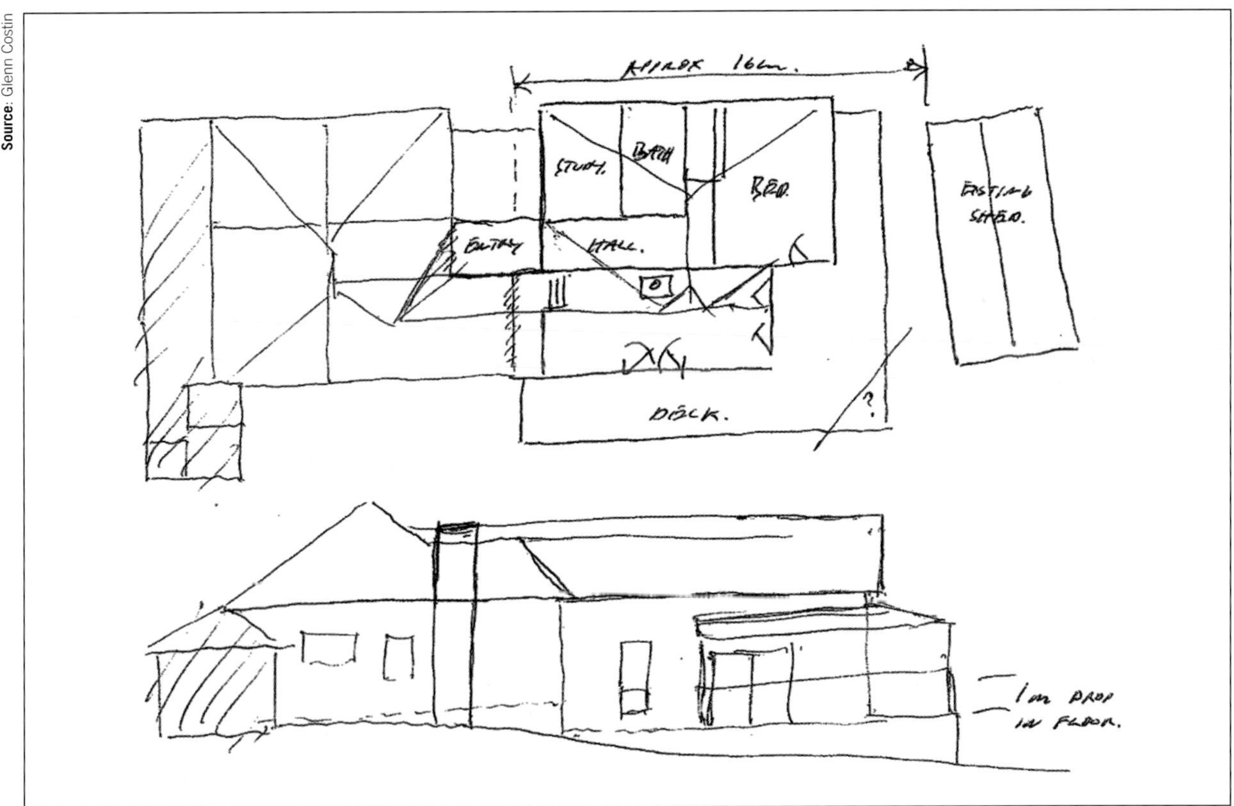

FIGURE 5.5 Example of an initial sketch for a house

or developed on a computer (computer aided design, otherwise known as CAD) using a variety of different software programs currently available (see Figure 5.6).

Presentation drawings

Architects often present artistically 'rendered' sketch plans or presentation drawings to clients. They can do this using any one of the pictorial drawing techniques discussed earlier, but the most favoured is the two-point perspective. Presentation drawings help non-technical people understand what the project will look like when complete more easily than they might through flat plane orthographic representations. It also helps the designer visualise their ideas in the context of the surrounding landscape (Figure 5.7).

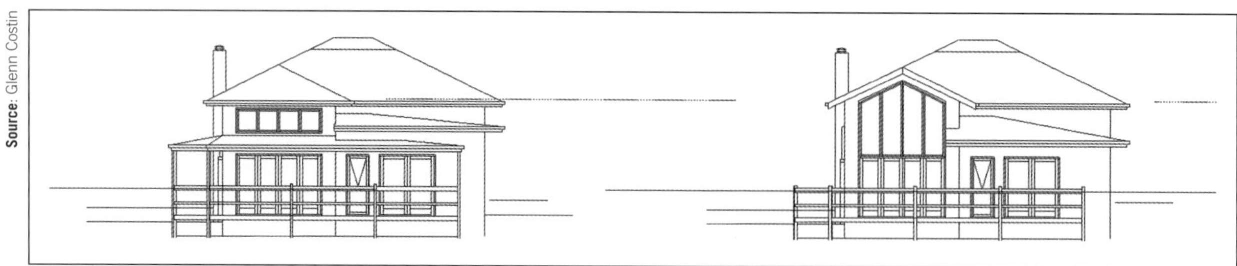

Source: Glenn Costin

FIGURE 5.6 Example of a preliminary drawing developed using CAD

Source: AAP Image/Universal Images Group

FIGURE 5.7 Melbourne Exhibition where the first Federal Parliament met on May 9, 1901. Although they had initially planned to hold the ceremony in Melbourne's Parliament building, these buildings were eventually chosen because of the amount of space they offered

Development plans

Sometimes referred to as 'block plans', these plans provide the location of the block or building site in relation to other blocks in the surrounding district, suburb, or new domestic or industrial development area. Figure 5.8 shows a typical housing development plan laying out roads, reserves, park spaces, block sizes and locations.

Figure 5.9, taken from the previous figure, shows that development plans also generally show the number, dimensions and area of a block. Block 34, for example, is a regular shaped block with an area of 448 m^2. Block 27, on the other hand, is quite irregular in shape. Note also that the outer boundary of the development has included the direction in degrees and minutes. The direction of a block boundary may also be provided as relevant to the development as a whole.

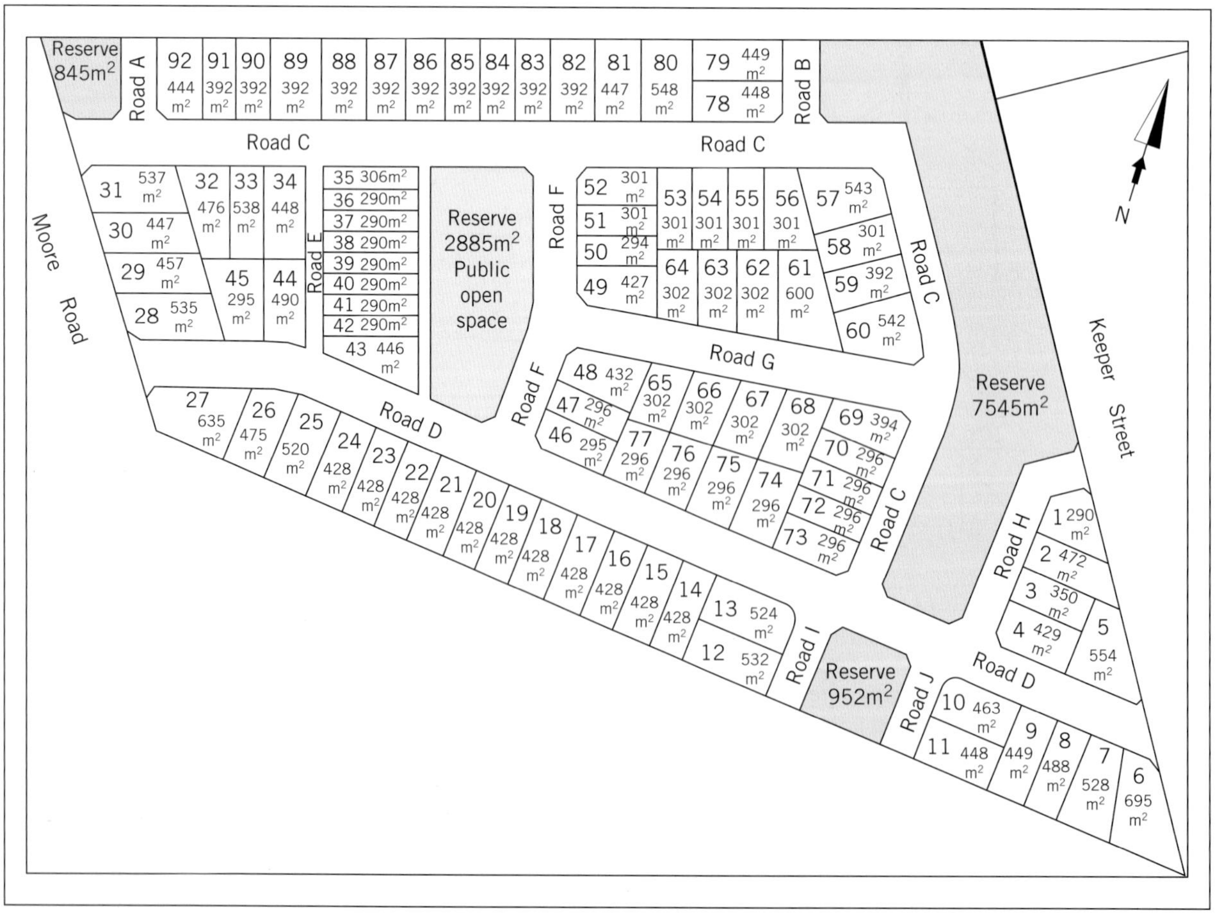

FIGURE 5.8 Example of a housing development plan

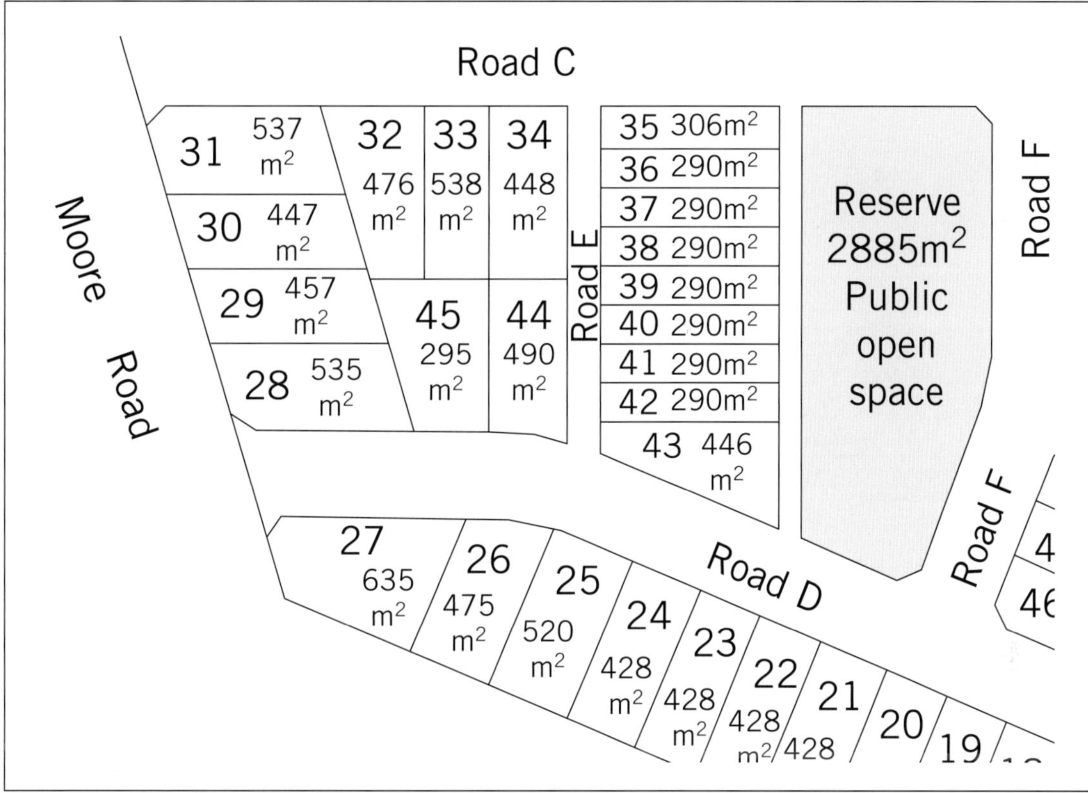

FIGURE 5.9 Development plan detail

Site and location plans

Site plans provide you with the location of the building with respect to the boundaries of the site as well as the location of other key features such as trees, and existing structures such as sheds or pools. They also offer an overview of the block in general, describing the fall of the land, direction and length of boundaries, creeks or streams and points of access and egress.

Although site plans are sometimes also referred to as ***location*** plans, location plans are generally used to describe the sites of proposed larger commercial, industrial, medical and/or educational facilities. Where site plans will aid the builder in selecting the right site, and position of the building within that site, location plans will tend to place more emphasis on the location of a facility in relation to other existing or proposed infrastructure that surrounds it. Figure 5.10 is an example of a domestic site plan with distances from the boundary, known as set-backs, clearly shown. Figure 5.11, on the other hand, shows a location plan for the Australian Technology Park (ATP) in central Sydney.

Working drawings

Also known as construction drawings, architectural drawings or blueprints, these are the 'plans' as most builders would understand them to be. Working drawings are a set of detailed technical drawings providing all parties, including all trades and subcontractors, with the requisite information to quote, approve and/or construct the building or structure.

Working drawings use a standard layout as defined in AS 1100.101 and are generally drawn to a scale of 1:100. Specific aspects of such plans may be at larger or smaller scales depending upon the detail required; i.e. the plan showing the location of the building on a block may be at 1:500, while details of how the subfloor is constructed might be as large as 1:5.

Working drawings should be, when combined with the specifications, the builders 'construction manual' – containing all of the relevant information required during the construction phase.

Typically, a full set of working drawings:

- is drawn orthographically and to scale
- documents the final design in complete detail, including all dimensions, structural and engineering details
- offers multiple views and details
- provides evidence of compliance to standards and codes
- shows the location of the construction on the site.

Working drawings may be used for quoting and tendering, and for off-site manufacturing. They also form part of the final submission to council for a construction certificate or building permit (the name varying from state to state within Australia). As such, the approved set of drawings, stamped by council, act as a legal document; providing evidence in the case of disputes or structural failure either during construction or in the future.

Source: Glenn Costin

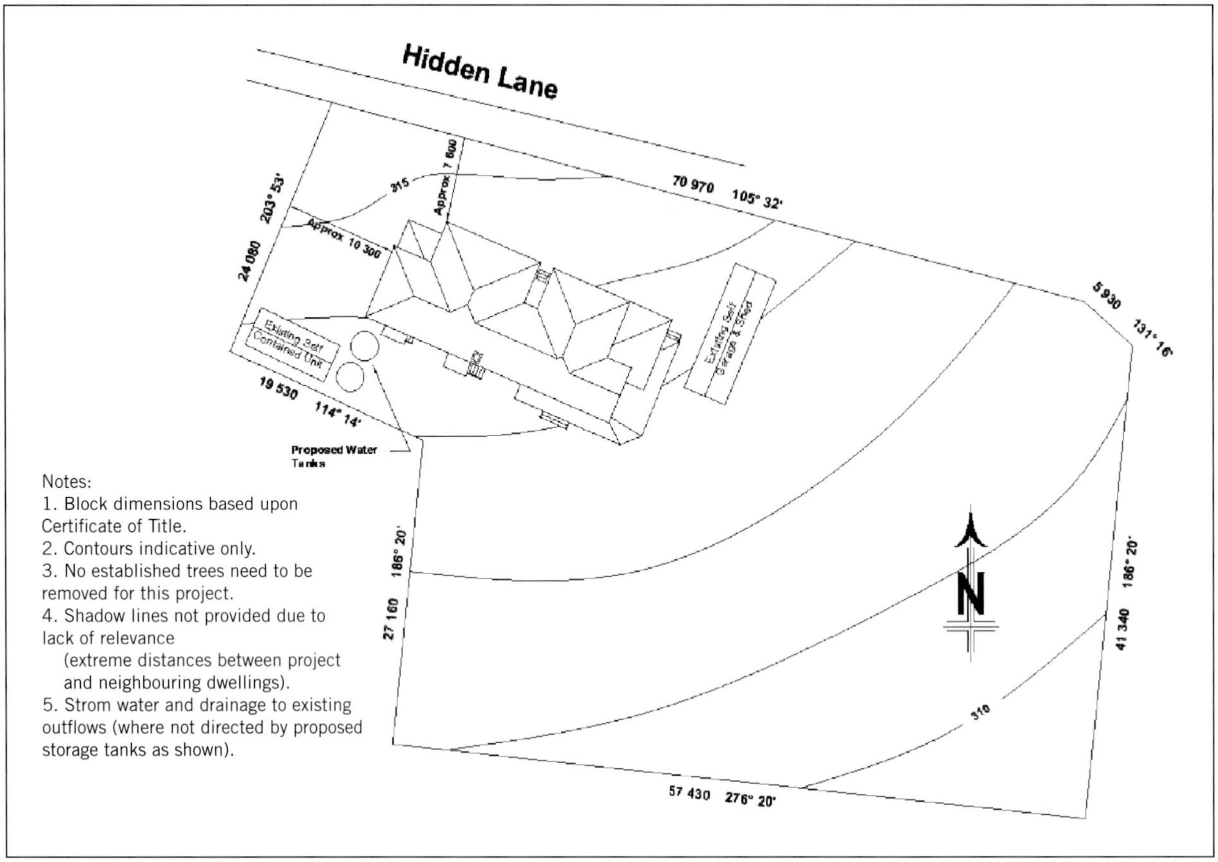

Hidden Lane

70 970 105° 32'

5 930 131° 16'

Approx 7 600

315

203° 53'

24 080

Approx 10 300

Existing Self Contained Unit

Existing Self Garage & Shed

19 530 114° 14'

Proposed Water Tanks

186° 20'

27 160

N

186° 20'

41 340

310

57 430 276° 20'

Notes:
1. Block dimensions based upon Certificate of Title.
2. Contours indicative only.
3. No established trees need to be removed for this project.
4. Shadow lines not provided due to lack of relevance
   (extreme distances between project and neighbouring dwellings).
5. Strom water and drainage to existing outflows (where not directed by proposed storage tanks as shown).

FIGURE 5.10 Domestic site plan with set-backs shown

Source: Mirvac

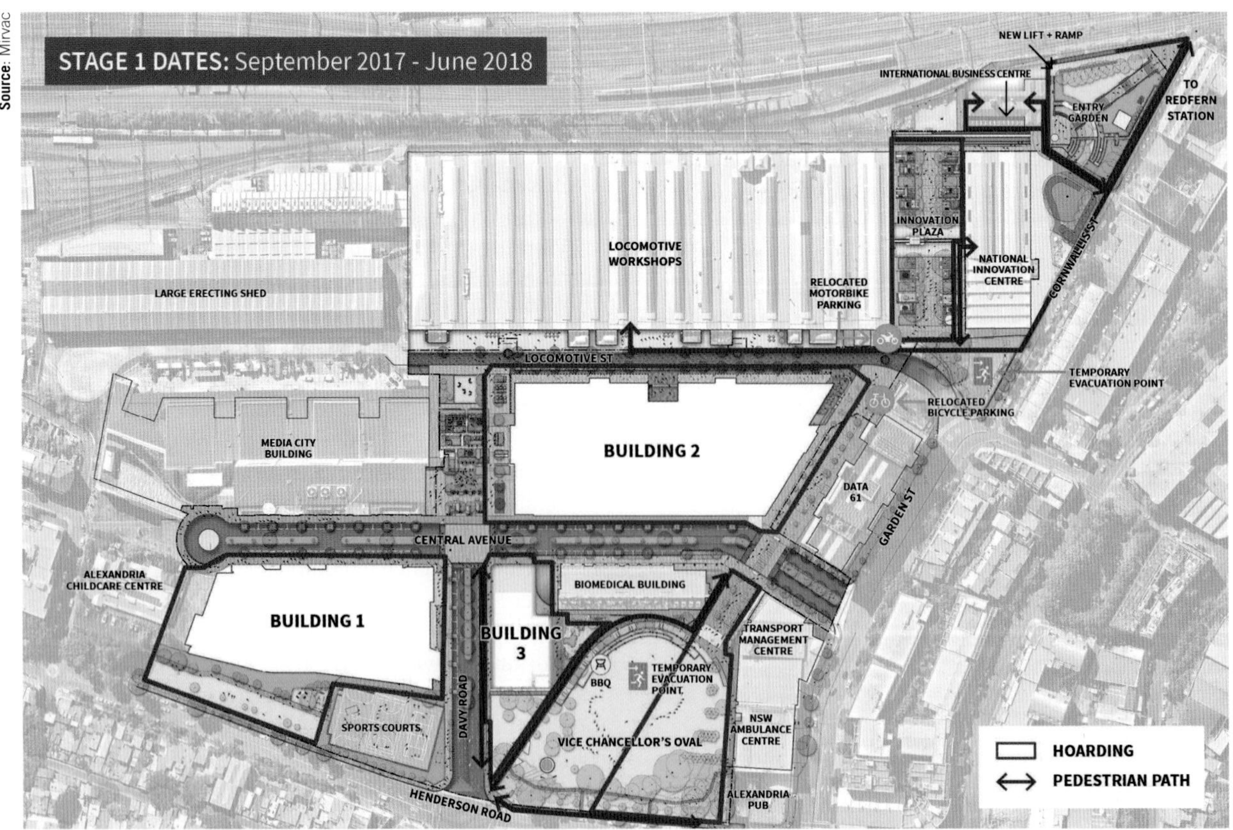

FIGURE 5.11 Location plan: Australian Technology Park, Sydney

CAD and BIM: computer aided drafting and modelling systems

Computer aided design, otherwise known as CAD, entered the construction industry many years ago. From the builder's and trades person's perspective its influence has been subtle; however, for architects, designers and draftspeople, and indeed clients, the shift has been substantial, allowing faster drawing times, and significantly greater stakeholder interactivity. CAD has also led to the introduction of a much more recent and less well-known acronym – BIM, building information modelling.

CAD

Computer aided drafting, or CAD, provides another variable in the types of drawings you might encounter as a builder. CAD drafting is the almost universal means of producing a set of working drawings for the contemporary construction industry. Indeed, the plans provided with this book have all been produced using CAD software. In itself, this does not make any huge difference in what will be provided to you as a printed set of working drawings. They are still orthographic, and they are still to the given scales – they are just drawn electronically rather than with a pencil or pen, traditional set squares and drawing boards (Figure 5.12).

However, the benefits of CAD drafting are multiple and significant; with the changes to industry still being worked through. Drafting on a computer allows for a single drawing to be copied, added to, subtracted from, or scaled up or down without repetition of work by the draftsperson themselves. This allows for very rapid changes to the design upon input from clients, engineers and other stakeholders upon their initial review of the first draft. It also allows for rapid dissemination of the drawing files to stakeholders by means of the internet or by any form of portable digital storage device (such as a USB flash drive). Likewise, any variations to the approved set of drawings can also be quickly made, disseminated and approvals sought – even to the point of being digitally signed by the parties concerned.

Though the printed drawings derived from CAD programs retain their 2D orthographic form, most architects are now drawing using 3D modelling systems. This approach allows the architect to create a three-dimensional model first, which may be rotated to provide views from any direction. The programs can even deliver a virtual tour of the proposed structure. Elements within the model may be drawn as different layers, which may be turned on or off as required to provide the information needed for any particular purpose. The software can then derive 2D drawings

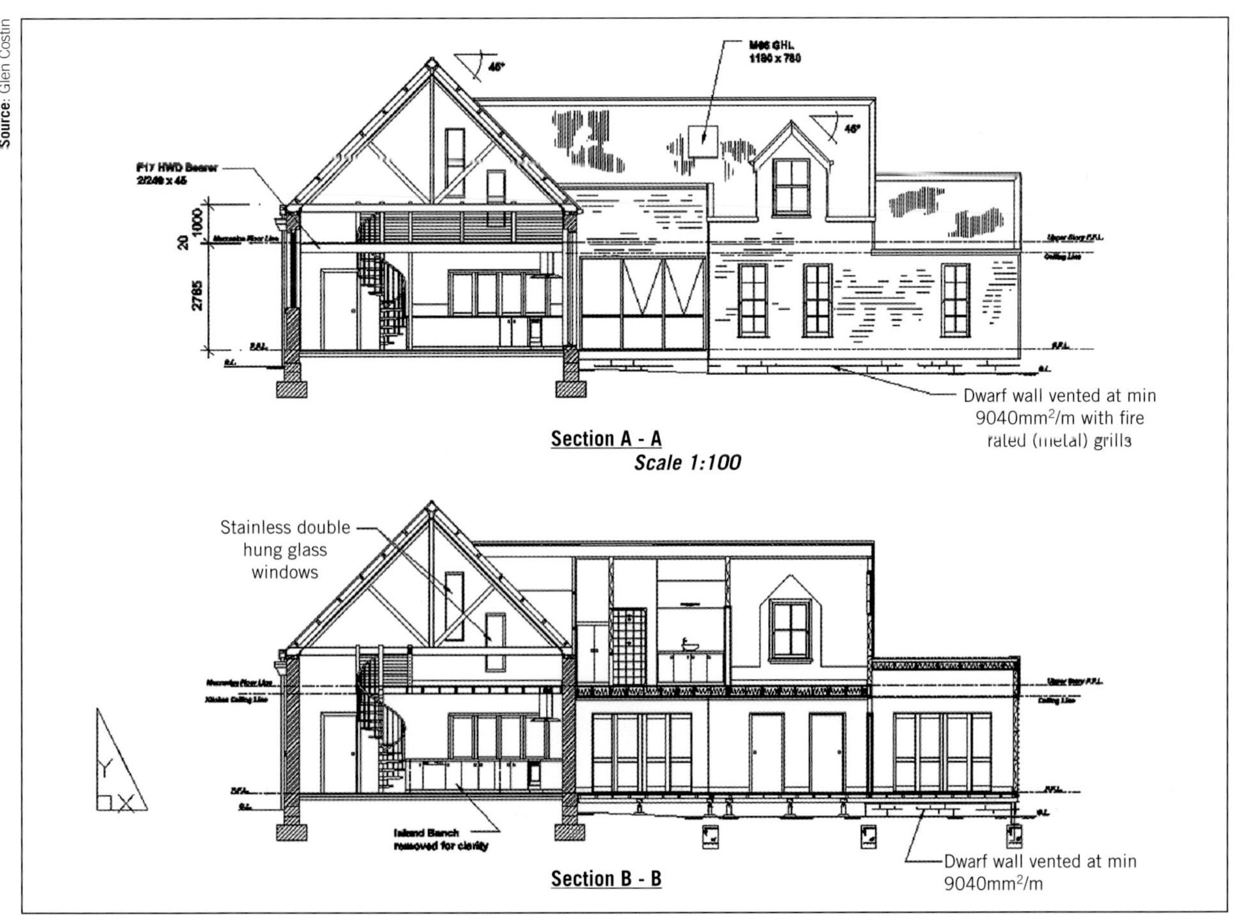

FIGURE 5.12 Example of a CAD drawing

from any particular perspective, or any chosen stage of work, to provide the orthographic renditions needed for the working drawings.

A further element, particularly with regards to commercial projects, is the capacity to quickly produce a final, as built, model (and hence set of plans), that takes into account any changes that might have been introduced by circumstance or last-minute material or fixture changes. The benefit here is in future maintenance and/or renovations. And so, we come to the concept of BIM.

BIM

All of the above benefits, excepting perhaps the last, are now very much commonplace in both domestic and commercial sectors of the Australian construction industry. Only in its early stages, however, is the natural extension of CAD modelling – 'building information modelling' systems or BIM (an acronym sometimes interpreted as 'building information management').

BIM is effectively still in its infancy, and as such is poorly defined – evident in the 'M' remaining disputed as either 'modelling' or 'management'. BIM has developed from 3D CAD systems as a means of managing the complexity of and improving performance in construction projects from conception to deconstruction. It is a way of managing and linking *all* of the digital material associated with any new structure within our built environment, from drawings to simple word documents, spreadsheets, photographs and the like. BIM applications (and there are many) link the 3D capacity of CAD to:

- spreadsheets
- specifications
- material and finishes schedules (indeed all schedules)
- tendering documents
- photographs
- supervisor diaries
- hand-written notes (scanned or digitally written on tablets or phones)
- communication devices (phones, iPads, laptops and the like)
- energy consumption monitoring systems
- maintenance schedules.

The list above is far from complete, but it offers some idea of the power of the system. Change management is hence one of the obvious benefits. Any change, on any element of the proposed or ongoing works, once fed into the system is disseminated to all parts of the system and the influence of that change may be evaluated for cost, time, safety (in construction or end use), energy efficiency, maintenance implications or aesthetics. And because all stakeholders can be quickly advised of the change, input into the evaluation from their various perspectives may be rapidly obtained. This information is constantly filed and backed up, and the working drawings or model becomes an 'as built' model, from which drawings depicting exactly the finished structure can be printed or electronically accessed. Future similar developments can now draw upon the data, as can any future works, maintenance, or emergency activities regarding the existing structure.

Aspects of working drawings

A set of construction plans will consist of a number of aspects, such as the floor plan, elevations and sectional details. Each of these aspects provides a view of the structure from which important information may be taken or interpreted. A full set of working drawings will generally contain the following aspects or drawings:

- site plan and/or location plan
- floor plan
- elevations
- sections
- footing plan
- bracing plan
- details (enlargements)
- overlay (new works to existing).

The recommended scales for each of the above may be found on page 19 of AS 1100.301 – 2008.

Each of these will be discussed later in this chapter; for the moment, a brief description will suffice. As the site and location plans have already been covered to some degree, we will begin with the floor plan.

The floor plan

A floor plan is a horizontal cross-section through a building as viewed from above. Think of it as if the roof was lifted off so that you can see all the walls and rooms. The purpose of a floor plan is much greater than that, however, for it details all horizontal construction dimensions such as:

- the size and shape of the building
- thickness of all walls
- width of eaves
- size and name of all rooms
- size of verandas, patios, porches.

The floor plan also provides information on:

- position of fixtures such as baths, showers, toilets and basins
- cupboards, halls and other spaces
- the roof outline
- the location and direction of sectional views.

As may be seen from Figure 5.13, there is a lot of information on floor plans and they should be interpreted with care. Usually drawn to a scale of 1:100, they are generally the first document or page of the working drawings you will want to review as you begin to understand the building. It is also generally among the first pages in a working drawing set, preceded usually by only the site or location plan.

Note: Dimensions to all construction drawings in Australia are in millimetres only. Metres or imperial measurements are not used unless to specify a particular product that is traditionally described by alternative dimensioning rules (such as some piping, for example). Also note that internal room dimensions are to the framework, *not* to the finished wall surface such as the plasterboard.

Elevations

Elevations show the side views of a building or structure and therefore show heights or vertical distances: hence their name 'elevations'.

In construction drawings, exterior elevations are drawn, showing the finished appearance (see Figure 5.14).

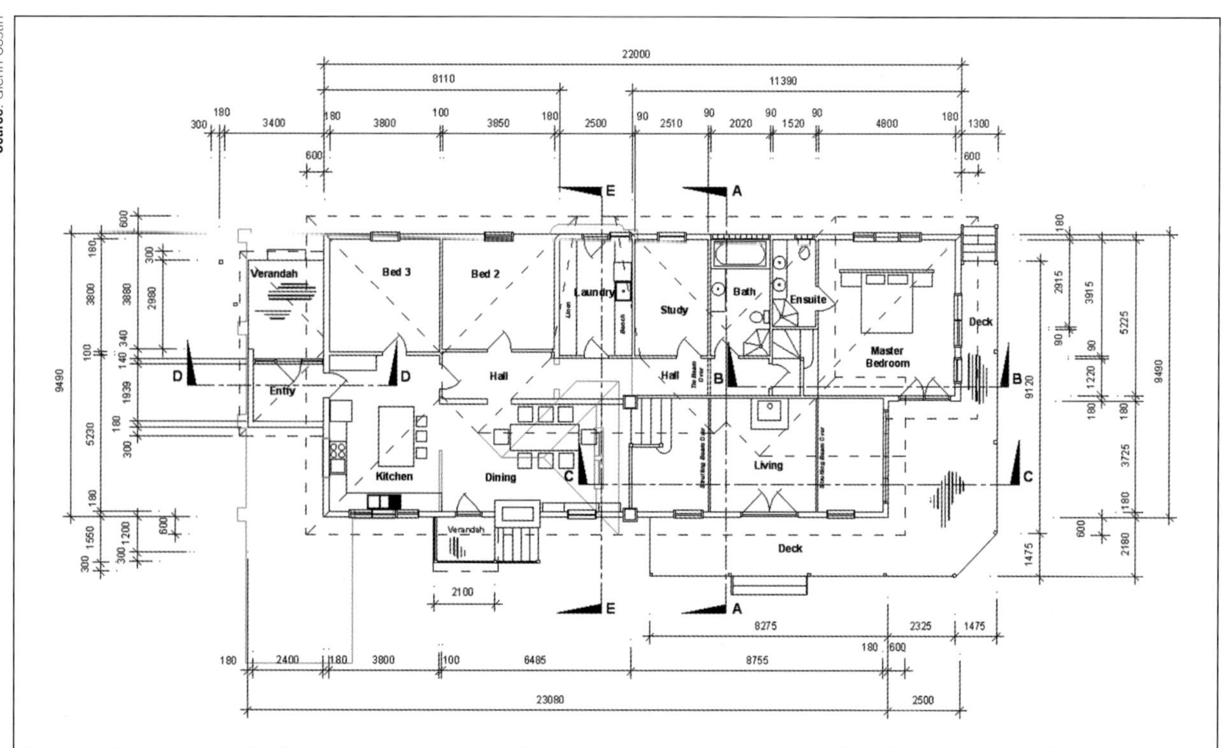

FIGURE 5.13 A floor plan

Source: Glenn Costin

FIGURE 5.14 East elevation of a building

Each elevation is named according to the direction it faces; i.e. NORTH, SOUTH, EAST, or WEST elevation. The elevations may show a lot of detail, but seldom much by way of detailed measurements. Only height (elevation) measurements are offered along with ground lines, finished floor and ceiling lines, and, where relevant, the pitch (angle) of roof surfaces.

The purpose of elevations, aside from showing particular heights and angles, is to offer the reader (you 'read' plans, not just look at them ...) a visual image of the final structure. From this you get the approximate location of windows and external doors, the desired approach to decking and trims, as well as some aspect of its 'fit' to the topography when viewed from a particular direction.

Sections and construction details

In reviewing a set of working drawings, you will find a number of views seeming to show the inside of a structure, or enlarged parts of a structure. These drawings may be labelled something along the likes of Section B–B or Detail 3–12 or the like. These will be explored in depth later, but a brief overview is relevant here.

Sections

Sections, also known as cross-sections, sectional elevations or sectional views, are drawings used to show detail that otherwise could not be seen without this alternative perspective. They are produced by making a virtual 'cut', as if by a large blade, through a structure (or a portion of a structure) and then looking at the cut face exposed. To see the detail with clarity, sectional views are generally drawn to larger scales than the floor plan; 1:50 being a common scale for domestic work. Sections generally cut vertically through a building, but other perspectives (horizontal or even angled sections) may be offered depending upon the nature of the design.

The position of a sectional view is shown on the floor plan by broken lines that include arrow heads indicating the direction in which the section is viewed. An alpha or numeric code by way of identification is also provided so that you can find the correct drawing with the working drawings for that particular view; for example, A–A (see Figure 5.13 floor plan, and example section A–A in Figure 5.15 below).

The sections may show such information as:

- height of floor above ground level (G.L.)
- ceiling heights, false ceilings, beams, trusses and eave types
- joinery fitments such as stairs, handrails, kitchens and wardrobes
- insulation, ducting, bathroom layouts and the like
- depth of footings, slab or subfloor details, retaining walls and areas of cut or fill
- sill or head heights to windows, doors, mirrors and many other features.

Sectional views may also depict the location areas for which more in-depth detail has been provided elsewhere. This will be done by circling, boxing or otherwise highlighting the chosen area and labelling it with an alphanumeric code such as 2–12, or 3–11, as exampled in Figure 5.15. These numbers or letters represent the detail number, in this case detail 2 and 3; and the page or drawing number on which you will find the detail, in this case drawings 12 and 11.

Construction details

Detail drawings are enlargements of key areas of a structure. In drawing them, the designer gets a clearer picture in their own mind of how a particular element of a structure might need to be built, or how it interacts with other elements and/or the structure as a whole. In reading them, builders and/or clients get a clearer understanding of how work that may be subtly or significantly different to normal practice needs to be approached, how it will look, or what materials will be used.

Source: Glenn Costin

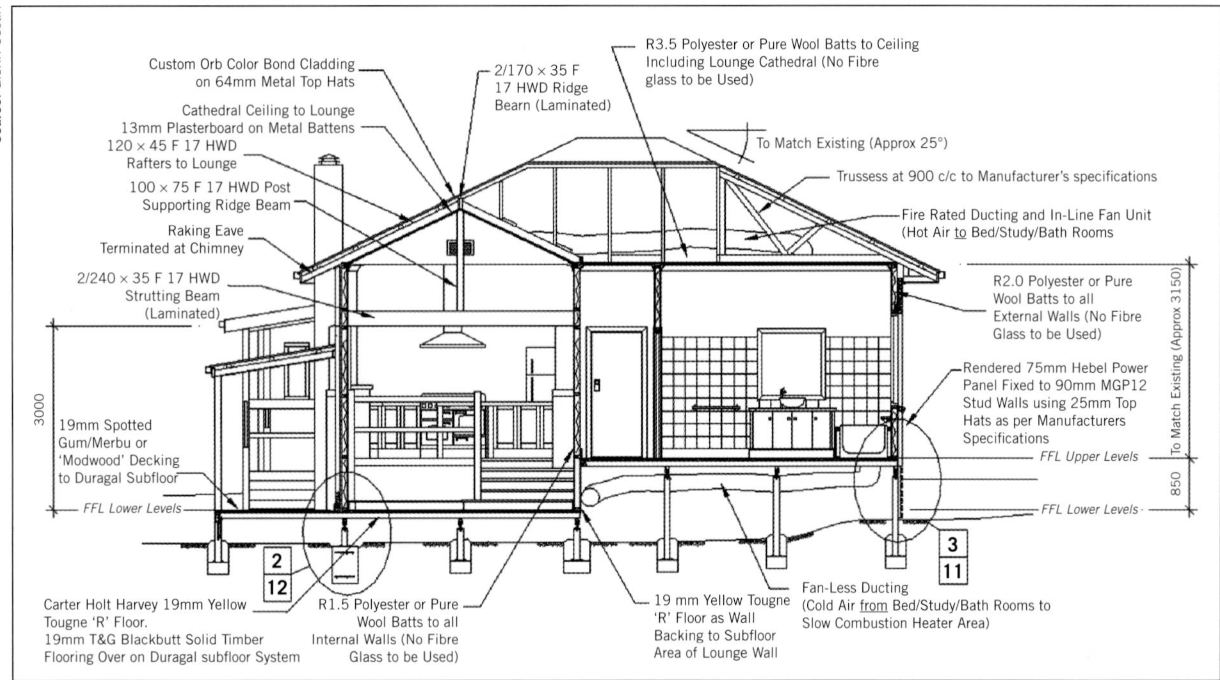

FIGURE 5.15 Section A–A

Details are critical elements of a drawing set. They not only hold important information, they also allow you, the builder, to look carefully at the designer's approach and perhaps offer alternatives that will be cheaper, faster, or stronger. Likewise, you might identify elements that your experience suggests will become issues in the future, such as termite access points.

Common areas where detail drawings will be required are:

- tie down requirements
- specific fixing details for new or uncommon materials
- retaining walls
- key loadbearing points
- tie in to existing footings or walls
- footings
- flashing details.

The above is not an exclusive listing, there a literally hundreds of areas of a structure that may need detail drawings before you can build what is desired, or for other trades to understand their particular role, or before you can even quote or develop a tender. It is also not uncommon for builders and subcontracted trades to call upon architects and or draftspeople to produce further details due to insufficient information in the existing drawing set.

Detail 2–12 is shown below (Figure 5.16) and from this you can see the level of detail that must be included in drawing plan sets. As was discussed in the section on CAD and BIM, this detail can be put into a drawing at the start, and then just 'zoomed' in on using the computer to generate the detailed view required. Details may be drawn to scales as large as 1:2; however, most commonly they will be at 1:20, 1: 10, and 1:5.

Other views and drawings

It is not that long ago when all that was required by way of working drawings for a domestic dwelling was the plan view, the elevations and maybe a section or a footing layout. Australia has since moved to far more stringent building codes and regulations, requiring significantly more detailed descriptions of the proposed structure. Working drawings now require detailed footing plans, bracing layouts, and shadow diagrams.

Other drawings that may also be included within the set are:

- service layouts – electrical, gas, water
- landscape plans
- joinery details
- internal elevations (differing from sectional views in that they only show the walls)
- window and door schedules (repeated in the specifications even if included in the plan set)
- materials schedules (also repeated in the specifications)
- reflected ceiling plans
- title page and general notes.

It's worth having a brief look at the more important of these before moving on.

Footings

One of the two more critical drawings required in today's working drawing set – the other being bracing – this document holds key engineering details. These details specifically cover the dimensions and reinforcing

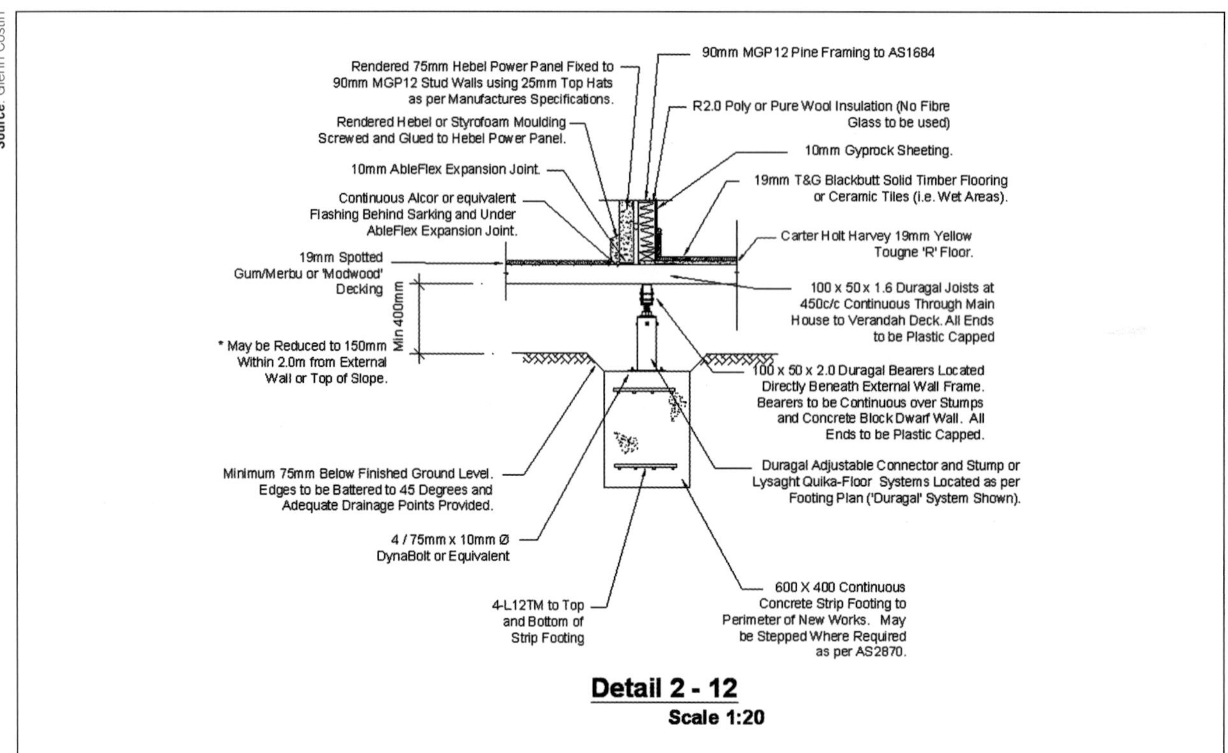

FIGURE 5.16 Detail 2–12

required to ensure the footings are adequate for the loads of the structure and the capacities and type of foundation material on which they rest. Without this drawing, even a basic house, garage or domestic shed will not be approved – this includes renovations and additions. On occasions, the subfloor layout of bearers and joists may be included in this diagram so that spans may be assessed by the relevant building authority. See Figure 5.17.

Bracing

Diagrams showing bracings are the other of the more critical drawings required for building approvals. An example is shown in Figure 5.18. This drawing shows the location and type of each brace required to ensure

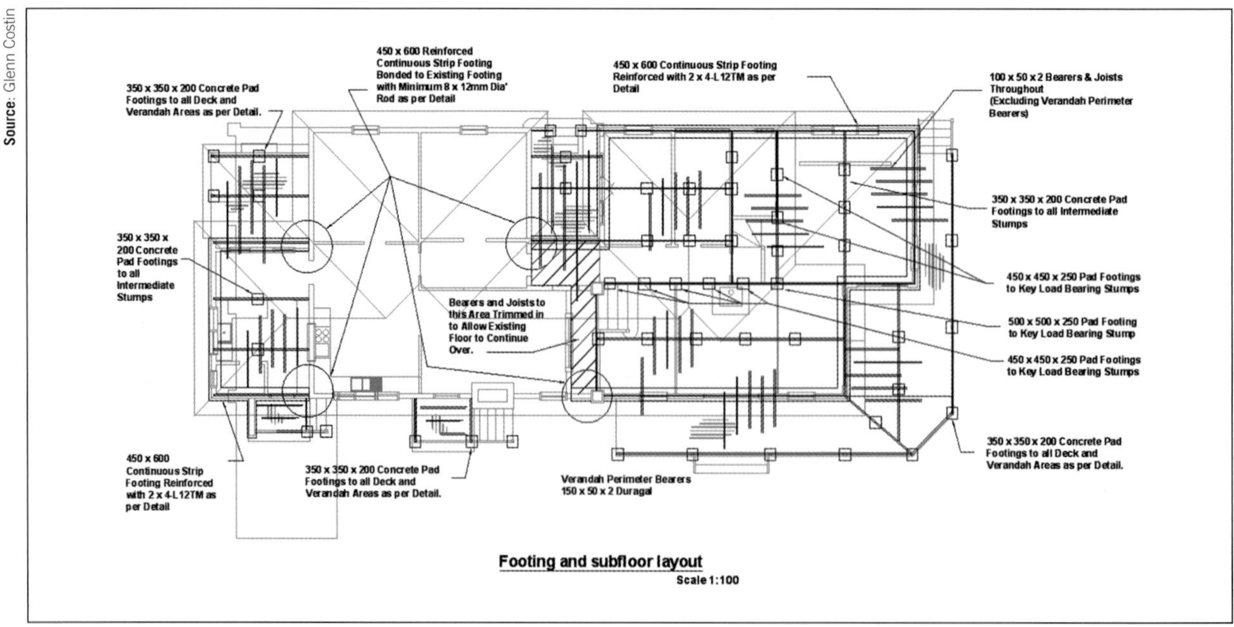

FIGURE 5.17 Footing and subfloor layout

Source: Glenn Costin

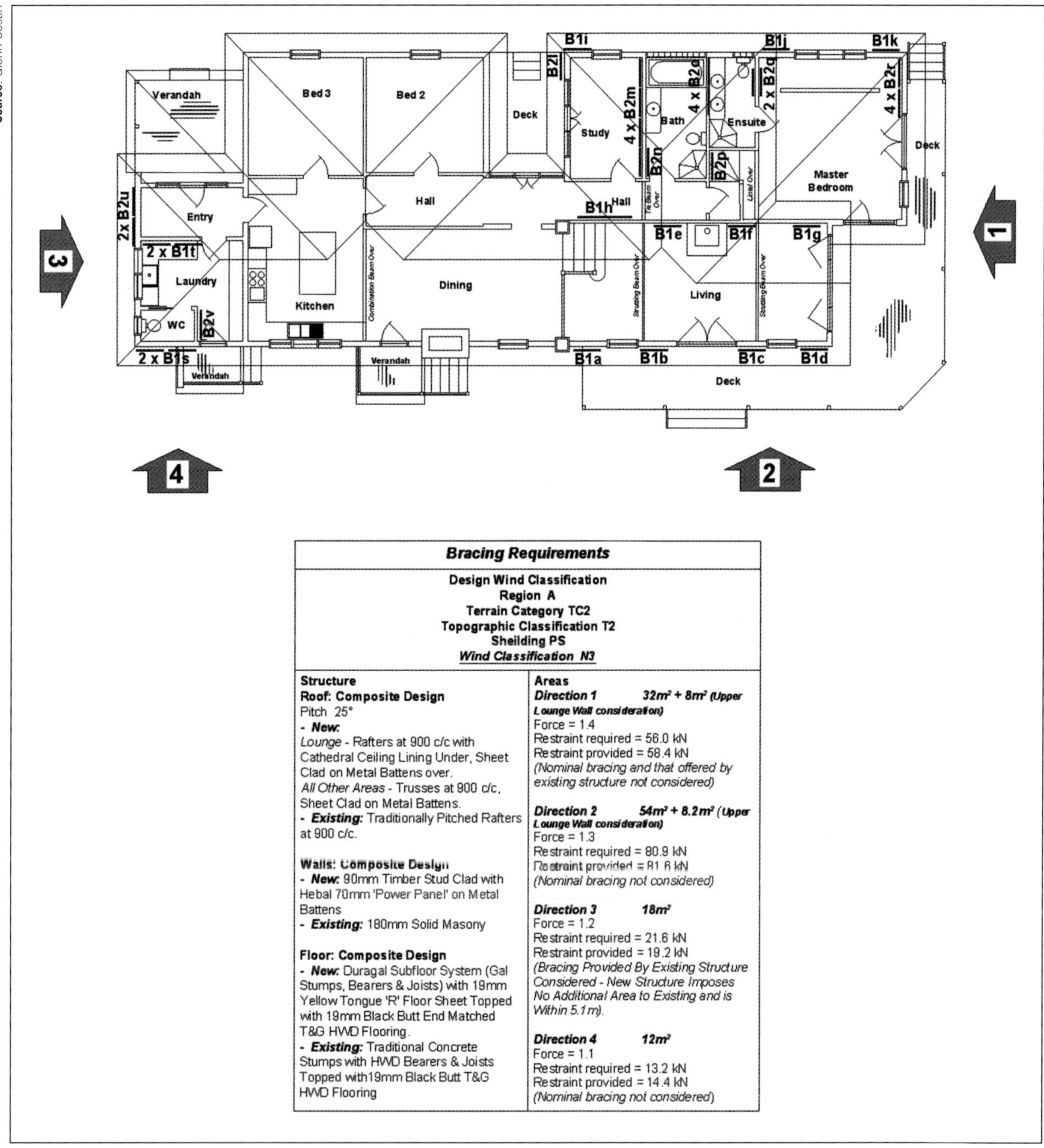

Bracing Requirements

Design Wind Classification
Region A
Terrain Category TC2
Topographic Classification T2
Sheilding PS
Wind Classification N3

Structure	Areas
Roof: Composite Design Pitch 25° - ***New:*** *Lounge* - Rafters at 900 c/c with Cathedral Ceiling Lining Under, Sheet Clad on Metal Battens over. *All Other Areas* - Trusses at 900 c/c, Sheet Clad on Metal Battens. - ***Existing:*** Traditionally Pitched Rafters at 900 c/c. **Walls: Composite Design** - ***New:*** 90mm Timber Stud Clad with Hebal 70mm 'Power Panel' on Metal Battens - ***Existing:*** 180mm Solid Masony **Floor: Composite Design** - ***New:*** Duragal Subfloor System (Gal Stumps, Bearers & Joists) with 19mm Yellow Tongue 'R' Floor Sheet Topped with 19mm Black Butt End Matched T&G HWD Flooring. - ***Existing:*** Traditional Concrete Stumps with HWD Bearers & Joists Topped with19mm Black Butt T&G HWD Flooring	***Direction 1*** ***32m² + 8m² (Upper Lounge Wall consideration)*** Force = 1.4 Restraint required = 56.0 kN Restraint provided = 58.4 kN *(Nominal bracing and that offered by existing structure not considered)* ***Direction 2*** ***54m² + 8.2m² (Upper Lounge Wall consideration)*** Force = 1.3 Restraint required = 80.9 kN Restraint provided = 81.6 kN *(Nominal bracing not considered)* ***Direction 3*** ***18m²*** Force = 1.2 Restraint required = 21.6 kN Restraint provided = 19.2 kN *(Bracing Provided By Existing Structure Considered - New Structure Imposes No Additional Area to Existing and is Within 5.1m).* ***Direction 4*** ***12m²*** Force = 1.1 Restraint required = 13.2 kN Restraint provided = 14.4 kN *(Nominal bracing not considered)*

FIGURE 5.18 Bracing diagram and associated specification or table

the structure can withstand the racking forces derived from wind loads. The drawing must be accompanied by the wind load calculations which are usually provided in a table or schedule. This table will also define the wind load zoning for that particular structure developed from AS 4055 Wind loads for housing, or AS 1170.2 Structural design actions for other classes of buildings. The calculations will be based upon the two worst-case directions for the building. In the case of extensions and the like, this may involve three or four directions if the extensions are on different ends of the house.

Many builders are still not following the bracing drawings provided and do so at a significant risk to themselves and their clients. Bracing forms part of the engineering behind a structural design and so to make unapproved changes to location, type or materials within a bracing unit can leave the structure at risk of failure.

Shadow diagrams

Shadow diagrams (as exampled in Figure 5.19) are a reasonably new requirement by councils and not all councils demand them unless they feel there is risk to the amenity of neighbouring properties. Shadow

Source: Amores

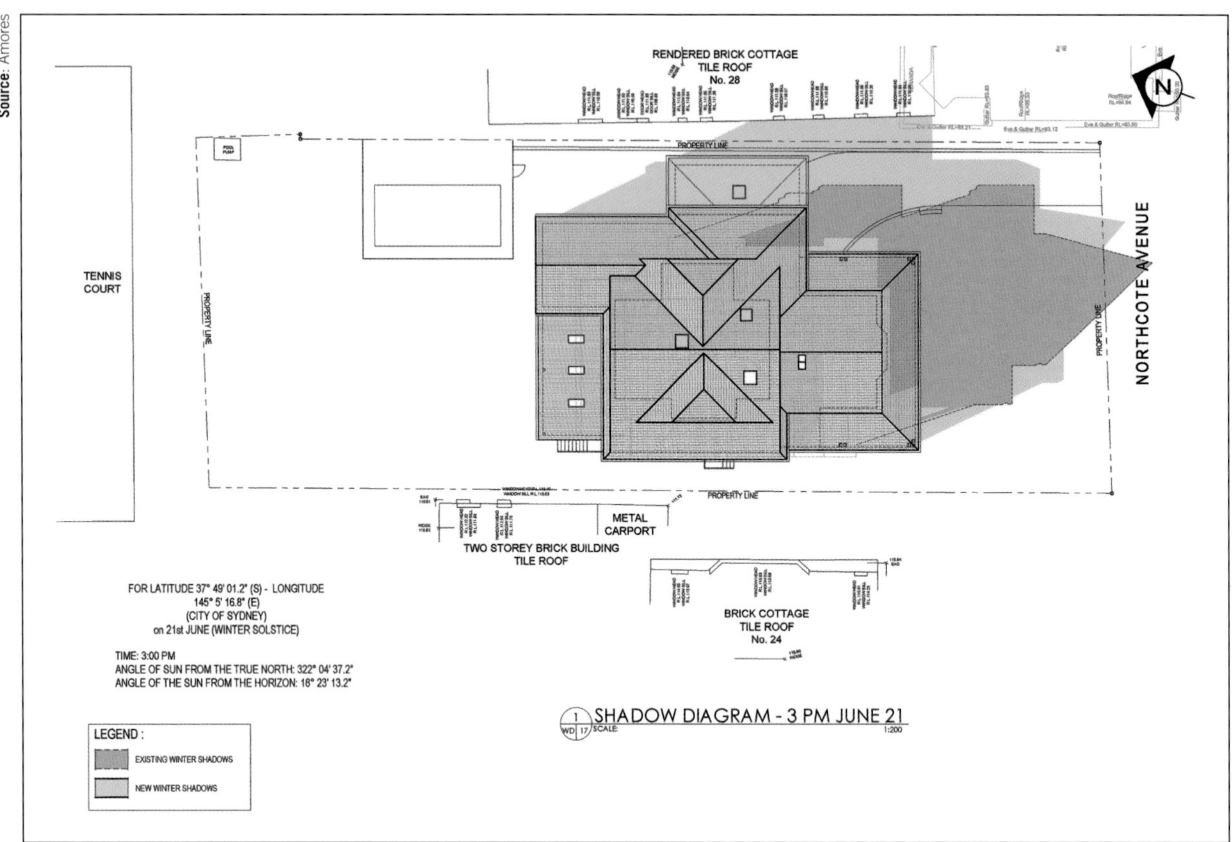

FIGURE 5.19 Shadow diagram

diagrams are just that, diagrams that show the shadow cast by a structure blocking the sun. The shadow will naturally be different for each day and time of the year; i.e. it is dependent upon the relative position of the sun in the sky. This is factored in by the regulations governing the development of the diagrams but significantly varies from state to state. Victoria, for example, requires that the shadow be depicted for the spring equinox (22 September) at 9am, 12pm and 3pm. NSW, on the other hand, requires diagrams for the same hours, but for 21 June (though some Sydney councils also require the equinoxes – 19 March and 22 September). Perth requires all three dates and the same times. Adelaide only requires 21 June between 10am and 3pm. Tasmania doesn't yet ask for shadow diagrams unless there is a dispute.

Services

Services such as electrical, water and gas supply and fixtures (see Figure 5.20) are sometimes shown on the one plan, while at other times each service will be depicted on its own: the complexity of the layout being the driving factor. These diagrams are important from quoting or tendering perspectives as invariably they are subcontracted both for pricing and installation. Their purpose is less regulatory (though regulations do apply) but more informative, both for the subcontractors and the clients. On review of the initial draft set it is common for clients to request variations to the initial proposal based upon their own desired end use of a room or space.

Landscaping

Landscape plans or layouts are usually prepared by, or with direct input from, landscape designers. Their purpose on the one hand is fairly obvious, they show the client and the contractors how the finished grounds are to look. On the other hand, their import is more subtle, and as the builder you should make yourself familiar with their content. Things that will be of importance to your build will be the location of existing trees and shrubs that must be retained, the final contours of the land (and hence where you would best relocate spoil and topsoil) and how access and egress is to be determined.

Landscape plans can be very simple, or highly detailed and specific, and include other works such as swimming pools, fish ponds, paths, fountains and flowing water courses, as well as isolated decks, pergolas, gazebos and shade sails (see Figure 5.21). They may also include information on drainage and the relationship to septic dispersion trenches or waste water recycling systems.

Source: Angela Elliss Design

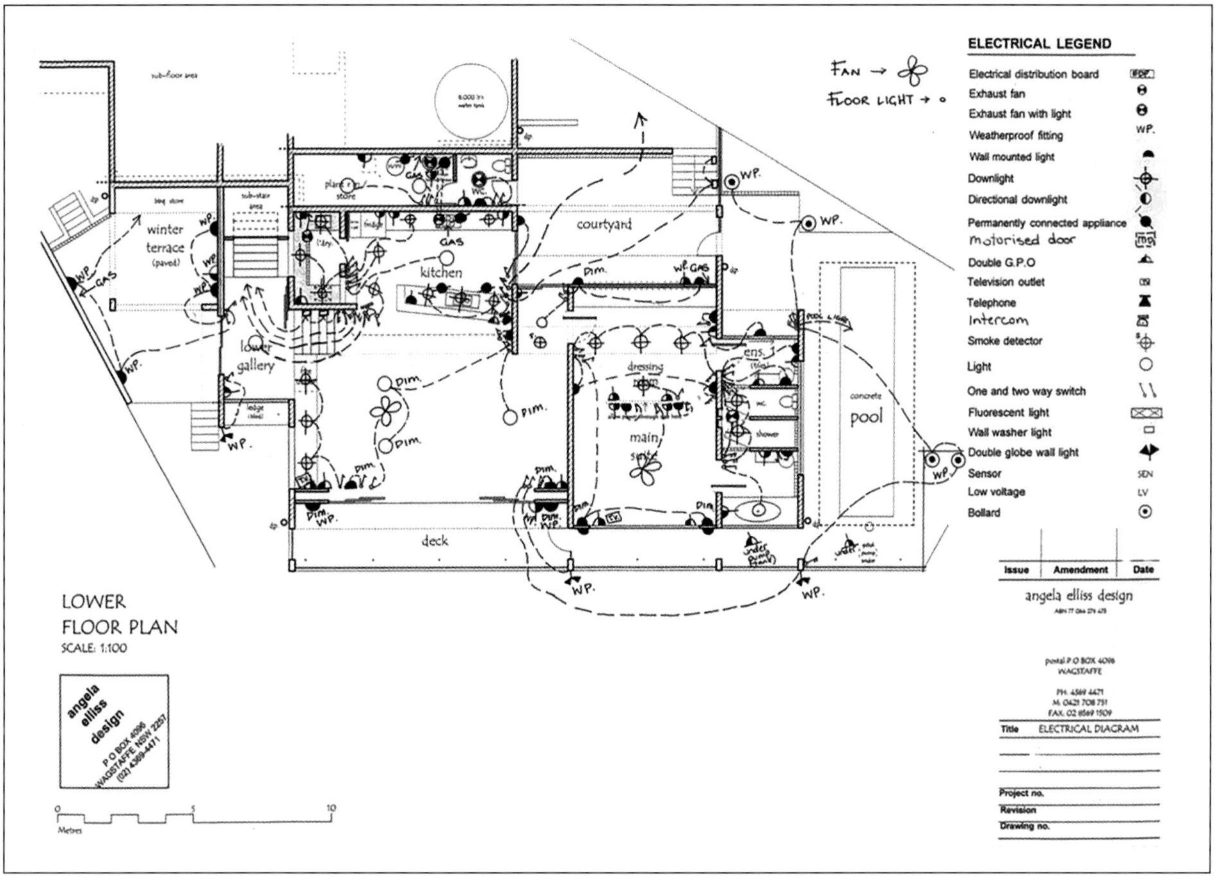

ELECTRICAL LEGEND

Electrical distribution board
Exhaust fan
Exhaust fan with light
Weatherproof fitting WP.
Wall mounted light
Downlight
Directional downlight
Permanently connected appliance
motorised door
Double G.P.O
Television outlet
Telephone
Intercom
Smoke detector
Light
One and two way switch
Fluorescent light
Wall washer light
Double globe wall light
Sensor SEN
Low voltage LV
Bollard

FAN →
FLOOR LIGHT →

winter terrace (paved)
lower gallery
kitchen
courtyard
dressing room
main suite
ens.
shower
pool
concrete
deck

LOWER FLOOR PLAN
SCALE: 1:100

angela elliss design
P O BOX 4096 WAGSTAFFE NSW 2257 (02) 4369-4471

0 5 10
Metres

Issue	Amendment	Date

angela elliss design

Title ELECTRICAL DIAGRAM
Project no.
Revision
Drawing no.

FIGURE 5.20 Electrical wiring diagram

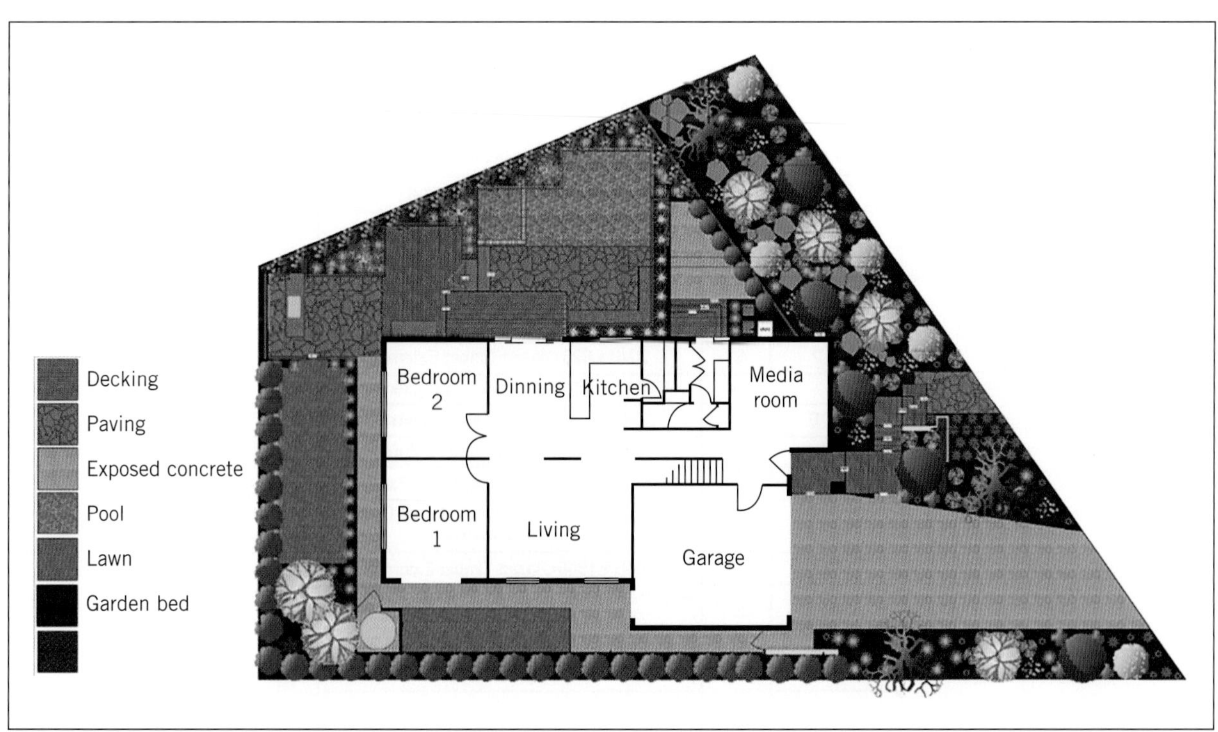

FIGURE 5.21 Example of a landscape plan

Window and door schedules

These are generally to be found in the specifications; however, they are frequently to be found within the working drawings set as well. This is because their content is often reflected in the floor plan, or a specially drawn floor plan, making for easy identification of location. These schedules will be in a table form and give the size, type and location of

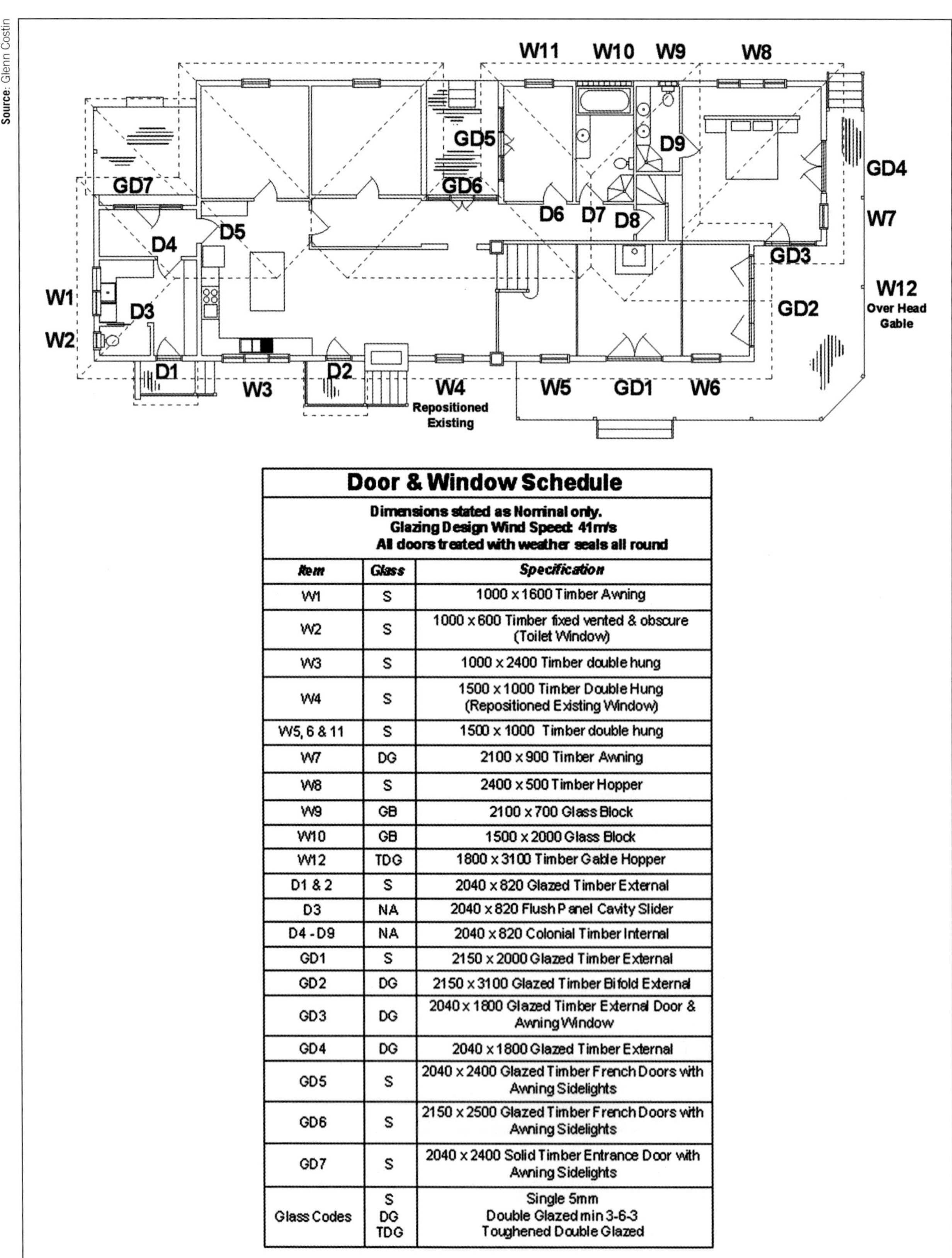

Door & Window Schedule		
Dimensions stated as Nominal only. Glazing Design Wind Speed: 41m/s All doors treated with weather seals all round		
Item	***Glass***	***Specification***
W1	S	1000 x 1600 Timber Awning
W2	S	1000 x 600 Timber fixed vented & obscure (Toilet Window)
W3	S	1000 x 2400 Timber double hung
W4	S	1500 x 1000 Timber Double Hung (Repositioned Existing Window)
W5, 6 & 11	S	1500 x 1000 Timber double hung
W7	DG	2100 x 900 Timber Awning
W8	S	2400 x 500 Timber Hopper
W9	GB	2100 x 700 Glass Block
W10	GB	1500 x 2000 Glass Block
W12	TDG	1800 x 3100 Timber Gable Hopper
D1 & 2	S	2040 x 820 Glazed Timber External
D3	NA	2040 x 820 Flush Panel Cavity Slider
D4 - D9	NA	2040 x 820 Colonial Timber Internal
GD1	S	2150 x 2000 Glazed Timber External
GD2	DG	2150 x 3100 Glazed Timber Bifold External
GD3	DG	2040 x 1800 Glazed Timber External Door & Awning Window
GD4	DG	2040 x 1800 Glazed Timber External
GD5	S	2040 x 2400 Glazed Timber French Doors with Awning Sidelights
GD6	S	2150 x 2500 Glazed Timber French Doors with Awning Sidelights
GD7	S	2040 x 2400 Solid Timber Entrance Door with Awning Sidelights
Glass Codes	S DG TDG	Single 5mm Double Glazed min 3-6-3 Toughened Double Glazed

FIGURE 5.22 Example of a door and window schedule

each door. The sizes given will be what is referred to as 'nominal'. A window, for example, may be listed as 2100 × 2400; this is nominal in not being the exact size that the manufacturer will supply. The actual manufactured size of the window will more likely be 2058 × 2410; the actual frame opening sizes and brick opening sizes will need to be gained from the supplier and will differ again.

Internal elevations

Rarely seen on standard residential plans, but common to commercial projects, internal elevations show you the features of walls as viewed as if you are standing in the room. These are useful when an internal wall has openings, wardrobes, shelving or other features that you cannot interpret clearly from any of the more traditional views. Internal elevations only show the walls, they do not show floor or roof details – that information you would look for in the relevant sectional drawings.

Internal elevations are usually identified on the floor plan, or another floor plan developed specifically for the purpose, using an arrow-like indicator with an alphanumerical code such as that shown in Figure 5.23.

Source: Glenn Costin

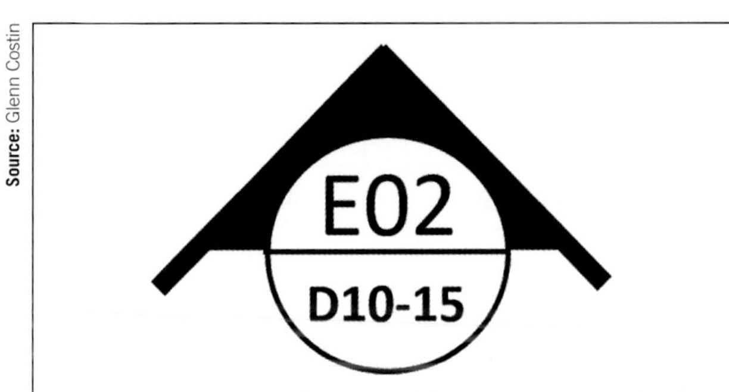

FIGURE 5.23 Internal elevation symbol

The E02 denotes the elevation number (E for elevation, 02 the number), while the D10-15 is the drawing number; indicating that this elevation will be found on drawing number 10 of a drawing set of 15 pages.

Reflected ceiling plans

The last of the drawing we will deal with in this section again are rarely seen on standard residential plans, and also are not overly common to commercial projects either. Reflected ceiling plans are not dissimilar to floor plans except they are viewed as if you are lying on the floor and looking upwards. Their purpose is to allow for measurements and locations that relate to the view you see if you are trying to map them on the ceiling while looking up. Electrical plans are occasionally laid out this way as are some more intricate plastered or panelled ceilings. Again, it allows for the measurements to relate to the way in which you would see them when attempting to apply them to the ceiling.

Reflected ceiling plans are often used because there is simply no better way by which to show the detail and information required to interpret the designer's concepts for plaster relief, raised ceiling features, shadow lines and the like.

LEARNING TASK 5.1 READING DRAWINGS

1 Looking at section A–A (Figure 5.16), on what page of the working drawing set would you find detail 3?
2 Detail 2–12 (Figure 5.17) describes the subfloor detail supporting an external wall where it meets with the decking. Answer the following:
   a What must go continuously behind and under the Ableflex expansion joint at the base of the external cladding?
   b What alternative to the Duragal stump system is offered? Which is shown?
   Alternative:
   Shown:

Common symbols, abbreviations and terminology

Already you will have noticed that construction drawings are littered with symbols, abbreviations and terminology specific to the construction industry. It is very important that you are able to decipher these attributes so that you can determine an architect's, engineer's, or client's real intent.

Symbols and abbreviations

In construction drawing, symbols and abbreviations are frequently used to represent items or otherwise further clarify an image for the reader. Abbreviations have become standardised in Australia, and so should comply with AS 1100.301. Symbols, on the other hand, can vary significantly.

Symbols

In construction drawing, a symbol is an image or artefact in the drawing (or computer-generated 3D space) that represents an object or feature, or seeks to elicit a response or action from the reader/viewer (such as 'go to page xx' or 'see file yy'). A number of symbols have already been included in the example drawings within the previous sections, such as Figure 5.23 above which provided a direction of view, and the location within the plan set where that view may be obtained. Other types of symbols are more representational; i.e. simple drawings of fixtures such as baths, taps, power points; or of materials when viewed in sections, such as concrete, steel, plasterboard, earth or the end grain of timber.

According to AS 1100.301, symbols should only be used in plans when drawing to scale is impractical.

Source: Glenn Costin

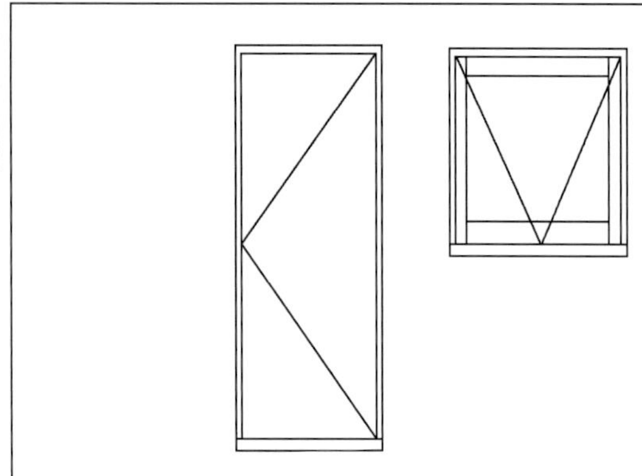

In contemporary Australian architecture the door shown would be hinged on the right-hand side. Likewise, the window described would be an awning sash hinged at the top (the symbolism being reflective of the flap on an envelope).

In the United Kingdom these symbols act as arrows *pointing* to the side that is hinged. That is, the door is hinged on the left, and the window is a hopper sash hinged at the bottom.

This was also the case in Australia until the late 1960s, and then for a brief period in 2008 when AS 1100.301 sought to realign Australia with some countries, before producing an amendment when that move was rejected.

As AS1100.301 – 2008 is still current, it is therefore important that your copy includes Amendment No.1.

FIGURE 5.24 Variations in expressing window and door opening directions

The standard provides examples of symbols depicting common architectural features, but recognises that there are numerous drawing templates and CAD programs with differing symbols. In addition, some site plan symbols are defined in AS 1100.401.

Note: It is important to understand that symbols are culturally agreed representations; i.e. while some symbols are internationally agreed, others may be nation specific. This means two countries may have different symbols for exactly the same item or concept, or, more confusingly, exactly the same symbol, but with a completely opposite meaning. For example, see Figure 5.24.

Symbols may appear or be used in all of the drawing types that have been discussed so far; however, they tend to fall into one of three categories:

- **fixtures** – sinks, basins, toilets and the like
- **drawing** – door and window representations, stairs, line types and material renderings
- **services** – power outlets, switches, lights, fans, gas meters and connections, etc.

Some examples of each of these categories are offered below, and further examples may be found in AS 1100.301–2008 (incorporating Amendment 1), which is available online. AS 1100 states that these symbols are recommendations only, and hence they do not have to look exactly as shown.

Symbols for fixtures

Fixtures are items that are fixed and so become part of a structure, such as doors, sinks, kitchens, wardrobes, cupboards, toilets and the like. Fittings are things that are easily removed and replaced, such as door and tap handles, TV antennas, washing machines, free-standing stoves and microwaves, etc. Fittings, such as door knobs and the like, may be shown within a plan set (such as in a detail, see Figure 5.15), however they are not standardised as generally they would only appear at a scale in which their form or purpose is self-evident (see Figure 5.25).

FIXTURE	SYMBOL	FIXTURE	SYMBOL
BASIN		SINK Single bowl Locate bowl as required	
BIDET		SINK Double bowl Locate bowls as required	
BATH		TUB Double	
DRINKING FOUNTAIN		TUB Single	
DRINKING TROUGH		URINAL STALL	
HOTPLATES		VANITY BASIN	
SHOWER		WATER CLOSET	

Source: Adapted from table 2.2 and table 2.3, AS1100.301—2008

FIGURE 5.25 Typical symbols for fixtures

Symbols for materials

Materials are typically shown by way of 'rendering'. Rendering is achieved using lines, patterns (hatching) and, where relevant, colours, such that the chosen materials are easily identified within a drawing that is otherwise heavy in detail. Rendering is standardised and is used in accordance with AS 1100.301. Some of the more common renderings are shown in Figure 5.26.

Line types

Line types are likewise standardised. The choice of line depends on what the line represents, such as: a solid part of the structure; something that has been removed for clarity (such as the roof in a plan view); a centre line; or the location of a cross-sectional view (see Figure 5.28). Although there are multiple line types, they typically fall into one of three categories:

	Brickwork Elev: Light Red Sect: Vermilion	Earth Sepia	
	Cement Render Terra Verte	Earth Fill Raw Umber	
	Concrete Hooker's Deep Green	Rock Vandyke Brown	
	Cur Stone Masonry Emerald Green	Hardcore	
	Partition Block Indian Red	Insulation Cerulean Blue	
	Concrete Block Green	Glass Cobalt Blue	
	Structural Steel Prussian Blue	Timber Yellow Ochre	
	Sanitary Fittings French Ultra-marine	Sawn Timber Chrome Yellow	
	Roof Tiles Cadmium Red	Dressed Timber Burnt Sienna	

Source: Adapted from Table 5.1, AS1100.301—2008

FIGURE 5.26 Common material rendering

continuous, dashed and centre lines, as shown in Figure 5.27.

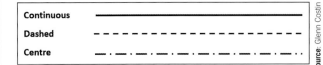

Continuous	————————————
Dashed	- - - - - - - - - - - - - -
Centre	— · — · — · — · — · — ·

Source: Glenn Costin

FIGURE 5.27 Line types

- **Continuous lines**: These are used to show the main elements of a drawing that may be seen directly in the specific view shown. AS 1100 applies different meanings to different thicknesses, though this is seldom strictly adhered to.
- **Dashed lines**: Generally used to draw hidden detail, or detail at levels different to the plan such as the roof and eave line. They may also be used in overlay plans or footing plans where the existing structure meets the proposed works. Various dashed lines may appear on the one drawing, yet represent different features. In such cases the dashes will vary in length.
- **Centre lines**: A form of dashed line, it differs in being a repeating pattern of short dashes (or dots) and long dashes. Centre lines are used for many things other than marking centres, such as ceiling and floor lines, and section lines (lines indicating the location of a sectional cut) as shown in Figure 5.28.

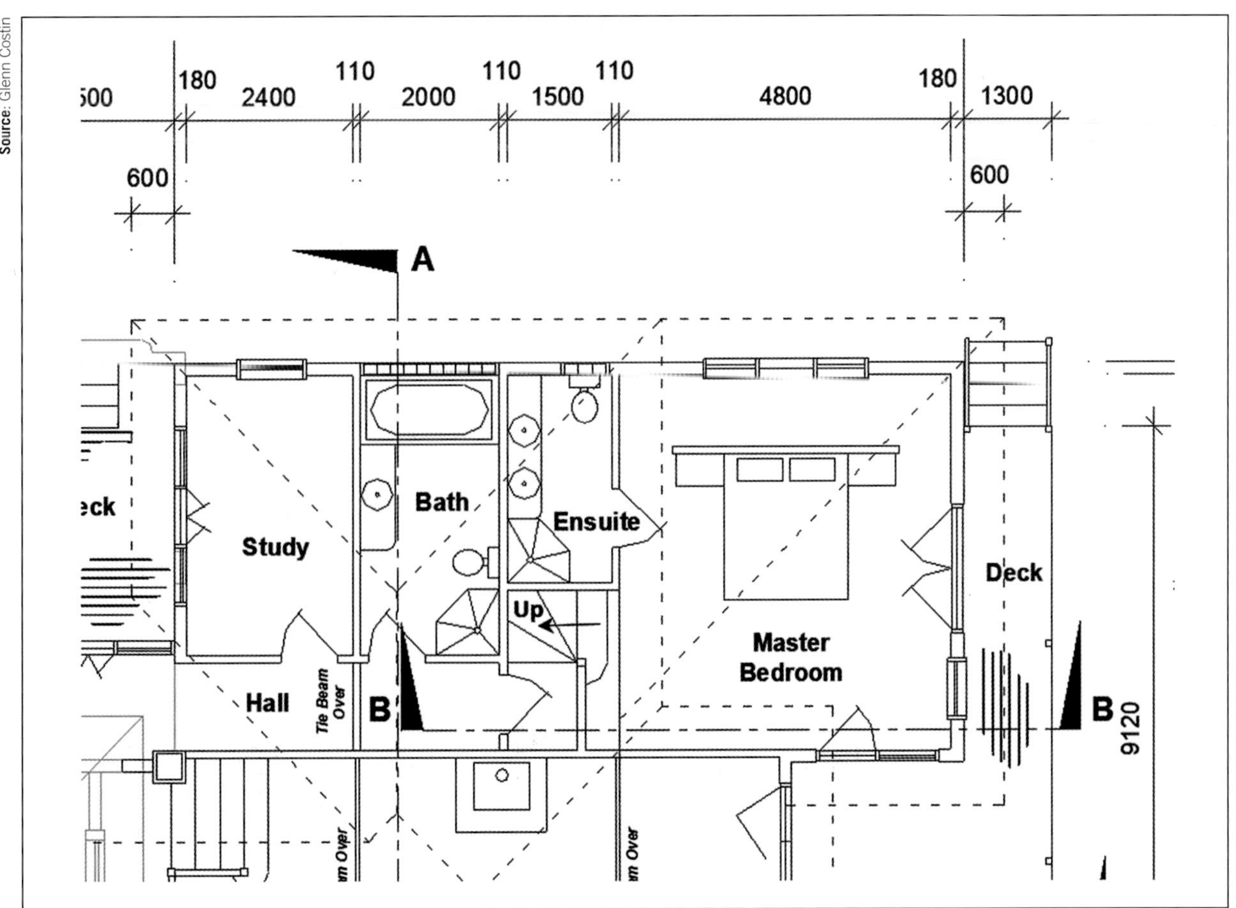

Source: Glenn Costin

FIGURE 5.28 Partial plan drawing showing location of sectional view B–B

Abbreviations

To the three-tiered list of symbols offered above (fixtures, lines, materials) may be added a fourth category – abbreviations. Technically not symbols as such, they often appear on drawings to help identify a symbol, material type, or some other aspect of the drawing that is in need of further description in order to be properly interpreted. While useful, AS 1100.301 recommends that abbreviations, such as WC (water closet or toilet) or SHR (shower), only be used if absolutely necessary, such as when similar objects are shown on the same drawing.

The following table (Figure 5.29) offers the more common abbreviations as derived from AS 1100.301-2008.

Construction terminology

Some construction information is presented on drawings and associated schedules as short notes. These notes convey details such as alternate dimensions, colour, materials, finishes, or will draw your attention to particular construction or design requirements. Notes may be annotated arrows, or cross-referenced lists within schedules. The abbreviations mentioned previously allow notes on drawings to be as concise and clear as possible. However, you will also come across a raft of terms and phrases that are specific to the construction industry – these too have become standardised to ensure concise explanatory notation.

Multiple glossaries of building terms are available on the internet; however, you must take care that they are Australian, not simply adaptations of American or English lists (this same issue can come into play when researching symbols). Standards Australia publishes a full glossary of building terms under the code HB 50 – 2004.

Common terms

This book has an extensive glossary which clarifies many terms you will see in sets of plans.

FIGURE 5.29 Common abbreviations

Abbreviation	Term	Abbreviation	Term	Abbreviation	Term
AL	Aluminium	CC	Concrete Ceiling	RL	Reduced Level
AS	Australian Standard	CF	Concrete Floor	RSC	Rolled Steel Channel
AUX	Auxiliary	CTR	Contour	RSJ	Rolled Steel Joist
B	Basin	CORR	Corrugated	RWH	Rain Water Head
BRR	Bearer	D	Door	S	Sink
BLK	Block	DAR	Dressed All Round	SD	Sewer Drain
BDYL	Boundary Line	DP	Down Pipe	SEW	Sewer
BT	Bath Tub	DW	Dish Washer	SF	Strip Footing
BRKT	Bracket	FC	Fibre Cement	SHR	Shower
BK	Brick	FFL	Finished Floor Line	SQ	Square
BV	Brick Veneer	FW	Floor Waste	SPR	Sprinkler
BWK	Brick Work	G	Gas	SWBD	Switchboard
BLDG	Building	GL	Ground Line	SWD	Stormwater Drain
BL	Building Line	HW	Hot Water Unit	T	Truss
BM	Bench Mark	HWD	Hard Wood	TC	Terra Cotta
CBL	Cable	KD	Kiln Dried	TM	Trench Mesh
CAB	Cabinet	MH	Man Hole	TR	Trench
CAN	Canopy	OUT	Outlet	TRH	Trough
CI	Cast Iron	OA	Over All	UB	Universal Beam
CW	Cavity Wall	OH	Overhead	U/C	Under Construction
CEM	Cement	P	Pier	U/G	Underground
CM	Cement Mortar	PBD	Plaster Board	UR	Urinal
CR	Cement Render	PBM	Permanent Bench Mark	V	Vent
CRS	Centres	PCC	Precast Concrete	VER	Version
CL	Centre Line	P/F	Plan of Subdivision	VERT	Vertical
CHY	Chimney	PF	Portal Frame	W	Window
CCT	Circuit	PM	Permanent Mark	WBD	Wallboard
CD	Clothes Dryer	RAD	Radius	WC	Water Closet (Toilet)
COL	Column	RF	Raft Footing	WRC	Western Red Cedar
C	Cooker	RHS	Rolled Hollow Section	WPM	Waterproof Membrane

Source: Adapted from Table 2.4 and Table 2.5, AS1100.301—2008

LEARNING TASK 5.2 INTERPRETING SYMBOLS, ABBREVIATIONS & TERMINOLOGY

Access AS 1101.301 and identify and/or draw the following symbols. All spaces in the table should be completed.

Symbol	Technical Name	Where usually found on a plan set
	Laundry sink	
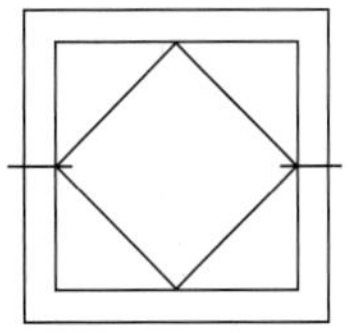		
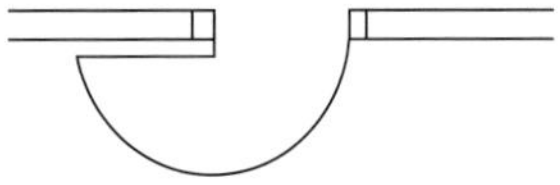		
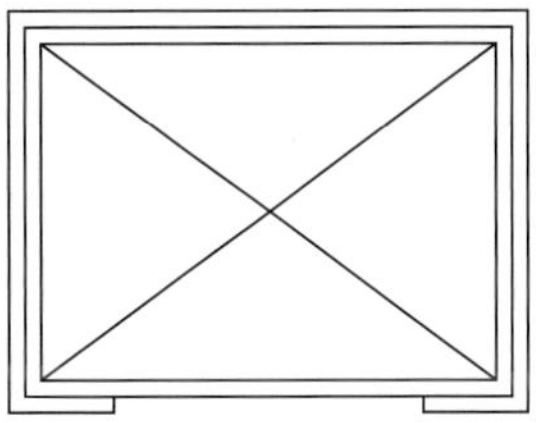		
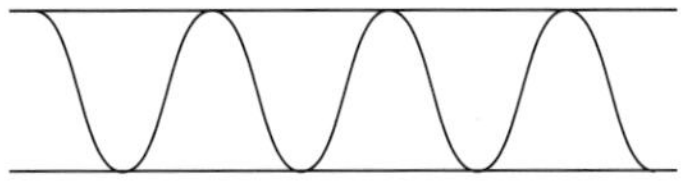		
	Datum	
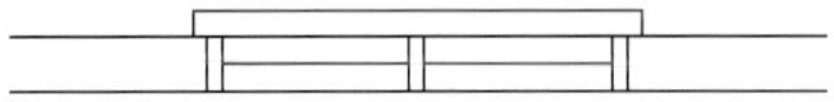		
	Spa bath	

	Brickwork	
	Window reference No.	
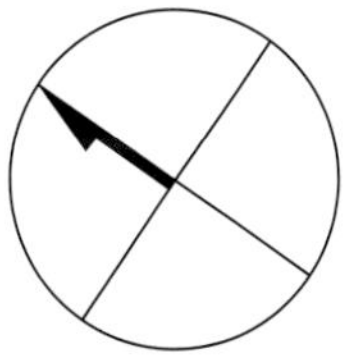		
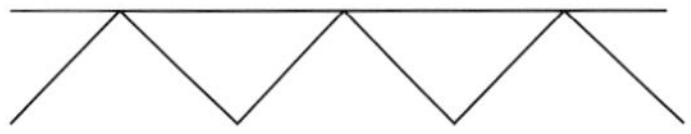		
	Stairway – arrow direction up	

Source: Adapted from Australian Standards - AS1100.301—2008. © Standards Australia Limited Reproduced by Cengage under licence by Standards Australia.

Key features of site plans

This section deals with the key features that you should be able to identify on a typical construction site plan. As outlined previously, a site plan is a detailed drawing of a building allotment from an overhead perspective. It identifies the position of the building or buildings on the site, the dimensions of the site boundaries, the distance that a structure is from those boundaries, and a number of other important features, such as:

- location, orientation and size of the construction site
- datum
- access and egress
- contours and slopes
- major geographical and topographical features
- existing dwellings, buildings or other structures
- retaining walls
- drainage lines
- septic tanks and dispersion trenches
- paving
- set-backs
- service connection points
- easements
- storm water disposal
- trees and vegetation.

Generally drawn to a scale of 1:200, site plans are usually oriented with north to the top of the drawing, though this is not critical. It is important, however, that all related drawings display a similar orientation: this is so they can be viewed as a coherent set of working drawings.

Identifying the building site and envelope

So far in this chapter you have been introduced to three types of plans that may be provided as a means of identifying the actual site of a proposed structure:

- development plans
- location plans
- site plans.

The first two are important, because from these plans you need to find your specific proposed building area from what can appear to be a number of seemingly identical locations (or building lots). Only once you have correctly located your site, can you begin even a preliminary evaluation of the task before you.

Identifying the building site

When seeking to identify the location of your site from a development plan you must take great care. In many new suburban developments, one lot (block of land) may look very much like another. Just going by lot numbers displayed on sticks or pegs is a trap for the novice builder as it is not unknown for young children to pull these out and mix them up just for fun (and the less scrupulous adult for material gain). When first identifying the site, you should therefore take note of as many identifying features as possible; e.g. the number of lots away from a street intersection, the length and direction of boundaries, the shape of the block and any other notable features in the vicinity. If you are in any way unsure of your correct identification of the block then you must get it identified by a licensed surveyor (see Chapter 10 Site surveys and set out procedures).

When identifying the construction zone or site on a location plan, care must also be taken to ensure your identification is sound. A location plan is often of a large site within which there are multiple existing structures. If these structures are still in use (such as a medical or educational facility) then you will generally be able to seek advice from a resident building services manager or similar. However, a location plan may also be of a large 'bare field' site on which multiple structures are proposed, but do not yet exist. In this latter instance you would engage a surveyor to assist in locating the exact area of your build.

In either case, you should take time to make your own assessment of the location based upon all of the identifying features available to you on the plan sets provided. These will include existing structures, trees, service lines, boundaries, paths, roads, fences and the like.

Identifying the building envelope

A building envelope is the sole area defined within a site that structures may be located on, with the only exclusions being boundary fencing. That is, no structures, not even sheds or effluent disposal facilities such as septic tanks or blackwater recycling systems, may be constructed outside of this area. While the description of a building envelope may appear on some commercial or industrial sites, it is more commonly applied to domestic housing lots (see Figure 5.30).

Building envelopes are defined by the local responsible authority (generally the local council) and should be clearly defined on the site plan, particularly when dealing with new developments. Frequently, however, this is not the case, and you are left to assume (a dangerous word in any endeavour) the envelope by distances known as set-backs and, on occasions, by easements. These two terms will be discussed more fully in the following sections; however, it should be noted that they cannot be used to fully define the building envelope. Figure 5.31 below demonstrates why this is so.

Such being the case, if you have any uncertainty in the location of your building you should seek clarification, in writing, from the council, architect, or client if they are the designer. In some cases, you may also have to get information from local service authorities such as water, power and sewerage.

Set-backs

Set-backs are distances from boundaries to the location of a structure. They form one element in defining the

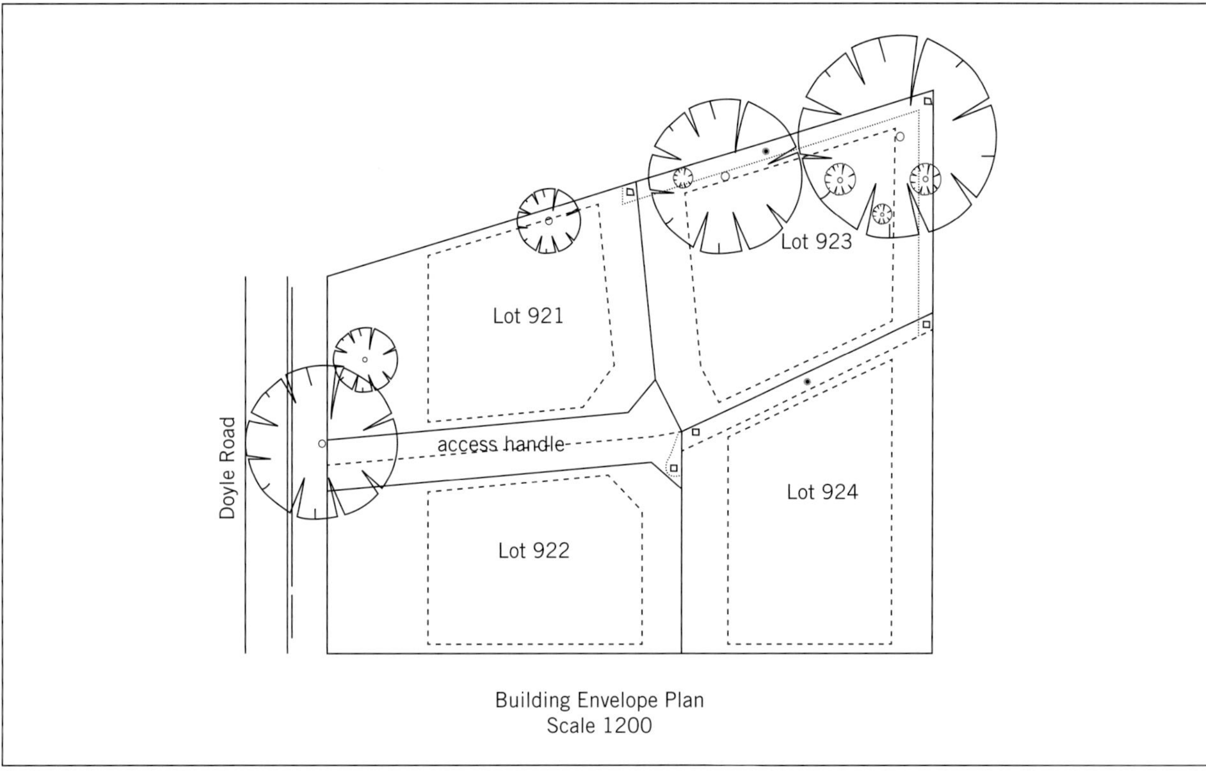

FIGURE 5.30 Building envelopes as defined on a development (subdivision) plan

Source: Penrith Local Environmental Plan 1998 (Urban Land), Historical version for 31 October 2008 to 14 December 2008, legislation.nsw.gov.au

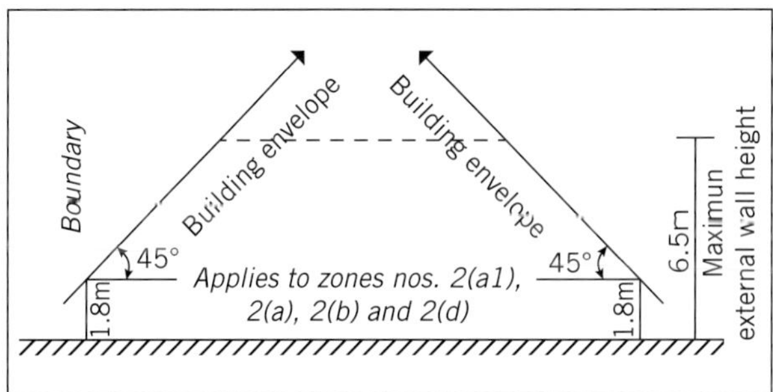

FIGURE 5.31 Example of vertical building envelope requirements

allowable building envelope but, as described above, do not totally capture the requirements of all envelopes. Councils develop set-backs for a number of reasons, such as providing:

- fire safety between adjacent dwellings
- similar street frontages
- access to services
- adequate light and ventilation to all dwellings in a development
- visual and noise privacy
- a buffer to hazardous and dangerous land usage.

Heritage locations will often have a set-back that is quite minimal (less than a metre) based upon the frontage presented by existing structures. New developments, depending upon the council intent for that zone, may likewise be very short for high-density housing, or very long (deep) for low-density rural zones (see Figure 5.32).

Source: Glenn Costin

FIGURE 5.32 Typical modern narrow set-back applied to all the houses in the street

Set-backs are typically a reference to the frontage of a block and defined by local councils; but set-backs also form part of, and are defined by, Volume two of the Building Code of Australia (BCA). As such, they form an integral part of the National Construction Code (NCC). Listed under fire safety, set-back distances are measured at 90° (right angles) from the boundary as shown in Figure 5.33.

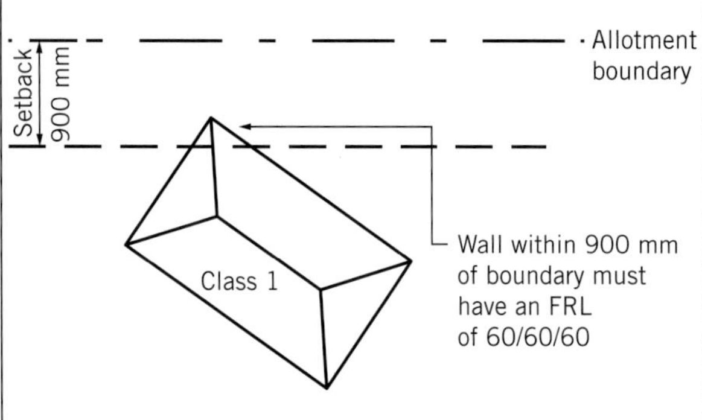

Source: NCC 2019 Building Code of Australia - Volume Two, © Commonwealth of Australia and the States and Territories of Australia 2019, published by the Australian Building Codes Board.

FIGURE 5.33 Set-backs must be measured at 90 degrees from the boundary

Easements

An easement is a section of a block over which parties other than the landowner hold certain rights, for certain purposes. As such, an easement can, and generally does, limit development over that particular area. Typical easements are:

- service or utility easements
- access or right of way easements.

Service easements are imposed for the purposes of allowing services such as power, water, sewerage, gas and telecommunications to pass through a property. An easement of this type is required to ensure two things: that such services are not damaged by future building works; and to allow access for maintenance, remedial, or future development of said infrastructure.

Access easements are sometimes imposed by a developer in conjunction with a service provider, or simply as being requisite to the development layout itself. An access easement is generally a strip of land that passes through a block, allowing vehicular access to a neighbouring parcel of land, or to a major services easement running along the rear of the property. Figure 5.34 offers examples of both service and access easements.

Each state and territory in Australia has different legal requirements regarding easements. Most states and territories, however, allow for easements to be created by and between individual land owners (private easements), as well as by public or local authorities (easements in gross [NSW] or regulatory easements [Vic]). Today, the latter formation is the most common as they are created during the development of subdivisions with full engagement from councils and service (water, power, etc.) authorities.

Note: In Figure 5.34, three easements are shown. If you look carefully you will see that lot 82, lot 93 and even the proposed development (unlabelled lot 81) all have sheds built over the easements; in the case of lot 82, even a carport. Small structures can be built over easements if prior permission is obtained from the relevant authority (in this case, most likely sewer). It must be understood, however, that such permission is only provisionally granted – in most cases the provision requires the removal of the structure should said authority require so for future works. The cost

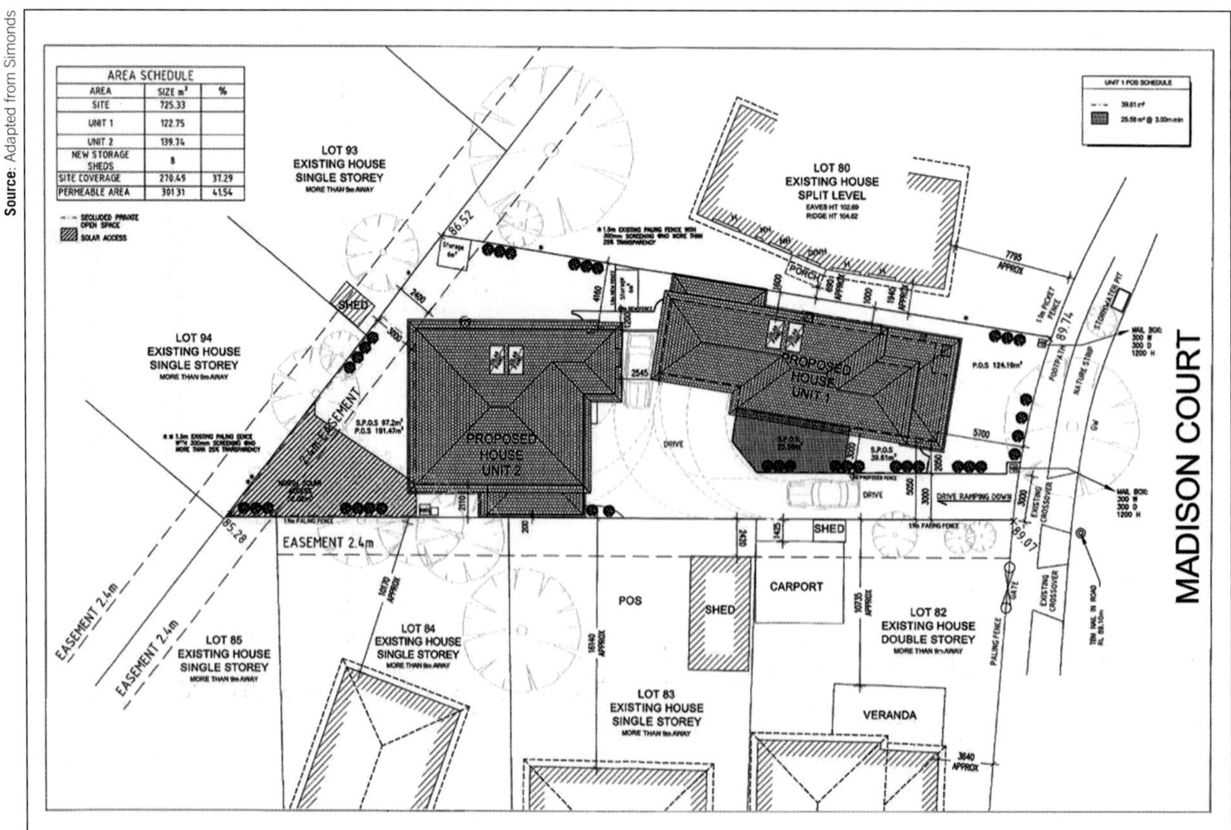

Source: Adapted from Simonds

FIGURE 5.34 Easement examples

of reinstatement of the structure is the burden of the property owner.

Orientation

As suggested previously, site plans are preferably oriented with north to the top, although this is neither critical, nor always practicable. What is definitely preferred is that all plans, and any similarly oriented images, be oriented the same way.

Which north is north?

A compass pointer (north) is always shown on a site plan, sometimes depicted in an unobtrusive location on the site plan itself, at other times in the title block. Unless specified otherwise, it may be assumed that True North, not Magnetic North, is represented. This allows the builder to orientate the building appropriately on the site. In addition, all boundary lines should have their alignment (in degrees and minutes from north) shown, along with their respective lengths.

True North may be described as a longitudinal (straight) line travelling directly from your location along the surface of the Earth to the geographic North Pole – the Earth's axis (see Figures 5.35 and 5.36). True North can differ significantly from Magnetic North depending upon the amount of magnetic deviation at any given location at any given time (at the time of writing, deviation for Albury Wodonga was approximately 11.5° east of north). This is caused by a number of factors including how the magnetic poles (north and south) themselves also vary over time, currently moving to the north-north west at about 55 km per year. The geographic (axis) pole also moves but so incrementally (about 170 mm east annually) it is not relevant for construction purposes.

You can check your own location's magnetic deviation by going to Geoscience Australia's website and inputting in your coordinates for latitude and longitude: http://www.ga.gov.au/oracle/geomag/agrfform.jsp

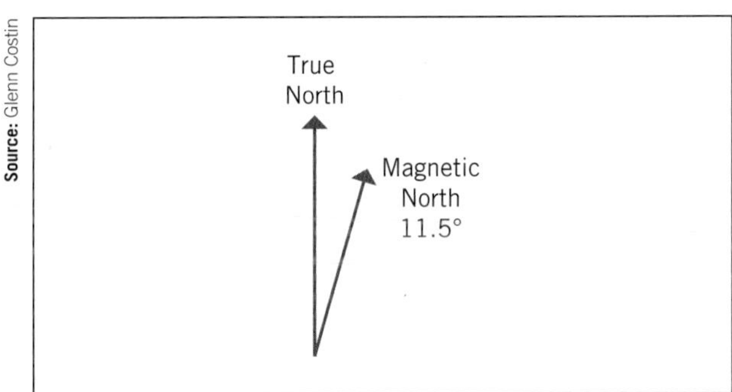

FIGURE 5.35 Magnetic North vs True North

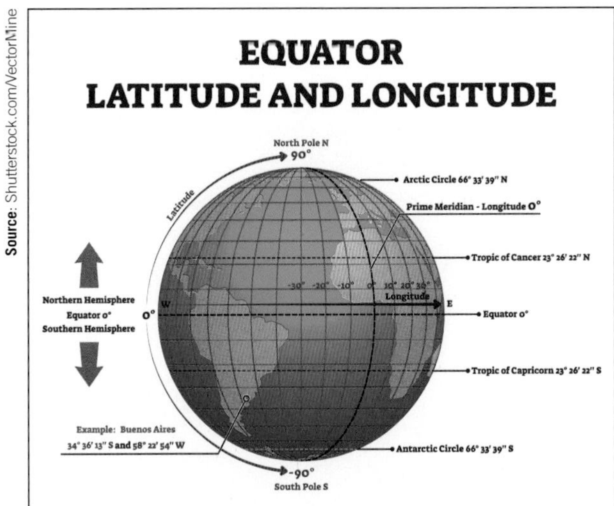

FIGURE 5.36 A line of True North runs longitudinally north and south as shown

The importance of correct orientation

Correct orientation of a building is critical to the fulfilment of the design intent of the architect. This is particularly so with regards to energy efficiency and solar access. Designers can plot the exact moment when the sun is to enter a room at a given time of year for any given location (see Figure 5.37) – but only if you construct and align the structure exactly as they have described. This shows the importance of knowing the difference between Magnetic and True North.

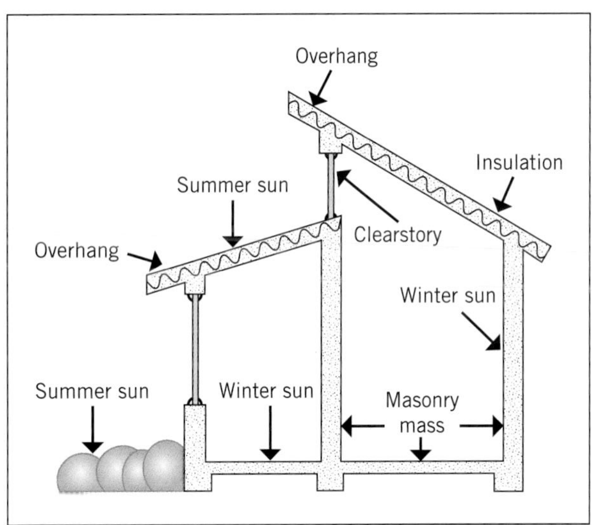

FIGURE 5.37 Solar access during the seasons

It is very important to have the correct information on the site plan regarding the true direction of north. You will note in the previously offered site plan (Figure 5.34) that north is shown point down at roughly the 7 o'clock position (see Figure 5.38). However, sheet TP.02, the ground floor plan, has been hand corrected to depict north at a very different orientation (see Figure 5.39).

In this instance, the shape of the block will tell you which is correct, and the nature of the alignment of the buildings in question mean that in reality this will have little to no influence upon your build. In other circumstances, such as on a large hectare or perhaps farm site, it may well be of great import. In such cases you would need to confer with the architect, designer, or surveyor for clarification.

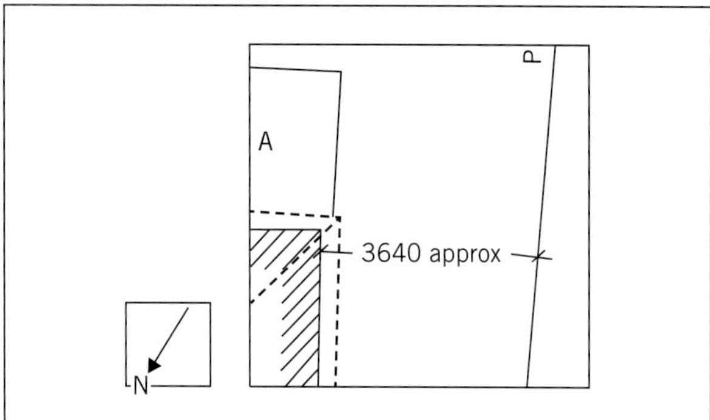

FIGURE 5.38 North at 7 o'clock position

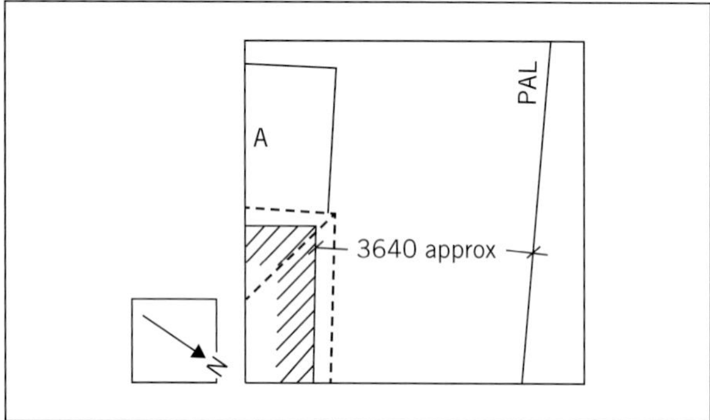

FIGURE 5.39 North at 4 o'clock position

Other key features

A site plan holds a number of other key features, many of which you will need to identify before you can correctly locate a structure or start your build. This information is not always easy to find, and on occasions it is omitted. Only in hunting for a piece of information, a need derived by the structure you need to build, will you realise if something is missing – in which case you must contact those responsible for producing the drawing.

The key features that you will generally need before you can start building, or even quoting for a job, are:

- access and egress
- datum
- benchmarks
- reduced levels
- contour lines and topography
- geology.

It is worth spending a moment developing an understanding of the relevance and importance of each of these features to both the quoting for, and building of, a structure.

Access and egress

One of the first features you will need to identify is how to get on and off a property. This is not always as straightforward as it sounds. Access for a person, or a light vehicle is one thing, access for a fully laden 7-tonne twin steer concrete truck, cranes, earthmoving equipment or the like, is quite another. Likewise, egress (exiting a site) is not always possible or desirable by the route taken to get in: one-way streets, difficult reversing areas, limited parking for workers, easements (rights of passage or carriage) all need to be considered. From the site plan you should be able to identify at least if access is readily available, and by which route. Frequently, however, a site visit will be required in order to ascertain the real nature of a location. In some cases, negotiations with owners of neighbouring properties may be required.

Datum and benchmarks

The datum is a point on or near a building site to which all other site levels are referenced. If on the mainland, the site datum is usually given as a known elevation relative to the Australian Height Datum (AHD), or the AHD (Tasmania) if in that state. These two datums are entirely separate and are based upon different sets of tidal gauges from which a mean sea level for Tasmania or the mainland has been developed. The mean sea level in each case has been assigned the value of zero.

On occasions, when working on rural sites, or when carrying out extensions or additions (domestic or commercial), the datum may be simply a peg in the ground, or nominated point on an existing floor surface to which a nominal value of 100.000 m is given. The value of 100 metres allows for rise and fall values to be taken without the resultant reduced levels (RLs – see following section) falling into negative figures.

On the site plan, the datum may be identified as a TBM (temporary benchmark), BM (benchmark), or PM (permanent mark), or simply 'datum' (see Figure 5.34). PMs are generally only to be found outside of the site boundary on a pavement or some other council serviceable location. All others are likely to be very close to the site, such as a nail and washer shot into the kerb, or on site, but sufficiently away from the proposed works so as not to be disturbed. Example datum marks are shown in Figures 5.40 and 5.41.

FIGURE 5.40 Example datum mark

Glenn Costin

FIGURE 5.41 Another datum mark

Reduced levels

In the previous section, the concept of a reduced level (RL) was referred to. The word 'reduced' in this case does not mean made smaller, but instead derives from one of its alternative definitions meaning 'to bring to a systematic form', or 'derived from'.

As stated above, the national or state (Tasmanian) datum has a fixed value of zero. All around the country Permanent Marks have been installed to which the height above (and on occasions, below) one or the other of these datums (as relevant) is known. That is, they all have an identified reduced level based upon the applicable AHD. This height above mean sea level is generally not recorded on the PM, but on a database accessible by councils, surveyors and other concerned parties, and is constantly updated to account for earth movement.

R.L.s will frequently be found on a site plan as a means of identifying known or desired heights on or around the building location. You need to be able to interpret these heights so that you can visualise the project, and to calculate quantities, be it for excavation or construction purposes (see Figure 5.42).

Contour lines and topography

The topography of any given geographical area is a description of the Earth's surface that includes the shape of the land such as hills, valleys, ridges and water courses, and existing structures such as buildings, sheds, bridges and the like. On a site plan, the detail can be more or less, depending upon the structure being built, and the features on the site. In which case it may include trees, rocks, drains, septic tanks, water tanks, fences, service lines and other similar features.

The key topographical features on most site plans are the contour lines (Figure 5.44). These provide you with a two-dimensional representation of the land, where it rises and falls, and hence insights into drainage patterns and excavation requirements. Contour lines represent a level line trace following the curvature of the Earth's surface. As may be seen in Figure 5.43, each contour line includes its RL (reduced

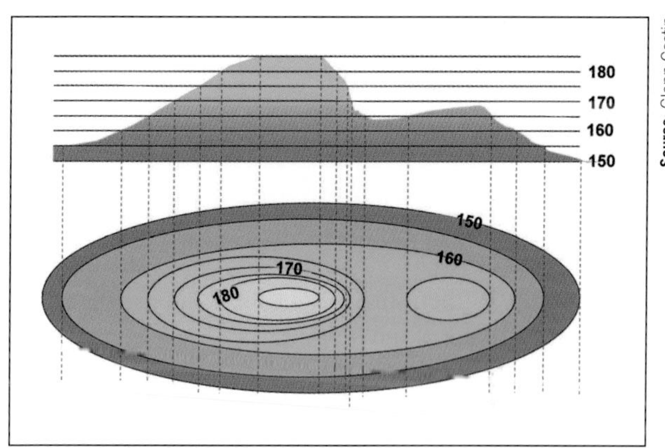

Source: Glenn Costin

FIGURE 5.43 Contour lines representing two small hills

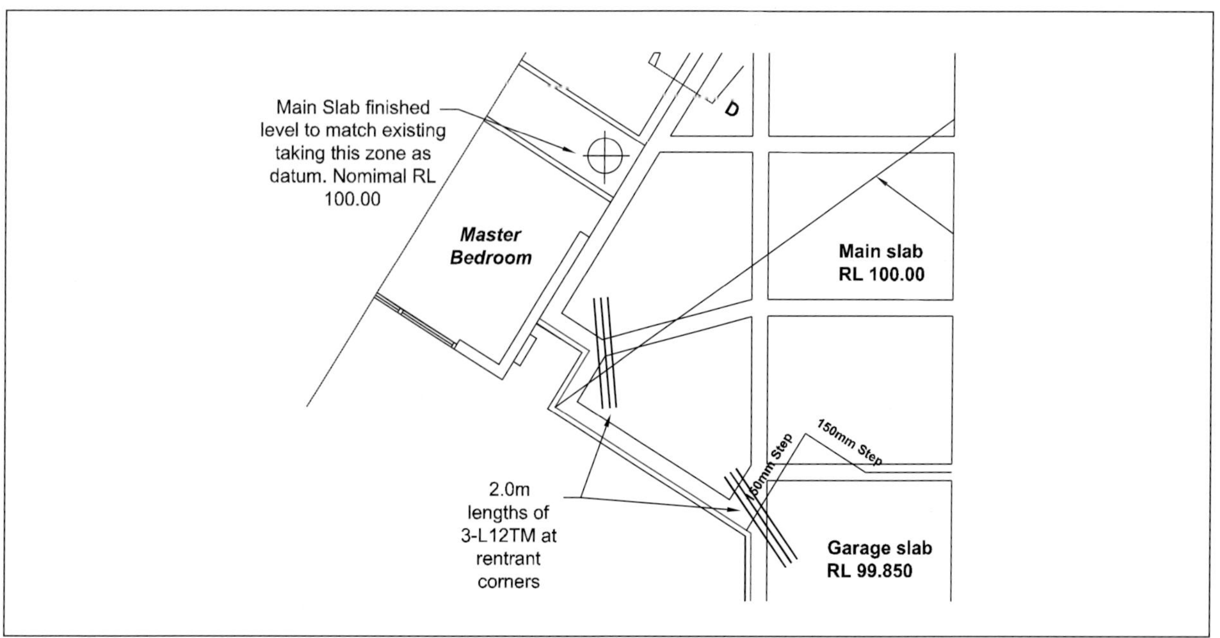

FIGURE 5.42 Reduced levels (R.L.s)

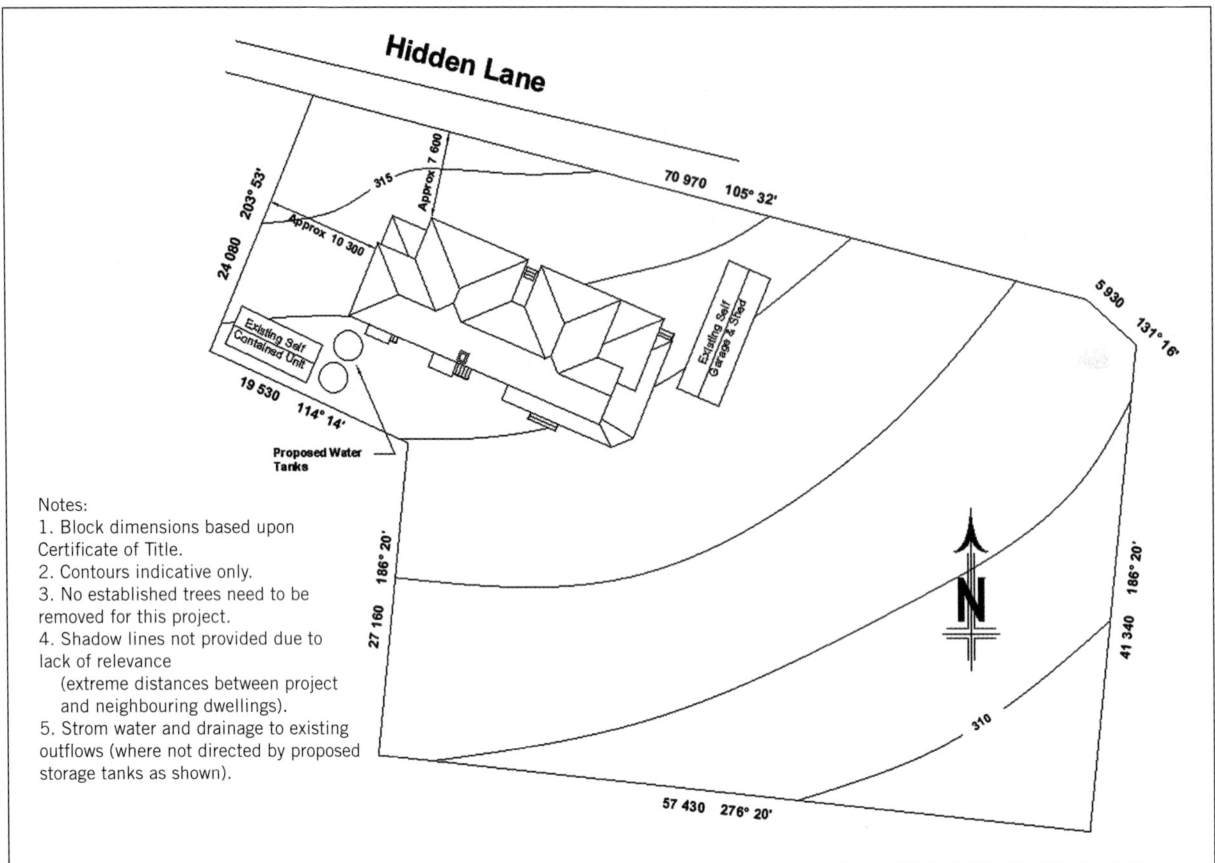

FIGURE 5.44 Contour lines on site plan

level) value. Without these values it is impossible to tell if you are looking at a hill or a pit.

Geology

The geology of a building lot is seldom to be found on the average domestic site plan; however, when it is, you can be assured it's there for good reason and you should take careful note. This is of particular import when initially developing a quote.

Geology refers to the study or science of the origin, history and structure of the Earth: rocks, minerals, stratification, oil, natural gas and fossils all inform this field of study. Pedology is the study of soils; their characteristics, origins and uses. For the purposes of this chapter, pedology will be subsumed into the one term 'geology' covering all elements of the foundation material.

The geology of a building site becomes important when it influences the engineering and excavation of the footings, the structure, or associated works such as retaining walls or subsoil drainage (see Figure 5.45). History is littered with structural failures that could have been avoided if greater attention to, or a greater understanding of, geology had been applied. Common geological failures include subsidence, slippage and compression of the foundation material. These failures can lead to fractures and on rare occasions, total collapse of a structure.

On a domestic site plan, common geological features may include identification of large boulders, protruding bedrock, filled in dams or creek lines, soils and gravels prone to compression, subsidence or slippage, springs, and references to soil tests such as notes on suitability for dispersion trenching associated with septic tanks, maximum allowable bearing pressures and/or minimum footing depths.

LEARNING TASK 5.3

UNDERSTANDING THE LANGUAGE OF KEY FEATURES OF SITE AND OTHER PLANS

In simple words, explain the following terms:

Named item	Name
access	
egress	
contour lines	
reduced levels (RLs)	
orientation	
solar access	
building envelope	
set-back	
datum	
benchmark	
easement	
right of way	
shadow diagram	
solar access	

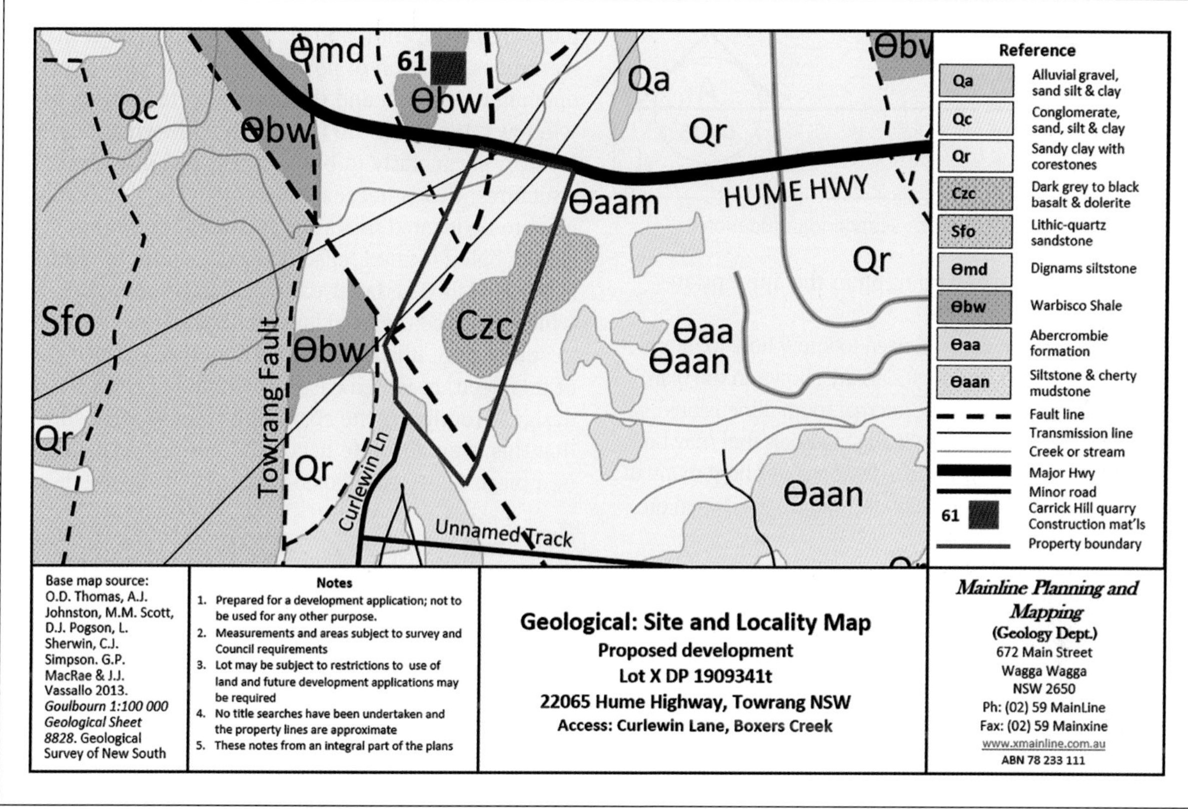

FIGURE 5.45 Example geology overlay on a site plan for a proposed development

Key features of working drawings

The previous sections have described the various aspects and types of construction drawings you are most likely to encounter. The purpose of this section is to focus in on the key features of the main drawing elements found in all working drawing plan sets; i.e. the floor plan, the elevations and sections. In addition, we will look at identifying approved variations within a plan set.

Plans, elevations and sections

The floor plan, elevations and sections are the most common aspects of drawings and each will be found in virtually all sets of working drawings. Of these, the first, and often most informative, is the floor plan.

The floor plan

As outlined earlier in this chapter, floor plans provide information on:

- the size and shape of the building
- thickness of all walls
- width of eaves
- size and name of all rooms
- size of verandahs, patios, porches
- position of fixtures such as baths, showers, toilets and basins
- cupboards, halls and other spaces
- the roof outline
- the location and direction of *sectional* views.

Your task as a builder is to find all this information, check for errors (e.g. comparing the summing of all internal measurements over the length of a structure to the stated overall length), and if errors are found, resolve them by contacting the relevant people (architect, designer, engineer, client).

A floor plan can also test your knowledge of construction terminology, acronyms, symbols and the like. In some cases, you many need to clarify terms with the person who drew the plans, or look them up online or in glossaries of building terms, such as the one at the end of this book.

Of particular importance in a floor plan, aside from the main measurements, is the identification of where a particular sectional view has been taken, and in which direction you are looking once you find it. A study of the partial floor plan offered in Figure 5.47 shows the location of three sectional views, with the direction of view being identified by the triangular black arrows. The letters beside the arrows tell you which section to look for in the plan set. Some draftspeople use a different arrow and identification system that includes the drawing page number (see Figure 5.46). Arrows of this type are usually applied to one end of the sectional cut line, with the other end showing a

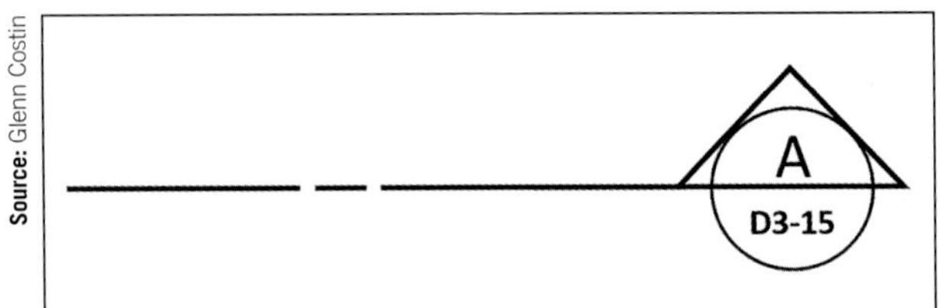

Source: Glenn Costin

FIGURE 5.46 Example of cross-section identification

simple triangular arrow pointing in the appropriate direction.

Another element you will need to know how to interpret is the way wall types and materials are shown in the plan. Again, you are reminded that internal room dimensions are to the framing, not to any lining materials that may be applied to that frame. Figure 5.48 demonstrates how many of these materials and construction techniques are drawn on contemporary floor plans.

While there will always be many more things that you will find on a floor plan, such as the floor coverings indicated in Figure 5.48, some important parts are the location and size of kitchens, built-in wardrobes, location of power board/fuse box, door-opening directions, and the approximate location of windows. Windows are frequently somewhat loosely located, particularly in brick veneer and solid brick structures; this is because the window will generally need to be located to suit brick sizes rather than simply being located 'exactly' in the centre of room's external wall. The latter case does occur, however, in which instance the bricklayer will need to cut bricks to suit. Again, it is the floor plan that will advise you of this need, in which case you should check with the designer (outlining the cost implications) to be sure that this is an absolute need, and not just a suggestion or a piece of easy dimensioning.

Elevations

As discussed earlier in this chapter, drawings of elevations generally show heights in the main, hence

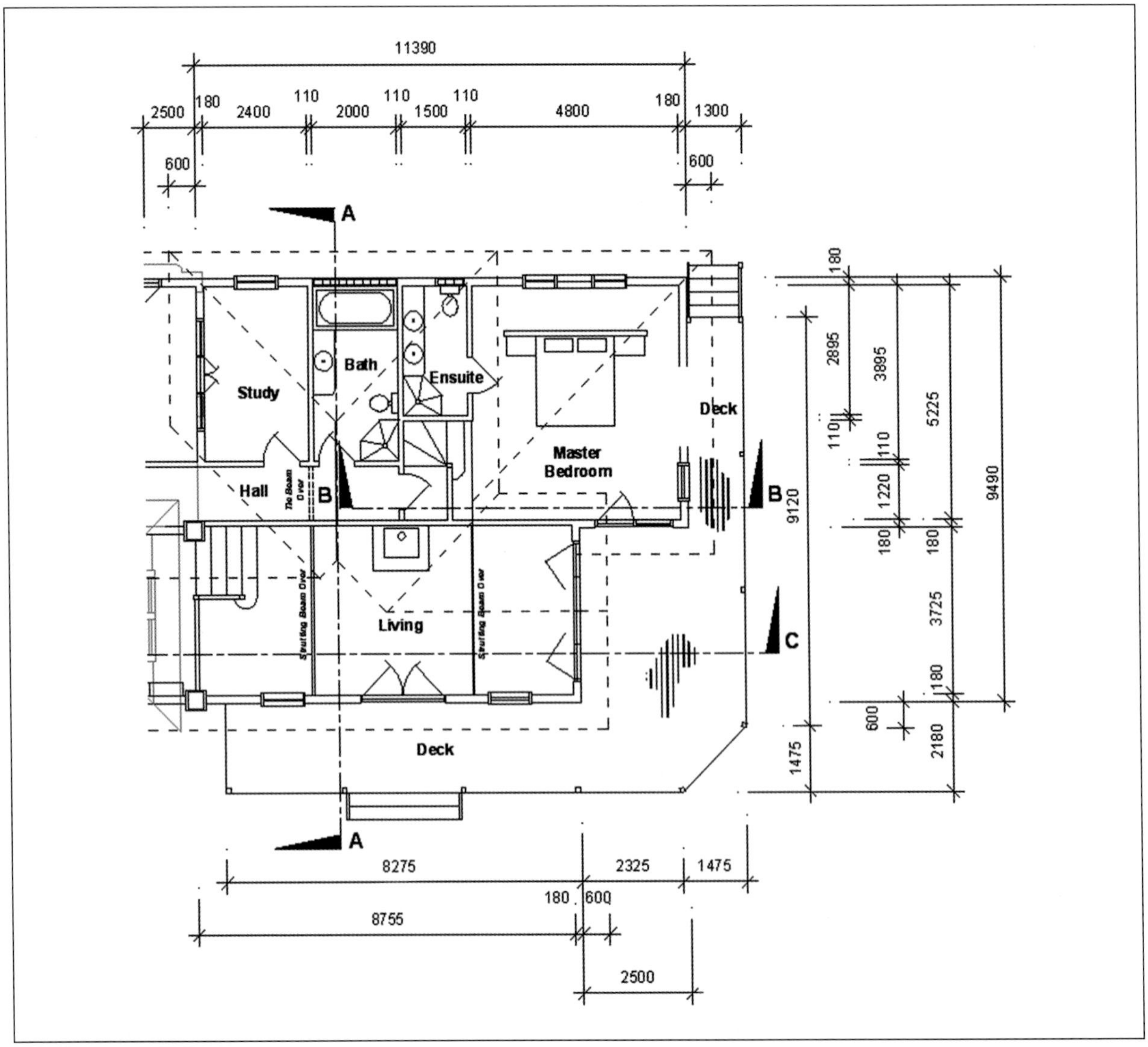

Source: Glenn Costin

FIGURE 5.47 Example floor plan detail

Source: Glenn Costin

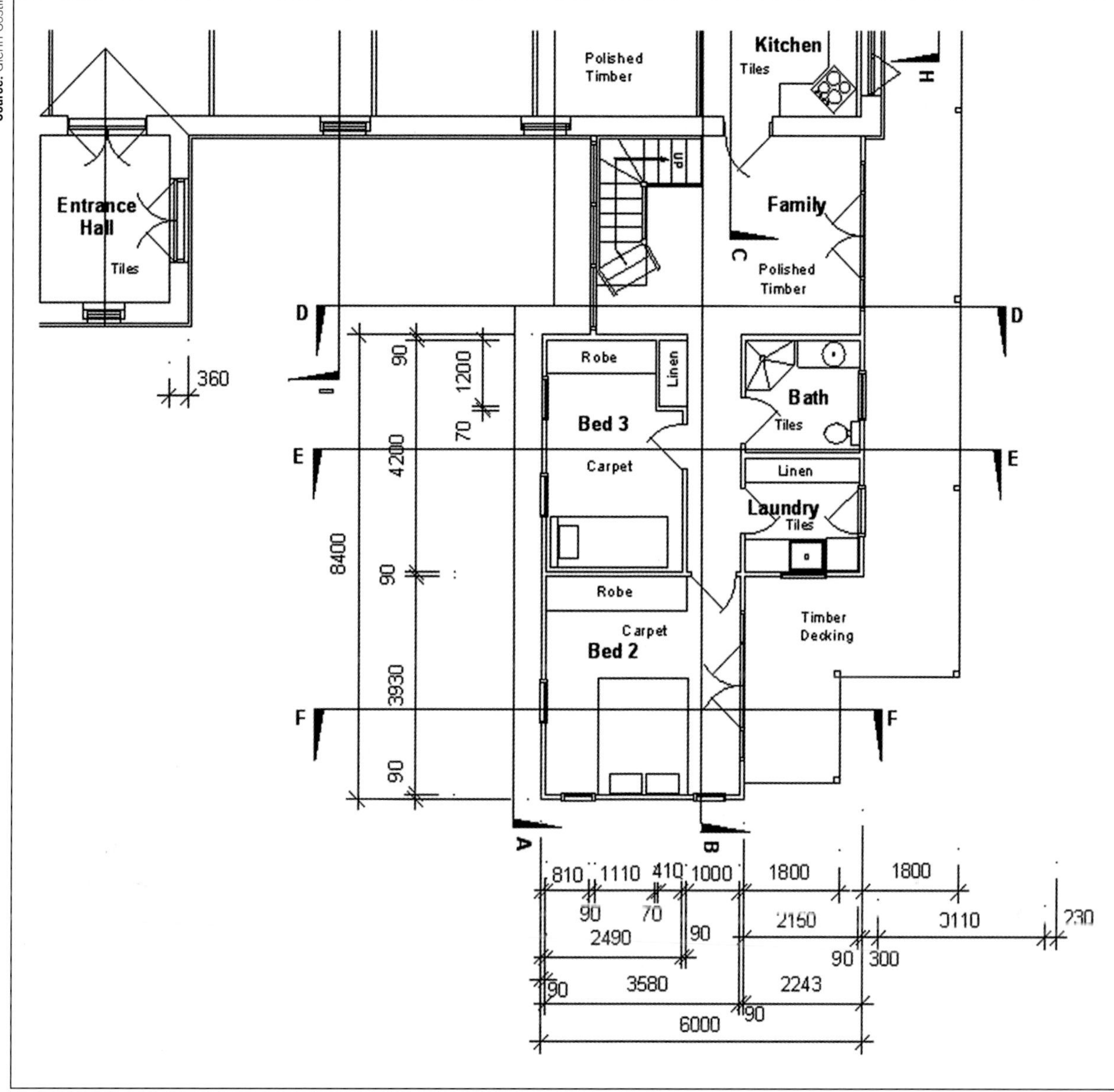

FIGURE 5.48 Example floor plan detail with material representations identified

their name. There are, however, a range of other pieces of information that you may glean from an elevation that will be important to your build, and before that, to your quoting. Figure 5.49 shows the elevation of an extension to an existing home. The floor plan (shown in Figure 5.48) cannot adequately represent the way that the end gable windows have been interpreted by the designer. The elevation can. Likewise, handrailing to the decks and roof pitch angles can all be understood in this view, but not from above.

Elevations may also be used to show the locations of particular features such as gable end claddings, roof and wall claddings, colour finishes, rendered sections and other external surface finishes. Figure 5.50 shows this use of an elevation.

Sections

Like the floor plans that tell you where and in which direction you are looking, sections are generally highly detailed and can take some practice to interpret correctly. The information sectional views contain is always important to your understanding of the designer's intent, otherwise they would not have bothered to draw them (they can be a lot of work to create). Take note of material types, heights and locations of key features. Often when you need to find something about how you are going to construct a particular element of the structure, it is in the sections that you will find the information.

Figure 5.51, section A–A, aligns with the floor plan given in Figure 5.48. You should always take a moment to compare the sectional view offered with the floor plan.

Source: Glenn Costin

FIGURE 5.49 Example elevation – note 'hopper' style windows (hinged at bottom) to gable end

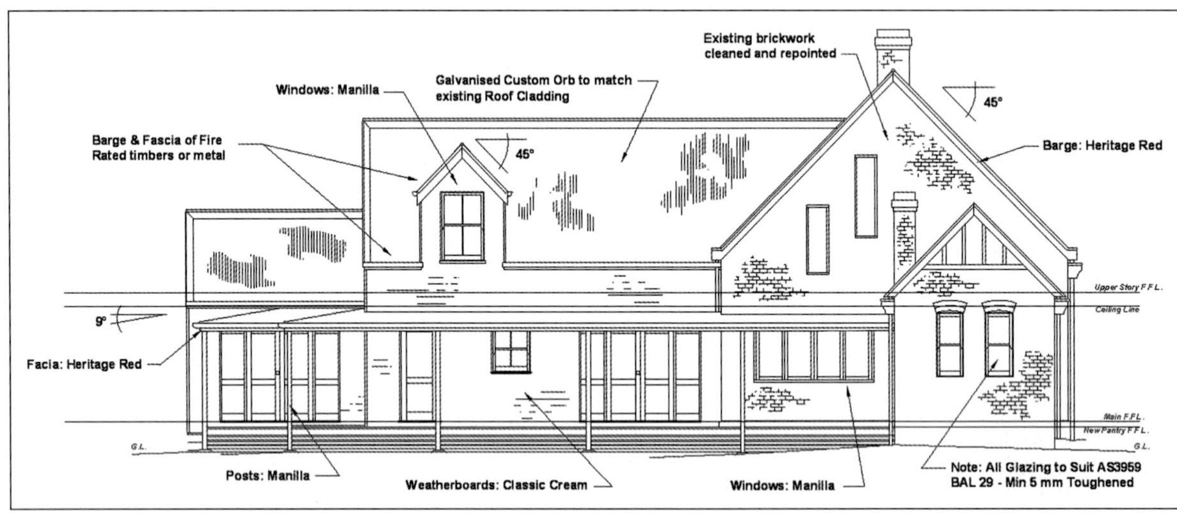

FIGURE 5.50 Example elevation – with finishes described. Note glazing thicknesses also described.

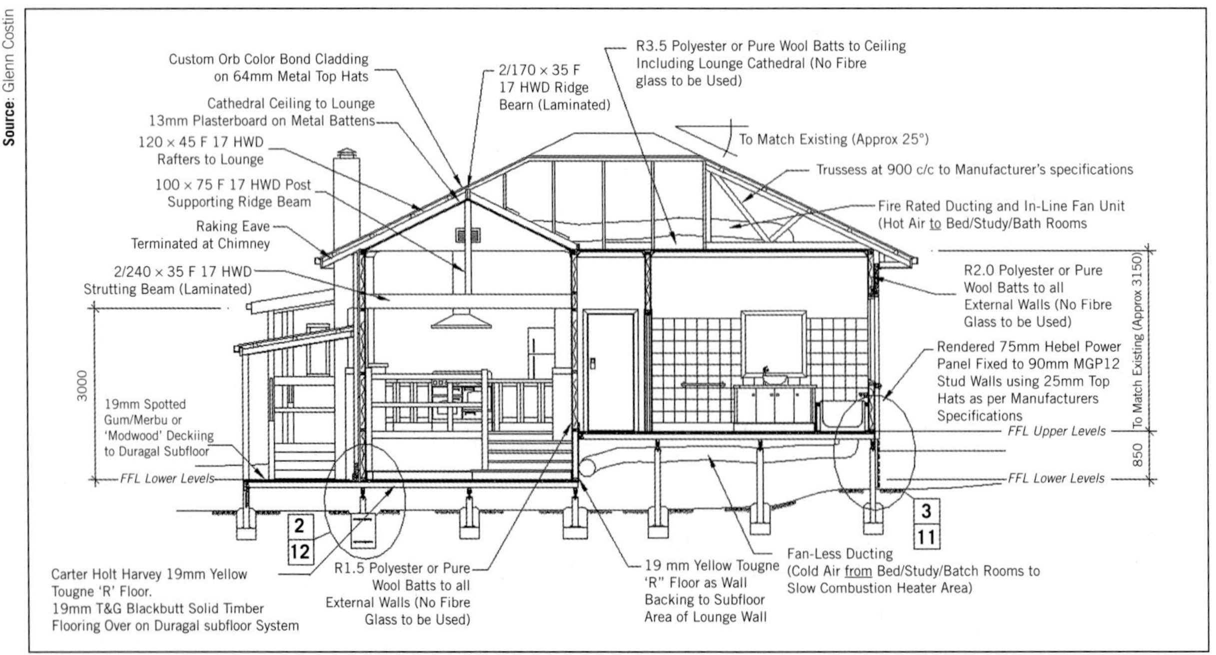

Source: Glenn Costin

FIGURE 5.51 Section A–A

This is necessary to orientate yourself with the view such that you can understand exactly what it is you are looking at, what it provides and what it does not. On occasions you may wish to go back to the design team (architect, engineer, etc.) and request further details to clarify something that appears to be hidden. This is a perfectly legitimate request.

Note that in section A–A the location of further information is also provided. These circled, shaded or otherwise highlighted areas are given a twin number system so that you not only know which detail to look for, but also which page of the working drawing set it is to be found upon. One feature important to this section drawing, and not easily identified on other drawings, is the height of the strutting beam over the lounge area. Other things of note are the internal handrail design, underfloor unfanned or convection ducting and the post supporting the ridge.

Reading and interpreting

The only way to really become familiar with interpreting floor plans is to study them, as many of them as you can. Only in this way can you get a feel for the common concepts that each hold, even though the draftspeople may have very different styles. To get you started on this journey, the following section is task oriented.

Variations

Clients frequently request changes to a structure after plans have been initially approved, and even after construction has begun. Likewise, architects, designers and engineers may also call for an alteration to the initial design at any time. And, on occasions, you, as the builder, may suggest, or need alternatives as well; this may be due to supply, practicality, an anomaly in the foundation material discovered during excavation, or an identified drafting error or structural concern. In either case, this requires a revised plan sheet, sheets, or complete set of working drawings needing to be approved and distributed. This is known as a variation, and the altered plan sets are known as revisions.

Note: Variations, and their influence upon contractual obligations, are discussed in greater depth in Chapters 3 and 6. This chapter only deals with variations with regards to their presentation within a set of working drawings.

Variations included in the approved documents

As the builder, your first task upon receiving a set of plans is to check that you have the latest revision. The revisions will be recorded within the working drawing set as a revision number appended to the page number in the title block (see Figure 5.52). Within the drawing itself, the changes should be clearly identified by what are known as revision triangles and 'rev clouds' highlighting the affected area on the particular drawing aspect (Figure 5.53). The revision triangle will be labelled with the revision number and any explanatory notes.

Client requested variations after commencement

Naturally, when a client requests a change to a structure after you have commenced the build, you will not find that change on the existing working drawing set. That request, however, must find its way in writing and in many cases, in drawing, into the contract. In the cases where a drawing needs to be produced, even if it is a simple on-site sketch, this must be transferred to the plans by way of a revision, and disseminated to the various stakeholders. Councils and the relevant building inspector will need this information, as will the design team. Only in this way can you be sure that all parties are in agreement.

On occasions, a change needs to be affected rapidly for site works to safely continue. In such cases every effort should be made to contact stakeholders and get their acceptance by text or email. These changes must still be transferred to the drawing set and a revision published. As mentioned earlier, this is where building information modelling (BIM) comes into its own on larger commercial works. This is because even minor requests for variations can be evaluated across the whole of the build, including the influence upon energy efficiency and post-construction maintenance.

Sometimes a drawing does not need to be made, not even a sketch: examples of this may be a change of tiles or fixture types, or paint colour.

Source: Glenn Costin

Detail 3 - 11 (Rev B)
Scale 1:20

B	Cladding Amendments	14/11/19
A	First issue	01/07/19
Rev.	**Amendments**	**Rev. B**

Project: Proposed Alterations to 1 Hidden Lane	**Contact information:** Ph:	Scale 1:20 or As Shown All Dimensions in mm	DWG 11-14

FIGURE 5.52 Revision identity in title box

Source: Glenn Costin

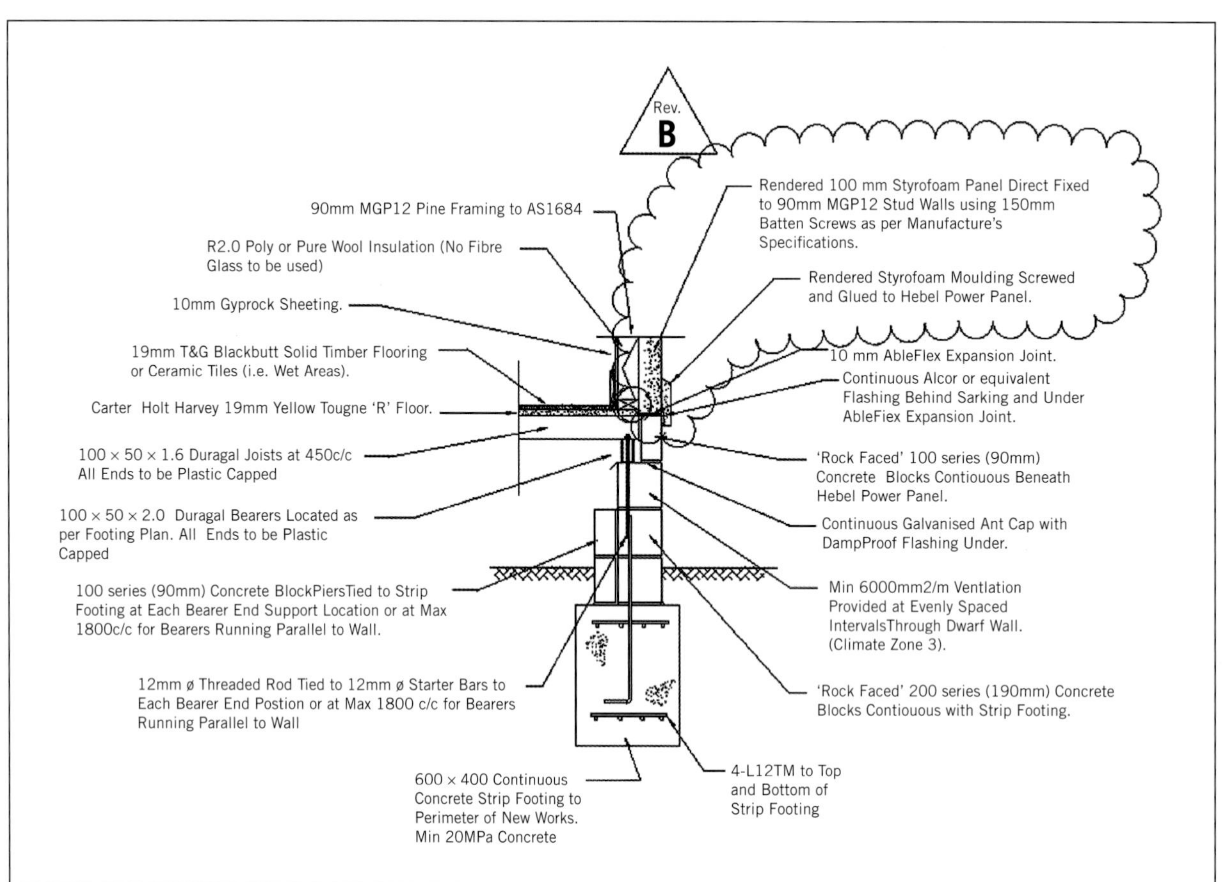

FIGURE 5.53 Rev cloud and triangle

LEARNING TASK 5.4 KEY FEATURES OF WORKING DRAWINGS

With reference to the plans provided in this section (Figures 5.47, 5.48, 5.49, 5.50 and 5.51), respond to the questions in the table below. In each case you are to state what you believe to be the correct answer, as well as the view or drawing aspect in which it was found, and the drawing sheet number.

Q	Question	Response	View	DWG No.
1	What is the width of the ensuite?			
2	What is the sectional size of the strutting beams over the living room?			
3	What is the difference in floor height between dining and living rooms?			
4	What is the eave width of the entry verandah?			
5	What is the floor to ceiling height for the master bedroom?			
6	What is the wall height to the living room?			
7	What type of glass is in W12?			
8	What forms of insulation are allowed in the walls?			
9	What is the approximate pitch of the existing roof?			
10	How is the bearer tied to the dwarf wall and footing?			
11	What is the total bracing restraint provided for wind direction 1?			
12	What is the minimum the top edge of the strip footing may be below ground level?			
13	What is the sectional size of the verandah perimeter bearers?			
14	What is the terrain category for this site?			
15	What must be done to the ends of the Duragal bearers and joists?			
16	Brace B2m offers how many kN of restraint?			

>>

>>

Q	Question	Response	View	DWG No.
17	In which room is brace B2m located?			
18	What is the ceiling height of the master bedroom entrance?			
19	What is the sectional size of the rafters to the lounge?			
20	What material are the lounge rafters to be made from?			
21	What scale is Detail 2–12 drawn at?			

With reference to the plans provided in this section (Figures 5.52 and 5.53), respond to be the questions in the table below. In each case you are to state what you believe to be the correct answer, as well as the view or drawing aspect in which it was found, and the drawing sheet number.

Q	Question	Response	View	DWG No.
1	What revision number is the drawing?			
2	What date was the revision made/published?			
3	What new material are to be used on the external walls?			
4	What change does this have on the fixings?			
5	What change does it make regarding external finishes?			

Specifications: how to read and interpret

Drawings and plans present construction information graphically in the main; however, there are types of information that will need to be conveyed in either written text, tables, or schedules. Some schedules may find their way onto the working drawing set; however, they will tend to be repeated in the document known as the specifications. Specifications hold information on a range of structural and non-structural matters such as: material choices, standards to which the work must comply, permits and application fees that may apply, the obligations of the client such as ensuring access, or works that the client intends to undertake themselves.

Specifications can be quite long, wordy and highly technical. In addition, they form an integral part (as do the working drawings) of the contract between builder and client. The working drawings and the specifications must be read and interpreted in tandem; without the one, the other is limited and easily misinterpreted. Both the specifications and working drawings are therefore used during the quoting and/or tendering process, as well as during construction.

The sections within the specifications are generally grouped by trades or subcontractor types. Likewise, they tend to be ordered sequentially to follow the way job activities flow on during construction. An **example specification** is found at the end of the book in appendix 2.

Builders specialising in domestic housing often use a ***standard specification***. These may be purchased online from entities such as NATSPEC (a government and industry not-for-profit organisation) or through building associations such as the Master Builders Association (MBA – a NATSPEC associate) or the Housing Industry Association (HIA). On government works (all states), a NATSPEC specification is a requirement for any tendering documentation.

This does not prohibit you from drawing up your own for minor domestic works, provided that you use a standard specification as an initial guide, or specifications from previous works in which you have been involved and have shown themselves to be adequate. On occasions the 'author' of the specifications maybe the designer or architect or even the client. Sometimes it is a collaborative effort of these and the builder (you). You should become very familiar with the required content of specifications before you try writing them completely yourself. Though it is almost a given that as a builder you will be involved in developing one at some stage in your career.

Whatever its source, a specification should detail requirements common to buildings of the type. The following table lists the work sections included in a standard NATSPEC domestic construction specification and are typical to most works of this type:

0121 Tender form	0451 Windows and glazed doors
0131 Preliminaries	0453 Doors and access panels
0171 General requirements	0454 Overhead doors
0184 Termite management	0455 Door hardware
0194 Door and window seals	0467 Glass components
0201 Demolition	0471 Thermal insulation and sarking membranes
0221 Site management	0511 Lining
0222 Earthwork	0551 Joinery
0223 Service trenching	0572 Miscellaneous appliances and fixtures
0241 Landscape – walling and edging	0611 Rendering and plastering
0242 Landscape – fences and barriers	0612 Cementitious toppings
0250 Landscape – gardening	0621 Waterproofing – wet areas
0271 Pavement base and sub-base	0631 Ceramic tiles
0274 Concrete pavement	0651 Resilient finishes
0276 Segmental pavers – sand bed	0652 Carpets
0310 Concrete	0654 Engineered panel floors
0331 Brick and block construction	0655 Timber flooring
0342 Light steel framing	0656 Floor sanding and finishing
0382 Light timber framing	0671 Painting
0383 Sheet flooring and decking	0702 Mechanical design and install
0411 Waterproofing	0802 Hydraulic design and install
0421 Roofing	0902 Electrical design and install

Source: Building Domestic by NATSPEC, https://natspec.com.au/. NATSPEC is a national not-for-profit organisation, owned by Government and industry, whose objective is to improve the construction quality and productivity of the built environment through leadership of information

In commercial construction, further information of an engineering nature may be included. This can be in the order of expected loads and the bearing capacity of members or elements of the structure (slabs and the like). Extended preliminaries, fire-stop and fire protection sections are also common to commercial works, as is a section on environmental management (though this may also be a requirement of domestic works in some locations).

In domestic renovation works the 'specification' is sometimes nothing more than a couple of schedules in the working drawings. Aside from a call for adherence to Australian standards, nothing else may exist. It is preferable, however, from both client and builder perspectives, to have something 'tighter'. Again, this maybe something written by yourself or, preferably, a standard specification purchased from a reputable agency.

Provisional sums and prime cost items

With many projects there will be works needing to be undertaken, or fixtures or fittings yet to be identified by the client, for which a fixed or definite sum cannot be offered. In such cases, in order to produce a final quote or tender, an estimated figure is offered and clearly identified as such within the specifications. These are known as provisional sums and prime costs.

Provisional sums

Sometimes, in perusing the working drawings, or in discussion with the design team or client, an activity will be identified for which no exact cost can be derived. The most common example of this is in excavation when there is a high chance of encountering significant time delays. In such cases a figure is proposed 'provisionally'. This figure may be based on what experience (yours and/or that of your subcontracting excavator) suggests is likely or, more commonly today, using a standard costing reference such as Cordell's Housing and Building Cost Guide. Within the specifications and contract documentation this is known as a **provisional sum (PS)**.

Once the work is complete, the difference between the provisional sum and the actual costs incurred are consolidated in the final statement of accounts. That is, where the cost of the works was greater, the contracted total price owing by the client increases; when the costs were less, the final monies owed by the client decreases by that amount.

Typical PS items included in a standard specification are rock excavations, retaining walls, landscaping and inground pools. In domestic work, electrical fit out also tends to fall into this category due to clients often not being clear on what they want where until after the build has begun to take shape. The specifications included at the end of this book provide examples of how provisional sums are included within such a document.

Prime costs

An identified **prime cost (PC)** within a specification is an allowance for the purchase and installation of items that the client or architect has not yet selected, or clearly specified. Common examples include tiles, baths, taps, cooktops, door handles, and other fixtures and fittings. The PC sums may cover the cost of labour and materials, or materials only, though in either case it shall be clearly listed in the specifications. As with provisional sums, the differences between the PC sums given and actual costs are accounted for, positively or negatively, in the final statement of accounts. Common examples causing a difference between the PC sums provided and the actual amount expended can be the client finding some bargain-priced tapware, or a discounted bath. Alternatively, the specification may allow for a standard bath and tapware, whereas the client selects items twice the cost.

Identifying variations

As variations to the approved works lead to revised drawings, the specifications likewise require to be updated when a variation is requested. Unfortunately, it is all too common for specifications to remain unchanged, leading to disputes. Done correctly, the revisions will be shown in the new specifications by way of comments or annotations down the side margins (see Figure 5.53). The date of the revision should, as with the working drawing set, be clearly identified on the front cover of the specifications document (Figure 5.52). Again, as with the working drawings, it is you, the builder, who is responsible for ensuring that the most up-to-date revision is in use.

A specification may also vary in ways which are more subtle. When a builder becomes familiar with specifications it is easy to fall into complacency and miss key points in a document that looks familiar, but has changes to it. Clients can request all types of small things, which can become big things if left unnoticed and hence not acted upon. Disposing of old timbers from the demolition (in a renovation), for example, may be an expensive error if the client had identified within the specifications that they were to form part of the new works – such as stair components.

When first getting involved with a project, you should sit down with both the client and the designer(s) and read through the specifications carefully together. Only in this way can you be assured that all aspects to the works have been fully covered and understood by all parties.

Communicating changes

Once more, it is in variations that BIM is such a valuable tool for communicating those changes through all documentation and to all stakeholders. The reality, however, is that BIM is still in its infancy in Australia and remains rare in the low-rise commercial sector, let alone the domestic housing market. Such being the case, it is up to the builder to ensure that variations requested by the client are documented in the specifications. This is particularly so post-commencement; i.e. after the documentation has been approved by the relevant agencies (councils and the like). In this latter instance, it is also required of the builder to make contact with said authorities and ensure the changes are acceptable and that authorisation is given.

Once the changes to the specification have been made or identified, these changes must be confirmed with the client. Upon receiving that confirmation – generally by way of dated initials signature to the relevant section(s) – the changes must now be conveyed to all relevant personnel and subcontractors. Depending upon the project, this is best handled by a site meeting and again getting writing acknowledgement from all parties concerned. Remember, it is your task to determine who those concerned parties are.

Interpreting essential elements for estimation, planning and supervision

A specification document is broken up into multiple sections, as described earlier. Each section relates to a particular part of the contracted works. The purpose of the sections is to clarify, often with intense detail, the builders' responsibilities, the standards and codes relevant for that element of the works, and any other matters that will ensure that the works undertaken are of a quality and type acceptable to the designers and the client.

The exact breakdown of the project, and the section names themselves, vary from specification to specification; however, there are similarities within this variance. The standard NATSPEC sections shown on page 126 at the beginning of this section are the common low-rise residential specifications.

Each of these sections is in turn broken down again as is required to provide the information necessary for a given project. In some cases, this will be a few numbered parts for clauses with very short descriptions. In other cases, it will include numerous clauses and subclauses that clearly spell out the needs and expectations of that area of the work. Section 0431 Cladding, for example, might be broken up into four clauses, with any one of those being broken down further into subclauses. Clause 3 of this section (execution of cladding) demonstrates a small part of this breakdown:

3 Execution

3.1 PREPARATION

Substrates or framing

Preparation: Before fixing cladding, check the alignment of substrates or framing and adjust if required.

Flexible underlay. Check that the underlay is restrained.

3.2 INSTALLATION

General

Requirement: Fix sheeting firmly against framing to the manufacturer's recommendations.

Fixing method: As documented or to one of the following fixing methods to manufacturer's recommendations:

- Steel framing: Screw.
- Timber framing: Nail or screw.
- Minimum penetration for profiled metal sheets: 30 mm for timber framing.

Accessories and trim

Requirement: Provide accessories and trim required to complete the installation with the same finish as the cladding sheets.

Corner flashing: Finish off at corners with purpose-made folded flashing strips.

Metal separation

Design for compatibility or detail separation.

Requirement: Prevent direct contact between incompatible metals, and between green hardwood or chemically treated timber and aluminium or coated steel, by either of the following methods:

- Apply an anti-corrosion, low moisture transmission coating to contact surfaces.
- Insert a separation layer.

Incompatible metal fixings: Do not use.

Fixing eaves and soffit lining

Nailing: 150 mm centres to bearers at maximum 450 mm centres.

Louvre sunscreens

Installation: Fix sunscreen systems in accordance with the current written recommendations and instructions of the manufacturer or supplier.

Proprietary systems or products

Product fixing: Fix proprietary systems to the manufacturer's recommendations.

Source: 436P Colorbond® Steel And Zincalume® Steel in Cladding from Branded Worksections by NATSPEC, https://natspec.com.au/. NATSPEC is a national not-for-profit organisation, owned by Government and industry, whose objective is to improve the construction quality and productivity of the built environment through leadership of information

This clause goes on for a further three subclauses covering cladding types and completion. These are extensive and detail fixing spacings, gaps, clearances, tolerances, supports and flashings.

However, it is the preliminaries section, the first part of the document, that has the critical information that helps you to both interpret the document and to see how it fits with the other documentation. The preliminaries will also include information on:

- the purpose and scope of the specification
- the specification's relationship to other documents
- what documents are included as part of the specification and should be read in conjunction with the specifications, such as the working drawings, addenda and schedules
- the legal precedence of the above; i.e. which document is to be considered correct in cases where documents are inconsistent
- units of measurement; in Australia this will generally be millimetres
- definition of terms
- standards referred to in the specification, such as the NCC/BCA and the relevant Australian standards.

It is important to read the preliminaries thoroughly, and acquaint yourself with any standards with which you are not familiar. Further, the section generally holds important information on the need for any permits, fees and the like that may be required before the site may be accessed, or the works commenced.

You must check for inconsistencies within the documents, and although the order of precedence is generally offered, attempt to clarify these issues with the design team, client or relevant authorities first.

Informing the team

As the builder, you are the conduit for all contracted works to the subcontractors and your own work teams and apprentices. It is therefore important, while obtaining quotes and during construction, to ensure that all subtrades are aware of the various sections, clauses and subclauses relevant to their area of works or associated works. This is crucial, as the specified works may differ significantly to those they generally undertake, be it in process, materials, or standard of finish.

Depending upon the work that your subcontractors are used to engaging in, they may or may not be familiar with the Australian standards applicable to or otherwise called up in the specifications. While it may not seem to be your job to furnish subtrades with these standards, it is in your interests, and the client's, to ensure that they do have access to and have read and understood the relevant parts. This can be a big call on some domestic projects in regional areas where such 'niceties' are not always adhered to, but it should be done nonetheless.

Identifying relevant codes and standards

As you read in the chapter on building codes (Chapter 1), in general application, Australian standards only have legal standing (with regards to construction) if called up within the NCC. However, they have equal legal standing if called up within the specifications for a particular project, even if they are not directly mentioned in the NCC or relevant volume of the BCA. Hence, identifying the standards applicable to a particular project becomes critical to both you and your subcontractors.

Generally, the applicable Australian standards will be listed in the specifications against a specific work item. An example is in Clause 1 General, of the work section 0184 Termite prevention, shown below:

PART 1 5

0184P TERMGUARD TERMITE MANAGEMENT

1 General

1.1 RESPONSIBILITIES

General

Requirement: Provide Termguard and Granitgard termite management materials and systems, as documented.

Performance

Objective: Protection of building from damage caused by termite attack.

1.2 COMPANY CONTACTS

TERMGUARD technical contacts

Website: termguard.com.au/your-nearest-termguard-installer

1.3 CROSS REFERENCES

General

Requirement: Conform to the following:

- 0171 General requirements

1.4 STANDARDS

General

Termite management systems: To AS 3660.1.

1.5 MANUFACTURER'S DOCUMENTS

Technical manuals

Website: termguard.com.au/contact-us

1.6 SUBMISSIONS

Certification

Certificate of installation: Submit evidence from an accredited technician that the system conforms to AS 3660.1 Appendix A and documented requirements.

Records

Completion: Submit record drawings identifying the locations of the installed system.

Tests

Site tests: Submit test results as follows:

- Chemical termite management systems: To AS 3660.1 Appendix E, and as required for the Codemark Certificate of Conformance.

Warranties

Requirement: Submit the following:

- 50 year warranty: For installations conforming to Termguard and Granitgard Marketing specifications.

1.7 INSPECTION

Notice

Inspection: Give notice so that inspection may be made of the following:

- Completed earthworks or substrate preparation before system application/installation.
- The completed termite management system

Source: 0184P Termguard Terminte Maangement by NATSPEC, https://natspec.com.au/. NATSPEC is a national not-for-profit organisation, owned by Government and industry, whose objective is to improve the construction quality and productivity of the built environment through leadership of information

National Construction Code (NCC)

The National Construction Code (NCC) and the relevant volume of the Building Code of Australia (BCA) will also be referenced, but generally in the preliminaries as an overarching standard. The purpose of this reference within the specifications reflects the purpose of the NCC itself; i.e. to achieve a nationally consistent minimum standard of structure inclusive of health and safety, amenity and sustainability.

On occasions either of these two documents may show up directly within a work section clause when no other standard is applicable, or in conjunction with those standards. A look at the first subclause of work section 0194 Door and window seals shows how this might look (BCA element in bold):

1.1 STANDARDS

Seals general

Quality management for manufacture: To ISO 9001.

Acoustic applications: To AS 1191 or AS/NZS ISO 717.1.

The BCA cites ISO 717-1:1996 and AS/NZS 1276.1 for testing of construction required to have a certain R_w rating.

Fire door assemblies: To AS 1530.4 and AS 1905.1.

Smoke door assemblies: To *BCA Spec C3.4*, AS 1530.7 and AS 3959 for silicon flame retardant PVC and TPE weather seals with a Flammability Index not more than 5 to AS 1530.2 providing BAL 40.

Combined fire and smoke door assemblies: To *BCA Spec C3.4*, AS 1530.4, AS 1905.1, AS 1530.7 and AS 3959 for weather seals providing BAL FZ.

Weather and energy saving seals for proprietary windows and door assemblies: To AS/NZS 4420.1 clause 5 and clause 6, and AS 2047.

Door bottom and perimeter seals for glazed external doors: To AS 2047.

Threshold plates: To AS 1428.1.

Source: 0194P Raven Door Seals and Window Seals by NATSPEC, https://natspec.com.au/

Specifications: the non-structural elements

A structure, be it domestic or commercial, large or small, may have its component elements categorised in two ways: structural or non-structural. The structural elements are those that withstand loads in some manner or form; i.e. either holding the structure up, holding it down, or bracing it against lateral forces. These are discussed in great detail in Chapters 12 and 13 of this book. The focus of the remaining section of this chapter is therefore about the non-structural components.

While non-structural elements may not play a role in either supporting dead loads or bracing against the wind, they have a role nonetheless: this may be pure aesthetics, or influence the sustainability or energy

LEARNING TASK 5.5 READING AND INTERPRETING SPECIFICATIONS

Complete this activity by accessing a plan set that you have been working on, or one provided by your instructor.

Separately document and provide a listing of the requirements on your plan set for the following, using the table headings below.

1 Provisional sums – list up to 5 items:

Provisional Sums			
Item	Found on Drawing No/s	Basis of work under the contract	Allowance for additional work 'over & above' contract basis

2 Prime cost items – list up to 10 items:

Prime Cost/s					
Item – full description to include product name, colour, serial no./ code & other characteristics	Found on Drawing No/s	Supplier name & details / date price obtained.	Item cost	Allowance cost for installation / commissioning of item	Total cost – inclusive of any GST.

efficiency of a structure. Alternatively, they may be critical elements integral to the structure's purpose, such as plumbing and electrical services, appliances, fittings, tiles and the like. Or even more crucially, they may protect those structural elements from environmental damage such as wind or rain.

Non-structural features and their implications

Non-structural elements of a building feature prominently in a specification as they are critical elements in the designer's and client's vision on aesthetics, end use, and maintenance and longevity. This being the case, you will find non-structural requirements in almost all work sections of the specifications, but particularly in those where products are mentioned, such as:

- masonry and pavements
- cladding and lining
- joinery
- flooring finishes and tiling
- electrical
- plumbing
- painting
- hardware
- insulation
- appliances
- landscaping.

Purpose and design: not just a question of aesthetics

Generally, the standard specification will allow for the client to have input into most of those items mentioned above. The choices made will frequently be based upon aesthetics and cost, but also on service value; to which the client will often seek a builder's opinion or advice. However, the specifications may list some constraints upon choice due to energy efficiency requirements precluding dark colours, or window frames requiring thermal breaks. With regards to other materials it may be an issue of structural integrity, such as loadbearing masonry, or the capacity of glazing to withstand wind forces.

It is important that as the builder, you take careful note of the non-structural elements described within the specifications and carry them through accordingly. There will be times when your experience base, or that of your subcontractors, contradicts the actions, materials or finishes depicted in the specifications. As with any section of this document, you must make contact with the designers and clients (and relevant authorising bodies) before enacting a change. In a heritage zone, for example, even an apparent slight shift in colour can mean wasted time and materials when upon inspection you are called to revert. Likewise putting horns on a double-hung bottom sash when historically, for the era being replicated, horns did not exist on bottom sashes.

LEARNING TASK 5.6 SPECIFICATION OF NON-STRUCTURAL ELEMENTS

Complete this activity by accessing a plan set that you have been working on, or one provided by your instructor.

Clearly identify the non-structural elements that would form part of the materials and items that would need to be priced as part of arriving at a price for a construction contract. An example template is provided for you.

Fill this part in for your response to question.
ADD more lines as required – use an Excel or similar spreadsheet

NOT REQUIRED – But will be used in future Study Units

Item no.	Item description	Quantity	Unit	Rate	GST – if applicable	Total amount

SUMMARY

This chapter introduced, and discussed in depth, the reading and interpreting of those plans and specifications associated with domestic and commercial low-rise structures. You should now be familiar with the types of drawings and their purposes, as well as the common symbols, key features and abbreviations associated with them. Likewise, you should now be able to engage confidently with a specifications document and draw from it the information you will need for estimating, quoting and supervisory purposes. What you should now do is access as many of these documents as you can, developed from as many different sources as are available to you. Compare them. Find their similarities and variances. Make yourself familiar with them so that you can easily find not only what you need for quoting, but also errors and contradictions so that these may be dealt with prior to commencement, rather than after the fact. Learn to enjoy picking up a new set of working drawings and comparing them to the specifications. They are, after all, a reflection of the vision behind a structure; that very structure that has been placed in your hands to bring into being.

REFERENCES AND FURTHER READING

NATSPEC is the government and industry supported producer of standard specifications for Australia. Free example specifications and sections of specifications can be downloaded, go to: **http://natspec.com.au/**

SAI Global. This is the portal through which you can obtain Australian standards. There is a fee payable for each standard unless you access them through your RTO library. **https://www.saiglobal.com/en-au/standards_and_content/effective_standards_management/**

ON-SITE SUPERVISION

6

Chapter overview

Site supervision is a multifaceted role where your knowledge of contracts and their administration is as important as your knowledge of the construction process. This role requires you to ensure that all those involved in the build, including yourself, comply with the company's quality control systems, record-keeping policies, and WHS and other regulatory requirements; all while following communication best practice and administering payments and claims.

As a resource aimed at enabling you to achieve these goals, this chapter is framed around the following elements.

Elements

This chapter provides knowledge and skill development materials on how to:

1 supervise the administration of claims and payment processes
2 supervise and maintain on-site communications
3 conduct on-site inspections
4 complete all project administration processes.

Introduction

Supervision is not just about observing to ensure that certain activities are undertaken in a compliant manner, it is also about communication: communication between those on the ground doing the work, those for whom the work is being undertaken, and to those who may need to look upon certain activities historically. It is about reading and following information contained in contracts, following specified procedures and communicating to others that those procedures are being followed. It is therefore as much about being seen to be doing as the doing itself.

In the following pages you will learn about general contract administration; more specifically, about the claims and payment processes contained within those contracts. You will also learn about sound communication practices, quality control procedures and general project administration. In many cases the topics have been discussed at some level in other chapters of this book; however, in this chapter the focus is upon appropriate supervision and the means of expressing and recording that this supervision has indeed occurred.

Payments, claims and general contract administration

In the construction industry, management is particularly focused upon the administration of contracts. You will need to engage with a range of key stakeholders that may include councils, designers and architects, subcontracting trades and businesses and, of course, the client.

One major element of this task is the negotiation and administration of claims and payments to ensure the construction process proceeds smoothly and to the satisfaction of all parties. It is the contract between you (as the builder) and your client that is the rule book by which you must shape your actions.

Payments: contractual orders and allowances

As a book of rules, the contract specifies the how, when, and why of payments by way of schedules, clauses, allowances and exclusions. Contracts and tender documentation are dealt with in detail in Chapters 3 and 8 respectively within this book; however, it is worth revisiting the basics here as contract type may influence both the expression of a claim and the manner of its payment.

Contracts can differ depending upon the state or territory you are operating in due to variations in legislation. However, there are some basic types of contacts that are consistent throughout the country, the most common being variously expressed as a fixed price, fixed sum or lump sum contract. Others, used only in the commercial sector, include:

- lump sum with rise and fall
- cost plus:
  - cost plus a fee
  - cost plus a percentage
- guaranteed maximum price
- bespoke or custom contracts.

Of these, only the 'bespoke' or custom contracts should be used in the domestic sector except in exceptional circumstances. In some states, cost plus contracts are illegal for domestic work under a certain project value. While you are advised to review Chapter 3 regarding the more specific elements of each of these contract types, the basics of each are summarised below.

Fixed price or lump sum contracts

This form of contract is captured in its name – 'fixed price'. This means that a set sum is established before the document is signed and both parties know what the expectations are with regards to outcome and cost. The main contract type used in the domestic sector, it is also common to smaller commercial works where the project is easily scoped and the likelihood of significant variation is low. That stated, in both domestic and commercial works variations to the 'fixed' price do commonly occur. This is due to contractually agreed areas where changes may be made, such as:

- prime cost items
- preliminary sums
- variations (design, structural, material)
- changes to taxes and duties.

Rise and fall contracts

Designed to reduce the financial risks to the builder, this form of contract is prohibited from use in the domestic sector in some states and territories, but not all. The basis behind this type of agreement is an allowance for changes in price due to a rise or fall in material, transport or labour prices. So, while there is the potential for a lower price to the client, the client carries all the risk for fluctuations, whereas the contractor will always be paid.

Cost plus

There are two forms of this contract: cost plus an agreed fee, and cost plus a percentage of those costs. This is a high-risk contract for clients as there is no limit to what may have to be paid. Project costs may expand for all sorts of reasons and the client must pay for them by the terms of the contract.

Guaranteed maximum price

Similar to cost plus, except that there is a limit to the overall costs. A cost 'ceiling' or maximum is imposed within the contract that cannot be exceeded. The problem with this contract is related to two issues. The first is that the builder must now bear the brunt of any cost overruns. Being aware of this may cause the builder to bring into being the second issue; that of reduced quality and the introduction of short cuts to prevent cost overruns from occurring.

Custom contracts

The previously discussed contracts are all fairly common and 'standard' versions of them are readily available throughout most states and territories. Custom contracts are ones that may be modelled on these standard versions (of whatever contract type deemed most applicable), but will be adapted to spell out some specific element or actions that the client or builder wants. In most cases, this is an attempt by one party or the other to offset risk and to impose that risk, generally financial, upon the other party.

Standard contracts

The bulk of all building and construction contracts use standard forms developed by industry organisations such as the Master Builders Association (MBA), Housing Industry Association (HIA), or Standards Australia. Other sources can be state or territory building authorities or consumer protection departments such as NSW's Department of Fair Trading. Most commercially used contracts are known in industry by some form of alphanumerical code such as AS 4000, GC21, or ABIC MW – 2008. The more domestically aligned contracts tend to use names targeting their use, such as 'Home building contract for work over $20,000'.

Tenders

Tenders will be covered in more detail in Chapter 8. A tender is a formal offer in writing to carry out works of a specified scope for a particular, generally financial, return. Tenders differ from 'quotes' or 'quotations' in being more formally composed, and offered in response to a 'call' for tenders from the 'principal': the principal being an individual, group, company or even a government agency. In responding to a tender, a company will use the tender package and forms provided by the principal; these dictate the structure or format of the tender application. As the prospective supervisor of what is only a potential project, you may be included in making the tender application, or you may simply be brought in after the tender is won. Upon winning a tender, contracts (as above) are then drawn up and signed by both parties.

Tenders may also be called for by your company for the completion of both large and small elements of the works to be undertaken. These sub-tenders are again more than simple quotes by subcontractors for the completion of minor plumbing or electrical works. For example, you might get a quote for the manufacture and installation of a single bathroom cabinet, but you would call for tenders to furnish and fit out all the bathrooms of a major multi-residential apartment or conference facility.

Other key terms applicable to contracts

The following terms must be understood before we begin discussing payments, as each of these has an influence upon how payments are legitimately claimed or contract amounts allocated.

Deposit

Generally, a contract will include a call for a deposit prior to the commencement of work. How much is allowed to be taken as a deposit is limited by law, which can vary between states and territories. The common figure is currently 5% of the total contract sum (inclusive of GST and all labour and materials). In some minor works the deposit can be increased to 10%.

Whatever the amount, this sum is considered as part of the overall payment due and so at the completion of the works you would have progressively brought in the remaining 90% or 95% only to achieve the total value of the works.

Allowances

An allowance in a contract is a cash amount that is a fair estimate of what a client may need to pay but cannot be calculated exactly for specific reasons. Excavation, for example, can be very difficult to price given the range of unknowns that may come to light once the ground is broken. Alternatively, tap fittings, tiles, kitchen appliances and the like can be priced precisely; however, the client may not have settled on a specific colour or brand. The price may therefore go up or down quite dramatically. These have been alluded to previously under fixed price or lump sum contracts and are known as prime cost and provisional sum items (see Chapter 5).

- Prime cost (PC): A prime cost allowance refers to items that may change as the project nears completion, such as tiles, plumbing fixtures, door handles, appliances and the like.
- Provisional sums (PS): These amounts can often be quite large, the excavation example being a case in point. Should the excavation run smoothly the price may be lowered; in the event of large boulders or a need for deeper excavations, the price may need to be adjusted upwards.

In each case, be it provisional sums or prime costs, any changes will reflect the real cost of the works undertaken or items purchased. If the allowance exceeds this actual expenditure, the final contract sum is lowered accordingly. When the reverse applies; i.e. the actual costs were higher than the allowance, then the total project sum increases by the difference owing.

Retention

Retention refers to money withheld by the principal or owner. It is retained as a form of surety should works fail to be finalised, the quality is subsequently identified as substandard, or to ensure that the defects period is fulfilled. The total retention monies allowed to be retained is generally a maximum of 5% of the total contract price. At handover of the project this amount is dropped to 2.5% for the duration of the defects liability period.

Implications for payments and claims

As the main contractor, or more likely their representative site supervisor, you must become expert in the reading and interpretation of contract documents. Once signed, a contract is extremely hard to change without clear agreement between both parties. You are both effectively bound by its terms. In most commercial or architect-led domestic projects, it is the client who selects or drives the contract process. As the site supervisor you are likely to have had very limited input into that selection; however, you must make yourself very aware of all clauses and if any are not understood you must have them clarified.

Of particular import are those areas where 'allowances' have been introduced. These are mainly prime cost items, and/or areas for where a preliminary sum has been nominated – such as in excavation where the likelihood of a variation in cost is highly probable.

In the latter case, this may not only lead to extra cost, but also extra expenditure of time. There will be a clause within the contract covering extensions of time and you should make yourself familiar with it, and also the any financial implications of running over schedule without application for extensions of time.

Claims against the principal: progress payments

Claims where the principal is the respondent, arise, or should arise, from the contractual conditions. Progress claims, for example, are ones that are authorised according to a schedule within the contract.

A common domestic housing contract progress payment schedule is shown in Table 6.1 – taken from Clause 12, p. 14 of the NSW Fair Trading Home Building Contract. The payment breakdown is typical for an Australian suburban project home with a contract price of $275,000. Note that the final sum is $13,750 short. This matches the deposit of 5%. Combined, they match the total contracted agreed price.

Each progress claim is made against either an agreed amount of work completed, a stage of completion, or a time schedule. In the latter case the amount owing is based upon a review of the to-date completed works, and costs calculated accordingly. Frequently it is the architect who, in discussion with you as the supervisor, will act as the administrator in such cases. The claim process is fairly straightforward and similar in each of these instances.

- As the supervisor, you submit the agreed claim based upon the architect's summations and/or the progress payment schedule within the contract.

TABLE 6.1 Progress payment schedule

1	Base stage 20% (Less 5% deposit)		
	(less deposit: $ 13,750.00)		$ 41,250.00
2	Framing stage 20%		$ 55,000.00
3	Lockup stage 25%		$ 68,750.00
4	Fixout stage 20%		$ 55,000.00
5	Practical completion 15%		$ 41,250.00
6			$
7			$
8			$
(If space is insufficient, attach a sheet referring to this schedule)		**Total**	$ 261,250.00

- The architect, the client, or the client's representative (may be the client's bank) will check the veracity of the claim before promoting approval by the client. This is generally in the form of a progress certificate with defined amount to be paid.
- The claim is converted to an invoice (if not already in that form).
- Payment is made within the contractually agreed timeframe.

The purpose of this system is to ensure that works are indeed completed within the dictates of the contract prior to payments being made. In submitting a claim (see Figure 6.1), you should take careful note of the following:

- **Reference dates**: These are the dates set down in the contract for when a progress payment may be made, or the date to which the amount of works completed (labour and materials expended) should be calculated as an amount payable.

  When a contract does not specify reference dates, you should check with the building authority in your state or territory. The Victorian Building Authority, for example, gives 20 business days from when work began, and between each subsequent progress payment thereafter. However, this can differ from state to state.
- **Claim time limits**: The contract will generally give a period after the reference date within which a claim may be made. If no such period is stated then you must check with your state or territory building authority. In most instances this will be three months.
- **Required information on claims**: A claim must:
  - identify your company including ABN number, contact details and the like
  - identify the respondent's (principal's) details
  - identify the project for which the claim is being made
  - some form of payment claim number or identifier

Ideal Pitch Design
and Construct
Contextulised
Home Design

Ideal Pitch Design & Construct
16 Curvature Avenue
Fremantle WA
6160
Ph: 08 1100 9222
Email: info@idpconstruct.com.au

PAYMENT CLAIM – SAMPLE FORM

This is a payment claim under the Building and Construction Industry Security of Payment Act 2002.

1 Claimant's Details

Company	
Contact Person	
Address	
Phone	Fax

2 Respondent's Details

Company	
Contact Person	
Address	
Phone	Fax

3 Project/Site/Job Description

Project/Site/Job Description	
Contact Number (if applicable)	
Date of Contract	DD/MM/YYYY

4 Payment Claim Number

Payment Claim Number

5 Payment Type

□ Payment based on a reference date
Claim Period DD/MM/YYYY to DD/MM/YYYY
□ Milestone payment
□ Single or one-off payment
□ Final Payment

6 Due Date

Due Date	DD/MM/YYYY

7 Payment Claim Summary

Base contract	$
Claimable variations	$
Total (the 'claimed amount')	$

8 Construction work done or related goods and services for which this claim is made

Item No.	Description of work, goods or services	Qty	Rate	Amount claimed
				$
				$
				$
Total				$

9 Claimable Variations

Item No.	Description of work, goods or services	Qty	Rate	Amount claimed
				$
				$
				$
Total				$

Ideal Pitch Design & Construct Progress Payment Claim Form Page **4** of **6**

FIGURE 6.1 Example of a typical progress claim form

- state the work completed, and the progress payment period that the claim is made upon
- clearly indicate the amount of money being claimed
- state when the money is due and any penalties for late payment – this varies between commercial and residential works.

In some states and territories there will be a formal statement identifying that the claim is being made under a particular regulation or Act provision. In Victoria, for example, the claim must include the 'Security of Payment Statement'; i.e.:

This is a payment claim under the Building and Construction Industry Security of Payment Act 2002.

This piece of legislation, the *Security of Payment Act* (the SOP Act), will be discussed later in this section as it has an important role in ensuring fair trading between all parties to a construction contract.

- **Excluded amounts**: In addition to the above listing of what must be shown on the claim, there are a few things that cannot be. Again, you should check for state/territory variations; however, in the main this comes down to amounts of money owed that are outside the contract progress payment description. Known as 'excluded amounts', typical examples include:
  - amounts for variations that are in dispute
  - changes in regulations
  - costs related to excess time expenditure (rain, storms, delays)
  - claims for damages for breach of contract.

  The SOP Act does allow for some disputed variations which are listed within the various state and territory versions.
- **Payment**: Despite a contract holding that payment of progress payments should be within a specified period of time, there may be an exclusion clause or paragraph that references banks or other lending authorities. Such a clause will generally require that you, as the contractor or their representative, accept that authority's normal payment terms. This may include the need for site inspections and the provision by you of certifications applicable to the works covered (e.g. plumbing, electrical).

  Upon receipt of payment you will generally notice three things:
  1. A statement clearly indicating that payment by the client is not evidence that the client accepts the quality or standard of the works completed – only that they have been completed. That is, the client can still inspect the works at a later date and declare them defective.
  2. That the sum for the first progress payment is less than the amount owing. This is generally due to the deposit amount being subtracted from that first payment.
  3. That the sum paid is still less than the amount owed on the following payments. This is generally about 5% and only on the first few invoices or until a specified percentage of the contracted sum is reached (generally 5%).

  The latter is known as 'retention' monies, and was defined earlier in this chapter and will be discussed further later in this section. Before doing so we need to look at the other side of payments and claims.

Claims against your company

Aside from sending out claims to the client for progress payments, the supervisor's role usually includes the receipt of claims or invoices for works done by subcontractors, or materials delivered by suppliers. While in many instances these will be handled by an accounts department within the company proper, it will generally fall upon you to approve, or prove, that these claims should be paid; i.e. that the materials have indeed been supplied, or works completed. In both instances, unlike the client paying a progress payment, the works or materials should be checked to ensure they do meet the quality standards specified before payment is approved. Payment is then made through your company's accounts department and copies of these transactions installed in both your files and those of the head office – today, that may mean a mutually accessible online database.

These claims require information similar to that of progress payments. On occasions, when they are part of a sub-tender or large contract, they may actually be progress payment claims and must be treated as such. On other occasions they may come in the form of a simple invoice for materials supplied, works undertaken, or a combination of both.

The *Security of Payment Act* (SOP Act)

The *Security of Payment Act* (SOP Act) has a range of formal titles depending upon the state or territory in which you are conducting your project. The Western Australian version, for example, is called the *Construction Contracts Act 2004* or more simply the WA Act. However it is named, in each case the intent of the Act remains the same – to provide a quick and easy path by which any person carrying out construction work, including contractors, subcontractors and those who supply related goods and services, may be paid for works undertaken, or to otherwise recover money owed to them.

Application

The SOP Act applies across construction contracts whether they are written, oral or a combination of both these forms. Notably, in Victoria, the SOP Act does not apply to the contract between a domestic homeowner and the builder, but it does between the builder and any subcontractors or suppliers carrying out work for that builder on a domestic project. In Victoria the builder/homeowner contract specifically remains under the older *Domestic Building Contracts Act 1995*. In all other states and territories, the SOP act applies equally to all forms of construction work based upon their given definition of such works.

The Act covers progress payments specifically, which includes the final payment claim and any single or milestone claims put forth by the contractor. The Act confers a statutory right to the contractor to receive these payments as part of the project, even when no written contract exists to stipulate these payments or the timing of them.

This Act also limits or excludes certain clauses that may be in a contract which would otherwise reduce a contractor's right to payment. These include terms like 'paid when paid' or 'paid if paid', which effectively means that the principal or client will only pay the contractor if and/or when they themselves are paid in full by some other agency. The SOP Act effectively nullifies such clauses, requiring the progress payments to be made anyway.

Operation

The SOP Act covers the whole hierarchy of the construction process. That is, from the principal, through the lead contractor to all subcontractors, suppliers and even transport operations bringing supplies to the site. The '***claimant***' can thus be seeking payment for works undertaken, partial works, goods or services.

Under the Act, the claimant submits a written payment claim – an invoice – to the relevant purchaser, contractor, principal or client. This individual or company is known as the '***respondent***'. The content of this claim is as previously described and may include elements such as prime cost and preliminary sum adjustments.

Disputes to the amount claimed must also be made in writing by way of a payment schedule specifying exactly the amount disputed and why this is the case. This must be done within the limited time framed by the SOP Act applicable to the state or territory in which the work was conducted.

Should the disputed sum remain unresolved, the SOP Act allows for an adjudication process to be enacted. The Act will provide a pathway by which an adjudicator may be selected and agreed upon, as well as a timeframe within which the person must make a decision. In some instances, when the amount in dispute is very large, an adjudication review is allowed for.

The alternative to the adjudication process, and on occasions as a back-up, is the court system. This is expensive in terms of time, money and emotional stress. However, the claimant can be assisted through this process by reference to the Act. Likewise, the claimant may appeal to the court when a respondent fails to pay an adjudicated amount. In some instances, this may lead to the claimant seeking payment from the respondent's principal. For example, the supplier to a plumbing subcontractor may seek payment from your company, and so through you as the supervisor of the project, as you are deemed the 'principal' who engaged the plumber.

Drawing against allowances: policy and procedure

Briefly discussed earlier, allowances are a means of nominating an amount against areas of work that cannot be accurately priced when preparing the tender or contract. The amount is a fair estimate based upon experience and information from skilled sources such as excavating contractors or material suppliers. The two main allowances discussed were prime cost and preliminary sums.

Drawing against the prime cost allowance

When drawing up the final accounts, or in making claims against these allowances in progress payment claims, you will need to reconcile the difference between the allowances made for each of the items and the actual expenditure. It is important to hold clear records showing an adjustment for each item, or cluster of items as they are purchased. For example, a single prime cost item might be a free-standing bath, a cluster may be door knobs/handles, or a set of special joinery windows. These allowances will be shown in the contract, but also in the specifications that accompany the plans; in the latter case, frequently with more detail, as shown in Figure 6.2.

In 'drawing against', to claim some, all, or more of the money in these allowances, you will be guided by the conditions within the contract. The NSW Fair Trading contract for works over $20,000, for example, provides this clause:

> If the actual cost to the contractor is greater than the prime cost amount allowed, the excess amount together with the contractor's margin of ________% on the excess, to cover overheads, supervision and profit shall be added to the contract price, along with the additional GST. If the actual cost to the contractor is less than the amount allowed, the contract price will be reduced by the difference between the amount allowed and the actual cost.
>
> Any such addition or deduction will be taken into account in the next progress payment or as agreed between the parties.
>
> The contractor must provide a copy of any relevant invoice, receipt or other document evidencing the actual cost of the work included in the provisional sums schedule at the time payment is requested.

Source: NSW Fair Trading, Contract for Works over $20,000.

Source: Glenn Costin

Lifestyle Constructions
Project Specification for
1587 Wanderrie Rd, Burnt Gully Creek, Victoria

20. **PRIME COST ALLOWANCES**
Prime Cost allowances are based on a reasonable cost for supply of a fitting or fixture where the exact requirements or product has not yet been specified by the Owner or is unknown at the time of contracting. The following allowances have been included in the Contract Price and include GST.

20.1 **Joinery** (*Refer Clauses 16)*
Allowance of $1200.00
To supply and install: and bathroom vanity with one bank of drawers.
Allowance of $3500.00
To supply only Hwd staircase to games room
Allowance of $3030.00 for 3 only Velux windows (note, plan shows 4 but only three required. Location of third window pending heritage approval for window on South Elevation)
Allowance of $6000.00 for 6 only Double Hung Windows.

20.2 **Floor Tiling** (*Refer Clause 18.3)*
Floor Tiles: Allowance of $35.00 per m2 for purchase of tiles.

20.3 **Wall Tiling** *(Refer Clause 19.2)*
Wall Tiles: Allowance of $35.00 per m2 for purchase of tiles.

21 **PROVISIONAL SUM ALLOWANCES**
Provisional Sum allowances are based on a reasonable cost for supply of work (including materials) required by the Working Drawings and Project Specification, where the extent of the work has not been ascertained. The following allowances have been included in the contract price and include GST.

21.1 **Electrical** – *(refer Clauses 11)*
Allowance of $2000.00 general electrical
Allowance of $1500.00 for relocation of solar panel array.

FIGURE 6.2 Prime cost and provisional sum allowances

You will note that this allows you to add a percentage of margin (profit) for the purchase and handling of the items involved. Also, that GST must be included.

Drawing against provisional sums

As stated previously, sometimes work will be identified for which no exact cost can be derived. The example given was excavation, an activity where there is a high chance of significant time delays. As the extract from a typical specification document shows, electrical works also often fit into this category of pricing. In such cases a figure is proposed 'provisionally'. This amount may be based on experience or, more commonly, using a costing reference like Cordell's Housing and Building Cost Guide.

As with prime cost allowances, once work is complete, the difference between the provisional sum and the actual costs incurred are consolidated and any differences accounted for accurately. Again, a profit margin is allowed to be added on any excess amount owing, along with GST. When actual costs incurred are lower, then the contract price is lowered by this difference.

One means of ensuring that the client or lending authority is satisfied with the actual cost claimed is to offer the opportunity for these stakeholders or their representatives to be present as the works are conducted. This is often written into the contract, in which case you must make the invitation in writing within the specified timeframe stated prior to works commencing.

Once the work has been concluded on that specific provisional sum task, you may claim these amounts within the appropriate progress payment invoice.

Variations: authorising and procedure

On occasions the contracted works will need to be varied due to unforeseen circumstances. This can be as simple as a change of mind by the owner or principal, or more complex issues involving the lack of access to particular equipment or materials leading to changes in the design.

Authorising

Whatever the cause, a variation must be agreed to by both parties for it to become part of the contractual process. This agreement must be:

- in writing
- dated and signed by both parties or their representatives
- fully costed prior to signing.

Note: in exceptional circumstances, such as urgent works to ensure stability of the structure, it is possible for the builder to proceed before the variation documentation is finalised.

Figure 6.3 provides an example of a typical variation document from the Queensland Building and

Variation Document No. ______________

FORM 5
VARIATION DOCUMENT

(**Condition 21** of the General Conditions of **QBCC New Home Construction Contract**)

NOTE TO CONTRACTOR: This form, which may be copied for multiple use, must be presented to and signed by the Owner **before** you start any work described in the variation.

To: (Owner/s) ______________________________
(insert name and postal address of Owner/s)

From: (Contractor) ______________________________

Regarding construction at: ______________________________
(insert Site address)

This document is for a variation:
(tick whichever is applicable)

- required by law
- for extra excavation and foundations
- requested by the Owner/Owner's Representative
- requested by the Contractor/Contractor's Representative for the following reasons: ______________________________

____
(insert reasons)

The change to the Works is as follows: ______________________________

____
(insert description of the variation including any change to the work or materials required by reason of the variation)

The Contractor's/Contractor's Representative's reasonable estimate of the period of any delay in the Date for Practical Completion that will result from the variation is: ________ business days.

NOTE TO CONTRACTOR/CONTRACTOR'S REPRESENTATIVE

If the variation causes you actual delay you must also submit a QBCC Form 2 - *Extension of Time Claim and Owner's Response to Claim* within 10 business days of the earlier of when you became aware of the cause and extent of the delay, or when you reasonably ought to have become aware of the cause and extent of the delay **(see Condition 23)**.

The variation will change the price payable by the Owner as follows: *(tick whichever is applicable)*

- **increase** the price by: $ __________ *(incl. GST)*
- **no change** to price
- **decrease** the price by: $ __________ *(incl. GST)*
- **increase/decrease** *(delete whichever is not applicable)* the price by an amount that will be calculated as follows:

____
(state how the increase/decrease will be calculated)

The increase or decrease (if any change) in the Contract Price payable by the Owner as a result of the variation will be taken into account in the Contractor's progress claim for the following Stage described in Schedule Item 8A or 8B:

____

____
(insert description of Stage from Schedule Item 8A or 8B)

____
(Owner/Owner's Representative to initial here)

NOTE: The Contractor cannot require payment for an increase in the Contract Price due to a variation before the variation work has been completed.

SIGNED: ______________________________
(Owner/Owner's Representative to sign)

DATED: ____ /____ /______
(day) (month) (year)

SIGNED: ______________________________
(Contractor/Contractor's Representative to sign)

DATED: ____ /____ /______
(day) (month) (year)

When form completed, Contractor to retain the original and give 2 legible copies to Owner.

QBCC NEW HOME CONSTRUCTION CONTRACT_JULY 2018

FIGURE 6.3 Example variation form

Construction Commission. Other examples may include a list of past approved variations as well as their cumulative costing.

Most contracts will give a clear indication of who may request a variation, and a procedure for their authorisation or formal acceptance into the contracted works. A variation, generally, can be requested by:

- the owner/client/principal
- the contactor – i.e. you, or you as the contractor's representative, or the contractor themselves. This is limited to the reason for the variation not being a fault of you or your company. In such cases no costs may be applied to the principal.

Variations may also come about due to:

- regulatory changes, council or other statutory authority requirements that could not be foreseen at the time the contract was agreed to
- other matters that could not have been foreseen by a competent contractor prior to the contract being agreed upon and are necessary for the completion of the work.

Procedure (corrective action)

As mentioned above, generally before you commence work associated with a variation you must provide written evidence of the need and cost. This must be supplied to the client or principal and their approval gained in writing in return. The procedure is frequently clearly spelled out in the contract documentation. The variation document shown in Figure 6.3 is often referred to as a 'notice of variation'.

Once this notice of variation is returned noting the approval by the client of all its elements, you may proceed with the variation as described.

Note: You should take careful consideration of the extra time that a variation may impose on the project. This should be factored into the variation and the overall timeline of the project, and the client should be duly informed of this change.

Potential work health and safety (WHS) or property damage issues

There are two instances where written consent is not required prior to commencing a variation. This when there is a clear risk to:

- the health or safety of any person, be they workers or the general public
- property, be it the client's or a neighbouring structure.

Generally, there will be a clause within the contract that makes this stipulation a legal part of the agreement. In such cases you should commence immediately with the works to ensure safety on the site has been re-established. Only after this will you document the actions taken and write up a variation. It is best practice to inform the principal or client at the earliest available moment and, if possible, have them attend the site as the work is undertaken or immediately thereafter. The variation may then be co-written and disputes generally avoided.

Adjustment of the contract price

While seeking agreement on a variation and its price, you will also be addressing the implications of this task to the whole project. Previously, the project duration was mentioned; i.e. it may now take longer. However, the overall contract price now also shifts. It will shift due to the following reasons:

- the cost of the materials and labour to carry out the variation including any overheads, profit margins and GST
- the changes to any insurance premiums needing to be paid due to the change in overall contract price.

As per the second point, depending upon the nature of the variation(s) involved, changes in the insurances covering the works as a whole may be incurred. This may happen whether the variation increases or decreases the value of the contract overall. Changes to policies will require a document signed by all parties detailing the changes and the cost. Usually you will also need to include a copy of the contract and the variation forms. In most cases where the works increase the value of the project, an additional premium will need to be paid to the insurer. Lowering the project price will generally lower the existing premiums payable.

It is very important to keep track of the total value of the variations as they come about and keep the insurer informed as penalties can apply if the correct premiums are not paid.

Notice of this change is provided in the notice of variation document. In some instances, this will be a lowering of the contract price; however, in most instances it is going to increase it. In either case the price variation must be signed and dated by both parties.

Any change to the total amount that is a result of changes to insurance premiums must be disclosed as a separate item and not hidden within that total price.

The final adjustments to the price, once agreed upon, may be taken up in the next progress payment or paid as agreed by both parties in a separate payment schedule.

Back charges: acting within policy

Back charge is an industry term that, in itself, has no legal basis; however, it is a term used industry wide to express what is legally referred to as the 'right to set off'. Effectively, what both mean is that one party, usually the client or principal, claims a reduction in payment due to money that they consider you owe them.

Back charges claimed by the client

This money you are considered to owe could be for some damage that your company has done to the client's property, or a defect that the client has identified in your work. For example: one of your subcontractors continuously parked their cars/trucks all over a lawn area that will have to be relevelled and resown. Your client feels that you should pay for this. Likewise, you feel the subcontractor should pay for it.

When you put in the next progress payment claim, or more usually the final account, the client may attempt to 'charge you back' that which they feel you owe them for the re-establishment of their lawn. In so doing they might submit to you a counter invoice or simply notice with their reduced payment as to why they are 'back charging' you and by how much.

This seems reasonable on the basis that you might wish to do the same to the subcontractor. The reality, legally, is somewhat different. Back charging can only occur if the contract expressly states that it can. This statement will be found in either the payment clause or as a separate clause within the document. It may be phrased as either 'Right to set off' or quite possibly 'Back Charges'. If this clause or expression is in the document and it expressly states that such is available to the client, then your company is open to such claims.

If, however, no such clause exists, then the client has no automatic right to back charge. To get compensation for the damaged lawn they would have to either appeal to your, or the subcontractor's, sense of moral responsibility, or take their claim through some higher agency such as the state's building commission, consumer authority (NSW Fair Trading, for example) or, failing that, through the courts.

Yet you should still carefully consider back charges should they occur even when no automatic right exists within the contract. This is because to refute the claim is to formally begin a dispute, and for minor amounts it is your company's reputation that you are considering – and hence future earnings – rather than the specific amount being claimed on any one occasion.

Back charges claimed by you against subcontractors

As with the previous situation, you can only legally make back charge claims against a subcontractor if the contract you hold with them expressly states that such may occur. If, as is occasionally the case, you have accepted a quote and there is no signed contract as such, it is highly unlikely that the required clause will be in that document. You therefore have no right to back charge for reparations to the client's lawn. Technically, legally, your company will have to wear the cost of those works.

Back charges and payment schedules

Whether you or your client have the right by contract clause or not, dispute of payment should be done by the submission of a payment schedule. A payment schedule will state how much your client is willing to pay, or you are willing to pay your subcontractor, and the reason for the difference; i.e. the claimed back charge. Under the *Security of Payment Act* in Victoria, for example, the respondent (the one who is being asked to pay) has 10 days within which to submit this schedule. After that period of time the client (or you in the case of dealings with a subcontractor) must pay the full amount. Each state and territory has their own variant of the *Security of Payment Act* so you must check this clause for the jurisdiction in which your project is conducted.

Invoices and payment of materials

A construction project is generally frequented by deliveries of various forms: concrete, steel reinforcement, timber, precast components or simply a box of screws may all appear at the gate. In each case, this material will have been ordered from, and supplied by, some party that expects payment; preferably as soon as possible. For the sake of the project's timeline, it is important that this transaction is completed promptly; however, there are also risks if such payments are made in too much haste. If you are too quick you may find yourself trying to get monies back for materials that never arrived or may never have been ordered in the first place.

Making the order

The best practice with materials is to establish a purchase order form system, be it hard copy or electronic. This becomes the first record of what it is that you are ordering. It will allow you to record the quantity of material, the specifics of the material, the date that the order was put through, and to whom it was put through. Depending upon the sophistication of the system established, it may also include the expected delivery date and quite possibly tracking.

A purchase order is effectively an offer to buy a specific quantity of particular materials at a stated price. Once accepted by the supplier, it becomes a contractual agreement between the parties involved. The order is generally developed in triplicate so that one is retained on site, one goes to the supplier and one goes into the company accounts. While electronic systems are generally centralised, allowing access from multiple points, hard copies are still preferred as a record of the transaction. This is the first document in the record-keeping system

Taking delivery

In taking delivery of the material, that which has been supplied should be checked against the original order form and the delivery docket. The delivery docket, the order form and the materials delivered should all match. The delivery docket then becomes the second document in the paper trail supporting the purchase and receipt of the materials on the site. One copy of this document will be signed and retained on site, another is signed and returned to the supplier.

Payment

The next document in the trail is the invoice. This will generally be sent directly to the accounts department, though on occasions it may come with the materials themselves. The invoice holds a range of information including the supplier's details, the amount owed and the period of time within which it should be paid. Payment is made against this invoice once proof is obtained from you as the site supervisor that the materials have been delivered and that the delivery matched that which was requested on the order form.

Payment should then be executed as soon as possible, at the very least within the terms stated on the invoice. Disputes can quickly arise when payment is delayed, which in turn can lead to the non-supply of materials critical to the project maintaining its schedule. Refusing to pay promptly also leads to your company gaining a poor reputation, making it harder to obtain materials from other suppliers in the region. From a legal perspective it is also in breach of the *Security of Payments Act* described earlier.

The final document is the receipt for payment which will be sent out by the supplier and will have sufficient information to associate it with the invoice you have paid against. This will usually go directly to the accounts department. Where an online record-keeping system is used it may be uploaded as a scanned image or simply noted that it has been received and the hard copy filed appropriately against the order form and associated documents.

Insurance claims

All registered building practitioners require a range of insurances covering various aspects of their operations. The exact insurances, and when they are required, differ marginally between Australia's states and territories, lending bodies (banks and the like), and the type of work being undertaken. Despite this, you or the company you represent, will most likely hold:

- domestic building insurance (home warranty) – domestic construction only
- public liability insurance (includes protection works)
- construction works or contract works insurance
- workers compensation insurance
- general property and tools of trade insurance
- commercial vehicle insurance
- commercial structural defects insurance (currently Victoria & Tasmania only).

Each of these has a defined range of cover which in some instances are fairly obvious from their names alone, such as workers compensation and commercial vehicle insurances. Others are less clear: public liability insurance, for example, covers injury or damage to 'third' parties – people or property other than those associated with your company or client. Domestic building (home warranty) insurance only covers the homeowner if the builder disappears, dies, becomes insolvent, or otherwise cannot be made to finish a home once started, or rectify problems with the home after its completed. All these are important; however, the insurance of greatest interest to the construction process, and so to you as the builder or their representative, is construction works insurance.

Construction works insurance

Construction works insurance, sometimes known as contract works insurance, protects you against damage to the works under construction. In some instances, it is coupled with public liability insurance and so covers damage or injury to third parties and/or their property as well.

This insurance will generally cover the following forms of material or property damage:

- fire
- theft
- water damage
- flood
- malicious damage (vandalism or other deliberate damage)
- storms and cyclones
- landslip and earthquake.

Public liability cover for construction work generally needs to be very high so most insurances will have allowances in the tens of millions of dollars.

Knowing precisely what is covered by an insurance policy is critical. You may think you have cover because it says fire, only to find that it covers the new works on an extension, but not the existing premises. Or it may cover your negligence in causing the fire, but not a subcontractor's.

Things you should check are included in a policy of this type are:

- materials storage and transport – off or on site
- professional fees
- subcontractor negligence
- mitigation – expenses incurred in preventing imminent damage or mitigating further damage
- site restitution – demolition and disposal of damaged works and restoring access including dewatering after floods
- business premises – your main office building and/or contents
- tools of trade
- cover over the maintenance period.

You should also be aware of when a policy starts and ends. In some instances, a policy may be for a single nominated project, ending at 'hand over', and so not include the maintenance period. In other instances, it may be for any works undertaken by your company on any site at any time of the year; i.e. an 'annual turnover' policy. This latter type is highly beneficial to volume builders or larger commercial builders who may have multiple active sites and/or display homes in operation. It will also cover any damages incurred while out quoting a job.

Making a claim

In making a claim your first step is checking that what you are claiming is actually covered by the insurance policy you are claiming against. If you are unsure, the insurance broker you purchased the insurance through is generally your best guide.

Having satisfied yourself, or it's the broker's opinion, that the event is covered, you need to determine what type of event it is and what is expected from your insurers for such an event. In the first instance, unless there are ongoing risks to people or property, you should not repair or replace anything until the insurer gives you authority to do so. This generally means after they have inspected the site. If there are risks that you have to mitigate, then do so and make a report explaining this action. Photographs can help in such cases.

When it is clear, or you suspect, that the damage was malicious or that a theft has occurred, then you must notify the police and obtain a police report and/or the number of that report, which will be forwarded to the insurer.

Although you may feel morally obliged to 'own up' to a particular action by you or your company with regards to losses to a third party, your insurer requires that you make no admissions unless as part of a police investigation.

The claim form

With most contemporary insurance firms, claim forms are easily accessed online from their websites. In some instances, you may wish to contact the insurer to determine the correct form for a particular claim, or how to fill out the form in its specifics.

Claim forms tend to be large and at times confusing or even conflicting in the information that is required to be input. If it is an online form you may be able to fill it out electronically as a downloadable PDF. You will likely still need to print the form for your records and/or to gain some signatures. The general information section is usually straightforward, seeking a description and location of the project, date and time of the incident, police report number and name of officers assisting if the police were involved.

The next sections of the claim form are about identifying the type of loss, describing the incident if possible, what exactly has been lost or damaged and who owned it. You may also be asked who you believe to be responsible. At some point you will need to offer an estimated cost of the reparation works. In making this estimate, be sure to include any demolition and disposal costs.

One of the final sections of these forms will ask about your insurance claim history, not just for this project, nor with this insurer, but over a stated period – generally the last three to five years. Any personal criminal history may also need to be declared.

Duty of disclosure

In all these declarations you must be open, honest and operate in good faith. Doing anything other will lead to either a rejection of the claim or the claim being delayed, which may cost you time and money in the end. In the worst case, you may be charged with insurance fraud if what you are claiming is untrue.

The duty of disclosure also applies when any insurance is initially sought. In making a claim, it may be rejected if the insurer feels that you or your company were not completely open in providing key information that may have caused the insurer not to accept the proposal initially. In all your dealings with the insurer during the claim process you must remain open and cooperative, providing any information that may influence the outcome, even when you may feel that keeping 'that bit' to yourself is in your best interests – because frequently it's not …

Appeals

It may come as a surprise to many, but insurance claims are granted more often than not. So, although the committee governing insurances in Australia admits that denied claim reporting is patchy, and most likely under-reported, the approximate figure of denied claims remains less than 4%. Despite this, it remains possible that a claim you submit may in fact be turned down. In such cases there are avenues of appeal that you can try. These are best handled through your insurance broker who can advise, firstly, if you were truly covered for that specific event, and then on how to pursue the matter.

Generally, a claim, if rejected, will have failed due to one of two factors. The first and most common is that it is not within the scope, or an 'operative clause', of the insurance policy. That, or there is an exclusion that you were unaware of. The second is a point of fact; i.e. some element of the factual account or site report suggests to the insurer that this is not an event covered by the policy. There are four steps you can take in either issue.

- advocacy through your broker
- dispute resolution – internal to the insurer
- dispute resolution – external to the insurer
- the courts.

In the first instance your broker is looking to confirm or challenge the exclusion clause or otherwise show that the claim is covered by the insurance policy. Where a fact or facts are at issue, the broker can help you to find competing facts that again demonstrate that the policy should cover the event. That is, your broker 'advocates' on your behalf.

Dispute resolution is a right of anyone. With regards to insurance companies, they are bound by regulation to conduct an internal review when requested by a disputing client. This is designed to test the decision they have made. If this still falls against you, you may call for an external review by a third party: for construction firms this remains the jurisdiction of the Financial Ombudsman Service Australia (the Australian Financial Complaints Authority does not include construction insurance complaints within its scope).

The final option is the courts. This is long, expensive and with little guarantee of success. It is to be avoided if at all possible.

Administration: regulatory and organisational requirements

The administration of a site is, as has already been seen, a multifaceted task. It is to the site supervisor that the responsibility of all this administration falls. Failure to successfully administer any one element of the task can lead to lost time, money, or in the worst of incidences, life. In addition to the activities previously discussed, the various key administrative tasks may be clustered under the following headings:

- compliance with federal and state regulations
- the building approval process
- contracts (covered earlier in this chapter)
- documenting, interpreting and communicating on-site consultancy and specialist reports
- updating and interpretation of plans and specifications
- planning and scheduling
- implementing and monitoring work health and safety
- implementing environmental controls
- mentoring and training
- financial reporting including wages and taxation.

Many of these areas have been covered independently in other chapters of this book. However, for the sake of clarifying the supervisor's role, it is worth revisiting each here, albeit briefly.

Regulatory compliance

Knowing which regulations you are supposed to comply with is always going to be your first step. This is highly dependent upon the nature of the work you are undertaking, and in some instances the client you are working for.

Domestic construction

In the domestic sector you will need to frame your administration and construction around the relevant state or territory building or home building Act and regulations. These Acts govern all elements of the domestic construction process including licensing and dispute resolution, while directing you to the relevant construction codes which is, of course, the National Construction Code (NCC). Through the NCC you are guided to various Australian standards.

Visit your state or territory's building regulator's website to access both the Act and the regulations, or you may find them at www.austlii.edu.au. It is important to note that if there is a difference between the regulations and the NCC, the regulations must be followed, not the NCC. This is the principle of legal hierarchy.

In addition to the construction regulations, your project must comply with the planning and zoning regulations which is part of the approval process discussed briefly below.

Commercial construction

In the commercial sector you are guided by the NCC and through this to a raft of Australian standards. In addition, when contracting with Commonwealth government agencies, you must comply with the Code for Tendering and Performance of Building Work 2016 – 'the Code'. This code requires particular actions to be carried out and standards of operation followed, which can include subcontractor compliance, on-site drug and alcohol testing, and freedom of association, among many others. It is important that you understand your obligations under this Code as failure to do so can lead to suspension from tendering for 12 months or longer.

In addition to the above, you again must go to your state or territory regulator to ensure you and your subcontractors hold the appropriate licences and qualifications for the works being undertaken. In some instances, this may require a diploma or bachelor degree-level qualification and/or a minimum number of years of experience. Insurance bodies can also drive the requirement for your workers and subcontractors to hold certain qualifications and/or proof of training. This may be said equally of your state or territory WHS authority and, to some degree, unions.

Building approvals

As a supervisor, you may well be brought into the project early and so be a part of the building approval process. Alternatively, you may enter the role only after this process is complete. It is still wise to have some knowledge of planning and zoning regulations, as well as overlays. For multi-storey construction this understanding can help in planning your site access paths and material transportation routes through urban and suburban areas.

Even after the approval process is complete the onus remains on the site supervisor to maintain accurate records of all required approvals including the planning (where required) and building permits. The approved, stamped and authorised plan set should also be available as hard copy. When changes occur to the plans you should be aware if these changes require further inspection by the relevant building surveyor. This applies as well when building information management (BIM) systems are being used. As the supervisor, you will frequently be involved in meetings around proposed changes, BIM-based or otherwise, and their influence upon site work flows particularly. Having an understanding of the approval process and requirements can be of benefit in these instances as well.

Planning approval process

The planning application can be time consuming and at times complex. Planning and environmental Acts, zoning, overlays and municipal strategic statements all interplay to determine whether or not a building is considered acceptable or not. Into this may be mixed public sentiment and cultural values. The decision, particularly with regards to commercial projects, may be made at local and/or state levels. Most of the key factors that may influence an outcome can be bracketed into one of two groups: economic drivers and social drivers.

- **Economic drivers**: population growth, transportation needs, infrastructure, industry or business development, agriculture, employment opportunities, tourism potential and the like.
- **Social drivers**: social and community housing, affordable housing, health planning, heritage and culture, housing density, liquor and gambling regulations, sport and recreation, perception of 'place'.

Applications generally follow fairly similar paths in most council regions throughout Australia, though some Indigenous areas will require additional actions to be completed.

- Design is conceptualised through basic elevations, plans and site drawings.
- Application is submitted including the above.
- Application is checked.
- Application is advertised for the required period for that jurisdiction (two weeks or longer). Advertising may be by letter and/or notices on site (see Figure 6.4).
- Objections are considered.
- Decision made, which may be yes, no, or yes with conditions attached.
- Review of decision if requested by court-like authority, such as Victoria's VCAT system.

As the site supervisor you may be involved in a range of discussions leading to planning approval, you may be the one who puts up the application for planning approval notice, or you may only be involved after the fact. You will be involved in any changes or challenges to those approvals so it pays to have an appreciation of how the system works.

Contract administration

The administration of contracts has been effectively covered previously in this chapter. The only addition

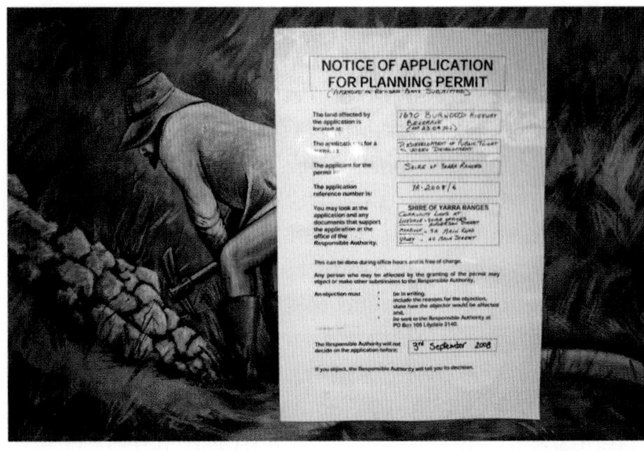

Source: Newspix/Steve Tanner

FIGURE 6.4 Notice of application for planning permit

here is that aside from operation within the contract, you must make sure you have a copy of that contract, and any relevant documents, immediately to hand on site as well as at the main office. You must also be aware of any changes to the contract. This can sometimes be complex, as the contract will itself state the how, why and by whom the contract may be changed. As has been discussed, variations are effectively changes to the contract which remain within the existing contract scope – extending or reducing it as may be. That is, variations, and how they are to be handled, are framed within the clauses of the contract.

Complete changes; i.e. the drafting of a new contract, are rare but do occur. Likewise, additional contracts can sometimes be required for works that are associated with, but not part of, the original project.

Your role is to understand the implications of each change, maintain copies of the contracts and communicate the implications to workers, subcontractors and material suppliers. You must also communicate the implications, particularly with respect to project timelines, to management and clients. All of these changes to projected timelines should then be reflected in your Gantt chart and the critical path re-analysed. In addition, you will need to review the specifications documents to see what the impact will be regarding nominated materials and any changes to specific prime cost items.

Engaging with consultants and specialists

On occasions you, or your client, will need to refer to consultants and/or specialists with regards to certain aspects of the project. This could include unexpected findings during the excavation, planning for excavation, or particularly complex engineering issues around post-tensioning of a slab. When conducting these meetings, you must make sure you follow all WHS procedures, including site inductions. You must also make sure that the meetings are well documented and reports not only filed away, but communicated to all relevant parties to whom the findings may be of importance. In many instances that means everyone.

Maintaining plans and specifications

This runs in parallel with maintenance of the contract documentation as your contract is effectively a reflection of the plans and specifications. Some commercial projects particularly can have multiple changes to the plans as the work progresses; this means you must have a system in place to ensure that you are working off the most current version. With contemporary BIM systems this is less problematic as the changes are effectively done live (as in Figure 6.5).

The complexity for you with BIM is not in the reading of new data provided to you, but in providing data back to the draftspeople and/or architect. This is because a fully-fledged BIM system is not just about plans for the build, it is also a record of 'as built'. This means that you must provide evidence back to the drafting team on any subtle changes that may have taken place in the heat of practice, so they can draw what was actually built. Ultimately, the client should be in receipt of a set of 'as built' plans that may then be used for maintenance purposes, emergency situations and any future adaption of the structure.

Implementing and monitoring WHS

You or your company will have a WHS policy in place that reflects the position they take with regards to this regulatory requirement under the relevant state or territory Act and associated codes of practice and Australian standards. This policy should be among your records on site along with the documents that this policy will require you to have in place.

The basics of the policy will be the aim for a hazard and risk-free workplace, safe work practices and systems, and regular reviews of those practices. In addition, depending upon the number of workers on the site, you may be required to have a WHS committee, or at the very least safety officers with nominated duties. Technically, for example, you should have a nominated laser safety officer whenever Class 2 lasers are in use.

Source: Courtesy of the Branch Group and G.J. Hopkins

FIGURE 6.5 BIM meeting on site

If committees are required, WHS committee training must be conducted and records kept of those qualified. Also in your records will be minutes or other records of 'tool box' safety meetings where tasks are discussed and safe work practices developed. Safe work method statements (SWMS) and job safety analysis (JSA) documents must all be held in your records and acted upon as per the state or territory regulations and the relevant WHS authority.

First-aid facilities, first-aid officers, emergency evacuation and emergency access routes should all be planned as part of your site establishment. This should be coupled with appropriate signage and traffic management, both pedestrian and vehicular, strategies. Included in this should be information on how incidents are to be communicated both within the site and to off-site emergency authorities.

Certain activities, such as scaffolding, must be under constant inspection regimes by qualified scaffolders where required. This is a high-risk area and it is not unknown for subcontractors (and your own workers) to remove components that are interrupting their work, doing so without authority or the requisite knowledge of the impact this may have on structural integrity.

However, WHS does not stop at the practical aspects of your project. You must also be aware of your duties regarding harassment, discrimination and bullying in the workplace and act to prevent it. Worksite drug and alcohol testing may be required, again depending upon the client. In addition, personal protective equipment (PPE) must be supplied by you, or the company you represent, for any workplace related risks that have been identified; this will include simple things like sunscreen.

A final part of any WHS policy is what will happen should the policy, or any part of it, be breached. Some form of disciplinary or performance management procedure should be included.

Environmental responsibilities

This has become a major factor in both the design and the construction of all buildings, be they domestic homes or major commercial projects. The legislation governing your actions is multi-layered and at times complex. The first level of legislation is the Commonwealth *Environmental Protection and Biodiversity Conservation Act 1999*. This Act works with state and territory laws and regulations without limiting those laws except in certain instances where the intent of the federal Act is superior to that of the state or territory.

State and territory Acts are currently in flux in some instances. Victoria, for example, is currently under an '*Amendment Act 2018*': the new legislation intended to take effect in July 2020. The purpose of these Acts is to provide preventative strategies to ensure our environment is not polluted or otherwise degraded. The new Victorian Act provides for tiered licensing, greater investigative and enforcement powers, along with increased penalties and sanctions.

Frequently, as part of the planning application process, an environmental audit will need to be undertaken on the proposed site, and an environmental risk management plan developed for the project. You should maintain a copy of these documents and match your actions to those prescribed or outlined in the plan. Remember that your environmental responsibilities do not stop at the gate. Wash-down pits are installed to prevent the carrying of mud and silt into the surrounding streets and therefore into the drains when it rains. The same may be said for the prevention of runoff from the site into drains, creeks and rivers (see Figure 6.6).

The onus is very much on you as the supervisor to ensure not only are your workers doing the right thing with waste disposal, air pollution, dust limitation, runoff and the like; but also that your subcontractors are doing so too. Where appropriate you may want to consider choosing contractors on the basis of their environmental policies and reputation. In such instances their reputation feeds into yours which can benefit your interactions with local councils and residents.

Financial reporting

It is a common misconception with those entering the supervisor's role that someone else looks after all the financial 'stuff'. Unfortunately, this is not the case and you will find that you must report not only to the head office – assuming you are not working for yourself – but also the Australian Taxation Office (ATO). If you are running your own business then this is obviously an area critical to your ongoing success as a builder.

General financial reporting

What you need to report to the government or to the ATO depends upon what type of 'entity' your company is. Are you a company limited by shares, by guarantee, or unlimited? Your accountant is the first person to whom you should turn for advice on your record-keeping requirements in either case. The information you capture and retain will be forwarded to your accountant who will then act in accordance with the Australian Securities and Investments Commission (ASIC).

Source: Aussie Environmental/Steve Berry

FIGURE 6.6 Silt fencing near waterways

The sort of information you need to retain includes:

- payments to employees
- payments to contractors
- business ABNs
- GST amounts
- withheld tax when ABNs are not provided
- superannuation payments
- asset purchases and disposal
- use of private vehicles for work purposes
- training and education.

Always check that the ABN on an invoice matches that within your records for a contractor in case they have changed at some point during your ongoing transactions with them.

Final accounts

To facilitate the above, you will need to organise carefully structured final accounts for a project. This is the last invoice, bill or account you send to a principal or client for payment. You will prepare and submit this at the time of practical completion. This document shows the entire financial history of the project from the original contacted sum to the final amount payable less retention money.

It is important that you prepare for this final account as you work through the project. This is the last claim you can make on the client, aside from the call for the return of the retention money at the end of the defects liability period. Everything you are owed, and any credits owed to the client must be tallied on this document, and must be done so without error. In completing the final account, you will generally need to acknowledge the following items:

- the contract sum (original price at contract sign-off)
- variations, be they positive sums or negative
- the original deposit
- prime cost consolidation; the balance between actual expenditure and the contracted allowance is determined across the whole of project and either claimed or credited as the case may be
- preliminary sum consolidation; as with the PC items above, the difference between the allowance and the actuals is stated as either a claim upon, or a credit to, the client
- all progress claims need be acknowledged
- the retention money will need to be listed as an outstanding amount – usually this is an adjustment down from 5% held until this point, to 2.5% of the final project total
- extended site facilities – when a project is extended in time through agreed variations you may claim for the extended hire of facilities such as shedding, toilets, fencing and the like; included in this may be the wages of supervisors and forepersons
- the balance payable and timeline for payment. This is the total owed by the client at handover, less the 2.5% retention money.

Taxation

All of the above information will help shape what is known as a taxable payments annual report (TPAR) which, given that you will have been making payments to contractors for construction services, is mandatory for businesses in this sector. At the time of writing, the TPAR must be submitted annually on 28 August. Employee payments, however, are not included in the TPAR. When reporting these payments, it is important to include both labour and materials as one single combined amount. The purpose of the reporting is so that the ATO can check that all of a contractor's earnings have been represented in their tax returns.

What needs to be reported in a TPAR is effectively a repeat of the list previously offered on what information needs to be retained. In particular, the GST amounts paid and or retained. What you *don't* need to report are payments for goods or materials only.

LEARNING TASK 6.1

PAYMENTS, CLAIMS AND CONTRACT ADMINISTRATION

You are about to request a progress claim from your client. Using the information in AS 4000 handbook HB140, produce a flowchart that shows the systematic method that should be used to properly coordinate and complete a progress payment request from the client.

Communications: the supervisor's role

As seen in the previous section, contract administration hinges upon sound communication between all the stakeholders to a construction project. Failure here is most likely to lead to the project failing as a whole: or at least bring about costly disputes which may be challenged without a suitable defence from yourself. The basis of good communication is openness and clarity. The support for good communications is sound record keeping. With regards to your role as a supervisor, this means sound filing practices, detailed site reports, a clear and visible project plan calendar and, most importantly, a very accurate and concise site diary.

The site diary

A **site diary** is a daily record of all activities that take place in relation to the project. Although it is called a 'site' diary, it will also include anything of relevance that is off site as well. The site diary is an important record that is frequently used as evidence in disputes and should be supplemented with photographic records where possible or relevant.

The sort of information held in a diary includes:

- targeted activities for the day/week
- workers on site:
  - number of company employees; include absentees
  - activities workers/work teams are doing including location
  - subcontractors – work being undertaken and location of activities
- non-working personnel on site – visitors, clients, architects, engineers, etc.
- meetings and minutes/notes of meetings
- instructions given or received
- telephone conversations and times
- material usage and prospective shortfalls (re-ordering needs)
- plant and equipment usage, location, hire, despatch, maintenance requests
- supplier interactions and orders
- deliveries
- weather conditions on site (twice daily)
- WHS incidents including near misses
- union engagement, matters raised.

Diary format: computer-based, online or hard-copy diaries

Interestingly, site diaries are not actually a legal requirement of the construction process. They can be used in a court as evidence, but neither they, nor the format or content within them, are mandatory. This being the case, site diaries come in a range of formats from basic paper-based booklets to extremely powerful 'cloud'-based tools. What is in them, and how they may streamline your project, tends to be a case of individual preference coupled with the scope of the works undertaken. Site diaries are now readily available as pre-printed hard copies with carefully considered layouts covering all the aspects listed above. However, there are also stand-alone computer-based programs and network-based programs (i.e. ones that allow multi-computer access from within your company's network). And then there are online or 'cloud'-based systems which hold all the records in independent secure databases. Understanding the pros and cons of each of these is important in making a decision on which to use.

The 'why' of having a site diary, however, is never in dispute. Such being the case, many clients or the larger main contractors, by whom the supervisor is employed, will require a site diary and dictate the format and content that is required. See Figure 6.7 and Figure 6.8 (on p. 152).

Hard-copy diaries

The advantages of these are fairly straightforward. They never run out of power and they are easy to purchase and start using. They can be extended simply by buying another one. And you just write

FIGURE 6.7 Site diaries

in them, and clients can sign off site instructions if that is how you have set it up, or the diary developer has arranged them. The downside is that they can become scrappy with messy writing, and installing photographs in them is problematic at best; though you can electronically file the photos and other documents and indicate within the diary where they are to be found.

Computer-based diaries

These are now fairly common for the reason that the writing is clear and easily copied. Copy and paste capacity allows for multiple copies of instructions to be distributed quickly by email or other electronic means. They can be readily searched by word or document title and may be linked to any number of other document control programs. Likewise, photographs, invoices, receipts, delivery dockets, site instructions and the like can all be directly linked to the relevant parts of the diary. Further, the planner or Gantt chart plan can also be directly linked and any implications within the diary linked to the changes in that document.

Yet another advantage is that the program can be installed on a central company computer or server, and then accessed from multiple secure points on the network with password-protected limitations on who can alter what. This access can include phones and small notepads.

The downside of these diaries is that they are open to hacking from malicious external parties who can either delete, damage, steal or 'kidnap' the information and hold it to ransom. Virus attacks and simple computer failure can all lead to issues if content is not securely backed up. And they rely on a secure source of electrical power. There can be significant down time when using these programs and training is critical to their successful implementation.

'Cloud'-based diaries

These have also started to become common. They are basically the same as the computer-based diaries except they are arguably more secure and well backed by security and backup systems. They must be paid for on an annual or monthly basis to maintain access and this can be an issue at times. Generally, however, these are cutting-edge programs that are easily expanded and improved live, as against having to change whole programs for the stand-alone systems (see Figure 6.9).

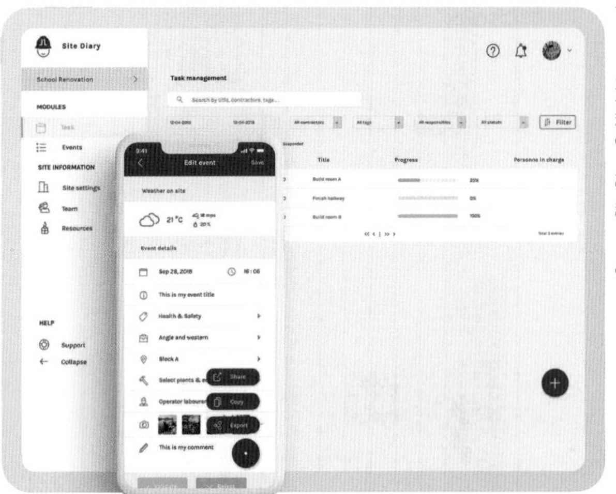

FIGURE 6.9 Electronic and 'cloud'-based diaries

Again, everything can be linked and searched, and the diary becomes the focus point of just about all the project management requirements discussed throughout this chapter.

File notes

As outlined in the previous section about site diaries, there will be a range of interactions with personnel, clients and the like that must be recorded in the diary. These records may sometimes need to be extended into other files and other documents; documents needing to be read and responded to by others. Two common examples are site instructions and requests for information.

Site instructions

You will note in the list of what should be in a site diary a number of records about conversations: on site, by telephone and/or in person, with visitors, the client or those involved in designing or engineering. These interactions can lead to instructions being given, to you or by you, that alter the project the given plans or specifications. All these changes need to be formalised.

When given an instruction to make a change to that which is in the existing documentation you might at first think that this is a variation. However, not all changes are variations, as not all changes have a cost implication. Changes of colour, texture, or even some materials do not necessarily change the cost involved. In such instances, it's easy to fall into the trap of taking a verbal

Source: Contractors Debt Recovery

TODAY'S DATE: / /20 Circle Monday Tuesday Wednesday Thursday Friday Saturday Sunday

PROJECT DETAILS

Site Address	
Client/Customer	Client Rep today was:

WEATHER

☐ Did not affect rate of work	☐ Slowed the rate of work	☐ Halted work for a while*
☐ Extreme heat	☐ Extreme Cold	☐ Extreme Winds
☐ Heavy Rain	☐ Moderate Rain	☐ Light Rain

* Work stopped between the following times:

EXTERNAL AUTHORITIES: *RTA, COUNCIL, UNION, TELSTRA,*

The following things occurred due to a direction given by the following external authority:

PHOTOS TAKEN *(Description)*

1
2
3
4
5
6
7
8
9
10

DELAYS *Today work was delayed due to:*

1 ☐ Other Trades [specify]
2 ☐ External Authorities 6 ☐ Site Meeting
3 ☐ Industrial Action 8 ☐ No access to crane/elevators
4 ☐ The client 8 ☐ Rework
5 ☐ Our Suppliers 9 ☐ Other:

Details & Times of Delay Events*:

* Is an EOT Notice required for these delays?

STAFF ONSITE TODAY	Hrs
1	
2	
3	
4	
5	
6	
7	
8	
9	
10	

OTHER EVENTS: ☐ Site meeting ☐ Payment Dispute Meeting ☐ Offsite meeting

This diary entry was completed by :

[PRINT] .... Sign ....

CONTRACT WORK DONE TODAY

Description: Refer to Scope of Works and detail what scoped works were done	References:

VARIATIONS & ADDITIONAL WORK DONE TODAY

Description	How requested?	References:
	☐ Written ☐ Oral	
	☐ Written ☐ Oral	
	☐ Written ☐ Oral	
	☐ Written ☐ Oral	
	☐ Written ☐ Oral	
	☐ Written ☐ Oral	
	☐ Written ☐ Oral	
	☐ Written ☐ Oral	

ORAL DIRECTIONS RECEIVED THAT CREATE A VARIATION*

We were directed to...	Direction was given by
1	
2	
3	

* Do you need to issue a Notice that the instruction constitutes a variation? Do you need to issue an EOT Notice?

NEW PLANS/DRAWINGS ISSUED TODAY
[To replace those in the contract documents]

Contract Plan Number/Version	Replacement Plan Number/Version

AGREEMENTS MADE
Today the client's representative........................ agreed to do the following

FIGURE 6.8 Example site diary page from Tenderfield

instruction without sufficient record of the request for that change, should it be disputed in the future.

This is where a document called a '**site instruction**' comes into play (not to be confused with a client's main site instructions sometimes given to the lead contractor covering all of that contractor's actions throughout the project). The site instruction is a brief document written usually by the supervisor (you) outlining the exact nature of the change, dated and then signed by both yourself and those who gave the instruction – usually the client, architect, or occasionally the engineer (see Figure 6.10).

Site instructions can also be issued by you as the lead contractor's representative to subcontractors undertaking work on site. Again, this will be for some change that is not instrumental in changing the cost of the works, but some form that the works will take.

As with any documentation, it is vital that site instructions be recorded and filed in a manner that makes them readily accessible should a dispute occur. This is where the electronic diaries are significantly superior to the hard-copy systems: documents can be electronically generated and so electronically linked to the diary, the date of the original conversations, records of phone conversations and photographs where relevant. They can also tie into BIM systems so that the change made can become part of the final 'as built' plan set.

Requests for information

Some forms of site instruction documents are multi-format, in that through a mere tick in a box they can be a site instruction, or a request for information (RFI). Requests for information arise when the plans or specifications need clarification over some aspect of detail. RFIs can be generated by you to the client, architect, or engineer; or they may come to you from a subcontractor. As with site instructions, these need to be suitably filed and linked to the site diary in a manner that allows for easy access when required.

RFIs are fairly frequent in the early tendering or quoting period of the project. The more that can be clarified prior to construction starting, the fewer delays there are likely to be later. RFIs issued during the construction phase can be onerous, sometimes taking weeks to be responded to; on occasions there may be no initial response, and so must be followed up. All of which leads to uncertainty, down time and thus cost: one researched estimate suggesting as much as US$1080 per RFI issued.

Other files

A supervisor may find themselves issuing a range of instructions that do not conflict with the existing plans or specifications, but are simply in support of those existing plans. Common instructions may include:

- instructions requiring workers to move from one area of the project to another to support activities in that area
- WHS decrees reminding all parties of their responsibilities in this area
- changed timelines on one activity or another
- memos and notes back to head office regarding deliveries or the like
- instructions to suppliers that alters a planned delivery schedule
- instructions to suppliers regarding specific handling requirements of materials.

These instructions all need to be kept track of and preferably linked into the site diary system, again supporting the use of electronic diaries over the traditional hard-copy form. Site reports are also intimately linked to the site diary, or should be, which are discussed in the next section.

Site reports

Site reports are a key level of communication between yourself as the supervisor and company management. Your company will use the content of these when reporting progress of the project as a whole to the client. The clarity, accuracy and detail of these reports is therefore of the utmost importance. There are a range of different types of reports that you may be required to submit dependent upon the scale and type of project you are engaged in. These include:

- site evaluations
- progress reports
- performance appraisals
- recruitment requests
- training outcomes
- job specifications
- job instructions (as against site instructions).

We will look at each of these briefly in turn.

Site evaluation reports

When you first walk onto a site prior to construction beginning, you will often do so with the relevant building surveyor, the client and/or the architect, and quite possibly utility representatives from water, sewage, electricity, gas and telecommunications organisations. The value of doing so is that it allows for a general overview of the implications of any one of these utilities to the project's practicality, buildability and timelines and hence budget.

In making a site evaluation report, you will be drawing input not just from the conversations, but frequently from reports that these organisations or individuals will subsequently provide to you. The relevant building surveyor may also provide a report from which you may draw valuable information, and then later receive instructions from management based upon this input. This may include the need for other reports such as a bushfire analysis, an expanded environmental audit, or updated engineering specifications.

Site evaluations are not exclusive to the pre-construction conditions of the site. These evaluations may be required at various stages of the project. In addition, they frequently include reviews of neighbouring sites for the purposes of assessing the need to protect other structures and property from the proposed works. In some instances, the site evaluation report may lead to a protection notice (also called ***protection work notices***), notices to the owners of those properties that their land will need to be accessed for the purposes of work required to protect their assets from potential harm.

Site evaluation reports should always be coupled with photographs where possible, and in the contemporary climate of digitisation there is seldom any excuse for not having them readily available.

Managers Light Industrial & Commercial

Site Instruction

Site Instructions are issued to clarify contract documents or to give direction on problems resulting from site conditions. These instructions are subject to provisions of the contract documents and unless stated herein and specifically co-authorized by the client, will not affect the contract.

If it is contemplated that costs are associated with the work described herein, YOU ARE REQUIRED TO SUBMIT AN ITEMIZED BREAK DOWN OF COST WITHIN 5 WORKING DAYS OF RECEIVING THIS INSTRUCTION.

The work is not authorized until a purchase order has been issued.

PROJECT: 10 Sacha Street Albury - 20 Units DATE: 20th February 2020

PROJECT#: 20—P4

SITE INSTRUCTION# 3

Revised Footing

The footing has been revised to accommodate identified changes in soil conditions (see Geotechnical Engineering notes GE 516, dated 10 February 2020). Construction drawing page A-32 (Rev B), also the footing specification on A-45, and all details relating to these pages are to be disregarded. Refer to Rolland Engineering structural drawings S1, S2, S3 & S4 for the new footing design. These are also reflected in the new construction drawing pages A-32 (Rev C) and A-45 (Rev A)

Issued By: Terry Nollan

Reason for SI: Revised Footing

Distribution List:

Reviewed & Accepted: Name:

Date:

Firm:

42 Coyte Street, Albury NSW 2640
Phone: (02) 34215984 Fax: (02) 34215985
www.tgcconstruction.com
admin@tgc.com

FIGURE 6.10 Example site instruction

Progress reports

Depending upon the requirements of either your client or the policies of your company, you may need to report on the progress of the site every week, or more usually, every month. This report must be compiled by you as the supervisor and accurately depict the project as it stands at that point in time. It should also include projections for the following month.

The content of this report will often be drawn from a range of site progress meetings and reports from employees, subcontractors and suppliers. You should frame the report around the following:

- summarised progress of each main element of the project
- an evaluation of the progress in relation to the planed schedule (Gantt chart)
- summary of delays, their causes, influence upon the schedule and means for rectifying
- weather conditions and influence on progress
- relationship to key performance indicators
- summary of subcontractor performance
- summary of interactions with client, architect and/or engineering teams
- identification and description of issues around quality (positive or negative)
- WHS report identifying any underlying or potential issues
- neighbour and site contextual issues if any have arisen, including access, noise, traffic and the like
- materials access and delivery including off-site manufacture
- a photo summary to support the report's contents
- projections for the following month.

Throughout the month leading to the report you will have had a range of meetings with subcontractors with regards to progress, some of which may have been no more than brief on-site encounters. These encounters are an important way of maintaining your connection with both the project's performance and the subcontractors themselves. They are also your way of gathering evidence and imparting your perspective on performance.

Performance appraisals

Performance appraisals form part of the progress report that you generate each month. The purpose is not to be heavy-handed with either your own personnel or the subcontracting teams you have in place. Rather, it is to determine if the work being produced, or the manner by which it is being produced, fits the standards required by your organisation.

This means you are reporting on issues of quality, quantity, and safe and efficient working practices. The latter being inclusive of your environmental responsibilities, which in turn covers areas such as air, ground, water and noise pollution, as well as sound waste management practices.

These performance appraisals are then fed back to the various parties concerned, which can be a way of rewarding quality workmanship or other sound contracting practices. Alternatively, they can be used as evidence in discussions leading to improved work practices and quality outcomes.

As with any other reporting element on the contemporary worksite, you should back-up the report with photographic evidence where possible. This is generally easily achieved with cameras in virtually all phones and notepad devices. Once more, this data, including the dates and times of these encounters, should all be linked into the site diary.

Recruitment requests

Also linked into the diary will be any requests you put forward for recruitment of personnel, be they specialists, trained, or trainees. These types of requests are usually framed as part of a site report that has identified the need for the skills or labour for specific reasons. Sometimes those reasons are opportunities for community returns by the company; i.e. chances to support trainees or sponsor internships at either vocational education and training (VET), or university levels. On other occasions you will need more labour, or it may be you need some skilled workers such as welders proficient at overhead seam welding that your current workers can't complete.

Coupled with or running alongside this form of request can be the identification of training needs for the existing personnel. This may be on any number of a range of skills or knowledge levels to ensure safe work practices or simply on the installation methods needed for novel materials or components.

Training outcomes

Once training is completed, you will need to report on the outcomes. This may require more than one report; for example, the first report would include the basic quantity data of who did what, when, where, at what cost and over what timeframe. It may also include any certificates or other qualifications gained through the training. A second report may need to be generated, however, to cover what was the 'real' outcomes of the training; i.e. what influence did this training have to the outputs of the teams involved and the project outcomes as a whole?

Job specifications

Reporting on the specifications of the job tends to be coupled with RFIs and prime cost reconciliation, but not always. The target specifications of the project can sometimes be called into question by either yourself, your employees, subcontractors, or suppliers. Sometimes the specifications are considered inadequate based upon experience. At other times they may be considered onerous or excessive based upon the stated or perceived project outcomes. These sorts of reports may also be the outcome of specialist or consultant engagement. Clarification of these matters is important when WHS issues are concerned; or where the client may not realise the cost implications of the existing requirements, when more cost-effective alternatives exist that may even provide a superior outcome.

Job instructions

These differ from site instructions in that they are effectively your orders for the day or week. They offer direction for personnel and subcontractors on what is expected over the next period of work, or how a work activity is to be conducted. They can include directions for appropriate WHS responses, the make-up of a given work team, or the timing of activities based on expected material delivery schedules.

Variations: the communications trail

Variations have been discussed at some length in the preceding parts of this chapter. Further to these discussions, we need to look at the communications trail and how this should best be implemented and recorded.

Recording and communicating requests and authorisations

As outlined previously, before variations can be acted upon, they must be authorised. For this to happen the variation must be accurately described, costed and communicated to both parties. For you, as the supervisor, you need to record a series of events in the site diary about the variation, starting from when it was first identified and by whom, and what documents you have to hand that gives evidence for that request being made. Photographic evidence of the existing situation should be included where possible, and these linked via the diary to the description of a proposed variation and the existing plan set.

When WHS issues preclude the gaining of prior authorisation, you should have this recorded in the site diary very clearly as a significant event that requires immediate action – again, coupled with photographic evidence.

Aside from WHS issues, there may be any number of reasons for a variation request being made, so this too needs to be recorded. In long-running projects, for example, it may be a technological change that gave rise to the request; i.e. a new material or technique has been identified and this is thought to be better, faster, cheaper or more aesthetically pleasing. Alternatively, a statutory change may have come about. Be sure to record these facts, whichever applies.

Your diary should also hold the time when you sent the variation request for authorisation and to whom. Upon receipt of the authority to proceed, record that. Be sure to link all of the above with the various documents through the diary, even if only by noting the file location if using a hard-copy diary.

Depending upon how advanced (electronically) your diary system is, you must ensure that the various parties involved, including those workers and subcontractors affected by the change, know of the change. This will take the form of one of the previously outlined communication systems dealt with earlier; i.e. site instructions, job instructions, and/or variation documents between you and the subcontractors. With all of this you must check the procedures within the contract to be sure your communication trail aligns with and is supported by the contract clauses.

Recording cost and time implications

You must also communicate these changes to your company management and accounts departments as there may be alterations to the total contract price and the project timeline. Your project planner now needs to be altered if the timeline has shifted. This is particularly important when your planner is being used by others, including subcontractors, to determine how and when they will interact with specific elements of the project.

Requesting time extensions

Most contemporary construction contracts have a fixed completion date for the project. Attached to this timeline can be clauses allowing the client or principal to receive compensation for work that goes beyond this date. Should you perceive a situation where the project may run overtime, you need to respond quickly to avoid these penalties.

Reasons for time extensions

As noted in the previous sections, variations frequently bring about a change in the timelines of a project. However, this should be communicated to the client during the initial negotiations over the variation, and agreed to in the signed documents supporting and giving authority to carry out each variation. But there are many other reasons that might cause you to request an extension in time.

Underestimation

In developing the original tender bid or quote you or your company may have underestimated the time required to complete the project. This is common on complex jobs where multiple subcontractors are involved or the design involves processes and/or materials which are uncommon or perhaps completely new to the industry or your firm. In such cases you may have begun to notice that the project is 'slipping away' from your planned schedule despite your best efforts in strategising the workloads, labour and available skills base.

Making a request to extend the project timeline due to what is effectively an error of judgement on your company's part is a test of your negotiation skills. You may find you will initially be opposed by the client, particularly when some level of loss is likely to be incurred by them. You need to be very clear on the reasons for the extension and may have to wear a bit of that loss.

Weather

Inclement weather is a frequent cause of project delays, particularly in the early phases where excavation and footings are challenging to complete in wet conditions. Depending upon when the job was projected to start, an allowance will most likely have been made for this contingency; however, at times this is not enough. In these instances, it is it is perfectly valid to request an extension of time.

Subcontractor performance

Subcontractors can speed up or slow down a project based upon their performance. This performance, or lack of, may be outside of their control in the same way that the weather is outside of yours. At other times you may have had to change subcontractors on the basis of poor performance and this has its own consequences on the schedule. Requests for time extensions on such matters is also a matter of negotiation. However, your main justification is the need to ensure the quality of the job and reduce the amount of time lost – all of which are in the interests of the client.

Making the request

A project that does not come in on time is likely to incur significant penalties that are clearly framed within the contract. You need to balance these penalties with any negotiated losses you might incur in seeking an extension. Before making the request, you should review the contract for those clauses that allow for extensions of time to be made. You should then frame your request around these clauses if they are legitimate. In such instances no penalties should be suffered. AS 4300 – a common commercial contract for Australian projects – offers some 13 clauses allowing requests for extensions of time. Some of these have been discussed already; however, others include:

- industrial conditions
- delays or disruptions caused by the principal or their agents
- quantities of work being greater than that specified in the schedule of rates
- changes in legislation
- directions or delays by councils or statutory authorities
- breaches of the contract by the principal or their agents.

On some occasions there may be no fixed date for completion, but the client may still expect a finished project within a 'reasonable' time. Again, if you perceive the project running in to what is clearly an extended period of time you should open communications with the client to ensure that both parties still find the projected timeframe acceptable.

Any request for an extension of time must be made in writing. The request should outline the following:

- the project details
- the current location of the project on the planning schedule
- the causes of the delay
- the planned actions on how you intend to limit any further delays
- the proposed new final project completion date.

If you are a supervisor employed by a large company, you should check the appropriate channels through which such a request should be made. This is important given that such requests can have a significant effect upon the project's bottom line. You will still, most likely, be the one who must frame the request; however, there may need to be a meeting with management prior to that request being forwarded to the principal.

Dealing with unsatisfactory work

As stated previously, subcontractor performance can have a significant influence upon the project's timeline and quality. Poor performance interrupts workflows and may challenge the contractual agreement you hold with the principal. One element of the progress reports discussed above is a summary of subcontractor performance, and at times this may not be positive. Your role is to ensure that this work is improved and brought up to the expected standards of both your company and the client or principal.

As a supervisor hiring in subcontractors there are ways by which you can reduce the risk of underperformance by systems such as prequalifying. By this process you select your subcontractors not just on price, but also their mental, physical and underpinning financial capacities. In precluding them from the project you should do so by informing them in a manner that helps them to perform better and so possibly be a part of future projects. The NSW government's website 'Procure Point' provides a best-practice guide to prequalification which is worth studying. Though aimed at works over A$1 million, it has sound messages for any level of construction including the domestic sector.

Warning signs

Poor work seldom comes unheralded. There are frequently a range of warning signs that indicate something is not right, or that the work is likely to be substandard. Some indicators can be the subcontractor frequently changing workers, failing to pay suppliers, poor time management or frequent down time due to constantly delayed deliveries of their materials. Being aware of these signs allows you time to make moves to mitigate underperformance before it becomes a significant issue.

Where possible this will be by way of considered, discreet discussions whereby the subcontractor's challenges are addressed and paths for improvement considered amicably. In these discussions you should keep in mind that subcontractors, as a general rule, do not want to underperform. Doing so is against their own best interests given they will have had significant outlays prior to actually starting their part of your project. Setting out a plan of action with the subcontractor is often all that is required. Helping them to understand and plan out what your expectations of them are can give them the tools for success rather than ongoing failure. This is rewarding for both you and them, leading to improved commitment to you and future projects.

Unsatisfactory work notices

Where work is unsatisfactory and no improvement is evident on behalf of the subcontractor then you should inform them in writing at the earliest moment. This is known as an unsatisfactory work notice. You should include in the notice the following as a minimum:

- the details of the work that has been contracted, and the date of the contract
- the contract price, or the amount relevant to the area of work in question
- the faults in the work
- the actions needed to remedy the problem
- the timeline for a written response
- a statement outlining the implications for failing to respond and/or improve performance.

Unsatisfactory work notices are not an end point. They are preferably the start of improved performance on the understanding that ongoing failure to perform will lead to termination of their contract. Termination is not to be considered lightly as while there may be significant cost implications in keeping a subcontractor on, there are always costs in replacing them. This must be balanced against the greater backdrop of the project's overall performance in both quality and time.

When termination is considered the only option left, be sure to check your contract with the subcontractor. The contract may have a termination clause and you must follow that procedure accurately. In addition, it is worth checking the relevant state or territory regulations regarding dismissal of a subcontractor, and the Fair Work Ombudsman, which is a federal authority.

As with all other documentation, you need to link all communication with the site diary so that you have a record of what was sent to whom and when. In the diary you should also note any verbal conversations you may have had with the subcontractors so as to provide a back story leading to unsatisfactory work notices or terminations.

Regulatory and organisational requirements: further commentary

As has been discussed, it is important that the communication trails in the construction sector be clear and well documented. This is essential for a raft of reasons already covered. Mixed into this discussion has been the hint of regulatory and organisational requirements, often around contractual issues. Not a lot more needs to be covered on this point aside from the recommendation that you ensure:

- Sound record-keeping practices are maintained with clear cross-referencing wherever possible. This is most easily achieved through electronic filing applications, but hard-copy files remain a strong fallback that should not be forgotten.
- That the most recent regulations applicable to your state or territory are frequently checked with regards to your obligations both in recording interactions, and in how you engage with issues such as termination, protection works, and education and training requirements.
- That in keeping with the above, you are also following your organisation's required practice.

When using digital formats, you should also check that your organisation, and/or yourself, have adequate backup systems. This is one reason why many major firms have moved to cloud-based applications which generally offer the greatest security.

LEARNING TASK 6.2

COMMUNICATIONS: SUPERVISOR'S ROLE

1. What are the main components that should be contained in a works progress report?
2. What are the main reasons that might lead to requesting an extension of time for your current worksite?
3. You are unhappy with some of the work that has been claimed by a subcontractor. Provide a detailed list of what should be included in any unsatisfactory work notice that you might issue to that subcontractor.
4. What are the main elements to be included in a 'site instruction' document, and what should you do with it other than giving a copy to the relevant workers/contractors?

Quality control

Quality control is a frequently misconceived concept in construction firms, mainly because managers and supervisors use that term only. A far more thoughtful and inclusive term is quality management systems (QMS). When in place in the construction industry it refers to quality planning, quality assurance and quality control. This more fully captures the goals of our industry in that we seek to bring a project in safely, on time, to the best possible quality and at the least cost to the client that the previous constraints allow.

Identifying quality control procedures

When quality management systems are absent, the development or identification of quality control procedures becomes problematic. Problematic because without the planning that QMS provides, it is hard to determine exactly what should be tested, when, how often and by whom. Quality Planning gives that certainty and direction of action that quality control needs.

In the contemporary construction climate, it is most likely your company or organisation has QMS in place. As the supervisor, your task is in part to identify the quality control procedures within that framework, and demonstrate the commitment and effective leadership needed to ensure those procedures are followed through. This means ensuring that personnel and subcontractors engaged in the project:

- are trained in the quality system being used
- understand the importance of a quality system more generally
- can use the system to the benefit of both their contracted works and the project as a whole.

At the heart of the system is the QMS manual. This manual will be explored in greater depth later in this chapter; for the moment what must be understood is that your procedures are to be found within this manual as a series of well-defined actions tied to specific documents.

What you also need to understand is that there may be large tracts of the QMS manual requiring your input into the development of both procedures and documentation. This is because the documents must be framed on the specifics of the project, and the various elements of that project. In addition, the procedures, particularly checklists and inspection test plans, should be developed in unison with those who actually carry out the tasks.

The next sections explore the various means by which we can clearly identify these procedures, and how they are developed.

Inspection checklists

The Western Australian government provides valuable resources for quality control via their Commerce website. The builders' technical quality assurance checklist shows how a simple form can be used to ensure you know what you are looking for, and by which standard you should evaluate performance. As an example, Figure 6.11 shows a portion of this checklist covering brickwork from the perspective of workmanship and technical adherence.

Using this as a template, you could add in your own references, such as a tolerances and standards text, along with any additional commentary you or your team feel relevant for a specific task.

The latter point is of particular importance. When developing breakdowns of tasks and activities for the purposes of quality assurance documentation, these should be created by those who would normally do these activities. This is so that the work can be fully scoped and those elements that need particular attention receive that attention at the most appropriate times.

Inspection and test plans (ITPs)

Inspection and test plans (ITPs) state when, what and how you will inspect works being undertaken for their compliance with the required project standards. Their value lies in many areas, not just in the promotion of quality. For example, ITPs can be useful in proving that your project has been performing to standards at given points in time. In addition, just the presence of an ongoing inspection system can be enough to ensure that the project performs better than if none existed at all. Further, dealing with unsatisfactory work is naturally part of any approach to quality control, and ITPs can form a valuable role in evidence-gathering or in the attempt to promote the improved performance of a subcontractor.

In developing ITPs, you should look for definable features of the work; staging points in particular elements of the project that allow for clear outcomes to be inspected at a particular standard at given points in time. On occasions this will reflect a safe work method statement (SWMS), but more frequently the element is larger, such as pouring and finishing of a concrete slab. In itself, this project too could be split into two elements, the formwork and placement of membrane and reinforcement, with the pouring and finishing of the concrete as a separate element. A typical ITP might look like that shown in Figure 6.12.

As mentioned previously, the NSW government's website 'Procure Point' provides documentation about many elements of quality control. Though this information is framed around ISO 9001, a standard that does not include ITPs, the website includes some powerful tools for their development: Appendix E, for example, offering some 13 steps for the documentation of test plans for a construction contract. This same document then covers a range of key elements in depth, including:

- Describing the work requiring inspection and/or testing (be it a stage or process) covering:
  - identification of logical stages: after multiple activities but before work is hidden or covered up
  - characteristics of the inspected element or process (influences timing, such as reinforcement in concrete)

Brickwork BCA 3.3 – Workmanship

Requirement	Compliant		Reference	Comment
	Yes	No		
Built in frames – alignment				
Built in frames – attachments				
Built in frames – other				
Weep holes			AS 3700	
Lintel – coating and thickness			AS/NZS 2699.3	
Utility			AS 3700	
Coarse/openings				
Perpends and joints			AS 4773.1	
Bonding				
Workmanship – face				
Workmanship – other				

Brickwork BCA 3.3 – Technical

Requirement	Compliant		Reference	Comment
	Yes	No		
Structure				
Cavity – cavity size			AS 3700	
Cavity – clean				
Cavity – other				
Insulation				
Damp-proof course (DPC) – liquid				
DPC – physical				
DPC – other				
Flashings – above openings			AS 4773.2	
Flashings – below openings			AS 4773.2	
Flashings – other				
Wire ties – spacing			AS 4773.2	
Wire ties – coating			AS/NZS 2699.1	
Wire ties – other				

FIGURE 6.11 The Western Australian builders' technical quality assurance checklist

INSPECTION & TEST PLAN TEMPLATE & CHECKLIST (Concrete Example)

Compliance Council

Inspection Test Plan

PROJECT:		CONTRACT NUMBER:			CONTRACTOR: (Insert logo)		
Inspection Reference	**Work Activity**	**Inspection Method**	**Inspection Frequency**	**Inspected By**	**Acceptance Criteria**	**Record**	**Date Completed**
#1	Collect pre-work documents	Visual check	Once prior to ...	(Insert inspector role)	Hard and soft copies of documents	Checklist #1	
#2	Prepare site	Visual check	Once prior to ...	(Insert inspector role)	Completed items 1 - 10	Checklist #1	
#3	Delivery	Visual check	Every truck	(Insert inspector role)	As per estimate	Checklist #1	
#4	Excavation	Dimensions check	Once prior to ...	(Insert inspector role)	As per spec.	Checklist #1	
#5	Pre-pour	Visual check	Once prior to ...	(Insert inspector role)	As per site drawing	Checklist #1	
#6	Foundation pour	Slump test	Once prior to ...	(Insert inspector role)	60mm +/- 20mm	Delivery record	
#7	Foundation pour	Strength check	Once prior to ...	(Insert inspector role)	As per spec. Must be over 40 MPa	NATA test report	

Checklist

PROJECT:		CONTRACT NUMBER:			CONTRACTOR: (Insert logo)		
Checklist Reference	**Work Activity**	**Inspection Reference**	**Inspected By**	**Inspector Sign**	**Pass/Fail**	**Comments/Rectification Information**	**Date**
#1	Collect pre-work documents	#1	(Name)	(Signature)	(Y/N)		
#2	Prepare site	#2	(Name)	(Signature)	(Y/N)		
#3	Delivery	#3	(Name)	(Signature)	(Y/N)		
#4	Excavation	#4	(Name)	(Signature)	(Y/N)		
#5	Pre-pour	#5	(Name)	(Signature)	(Y/N)		
#6	Foundation pour	#6	(Name)	(Signature)	(Y/N)		
#7	Foundation pour	#7	(Name)	(Signature)	(Y/N)		

FIGURE 6.12 Example inspection test plan

- type and extent of inspections and tests
- consequences of failure (the importance of the test/inspection).
- Stage and frequency, covering:
  - testing/inspection frequency and sampling processes
  - relevance of testing/inspection to contractual obligations.
- Record keeping: noting that this will be in various forms depending upon both the test/inspection, and that which is inspected.
- Specifications and standards being referenced including:
  - contracted specifications and drawings
  - Australian and international standards
  - regulatory requirements
  - manufacturer's recommendations and standards.
- Criteria for acceptance: when not in contract documents, may need to be defined in partnership with the client or principal, or local regulatory authorities.
- Inspection or test procedures, including:
  - context and conditions of testing/inspection
  - qualifications and experience of testing personnel or agents
  - equipment specifications and calibration requirements
  - forms of documentation required.
- Hold and witness points: these are points in time or workflow that require an inspection to take place before work progresses further.
- Checklists: these are not the test plans, but rather checklists that the subcontractor, service provider, or employees on that element can work through knowing that they will be inspected on those identified elements, products or outcomes. A portion of the checklist for painting is offered in Figure 6.13.

Quality manuals

Any building firm, large or small, should have some form of quality management manual with which to guide employees and subcontractors on their expected performance. The manual is also used to inform the principal or client and subcontractors of how quality is to be evaluated. As a supervisor, this is your guidebook; it tells you the components of the system and your role in making that system work. Those components are, or hold within them, your procedures.

The manual, as an expression of a quality control system, will generally begin with a statement of the company's philosophy and strategy towards quality in general terms. This will then be followed by:

- the firm's goals, both primary (the big picture) and secondary, which may include social and environmental aspects as part of a triple bottom line (TBL) accounting approach; i.e. social and environmental as well as profitability
- the organisational structure, including management, owner, staff and location or locations of the firm, its offices and projects
- the target of the firm's services; i.e. what the company does and generally who for.

The specifics of what follows after these is very much based upon what each of the above identifies. In general, it will introduce procedures for the development of a project quality plan. In so doing,

Checklist

Inspection and test plan checklist for: ***painting*** *(To be completed by the person(s) directly responsible for the work)*				
Contract number: **Contract/Project name:** **Contractor:** **Subcontractor:**	**Work area:**		**Checklist number:**	
Work	**Items/activities to be verified**	**Reference**	**Initialled/OK**	**Comments**
Preliminary activities	Access permission obtained			
	Access obtained			
	Equipment approved/on site -			
	Scaffold/ladders			
	Signage/barricades			
	Brushes/rollers/drop sheets			
	Materials approved/on site -			
	Filler/thinners			
	Paints/colours			
	Repairs completed			
I have carried out all necessary inspections and verity that the above items/activities conform to the contract specification/documents			**Name:** **Signature:** **Date:**	

Source: ITP Preliminary checklist from 'Appendix E - Guideline for Inspection and Test Plans', Construction procurement policy, ProcurePoint

FIGURE 6.13 ITP Preliminary checklist

it may reference either ISO 9001, the international standard on quality assurance, or APES 320, a standard for quality control that deals more with the financial aspects of the company. If seeking to do work for the NSW government, for example, ISO 9001 is required to become a preferred supplier. Their website 'Procure Point' (see websites list at the end of the chapter) provides a wealth of checklist documents for ensuring various elements of quality control are addressed appropriately, quality planning being but one of them; others having already been explored previously.

The Tasmanian government's procurement site, www.purchasing.tas.gov.au, offers an excellent example of a QMS manual for contractors as part of their prequalification system. This manual begins much as stated above, but then moves straight to one of the most important aspects of quality management (QM), and critical to your role as a site supervisor: leadership. It is very important to understand that a QMS is not simply a manual and a bunch of documents that you tick your way through. It is about guided advancement of the entire project site, off-site manufacturing, suppliers and management itself, towards the highest level of performance achievable within, and despite, all the contextual constraints.

The breakdown of the various elements of the manual help you to play this role. The section on leadership, for example, provides you with guidance on identifying QM roles within the organisation, along with the responsibilities and authorities of each. A sound QMS manual is structured so that this guidance flows through its entire make-up. In the Tasmanian example for contractors, leadership is followed by the following sections:

- planning, including:
  - actions to address risks and opportunities
  - quality/project/construction objectives
  - planning to achieve the above
- support and resources, covering:
  - people and competence
  - communication, records and document control.

The last major section, excluding management review and the project management plans themselves, is operation. This is the management of the project, and the processes involved in its construction. In itself, this is broken down into a range of clear and informative parts, including:

- the requirements for planning a project – from a construction management perspective
- determining, reviewing and changes to a project – looking at the construction requirements
- control and management of the project and the construction activities
- control of externally provided processes, products and services
- control of non-conforming outputs and corrective actions.

You can see that the last three sections particularly are focused on how to maintain leadership (control) over the project as a whole, as well as all its various elements and processes, including those occurring off site. The last section deals with underperformance and how to get these activities back on track.

Review of quality management

The manual itself is therefore not a fixed entity. The manual must be updated and procedures reviewed both between projects and within projects. Where a system or approach worked well in one context, it may not do so well in another, and this must be recognised. Companies that work across multiple states or territories will frequently find this to be the case, coupled with regulatory inconsistencies that must be addressed. When a process or element of the project is underperforming it will not always be the footing, or the electrician, or a framing team. This means some ITPs will not be 'tested', challenged, by underperformance – so their inadequacies as a quality control document may remain unknown if not compared with those that have exposed issues.

The last parts of the manual therefore act to review the success or otherwise of the QMS on that specific project or cluster of projects. In so doing you seek two types of information, known as qualitative and quantitative.

The quantitative is effectively numbers – financials, time, quantities and the like. This is information we can look at fairly objectively because we can hold up numbers for a fairly straightforward comparison between one project and another.

The qualitative is about what people think about why something happened, about how did 'this' action influence 'that' outcome, etc. It can be highly subjective, but it gives reason to the numbers and asks the more complex question of 'why'. This is where interviews and meetings with the client or principal are important – were they satisfied? What do they suggest could be improved?

Through these two levels of data or information you review the project performance from a quality perspective overall. This will include all the sections listed previously but also, and most importantly, identifying opportunities for improvement: evaluating the effectiveness of corrective actions, for example, and the actions taken to address risks and opportunities as they presented themselves before and during the project. Likewise, you are looking at the participants in the process and seeing where they feel their strengths were, where they could improve, or if someone from a different position within or external to the company should play that role.

Inspections: local authorities, building surveyors, inspectors and certifiers

Any building or construction work in Australia, including subdivisions, will require a number of inspections to take place. These are required by the local authority or council prior to occupation by the client or principal. In the past, these inspections were performed by the local authorities exclusively. Today, you can still engage the local authority, but you may also choose to engage an independent agent to carry out these inspections and certify the work. The name of these inspectors varies between states and territories with titles such as: Certifiers ('Principal' NSW, 'Building' QLD & ACT, 'Private' SA, 'Registered' NT), Building Surveyors ('Relevant' VIC & TAS, 'Private' WA and 'Council' SA – note that SA uses a different title depending upon their being a council or private certifier). In any case the inspectors must be qualified and authorised to act in the particular state or territory in which the project is sited.

Required inspections

Despite this variation of titles across states and territories, the inspections themselves tend to fall into much the same categories and timings. As this is not exactly the case, you should check with your chosen inspection authority to determine the 'hold points' for your particular state or territory. However, while the required inspections tend to be similar across states, they vary between building classes as defined by the National Construction Code (NCC). The typical mandatory inspections (as required by the state building Act) for each classification are as follows:

Residential work – Class 1 and 10 buildings:

Houses, alterations and additions, garages, carports and swimming pools.

- Footings prior to concrete pour*
- Slab and other steel reinforcement prior to being hidden by the concrete*
- Framing (including subfloor and roof framing)
- Wet area waterproofing
- Stormwater
- Final / completion

***Note**: these may be inspected simultaneously

Multiple dwellings – Class 2, 3 and 4 buildings

Multi-unit developments, residential components within commercial or industrial buildings.

- First footing inspection
- Wet area waterproofing
- Stormwater
- Final / completion

Commercial and Industrial Buildings – Class 5, 6, 7, 8 or 9 buildings

Examples include offices, shops, factories and commercial buildings.

- First footing inspection
- Stormwater
- Final / completion

To these may be added, depending upon the state or territory, inspections of:

- Block work reinforcement
- Fire separation
- Wet area waterproofing
- Fire control systems (hose reels, hydrants, detection, etc.)
- Intent to fill a pool

The inspections listed above are those handled by the certifiers/surveyors mentioned previously. To these there may be added inspections for certain activities such as plumbing and electrical – which may require inspections before the work is covered – such as in ground sewage connections and/or in wall applications. However, in most states and territories both plumbing and electrical are self-certifying trades. That is, having completed the work, the plumber or electrician provides the principal via the building supervisor a certificate stating that the work has been completed to the standard required by the relevant codes and standards.

It should be noted that although each inspection is officially a 'hold point'; i.e. no work can proceed until the required inspection(s) have been conducted and authority given to continue, there is a time limit to which inspectors must comply. Within each state and or territory building Act there is a prescribed period, after which works may continue as if the inspection has been performed. In certain plumbing instances, such as the inspection of a domestic sewer connection prior to backfilling, this might be as little as a few hours.

Documenting inspections

Even on small domestic projects, a number of certificates and/or approvals will be generated. On larger commercial projects there will be many more. These are very important documents that must be filed appropriately. In addition, the dates and times of the inspections should be recorded in the site diary and the file location of the certificates noted. Copies of the certificates must be forwarded to the local authority and the principal or client.

The final inspection must be completed and forwarded to the local authority before occupancy of the building, whatever its classification, can occur.

What building surveyors and certifiers are not

While the building surveyor/certifier provides inspections, and certificates relating to these inspections – all of which form a part of the evidence trail within your QMS – the inspectors themselves are not responsible in any way for quality control. They just inspect and approve a particular element. Likewise, they are not in any way involved in site supervision, nor do they determine if you are complying with your contract. These actions are within your role as the site supervisor and remain part of your responsibilities.

Communicating and assessing quality requirements: construction standards

Within your QMS manual there will be continuous references to standards, codes and possibly codes of practice. There will also be approved ways of doing particular tasks modelled around safe work method statements (SWMS) and similar documents. These describe how you want a task to be completed. Further to these are the inspection test plans (ITPs), which help you determine if a task has been completed to a particular standard or level.

While these are all excellent resources and help guide you and your workers along, it is up to you to communicate this information to those workers and subcontractors. It is up to you to promote the way of thinking behind their formulation, to help workers and subcontractors to be a part of the QMS and not see themselves as simply workers constantly under some protracted and threatening review.

The standards

Most subcontractors will be well aware of the Australian standards and/or NCC requirements to which they must work. However, many only know these standards superficially, and many of their work teams or employees will know very little. It is important that you identify the standards listed in the NCC that are relevant to each task and promote these to both your workers and subcontractors.

If you have developed the QMS manual appropriately, you will have already identified these standards within the inspection test plans (ITPs) explored previously. Because you want your work teams on your side and working with you on this, you should 'test' the ITPs with the subcontractors and workers. They will be able to tell you if certain standards are more relevant than others, or not relevant at all, and if anything important is missing.

Codes of practice

Generally part of WHS management, codes of practice are useful in breaking down a task safely, and aid in identifying where ITPs should be applied. In addition, working to a code of practice means that the workers are more secure, safer, in their activities, allowing them to focus more on quality output due to the reduction in risks and hazards. This is particularly so through the use of appropriate scaffolding instead of ladders, for example.

Tolerance and standards guides

Each state and territory publishes some form of standards and tolerances guide, which is updated as the industry moves forward in its efforts to improve outcomes. These guides provide the bare minimum required qualities for construction or installation of an extensive range of building elements. Though backed by reference to many Australian standards, you should negotiate with your client or the principal on what they perceive to be acceptable. For example, paving elements, according to one such standard, may be 'stepped' where they meet by as much as 5 mm when the aim is for them to be at the one level. Five mm is easily tripped over by an elderly person so this may not be 'tolerable' to the client. Likewise, brickwork, supposed to have 10-mm perpends (vertical joints) are allowed to be as little as 5 mm and as great as 15 mm. That is, seemingly allowing 10 mm difference between any two joints. 'Seemingly' because a more thorough study of the document will show that in any one wall the maximum difference allowed is only 8 mm – still a visually significant amount.

These standards and tolerances should be reviewed and discussed with your work teams and clients to determine what will be acceptable for this particular project, client, or end use of the structure.

Communication settings

There are a number of well-respected avenues for 'connecting' with your workforce on a construction site. What are whimsically referred to as 'tool box' meetings – sitting down with a work team on their tool boxes, backs of utes, material stacks, or any convenient location around the work area – are particularly effective. This is because you are in 'their' space, their familiar context and surroundings. They know who they are in these settings and are more likely to speak freely and openly about a task. This allows you also to use the task and worksite as your story board, and point to areas or materials as you explain the importance of a particular factor in relation to quality control. Minutes of such meetings can still be recorded, formally or informally, and entered into the site diary and files.

On occasions you may need to make meetings more formal. These may occur in site offices dedicated as meeting spaces where data projectors and screens are available. Sometimes these spaces are important so as to 'extract' your workers from the 'hot action' of the site, providing shelter from wind, noise, rain, or sun. In such settings, documentation can be reviewed either in hard copy or digital and details made clearer. Be aware though that in such settings some workers feel out of place and less likely to speak. Also, in these settings it is important, when it is feedback you seek, to not present your information or planned actions as a finished product: give your work teams openings for input. On other occasions these are settings where you can lay out an agreed or necessary plan with some force and conviction, ensuring that it will be followed.

Follow up and support

Meetings, no matter their context, are only one path for communication of the QMS. After a meeting you need to follow up plans and the agreed or required actions with formal work instructions documenting procedures or outcomes. Never assume that the message has been received, no matter its form – email, phone calls, texts – go and see the work as it progresses to be sure it remains on track. Have the team leaders report on progress regularly – but still review the work between ITPs.

Most importantly, listen! Your work teams must be supported in their efforts to undertake the work that you require of them. Materials and equipment must be to hand at the right time. Communication tools, such as mobile phones, two-way radios, whistles, signage, iPads and the like must all be available. In addition, training may be required in the use, application, or installation of any of the above. On occasions you may need to have a supplier provide on-site instruction on either a new material, such as an adhesive or waterproofing membrane; or equipment, such as a new model of powder actuated tool. Your work teams, given the appropriate context to communicate, will let you know what they need and when – listen to, and action those requests as matters of priority.

Supervising on-site work: regulatory and contractual standards

Constant engagement with your work teams and their outputs has been an underlying message in much of the previous sections. What you need uppermost in your mind, from a quality control perspective, is the contract. Within this document, and the documents it references, you find the bulk of the information you need to ensure the project comes in on specification or better. To this may be added documents listed previously, such as the guide to standards and tolerances, the NCC, and the applicable Australian standards.

It is also beneficial to review these documents at each phase of the construction process.

Phases of construction

The construction of a project may generally be broken into four phases, whereby each phase has somewhat differing requirements regarding inspections and oversight procedures. These may be defined as the:

- preparation phase
- site establishment phase
- construction phase
- handover and defects liability phase.

We will look briefly at each of these in turn.

Preparation phase

This phase is the one where you are preparing to engage with the project and includes the tendering process if such exists, or the generation and offering of a quote. Tendering is a significant part the construction sector, and is important enough for all companies to have set processes by which a tender is generated and offered. These processes and systems are also encompassed by the QMS of a company and so must be monitored accordingly. While it is true that in most cases the tender is not within your scope as the site supervisor, in this phase you should make a point of familiarising yourself with the document. In so doing it helps you to place the project in context and understand the architectural direction (of some projects) and the intent of the client or principal regarding end use.

Beyond the tender there lies the contract, plans and specifications. It is from these that you will determine the project needs regarding:

- equipment
- materials and appropriate/approved suppliers
- skills base
- training
- regulatory, certification and inspection requirements
- quality requirements and standards to be followed/achieved.

There are a number of activities that may need to be conducted in this phase and each must be supervised to ensure they are being done to the standards required. This does not mean that you must stand and watch each activity, but you must inspect and/or check inspection/certification documentation once it becomes available. Some common activities in this period include:

- testing of materials and equipment for suitability and quality output/finish and regulatory compliance
- surveying and geotechnical evaluations
- training, upskilling and evaluation of personnel
- identification and hiring or contracting of specialist skills personnel
- site inspections and existing condition reports generated.

Once conducted, the outcomes of the activities should be documented and appropriately filed – linking to the site diary as required by date and file location. Included in this phase will be the identification and procuring of requisite insurances.

Site establishment phase

Sometimes referred to as the 'start-up' phase, this is where the work on the ground begins. Again, it is a case of ensuring that appropriate regulations have been checked and that the requirements can be met. The individual subcontracts that have been established in the preparation phase should be checked for surety that the requirements of the greater project will be satisfied and the potential for failure is low.

Some key activities that are generally part of this phase include:

- establishment of site access and egress points
- traffic management shifts established
- site security – fencing, hoarding, signage and other barriers installed
- preliminary excavation and on-site traffic routes where required
- site facilities installation – offices, gate checkpoints, vehicular wash down beds, toilets and the like
- services connections – potable (drinking) water, power, telecommunications.

Each of these need to be inspected and that inspection matched against the regulatory requirements. In some instances, this will be the first applied use of ITPs. Though seemingly basic, poorly installed hoarding has killed pedestrians in the past, and so nothing at any part of this phase should be taken for granted. As with the previous phase, any documentation, certificates, or the like should be carefully filed and event details entered into the site diary.

Construction phase

The quality assurance aspects of this phase have been fairly well covered in previous sections. In summary, this where the structure, whatever it may be, gets built. It is done so through the carefully coordinated activities of a range of subcontractors and on-site personnel. Most importantly, the oversight and communication trails covering all the various QMS requirements must be carefully documented, as described earlier, to ensure that the project performance meets specification.

One of the areas that does tend to get forgotten in this phase, however, is the feedback on completed process. As the supervisor you should not only record what has been done, but also any lessons learnt from that doing. In many instances the project can be depicted as running smoothly and to contracted specifications. While true, this does not mean that some things could not have been done better, faster, cheaper; that some materials were problematic; or some processes overly complex or understated. From an ongoing, project to project, QMS approach this should be documented and fed forward into new projects.

Handover and defects liability phase

This is often the 'forgotten' phase when it comes to quality control. As a phase it will appear (by this title or something similar) within most QMS manuals; however, it is sometimes forgotten by the site supervisor. This is because many site supervisors move on to other projects once handover is complete. Yet, as has been discussed previously, preparation of handover is a critical aspect of the accounting side of the role, and this too must be completed with due regard to quality assurance.

It is at this point that all the previous work on ensuring document control and diarising comes into play. Inspections and any associated certifications are now gathered and copies passed on to the principal. Various activities involved with preparing the site for handover now take place, including:

- removal of hoarding and/or fencing and signage
- final landscaping completed
- decommissioning of temporary services
- removal of waste services and site facilities, if still present
- access and egress points that crossed over street pavements restored to original condition or better based upon existing site condition reports
- a final report on the quality of the building is generated and in some instances a BIM model of 'as built' is supplied along with a location survey confirming the position of the structure or structures on the site.

Upon successful handover, the defects liability period is entered. The duration of this will be stated in the contract. In many instances this will not fall into the responsibilities of the site supervisor; however, you may well find yourself called upon to assist in rectification processes due to your intimate knowledge of the project. When this is the case you should again review the contract details, the contracts of those involved with that particular field of work, insurance documents and any ITPs that were conducted on that element. This is important as it is only through this information, coupled with the original plans and specifications, that you will know if what did fail needs to be addressed in a different manner to ensure the failure is not simply repeated.

Once more, any new documentation, such as altered processes or materials to render defects, should be filed in a manner that allows these lessons to be fed forward to new projects.

LEARNING TASK 6.3

QUALITY CONTROL VS QUALITY ASSURANCE

Complete this activity by accessing a plan set that you have been working on, or one provided by your instructor.

You have been asked to prepare an inspection test plan (ITP) for the bathroom finishing works detailed in the plan.

The room is now ready to be completed as the building has reached 'lockup stage'. Internal doors and windows are all fixed in place. Walls and ceilings are not sheeted as plumbing and electrical rough-ins are still to be done and the regulatory authorities still need to do their inspections before the wall sheets can be installed.

Final administration: towards the completed project

Keeping abreast of the documentation has been the key theme of this chapter, be it from a financial perspective, communications, or quality management systems approach. As the project draws to completion you need to begin organising all this material in a way that allows it to be reviewed and commented upon. This commentary may be upon the lessons learnt, or it could be for the development of handover documentation demonstrating the alignment with the contract originally entered into. How the documentation as a whole should be handled will be down to company policy, regulatory requirements and specific clauses within the contract.

Towards practical completion: contractual and regulatory requirements

The section on financial reporting earlier in this chapter refers to a project's final accounts. This summarises all the various costs, payments and claims so as to provide a final figure owed by the principal to you or your firm. This final accounting gathers all the evidence together to support this final claim. Preparing for practical completion from a contractual and regulatory sense is very similar. If you have been following the procedures suggested in this text so far, then you have effectively been preparing for this moment from the first day on site.

Practical completion

Practical completion does not require the building to be 100% complete in all aspects, nor that all identified defects have been resolved. Practical completion means that the structure is fit to be occupied and used for its intended purpose. In addition, it means that:

- excepting minor defects, all works have been completed
- all tests and required inspections have been conducted and passed
- all required documentation and information have been supplied and are ready to be passed to the client.

Practical completion may also apply only to parts of the structure. In this instance, the client may have a contractual agreement in place that requires you to allow them to take possession of a portion of the building, such as tenancy for shops, or sole occupancy units (SOUs) in a multi-residential building. In so doing, the project has been designed to take account of this requirement, allowing said tenancy to occur without them being interrupted by ongoing construction; and that this ongoing construction is not inhibited by the tenants.

In moving towards the practical completion or handover date you should meet with your client to ensure they understand what 'practical completion' actually means, and prepare them for the handover. Most importantly, you should review the contract as to how it defines practical completion. Under AS 4000 series contracts, for example, practical completion, and the process of moving towards it, is clearly defined. The path is mapped and looks much like that shown in Figure 6.14.

Generally, before handover occurs, there will be an on-site meeting with you as the site supervisor and the client whereby a list of items – incomplete items or those considered a defect – will be agreed to. Handover itself, assuming there is nothing still outstanding inhibiting occupation, will then usually take place about two weeks later.

Regulatory requirements

As noted previously, the exact regulatory requirements you face on any given project depend significantly upon the state or territory in which the project is sited. You will have reviewed these requirements prior to starting the job, as your company would have done prior to submitting the initial tender bid.

In the meantime, you should be making sure you have all the:

- relevant regulatory inspection certificates, including the 'final' inspection
- 'self-certification' documents from such trades as electricians and plumbers (including roofing plumbers)
- certifications
- reports and connection notices from service providers or utilities such as:
  - gas, electricity, telecommunications, water and sewage
  - maps of where these lines have entered to the property and to where they connect
- warranties and user manuals for appliances and other relevant products.

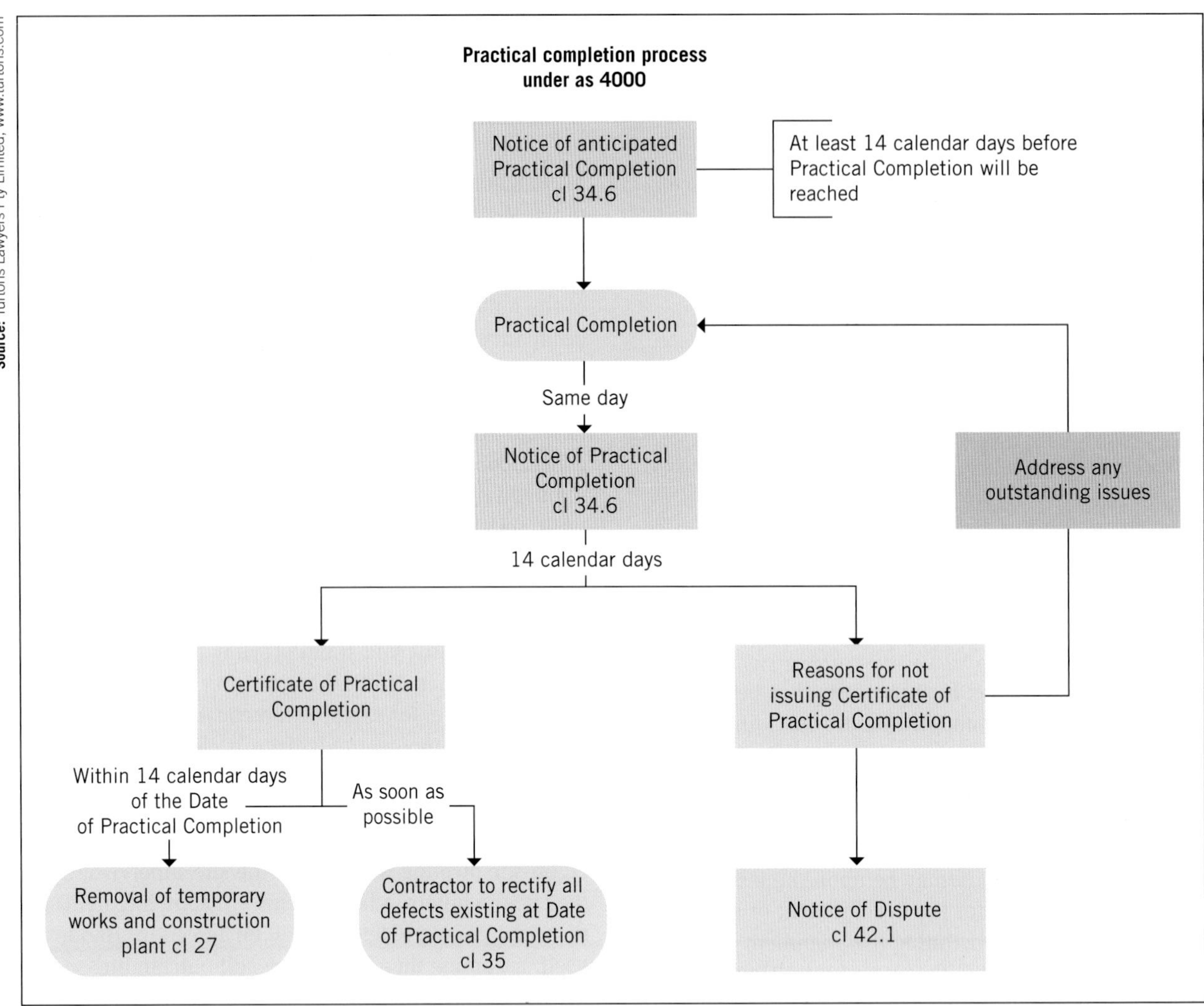

FIGURE 6.14 Path to practical completion – AS 4000

This means, in short, that the building has met all the required statutory approvals.

Once practical completion has been satisfied and handover completed, the client becomes the owner of the building. They become responsible for all insurances, security and maintenance of the property. All liability for liquidated damages also ends. At this point a portion, usually 50%, of the retention monies passes to the builder, and the defects liability period (generally 12 months) begins.

Towards practical completion: inspection procedure

The contract will outline how much prior notice you must give to your client of an impending practical completion inspection (PCI). This may be as short as one week, but is generally two weeks or 14 days. At the same time, you will advise, or re-acquaint, your client with the handover date (generally in the contract). Your procedure should be framed around a two-part inspection process:

- a first inspection where any defects or minor rectifications will be identified and a list made and signed by both parties
- a second inspection just prior to handover to check that each item has been addressed – acknowledging that handover is still possible if non-critical aspects have not been fully completed at that point.

The initial inspection

This is best done with a checklist. Do not be surprised if the client requests that an external inspector of their own nomination comes as well – you can guarantee that this person will have their own checklist.

In conducting the inspection, you will look at all aspects of both the interior and exterior of the building. How you define work that is incomplete is framed around the usability of the building; i.e. does the incomplete work prevent the building from being used for its intended purpose.

To define what is a defect, you should refer back to that document or system to which both parties agreed at the outset. Frequently, this will be the guide to standards and tolerances published in your state or territory, or an agreed adaptation. This means you will look at the whole of the building through the lens of acceptable levels of error. In creating your checklist, you can either include these tolerances, or reference them, beside the element in question. Such a list is too long for inclusion in this chapter; however, that which follows is suggestive of just some of the things you should include in just a portion. This listing is still quite limited in scope, and you should consider it a guide only; for example, the single point on 'landscaping' could easily be expanded into a one- or two-page document in its own right.

Example check list content (limited exterior only):

- the exterior for:
  - paint issues
  - footpaths and driveways for cracks in concrete or uneven paving
  - retaining walls
  - fencing (if included) and gates including latching
  - landscaping works incomplete, gravel, plants, edging and the like
  - poor brickwork – cleaning, missing mortar, inconsistent laying
  - other cladding inconsistencies and poor weather sealing
  - expansion joints incomplete
  - inadequate ventilation and blocked weep holes
  - termite prevention and damp-proof course
- roofing:
  - damaged or missing fascia
  - damaged or missing gutter
  - damaged or missing downpipes
  - dinted or scratched roof sheeting
  - cracked or misaligned tiles
  - pointing missing on tiles
  - poor flashing installation
  - inadequate or inappropriate fixings/hold down
- windows and doors:
  - glass cracked, chipped, or not clean
  - chipped, cracked, or dinted frames
  - gaps around frames
  - poor flashing practices
  - doors poorly hinged or hung
  - missing weather seals
  - missing or poorly fitted furniture.

How this content might be mapped into a checklist is up to you; however, where possible, provide it in a table format that allows you to put in notes, references to standards and tolerances, as well as being able to attach photographs as required. This means that the digital format, accessed on a small tablet computer (iPad or the like) with a built-in camera is by far the most appropriate and adaptable.

An example of how this might look in a commonly used spreadsheet program is shown in Figure 6.15.

Source: Glenn Costin

PCI Checklist

Element: **Exterior cladding**
Type: ***Brickwork***

Date of inspection

Item:	Quality	Standards & Tolerances Ref NSW	Notes	Photographs
Bricks				
	Chipping	3.10		
	Colour discrepancy	3.6/3.7		
	Poor cleaning	3.11		
	Blocked weep holes	3.9		
	Excesive variation in perpends	3.13/Table 3.04		
	Excesive variation in bed joints	3.14/Table 3.04		
	Expansion/control joints not filled	3.3		
	Cracking	Table 3.02		
	Damproof joints poorly laid	3.17		
	Verticality (Plumb)	Table 3.04		
	Level	Table 3.04		
Mortar				
	Missing	3.9		
	Inconsistent colour	3.8		
	Inconsistent raking	3.18		
	Soft: inappropriate formulation	3.8		

FIGURE 6.15 Spreadsheet version of a PCI checklist

The follow-up inspection

This inspection will use the same checklist from the initial inspection, paying particular attention to the attached notes and photographic evidence. The purpose of this second inspection is to satisfy both parties that adequate progress has been made to those issues identified previously. 'Adequate' in that they do not all have to have been completed, but that those that would have inhibited occupancy have been dealt with. Of the others, the client needs to be comfortable that sufficient work has been undertaken to demonstrate good faith and that they will be completed without inconvenience to themselves or their tenants. The latter case is important to the client, for once handover is complete, they have no immediate recourse to liquidated damages (loss of return from rent, for example) through the contract – though they can still pursue this through the courts.

Handover procedures

Handover takes place after the client approves of the final inspection, final payment is made (less the limited retention monies) and the building is deemed suitable for its intended purpose – be that residential occupancy or industrial use. What needs to happen now is dependent upon that intended use.

Residential occupancy

On the day of handover, you must supply your client with all the documentation discussed previously. This includes:

- the practical completion certificate
- all certificates of hold point inspections
- 'self-certification' documents from such trades as electricians and plumbers (including roofing plumbers)
- engineering certifications
- product warranties and manuals
- reports and connection notices from service providers or utilities such as:
  - gas, electricity, telecommunications, water and sewage
  - maps of where these lines have entered to the property and to where they connect.

Although this is not mandatory by any means, you should consider giving your client a guided tour of the home, and demonstrate any features that influence the efficiency of the building. That is, you should offer the homeowner a 'user's manual' and personalised instruction. This is the failure in many handovers of homes designed to be energy efficient. Most such homes can only achieve the targeted efficiencies if windows, blinds, ventilation and any mechanical heating and cooling is used in a coordinated manner.

Disputes

Although any disputes should be settled prior to agreed handover, it is not impossible that there may be something that continues to be an issue. Handover can still occur with the agreement of both parties as the defects liability period will continue to run generally for 12 months afterwards. During this period the client still has control of 2.5% of the contracted sum ensuring your compliance to their requests, within reason.

Commercial and industrial handover

The documentation trail at handover is very similar in this sector as in the residential, at least in name. Product warranties and manuals need to be provided, as do certificates of inspections and connection notices. The shift is in the scope of these documents; i.e. there are more of them, and they will be for commercial connections (more expensive) than for domestic. The manuals for installed equipment may well be significantly more important and the use of such equipment may have to be very clearly explained. Operators may have to be 'checked out' on high-grade equipment, power supply back-up generators, water softeners, pumps and the like. This will form part of a 'users' training program.

Just prior to handover you will have had your contractors and design teams review the operation of specialised equipment, and any necessary commentary inserted into the manuals. Houses rarely have 'as built' plan sets provided to the owner; commercial and industrial projects regularly seek this information today as it is the best source of information on which to base maintenance programs. If your project uses a BIM system then this will have tracked the construction to completion. Such being the case, you will be able to provide your client with a full set of as built plans, ensuring all contractual requirements of the structure have been demonstrably met.

A program needs to be set up to run over the defects liability period to ensure correct operation of the building and contractually included services or equipment: included will be a compliance report submitted at the end of this period.

Certificates and other client documentation

The past few sections have made clear reference to a range of documents that need to be supplied to the client or principal at handover. These included:

- relevant regulatory inspection certificates, including the 'final' inspection
- 'self-certification' documents from such trades as electricians and plumbers (including roofing plumbers)
- engineering certifications
- reports and connection notices from service providers or utilities such as:
  - gas, electricity, telecommunications, water and sewage
  - maps of where these lines have entered to property and to where they connect

- warranties and user manuals for appliances and other relevant products
- the occupancy permit.

In addition, for commercial and industrial facilities, there may be required annotated user manuals from preservice testing and preliminary in-use observations.

For residential works, an occupancy permit will only be issued once all factors concerning amenity, health and/or safety have been satisfied. This includes things like laundry taps, smoke detectors/alarms, handrails and the like. In some instances, this may not even include painting. However, strong arguments can be made for anything that is fume-generative (such as paint) to be completed prior to occupancy by those who are sensitive to such, particularly children. Likewise, calls for loss of amenity may be made if carpet is not laid or timber floors are not polished.

In many states and territories, the occupancy certificate is not a single document, but must be chosen from among several based upon the structure in question. In other cases, more than one form must be supplied. In Western Australia, for example, there are four documents to choose from: two referring to construction or building 'compliance', one of which must accompany either of the two others – being the applications for occupancy permits (one for strata-titled premises). In the compliance forms there is a declaration that clearly spells out your responsibilities regarding the ongoing health and safety of the occupants. In the case of the BA18 form (see Figure 6.17) this reads as:

> Occupying or using this building or incidental structure in its current state in the way proposed would not adversely affect the safety and health of its occupants or other users.

Source: Department of Mines, Industry Regulation and Safety

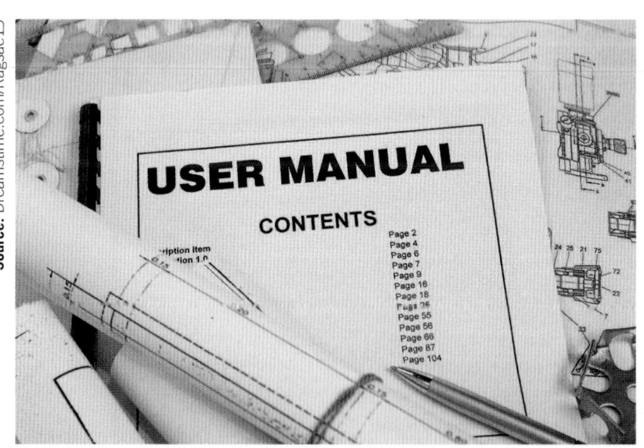

Source: Dreamstime.com/Ragsac13

FIGURE 6.16 User manual

Property sewerage diagrams (sewer diagrams) are also required, and this should be supplied by your contracted plumber. Diagrams such as that shown in Figure 6.18 are the norm for structures within the Sydney region.

Similar diagrams are required for electrical and gas supply lines. It is essential that these be provided to the client for future landscaping, maintenance and building works to reduce the risks of accidental severance or rupture.

Insurance documentation

Under Australian law there are a range of implied warranties associated with domestic building insurance. These implied warranties require the builder to carry out their work in a proper manner in accordance with the plans, law and the contract documents; also, that the materials used are suitable and new unless specifically requested within the contract to be otherwise.

These warranties transfer automatically to the owner, and any subsequent owner for up to 10 years. Homeowner warranty insurance was always with the client, but only gives limited cover in the event of a builder becoming insolvent, dying, or otherwise being unable to pay off the reparation works themselves. At handover you should explain these facts to the client again.

You should also explain that as of handover, the onus of insuring the property against damage from fire, flood, rain and any other event for which a home is normally insured falls upon the owner. As of handover the builder is no longer responsible for any damage to the structure, or to secure or protect it in any way. This applies to commercial structures as well. Such information should be supplied to the owner in a timely manner, allowing them to secure the appropriate insurances.

CERTIFICATE FORM BA18

Certificate of building compliance

Building Act 2011, section 57
Building Regulations 2012, regulation 4, 36

PERMIT AUTHORITY USE ONLY	Reference number
BUILDING SURVEYING USE ONLY	Reference number

Source: Department of Mines, Industry Regulation and Safety

FIGURE 6.17 Form BA18

Source: NSW Fair Trading

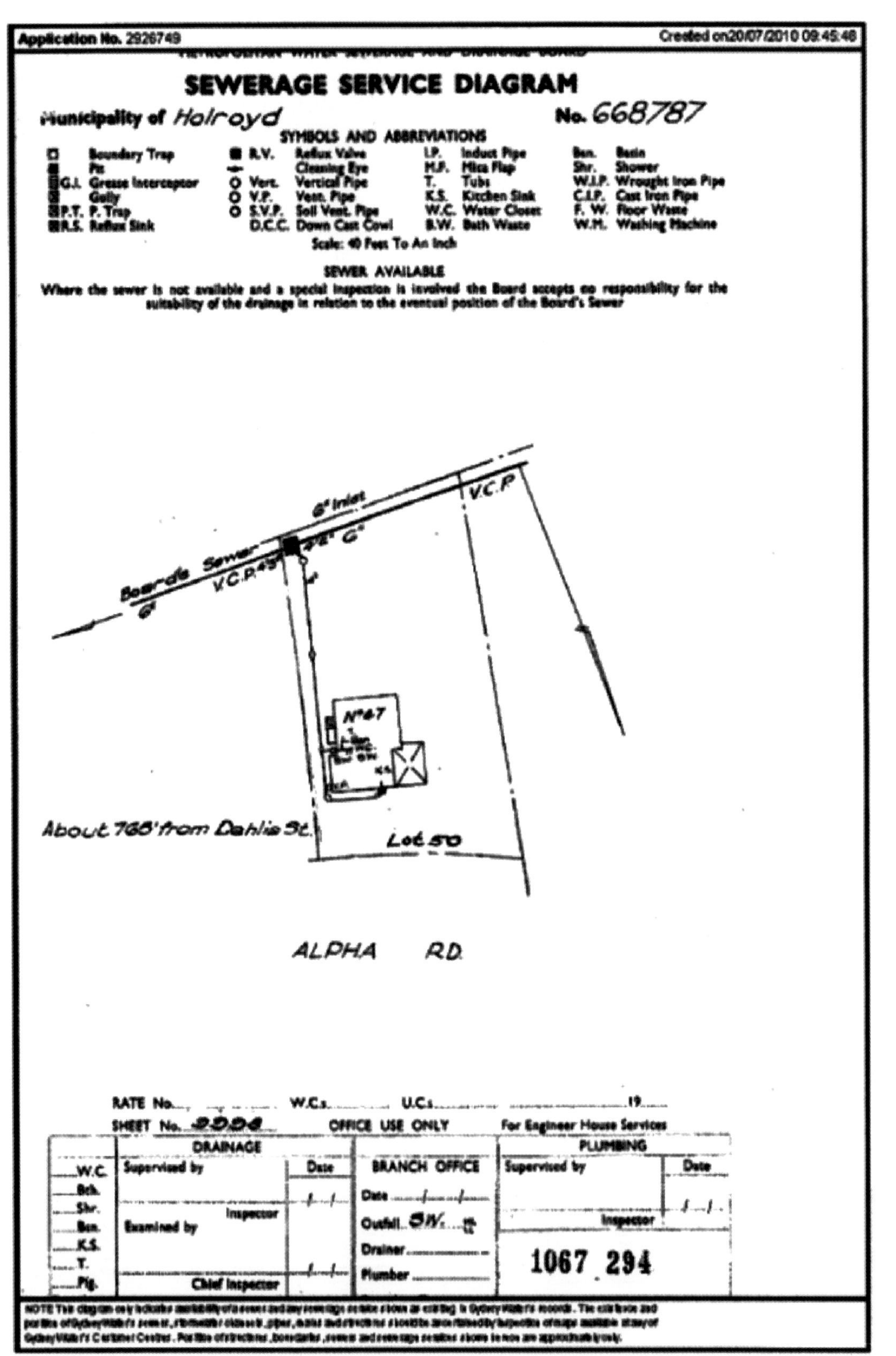

FIGURE 6.18 Sewerage service diagram, Sydney, NSW

Client identified defects

Defects found during the practical completion inspection (PCI) will have been noted in some form of checklist-like document and any required reparation works put underway before handover – even though some may not be complete. These defects are, however, known to both parties.

Once the client is in occupation, or their tenants are, further defects may be found. Provided that they are found within the defects liability period, and that they are genuine defects in workmanship or materials, then you will be bound to cover them. The defects liability period is given to the start at the time when the certificate of occupancy was granted, and runs for 12 to 24 months for most domestic contracts; sometimes more, sometimes less, with respect to commercial contracts.

Statutory and implied warranties

The caveat to this is the statutory or implied warranties mentioned previously. The existence of a specific defects warranty period within a contract cannot limit or remove a client's rights to these warranties. This can mean that for some defects, you may remain liable for a much longer period of time.

The timeframe within which claims may be made under statutory warranties is dependent upon the state or territory in which the work was conducted. Within these states there may be further complexity around the applicable claim period due to changes in regulations or definitions around 'major' or 'structural' defects and 'non-major' or 'non-structural' defects. In NSW, for example, major defects have a claim period of six years (plus six months if the defect found in the last six-month period), whereas 'other' defects receive only a two-year (plus six months) period within which a claim may be lodged.

Further complexities arise when a statutory warranty is said to have begun. Under the NSW *Home Building Act*, for example, a warranty period is said to have begun '... on completion of the work to which it relates ...' (*Home Building Act 1989* No 147: 18E). This means that the work could have been done very early on in the project, with handover not occurring until 12 months later. If it was a non-major defect, this means that the owner may only have 12 months left within which to identify and claim against it.

Statutory and implied warranties are complex and must not be dismissed by the builder, and limitation periods can vary depending upon which cause of action applies – breach of contract, negligence, Australian consumer law or state building Acts. If you intend to dispute a defect claim, it is best to get legal advice pertinent to your state or territory.

Types of defects

The types of defects a client may identify over time vary markedly. Some clients will not see a major defect at all, when it is patently obvious to others. Other clients will pick up every little thing that is not absolutely perfect, things that are not technically defects at all. In the previous section the terms 'major' and 'non-major' were used to express types of defects and their implications around statutory warranty periods. These terms need to be understood more fully.

Major or structural defects

Each state and territory has its own definitions on this point. The Queensland Building and Construction Commission (QBCC) frames their wording around 'structural' and is quite simply stated:

- Structural – e.g. leaking roof, leaking shower, health and safety issues

Whereas the NSW *Home Building Act 1989* offers one of the more complex definitions.

(a) a defect in a major element of a building that is attributable to defective design, defective or faulty workmanship, defective materials, or a failure to comply with the structural performance requirements of the National Construction Code (or any combination of these), and that causes, or is likely to cause:

  (i) the inability to inhabit or use the building (or part of the building) for its intended purpose, or

  (ii) the destruction of the building or any part of the building, or

  (iii) a threat of collapse of the building or any part of the building, or

(b) a defect of a kind that is prescribed by the regulations as a major defect, or

(c) the use of a building product (within the meaning of the Building Products (Safety) Act 2017) in contravention of that Act.

A major element of a building meaning:

(a) an internal or external load-bearing component of a building that is essential to the stability of the building, or any part of it (including but not limited to foundations and footings, floors, walls, roofs, columns and beams), or

(b) a fire safety system, or

(c) waterproofing, or

(d) any other element that is prescribed by the regulations as a major element of a building.

Source: Legislation NSW

Non-structural or 'other' defects

Again, the QBCC provides the simpler definition:

- Non-structural – e.g. sticking doors or windows, minor cracking of plasterboard

Whereas NSW actually offers no definitive comment at all on what 'other' defects might mean, allowing the definition of 'major' defects to stand as the counterpoint to all 'other' defects that may arise. However, given that it is common within the state and territory Acts that no real definition is provided for defects, it is helpful to find at least this in the NSW version.

Note: Not all states and territories reflect on defective work in either of the manners stated above. The Northern Territory, for example, does define defective work within its *Building Act 1993*. In so doing, however, it does not differentiate between structural or non-structural or major and 'other'. A defect is simply defined as building work:

(a) carried out by a residential builder in a way that contravenes a consumer guarantee; or

(b) for which the residential builder has supplied materials otherwise than as required by a consumer guarantee.

Once again it is important to understand the concept of 'defect' from the legal standing of the state or territory in which you have conducted your work.

Guides to standards and tolerances

With regards to commercial projects, refer to your contract for definition of defects. In most instances this will then refer you to another agreed document such as an adopted guide to standards and tolerances. So, while there may be structural and non-structural, major or 'other', defects, it is this document that determines what is acceptable, and what is actually a 'defect'.

This same document is equally applicable to domestic structures, provided that it has been mutually agreed to. If there is no evidence of this agreement within the contract then you, or your client, may still reference it as a 'guide' to what is acceptable or not – and, if necessary, it could be used in any dispute resolution process.

Defect rectification

Should defects be found, and they are not uncommon on all types of structures be they commercial or domestic, then you should rectify them at the earliest convenience to both parties. Allowing a defect to remain unresolved on a project leads to a lowering of goodwill between you and the client, and can affect your potential to gain other work in the future.

Defect identification

Whether the defect was identified in the practical completion inspection (PCI) or by the client months after handover, you need to determine its level of importance. In identifying the type of defect, you are then in a position to act in a manner that is in the best interests of both parties. For example:

- **Issue identified by client**: A number of doors have begun to stick or jam. These doors are all in the one area of the house. The house has been built on a waffle raft slab and is brick veneer clad.
- **Inspection by builder/supervisor**: Upon inspection it becomes clear that the footings have moved disproportionally in one portion of the building due to excessive moisture in the clay foundation material beneath the home. This in turn leads to the identification of a leak in the main water line feeding into the home. That is, it is a major or structural defect that has been identified and it must be dealt with quickly.

The rectification is therefore not a matter of fixing doors to stop them sticking (a non-structural or 'other' defect), it is about eliminating a major issue that can ruin the home's footings, with major inferences for the longevity of the home itself.

Rectification

Having identified the true nature of a defect; i.e. the cause not the symptom, it is now a matter of determining who is best positioned to do the work and who pays.

In the example offered previously, the first instinct is to blame the plumber. Blame is not your best approach as it can lead to the undermining of what were otherwise good relationships between yourself and your subcontractors. Instead, contact the plumber, state the problem calmly and rationally and determine the earliest time that they can attend the site.

It is now for the plumber to determine what has gone wrong. This may be a faulty connection, a poorly installed connection, a rupturing of the line due to a failure to provide appropriate lagging at a penetration, or a landscaping error where the pipe has been penetrated by a metal stake.

The 'fault' is therefore not always with the subcontractor who will need to rectify the problem. Even when it is, treat them with respect as next time it may be your error of judgement that comes under scrutiny. In addition, you need the work done and quickly so it's best to work cooperatively.

Who pays?

The insurance broker is always your first point of call in these instances. If insurance covers the work then this is your problem solved. It may not solve the plumber's, however, as the insurance company will seek to gain recompense from them. The plumber will most likely have their own insurance. This is where discussions may be had between insurers as to who pays for what. It is best for both you and the plumber to allow your mutual insurance brokers deal with the problem on your behalf, which keeps the pair of you in good relations generally.

If there was a third party involved, such as the concreter or a landscaper, the same approach applies. Allow the insurance brokers to deal with the issue.

For smaller defects, such as a door that was simply poorly hung, sometimes its simpler just to fix it and move on. It may cost you or your company more to deal with the paperwork than to pay a tradesperson to attend the site and do the work.

Regulations and organisational requirements: a final review

With the project nearing completion, you are into the final administrative phase. It is important not to rush towards the finish line but rather to slow down and take your time. In so doing you can ensure that all contractual, regulatory and organisational requirements are being met. Most, if not all, of these requirements have been covered in the preceding sections; however, it is worth revisiting them, albeit briefly.

The administrative process: practical completion

As you approach the date of practical completion or handover, there are a few things that may come to light: first, that there are a raft of defects or minor incomplete works that still need to be rectified; second your client does not want to take possession 'just yet'. The first we have dealt with at some length previously. The latter, however, is of some import.

Why practical completion is sought by both parties

As the lead contractor, your firm wants practical completion and handover for a range of very clear reasons:

- responsibility for the security and insurance of the building is no longer yours
- site staff are no longer required which saves on wages
- the defects liability and, in some instances, statutory warranty, period begins – allowing your company to count down towards payment of the final retention monies owed
- exposure to liquidated damages is limited.

A client generally wants handover to take place because:

- they, or their tenants, want to move in, saving on rent paid out, gaining rental paid in
- it is their home; they have an emotional attachment to gaining entry and setting it up
- it is the premises of their business and they need occupancy to begin their work

As a general rule then, handover is in the interests of both parties. However, there will be times when this is not the case.

Why your client may not want practical completion

Your client may not want to take possession for a number of reasons that they perceive to be in their best interests:

- the project is simply not complete and their amenity would be compromised while you continued to work around them
- the work may not be to the designs they hold
- the work may be, in their eyes, unsatisfactory
- if the building is to be rented out, they may not have any tenants ready to move in; they will be paying for the security and insurance of an empty building
- on commercial jobs particularly, you may not have been able to demonstrate or prove that the equipment or systems installed actually work to specifications.

In these instances, you need to look closely at what the client is telling you and determine if they are holding up practical completion for reasons that are not compatible with practical completion as it is defined in the contract. If the work is incomplete, then the onus is on you to complete it in a timely manner. If the client is pointing at minor defects that do not disallow the building to be used for its intended purpose, then you can challenge their position on the matter.

Putting the administration into place

In summarising the previous sections of this chapter, and covering the requirements of both domestic and commercial projects, you should ensure the following are undertaken:

- all defects are attended to or programmed to be attended to
- the WHS files are completed
- operation and maintenance manuals are ready for handover, including:
  - service testing certificates and data
  - training records
  - notes on observed operation
- testing of systems is completed and observations recorded
- consumables such as filters and the like are replaced with new before or at handover
- all keys handed to client
- maintenance periods advised and record books are handed over
- certificates of regulatory hold point inspections are supplied
- certificates from self-certifying trades are provided
- sewage service diagram is provided
- site diagrams for other services (electrical, gas, telecommunications) are provided
- 'as built' plans are handed over and any variations from the original are ratified by both parties
- insurances change from builder to client
- a receipt is provided for final payment made by the client (except 50% of the retention monies)

In completing all of the above, you should make a record that captures the signature of the client as being in receipt of each item, and their understanding of any information presented to them with regards to maintenance, for example. This should be additional to the site diary entry that will record the date and time of this transaction, and what documents and articles were transferred.

In some instances, the climate will not allow for the testing of some systems. In such cases you need this recorded along with a program for when such systems will be tested. This should also be sighted and signed as understood by the client.

As part of the handover, a locations diagram coupled with a guided tour should be provided for the client to show them where important elements of their building are located. This would include basic things such as those in the following list.

Domestic:

- electrical metre box and any sub-boards
- mains water shut-off valve
- in-cupboard or wall-mounted water valves
- mail box
- hot water system
- heating and cooling systems
- boundary lines and survey pegs where relevant (on new subdivisions, for example)
- smoke detectors and back-up battery exchange process, where required
- hoses and bushfire protection measures, where relevant.

Commercial:

- the above plus ...
- fire hydrants
- emergency exits and back-up lighting systems
- emergency power cut-off switches, if installed
- specialised equipment such as retracting bollards and similar barriers
- security camera locations and operation.

The administrative process: defects liability period

In this phase you have options. You can stand back and wait, or be active – but not intrusive – in checking up on your client's satisfaction and the building's performance. During the lead up to handover you will have conducted at least one, preferably two, inspections of the building to check for defects. You have a checklist and this will now be a record of what was found, what has been rectified and that which remains to be, or is in the process of being, rectified.

It is good practice to provide your client with a copy of this document. This may be in hard copy, or it could be digital if the client has access to the appropriate hardware and software to run it. This allows you to inform your client of what is, and is not, a defect, and in some instances the importance of some types of defects. Coupled with this document can be a diary of when you would attend the building and carry out any further inspections with them. You may only choose to do the first inspection with them as a means of helping them to understand the process when they do it on their own, and for you to understand their concerns or perspective.

You also need to have a process in place for receiving and responding to the defects identified by the client. This process should ensure that the client feels their concerns are of a priority to your company. This means not only recording their phone calls, but back-up responses, clear timeframes for on-site visits and avenues for recording proposed actions. Included should be dedicated files for photographic evidence of before and after rectification works and reports – these should tie into the 'as built' plan set if using a BIM system and the works are on major defects.

LEARNING TASK 6.4

FINAL ADMINISTRATION – COMPLETING THE PROJECT

Develop a 'handover' list of documents and other relevant information to which the client should have access, once the works have been completed for the 'most recent' plan set you have been working from.

PART 1

6

SUMMARY

This chapter set out to show you that your role as a site supervisor is a multifaceted one: where your knowledge of contract administration is just as relevant as your knowledge of construction. In so doing, you have been guided through quality control systems, record keeping policies, WHS and other regulatory requirements, the administering of payments and claims, and communication best practice.

This last point is critical: you should now know about the importance of communication between those doing the work, the client and future readers of the project's history. The role cannot be conducted effectively if you cannot also interpret the communication of others, in all its forms: be it contracts, specific procedures, or communicating to others your interpretations so that those procedures may be followed. As stated at the outset of the chapter, the role of the site supervisor is as much about being seen to be doing as the doing itself.

REFERENCES AND FURTHER READING

Weblinks

https://www.fos.org.au/custom/files/docs/cgc-report-general-insurance-in-australia-201617.pdf, p. 20

https://www.cmaanet.org/sites/default/files/resource/Impact%20%26%20Control%20of%20RFIs%20on%20Construction%20Projects.pdf

NSW government: **https://www.procurepoint.nsw.gov.au/policy-and-reform/construction-procurement-policy**

Tasmanian government purchasing: **https://www.purchasing.tas.gov.au/**

WA government: **http://www.commerce.wa.gov.au/publications/builders-technical-quality-assurance-checklist**

7 BUSINESS DISPUTES

Chapter overview

You may be the most competent builder in history, and your client the most amenable and rational you have ever dealt with, but a dispute can still arise on the construction site. In the main this is due to the scale of financial risk involved by both parties, but there can be many causes. This chapter discusses these causes, and the processes for their resolution. Covering issues from the development of contractual resolution procedures through to court determinations, the chapter also provides the most important guidance of all – paths for avoiding disputes in the first place.

Elements

This chapter provides knowledge and skill development materials on how to:

1. evaluate dispute information and develop and implement resolution procedures
2. negotiate with parties: developing resolution strategies
3. identify opportunities for dispute resolution.

Introduction

As stated above, the aim of this chapter is not only to provide the knowledge of how to resolve a dispute, it is also about how to avoid them. As we will discuss, disputes arise from a range of issues, some seemingly minor, some clearly significant. Mostly they arise from misunderstandings; seldom, if ever, from one party trying to cheat the other. Either way, if you are not appropriately prepared, the cost in time and money can be so great that it may exceed the original project contract price.

Knowing this, governments have established a range of mechanisms by which construction conflicts can be resolved: mediation and conciliation being preferred. State-backed tribunals have also been developed over the years which remain comparatively cost-effective compared to the courts. The most preferred solution, however, is to have effective resolution procedures in place prior to a dispute arising. Therefore, we will discuss the development of such procedures first.

Evaluating the information and developing dispute resolution procedures

Irrespective of their nature or cause, you can go a long way towards avoiding or resolving disputes quickly and painlessly simply by acknowledging their possibility as you develop the contract. If you are a small contractor, dispute resolution procedures may be new to you, or at least you may never have had to deal with them as yet; or perhaps you have simply used standard contracts and never had to concern yourself with the fine print. Alternatively, you may be part of a large commercial construction company where such things tend to be handled by others. No matter your role, you should make yourself aware of any existing dispute processes, and at least have a passing understanding of their limitations. It's through this understanding that you can avoid or at least mitigate disputes as they arise.

What constitutes a dispute?

Disputes are many and varied in form. They may be a simple question of acceptable quality over which no agreement is immediately made. Or a difference of opinion over the work output or the value of that work, or the speed at which the work is being conducted. In many cases these differences can be settled quickly through face-to-face conversations. If you treat every 'complaint' to your company as you would expect to be treated when making a complaint to any company you have dealings with, they may well be resolved without ever becoming a formal 'dispute'. Courtesy, politeness and a clear willingness to listen can go a long way to defusing a situation, even if the complainant begins loudly and aggressively.

At the end of this chapter there is a weblink to a valuable resource for any company. Produced by the Commonwealth Ombudsman, it is titled '*Better Practice Guide to Complaint Handling*' and has advice that should be followed by companies of any size.

In addition to the this resource is a link to The Ethics Centre. Ethics in business has become an important topic in recent times, particularly in the banking sector. Yet it is just as important to the building sector, a sector that has had its own share of bad publicity surrounding unethical behaviour. Understanding ethics is not about the blind acceptance of a particular doctrine or set of rules; rather, it is about accepting that other people matter. In acknowledging that people in general deserve respect and dignity, we can frequently step back from a dispute and find mutual ground from which to find an appropriate resolution.

Reviewing established dispute procedures

As you learnt in Chapter 3 Construction contracts, even the most basic of standard contracts will have a clause concerning disputes. This clause will outline the steps to be taken should a dispute arise. As a quick refresher, this will generally look something like this:

- document the issues for each party to study in writing
- talk to each other honestly and in a timely fashion
- continue with your obligations under the contract
- only if the above fails, then lodge a complaint with the building complaints authority of your state or territory.

In standard domestic contracts a cluster of associated forms will usually be nominated for use in these instances, along with timelines for issue and response.

In the commercial sector, contracts generally focus upon the role of the superintendent as the first level of dispute resolution. **Clause 42** of the Australian standard contract **AS 4300** (pp. 38–9) is an example of this approach for commercial construction. In this clause you will notice that despite the superintendent themselves being cited as a possible cause of dispute, this officer is still the first point of mutual contact. Hence the system flows something like this:

- dispute noted by either party
- superintendent notified
- conference(s) to resolve issues undertaken within a limited period (usually 14 days)
- arbitration, if not resolved after 28 days
- court proceedings not eliminated if determination by arbitration is not accepted.

Note: Arbitration is a form of dispute resolution whereby the decision is made by an external professional party. This concept we will discuss in detail in the next section.

The purpose in your reviewing the existing procedures is three-fold:

- first, to ensure that you understand the procedures yourself, and that they are appropriate to your project
- second, so that you may informatively discuss the procedures with your client and gain their input and perspective on their appropriateness to the project
- third, so that both parties are aware of what these processes entail. Such as: who is the arbitrator and how they are accessed; where conferences should be held; what role state tribunals have in the domestic process and how they are accessed.

It is not uncommon that either you, or your client, may decide that these existing procedures are inappropriate for the project, or that they are unclear. This may be because they are too commercially or domestically oriented; they could move too quickly to being outside of your joint (yours and the client's) control; or you could jointly decide to have access to different mechanisms that have been successful for others.

In such cases, you may wish to change the contract and insert your own agreed procedures. We will discuss this shortly. For now, you should complete the first exercise of this unit as outlined below.

Developing dispute resolution procedures

It must be stated up front that this is not a task that you should do without expert advice. In most cases, if not all, you should engage a construction lawyer to phrase any dispute resolution clause to the mutual satisfaction of both you and your client. That said, it is wise for you to know what an effective clause should contain. To know this, you need to know the dispute resolution options. You should also understand the purpose of such a clause from the legal perspective: put simply, this is to bring about a final, mutually adhered to, resolution. Correctly written, such a clause is not a path that 'might' find a solution, it is a path that 'will'. If it does not, then the process may be challenged in court.

The options

Alternative dispute resolutions (ADRs) are becoming highly favoured in all sectors of Australian industry, not just in construction. Yet despite their popularity, and high success rates, many are not legally defined as a method of resolving disputes because they do not guarantee a solution, nor are they binding. This is particularly important when they are included in contracts of commercial projects where the stakes are high. This does not mean they cannot be used as stepping stones towards a binding solution, it just means they must be backed up by other means. This being the case, it is important that you take note of the ways by which such may be done as outlined below.

Each option is presented in order of preference, complexity, timeliness and cost. They are also discussed in view of their relevance and applicability to the domestic or commercial sectors.

Negotiation

Clearly, the most preferred and cheapest option is to talk it through. This path to resolution is known as negotiation. Negotiations need to be done in good faith, calmly and for the best interests of both parties and the project outcomes as a whole. The problem is that when money and project deadlines are involved, emotions can run high quickly, and clear communications can be interrupted. Hence, while negotiation may be included in the clause as a first step, it needs to be backed up by some clear, predetermined pathways by which the relationship may be reconciled and the project brought back on track. Such paths include mediation and conciliation.

Mediation

This is one of the most successful and cost-effective ADR approaches used today. Mediation is a process whereby an independent third party, trained in mediation, helps the two parties in:

- identifying the key issues underlying their dispute
- guiding them towards developing options and alternatives
- facilitating their reaching an agreement.

It is important to note that mediators are not industry experts, nor do they play a part in advising on the dispute or in determining its outcome. The mediator's expertise is in guiding the two parties though the process with the least level of unnecessary emotional conflict, clearing the way for them to come up with the solution themselves by focusing upon the facts, and the outcomes. Mediators therefore do not:

- take sides or make decisions
- give legal advice
- provide counselling.

Like negotiation and conciliation, mediation is not a certainty for resolving a dispute under the Victorian *Building and Construction Industry Security of Payments Act 2000*. This act requires that a contractual dispute resolution clause must include a method leading to an 'actual resolution', whereas mediation is seen only as a means by which a solution 'might' be found. And so, like negotiation, it too must be backed up in the dispute resolution clause by other paths. That said, a solution to a dispute found through mediation remains the quickest and most cost-effective approach you can take. In addition, solutions can be made enforceable through agreement by both parties to have the resolution endorsed by a court or tribunal order.

On projects where the dispute may involve many thousands of dollars, the presence of lawyers is common, and their input important. They can also ensure that the final mutual determination is correctly worded for conversion to a court order.

Conciliation

Similar to mediation, conciliation involves meeting with an independent third party to help resolve a dispute. A conciliator has many of the skills of the mediator but differs in having knowledge of the construction industry and so is able to assist in developing options, alternatives and in clarifying the key issues of the dispute. Conciliators are therefore more involved in the decision-making process. However, like mediators, they do not make the decisions and they do not make any final determination.

Also, like mediation, the mutually agreed decision can be made enforceable through a court or tribunal order if both parties agree. Again, since it is not a given that a solution shall be found through this path, conciliation must be backed up by something more definitive.

Arbitration

In the previous paths the power of decision making remained in the hands of the two parties involved. Though still seen as an alternative dispute resolution (ADR) process, arbitration takes this power away. Arbitration is more court-like in that both parties present arguments, evidence, and documents to an independent third party (the arbitrator) who makes a decision. Like a court, this decision is binding.

The benefits of arbitration are that the arbitrator (or arbitrators – there can be more than one; i.e. a panel) is also knowledgeable of your industry; and though the process is formal, it is also private. This means the dispute is not open to public scrutiny as it would be in a court. It does, however, hinge much more on facts and that which can be proved. Arbitration is seen as a definitive resolution process and so can be an end point in a contractual dispute resolution clause. Despite this, a resolution so found can be challenged in a court, but it would be hard to fight unless some form of bias or error of fact on behalf of arbitrator(s) could be demonstrated.

It is important to note that arbitration clauses in domestic contracts are prohibited (made void) in most states and territories. These states will require instead that you take the path of a tribunal.

State tribunal

In most states and territories some form of state tribunal is in place for the hearing of disputes arising from construction projects. Generally focused upon the domestic sector, they can and do make determinations on commercial projects as well. These tribunals are:

- Victorian Civil and Administrative Tribunal (VCAT)
- QCAT (Queensland)
- NCAT (NSW)
- SACAT (South Australia)
- NTCAT (Northern Territory)
- ACAT (Australian Capital Territory)
- SAT (State Administrative Tribunal of Western Australia)

Note: Tasmania works slightly differently through its Department of Justice, deploying an adjudication panel should mediation fail.

Technically not courts, state tribunals are the main avenue for the final determination of most domestic construction disputes and, to a lesser degree, commercial construction issues as well. Determinations made in these tribunals can be ratified (made legally enforceable) by a court, if both parties agree, without further costs being incurred. In most states, even if you or your client decide to go directly to court, the dispute is frequently referred to the tribunal for determination. In addition, once in a court, either party of the dispute can call for the matter to be referred to the tribunal where it will proceed as if it had begun there.

Once at a tribunal, you may be required to go through mediation or conferencing (negotiation), even if you have already done so. You will also be required to provide a comprehensive list of all documents that either party intends to use in the hearing.

Court hearing

It is to be noted that none of the above processes eliminate your right, or that of the other parties, to pursue a matter through the courts when unsatisfied with the outcome. This is why in each of the previous cases it is possible, indeed preferable in instances where the issues are significant, to have the resolutions ratified by a court order. A court hearing is a lengthy affair that can take years to be 'heard', and can sometimes cost many times more than the total contracted sum of the project you were working on. Before even considering this path, you should seek formal, experienced, advice from construction lawyers. In many cases they will advise you against it. Some things they will ask you to consider are:

- high financial costs (to both parties no matter who 'wins')
- both barristers and solicitors are frequently required (more cost)
- expert witnesses, if required, must be paid for by you
- time (they can run for weeks)
- the awarding of costs to the 'winning' party seldom cover the actual costs
- the other party may not be able to pay the costs anyway
- the decision can still be appealed, leading to more cost and time
- the stress on you, family and employees.

You will also be advised that it is possible to make a negotiated decision at any point prior to the court handing down its ruling.

Developing a procedure

In developing a procedure, it is not suggested that each of the options should be undertaken one after the other until a resolution is found: though this can be done. Rather, it is about jointly (you and your client) selecting those paths which seem most likely to succeed in the first instance, given the nature of the project and the types of disputes that might arise.

Heed should also be given to the recent Victorian determination that a dispute resolution clause must lead to a definitive resolution, not just a possible resolution. Such being the case, irrespective of your state or territory, you should ensure that this is the case in any process you are considering to document. The first step in this direction may be a provision requiring any resolution found through an ADR path, such as mediation or conciliation, to be ratified by a court order to which both parties shall be bound. If not a court order, at least a pre-stated agreement that both parties will bind themselves to the resolution.

The second step in the wording is what must happen next should an ADR path fail. What must be clear in the procedure is an end point at which a determination 'will' be found and adhered to. This end point will generally be a tribunal for the domestic sector, or arbitration in the commercial.

The third step is timing. As mentioned previously, disputes can delay a project even if work continues. Through distraction, errors can be made, leading to more disputes. Timelines for each action in the dispute clause must therefore be inserted and made clear. The time allowed for ADR pathways before the process is taken over by arbitration or a tribunal will depend to some degree upon the scale of the project.

You should be aware that the above are not the only options available, though they are the more common. There are many more, such as: expert determination, adjudication and hybrid or combined approaches. For information on these and other paths, go to the Australian Disputes Centre website at www.disputescentre.com.au.

Record keeping

Should a dispute arise, you will need good, clear records of your activities to date; this is particularly so if you end up before a state tribunal. A tribunal, court or even an arbitrator may make an order for 'discovery'. This requires both parties to develop a comprehensive and accurate list of all documents that they may rely on in making their presentations. You cannot be selective in determining what documents you wish to include, and you most certainly cannot destroy any. 'Discovery' is a time-consuming process, often needing to be done by a disinterested third party. The easier you can make it, the quicker and cheaper it will be. Sound record keeping is therefore as essential to dispute resolution as it is to any other element of the construction process.

Types of records

With regards to disputes, the types of records required may differ with the nature of the dispute. In the main, however, you will find there are commonalities, such as:

- the contract documents – plans, specifications, council approvals, etc.
- all subcontractor documents
- any contract variations, signed and dated
- all communications, letters, emails, texts, phone call logs and the like with the client, their representatives, subcontractors and suppliers
- procurement – all purchase orders, deliveries, times, dates, etc. included
- WHS documentation including licences, qualifications, safety data sheets, safe work method statements, risk controls, plant and equipment maintenance records and the like
- insurance documents and business financials.

You need to ensure that all these documents are safely filed for easy retrieval. In some cases, such as the main contract, a reduction or consolidation of the project is included. This is referred to in Australian standard contracts as 'Works under Contract' or WUC; on other occasions it may be known as a scope of works or statement of works (SOW). This outlines:

- an overview of the project
- the identifiable milestones and dates
- the overall schedule including all tasks, critical path, materials supply and labour provisions
- progress payments.

Though not as common in the domestic sector, a WUC is of equal value there, and comes into its own when multiple projects are being undertaken simultaneously.

Other documents

In the contemporary world of electronic documents and rapid communications, is equally essential to back-up all documentation using online servers or 'cloud'-based document storage. It is also useful to scan hard-copy documents for similar archiving. That which follows is a non-inclusive listing of the other main documents that you should be storing in a structured manner.

Site diary

The site diary is one of the more critical documents you may need in a dispute. This may be a hard-copy diary, or electronic document linked to a major construction software package. It will be filled in by yourself, a supervisor, foreperson, or superintendent. If it is a hard-copy diary it must be kept secure when not in use.

Photographs

Photographs and video material should likewise be stored in multiple locations, filed with their date and location details (site, stage of project). Given that these photographs will be, almost without variation, digital in form, back-up is quick. Photographs should be taken at each stage of the works, and at key points in progressing through those stages, such as at the completion of excavation, steel work prior to pouring, preliminary service connections and the like. Photographs of footings and their preliminaries are particularly important given that they are inaccessible later in the project.

Certificates

There are a range of certificates that may be compiled over the period of construction. In NSW the obvious one is the initial Construction Certificate obtained after development approval has been granted. In most states and territories this is referred to as a Building Permit. Other certificates may include:

- scaffolding handover certificates (tags should be on the scaffolds themselves)
- completion certificates (may be for stages or specific buildings on a large project)
- plumbing and electrical certificates
- installation certificates (e.g. solar panels, major plant and the like).

On major commercial projects, or if you are part of a large domestic house-building firm with administration support, these documents may be scanned and stored digitally. However, the originals must still be securely filed for easy access.

Variations

Any changes to the original contracted works are described as a variation. These must be fully documented and signed by both parties before such work can commence. These too can be in digital or paper form.

Defect notices

Defects are not just issues to be solved after completion. Defects may be identified by anyone (including yourself) at any time during construction. Informal and formal notes and notices, photographs and videos may be received at any time: and aside from acting on them, you should record and store them, along with a record of the actions taken.

Other communication documentation

All communication between yourself and subcontractors, the principal or client, banks or other lending agencies, councils and authorities, no matter how seemingly insignificant, should be kept on file. Note phone conversations, and store text messages. These will all be required if a 'discovery order' is issued by an arbitrator, tribunal or court.

Record management systems

When dealing with hard-copy file systems, each administration team will tend to have their own system. Common to all, however, will be some form of central file or register that logs all the document types, where they are filed and who is responsible. Larger projects now tend to favour major software systems that include cloud-based storage for back-up.

The value of such systems, even for minor domestic housing projects, cannot be overstated. In addition, these systems either are, or will tie in directly with, building information modelling (BIM) software. This means that documents can be easily cross-referenced with multiple parts of a project for easy access when a dispute is about a particular element or stage. Further, three of the leading systems, Aconex, Buildertrend and Procore, have 'no delete' document tracking facilities for fast document discovery in times of dispute.

Nominating a mediator, conciliator or arbitrator

As outlined previously, developing a dispute resolution procedure will tend to include mediation, conciliation and/or arbitration. Nominating who or what agency will fulfil these roles, and having them listed within the contract, saves time should a dispute arise.

Mediators

Mediation services abound in all states and territories so finding one is never an issue. The question is, which is the most appropriate and mutually acceptable? Mediation services are available through government agencies and the courts of each state and territory of Australia, but you and your client may favour a private agency such as a construction law firm or registered individual. In making the nomination, you should inquire as to the wait time (court mediations can be three to four weeks), the process for access and, in the case of private agencies, if they are agreeable to taking on disputes of your type.

Conciliators

In some states and territories, conciliation is a required first step that must be undertaken prior to domestic housing disputes being heard by tribunal. An example is Domestic Building Dispute Resolution Victoria (DBDRV) which uses conciliation instead of mediation. As conciliation is also a valuable ADR path in commercial projects, there are numerous private agencies also available. Some of these are offshoots of legal firms, others are more dedicated ADR organisations listed for deployment by the Supreme Court and state and federal governments. Examples of the latter are the Resolution Institute, The Institute of Arbitrators and Mediators Australia, and the Australian Disputes Centre.

Arbitrators

Finding and nominating an arbitrator is similar to the above, as the agencies generally can provide either of the ADR services.

Making the nomination

With regards to arbitrators and conciliators you should consider the following points:

- the type of construction project involved
- the areas of expertise and experience of the agency or individual
- the qualifications of individuals
- the costs
- timelines and accessibility.

You must remember that this is a mutual choice as the agreement will form part of your contract. If you are choosing an individual, rather than an organisation or agency, then you should produce a list of alternatives or back-ups. In this way you can ensure that you do not lose time if your nominated person is unavailable through other commitments or is on leave.

LEARNING TASK 7.1

Based on what you now know, are the existing dispute resolution procedures in the domestic housing contract of your state or territory definitely going to bring about a resolution?

- Find the existing dispute resolution procedure in the Office of Fair Trading (OFT) domestic housing contract of your state or territory for work over $20,000.
- Find the dispute resolution agency in your state or territory – determine their timelines for access and hearings.
- What is the process for having an issue heard in your state or territory?
- Nominate a mediator, a conciliator and/or arbitrator accessible in your local area/region.

Investigate disputes and resolution strategies: negotiating with parties

Disputes expose your organisation to negative publicity and can damage the relationship you have with the client or principal. Neither are desirable as they can ultimately affect your bottom line through direct and indirect loss of earnings. Such being the case, you need to get on top of the problem early by identifying the nature and cause of the dispute and developing a resolution as quickly and possible. This is also in the best interest of your client as, they too, can lose financially as the project lingers.

Identifying the nature and cause

In the construction industry a dispute cannot be said to exist until a request for action has been received and rejected. Formally, a 'dispute' is not a conflict. Conflicts – disagreements, arguments – can lead to a dispute, but are not a dispute as such. Likewise, a '**claim**' is not a dispute, but will lead to one should one party reject the claim of the other. Hence, the action requested could be a claim for payment, a demand to accelerate the work, a call for improved standards, or the rectification of perceived errors or damage. This is the '**nature**' of a dispute. The **causes** are, of course, many and varied and are what initiated the request in the first place. Commonly cited examples include:

- client-initiated post-contract variations
- poor supervision
- poor workmanship
- extension of time (EOT) claims by the builder/contractor
- site access
- poor communications
- unrealistic expectations
- misinterpretation of contracts, plans, specifications and schedules
- unanticipated site conditions
- inflexibility
- loss of trust.

You will notice that many of the above causes could, and do, apply equally to either party. Because of this it is not always easy to determine a specific, singular cause for any given dispute. However, it is important to acknowledge the perceived cause(s): the issue(s) as they are held to be by each party. Only then is it possible to reduce the influence of this and any underlying factors that have brought the issue to the point of one party refusing the other's request for action.

Identifying the cause as it is perceived by one party or the other is best conducted through documentation. The first steps are:

- gain from, or give to, your client a written outline of the issue to hand; i.e. confirm and clarify the **request**
- you or they then respond in kind; i.e. clarify the grounds for the **refusal**.

You now have at least the two perspectives of the issue and the reasons stated clearly as to why a request has been made, and why the request was rejected originally. From these two documents a preliminary perception of the dispute's cause can be developed. Note that this is preliminary: the actual cause(s) may be underlying and not exposed until later in the path towards resolution. Note also that there may be more than two parties to a dispute, particularly in the commercial sector. Such being the case, you may have more than one perspective on the cause of the dispute.

The parties and their perceptions

In the domestic sector, generally the dispute is between the builder and the client, or the builder and a subcontractor or supplier. On occasions it may be between the client and the council or a services authority such as power, water or sewage. In the commercial sector it is similar, though there can be a greater range of stakeholders, such as:

- architects/designers/landscape architects
- engineers (structural, mechanical, geotechnical, electrical)
- surveyors
- superintendents
- the principal's representatives
- site, project and construction managers/foreperson
- subcontractors/contractors
- unions.

As stated earlier, when seeking to identify the nature and cause of a dispute, you need to communicate with the relevant parties and have them state their position on paper. This is crucial, because once it's on paper you have greater clarity – clarity in your mind, and in the mind of the person writing the document. It may not be what you want to read, but it will be clearer.

The prime purpose in documenting both perspectives is to ascertain the facts – as each party holds them to be. Facts are sometimes less easily defined than what you may at first think. Memories of events can differ, interpretation of a clause or a set of plans and specifications may differ. What you seek is these areas of divergence – where the two parties, e.g. you and your client, see a point of seeming fact differently.

Sometimes you may want to repeat this step, though this is not always effective and can lead to increased tension if you are not clear in what it is you want clarified. On other occasions you may feel the need to clarify a point by verbal communication. In so doing think carefully about what it is you need to know, and ask questions accordingly. Later, we will look at how this might be done. Before that, however, you need to look to your contract and determine just what you are obliged to do with regards to a dispute arising, and what solutions you may present.

Contracts and solutions

In the first two sections of this chapter, we looked at dispute resolution procedures as they might be developed in a contract. In Chapter 3 we discussed contracts more fully. It is only when a dispute arises on a project (and one will, one day) will you find out how effective your process really is.

As stated previously, the purpose of a contractual resolution procedure is to bring about a final, mutually adhered to, resolution. Correctly written, it is not a path that 'might' find a solution, it is a path that 'will'. In reviewing the procedure, you will most likely find that the first step after any initial negotiation meeting is mediation or conciliation. As you have learnt, both of these involve an independent, neutral third party. The difference being that mediators will know little to nothing of the construction field, and hence play no part in developing a solution; conciliators, on the other hand, will know the field and so can aid in suggesting a solution. Neither, however, will make decisions. This is the first step that 'might' find a solution, but it is not a given.

Because of this, it is important to ensure you enter such sessions with an open mind, an array of possible solutions and the 'intent' to achieve a resolution. In so doing, you want to hold clear in your mind which points you are willing to compromise on, and those you would rather not. You should likewise keep in mind what the next step is should this first one fail, and the cost of such a step in time and money. This latter point is likely to be raised by the mediator or conciliator during the session, particularly if the session is not being productive.

Developing preliminary solutions

Normally you will have had at least one negotiation meeting prior to a mediation or conciliation session. From this first meeting you will usually have gained some insight into the other party's stance on the issue. You will need this information to develop your preliminary solution.

In developing possible solutions to take into a session there are a number of strategies open to you:

- clarifying the problem
- defining goals
- brainstorming (individual or group)
- comparing previous examples (within the industry or outside of it)
- empathy.

Clarifying the problem

This step will need to be repeated, possibly multiple times throughout the resolution process, due to the issue raised previously; i.e. individual perspective on 'facts'. In this preliminary stage you are only clarifying your perspective – how you and your team hold the matter to be. It is nonetheless an important action to take. Without it you cannot possibly come up with any solutions that are likely to satisfy you or anyone else.

Defining goals

This strategy is about determining what elements are needed in a solution that would satisfy yourself or your company. Likewise, it can be used to determine the goals you will be seeking at the end of the resolution process.

Brainstorming

You can do this by yourself, or preferably with a group. It's a process whereby ideas are put up without judgement no matter how crazy or outlandish they appear upon first hearing. No judgements are made until later, when they are workshopped and consolidated. The advantage of this process is that a suitable solution may be formulated by combining, or adopting part of, several ideas.

Previous examples

This involves researching past examples of disputes similar to your own situation. Accessing this information can sometimes be problematic as neither mediators nor conciliators can give out details of disputes they have handled. Both services can, however, offer examples in a more general sense that may be applicable. Lawyers and solicitors who are experienced in building and construction matters may also be a source, but with an associated cost. The best sources are from fellow builders and subcontractors who have had to go through something similar in the past. Their experience will be invaluable.

Empathy

This is where you change hats as the saying goes. You put on the 'hat' of the other party and attempt to look at the issue from their perspective. This is harder, yet easier, than you think. Harder, because if it is not your natural inclination to imagine alternative ways of looking at things then you will initially struggle. Easier, because once you master the concept, it is remarkably easy to find holes in your own arguments this way. The reality is, of course, that you can never fully appreciate another's perspective: particularly if the other party is from a very different culture to your own. Even so, it is well worth the time making the attempt.

Once you have identified at least one preliminary solution, coupled with notes on what you are willing to compromise on, and that which you feel must be retained, you are ready for your first formal meeting, be it mediation or conciliation. As stated earlier, you enter this meeting informed by the contract of the stages that must come next, should these sessions fail. These should be spelled out to the other party(s) so that each knows the possible costs of not reaching an early settlement.

Resolution: identifying the opportunities

Clearly, disputes are not desirable on any construction project. Time and money are lost all too quickly if a resolution is not found in the initial meetings. This is why you must look upon each meeting as an opportunity to refocus both yourself and the other party back upon that which is important – the project. No matter how distant your individual positions appear to be, you must approach each opportunity in good faith and show that you are willing to both listen and seek an appropriate resolution. This is particularly important in your first meeting – the informal meeting of negotiation.

First avenue: meeting and negotiating

This is your first opportunity to discuss the issue and explore the reasoning behind each other's stance. Questioning techniques are crucial here. How you frame your questions will dictate the quality and value of the information you obtain. Likewise, your questioning technique, and your responses to theirs, can have a big influence upon the outcome of this initial session.

LEARNING TASK 7.2 INVESTIGATE DISPUTES AND RESOLUTION STRATEGIES

Consider the options discussed in the text. How might you approach the following issue, which is very likely to lead to a dispute?

Scenario:

You have ordered roof sheets, barge flashings and guttering materials for your latest job and taken up the supplier (which you regularly use) on his offer to deliver the materials at no extra charge.

The materials arrive and upon inspection as the materials are coming off the truck you notice that most of the barge flashings are bent and/or damaged in various places.

The truck driver is adamant that you need to sign the delivery docket and take up the damages issue with the supplier.

You refuse to sign – telling the truck driver to take all the materials back to the supplier as you are not prepared to sign for the delivery on the basis that there are damaged goods in the delivery package.

The driver returns two days later with the supplies order in good condition and you accept that delivery.

You are now two days behind schedule and have needed to re-arrange your subcontractors to attend the site.

This has cost you an anticipated additional $500 in reorganising the on-site work around the materials delay.

Your invoice from the supplier in the following weeks shows $3600 for the 'agreed' ordered materials and additional charges for that order – 2 x lines indicating:

- additional charge of $280 for replacement of damaged flashings
- additional charge of $115 for re-delivery of original order
- total invoice reads $3995 (inclusive of GST).

1 What are your options in regards to the invoiced amount?
2 What 'proactive' strategies should you take if you believe the invoice is inaccurate?
3 Provide legitimate/legal reasons for your answers.

Questioning techniques

There are many types of question; however, most fall into one of the following three categories:

- open questions
- closed questions
- reflective questions.

Open questions

An open question is one that allows the other party to express themselves fully upon a subject. You use these questions to find out details of the other party's stance on the issues at hand and to invite them into the discussion. You will typically use open questions to start proceedings, but they can be used at any point to encourage the other to express their opinions, possible solutions, or the like. Such questions might begin with: 'What do you think ...'; 'Explain to me ...'; 'What's your opinion on ...' and so forth.

Closed questions

A closed question seeks a short, concise response: such as 'yes' or 'no', a choice between two or more options, or at most a very short affirmative or negative sentence. A typical closed question may be 'Do you agree with ...'; 'Which do you choose ...'; or '... is this true or false?'. While these are handy for getting quick answers, they should be used with discretion and mixed with open and/or reflective questions. Too many closed questions may lead the other person to feel pressured and without any level of control or input. Ultimately, these questions may become dictatorial and offensive if used excessively.

Reflective questions

Reflective questioning is you feeding back to the other party their own statements in a clear and concise manner, paraphrasing their statements, while seeking confirmation that you have their position clear. Such statements might start with 'So what you mean is ...' and end with '... have I got that right?'

This is a particularly important technique when dealing with those from a different cultural background, those for whom English is not their first language, or those who have become emotional about the issue or otherwise find it difficult to express their opinions clearly. Likewise, it is useful when dealing with someone who does not know the industry and is battling with construction terminology or concepts. Again, you must not overuse this form of questioning as it can begin to sound condescending.

The negotiation

There is a high probability that both yourself and the other party will enter the negotiation with a possible solution to which you will each be seeking the other's agreement. The problem is that these two solutions may be diametrically opposed. What happens next is down to the skill of each of you as negotiators.

Negotiation is not a linear procedure that may be mapped and followed identically in every case. It tends to be a fluid and dynamic process rather than a formula and will depend greatly upon the context, the skills and the backgrounds of each of you. Aside from questioning techniques, there are a few actions you should consider at various times in the meeting that may assist both parties in reaching your goal:

- Trust building – share and explore each other's ideas and possible solutions. Don't be afraid to try brainstorming (see p. 186) with each other. This shows a willingness to try anything that may bring about a resolution appropriate to the situation and agreeable to each. Indeed, it may lead to a solution derived by a part of each other's ideas.
- Knowledge sharing – don't try to be secretive, freely offer information and explain issues with clarity.
- Don't lay blame on the other – this may or may not be the case, but hold back on 'pointing the finger' at the other party. Lay out the issues clearly: if the other party accepts a level of fault or error, good; if they do not, let it go. Likewise, you avoid being defensive in responding to their questions, respond as openly as you are able.
- Take breaks – sometimes the discussion might become difficult and stressful. Anticipate these moments if you can, or at least recognise them as they evolve, and take a break.
- Be true to yourself – be willing to speak your own mind calmly and with authority. Your opinions and needs are just as valid as the other party's.

Telephone and online communication

When a relationship is sound, despite the dispute, a well-thought-out telephone call may be sufficient to solve the dilemma. Well thought-out, in that you should be just as prepared for the conversation as you would be if it were face to face. Well thought-out, in that you must make your points clearly – and lacking the visual cues of face-to-face meetings – listen carefully without interruption to the other's ideas, statements, responses and questions.

Online video chat platforms such as Skype, Viber and WhatsApp are one up on the phone call in that you regain the missing visual cues. How effective these platforms will be depends much upon the quality of the internet linkage at both ends, but they can be very useful. Again, you must be well prepared prior to the event, and take notes during.

In both cases, phone or online chat, whatever resolutions or ideas you come up with should be put down in writing, sent electronically or in hard copy to both parties and signed by both parties as a true account of the meeting. If it is an agreed resolution then this will be signed and filed alongside the contract with copies to both parties.

Moving forward

It's not impossible, indeed it is hoped, that this session concludes with a resolution and the dispute goes no further. If such is the case, then you should ensure that the minutes to the meeting are written up and the resolution signed off by both yourself and the other party. With this agreement in hand the work may proceed and you take what lessons you can from the experience.

Occasionally, the other party may enter negotiations from a position of power and wealth, capable and willing to take their time to get what they want. You must recognise such actions early and weigh up your options – sometimes it may be appropriate to see a wall for what it is and simply follow the procedural steps of the contract towards arbitration or tribunal as quickly as possible. In so doing, you should continue to present a willingness to negotiate should they respond accordingly.

On other occasions, despite a genuine willingness to negotiate, things may not play out as you both hoped. Such being the case you must look to the contract and move to the next level. This should have been spelled out at the beginning of the session, but agreement at least on the necessity of the next step, possible dates, time limits, and who will be the third party (mediator, conciliator) must be gained at the meeting's close.

Second avenue: mediators, conciliators and arbitrators

At the start of this chapter the roles of mediators, conciliators and arbitrators were outlined. You should therefore be aware of the differences between each and where they might fit in a domestic or commercial contract. In most domestic contracts it will be mediators who will be the first option after a failed initial negotiation. Remember, with mediators and conciliators the power of decision making remains in the hands of you and the other party. Arbitration, on the other hand, removes this power.

The mediation session

Mediators come from a variety of different backgrounds, none of which need to be aligned with the construction field. Their skill is in helping two divergent parties reach an appropriate resolution of their own making without any legal or technical input from the mediator themselves. It is a significant skill that should not be underestimated.

Early in the session, mediators will stress the value of an early settlement, and the risks and potential costs of litigation through the courts, or the lack of power in the determination of a resolution by arbitration or tribunal. They will not give any insights into legal processes, however, nor will they offer any suggestion upon the outcome of any of these court-like proceedings.

What will also be stressed is that a willingness to compromise is the key to success in a mediation session, and that both parties must enter the discussions in good faith; i.e. with an intent to find a solution.

The conciliation session

Unlike mediators, conciliators come with a strong knowledge of the construction field. Their skill is in guiding parties to reach an appropriate resolution with experiential and technical input. Like mediators, they will not provide legal advice, nor comment on a likely outcome if the issue went to arbitration, tribunal or court.

Preparation

You should approach a mediation or conciliation session well prepared; in some cases, both yourself and the other party may decide to bring a legal advisor to the session. It is preferred, however, that either both parties do, or both parties do not, bring legal support. Indeed, most mediators prefer that no legal representatives are involved at all as they can tend to try and dominate the sessions with what can amount to legal threats rather than assistance towards a resolution. That stated, you should get as much legal and construction advice as possible prior to the session starting. This will mainly be in the form of documentation, including:

- contracts
- plans
- variations
- text messages, emails and site notes
- photographs
- minutes of the negotiation session(s)
- site diaries, phone records, notes on past conversations leading up to the dispute
- engineer's reports and/or expert witness reports.

Your approach to and operation within the session should be similar to that of negotiation sessions: open, honest, with clarity of speech, polite and courteous without rancour, and a willingness to listen and follow the guidance of the mediator and the principles of mediation they lay before you.

In cases where the project is effectively finished and the dispute is about some level of required rectification or restoration, be prepared to accept a monetary solution. Returning to a job to do something can actually end up with further disputes arising.

When it is a matter that has arisen prior to the project's completion, it is important that the work continues. The pay now, argue later, philosophy is often best kept in mind here; likewise, the contractual requirements that in most cases demand work proceed while the dispute is settled.

Most lawyers will strongly advise both parties towards settlement at this session, and provide evidence that all other future avenues will be expensive and most likely in excess of the value of 'winning'.

Arbitration

This is where you hand over control and await an outcome. You and the other party have decided that you each believe strongly in your stated positions and do not feel it appropriate to compromise. Such being the case, you both have determined to place the matter before someone, or a panel, that will make a determination based upon the facts placed before them – each hoping that the process will find in their favour. You are reminded that in most states and territories domestic housing disputes cannot go to arbitration. Instead you will be guided into the state tribunal. This will be clear in your contract.

As with mediators and conciliators, you should have a nominated arbitrator (and back-ups) as part of that contract. Prior to the sessions, the arbitrator will most likely call for a discovery of documents from each party (see p. 182). This will take time, and probably some cost. You will then be called to the arbitration meeting where you may verbally present your case and so argue the merits of your position – as, of course, will the other party. There may be more than one session as the arbiter seeks clarification on one issue or another. These sessions are private and can be confidential in their findings if both parties agree.

In entering arbitration, it will be critical that you have strong legal support or advice. This is as court-like as it gets without it actually being a court. Facts are key elements, even though arbitration is more concerned with equity than law. This is an important difference. As you will read in the next section of this chapter, courts are about the law, not what is fair or seemingly reasonable, despite what a judge themselves may think of the issue. An arbitrator, on the other hand, keeps equity (and therefore fairness) in mind. This means that the determination should, on balance, give a result that is fair and reasonable when looked upon dispassionately – even though, if it goes against you, you might feel otherwise.

Once the arbitrator has made a determination, it is final and any directions must be adhered to within the timelines specified. Should one or the other party renege on their part of the determination – i.e. fail to undertake some action or pay some monetary compensation – then the matter may be brought before the courts. The party that has reneged is then in a very poor position regarding any potential court hearing falling in their favour unless some legal issue surrounding the arbitration process can be demonstrated.

Third avenue: the statutes

This is not a separate avenue as such, but more of a strategy or avenue within the others. Knowing the statutory law surrounding contracts and the construction industry in general can be of vital import to your success in any dispute resolution process, including the initial negotiations.

In researching the various applicable Acts and regulations you will be seeking to determine:

- the compliance requirements of the building, and contractual legislation of your state or territory
- requirements of industry codes and standards called up by the legislation as they apply to your state or territory.

You will need to take particular note of how the above applies to the nature of your particular dispute.

Legislation is the written law as dictated by parliament and formulated as an 'Act'. There are federal and state Acts as there are federal and state parliaments. These Acts are then defined or detailed through regulations. In addition, both federal and state Acts may call up codes and standards, such as the National Construction Code (NCC) which may include state variations that need to be considered. In some instances, laws are made by local councils that are specific to that shire or area.

Contract law was covered in Chapter 3, so only a cursory reminder needs be provided here that the Acts concerning contracts are specific to each state or territory. Sources for information on these and other Acts may be found online through the state and federal government websites. In addition, you may find the Acts, the regulations, and supporting information through state-based government agencies such as those listed in Table 7.1.

TABLE 7.1 State government building and planning agencies

State or territory	Government organisation
Australian Capital Territory	Environment and Planning Directorate
New South Wales	NSW Fair Trading
Northern Territory	Building Practitioners Board
Queensland	Queensland Building and Construction Commission
South Australia	South Australia: Housing, Property and Land
Tasmania	Department of Justice
Victoria	Consumer Affairs Victoria
Western Australia	Building Commission

Having found the required regulations and/or applicable codes, you should study them to determine if or how you may have breached them. If you are convinced that you have not, then this information will logically form a significant part of your rebuttal of the other party's position. If you have breached them, then you will be more informed on the likelihood of any success you may have in a court of law, a tribunal or even in arbitration. Likewise, it informs your stance in mediation and conciliation sessions and how concessional you may need to be.

Fourth avenue: tribunals, courts and common law

In most states and territories, you will have access to the courts for the adjudication of a construction dispute irrespective of the size or nature of the dispute; i.e. whether it is a domestic housing issue or a large commercial issue. However, in most states and territories there is some form of tribunal such as the Victorian Civil and Administrative Tribunal (VCAT), QCAT (Queensland version), NCAT (NSW), or State Administrative Tribunal (SAT) of Western Australia. It is to these that you will invariably be guided first.

State tribunals

Technically, these are not courts, although recently VCAT has been deemed to be effectively so in limited circumstances. Irrespective of this, state tribunals are the main avenue for the determination of most domestic construction disputes and, to a lesser degree, commercial construction issues as well. Determinations made in these tribunals can be ratified (made legally enforceable) by a court, if both parties agree, with some further costs being incurred. In addition, either party may request that the resolution be enforced by court order should the other party not respond to the tribunal's directives within the timeframe given. The cost of the order is then applied to the offending party.

With regards to housing disputes, if there is a state tribunal (VCAT, QCAT, or the like) then you will be strongly advised or directed to that institution first. In addition, if you do go to court and during the proceedings you decide that the matter would be better served through a tribunal process, then you can have the matter transferred. In most states it shall be taken up by the tribunal as if it had begun there in the first instance.

The courts

Just to be clear, don't go here. The courts don't want you wasting their time, the lawyers do want your money (though even they will advise you to stay clear), and time is on everyone's side but yours. It is expensive, and the law is about the law, not fairness: you can be in the moral right, but by law, you can be in the wrong. And it will cost you: in money, time and stress. It is this latter point that many do not understand until it is too late. Long legal processes can be emotionally debilitating: the stress over time, and the impact that this period can have on your ability to retain focus upon other projects, can mean further construction issues may arise and/or money may not be coming in. And because the stakes can be high, rational thought – on both sides – can be lost, and the fight taken far beyond what either had intended at the outset: and so you both lose.

There are times, however, when you may be taken to the court by the actions of the other party. In such cases you will need to be prepared and this will mean sound and experienced legal advice. Each state and territory has its own specific systems for conducting construction disputes under common law. However, the basic structure of our legal framework is similar in each instance and is based upon a separation of powers, as shown in Figure 7.1.

What we know commonly as the Australian government is actually made up of two parts with regards to the making and administration of law:

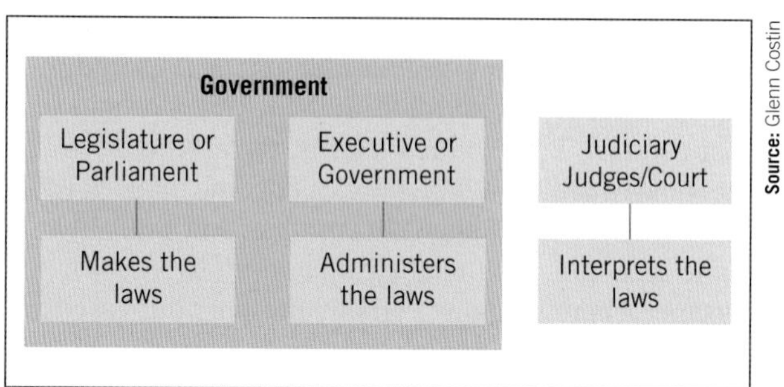

FIGURE 7.1 Administering the law in Australia

- the parliament, or legislature, makes laws applicable to the country at large
- the government is the executive: this is the body that enacts and upholds the laws established by the legislature.

You can see that these two bodies are framed collectively under the one heading 'Government'; this is because the same people can be both ministers of government, and members of the legislature.

The judiciary stands alone. They enforce the law, as developed by government, through the court system. They also review the two arms of government – legislature and executive – to ensure that they are acting within their powers. It is the judiciary with which we are most concerned in this section.

The judiciary – the judges, the courts, the lawyers, bailiffs and solicitors that make up this arm – are the body of people who interpret the laws passed by government. They make the determinations that affect those who appear before them. This is the key point. The law is open to interpretation. It is an argument between two parties backed by their understanding of the written law: the better they understand it, the better their argument 'appears' to be. Arguments are supported in the main by previous cases. New interpretations are also made as community expectations and the peculiarities of specific cases challenge older understandings. It is this process that is known as common law or case law.

Common law

Based historically on English law, common law in Australia is now based firmly on the principle of precedent – that which went before. This principle applies across all Australian courts irrespective of state or territory systems. When an appeal (case) is heard in court, the findings, once published, form part of this common law history of past cases (precedents) and so can influence new cases. The law in this regard is not static: it is interpreted. However, the law itself – the wording or 'letter' of the law – remains fixed. Changes to common law is the role of government, or more correctly, of the legislature within government. The courts can recommend a change, but they cannot enact it.

This is the key point to understand when considering pursuing your case through the courts. It is about interpretation of law, not about fairness or equity. Often a case is 'won' not on the basis of what is 'right' but on the capacity of one party's legal team to convince the court through argument that their position is the stronger 'in law'; i.e. as per the 'letter' of the law, rather than the purpose of that law.

Preparation

As with all previous levels of contractual resolution pathways, preparation is key to success in both these arenas. Indeed, success itself may be a matter of perspective. That is to suggest that you may not win, but you may defend yourself such that your losses are less than they may have been.

To this end, you must ensure that you have access to very good legal representation who have expertise and experience in construction law. You may need independent technical experts or witnesses who may support your arguments in matters of structure, code adherence and the like. They can also give estimates of costs associated with delays, rectification of defects and other construction details.

Your legal representatives will also have advised you on your need to ensure the other party has the capacity to pay should the verdict fall in your favour. It often can happen that despite you winning both the case and the costs, the other party is not in a position to pay anywhere near that which has been awarded to you. Likewise, they will advise you on the potential costs you may incur should you lose – you need to be sure not only that you can pay them, but that you can afford to lose such sums and still be in a position to conduct business.

Hence, no matter which avenue you end up taking, it is vital that you gain advice from experienced construction lawyers both beforehand and during either a tribunal or court session. And if they advise you to stay out of either, stay out.

LEARNING TASK 7.3

RESOLUTION: IDENTIFYING THE OPPORTUNITIES

Develop a list/flowchart of the systematic steps that should be worked through in attempting to resolve a business dispute before it ends up in a civil court.

SUMMARY

The purpose of this chapter was to provide the knowledge of how to resolve a dispute, and more importantly, how not to get into one in the first instance. With disputes arising from a range of issues, frequently minor misunderstandings, it was shown that rarely is one party attempting to cheat the other. Underlying this message was that touched upon at the very outset of the chapter – ethics and customer complaint handling. Respect and dignity towards others are your greatest assets. At the same time, the chapter showed that if you approach a dispute without a willingness for open and calm discussion, the cost in time, money and emotional energy can be significantly greater than the value of the project itself. A range of government-supported mechanisms were therefore discussed. These included mediation and conciliation as the preferred options. In addition, it was shown that to ensure access to these options as a first action, it was important to have them as clear clause directives within your contract. Ethical conduct, preplanning and clear procedures – these are your best friends in business, and in defusing disputes this remains true.

REFERENCES AND FURTHER READING

Weblinks

https://www.disputes.vic.gov.au/about-us/mediation

https://www.ethics.org.au/consulting

https://www.fairtrading.nsw.gov.au/trades-and-businesses/construction-and-trade-essentials/resolving-a-dispute

https://www.fairwork.gov.au/how-we-will-help/templates-and-guides/best-practice-guides/effective-dispute-resolution

https://www.ombudsman.gov.au/__data/assets/pdf_file/0020/35615/Better-practice-guide-to-complaint-handling.pdf

https://www.sat.justice.wa.gov.au/C/commercial_and_civil.aspx

PART 2

BUSINESS PREPARATION

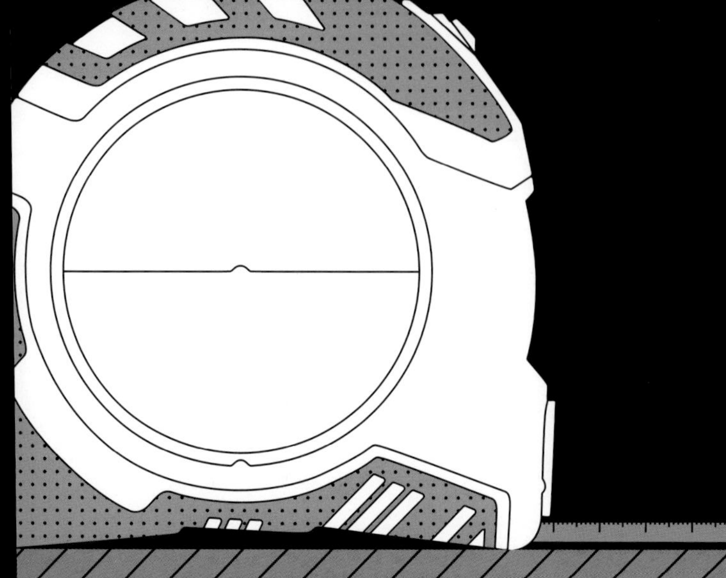

8 TENDER DOCUMENTATION

Chapter overview

Tendering in the construction industry is a deceptively simple, often multi-layered process used by clients and builders alike when undertaking a project. It is one of many paths in what is known as procurement, the process of acquiring goods and services in the construction industry. Tendering has been identified by the Australian government as being one of the most crucial yet most difficult areas with regards to contracting. Undertaken appropriately, with due fairness and honesty, it will provide access to the best capability, quality and cost. Inappropriate tendering practices can impose excessive costs on not only the initiating client, but also on potential subcontractors as well. Hence, tendering for government projects particularly – federal, state or territory – is subject to strict rules and codes of practice.

The purpose of this chapter is to introduce you to the subject in both a general and an applied manner. Note that the focus of this chapter is the commercial sphere on the basis that occurrences of tendering in the domestic housing market are rare, and when they do occur, they commonly use commercial contracting practices.

Elements

This chapter provides knowledge and skill development materials on how to:

1 evaluate contract risk
2 prepare tender documentation
3 identify and attach appropriate supporting documentation
4 evaluate completed tender documentation
5 obtain tender approval or endorsement.

The Australian standard that relates to this chapter is:

- AS 4120 – 1994 Code of tendering.

Introduction

We will begin this chapter with a brief introduction to the terminology. Without a clear understanding of at least a few basic terms it will be hard for us to move forward in an informative manner. There are just three terms you need to know up front, two of which define the key roles involved in the tendering process:

- Tender: This is a formal offer in writing to undertake a project, supply goods, or carry out works in accordance with a prescribed set of documentation. It includes a definitive price or expected remuneration.
- Tenderer: This is the party that submits the tender. These may be contractors, subcontractors, or suppliers of materials or services such as labour (a non-exclusive listing). They are the respondents in the tender process.
- Principal: The individual, group, or company that has called for the tender, and will receive the tender application. They are the initiator of the tender process.

Within the construction industry, many people can play these roles. In fact, as a builder it is not impossible that you will find yourself playing both of the key roles at some point, possibly within the one project. This may come about in the following manner:

- As the respondent to a call for tenders. You are the 'tenderer' making an offer to do that work for which the principal has requested prices and timelines.
- In the preparation of a tender, or having gained a contract, you call for tenders from other trades to complete parts of the project. In such cases you are the 'principal' seeking quotes and timelines.

The tendering process refines this calling for, or offering to undertake, work in a number of ways. The most common approaches being:

- open tender – effectively allowing anyone to make an offer, which is then judged on capability and experience as well as cost
- prequalified tender – generally open only to those already qualified to be on a listing
- limited tender – by which predetermined prospective contractors are approached directly.

Such paths are heavily backed by best-practice methods developed by various agencies and bodies including governments, building organisations and architectural groups.

The competency unit around which this chapter is framed is focused upon the role of the contractor or 'tenderer' – the one making the submission – as opposed to the 'principal' or client – those calling for tenders. Despite this, we will at times take up the role of the principal. There are two very good reasons for doing so: first because, as has been pointed out previously, large and small construction firms frequently call for tenders from subcontractors and so you may well find yourself in this role at some point; second, to improve your appreciation of the tendering process – particularly with regards to risk management.

The purpose of tendering is to identify the best party to undertake some required works, leading ultimately to a binding contract. Given the legality of these contracts, there are significant risks surrounding tenders on both sides; either in making an offer, or in the acceptance of that offer. It is, therefore, the evaluation of these risks that we will look at first.

Contracts and risks

There are risks in signing any document involving a transaction of some form. However, the risks in tendering are higher due to the processes and possible cost implications involved. Further, the nature of these risks changes depending upon whether you are acting as the principal or the tenderer. Both risk management and quality management systems play a key part in the tendering process no matter which role is being played.

Risk management is the coordinated processes and actions undertaken by an organisation to ensure the risks associated with that organisation's activities are identified and understood. It allows the organisation to manage these risks informatively, as well as identifying potential areas of opportunity.

A quality management system, on the other hand, involves the development of policies and procedures governing the planning and execution of an organisation's core business. It includes both quality assurance and quality control elements (see Chapter 6). It is the hub of an organisation's documentation and policies designed to ensure that the company meets its own quality objectives, as well as those of the client, and any regulations, codes or standards including work health and safety (WHS) requirements.

In the case of tendering, the first part of the risk management process is to identify the type of contracts that may be involved. From these, the risks associated, from a tendering perspective, can then be identified and means for their appropriate management developed; assuming it is considered that submission or acceptance of a tender should proceed.

Selecting the contract

As stated above, effective quality management in any company includes some statements that define its core business and policies outlining how it conducts or engages in that business. This informs you on what sort of projects your company is willing to get involved with, quote for or tender for. All companies are 'selective' in their willingness to tender, this being their first line of risk management: being choosy as to where they will risk their time, money and reputation.

For example, as part of small rural building firm with an experience base in domestic renovations and basic maintenance works, you might consider taking up the contract for maintenance of the local secondary school. However, it would not be wise – i.e. the risks would be too great – to take on the contract to build a multi-storey apartment block in a neighbouring city.

Even if your firm has developed a quality management strategy, such a contract would be well outside your capacity and experience base. Equally logically, it is unlikely to be received with any favour by the principal.

Alternatively, as an employee of a major commercial construction firm your selection of tendering contracts may be much broader. A call for tenders for a multi-storey project would most likely fit your company profile, and be possibly well received. However, you would still need to assess the project against your company's defined policies. To do this in an informed manner, you must have a clear understanding of the tendering process.

The tendering process and best practice

To tender is to enter into a competitive process. Tendering is the means by which a company, or consortium of companies, will be selected to do the work which the principal has outlined in the original tender documents. The selection process will include the overall cost, but will not be limited to that (or should not be). Tendering is generally considered a fair approach because everyone knows the exact nature and scope of the works to be conducted, hence what they are pricing on. Likewise, they know what information the tender application must include if it is to be considered.

The tendering process may be reduced to five basic steps:

- call for expressions of interest (not always included)
- call for tender
- submission of tender
- evaluation
- negotiation.

When deployed, the call for expressions of interest is used to identify those agencies or organisations most capable of carrying out the works. The principal can then check, and the interested party can check, for those factors that might make an organisation eligible or ineligible for consideration – such as qualification, experience, or knowledge base.

However, while tendering should be a fair and equitable process, this is not how it always plays out. Likewise, it does not always result in the most appropriate or capable contractors receiving the contract.

Best practice

In response to these issues, in 2008 the Municipal Association of Victoria, the Institute of Public Works Engineering Australia and the Civil Contractors Federation collectively published a document referred to as the *Victorian Civil Construction Industry Best Practice Guide for Tendering and Contract Management*. Similar documents now exist in other states and territories, though mostly focused upon government (local, state, or federal) tendering guidelines. On 1 January 2018, the federal government commenced application of the *Commonwealth Procurement Rules*. Again, these are focused upon government procurement practices, which includes tendering. Such documents offer very clear and concise advice on how the tendering process 'should' work if managed under a best-practice model.

Although AS 4120 Code of tendering predates all of these documents, these new guides and rules are more detailed and are more aligned with contemporary procurement practices and the identified issues.

You are advised to access and read these documents for the details; however, the basis of each with regards to tendering best practice is a strong emphasis upon ethical behaviour. The key points of their principles are outlined in Table 8.1.

It is with these in mind that you begin to frame the tender documentation. If acting as the principal; i.e. you are calling for subcontractors to undertake works on a project for which you have been awarded the main contract, then the selection of which contract is most applicable must now be considered.

Choosing contracts: past, present and future considerations

As the manager or employee of a construction firm, your task in selecting those contracts or calls for tender for which you will make a submission requires strategic thinking: your choice is based upon not only what has gone before, but also on those projects you are currently undertaking, and where your company would like to be in the future.

The actual type of contract framed by the tender is invariably the decision of the principal party, though there are instances where early engagement allows for discussion on this point. It is chosen on the basis of the

TABLE 8.1 Tendering principles

Subject	Principle
Ethical business practices	covering all aspects of the project's management
Probity and transparency	fairness, impartiality, consistency and transparency
Value for money	having regard to policy, performance, risk management and life-cycle costs
Open and fair competition	maximising the opportunity for firms and individuals to compete for business
Accountability	allocating responsibility for compliance with policy and adoption of best practice
Risk management	risk should be managed by the party best able to manage the specific risk
Local industry participation	using local suppliers whenever legitimately possible
Minimisation of tendering costs	consider costs of tendering to both principals and contractors
Confidentiality	tenderer's submissions remain confidential and not disclosed to other tenderers
Intellectual property	intellectual property of one party's tender not to be used to obtain alternative prices
Open market	no discrimination against tenderers who have previously declined invites to tender
Open evaluation	all tenderers informed of the process by which tenders will be considered
Selective tendering numbers disclosed	number of tenderers invited to tender is disclosed to all
Invite documentation	same conditions of invitation, selection and engagement to all tenderers
Collusion	parties shall not engage in collusive tendering or anti-competitive practices
Complaints	establishment of a complaints process which is included in invite documentation
Timeframes	sufficient and appropriate to tender requirements
Advice	contracted and non-contracted tenderers to be advised as soon as possible

type of project and scope of works involved. It is also chosen on what is known as the 'procurement strategy or model' or 'delivery method'. All of these factors will have an influence upon your decision to submit a tender or not.

Procurement strategies

This is a statement of how the client or principal wants to approach the construction of a particular project; e.g. do they want someone to build something for which they have a design; or do they want the contractor to design and build it? Even with invites to subcontractors, these strategies may apply. Common strategies or methods for which tenders are called include:

- construct only
- design and construct
- construction management
- managing contractor.

If contracting with the public or government sector, there is also the possibility of being involved in public-private partnerships (PPP), project alliancing, and early contractor involvement (ECI). As these are generally for major infrastructure projects, they have not been discussed previously in Chapter 3 Construction contracts; nor will they receive more than passing commentary here. However, they are procurement models that are gaining importance in Australia. For further information on these models, access the *Building and Construction Procurement Guide: Principles and Options*, which may be found at: https://www.apcc.gov.au/publications

and

https://infrastructure.gov.au/infrastructure/ngpd/files/Volume-1-Procurement-Options-Analysis-Dec-2008-FA.pdf

There is then the manner of pricing to be considered: In what form is the project required to be costed?

Costing models

In most cases, construction contracts in the commercial sector tend to fall into one of the following categories:

- fixed price or lump sum
- lump sum with rise and fall
- cost plus
  - cost plus a fee
  - cost plus a percentage
- guaranteed maximum price
- bespoke or custom contracts – into which PPP, alliancing and ECI will fall.

You need to identify which of these formats has been chosen as they have obvious restrictions, as well as a range of other implications on cost flows such as profit margins and financial risk.

Contract types

As Chapter 3 described, most commercial construction works will be conducted through one of the various types of 'standard' contracts listed below. This is despite the fact that in over 80% of cases these contracts will need to be amended from their original published form, generally in order to shift risk.

- Australian standard contracts; e.g. AS 2124, AS 4300 or AS 4000
- Australian Building Industry Contracts or ABIC Suite; e.g. ABIC MW – 2008

- GC21 – mostly NSW but also Queensland
- National Public Works Contracts or NPWC forms.

Each of these will have inclusions on how this particular document is to be used with regards to tendering. Some will include the whole of an additional Australian standard, or will have a clause covering the 'general conditions of tendering' as it applies in that particular case.

In-house policies

Quality management systems develop policies and procedures governing the planning and execution of an organisation's core business. It is these policies that you must review to ensure that the proposed tender aligns with the company's objectives. In some cases, these policies will dictate the types of projects and contract types your company is willing to enter into in a very finite manner; on other occasions there will be room for discussion on opportunities to expand or otherwise explore projects of a different nature to the past.

Most companies of any size will have policies about contracts and projects that reflect the following:

- maximum (and minimum) price of contracts for each contract type available
- scope of works – governing the nature and size of work to be undertaken within a specific contract type
- clarity and division of scope of works – how well a project is to be defined for the purpose of calling for sub-tenders
- timelines – how to estimate the period of time within which the works must be completed, as well as start/end dates
- operational influences – ensuring a capacity to succeed (outlines of internal client checks to be undertaken prior to furthering a consideration); conflict with other projects or tenders
- relevance to company expertise base (or capacity to outsource or subcontract)
- clarity of risks (type of risks, the capacity of the company to absorb or offset those risks)
- the 'fit' of the proposed project to the profile of the company
- the capacity to profitably advertise engagement with the project (during the works), and after completion
- the risks to successful completion if the project involves partnerships with other tenderers.

These policies will also have commentary on the accepted practices for the amendment of standard contracts to fit more closely with the project requirements. This point is raised for two reasons: first, because you may be in the position of principal for the purposes of tendering out to subcontractors for parts of a project; and second, these policies can guide you when assessing any changes or amendments that the principal may have made to the tender contract you are considering.

Amended contracts

Amendments to contracts must be done with great caution. Re-wording a standard contract can lead to a number of unforeseen complications and legal ramifications. Any changes to wording should be checked by a lawyer experienced in construction contracts. Likewise, bespoke or custom contracts must be assembled with a legal firm of a similar experience base. There are risks associated whichever path you or the principal might make.

- **Using a standard contract without amendments**: this can lead to the contract framing, and thereby limiting, the project possibilities, instead of the project framing the contract.
- **Amending a standard contract**: can lead to a raft of 'knock-on' affects, with a new clause conflicting or annulling an existing clause. In any dispute, arbitrators and administrators must go by what is written, which can lead to unwanted and unfair outcomes.
- **Using a custom or bespoke contract**: this has similar risks to amended contracts in that without very thorough checking by more than one legal entity, it can lead to poorly phrased clauses, conflicting clauses, and hence unwanted outcomes.

Contract selection, in all its aspects, is therefore a matter of careful consideration of risk at all levels, even within the document chosen to limit those risks – the contract document itself. As always, you are strongly advised to seek to identify and then allocate risks to those best positioned to manage those risks. The *Building and Construction Procurement Guide: Principles and Options* mentioned earlier provides a comprehensive (six-page) table of where risks should best be allocated based upon the type of delivery model being used (Appendix A: Delivery model profiles); it is strongly advised that you review it.

Some final tips

The first is simple: if you don't believe you can deliver, or there is a question mark on your capacity to deliver, don't put in a bid – it's that simple. Make no assumptions about your position to win a tender; most certainly do not place yourself in a financially dangerous position framed on this assumption. Monitor your existing resources (financials, labour) prior to putting in bids; winning new tenders can put a strain upon your capacity to deliver existing contracts.

The tender documents and risk consideration

In this section, we will focus upon the role of you as the prospective tenderer. That is, you are in receipt of a tender package or packages (if reviewing multiple possible opportunities to make submissions) and must evaluate them – particularly from the perspective of risk management.

As outlined previously, for the tender to be fair and equitable, the documentation must be completely closed. Closed meaning that anyone seeking to submit a tender will obtain the same package of material. No additional information or documents may be added to any individual's package, even verbally.

However, if upon receiving and reviewing the documents, one or another prospective tendering party considers that further information or details are required, then a formal request may be submitted to the principal to that end. Anyone is entitled to do this, and it is not in breach of any codes or standards. What occurs after this request is made is then at the discretion of the principal; they can choose to either:

- decline to expand on the existing information and details
  or
- amend the package as a whole for all prospective tenderers.

The principal cannot inform you at the exclusion of others, verbally or otherwise. Each party thus remains in a fair and even contest, and can review the risks in an informed manner without disadvantage.

The documents

A tender package is often extensive, and its contents will vary according to the nature of the project. Its purpose is to inform you of project size, complexity and delivery or procurement strategy proposed by the principal. Documents may be in hard copy, or more frequently today, in electronic form. It is from this, and the known history of the principal, that you must make your next levels of risk assessment.

Despite the variances associated with different project types, a typical package may be broken into six parts, the first three of which may be combined into one on small projects.

- **Notice to tenderers**: Project summary, document listing, key dates, period for which the tender remains open or valid, contact names and details, format of submission including where and when to submit.
- **Conditions of tendering**: Details the tender process, procurement strategy, probity and communication issues, selection criteria and evaluation process and timelines.
- **Tender form and schedules**: Require specific information from the tenderer with regards to the proposed works. This may include costing breakdowns, program of activities, details of personnel, plant and equipment requirements, subcontractors and suppliers.
- **Conditions of contract**: Sets out the general and specific conditions of the contract and any amendments to standard contract forms. This you won't sign, it's simply a sample so you know what the contract will look like and hold if you are awarded the tender.
- **Specification**: The contents of this section will differ depending on the type of procurement strategy being used.
  If design and construct, it will not have a complete specification as would normally be supplied with a set of plans; rather, it will have a design brief. If construct only, then a full specification and/or bill of quantities will be supplied as normal.
- **Design and engineering details**: Again, this will depend upon the procurement strategy. In may hold an artist's impression of a design concept, only site drawings and context, or full drawings including engineer's details, sectional details and the like. These may be supplied as hard-copy 2D drawings, computer aided drafting files (CAD), or, on major projects, a set of 3D images that are in a building information modelling (BIM) system.
- **Additional information**: May include environmental impact statements and studies, community studies and other information that may be relevant to the type of project and procurement strategy involved. In some cases, it may request that such studies be undertaken as part of the contract.

While the above outlines what is in a tender package, what we haven't discussed is how to access these packages in the first place.

Accessing tender documents: the modern world

How you access these documents is changing. Calls for tenders are still advertised in the national and local papers, in which case you must call or email a request for the documentation to be sent to you, or you go and pick the package up in person. More frequently, they are being advertised through online 'portals'. In many cases, particularly with regards to major projects and government contracts, the whole process in handled through these internet portals.

Each state and territory government has its own online tendering portal, as does the federal government. Only a few these are listed here. In addition, there are many private portals in use. These capture and direct you to a range of other portals including the government ones, as well as advertise and provide full tender processing for non-government organisations and private clients or principals. See Table 8.2.

TABLE 8.2 Tender portals

NSW Government: NSW eTendering and ProcurePoint	https://tenders.nsw.gov.au/
Queensland Government: eTenders	https://www.hpw.qld.gov.au/bas/eTender/Default.aspx
Tasmanian Government Tenders	https://www.tenders.tas.gov.au/#
Australian Tenders	https://www.australiantenders.com.au/
Federal Government: AusTender	https://www.tenders.gov.au/
Tenders Online (formally Cordell Tenders)	https://www.tendersonline.com.au/

Evaluating a tender for risks

Having obtained the tender package and confirmed the contents, you are in a position to evaluate the tender for risks. You should do this at two levels: both involve a study of the documents; however, the evaluation is done from two different perspectives.

First-level evaluation

At this stage you make a thorough study of the documents to ensure they are complete, that they outline the project appropriately and that there is no immediate requirement for further information. This first study will allow you to identify some of the more obvious risks, such as:

- WHS risks, particularly ones that might be beyond the norm due to the nature of the project to be undertaken.
- Project scope and its alignment with your company's experience and expertise base.
- Timelines: particularly contract clauses that impose costs on your company if there are delays.
- Timing: how does this project's schedule fit with projects for which you are already committed? This can cause cash flow, skill, labour and equipment shortfalls.
- Profit margin: is it sufficient to cover contingencies such as additional labour, equipment hire, interest rates and the like.
- Need for outsourced expertise, skill, trades, labour or equipment.
- Principal or client history: what is their credit and payment history? What is their general reputation?
- Stakeholder concerns: community, environmental, cultural or other concerns that might slow or halt the project, or otherwise damage your company's reputation.
- Guarantees, quality and maintenance period requirements.
- Demolition or earthwork risks: identified, and the possibility of unidentified works that could be associated with the project.

Though not an exhaustive list, it covers the common risks associated with most projects for which a call for tenders may be made. Each of these issues should be taken on their merits individually: alone none need cause alarm, but would need to be factored into your pricing structure. If, however, a number of these issues showed themselves to be problematic, then collectively they may cause a company financial issues, damage its reputation and/or interrupt other projects in which you are engaged.

Risk is part of being alive, and of being in business. Hence, risk management is not about elimination of exposure to risk, but rather limiting the extent of that exposure and any implications that might arise from that exposure. It is about ensuring that if the company was hit with whatever burden arising from a risk becoming a reality, it is in a position to carry that burden, continue to operate and ultimately deliver.

However, risk management is also about the apportioning of risk: sharing the risks associated with a project such that each stakeholder carries those risks for which they are best positioned to control and/or withstand.

Hence, in tendering you must not only identify the risks, but also clearly pronounce in your tender application those risks that you are prepared to carry, and those which you would expect the principal or subcontractors to carry. This applies equally in those instances where you are the principal and are calling for subcontractors to tender for elements of the project.

Second-level evaluation

The second level of evaluation requires more time, and greater experience in document interpretation. This is effectively a case of reading between the lines, and at times between documents. After this second evaluation, it is not impossible that you will need to call for further documentation, or at least clarification of particular phrases, clauses, or diagrams.

This second level of evaluation is important due to the wide-ranging nature of the documentation that must accompany any tender. In gathering the material, the principal is seeking to provide clear and accurate information on the proposed project. In addition, they must compose their invitation and preliminary material in a manner so that prospective tenderers are clear on what it is they are tendering for. Depending upon the project scope or nature, this can be a significantly complex exercise, and so flaws can exist.

To combat this, many agencies that commonly act as the principals provide their officers or staff with a clear risk checklist or tables for the creation of tender documentation, and the package as a whole – this is part of their risk management. As the tenderer, these tables and lists offer insights into areas of risk for your company. That is, where there is a risk for the principal, there may be an associated risk for the tenderer. Hence, these lists and tables are your guide to reading between the lines of the documents the principal has provided.

Such tables and lists typically focus upon the following key areas:

1. identifying the need and procurement strategy
2. developing specifications and/or design briefs
3. selecting the costing model
4. purchasing documentation, e.g. contracts, in preference to in-house development
5. procedures and processes for inviting, clarifying and closing offers
6. how tenders will be evaluated
7. selecting the successful tenderer

8 negotiations
9 contract management
10 evaluating the procurement process
11 disposals.

Other lists may work around phases or stages of the tendering process. In some cases, these documents may include a risk and consequences action table, not unlike that in the *Building and Construction Procurement Guide: Principles and Options* mentioned previously. The actions will include where risks should be allocated, generally away from the principal. It's not that you are going to try and foist such risks back upon the principal; rather, in identifying the risks, you can now make decisions about the project more informatively.

Loss leader tendering

Loss leader tendering is a considered risk that some companies will take in an endeavour to enter new markets. On occasions you will find that your company is in a position to make a strong tender submission, and most likely win the bid, for a particular tender. But it may also be clear that little or no profit will come from the project. Normally, sound business practice would discount making the application in the first place. However, it is possible that the tender provides an opportunity that ties in with the company's long-term strategy.

In such cases, provided that delivery will not cause clashes with existing projects, and that the business is in a position that it could withstand a minor loss if the project failed to break even, then it may be worthwhile. It can open up new markets, and build new business relationships, either of which may bring profits in the future.

LEARNING TASK 8.1

IDENTIFYING COMMERCIAL ETHICAL PRACTICES AND THEIR CONTRACT RISK

Access AS 4120 and compare the Australian standard to the online document nominated below:

- http://www.infrastructure.nsw.gov.au/expert-advice/nsw-government-construction-leadership-group-1/, see 'Practice Notes' at the bottom of that web page.

Or use the alternative (government) *Code for the Tendering and Performance of Building Work 2016*:

- https://www.legislation.gov.au/Details/F2019C00289

Document your 'comparison' findings using a template as shown. See earlier in the chapter for items you could include in the first column.

No.	AS 2140 says:	Nominated online document says:
1		
2		
3		
4		
5		
6		
etc.		

Tender preparation

Having identified a particular tender contract as a match to your company's profile and/or projected direction, and determined the risk profile is acceptable, you are in a position to create a submission. When making a submission you need to plan for it. The first steps of which have been covered in the previous section; i.e. analysing the tender request and confirming the tender aligns with your company's capacity to deliver.

The next stage is twofold: first, identifying the format that the tender package requires a submission to follow, and the information required to make such a response; second, ensuring that the approach to the submission follows any stated company policies and procedures.

Identifying and obtaining information

There are a number of sources from where you might obtain the information you need to make a submission. The bulk of the information, however, will stem from the following:

- the tender package or document set
- the pricing structures and/or quotes from suppliers and subcontractors
- in-house personnel
- your company procedures and tendering formats
- planning and development meetings.

The tender package

It would be erroneous to believe that the documents provided by the principal are sufficient in themselves to compose a tender submission; even assuming that the package is complete, or that the principal has been forthcoming with any additional information requested. What the package does provide, however, is the basis of your search for information, and the manner in which that information must be compiled for submission.

Most tender specifications will include a response template. In short, use it. It will have required formats, word limits and be of a particular digital file format. You may have the option to work on it in one format, but be required to submit it as a locked PDF. By following the presentation format requested, you are more likely to at least get before the review panel.

Depending upon the type of project or procurement strategy involved, you will also use these documents to inform your requests for quotes from suppliers and subcontractors. Likewise, the information contained will allow you to frame preliminary timelines and critical activities. Such information may include:

- types of trades and skills involved
- the scope of works
- the scale of works
- timelines and construction programs as required by the principal
- proposed payments schedules
- quality standards required
- qualifications required by the contractor (you), subcontractors and workers
- specific materials and/or technologies (common, new, or otherwise unfamiliar)
- specialised equipment requirements
- access limitations
- bill of quantities (or the information with which to create a bill of quantities).

In addition to this information, you will need to confirm the details of what constitutes a conforming tender; likewise, if a non-conforming tender may be submitted.

Conforming and non-conforming tenders

A **conforming tender** is a submission that meets all the criteria as laid out in the tender documentation. What constitutes a conforming tender will depend significantly on project type, the contract documents involved and all the other variables of the procurement process. In general, however, a conforming tender is one that acknowledges all the elements of the project as described in the tender documents and is completed in the manner required.

A **non-conforming tender** is effectively a submission that does not meet all such criteria.

However, even though a tender does not conform exactly to all elements of the tender package, this does necessarily preclude it from consideration by the principal. It depends upon whether or not, within the invite to tender or elsewhere within the documentation, there is a statement that allows non-conforming tenders to be submitted, and on what basis. The reason non-conforming tenders are sometimes allowed is the nature of certain projects, particularly ones that are design and construct. The tender may call for particular materials or processes to be used, but the principal is aware that alternatives might exist. In such cases the allowance for non-conforming tenders may arise.

The basis for the acceptance of non-conforming tenders is generally the inclusion of specific documentation that supports an alternative approach. In particular, those that demonstrate that such an approach will provide the principal with an equitable or improved outcome by way of time, cost and/or quality.

Suppliers, subcontractors and pricing structures

In almost all cases, the principal requires a total cost. In assembling the relevant documentation for the tender, your documents must show a clear breakdown of the costs leading to this total. Again, in most cases, this will need to include some form of **bill of quantities**.

On occasions a bill of quantities will be included in the tender documentation offered by the principal. You should not take this as being in any way accurate. It is generally only supplied as an approximation; acting as a guide to prospective tenderers and as an aid in making that initial decision of whether to make a submission or not.

For the purposes of quoting, you will generally need to determine those areas for which **subcontractors** will be required. On occasions, depending upon the nature of the project, you may even need to split the job up into parts, whereby certain specialty contractors are used for those whole elements, such as demolition, toxic material removal and the like. In such cases, you may choose to call for sub-tenders as against standard quotes. On these occasions you are then acting as the principal and supplying your own documentation packages.

In calling for either sub-tenders or quotes, you may have preferred suppliers, tradespeople, or contractors. However, you must check the original documentation to identify any required qualifications and memberships, or if the principal has specified their own preferred suppliers, etc.

On occasions, your suppliers and subcontractors will not provide a quote, with its associated costing breakdown, but rather a pricing structure. Common pricing structures may include a cost per metre rate for wall framing, or a cost per square metre for tiling, floor sanding, painting, plastering, or the like. You would then need to make your own calculations based on these figures. In such cases, you should note the time period for which such prices will hold, and provisions by which the prices may fluctuate.

These calls for quotes and/or sub-tender submissions must be disseminated, and responses received, in a timely manner. They will then need to be formatted and included in the pricing and documentation that forms your tender submission.

In-house personnel

In larger companies particularly, you will have departments and staff that are either experienced in the compiling of tenders, or in the particular type of project you are seeking to engage. You need to take advantage of these sources of knowledge and skill.

Departments

Policy and procedure will probably require that you access these departments at various stages in the process of compiling a tender. The tender may have to pass through them as points of checking (such as finance) before submission can be approved. However, in compiling the information in the first instance, these departments can inform you of the types of information that they will be looking for and so speed up your task. Hence, it is worth holding workshops or meetings with the broader group early in the process.

Staff

Depending upon the nature of the company or its structure, there may well be staff that are experienced with the project type under consideration. Such staff should not be underestimated in what they can bring to the project in the planning stages, including the tender submission. If they don't know, they may be able to advise you where such information or expertise may be found; i.e. identify prospective subcontractors or suppliers.

Company procedures

Quality and risk management policies, particularly in larger companies, frequently have fixed procedures, or at least a guiding structure, for the compilation of a tender. In developing a submission, you must access these procedures, forms or other documents, and ensure that they are used as such procedure dictates.

For example, your company may have a set format in which the tender should be laid out. This could be as simple as document formatting – such as headers and footers, page number styles and the like. Or it may be the manner in which a bill of quantities is presented and/or that the costings fit the accounting practices of your firm.

Following such procedures is not necessarily as easy as in sounds. As mentioned earlier, the tender package, as received by the principal, may also have clear guidelines for a submission. On occasions these may clash, in which case you must negotiate the company policy such that a conforming submission is viable. Generally, this will mean sticking with the tender's required format and file type, but ensuring compatibility with your company's preferred accounting and/or quantifying systems.

Planning and development meetings

Accessing either individual staff or departments is preferably handled in a professional and coordinated manner; this allows input to be recognised and documented. The best way for this to happen is that you develop a timeline into which you locate development and planning meetings at which calls for information may be made, key stakeholders identified and their concerns aired (see 'In-house stakeholders and clients'). Information, its relevance to the project and example documents may be discussed at these times, as well as being actioned for tabling at future meetings. These meetings are also one level of quality checking, ensuring that company procedures are being followed as the tender submission is staged out and a whole-of-company response to the project is developed. Figure 8.1 shows an example tender submission timeline.

Company procedures and instructions

Company procedural processes have been touched upon in the previous section, but only with regards to the appropriate use of form and format. What must now be addressed is equally important: the procedure by which a company may consider it is appropriate to 'approach' the compilation of a tender submission.

Compiling a tender: a corporate approach

When you work for yourself, such as being the owner–operator of small domestic building firm, you develop your own way of doing things. Provided these methods stay within the various regulations governing all the various aspects of construction, then these practices will work fine in most cases: you don't have to explain your actions to anyone. However, when compiling a tender as the employee of a larger firm, it's likely you will need to explain something to someone at some point.

Companies have their own in-house way of doing things and these will be laid out in a set of formal procedures. This is no longer an issue of basic forms and format styles but rather procedures for how you go about doing a particular activity. Such procedures are important for a number of reasons, in that they serve to ensure coordination between the various activities and departments of an organisation. These procedures may include:

- the identification of in-house stakeholders and clients
- steps to follow to ensure external pricing is accurate and current
- quality control procedures for ensuring tenders are appropriately compiled
- stages through which each element of a tender must pass (including nominated personnel) to ensure all foreseeable contingencies are accounted for (i.e. clashes with other project timings, cash flow constrictions and the like).

In-house stakeholders and clients

In-house stakeholders are people and departments within a company that will somehow be affected by the project that is being tendered for. They have an interest in the proposal because they will most likely have to do something if the tender is won. It is important that you identify these parties early in your process. Important because even compiling a tender may affect their capacity to perform other activities or the functions they operate within the company or for other departments.

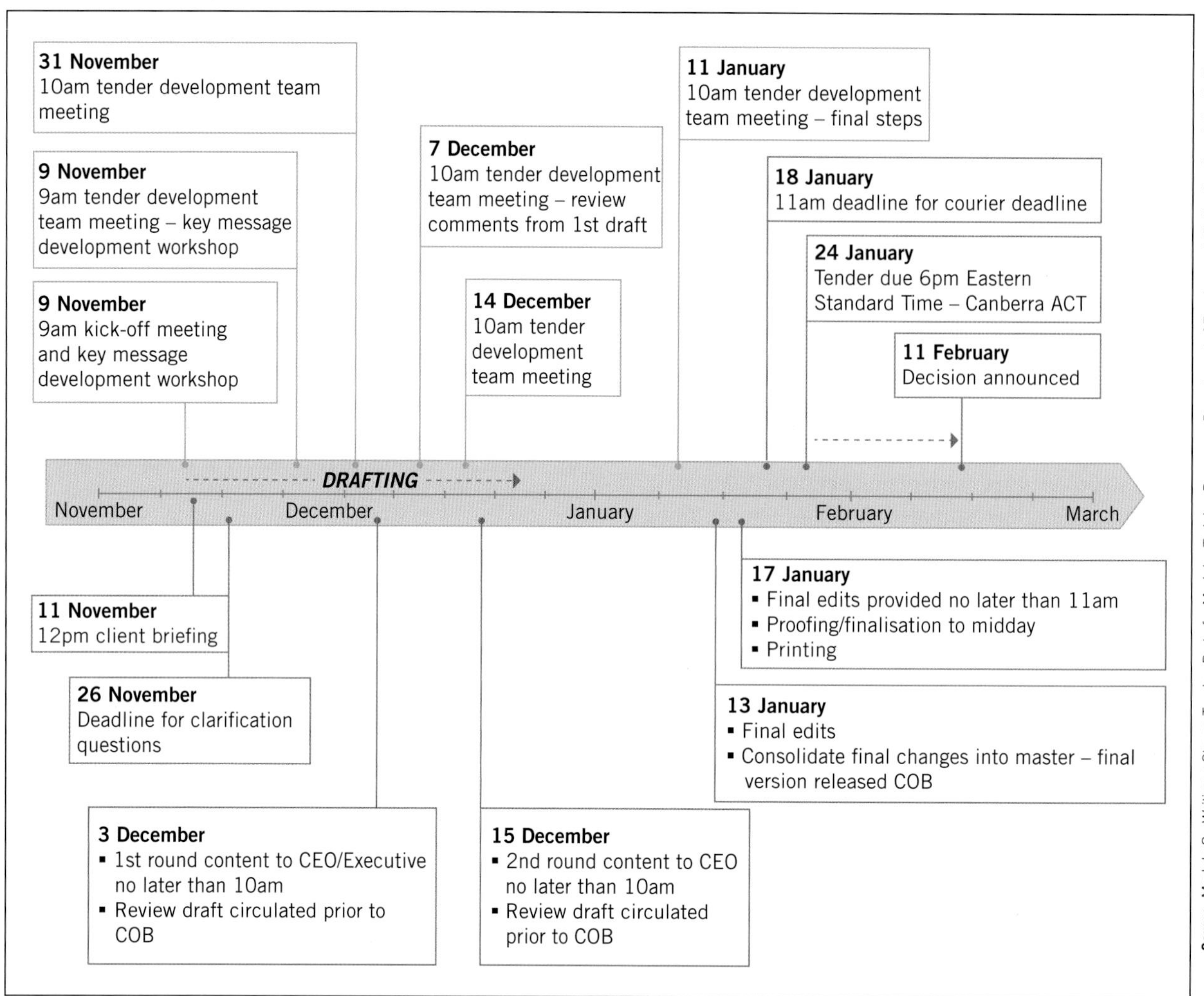

FIGURE 8.1 Typical tender submission timeline

In-house clients are people whom you serve effectively as customers. You are the provider of a service to them or they rely upon you to provide some form of product such as documents, quotes, or the like. Many of these people will need to engage with the tender process to ensure that their roles are not adversely impacted upon; that your service to them is not compromised, or that their capacity to perform as the project unfolds in the future is not jeopardised by the project.

The company's procedural guidelines are there to ensure that you contact each of these stakeholders at appropriate stages during the tendering process. In so doing, the risk of misinformation within the finished document is reduced, and it limits any adverse impact upon company performance should the tender be won.

Pricing procedures

Irrespective of the size of a project, pricing errors may creep into a quote or tender all too readily. Pricing procedures are put in place by companies to limit such errors and to ensure that last-minute changes to pricing have been included in the final documentation. They also ensure that pricing includes all identifiable parameters of the works to be undertaken; this is where a very careful analysis of the tender package is most crucial. Most importantly, the approach to pricing, its format and content breakdown, must align with the company's accounting procedures.

Once again, this will be a set of guidelines. The guidelines will inform you as to how you obtain your quotes, from what firms, or how you nominate new firms for the purposes of obtaining prices. It will also provide the procedures by which your pricing will be checked and at what points in the process these checks should take place.

Quality control

Quality control is that arm of quality management that checks procedures are being followed and to the standard required. With regards to the tendering documents, this will most likely be a set of steps through which your documents must pass. Each step ensures that company guidelines have been met, and that the final product, the tender submission, and all documents associated with it, are professional. Likewise, that they are presented in a manner reflective of the company's standards.

Stage checking

This is a natural extension of having identified in-house stakeholders and clients. Stage checking ensures that at each phase of the project, should it be won, there will be no clashes with existing or other proposed projects. This is critical to the tender submission's proposed timeline and/or progress payment schedule. Without these checks a company may encounter issues such as cash flow, labour or skills shortages, and equipment availability. Each of these problems will inhibit the firm's capacity to deliver on all projects, not just the new one.

Developing procedures

If part of a small firm, you may not have any procedures for tendering. Finding in-house procedural examples to replicate will generally not be possible as they are by definition 'in-house'; i.e. closed to outside scrutiny. Examples of government tendering or procurement policies are, on the other hand, fairly easy to find. Unfortunately, these will invariably to be from the perspective of the principal.

However, some government agencies provide excellent guides on their websites on how to approach a tender submission. Business Queensland's multi-webpage guide is highly informative; it may be found at: https://www.business.qld.gov.au/running-business/marketing-sales/tendering. It is worth examining this website, and many of the other documents previously mentioned, for information on appropriate approaches to tendering: either with the intent of creating your own policies and procedures, or if you are tendering for the first time.

LEARNING TASK 8.2

MANAGING THE TENDER PREPARATION PROCESS

Provide a flowchart of the activities you follow when initially obtaining access to an advertised tender and requesting that the tender details be forwarded to you (or your organisation). Also provide a systematic sequence of the in-house activities that would need to be carried out through to the point where your tender offer/proposal to complete the works is forwarded to the principal.

The supporting documentation

Within the tender package there will be a listing of the documentation that must be included in your tender submission. It is important that you take note of this listing and respond accordingly if you aim to submit a conforming tender. There are occasions, however, when you will need to add further information, the most common being when:

- tendering for the first time in a particular field of construction or type of project
- this is your first tender to this client
- submitting a non-conforming tender.

In any of these cases you are likely to need to offer further supporting documentation.

Identifying and selecting supporting documents

The operative word here is 'supporting'. In creating or compiling information for your tender you are seeking that information which supports any claims, prices or concepts you are putting forward. The tender package will request that you provide a price, but the principal will be looking for evidence that supports your claim to be able to deliver for that price, and on time. Areas of your tender that may require supporting information include:

- Pricing structures and quotes – evidence will be needed on who supplied the quotes, as well as the quotes themselves. Likewise, for pricing structures, evidence that these prices will be maintained over the life of the project. Evidence will also be required that demonstrates the capacity of the subcontractors and suppliers involved to supply as and when requested, and that what they supply will be of the requisite standard. In each case, evidence may be in the form of:
  - quality management credentials (such as ISO 9001)
  - past performance and histories
  - references, which can be useful for smaller enterprises.
- Technological capacity – here you must demonstrate that you and your subcontractors are positioned to handle the materials and equipment required to complete the project. In some cases, this may be data management systems such as BIM, or a new metal alloy or polymer that requires specialised jointing equipment. Alternatively, it might simply mean the context in which the project must be delivered, such as in Antarctica or underwater. Evidence may include:
  - past histories and reports of previous successful projects
  - evidence of training and experience
  - any qualifications for materials or equipment handling relevant to the project.
- Financial capacity and insurances – you need to demonstrate that the company can carry any cost burdens required between progress payments, and beyond if there are disputes that may delay such payments. In addition, you will need to confirm that you have the requisite insurances in place, such as:
  - insurance of the works
  - insurance against public liability

- workers compensation insurance
- professional indemnity insurance.

  Less common:
- marine insurance covering the operation of waterborne craft more than 8 m long used in carrying out the contract.

  Supporting documents may include:
- letters of surety from banks and lending bodies
- financial statements
- copies of insurance policies including public liability
- proof of WHS statutory liability insurance.

- Quality workforce – you need to show that you have access to workers with the requisite qualifications, skill and experience base. This is for all workers and administration staff, not just specialist areas particular to this project. Evidence may include:
  - listings of staff qualifications
  - quality management credentials (company; e.g. ISO 9001)
  - past histories and reports of previous successful projects
  - evidence of training and experience in areas of high risk.

Tendering for the first time

Clients like surety. They want to be certain that you have the capacity to carry through on the project, delivering it in a timely and cost-effective manner, and to the quality they require. If this is your first tender, your first project of this particular type – shopping mall, high-rise, multi-residential, or the like – or it is simply your first project with this client, then they made need more convincing.

Often the principal will have some criteria of historical performance that you will need to satisfy. The usual approach to this issue is through a careful study of such criteria, to which you would respond with:

- references
- past project case studies that may fit with elements of the proposed project
- examples of exemplary past projects (of any form) demonstrating high quality and timeliness
- a general history of quality performance in construction.

Non-conforming tenders

On occasions, in calling for tenders, a principal will allow tender submissions that do not conform to the exact requirements of the tender package (see p. 202). Likewise, on occasions, you or your company may be in a position to offer something that differs from the project as it has been described in this package. What you propose might be safer, requires less labour, reduces cost, or speeds up delivery of the project; and doing so while retaining or improving quality and maintaining the client's end purpose. This may be in the form of alternative materials, changes to design, or better equipment or technology.

In making such a submission, you will naturally have to be able to back-up your claims. Such documentation may include:

- technical information and/or material safety sheets from material manufacturers
- costings signed by subcontractors and suppliers
- building information modelling (BIM) data
- construction programs and schedules
- artists' impressions
- examples of previous projects where the suggested approaches have been successful
- alternative bill of quantities by a qualified building surveyor.

Principals allowing non-conforming tenders do so because they are aware that they may not have the best solution defined within the tender package and so are seeking ideas, options, alternatives. However, what the package contains is 'safe'. Doing it that way has been tried and tested over time and so they know it will work. It just may be that it could be done better. But you need to persuade them that the alternative proposed will succeed and that your company can deliver. Hence, your supporting material must be clear, evocative, precise, and above all, convincing.

Drawings and specifications

This is a deceptively simple task that often is the weak point of many tender submissions. Before submitting the tender, you must gather and append all the plans, specifications, bills of quantities and the like associated with the project. Depending upon the nature of the project, or the procurements method involved, the bulk of this material may have been in the original tender package, or the onus may have been on your company to produce it. In either case, you must make sure that all such material is included, particularly any changes, or additional material, to that of the original package.

The main documents

The documents may be clustered into five main groupings: plan sets, specifications, bill of quantities, documents supporting proposed amendments, completed forms and the like from the tender package.

- **Plan sets** are the most obvious documents required and the most open to potential errors of submission. These must be complete with all elevations, site plans, details, sections and the like. In particular, you must make sure that you are supplying the latest amendments and that they are signed and dated accordingly. Where changes have been made to any original drawings supplied by the principal you must make sure they are highlighted appropriately and clearly.

- **Specifications:** If they have been developed by your company then again you must make sure that they are complete, respond appropriately to the design, indicate where they align with Australian standards or other codes or requirements, and are complete in all details. Where the specification is an alteration of that supplied within the tender, you must be even more careful that it has remained a comprehensive document and that critical areas have not been inadvertently lost.
- **Bill of quantities:** As with the specification, this may have been part of the original package, or developed by your own quantity surveyor. In either case you must make sure that it is the most current version, and that it responds to the most current set of plans and specifications. Errors in correlation between any one of these three can lead to a failure of the bid, or financial risks should the tender be successful.
- **Documents supporting proposed amendments** should be concise and relevant, yet detailed. Such documents have been discussed previously, but in compiling the tender submission you must ensure that they cover all aspects of any alternative approaches. Be aware that other parties within the company may be responsible for these documents, as may be suppliers or subcontractors. In cases where documents are missing or incomplete, those people will need to be contacted for supply; allow time in compiling the submission for such occurrences.
- **Tender forms**: A tender package will contain a range of forms and documents that must be completed and signed by appropriate members of your organisation. You must make sure that these are complete. Likewise, there may be documents that only specific departments could reasonably respond to, or from whom advice will need to be sought for their completion. You must ensure that such information has been sought and the forms are correctly completed.

Building information modelling

BIM software is becoming an important tool in the development, planning, construction and ongoing maintenance of buildings in Australia. In some regions, BIM data files are the only approved means of plan submission. This being the case, it is not unlikely that you will be required at some point to submit BIM files as part of your tender submission.

In such instances, you must make sure from the outset that your file retains compatibility with the principal's: there being multiple BIM platforms, most of which will 'talk' to each other, but there can be errors. You also must make sure that all relevant files are included, and that any links within the system continue to function when the files are transferred to either another storage device or when uploaded through online (cloud) portals.

BIM systems are powerful tools requiring specialist training and input. When available, make sure you engage their full potential. For example, where an alternative approach is being applied to a glazing element, you can create links in the 3D model to supplier details, images and pricing schedules, as well as the bill of quantities and the total costing. You can also build in a variety of alternatives so that preferred approaches and their costs can be demonstrated.

LEARNING TASK 8.3

CHECKING AND RECORDING THE STATUS OF SUPPORTING TENDER DOCUMENTATION

Complete this activity by accessing a plan set that you have been working on, or one provided by your instructor.

Develop a template including the main headings that will be used to ensure a formal check that all supporting documentation and items contained in a tender package have been investigated in-house and properly processed, to ensure that they are included/excluded before your final tender submission occurs.

Alternatively, gain access to and supply information against that document in the format a response template requires.

The completed document

Ensuring that a tender submission is complete is something that should be checked very thoroughly. It is best approached using checklists that have been developed as the tender has progressed. Generalised checklists are an acceptable starting point, and have value in ensuring that some of the more basic inclusions are covered. However, more detailed and tender-specific lists, building from the general, will be required for a complete final analysis.

Preliminary evaluation

This is a fairly simple check that must be very thoroughly applied. As stated above, the best approach is by a checklist based upon general tendering principles and then expanded to include the known specifics of each particular tender. Part of this expanded checklist may be derived from the document package, or the principal's website. From these sources you can frequently obtain the submission checklist that the principal's own office will use for acceptance and conformity.

An example checklist is offered in Table 8.3. Though by no means comprehensive, it may act as the basis of a generalised approach.

TABLE 8.3 Tender checklist

Item	Included	Comment
Original tender package document listing checked against submission content listing		
Mandatory tender schedules are included		
Specified quality management requirements met		
Workplace health and safety (WHS) management system requirements met		
Full plan set or BIM file included		
Revised specifications included		
Licensing and registration requirements met		
Specified environmental management system requirements met		
Total tender price included		
Financial capacity demonstrated		
Technical capacity demonstrated		
Specialised subcontractors listed with qualifications		
Specialised suppliers listed with credentials		
Experience, nature and complexity of work, demonstrated		
The tender sum includes provisional sums, prime costs allowances and/or schedule or rates sums (if these are included in the tender)		
Construction methods outlined		
Project environmental management plan included		
Workplace health and safety (WHS) contractor management system demonstrated		
Construction program outlined		
Current commitments disclosed		
Technical capacity disclosed		
Management and personnel capacity disclosed		

Note the inclusion of a comments section. This allows for statements on partial completion or from where the outstanding document or information must be sourced.

Conformity to administrative guidelines

In-house policies and guidelines were referred to earlier in this chapter. Checks now need to be made to ensure that such policies and guidelines have been followed through and are evident in the tender submission. Summarising from pp. 203–205, the key points that you will be checking are:

- in-house stakeholders – check calculations align with the processes and procedures of their departments or systems
- external pricing – these also must be in a format and of appropriate detail that aligns to the in-house policies and procedures
- quality control – ensure these have been carried through with regards to the conduct of calculations and their format and presentation
- calculation accuracy – ensure that calculations have travelled through and/or are derived from the approved departments, and their accuracy confirmed before being included in final documents.

Most importantly, you must confirm that all calculations have been approved by the accounting department or, for smaller businesses, your accountant. In this way you can ensure that they follow a format that aligns with in-house accounting procedures. Failure to undertake this step can lead to missed inputs, and increased financial risk should the tender succeed.

A last check

It has been mentioned previously that the best path for ensuring that a tender is conforming with the requirements of a particular principal is to access their own tender acceptance policies, rules, or, better still, examples. This is not always an easy matter when dealing with private principals; however, with government contracts, examples are plentiful. It is recommended that you access and download the NSW government's Service Provider Tender Evaluation Plan – Worked Example Form from their ProcurePoint website: https://www.procurepoint.nsw.gov.au/ (scroll down to 'Service provider selection'). You can also find the full url at the end of this chapter.

In studying this document, you will quickly see the basic points by which a principal may automatically exclude a tender, and by which paths they are most likely to approve. It is invaluable reading.

LEARNING TASK 8.4

INVESTIGATE PRE-QUALIFICATION CRITERIA

Access the NSW government website listed in this chapter. After reviewing the document, provide a sample checklist to your trainer based on the pre-qualification criteria on that website.

Company endorsement

As the last stage prior to submitting a tender, this is in many ways a reprise of the previous checks and balances. It is also backtracks over the initial decision to take on the tender. What follows, therefore, is in many ways a revision of the previous sections.

Paths to endorsement

Endorsement from company management for the submission of a tender requires that you have obtained, followed and documented compliance with all the requirements of:

- the principal's tender package
- the company's policies and procedures.

Tender package compliance

The checklist you produced to ensure the tender submission's contents were complete (see Table 8.3) should front the package you have compiled. This acts a bit like a contents page – though you should have one of those also. This checklist should be slightly changed such that either the 'comments' or 'included' columns provide instead the location of a document or part within the submission package. The use of removable coloured tabs strategically located throughout the document can also assist in the reader identifying the required aspects.

Upon submission to the appropriate manager or director, the onus then falls upon them to determine if the tender should proceed, if it needs more input prior to submission, or that the risks outweigh the opportunity.

Company policy and procedural compliance

The purpose of establishing a timeline and staged meetings was to ensure that policy and procedures were being followed through right from the beginning. If this has not taken place, then checking the document for compliance at this stage becomes a laborious task: for now you must pass the document back through every department and individual (all the in-house stakeholders) and gain their endorsement.

The 'Company procedures and instructions' section (see p. 203) laid out the means for ensuring that policy and procedure was being met during the creation of the tender. Included in this section, and in most company procedures, is the requirement that you gained endorsement, by departmental signatures – electronic or otherwise – at each stage at which their input was gained or required. This is a part of their check, their departmental procedure.

Small construction firms

In smaller construction firms the process is simplified and it is easy to suggest that you simply sign off your own work. The risks here are obvious, and so it is best practice to have the documents studied by persons other than yourself prior to submission. Best practice suggests three steps of consideration:

- an **informed partner** – this person looks over the work first for basic errors and completeness according to an agreed checklist. An 'informed' partner is one who is knowledgeable of the tender process and your business practices.
- an **accountant** – also looks over the whole document checking calculations, sums and any errors where figures in one section may not align with those in others. Your accountant may be your 'informed' person if they have sufficient background in your business practices and tendering in general.
- a **solicitor** who can advise on the contract conditions. Even if you have already done so earlier when first considering the call for tender, a second look once the tender submission is complete allows the solicitor to check alignment and risks that you or they may not have identified earlier.

Readiness for submission

When the tender submission is complete in all its parts, and endorsement gained from company management, the submission needs to be appropriately packaged. And despite the adage that books should not be judged by their covers, they invariably are, which is why marketing companies are so sought after in all avenues of business, including construction (and books …).

The purpose of correct packaging is not, however, just to catch the eye; indeed, in construction, this often acts against you. Rather, it is to ensure that the principal can evaluate your tender easily, check quickly that it meets all the requirements for a conforming tender (assuming that is what you are submitting), and is easily handled, page to page, part to part. Simplicity and clarity are the major drivers in how a submission should be packaged.

Final cautions

Tendering, as you should now be fully aware, is a process bound by law. These laws and regulations are in place to ensure fair play – that no tenderer has an unfair advantage over another. To ensure this, the tender process is closed and private. What one tenderer submits is not disclosed to another, and two or more tenderers cannot collude with each other to bring about higher winning bids. Likewise, taking information or material from another tenderer without their consent – in effect, stealing – is a breach of the law.

In taking your documents to an external party; for example, a marketing firm for the purpose of improving its visual appeal, is to take a risk. For this reason, most businesses will conduct all packaging and tender preparation in-house where possible. Marketing teams, if not in-house, will never see the whole package. Such agencies will only be employed for front covers, format of page headers and footers perhaps, and maybe template designs. This is a policy that even small firms should consider when venturing upon larger tendering projects.

Insights to construction law and tendering

There are a number of resources that you may like to consider investigating with regards to the law as it applies to tendering. This weblink provides some excellent case studies as well as some very considered information about the legalities of tendering: http://www.constructionlawmadeeasy.com/managingthemainrisksoftendering. You will also find numerous links within this webpage that can expand upon the subject more fully. In light of your knowledge of tendering, you may also like to review what you studied in Chapter 2 Legal requirements for building and construction projects.

LEARNING TASK 8.5

MANAGEMENT DECISIONS ON FINAL TENDER ACCEPTABILITY

You have just completed 60 hours of work on preparing a tender submission which has been circulated to management for final approval. You are 'shattered' to find out that the tender submission you have submitted to them will not be forwarded to the principal.

Provide a list of 10 possible reasons for why this may have occurred.

SUMMARY

As stated at the outset, tendering in the construction industry is a deceptively simple process. You should now be aware of just how deceptive. It is indeed a multi-layered process and you may be acting as either principal or contractor, and frequently both, when undertaking a project.

You should also now be aware of the many types of procurement, and where tendering fits into these strategies. More importantly perhaps, you should now understand the implications these strategies have upon the final format of a tender response, and the risks these different strategies hold for each party. This is why the Australian government continues to identify tendering as one of the most crucial yet difficult areas in relation to contracting generally, not just in the construction industry. This is also why this is an area of construction that is surrounded by law and ever-improving descriptions of best practice.

The purpose and basis of all these rules and attempts at best practice is to ensure fair play, as tendering has been shown to be open to abuse. Throughout this chapter, the issue of fair play has been at the forefront, as to breach it not only leaves the principal open to excessive prices and possibly poor quality, but it also leaves you and your subcontractors just as open to financial ruin. Tendering is not for the faint-hearted, but it can be one of the most rewarding business strategies, both in profit and reputation, when conducted in good faith.

REFERENCES AND FURTHER READING

Australian government: AusTender, **https://www.tenders.gov.au/**

Australian Tenders, **https://www.australiantenders.com.au/**

Building and Construction Procurement Guide: Principles and Options, **https://9104f275-f216-4fd2-9506-720eb252b4fc.filesusr.com/ugd/473156_2142409e24444d7c981cb795a5010856.pdf**

Business Queensland, **https://www.business.qld.gov.au/running-business/marketing-sales/tendering**

Commonwealth Procurement Rules, **https://www.finance.gov.au/sites/default/files/commonwealth-procurement-rules-1-jan-18.pdf**

NSW government: NSW eTendering and ProcurePoint, **https://tenders.nsw.gov.au/**

Queensland government: eTenders, **https://www.hpw.qld.gov.au/bas/eTender/Default.aspx**

Tasmanian government tenders, **https://www.tenders.tas.gov.au/#**

Tendering Guidlines for NSW local government, **https://www.olg.nsw.gov.au/sites/default/files/Tendering-Guidelines-for-NSW-Local-Government.pdf**

Tenders Online (formally Cordell Tenders), **https://www.tendersonline.com.au/**

Victorian Civil Construction Industry Best Practice Guide for Tendering and Contract Management, **http://www.buloke.vic.gov.au/ArticleDocuments/148/Victorian%20Civil%20Construction%20Industry%20-%20Best_Practice_Guide_Final_May08.pdf.aspx**

Weblinks

https://www.business.qld.gov.au/running-business/marketing-sales/tendering

http://www.constructionlawmadeeasy.com/managingthemainrisksoftendering

http://www.constructionweekonline.com/article-1206-to-bid-or-not-to-bid-contractors-face-tough-tendering-decisions/

https://infrastructure.gov.au/infrastructure/ngpd/files/Volume-1-Procurement-Options-Analysis-Dec-2008-FA.pdf

https://www.procurepoint.nsw.gov.au/before-you-buy/construction/index-construction-documents

SMALL BUSINESS FINANCES

Chapter overview

This chapter covers the essential skills and knowledge required to effectively manage the financial operations of a small business. Specifically, this chapter considers the implementation of a financial plan and the ongoing monitoring of the financial performance of a small business operation.

Elements

This chapter provides knowledge and skill development materials on how to:

1. demonstrate your knowledge and understanding of the requirements of implementing a financial plan for a small business operation
2. undertake the required actions to monitor the financial performance of a small business operation.

Introduction

In this chapter, we will cover the essential skills and knowledge required to manage the financial operations of a small business. In particular, cash – which is a critical component in all businesses, but is particularly key when running a small construction business involving the purchasing of materials, business running costs and managing the receipt of payments from customers, while at the same time providing the quality service and workmanship that your business needs to succeed.

Implement a financial plan

A financial plan is a key element of a successful business, particularly when it comes to small business operations like those in the construction industry, as most often the financial management of small businesses is outsourced to accountants and bookkeepers. The value of having an understanding of financial concepts and the ability to apply them through the use of some basic skills in your own business is something that this chapter seeks to emphasise. Although it may seem at times that the work involved in managing the finances of your business is too great, the advantage of having a good working knowledge of your finances and to be involved in its management to the extent that you are able will provide a strong foundation for you to build your business into the future.

In this section of the chapter, we focus on the specific concepts and skills that are required to implement an existing financial plan, but first we will take a look at the information requirements for managing the finances of your business and the basic process that needs to be undertaken to prepare a financial plan for your business.

To help guide you through the process of implementing your financial plan and monitoring the financial performance of your business, we are going to use a case study scenario for Bob's Carpentry.

Financial information and the business plan

As we have already mentioned, there are specific information requirements for managing the financial operations of a small business, and it is this information that a small business owner will use to make decisions about the business, its direction and its goals. A **business plan** is the document that outlines the goals and objectives of your business, both in written statements and through the presentation of financial **projections**. By preparing and using a detailed business plan, you will have a useful tool for not only managing your finances but also for arranging finance for your business. It is the business plan that provides important information to potential financial backers and also shows that your business is properly organised and managed.

CASE STUDY

Scenario: Bob's Carpentry

Bob's Carpentry is a family-run business based in a suburban area of a large Australian city. Bob's business is a dynamic building company providing good old-fashioned honest service. Bob employs one apprentice and has several contract tradespeople that he calls on throughout the various projects that he takes on.

As well as organising and supervising all the trades associated with each project, Bob is on site every day, which means that he is focused on one project at a time, ensuring that his finished product is of the highest quality. This means that most of his work is generated through 'word of mouth' and recommendation, enabling him to maintain a steady income stream into the future.

Bob's main clients are homeowners who love where they live but have either outgrown their current home with a growing family or feel that it is looking tired and worn out. Bob provides his clients with endless possibilities for transforming their existing houses into their dream homes.

Maintaining financial records

Creating and maintaining a system of accurate and detailed financial information is the key to managing the finances of any small business and no more so than in the construction industry, where the demands on the small business operator are complex and constantly changing. The combination of managing various projects, contractors and clients while maintaining an eye on the finances can result in not only confusion but the possibility of losing track of legal requirements, particularly taxation matters, which may result in you and your business being penalised.

It is also important to remember that maintaing records of your financial information is affected by a range of legal requirements, such as taxation, and you need to be familiar with these legal obligations, particularly in relation to any reporting and compliance requirements applicable to your business.

Keeping financial records is not only a compulsory requirement for taxation purposes, but they also provide the basis for recording financial transactions, making them the foundation of producing financial reports that can be used to determine how well your business is performing according to your business plan.

Legal requirements

The legal requirements for all businesses, including small businesses, are continually being updated and modified, so it is critical for you to have some form of system to ensure that any changes are implemented into your record keeping. This is where the services of a bookkeeper or accountant can be invaluable.

In Australia, the following legal requirements apply:

- If your business generates more than $75,000 turnover annually, you are required to register for an Australian Business Number (ABN) and to register for GST.
- If you are registered for GST and/or have employees, you will be required to lodge a Business Activity Statement (BAS) disclosing relevant taxation information including GST, PAYG and other taxation-related monies.
- If you have employees you are required to administer the Superannuation Guarantee payments for your employees.
- If relevant, you will need to ensure you comply with any Capital Gains Tax (CGT) or Fringe Benefit Tax (FBT) obligations.

Record keeping

So, clearly one of the main legal requirements of maintaining records of financial information is taxation legislation, most significantly, GST and income tax law. It is a requirement that taxpayers take reasonable care to maintain their tax records through the use of an appropriate accounting system, providing assurance that the income and expenditure of the business is correctly recorded and classified for tax purposes. Those documents related to tax must be retained by the taxpayer for a minimum of five years after the year that they relate to.

The (ATO) provides a wide range of information for small businesses on their website, which includes the requirements for record keeping. The legal requirements for the way that financial information is kept are:

> By law your records must:
>
> - explain all transactions
> - be in writing (electronic or paper)
> - be in English or in a form that can be easily converted
> - be kept for five years (some records may need to be kept longer).
>
> If you don't keep the right tax records, you can incur penalties.
>
> **Source**: Australian Tax Office, https://www.ato.gov.au/General/Other-languages/In-detail/Information-in-other-languages/Record-keeping-for-small-businesses/

Record keeping, however, is not just for the purposes of meeting your legal requirements, there is a genuinely positive purpose for you to keep good financial records. By keeping records and ensuring that they are correctly classified not only for taxation purposes but also for managing your business generally, you have a powerful tool to assist you with identifying and avoiding any cash flow problems. Good records also provide you with the ability to make proactive decisions about the direction that you want your business to take by identifying opportunities to maximise your profits and even minimise your tax!

Accounting information

Financial information is also known as accounting information and is the basis of all reporting for a business. As we have previously considered, if it is not accurate, then the reports that are produced from the system will not be meaningful and can create significant difficulties when making decisions about the business.

Financial transactions or events are the trigger for information collection into the accounting system. Having a computerised accounting system can assist with this process but is not essential. A simple system can be established for less complex businesses on a spreadsheet if necessary, but the benefits of an accounting system are really only realised when a report is required to be generated.

Source documents, such as tax invoices, receipts, cheque butts, bank deposit slips and bank statements are used to create the record in the accounting system. The source document provides the evidence of the transaction occurring and will be required to be produced for taxation or other audit purposes. Having this information to hand will also mean that you have a detailed reference for a transaction should you require it.

Once these transactions are recorded into the accounting system, you will have a written record of your business operations, which you will then be able to use to produce reports to summarise the information. This allows you to get a clear overview of key decision-making facts, such as how much profit you are making, what your cash flow looks like and even how much GST credit you can claim back from the ATO. These reports enable you to make decisions about your business because they are able to provide you with the information you need in a format that is specific to the decision at hand. This is how the process of managing the finances of your business works and is shown in Figure 9.1.

Financial management

Managing your financial information effectively will enable you to produce reports to monitor and manage your business to ensure it is on the right track. The purpose of financial management is to manage the

FIGURE 9.1 Financial management process

financial resources of a business so that it can meet its desired goals and objectives. The process itself is a key component of the business plan, which ensures that the financial objectives set out by the business are met and that it remains viable.

While financial management actually has a focus on the cash management of the business, it really encompasses all activities of the business, as everything that the business does will have a financial impact. As we have seen in Figure 9.1, there are financial reports that are produced from the financial information, as listed below.

- Tax reports: these reports cover all aspects of the taxation liabilities of a business including GST, income tax, payroll tax, PAYG and FBT. The most significant report that is prepared from this information is the BAS, which is required to be submitted to the ATO at regular intervals.
- Profit reports: reports that provide an overview of the financial performance of the business can be called profit reports. In some cases, a formal Profit and Loss Statement or Income Statement will be prepared. These reports list all the expense and income amounts for a period (generally, a month or financial year, depending on the purpose) and calculate the amount of profit or loss. Sometimes a profit report may be called a *Projected Profit Statement* or a *Budgeted Income Statement*; these provide information relating to the anticipated income and expenditure of the business and are usually produced at the planning stage of financial management.
- Cash flow reports: these reports will help you to keep an eye on your cash position, whether it is the balance in your bank account or the projected amount of cash you will need to complete a specific project. Cash management is critical to the viability of the business and must be monitored frequently.
- Balance Sheet: this single report provides a picture of the financial position of a business at a point in time. It looks at the assets and liabilities of the business and provides a method of calculating the owner's investment and ongoing interest in the business. This gives a good picture of the ongoing viability of the business.

Accounting terminology

Understanding accounting terminology will assist with determining financial information needs and maintaining financial records for the purpose of generating financial reports. The following provides an overview of key accounting terms that will help you with working through the rest of this chapter.

- Assets: Resources that the business owns or utilises for a future benefit; e.g. cash, motor vehicle, plant and equipment, debts owed by customers.
- Liabilities: Amounts owed (debts) by a business to outsiders, which represent a commitment to pay cash at some point in the future; e.g. amounts owed to suppliers, borrowings from the bank or other financial institution.
- Income: Also referred to as revenue, the funds received or earned in the course of running your business; e.g. sales revenue, fee income, service income.
- Expenses: The costs incurred in the course of running your business, that is, earning income; e.g. wages, advertising, insurance.
- **Owners' Equity:** What the business owes to the owner(s) of the business. The owner(s) of the business are the beneficiaries of any profits and bear any losses incurred.

Obtaining specialist services

As mentioned previously, the financial management of a small business in the construction industry would rarely be done solely by the owner/operator of the small business. This means that there will always be a requirement for obtaining specialist services including professional advisers and other support services.

Professional advisers

Seeking the services of professional advisers is an important aspect of managing the finances of your construction business. Including a list of these in your business plan will mean that this information is always centrally located. Business advisers may include any or all of the following:

- qualified accountant
- certified bookkeeper and BAS agent
- solicitor
- insurance broker
- bank small business adviser.

Other specialist support services

While the support of professionals is a critical aspect of the specialist support services that will generally be required when operating your construction business, there are also a range of other support services that may be of benefit. These services provide support in specialised services and can include any or all of the following:

- 'cloud'-based services, such as online gateways
- other software services, such as training
- business brokers or consultants
- industry or trade associations
- mentors and other advisers.

The Learning Tasks in this chapter will use the information provided in the case study scenario for Bob's Carpentry on page 213 and the excerpt of the business plan in Appendix 3 at the back of this book.

> **LEARNING TASK 9.1**
>
> **IDENTIFYING FINANCIAL AND OTHER INFORMATION REQUIREMENTS**
>
> For Bob's Carpentry, identify the sections in the business plan excerpt that detail the financial information requirements and the specialist services that Bob uses for his small business.

Operating profitably in accordance with the business plan

We have already considered the concept of financial management and its significance in the managing of your business finances. The question remains, how do you know if your business is on the track that you had anticipated? With this in mind, it is clear that there is a need for a plan to be in place for you to understand what your goals are for your business and where you would like it to be at some point in the future.

As we have previously mentioned, the business plan provides a written statement of your intentions in relation to how you would like your business to develop over the medium to long term. So, operating in accordance with the business plan means managing the finances of your business effectively to make sure that the business achieves its objectives of operating profitably.

Benefits of business planning

There are significant benefits for small business operators in the construction industry in preparing a business plan, the main one being that you will have a way in which to establish and maintain your business's viability both now and into the future. Just as importantly, the business plan provides the detailed information that you will be required to provide to potential and current financial backers or lenders.

Another benefit of having a business plan is that you are able to monitor the performance of your business and manage the operational functions effectively. Managing the operational functions of your business requires planning and control of each operating function, including marketing, purchasing, personnel and finances. The success of your management of these functions is measured by your financial results and other measures of business performance.

Understanding the financial plan

Preparation of the financial plan occurs at the business planning stage and involves the following process:

- assessment of personal financial position to determine cash drawings required
- identification of overheads and an estimate of costs
- identifying the target net profit to cover these costs and to provide an adequate income
- calculation of an appropriate charge-out rate for labour
- preparation of a Projected Profit Statement to provide to potential financial backers
- planning the financial structure of your business to calculate capital requirements
- preparing a cash flow forecast to ensure adequate working capital.

In the case study scenario provided, Bob's Carpentry used this process to prepare the plan for the next three years, the result of which is shown in the business plan excerpt in Appendix 3.

Understanding business costs

It is important to understand the costs associated with operating your business and the market in which your business operates. To successfully manage the finances of your business, you will need to make sure that you cover these costs and generate a profit.

When considering the costs of your business and the potential for earning profit, you will need to consider the following:

- How much will it cost you to provide your service?
- What price can you charge?
- How many projects will you need to complete to make a living?
- Is there sufficient demand for this to occur?

Your estimated profit will depend on how much you can charge and working this out can be quite challenging. Before anything else, the amount that you charge must cover all of the costs of running your business.

The first thing you will need to do is make a list of all the costs you will face in providing your service. There are two types of cost to consider:

- operating costs: these costs are recurring, such as rent, insurance, fuel, etc.
- capital costs: this includes all items that your business will use to make a profit, such as your tools, your motor vehicle and any other significant purchases that will last for more than one year.

Projecting profit

A profit projection is one of the first calculations you will need to make when preparing your financial plan. To accurately calculate a profit projection that will cover the costs of operating your business as well as generate sufficient profit to give you your desired income, you will need to calculate both the break-even point and the income level that you will need. In these calculations, GST will be ignored.

As we previously mentioned, a Projected Profit Statement is usually prepared to estimate the expected financial performance of a business. It is a useful tool in the financial management process as it will allow you to regularly analyse the financial performance of your business. That is, you will be able to compare your actual expenses and income against those that you have estimated. In addition, the preparation of this report will give you a document that you can provide to creditors and lenders to assist them with estimating the financial viability of your business.

The concept of break-even

A useful calculation to determine the profitability of your business is the break-even point. The break-even point is where the income of your business equals the fixed expenses of your business; that is, those costs that you must pay regardless of whether you are earning an income or not, such as insurance and rent.

Once you have established how much you need to earn to cover your fixed costs, you can determine how much profit that your business will need to generate, both to cover its costs as well as any personal drawings (i.e. your income), and any other expenses related to the business such as income tax and loan repayments.

CASE STUDY

Conducting a break-even analysis

Bob is costing out a new job and estimates that his overheads (fixed costs) for this will be $8500. He is expecting the job to take around a month to complete. He usually works six days per week.

What daily rate would he need to charge for this job if he wants to simply break-even?

Answer:

Break-even in days = $8500/24 days = $354

Produce financial budgets and projections

You may already have an idea of what a budget is, you may have already used one in your business; the fact is, financial budgets are crucial to managing your small business finances because they will provide you with a plan. In fact, that is what a budget is, a financial plan, a road map to help you track how well your business is going and whether you are headed in the direction that you intended.

Understanding budgets

It is at this point that an overview of the key benefits of preparing financial budgets for your construction business will help you to get a handle on the business of budget preparation. If you have already been involved in managing projects, you will already understand the complexities involved in managing the purchasing of materials, paying contractors and collecting payments from your customers, among the other aspects of running your business, and how crucial it is for you to keep a close eye on all of these aspects. The tool that can help you to manage this is a budget. Budgeting is the process of planning a future action in quantitative terms (units or dollars).

Key benefits of budgeting

Budgeting for your construction business may seem to be a too complex and time-consuming activity without any real benefit to you and your business. Or, you may think it more important to focus on the actual day-to-day operations of running your business rather than going through a process that may seem to be just something else to add to your already overloaded schedule. Just as keeping records is essential to business success, so too is keeping tabs on how your business is performing and in order to do this you will need to have a budget.

LEARNING TASK 9.2

REVIEW THE BUDGET FOR BOB'S CARPENTRY

Using the business plan excerpt for Bob's Carpentry, review the budget and identify some of the key areas that he should be monitoring closely.

LEARNING TASK 9.3

WORKING CAPITAL FOR BOB'S CARPENTRY

Using the business plan excerpt for Bob's Carpentry, calculate the amount of working capital that he has projected.

Is the figure you have calculated positive or negative?

Understanding cash flow and its relationship to financial viability

Cash flow is the movement of money in and out of your business. By analysing your cash flow, you can see when and where money comes and goes within your business and it is useful for you and your financial backers to assess how well your business is doing. It also helps you to plan your cash flows so that you can maintain sufficient working capital to maintain the financial viability of your business.

Financial viability

The financial viability of your business is defined by how well the business meets its financial obligations and achieves its goals. A business that cannot pay its staff what they are owed and cannot generate enough income to pay its bills may be in serious trouble. In this case, financial backers may not want to or be able to bail the business out. Businesses like these are generally sold or closed down.

Usually, your business' financial viability will be measured and defined by its ability to:

- earn sufficient income to cover its operating expenses
- meet its financial obligations
- make a profit
- show potential for growth.

Working capital

Working capital is the amount of cash that you will need to meet the short-term financial obligations of your business, such as rent, loan repayments, wages, materials, tools, fuel and other ongoing operational expenses. Working capital can be taken from any assets that you have that can be easily converted into cash; that is, the assets that are most liquid. To determine how much cash you need, you must subtract your current liabilities from your current assets, or subtract the debts that you need to pay within the next 12 months from the amounts that are owed to you (including any cash) within the next 12 months.

working capital = current assets – current liabilities

If you calculate a positive working capital figure for your business, it is operating securely and efficiently because you have sufficient cash to pay all your short-term obligations with money left over. However, if you calculate a negative working capital figure this will show that you have more liabilities that need to be paid within the next 12 months than cash assets, suggesting inefficiency and poor short-term financial health.

Managing the working capital of your business is a key tool for ensuring that your business plan is implemented effectively, and is key to managing your cash flow; however, failing to manage this aspect of your business could lead to a cash flow crisis where you are unable to meet your obligations due to a lack of available funds. So, you will need to make sure that you have sufficient working capital for the duration of your construction projects, including enough cash to pay for materials, wages and other operating expenses, both relating to the project and the ongoing operations of your business.

Insufficient working capital or cash can lead to the failure of many small construction businesses, making it critical for you to factor in all of your possible financial requirements at the business planning stage to avoid any unexpected shortfalls when your business is up and running.

Some strategies that you can implement to make sure that you manage your working capital requirements include:

- a regular review of your working capital needs; this could be on a daily, weekly, monthly, quarterly and yearly basis
- make sure that you have effective credit collection procedures to minimise the length of time customers take to pay
- negotiate extended payment terms to maximise the length of time you have available to pay suppliers
- perform a regular stocktake to optimise inventory levels and minimise the costs tied up in excess inventory
- negotiate options with your lenders, such as debt financing or extending your overdraft facilities, to support your short- to medium-term liquidity issues.

Analysing cash flow

When reporting on cash flow, it is calculated as the difference between the cash you have on hand at the beginning of a period (e.g. month or year) and the closing balance on hand. If the closing balance is higher at the end of the period, this is a positive amount; if the closing balance is lower at the end, this is a negative amount and, over time, without cash to support its operation, a business can become insolvent.

With this in mind, you can see how important it is to analyse your cash flow regularly to make sure that your working capital goals are being achieved. Analysing cash flow can help you to identify where your money is being spent and where any overspending may be occurring, which will enable you to identify any insufficiencies and implement a strategy to manage your cash flow.

One way to use the analysis of your cash flow to help you manage your working capital is through identifying any trends that might occur. So, for example, you will know when your busy periods are by simply analysing the cash coming in across the year and at the same time you will know when your slower times are and be able to plan accordingly.

Cash flow statements

The cash flow statement shows the calculation described above and is used to explain how well your business is performing in terms of the movement of cash; that is, cash flows. Cash flows include both cash receipts and cash payments.

Cash flow statements are broken down into three areas that are useful when analysing cash flow trends. The three areas are:

- operating activity: this is all cash received, and cash paid out through the operation of the business
- investment activity: involves the purchase and return on investments, such as property, fixed assets and equipment
- financing activity: this is money used to finance the activities of your business, such as loans received and loan payments.

LEARNING TASK 9.4

REVIEW THE CASH FLOW FORECAST FOR BOB'S CARPENTRY

Using the business plan excerpt for Bob's Carpentry, review the cash position of the business and identify some of the key areas that he should be monitoring closely.

Business capital

For sole traders and partnerships, their financial resources are the capital introduced by the owners and any short-term or long-term borrowing from financial institutions and/or other private sources.

Business capital can be either debt or equity; that is, the funds used to finance the business are either through investment into equity or through borrowing. With this in mind, the implementation of the business plan needs to consider the requirements of these financial backers, which might include banks, leasing companies and providers of venture capital as well as others such as partners, owners, family and friends.

Sources of business capital

As with all major decisions relating to your business it is wise to seek advice from your legal advisers and accountants prior to entering into any legally binding financial contract. The following list provides an overview of some of the most common sources of business capital financing available.

- **Personal savings**: This is usually the easiest and cheapest way to finance your business and is a viable option for everyone. However, this will also involve a level of personal financial risk that you need to carefully consider.
- **Credit cards**: You may already use credit cards for your personal expenses, so you will probably know that credit cards are relatively easy to obtain. In terms of your business operations though, you should be aware of the high interest rates and other charges involved in borrowing through credit cards. It is generally recommended that credit cards are used for short-term access to a low level of finance.
- **Family, friends and relatives**: This can be quite a cheap and flexible alternative for sourcing finance for your business, but remember that this can get complicated and may end up affecting your personal relationships if your business runs into financial difficulty.
- Angel investors: Finance is provided in exchange for an equity stake or a share in your business, which means that your business will no longer be wholly owned by you. Basically, you are selling equity in your business to generate finance and this generally triggers various legal, taxation and accounting consequences.
- **Leases**: This involves a financier purchasing the equipment that you require and then leasing it back to you in return for regular rental payments for the duration of the lease term. At the completion of the lease term, you are offered the option of purchasing the equipment at an agreed residual value. Leasing can be a good option as you avoid high initial purchasing costs; however, they can become quite expensive and involve complex treatment for accounting and taxation purposes.

- **Bank loans**: These are the most common source of funding for small businesses and include term loans, mortgages and bank overdrafts. They can be more difficult to obtain than some other types of finance as you will have to meet various requirements; however, bank loans are generally very secure.
- Trade credit: Instead of paying cash for purchases on receipt of goods, payment can be delayed to a future date with the agreement of the seller; this means that the money payable on receipt of goods stays with your business until it is due for payment on the agreed date. Therefore, this money becomes a source of finance to the organisation during this period. The cost of this resource is nil.
- Overdraft facility: This is an arrangement with your bank to overdraw the business bank account to an agreed limit and it is a useful source of short-term finance. There is no need to use the entire facility if not required.

Maintain financial provision for taxation

It is important for the viability of your business that you maintain adequate financial provision for your taxation obligations. Previously, we have considered the taxation obligations of a small business operator from the perspective of maintaining financial records. We will now consider how these taxation obligations can impact on both the net profit and cash flow of your business and how monitoring these liabilities regularly will help you to maintain an effective financial provision to cover these taxation obligations.

Taxation for small businesses

All businesses in Australia, including small businesses who are registered for GST, must complete an Activity Statement to report their taxation obligations to the ATO. These statements may be issued monthly, quarterly, or annually depending on the set-up that has been made with the ATO.

Other taxation obligations include payroll tax on wages which may need to be paid when reaching a threshold amount.

Impacts of GST on cash flow

GST on sales is collected by the business and remitted to the ATO according to the timeframe provided. What this means is that in adding GST to your income earned, you are really only acting as a collection agent for GST, which means that the GST is not included in your budget. However, as you are holding the money on behalf of the ATO and will be required to remit this at the time of submitting your Activity Statements, it will have an impact on your cash flow. This is why keeping track of your cash flow and your financial performance is critical to ensure you have sufficient funds available to pay your tax liabilities.

On the other side, however, you as a business operator will pay GST for the purchases that you make and for this amount, you will receive a credit on your Activity Statement. So, the net of the GST paid and the GST collected is calculated and the difference between these two amounts is paid to, or received from, the ATO.

It is important to note that, depending on the way that you have set up your Activity Statements with the ATO, any amounts outstanding from customers would include GST and will be required to be included in the calculation of the GST amount collected on behalf of the ATO.

In terms of record keeping, you will need to maintain separate accounts in your system to record the amount owed to and from the ATO. This separation from the records of your income and expenses is essential to manage your cash flow and maintain financial provision for these taxation obligations.

Developing client credit policies

In the construction industry you may be paid in cash at the time of completing a job; however, on larger projects, you may be forced to provide credit, either formally by operating a 30-day account, or informally by invoicing the client and waiting to be paid. The result of offering credit in this way means that the cash won't come to you straightaway and is therefore tied up in what is called Accounts Receivable. This means that your working capital needs to be greater than your cash debts, to make allowances for the difference between when you pay cash for the materials and other costs of projects with when you receive the cash from the customer.

In addition, the managing of customer credit involves time in chasing up debts, which will have a cost impact either in your time or by paying your bookkeeper to do this. There is also the risk of the customer not paying; however, you can minimise this risk by developing a clear policy on how you will provide credit to your customer by making a clear statement of your credit terms and also having a documented procedure to follow up any overdue accounts.

Credit policy and terms

Your terms should be very clear and consistent. You will need to make sure that your customers have a good understanding of when they are expected to make payment by including your terms on all of your documentation as well as telling them at the time of quotation.

To keep control of Accounts Receivable, at the end of each month, you should obtain an Aged receivables list report which summarises the amount outstanding for each customer and whether the amounts are current (i.e. within trading terms) or overdue.

Collecting outstanding accounts

If your policy is clear and you have already decided what you will do if you have a late-paying customer, you may be tempted to delay following it up as it might be difficult and potentially unpleasant.

By documenting the steps that you will take at specific stages of an overdue invoice, you will have a process to follow that will remove any difficult decisions. The steps will generally take the form of timeframes. Table 9.1 provides an action plan that can be used to follow up individual customers and at the same time provide documented evidence of the process that could be used in the event of legal proceedings being required.

TABLE 9.1 Debt collection action plan

Action plan for following up outstanding debts			
Customer Name			
Amount outstanding		Date Due:	
Step	No. days past due date	Actions	Notes (include date of contact and details of communication
1	7 days	Send a reminder notice	
2	14 days	Follow up with a phone call. Remind customer of credit terms.	
3	30 days	Follow up with another phone call and perhaps a personal visit.	
4	60 days	Send a letter of demand, threatening legal action.	
5	14 days after letter of demand	Place in the hands of a debt collector.	

CASE STUDY

Review of client credit

At the beginning of the financial year, Bob had the following existing debt, which was proposed to be collected as follows:

July	$25,000
August	15,800
September	8,500
Total	

Bob's current collection policy is via progress payments where he collects one-third prior to the commencement of the job, one-third one month later and the final third three months later.

LEARNING TASK 9.5

DEVELOP A CLIENT CREDIT POLICY

For Bob's Carpentry, develop a credit policy which includes: trading term, credit limits (if applicable), and a procedure to follow up overdue accounts. Also identify how Bob can best communicate this policy to his customers; e.g. on all written quotes, on invoices, on monthly statements.

Contingencies for debtors in default

A contingency plan is a list of the actions you could take if a debtor does not pay. As we outlined above in Table 9.1, if all else fails then put the debt into the hands of a debt collection agency; however, how will you manage the shortfall of cash in the meantime?

It is extremely useful for you to have options in the event of any change in circumstances that result in an adverse impact on your cash flow. In the case of slow-paying debtors, you may wish to arrange for an overdraft facility with your bank. This will cover any short-term cashflow problems, including slow-paying debtors.

Selecting key performance indicators

When you write a business plan and subsequently quantify that plan into a financial plan, you would have identified some key areas that you wanted to focus on to achieve success. These key areas can also be called key performance indicators (KPIs) and can be used to help you focus on what you want your business to achieve and to make sure that you are on track to achieve this.

A performance indicator needs to be represented by a measurable statement that can be used by a business to evaluate its performance. KPIs are, as they imply, key to the objectives of the business and so are used as a way to measure the progress of a business against its goals and objectives, as set out in its business plan and its financial plan. In order to be able to use the KPIs effectively, they need to be measurable and expressed in either financial or non-financial terms while reflecting the nature of your business.

Identifying KPIs

The two main types of performance indicators used are: *financial indicators*, which use information from the financial records and are expressed in dollar terms; and *non-financial indicators*, which are commonly expressed in real terms (written descriptions of how the company is actually operating) and often make use of qualitative data. Performance indicators focus on specific areas of the business as determined by its goals and objectives and often include the comparison of figures, either across different years or within the same year. This comparison is an internal comparison; however, it is also possible to compare your business to other businesses in the same industry, which is known as **benchmarking**.

The types of analysis that are made using performance indicators can include:

- whether the business is performing as planned
- whether its performance has improved over time
- how its performance compares to that of similar businesses.

Financial performance indicators

Financial performance indicators generally focus on profit and are usually calculated as ratios or percentages, as they are expressed as a relationship between net profit (or gross profit) and sales, or between net profit (or gross profit) and cost of sales.

We will look at the detailed calculation and analysis of these ratios later in the chapter, such as:

- percentage of Net Profit to Sales
- percentage of Gross Profit to Sales
- percentage of Net Profit to Cost of Goods Sold
- percentage of Gross Profit to Cost of Goods Sold.

Non-financial performance indicators

As is the nature of non-financial performance indicators, they are not measured directly and depend on the area that you choose to focus on. Some examples of these indicators are:

- customer complaints: the number of complaints received from customers as a percentage of the total number of customers served
- employee turnover: the number of employees who have left the business as a percentage of the total employees.

Selecting KPIs

To select appropriate KPIs for your business you will need to consider what is most important to your business. The following are some questions that can help you with this decision.

- What are the most important areas in my business plan? Are they:
  - cutting costs
  - improving quality
  - improving employee productivity
  - improving customer satisfaction
  - lowering employee turnover
  - other?
- What would be the impact if something went wrong?
- How do these areas support where you want your business to go?

Once you have some answers to these questions you will have a basis to identify what areas of your business should be monitored. Remember that your measures should monitor those key areas that you have already identified, including those that support the direction that you want your business to take. Some key factors to consider when doing this are:

- make sure that the measures you choose are the most important activities for your business
- also keep an eye on your use of resources
- try to measure things that give you the information you need, quickly and easily.

Establishing targets for performance

Once the KPIs have been selected, it is important to effectively utilise them in monitoring the performance of your business. This is done by establishing targets for each KPI; such as identifying the planned or expected outcome for the KPI selected. One method of selecting a target is to benchmark your business performance; that is, to set a target based on information you already have about your own business' past performance or obtain information about the performance of other businesses within the construction industry.

It is possible to benchmark almost all the areas of your business that are important to your success, and the following lists some of the financial and non-financial data that can be used to measure the performance of your business.

- invoicing customers
  - Accounts Receivable days
  - percentage outstanding bad debts per time period
- paying suppliers
  - Accounts Payable days
- ordering and purchasing materials
  - average order value per month
- quoting jobs
  - percentage of quotes accepted per month
- staffing
  - staff turnover
  - cost of recruitment
- customer feedback
  - number of complaints
  - word of mouth jobs
- service and quality
  - average job lead time
  - overall job timeframes.

Using KPIs to monitor business performance

As we have already discussed, once a KPI is selected, an achievable target needs to be set in order for progress to be tracked. So, when monitoring the performance of your business against the target set for the KPI, if the numbers show that you have met the target, then you will know that you are on track. However, if the measure is not met then this will give you an early warning signal and enable you to take some action to rectify the situation. By having a set of KPIs that you can review regularly, you will not have to wait until the end-of-year figures tell you that your business might be in trouble.

Part of any monitoring process is identifying where actual performance is not in line with these indicators of planned performance. The measuring and monitoring of the performance of the business against the KPIs and targets set forms a critical part of this process. We will consider the use of KPIs in monitoring the performance of your business later in the chapter.

CASE STUDY

Selecting key performance indicators

Bob has selected the number of charge-out hours as his KPI and has set a target for the next financial year.

He will use this target to measure his daily, weekly and monthly hours. If his KPI measurement is on track he will know that he is on track to meet his sales target.

However, if the hours he is actually working are consistently below his forecast, this will indicate that he needs to take some further action such as reducing his costs and/or drawings so that he can continue to meet his financial commitments.

The indicator Bob has chosen is what is called a lag indicator; that is, it tells him what has already happened. On the other hand, a lead indicator can give Bob an early warning signal that will enable him to take corrective action to stay on track to meet his goals and objectives.

Identify a lead indicator for Bob's Carpentry.

Documenting financial procedures

It is important for you to have your procedures documented in some form to enable you to communicate to your bookkeeper or administration staff how your financial information is maintained and how you will comply with your legal obligations.

Another benefit in a small business situation is if there are tasks that are not performed on a regular basis, the documented procedures can serve as a memory jogger and will also assist in training new staff. Documented procedures, however, go further than just reminding you or providing you with the steps required to complete a task related to the financial management of your business. Through documenting the processes you use for managing the finances of your business, you will also have an opportunity to think about why you follow those particular processes.

To help you think about this, it is useful to gain an understanding of the differences between a *policy* and a *procedure*. In the online support provided by Business Victoria they have a section on *Developing good financial procedures* which discusses these differences.

A policy is a statement that outlines the principles and views of a business on each topic covered. It is an overview of certain rules that a business should operate by.

A policy should:

- align with business goals and plans,
- reflect the culture of the business,
- be flexible, and
- be easily interpreted and understood by everybody in the business.

Sometimes a policy will need a supporting procedure. Procedures are clear and concise instructions on how to abide by the policy. They detail the sequence of activities that are required to complete tasks and should include the 'how to' guidelines to achieve the necessary results.

Procedures should be:

- factual, simple to understand and succinct;
- written in a step-by-step style that shows people how to follow the procedure through from beginning to end;
- include references or links to any related documents and forms that need to be completed when following the procedure; and
- in the best format for their purpose e.g. a procedure could be presented as written steps, a flow chart or a checklist.

Source: Accounting and financial policies and procedures, Business Victoria, State Government of Victoria, © Copyright DJPR 2019 https://www.business.vic.gov.au/money-profit-and-accounting/financial-processes-and-procedures/accounting-and-financial-policies-and-procedures

Developing a policies and procedures manual

It would not be unreasonable for you to be thinking at this point that your business is not large enough to warrant the documentation of financial processes, let alone the preparation of a formalised policy and procedure manual – but rest assured, there are many benefits in doing this. The documentation of policy and procedure for your business is not only considered a good business practice but the benefits are great. In its publication called *Good practice checklist for small business*, CPA Australia provides a summary of the possible benefits as follows:

A business that follows good practice benefits in many ways, including:

- the business is more likely to be profitable, have better cash flow and operate with less financial risk
- the business may be easier to sell in the future, and possibly at a better price
- the business may find it easier to access external finance, including bank finance, if needed
- the business may be better placed to respond to future challenges and opportunities.

Source: CPA Australia, *Good practice checklist for small business*, June 2015, https://www.cpaaustralia.com.au/~/media/corporate/allfiles/document/professional-resources/business/good-practice-checklist-small-business.pdf. Retrieved 14 March 2017.

Even though the benefits may be clear, it may still seem like a daunting task for you to undertake the writing of a policies and procedures manual along with all of your other administrative tasks. However, this process does not have to be done all at once and, in fact, the progressive development of such a manual will allow you to closely review specific areas of your business management while continuing to focus on your day-to-day activities.

A good way to make this task more straightforward for businesses that may have five or less staff members, is to write up the steps required to complete a specific task as the task is being done. This means that the person currently responsible for the particular task being documented is responsible for writing it up. This can then be added to the policy and procedure manual once you as the business owner have reviewed and approved it.

So, for example, if your business has a new customer, your bookkeeper or administration assistant may be responsible for setting up the customer account and they could write up how the details will be recorded, where these will be kept, how the customer credit limit is set, etc.

This process of writing while doing is a great way to ensure that all the steps are captured, and using this method of documenting your procedures can add significant value to your business.

Choosing procedures to document

The process of deciding on which procedures you should choose is very much dependent on the size of your business, and the following key areas should be considered so that you can determine where the documentation of a policy and/or a procedure will add the most value. These key areas include:

- **banking authority**: a list of people in your business who have access to your financial information as well as those who have the authority to make payments from your business bank account. This document would also provide the processes for opening and closing accounts.
- **new suppliers**: provides the steps on how to choose them
- **new customers**: includes the provision of credit (credit policy) and other trading terms such as discounts as well as debt collection and management
- **buying and purchasing**: includes how to determine when stock, equipment and assets need to be purchased.

We have already considered a documented procedure for following up late-paying customers. Table 9.2 provides a quick overview of some of the other types of financial procedures that may be valuable to document, while Table 9.3 is an example of a document identifying the steps that need to be taken to successfully complete various procedures.

TABLE 9.2 Common financial tasks requiring documented procedures

Key area	Policies	Related procedures
Bank accounts	Bank account policy	• Opening bank accounts • Closing bank accounts • Bank account transactions • Bank account reconciliations
Customers	New customer policy Customer credit limit policy Customer debt collection policy	• Choosing customers and establishing credit terms • Review customer credit terms • Debt collection procedures
Suppliers	New supplier policy	• Selection of new suppliers • Supplier payment terms
Purchasing and stock control	Purchasing policy Stock control policy	• Purchasing requests • Purchasing stock • Receiving and managing stock
Employees	Payroll management policy	• Wages and salaries payments

Source: Adapted from: Business Victoria, *Accounting and financial policies and procedures*, 6 December 2016, http://www.business.vic.gov.au/money-profit-and-accounting/financial-processes-and-procedures/accounting-and-financial-policies-and-procedures

TABLE 9.3 Operational tasks for procedures

Operation	Tasks or steps involved in process or procedure
Purchasing and **procurement**	The tasks involved in this operation are: 1 Requisitioning for goods and services by individuals or departments requiring them 2 Identifying the potential supplier 3 Deciding on a supplier including competitive tendering or obtaining quotations 4 Placing the order 5 Receipting of goods 6 Recording the liability to the supplier.
Wages and salaries payments, and record keeping	The tasks involved in this operation are: 1 Gathering information concerning time worked by the employees and rate of pay 2 Calculating the gross amounts of pay and taxes payable by employees 3 Calculating the net amount payable by gathering information on deductions to be forwarded to outsiders under authorisation from employees 4 Payment or remittance of net amount to employees; and record keeping of employees' earnings.
Maintaining journals, ledgers and other record-keeping systems	The procedures for record-keeping systems including maintaining journals and ledgers are generally twofold: 1 Setting of limits such as time to record, levels of approval and levels of authority for recording, within which the system operates 2 List and define various codes such as accounting codes and cost centre codes which need to be used for recording. These codes are important for analysing the data for information.
Arranging for use of corporate credit cards	This procedure should include: 1 The purposes for which a corporate card could be used 2 The maximum monthly credit limit for each employee card holder 3 The authority level of the approving officer 4 The method and time limit within which the expenditure on the corporate card should be accounted for 5 Penalty for not conforming to the rules of accounting for expenditure and/or misusing the corporate card.
Banking	This procedure involves: 1 The rules as to the frequency of banking 2 The number and levels of employees involved in the transporting of cash to the bank 3 The mode of transport, such as using the organisation's car 4 The times and routes used 5 The cash-carrying equipment.
Debt collection	The procedure for debt collection includes: 1 The period of credit allowed for customers 2 The methods of reminding/demanding overdue payments such as telephone calls or letters and emails 3 Seeking the help of specialist debt collectors and/or legal personnel 4 Writing off bad debts.
Invoicing clients, customers and consumers	This task is part of the sales and marketing operations. The procedure includes: 1 When the invoices are to be prepared (e.g. together with the delivery note, after the delivery/service is confirmed, or before delivery/service is organised) 2 Any special prices or discounts applicable to which type of customers/consumers.
Maintaining a petty cash system	This procedure should include: 1 The method of appointing a petty cash custodian and that person's responsibilities 2 Deciding on the amount of the petty cash float 3 Reasons for increasing/decreasing or cancelling a petty cash float 4 Types of expenditure allowed to be made using petty cash 5 Maintaining a register of petty cash payments 6 Methods of reconciling and reimbursing 7 Changing a petty cash custodian.

>>

>>

Operation	Tasks or steps involved in process or procedure
Ensuring security, accuracy and currency of financial operations	The procedures for financial operations should incorporate: 1 The internal control methods such as segregation of duties and approval procedures to ensure security and accuracy of the system 2 The frequency and methods of auditing; and the reviewing intervals and methods of the system for its currency (i.e. the recording conforms to the current legislative needs) 3 Procedures for establishing online accounts, levels of approval, access to confidential information and requirements for issuing new passwords to replace forgotten passwords.

Source: Adapted from: Business Victoria, *Accounting and financial policies and procedures*, 6 December 2016, http://www.business.vic.gov.au/money-profit-and-accounting/financial-processes-and-procedures/accounting-and-financial-policies-and-procedures

Monitor financial performance

In the previous section, we looked at what is involved in implementing the financial plan; in particular, we looked at implementing the plan for our case study scenario Bob's Carpentry. We have also looked at how selecting and monitoring KPIs is a useful way to keep your finger on the pulse of your business by monitoring these on a regular basis; thereby providing you with the opportunity to be alerted to any potential issues in the implementation of your financial plan. Effectively, KPIs act as an early warning system allowing you to act should the circumstances of your business change.

However, in creating a financial plan, you forecast a series of budgets for sales, cost of sales, projected profit and cash flow. These budgets give you clear financial targets to aim for. But they also give you a yardstick by which to judge your performance on a monthly and year-to-year basis.

Report on financial performance and analyse data

In order to measure the performance of your business it is important to create a set of reports that can easily be generated from your accounting system. Most accounting software packages provide standard reports, or you may be able to access your professional support services to have reports developed to your needs. In any event, your reports will need to compare actual results for the month or year against the budgeted or planned figures from your financial plan.

Financial performance reports

A financial performance report is a key tool for you to be able to monitor how well your business is achieving its goals and objectives. As you will recall, previously we emphasised the importance of maintaining sound financial records and, as performance reporting involves comparing what has actually happened (i.e. what has been recorded in the accounting system) with the budgeted performance, we are relying on the accuracy of that information to make decisions about the performance of the business.

There are a number of reports that can be generated from an accounting system that can be used to measure the performance of your business. However, they all do the same thing, which is to take the actual data from the accounting system and compare it to the budgeted estimate that you have prepared as part of your business plan implementation.

Monthly performance report

By preparing a monthly **performance report**, you will be able to determine any significant issues through your analysis, and as this is done on a monthly basis, you will keep on top of any issues as and when they arise.

By comparing actual results as they come to hand with your planned performance you can identify the key features that cause the difference between your estimated and actual cash profit position. What you are looking for in performance control are:

- significant variations from budget
- trends in overall variances.

Annual Profit and Loss report

You may need to understand why your business turned a profit or a loss as a whole, or in certain areas. For example, in the case of a profit you may want to look at areas that may have contributed to this success, such as your pricing, the quality of your work and the process you use to purchase your materials. In the case of a loss, however, you may need to consider these same elements but from a different perspective such as:

- Are your prices too high?
- Is your materials purchasing process complicated and/or inefficient?
- Are you providing quality workmanship and service?
- How are you managing your projects and your availability?

On a different note, your profit or loss could also be affected by factors that are beyond your control such as:

- a cheaper rival appears on the scene
- the business receives bad publicity or is the subject of a scandal and boycott
- there is a resource shortage
- there is an economic crisis

Understanding end-of-year financial reports

We have already mentioned the Profit and Loss Statement above; the Balance Sheet is the other key financial report that you will need to understand. The following discussion provides an overview of the information that is included in each of these key reports.

Profit and Loss Statement

The Profit and Loss Statement can also be called the Revenue Statement, or the Income Statement, and it calculates the profit or the loss of a business by first calculating the total income or earnings of the business and then deducting from that figure the total costs or expenses incurred in earning that income. This calculation is represented as follows:

Total Income – Total Expenses = Profit

It is necessary to prepare the Profit and Loss Statement using an accrual basis. This means that all income earned in that financial year, regardless of whether the cash has been received or not, will be reported; and, in the same way, all expenses that have been incurred during the financial year regardless of whether the cash has been paid out are also recorded. By doing this, you are ensuring that your reports are prepared in a consistent way, enabling you to use them for financial analysis.

We have already mentioned that the GST collected on your service revenue is not actually income to the business because you actually owe it to the ATO. Similarly, GST paid on business costs is not an expense of the business as you will be able to claim a GST credit. This means that the data used in the Profit and Loss Statement will need to be GST-exclusive figures. Any GST accrued or refundable is accounted for in the Balance Sheet.

The following list provides some specific detail about the sections included in the Profit and Loss Statement.

- Gross Profit: Gross Profit is calculated by subtracting the Cost of Sales from the Total Income you have earned, as follows:

  Total Income – Cost of Sales = Gross Profit
- Total Income: to calculate this you will need to consider which information to include. If your business operates solely on a cash basis, you will include all your cash sales as your Total Income earned throughout the year. However, if you offer credit facilities to your customers, you will need to calculate your total of cash sales plus the total of sales invoiced during the year; i.e. 1 July – 30 June, even if you have not yet received the cash from those customers.
- Cost of Sales: in the construction industry, this calculation represents the mark-up that you place on the materials included in your costing for a project.
- Net Profit: uses the Gross Profit calculated and subtracts all other Operating Expenses, or overhead costs and is represented by the following calculation:

  Gross Profit – Operating Expenses = Net Profit

Balance Sheet

The Balance Sheet calculates the financial position of the business by calculating what the business owns (assets) less what the business owes (liabilities) and the net worth of the business (Owners' Equity). It uses the following calculation:

Assets – Liabilities = Owners' Equity

Assets and Liabilities are also broken down into current (meaning utilised or consumed within a 12-month period) or non-current (longer than 12 months). The following list provides some specific detail about the sections included in the balance sheet:

- Current Assets: represents the working capital of the business and consists of the short-term assets that are used within a 12-month period to generate profit for the business and include:
  - **Accounts Receivable:** the amount of money owed by customers (debtors) at 30 June
  - Cash at Bank: the balance of the business bank account at 30 June
  - Materials Inventory: materials that you have on hand as at 30 June.
- Non-current Assets: assets such as your motor vehicle, tools and other fixed assets, which are purchased for use, not for sale.
- Current Liabilities: these are the amounts owed by the business that must be paid in 12 months or less and include:
  - **Accounts Payable:** represents the unpaid accounts of the business at 30 June for materials or any other expenses
  - **Employee entitlements:** represents the PAYG withholding tax, superannuation or workers compensation levies due but not yet paid at 30 June
  - **GST liability:** represents the balance of GST owing (GST collected on sales less GST credits available) as at 30 June.
- Non-current Liabilities: These are amounts owed by the business but not due for immediate payment. Non-current Liabilities usually consist of bank loans or other sources of long-term finance. For example, Super Plumbers owed $42,120 on a bank loan. This had decreased from $54,000 in the previous year, suggesting they are making consistent loan repayments.
- Net Assets: When the Total Liabilities are deducted from the Total Assets, the difference is the Net Assets or net worth of the business as at 30 June.

- Owners' Equity: The total of Owners Equity must equal the Net Assets of the business otherwise the report does not 'balance'.

Analysing performance reports

As we have already mentioned, it is important for you to analyse your performance reports on a regular basis to ensure that any variances are identified and actioned early to avoid any adverse impacts on the viability of your business.

To effectively analyse your performance reports, you will first need to identify any variances and the following steps provide a basic process to do this:

- **Check for variance**: At the end of each month use your performance report to compare your budget to the actuals.
- **Highlight variances:** Identify significant positive and/or negative variances for investigation and follow up.
- **Identify required actions:** What needs to be done, who will be doing it and when it needs to be completed.

Analysing variances

When variances have been identified, it is necessary to determine the reason for them. The variance may result from several causes, including incorrect information used to prepare the budget, better or poorer performance by people responsible for the activity, incorrect processes or uncontrollable events occurring outside the responsibility area.

An example of an uncontrollable event is the ongoing fluctuations in the price of fuel. In the construction industry, this may have an impact on you if you pay for the fuel consumption for your business. If this is a significant cost, then this fluctuation will make planning difficult and this could result in regular variances.

If a variation is identified it may become necessary for you to take corrective action which can include:

- **Interim action:** action that you take to solve the issue in the short term; usually you would implement this type of action while you decide on a more permanent solution. An example could be employing agency staff in the event of a full-time staff member leaving suddenly.
- **Adaptive action:** actions that are required when conditions change, such as a change in the demand for construction or a change in the supply of materials.
- **Corrective action:** these actions are implemented to rectify a negative situation, such as when there is a downturn in work. An example of an action could be running a marketing campaign to build up business again.
- **Preventative action:** this type of action is one that you might take to remove the cause of an issue. An example might be the management of a non-performing staff member, where they could be rotated to a different role in your business or performance managed.
- **Contingency action:** sometimes you might notice a negative trend in the performance of your business and you might then decide that if the negative trend continues you will need to act to reverse the trend and prevent any future impact.

Evaluating marketing and operational strategies

Your business plan needs to document your marketing and operational strategies. As with the managing of your financial plan, your marketing and operational strategies need to be reviewed and evaluated to ensure that they are effective. You need to remember that your financial plan hinges on these strategies and if they are not working effectively then there will be flow-on impacts onto your financial plan.

Marketing objectives and controls

Marketing is basically everything the business does to make itself known in the marketplace in which it operates. In the case of a small construction business, its strategies may include the placement of advertisements in local newspapers, displaying the name of the business on vehicles and perhaps even advertising on a regional radio or television station. While some businesses might do this without a real plan of action, it is useful to have an objective stated in your business plan.

A marketing objective is a statement of what the business expects to achieve through its marketing activities and will be focused toward the target market; that is, the intended group of customers. Marketing your business effectively will ensure that your business has the best chance to meet its operational and financial goals. A business that has no idea about who its customers are will not have effective marketing strategies.

A good marketing strategy should also be adaptable to change. This is an important element of reviewing and evaluating strategies on an ongoing basis as it allows for any changes in business conditions to be adapted to or new opportunities to be considered and included in the strategy.

LEARNING TASK 9.6

IDENTIFYING MARKETING STRATEGIES AND CONTROLS

For Bob's Carpentry, identify the sections in the business plan excerpt that detail Bob's marketing strategies and controls.

Calculate and evaluate financial ratios

The purpose of financial ratios is to allow you to compare the trading results of one year to another in a meaningful way, giving you the opportunity to think about how well your business has achieved its goals and objectives for the year.

Financial ratios provide a range of measures of financial performance which can be used to both benchmark your business against others within the same industry or as a key performance indicator to help you maintain focus on the goals for your business.

Financial ratios are calculated using figures from the Profit and Loss Statement and Balance Sheet to compare elements of the reports to focus on a particular measure (see Table 9.4). They include:

- profitability ratios measure efficiency and indicate the ability of a business to generate profits from sales, assets employed and owners' investment
- liquidity ratios measure the ability of a business to meet its short-term financial obligations
- activity ratios indicate the effectiveness with which financial resources are used
- leverage ratios indicate the extent to which debt funds are used in the business.

Calculating financial ratios

As the purpose of financial ratios is to compare data across different periods of time, it is important that the ratios are calculated in the same way. With this in mind, it is important to take the correct data from your financial reports when making the calculations. It is also important to be aware of different terminologies that may be in use: for example, the term *capital* refers to the Owner's Equity in the case of sole trader and partnerships or shareholders' funds in the case of companies. You will remember that Owner's Equity is the total of the owner's investment in the business and any retained profit.

Profitability ratios

- **Return on capital:** measures the performance of a business in terms of Net Profit generated. Dividing the Net Profit by the capital invested and expressing the result as a percentage for each year enables the business to compare its performance over a number of years and also compare it with the performance of similar businesses in the same industry.
- **Return on total assets:** shows the Net Profit for every dollar invested in assets, or in other words it shows the efficiency in the use of assets by a business.
- **Gross profit margin:** the difference between Sales and the Cost of Sales, so this ratio measures the profit earned by every dollar of sales before Operating Expenses are considered. By comparing the Gross Profit to Sales on a regular basis, a business is able to determine the causes for variation in the profit earned per dollar of Sales before Operating Expenses are considered.
- **Net profit margin**: measures the Net Profit earned by a dollar of Sales.

Liquidity ratios

- **Current ratio**: shows the short-term debt-paying ability and is calculated by dividing the Current Assets by Current Liabilities. This ratio will be at least greater than 1:1 if the business is able to meet all debt obligations as and when they fall due. As a rule of thumb, a ratio of at least 2:1 indicates the financial soundness of the business and that it is able to avoid liquidity problems.
- **Inventory turnover**: measures how many times per year the inventory is replenished. Inventory turnover is obtained by dividing the Cost of Goods Sold by the average Inventory. The average inventory is the average of the opening and closing inventories for the period.
- **Collection period (in days)**: this indicates efficiency of collections from credit customers. Collection from Accounts Receivable is a source of finance for a business. An efficient collection procedure is thus important for any business. Too much money tied up in Accounts Receivable for too long would make the business face the risk of bad debts.

TABLE 9.4 Commonly used ratios

Ratio	Purpose	Formula
Profitability ratios		
Gross Profit margin	Expresses the Gross Profit as a percentage of Total Sales	Gross Profit/Sales × 100%
Net Profit margin	Expresses the Net Profit as a percentage of Total Sales	Net Profit/Sales × 100%
Liquidity ratios		
Current ratio	Indicates the capacity of the organisation to meet its short-term debt	Current Assets/Current Liabilities

LEARNING TASK 9.7

EVALUATING FINANCIAL RATIOS

For Bob's Carpentry, review the key financial ratios (Gross Profit and Net Profit) provided in the business plan excerpt and comment on what these might mean for Bob's business.

Assess financial plan

It is necessary to assess the effectiveness of your financial plan.

It may help to consider the following questions as part of your review:

- What went well?
- What didn't go so well?
- Did your financial management processes help to detect problems before they became major issues?
- Did any major problems occur within the business that could have been prevented?
- Can the process adapt to match the organisation's needs as it grows and develops?
- Do the processes allow you to manage the business' finances effectively?
- Can the process be improved?

SUMMARY

This chapter provides guidance and direction on some of the key skills and knowledge required to manage the day-to-day financial operations of a small construction business operation. While there will always be a requirement for access by small business operators to external professional services in relation to financial management, such as an accountant or a bookkeeper, the need for you as the operator of the business to take prime responsibility for the financial management of your business is a critical success factor.

Specifically, this chapter covered the following key points:

- understanding the financial information required to operate the business profitably
- producing and evaluating budgets including, cash flow estimates for forward planning
- understanding the importance of managing business capital to meet the requirements of financial backers
- understanding the taxation requirements of the business to ensure that adequate financial provision is maintained
- maximising cash flow through the development of client credit policies
- identifying and selecting appropriate key performance indicators to monitor financial performance
- ensure documented procedures are recorded and communicated to key staff
- prepare regular financial performance reports to monitor financial performance
- monitoring the effects of relevant marketing and operational strategies
- demonstrating how to calculate and evaluate financial ratios
- assessing the implementation of the financial plan.

REFERENCES AND FURTHER READING

Birt I. 2016, *Writing Your Plan for Small Business Success* (6th edn), Pearson Education Australia, Frenchs Forest, NSW, Australia

Rumble S., Anandaraha A. and Aseervatham A. 2017, *Manage Budgets and Financial Plans* (5th edn), Cengage Learning Australia, Victoria, Australia

TAFE SA 2013, *Manage Small Business Finances*, TAFE SA, Adelaide City Campus, Small Business Training Centre, Adelaide, SA, Australia

The Australian Government provides a wide range of useful information for setting up and running a business, providing specific tools for business planning and budgeting: **https://www.business.gov.au/**

APPENDIX

See the Bob's Carpentry business plan (excerpt) in Appendix 3 at the end of the book.

PART 3

DESIGN CONSIDERATIONS

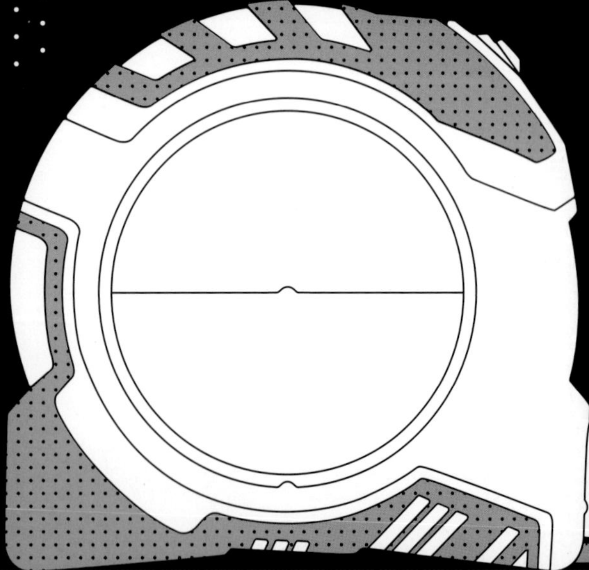

SITE SURVEYS AND SET OUT PROCEDURES

10

Chapter overview

Accurate set out is the basis of any built structure, be it a domestic home or a large commercial warehouse. Likewise, being able to conduct a basic survey of the land upon which that building is to be sited is necessary not only during the design of the building, but also in estimating and quantifying such things as excavation spoil, slopes for landscaping and fall for drainage.

This chapter introduces each of these areas. In so doing, the chapter covers basic levelling and measuring techniques, the use of surveying and levelling equipment, and the calculations necessary to the successful conduct of a typical site survey and the setting out of a low-rise building.

Elements

This chapter provides knowledge and skill development materials on how to:

1. set up and use levelling devices
2. perform setting out using appropriate measuring techniques and calculations
3. mark out and determine levels on a grid for contouring and volume calculations
4. construct longitudinal sections and determine associated grades and levels in typical drainage and pipeline situations.

The Australian standards that relate to this chapter are:

- Intergovernmental Committee on Surveying and Mapping (ICSM): Standard for the Australian Survey Control Network – Special Publication 1 (ICSM SP1)
- AS/NZS IEC 60825 Safety of laser products
- AS 2397 – 2015 Safe use of lasers in the building and construction industry.

Introduction

Surveying and set out are core activities in almost any construction project. Today, much of this activity is done professionally by highly skilled and experienced surveying companies. Contemporary professional surveyors use a range of high-precision equipment including drones, lasers and global navigation satellite systems (GNSS). Despite this, the basics of surveying and set out practice are still highly relevant to builders involved with small and medium-sized domestic or commercial projects.

The purpose of this chapter is therefore to provide the theoretical underpinnings, as well as practical guidance, into the how, what, why and when of both surveying and set out for construction purposes. The chapter begins with the basic instrumentation, their set up and use, along with how to test them for accuracy. This is followed by some of the more straightforward techniques for developing a set out, including common calculations, the identification and reduction of error, and quantifying cut and fill requirements.

The chapter will then move to the field of topography. Here the concept of contour plans is introduced, and how they are produced through basic grid surveying. Calculations for the determination of excavation volumes are again visited in this section. The final part deals with the development of longitudinal sections of the Earth derived from contour surveys and the various calculations associated with grades and batter levels.

Levelling devices: set up and use

The basic collection of levelling devices used by builders has expanded dramatically from the humble Cowley and dumpy levels of old. The Cowley is seldom, if ever, seen on site today – despite it not losing its appeal to some. The dumpy, or automatic level, however, retains its relevancy as a highly accurate optical levelling device and is still the workhorse for many builders. But it is the automatic laser that has captured the building market of today due to its speed and capacity to be a one-person operation.

Other laser devices have also found relevance in construction, particularly in electronic distance measuring – highly accurate yet remarkably inexpensive tools that again speed up on-site operations. These, among others, will be discussed in the following pages, along with how they should be used and their accuracy tested.

Levelling devices: the basics

Although the automatic laser level has taken top ranking for most standard forms of set out on site, we will begin this review of levelling devices with the optical automatic level. Despite generally requiring two people to use it with precision or speed, this instrument remains one of the best and most inexpensive ways to obtain accurate level readings. It is also an easy tool to check on site for accuracy through one simple testing procedure.

Optical automatic levels

Commonly referred to as 'dumpy' levels, these are basically a self-levelling telescope with cross-hairs for sighting. Incorporating high-quality optics (lenses), these instruments use prisms and a pendulum to automatically develop a level line of sight. A 'staff' – basically a long ruler – is used to measure distances from this level line of sight to the ground or other features as required (Figure 10.1). Correctly looked after, these tools are a lifetime investment that can easily be site checked for accuracy, though adjustment is generally best left to professionals who can bench-test the whole optical train.

Source: Glenn Costin

FIGURE 10.1 Automatic optical level 'dumpy' with staff in background

Setting up an optical level

Optical instruments may be set up on tripods with either flat or domed heads. The approach to setting up an optical level (dumpy) outlined below may be used to set up any form of levelling instrument that uses a tribrach – three screw adjustable base – system (Figures 10.1 and 10.2). This approach applies equally irrespective of the type of tripod the level is mounted upon.

HOW TO

SET UP AN OPTICAL LEVEL

Step 1	Set up tripod in a location that is out of harm's way, but is reasonably central to each element you need to sight. Extend the legs such that the optical level will be easy for you to view through. If a flat-topped tripod, ensure the top is reasonably level based upon horizon or known to be level features such as brickwork, house eaves, roof lines or the like. If a round-topped tripod, simply set up to the desired height. Make sure the bases of the legs are firmly pressed into the ground.
Step 2	***Round-topped tripod***: mount optical level on the tripod and swivel until the blister bubble on the instrument is close to the middle (Figure 10.3), then adjust the tribrach as per a flat-topped tripod (below). ***Flat-topped tripod***: Mount the optical level on the tripod and adjust tribrach as follows. The tribrach consists of three thumb screws for adjustment as shown in Figure 10.2. In addition, there is a blister or bubble level mounted into the main body of the instrument (Figure 10.3). Adjust these screws using the following technique:

Source: Glenn Costin

FIGURE 10.2 Tribrach adjustment screws

Source: Glenn Costin

FIGURE 10.3 Blister or bubble level

>>

>>

Step 3	• Position yourself so that you can look directly down onto the bubble level, or can see the bubble clearly using the mirror (Figure 10.3). • Picture in your mind the concept described in Figure 10.4. In this diagram the screws each sit under one arm of a 'T'.	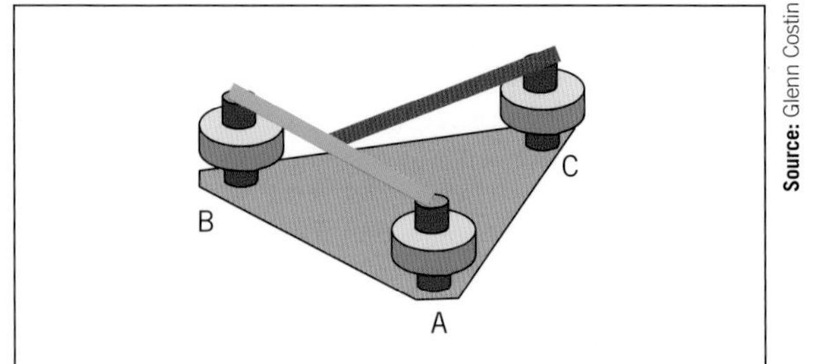 FIGURE 10.4 Visualising a 'T' over the tribrach
Step 4	• Hold two of the screws between fingers and thumbs as shown in Figure 10.5. These are the screws sitting under the head of the 'T' (green line – screws A and B in Figure 10.4). • You begin by adjusting these two screws simultaneously. This is done by moving your thumbs either towards each other, or away from each other as shown in Figures 10.5 and 10.6. This causes one screw to wind up, the other to wind down. As you do so, watch the bubble in the blister level. Your aim is to get it 'centred' *left to right*. **Note**: You are not trying for absolute centre at this point, just centre between left and right (Figure 10.7).  FIGURE 10.5 Rotate screws by moving thumbs towards each other or …  FIGURE 10.6 … by rotating thumbs away from each other 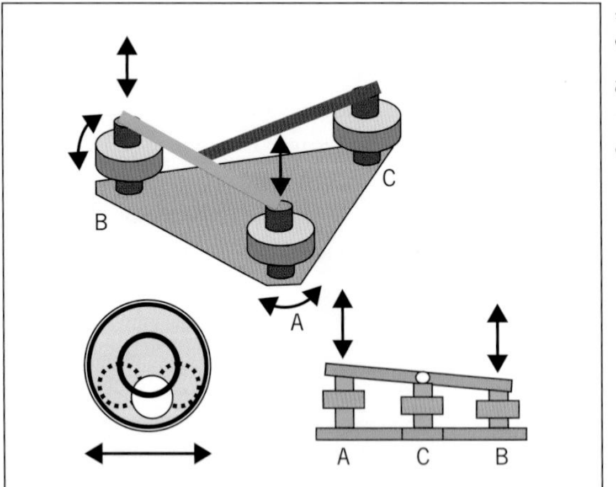 FIGURE 10.7 Adjusting the first two screws for level – by moving thumbs towards or away from each other, one screw will go up and the other down	

>>

>>

Step 5	• Now go to the third screw (C, Figure 10.8). ***Without*** touching the other screws, turn this single screw until the bubble is moved to absolute centre. **Note**: If the bubble remains slightly to the left or right of centre, go back to the first two screws and adjust them simultaneously again – ***never one at a time***. Learn to do this properly and setting up an optical level, or manual laser level (which uses the same tribrach-style base), is not more than a 30-second job.

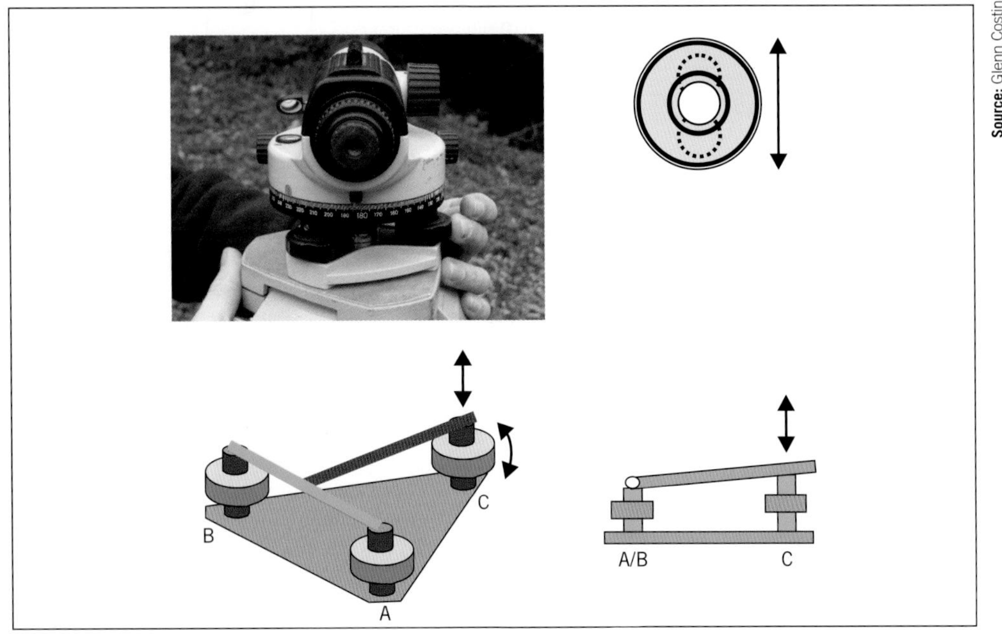

FIGURE 10.8 Adjusting the third screw only to bring the bubble to centre

The optical train and parallax error

The optical train is demonstrated in the diagrams shown below (Figures 10.9 and 10.10). It is important to understand this system, particularly with regards to eliminating a level of operator-induced inaccuracy known as 'parallax error'.

Parallax error occurs when the cross-hair or 'reticule' is in focus but seems to float in front or behind the image of the staff as seen through the eyepiece (Figure 10.10). This means that if you move your head up or down, the measurement you see on the staff will change. Only when the focal image is located precisely on the cross-hairs can this movement be eliminated (Figure 10.9). To achieve this, you carry out the following procedure.

Hold a blank, white, piece of card (the back of a white business card is fine) about 300 mm in front of the main (objective) lens of the instrument.

Look through the eyepiece, and with your eye relaxed; i.e. without trying to focus your eye upon anything, rotate the eyepiece focusing ring until the cross-hair (reticule) becomes clear and sharp. Now

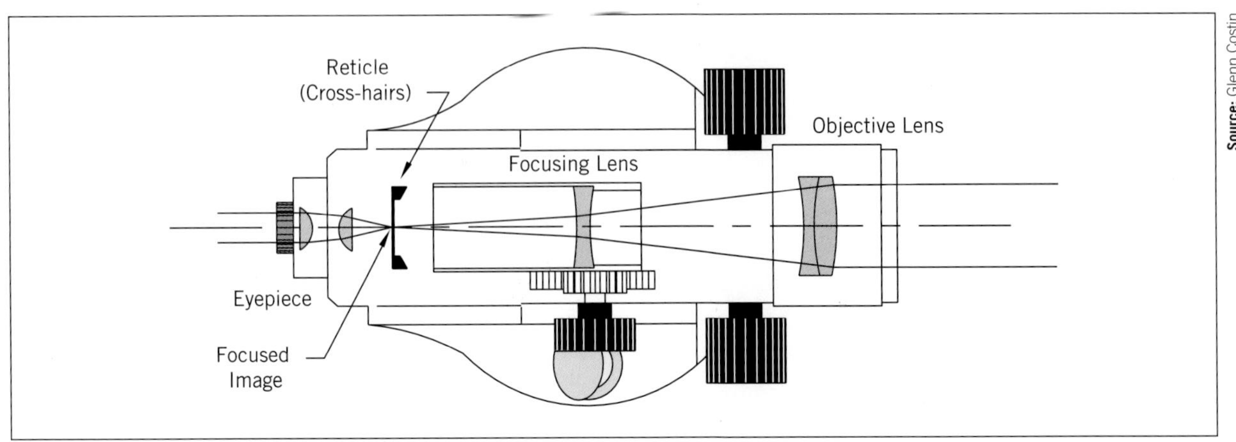

FIGURE 10.9 Parallax error eliminated by image being focused exactly upon the reticule

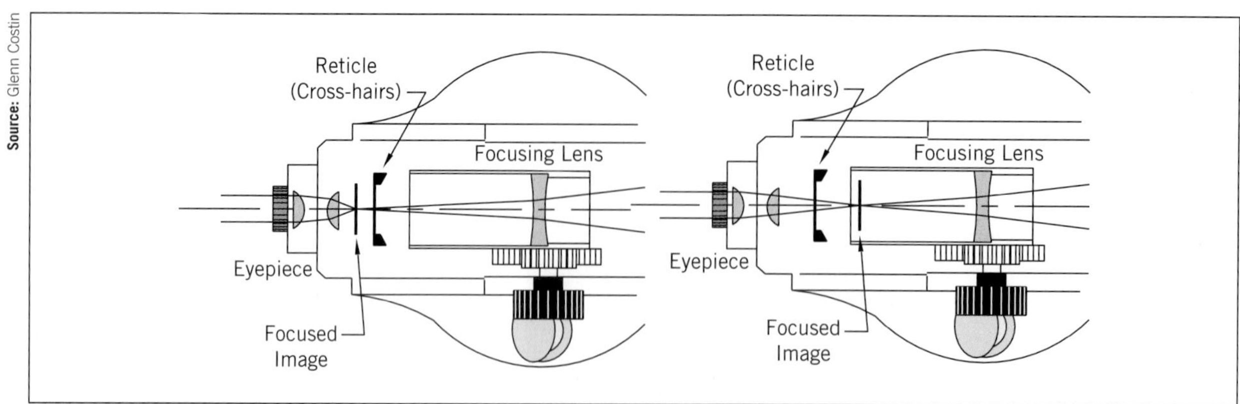

FIGURE 10.10 Parallax error present because focal image is displaced – in front or behind – from the reticule

when you use the main focusing knob to focus on the staff, the objective image and the reticule are in exactly the same spot and the potential for parallax error has been significantly reduced.

As each operator's eye is different, this procedure must be carried out whenever a new operator takes a reading.

The survey staff

This is effectively a very large, collapsible ruler for use with either laser or optical levels (Figure 10.11). When held plumb, the staff may be read by sighting through the optical level, or by a laser receiver. Being able to read a staff accurately is critical to many surveying tasks.

FIGURE 10.11 Survey staff

Taking a reading (optical levels)

The surveyor's staff (Figure 10.12) is marked in solid block increments of 10 mm. When reading any measuring tool, the most accurate one is one half of one graduation. With staffs of this type, therefore, you should only work to 5 mm increments. Trying to read finer is generally pointless as you are using them on rough ground or rough sawn pegs whereby the smallest stone is going to shift your reading by 2 or 3 mm anyway. When needing to read more accurately, and therefore over shorter distances, you might choose to use a staff graduated like a standard ruler – though experienced operators can, and sometimes do, take fine readings using the normal staff.

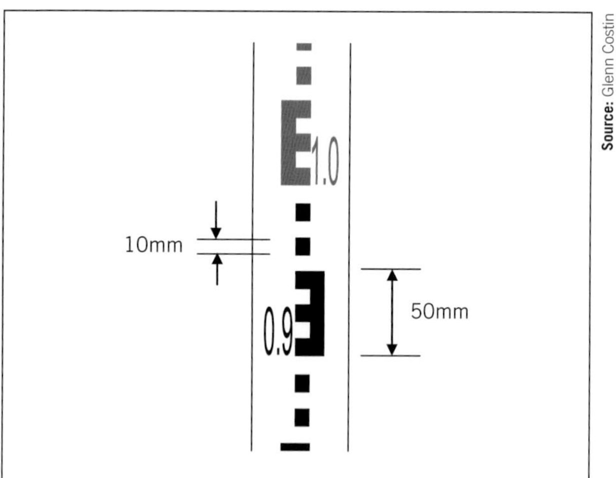

FIGURE 10.12 Increments on the traditional surveyor's staff

When looking through an optical level you will see the image as shown in Figure 10.13. You will note that, aside from the cross-hairs passing through the centre, there are two other lines. These are stadia lines, used on other occasions to calculate distance, and you must be wary of them. It is easy to accidentally take your reading off one of these lines, rather than the central cross-hair. Figure 10.13 shows a reading of 1.0 m.

Source: Glenn Costin

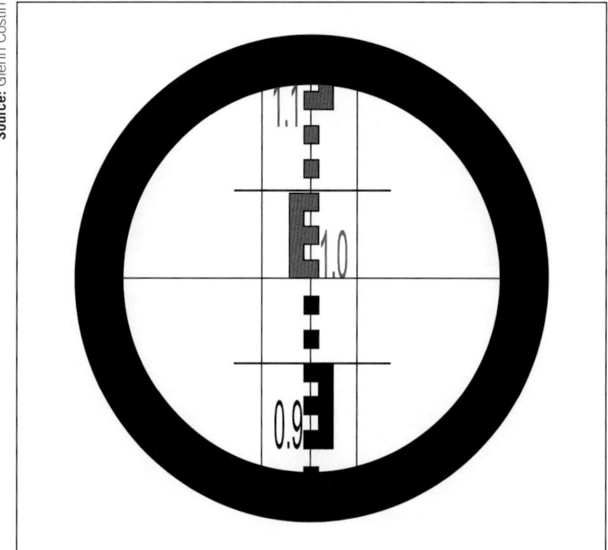

FIGURE 10.13 Sight of staff through automatic level

Laser levelling instruments

There are a range of levelling instruments available today that are based on laser-generating technology. Many of these, such as laser line generators, dot point lasers and the like, though useful at times, are incidental to survey and site set out work. Others, however, have become fundamental to our industry, bringing a level of accuracy and speed that was unthinkable only a few decades ago. It is the latter group that are described below.

Rotary laser levels

These are now the backbone of most domestic and commercial low-rise projects. The instrument casts a non-visible level line that rotates around the level's centre. The laser's beam is then read by means of a small receiver.

Source: iStock.com/roman023

FIGURE 10.14 Electronic self-levelling

There are several types of rotary laser levels, of which only two are considered suitable for professional trade craft.

- **Electronic self-levelling lasers** (Figure 10.14): These are now the most common laser levelling units and are to be found in use on most construction sites, commercial or domestic. Once the instrument is set up roughly level and turned on, servo motors self-align the unit to a level plane. Some of the higher end units will also tell you if they need service or recalibration. The best of these units are used by professional surveyors for large building works, engineering, mining and the like.
- **Pendulum automatic levels** (Figure 10.15): Once the main form of laser level found on domestic construction sites, the pendulum mechanism has now been relegated to the smaller line generator types as shown in Figure 10.15. The pendulum system is a cost-effective way to ensure the laser only casts a level beam or shuts off: the 'swing' of the pendulum generally being dampened by magnets. Most of these instruments will compensate for the unit sitting up to about 5° out of level. When not in use, and in order to protect the pendulum mechanism, there is a locking device that should be activated prior to packing and transporting. Some units will allow this lock to be used such that the beam may be cast at an angle.
- **Grade levels:** This is a form of automatic yet manually adjustable level that allows the instrument to cast a plane of light at an angle rather than level. This is highly effective in drainage and plumbing excavations where a continuous fall of a set gradient or angle is required. With most instruments of this type, the required angle is fed into a small on-board processor which then automatically tilts the laser beam through the compensators (Figure 10.16). Angles up to 15% from horizontal are achievable in many cases.

Electronic distance measuring (EDM) equipment

Most distances in surveying and set out work for domestic construction will be by direct measuring

Source: Imex

FIGURE 10.15 Pendulum

Source: Position Partners Pty Ltd

FIGURE 10.16 Grade levels

techniques using tape measures of a form. Optical instruments can also develop distances using tachometry or telemetry – both deriving distance through fine angle discernment measurement of one short length at the target end. In all the above cases, measurements are limited to relatively short distances, and easily defeated by obstacles.

Electronic distance measuring (EDM) overcomes these issues using electromagnetic waves. The waves are projected towards either a receiving device for amplification and return, or a simple reflector. In either case the waves may be counted and, knowing the length of each wave, the distance may be computed, a technique referred to as phase measurement. Alternatively, a pulse or time of flight measurement is used where the speed of the wave is matched to the time taken to send and receive and the distance so derived.

There are three basic types of EDM instruments (Figure 10.17), framed around the form of electromagnetic wave used:

- **Microwave instruments** – these can be used in day or night conditions over extreme distances up to 100 km. They can also be used in poor atmospheric conditions such as fog or rain. Accuracy, however, is limited to around +/– 5 mm – 15 mm per km.
- **Infrared instruments** (also known as electro-optical) – these are a shorter wavelength type, with moderate accuracy of the three at +/– 10 mm per km. Range, however, is the limiting factor: distances are limited to 3–5 km.
- **Viable light instrument** (another form of electro-optical) – effectively these are lasers using the visible light spectrum such as red or green. Extremely high accuracy, +/– 0.5 mm – 2 mm per km, is achievable with these units. Range is variable with some limited to as low as 3 km, while others claim as much as 70 km in clear conditions. Newer units, such as that shown in Figure 10.17 below, are capable of achieving accurate measurements without a reflector at distances up to 500 m and as close as 0.3 m.

Source: Position Partners Pty Ltd

FIGURE 10.17 Topcon GM-50 Geodetic Measurement Station

Theodolites

Traditionally, a theodolite was, is, an advanced optical instrument that can be used to determine angles and distances in both vertical and horizontal planes. Distances are calculated by tachometry and/or trigonometry, while angles were read off graduated analogue scales. Analogue instruments (Figure 10.18) are seldom, if ever, used today as their precise use is heavily operator-dependent and calculations are time consuming.

The contemporary theodolite (Figure 10.19) uses an inbuilt computer to digitally determine angles and calculate distances based upon the inputs received. The optical component remains much the same though even less-expensive units can achieve far greater optical quality today than many were capable of previously. Most theodolites are mounted on a tribrach similar to that found on a dumpy level, meaning the final adjustment process in setting up a theodolite is similar to that described previously. It is in the initial location of the instrument over a defined mark or peg that requires a far more protracted set up process. This is best demonstrated practically by an instructor; however, there are a number on online video or YouTube clips available from which the process may be learnt, such as: https://www.youtube.com/watch?v=QUX9_1fRnlo and https://www.youtube.com/watch?v=lp824ZRIWQs

The main advantage offered by theodolites when setting up over a specific point is the provision of a perpendicularly down viewer or 'optical plummet' by which to adjust the instrument to an exact location.

Source: Alamy Stock Photo/travelib prime

FIGURE 10.18 Standard analogue or Vernier theodolite

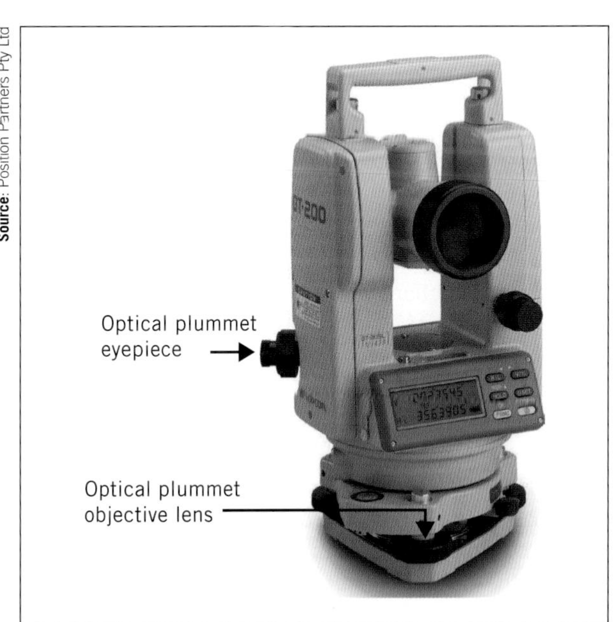

Source: Position Partners Pty Ltd

FIGURE 10.19 Modern digital theodolite

Optical and laser plummets

Optical and laser plummets (Figure 10.20) are tools for providing highly accurate positioning over a chosen mark or feature. Optical plummets use an eyepiece and prism to allow visual sighting of the mark, while lasers use intense visible light (generally red or green) to plot a visible position in the form of a bright dot. Plummets,

Source: Glenn Costin

FIGURE 10.20 Optical plummet (left), laser plummet (right)

optical or laser, are integrated into the tribrach and/or bodies of EDMs, theodolites and total stations to ensure their accurate positioning over markers, and hence their use forms part of the set-up procedure for these instruments (see Figure 10.19).

Total stations

The total station (Figure 10.21) looks, is set up and operates much like a modern digital theodolite.

Source: Alamy Stock Photo/Chris Pancewicz

FIGURE 10.21 Theodolite - Leica TCR1205 R100 TPS Total Station EDM. Reflectorless total stations

The main difference between these two instruments is the inclusion of distance measurement (EDM) capacity and, in some more advanced models, global navigation satellite systems (GNSS) and robotic control. GNSS units capture signals from orbiting satellite arrays or 'constellations' that enable the instrument to accurately locate itself in three dimensions on the Earth's surface, thus radically speeding up the survey process. There are currently four main GNSS constellations in place:

- BeiDou Navigation Satellite System – China
- Galileo GNSS – European Union
- GLObal Navigation Satellite System (GLONASS) – Russian Federation
- NavStar Global Positioning System (GPS) – United States of America

Depending upon the instrument's manufacturer and its location when in use, any one of these constellations may be used to locate the total station. In some cases, multiple constellations may be used to enhance accuracy. GNSS accuracy is further enhanced as the total station tends to remain static, and mobile satellite receivers are then used to plot outlying points that are referenced not just to satellites, but also back to the base station, mutually enhancing the accuracy of both positions. This is referred to as Real Time Kinematic positioning or RTK.

Accuracy

Total stations are currently the most accurate surveying and positioning systems available, though accuracy varies between manufacturers. GNSS units can currently achieve accuracy within 3 mm post processing within the instrument (Leica Viva GS16). As noted previously, for local measurements (within 1 km of the total station) the EDM element of the unit remains the most accurate with only 0.5–1.0 mm of error over distances of up to 1000 m.

Use in industry

It is highly unlikely that you will see these instruments on domestic construction sites for some years as they are complex and expensive pieces of equipment. They are, however, becoming common on larger low-rise commercial and residential projects, as well as in the initial phases of residential housing developments in the establishment of roads, services and block boundaries.

Surveying equipment maintenance: national, state and territory standards and regulations

Each state and territory has some form of regulation governing the practice of surveying. These regulations stipulate the levels of accuracy required and also the maintenance periods for specific types of equipment: maintenance having a significant influence upon accuracy. For example, the NSW Surveying and Spatial Information Regulation 2017 (under the 2002 Act of the same name) requires that EDMs and GNSS equipment must be tested against state control surveys at least once every year. They must likewise be tested after any repairs and even after a software update.

In addition to state and territory regulations, there exists an authoritative body known as the Intergovernmental Committee on Surveying and Mapping or ICSM. Australia's network of survey controls (permanent marks, positional locations and the like) is governed to a degree by the ICSM standard SP1. This document also provides advice on minimum equipment quality and control. Although the standard applies to work of a quality generally superior to domestic housing, many professional surveyors use this as a guide for all their contracted works.

Safe use of laser equipment

Lasers have now become extremely common on all types of low-rise construction, be it domestic or commercial. The strength of the lasers varies quite markedly between tool types and purpose. Laser line generators, those that produce a visible line, are generally the most powerful, though laser-based theodolites, EDMs and total stations are also of significant strength. Such powerful lasers create potential safety hazards and associated risks, mainly to the eyes, but also to the skin, which must be managed.

Refer to:

- AS/NZS IEC 60825 Safety of laser products
- AS 2397 – 2015: Safe use of lasers in the building and construction industry

Australian standards

The manufacture and use of lasers come under two Australian standards:

- AS/NZS IEC 60825 Safety of laser products
- AS 2397 - 2015 Safe use of lasers in the building and construction industry.

Both these standards describe eight laser classifications:

- Class 1: low power and low risk. Considered eye safe in general usage.
- Class 1M: Safe unless coupled with telescopic optics.
- Class 1C: Generally medical or cosmetic lasers not likely to be used in the construction industry.

- Class 2: Visible lasers but of lower power, blink response as natural aversion to bright light is considered sufficient eye protection.
- Class 2M: Visible lasers of moderate power offer little risk from momentary or accidental exposer, but high risk if deliberate, prolonged, or optically enhanced exposure occurs.
- Class 3R: These are up to five times more powerful than the previous classes and should not be used in dimly lit areas or where possible prolonged accidental exposure is possible.
- Class 3B: May be visible or invisible lasers of very high power from which significant injury may occur even with limited accidental exposure. Class 3B lasers are not permitted on construction sites under these standards.
- Class 4: High-powered visible or invisible lasers offering significant risk to skin or eyes even with momentary exposure durations. Class 4 lasers are also not permitted on construction sites.

For all but Class 1 lasers, when used on construction sites, administrative controls must be in place to ensure: operator training, exposure avoidance, appropriate use and, with regards to the 'M' and 'R' classifications, avoidance of exposure through optical instruments (e.g. theodolites and dumpy levels).

In addition, *all* operators of Class 2M and Class 3R lasers must be trained in their use, have access to all operational manuals and have a copy of AS 2397 kept on site. The operator (you) must ensure that the tool is used in accordance with the instructions in these manuals and guidelines in the standards.

Because Class 2 and stronger lasers have the capacity to cast a line beyond the immediate area of use, a warning sign (Figure 10.22) must be in place at all points of access.

Identifying the laser class

All laser equipment must be labelled with a warning stating the class of laser, such as shown in Figure 10.23.

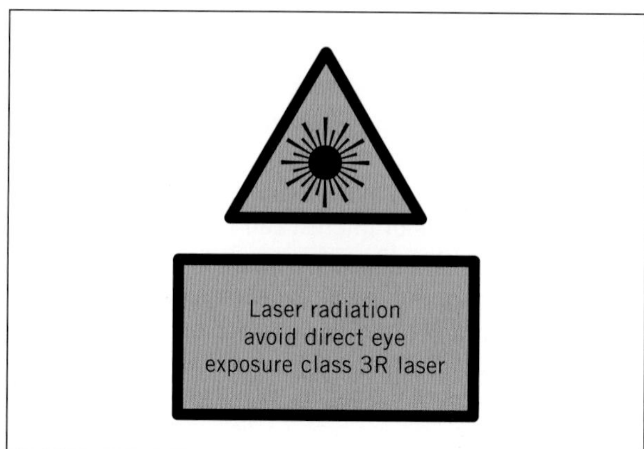

FIGURE 10.22 Laser warning sign

Source: Glenn Costin

FIGURE 10.23 Typical laser classification labelling

As standards have changed, so has the tightening of the labelling requirements, coupled with changes in the classification of lasers. This means that there remains in use laser equipment labelled to old standards, and those labelled under US Standards rather than Australian. Typical alternative labelling may be:

- Class 3A: treat as Class 1M if invisible, or 2M if visible
- Class 3B (restricted): treat as Class 3R.

US Standards use Roman numerals such as I, II, III; meaning 1, 2 and 3 respectively. Though slightly different in definition, US Class I and Class II may be treated as Australian Class 1 and Class 2, respectively. US Class IIIa should be treated as for Australian Class 3R.

Laser safety officers

AS 2397 requires that a laser safety officer (LSO) be in place at all sites where Class 1M, 2M and 3R lasers are used. It is this person's responsibility to ensure all tools are correctly maintained and that the beams of the more powerful lasers are confined to the site of use (not allowed to shoot across the street, for example). In small one- or two-person sites (tradesperson/employer and apprentice, for example) this responsibility lands on the employer.

Staff readings and reduced levels

The use of levelling instruments is not just about the determination or transfer of the one height or level across a site. It's not just about making things 'level'. Rather, surveying and levelling is frequently about the determination of the difference in height between one point and another, or the actual height of one location above another very specific location known as the *datum*. Such heights or levels are generally referred to as reduced levels or RLs: *reduced* in this case meaning 'to take from' or derived from. These two terms are discussed below – for a visual description, see Figure 10.28.

The datum

On a construction site or plan set where a datum is described, this becomes the point to which all other heights are referenced. The datum is your beginning point.

On or near a building site the datum is an easily identifiable peg or mark of known or nominated height: 'known' as in its height above the Australian Height Datum (AHD) or the AHD Tasmania (AHD-TAS83). Both these approximate the mean sea level height for those regions. Both tie in with Geocentric Datum of Australia 2020 (GDA2020) referenced by GNSS systems such as GPS and the like described earlier. On other occasions the datum may be offered a 'nominal' height, usually 100 m. This peg or mark is identified on a set of building plans as a TBM (temporary benchmark), BM (benchmark), or PM (permanent mark). The latter, however, is generally only to be found outside of the site boundary on a pavement or some council serviceable location. PMs are numbered pins set under a removable brass plate – the number is a reference to a council-held database of constantly adjusted values that cater for earth movement.

The purpose of a datum is to allow the accurate determination of important levels on and around a structure, such as floor levels, drainage lines, or landscaping features. A datum also allows a building to be sited relative to the height of with respect to known flood levels, or aspects (such as floor or window viewing levels) of other buildings. Alternatively, the datum may simply provide a point of reference to ensure that differing floor levels, window heights, tile heights, ceiling heights and the like, within the one building, are correctly maintained.

Reduced levels (RLs)

Reduced levels (RLs) are the height given to a point relative to, 'reduced from' or derived from the datum. So if our datum has a height of 100.00 m, and the point we are measuring is 1.000 m lower, then we would say that it has an RL of 99.00 m. Note that the datum itself is also always expressed as an RL: so our datum in this case is said to have an RL of 100.00 m.

The basic principle of levelling

The principle behind levelling, be it by laser or optical instrument, is the casting of a level line from which all other points of interest are referenced vertically. This is known as the 'line of collimation' in that it passes through the optical centre of the level, or generational centre of the laser. Figure 10.24 shows this line being cast around the instrument a bit like a flat, perfectly level, disc.

This is your level 'plane' (a flat horizontal surface passing through the air). With a laser this is exactly what's occurring and, in a foggy evening light, is actually visible. With an optical instrument, this is the line of sight you will have as you turn the instrument through 360°.

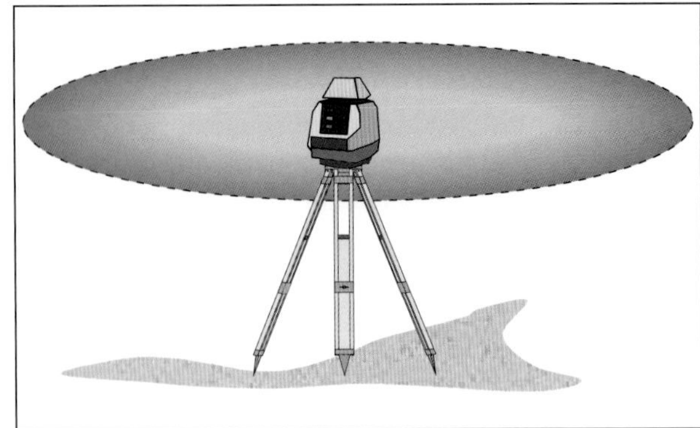

FIGURE 10.24 The line of collimation projects like a large, flat, level disc

To transfer a height from one point to another you only need to measure down or up from this level plane to the point you are referencing – a floor, door height, ground surface, formwork edge or the like – and then move to your chosen location or testing point and measure the same distance.

When trying to find the difference in height between just a couple of points, your task is almost as simple. Measure from this 'disc' down to the first peg or point and record the measurement. Then do the same for the other point location. Subtract one from the other and you have the difference in height between these two points. This is demonstrated in Figure 10.25:

- Peg A is measured as being 0.600 m below the level plane
- Peg B is measured as being 1.100 m below the level plane.
- => 1.100 – 0.600 = 0.500 m
- 0.500 m is then the difference in height between Pegs A and B.

Determining a reduced level (RL)

Figure 10.25 also demonstrates the development of RLs from a datum. In this case, Peg C describes a datum with an RL of 100.00 m. The line of collimation, reference plane, or line of sight is measured at being 0.800 m above this point. The line of collimation therefore provides a reference plane at 100.800 m. All other desired points of interest can now be referenced to this plane. In this example:

- Peg A measures at 0.600 m below this line.
- Peg B measures at 1.100 m below this line.

From these it can be determined that:

- The RL for Peg A is 100.200 (100.800 m – 0.600 m).
- The RL for Peg B is 99.700 m (100.800 m – 1.100 m).

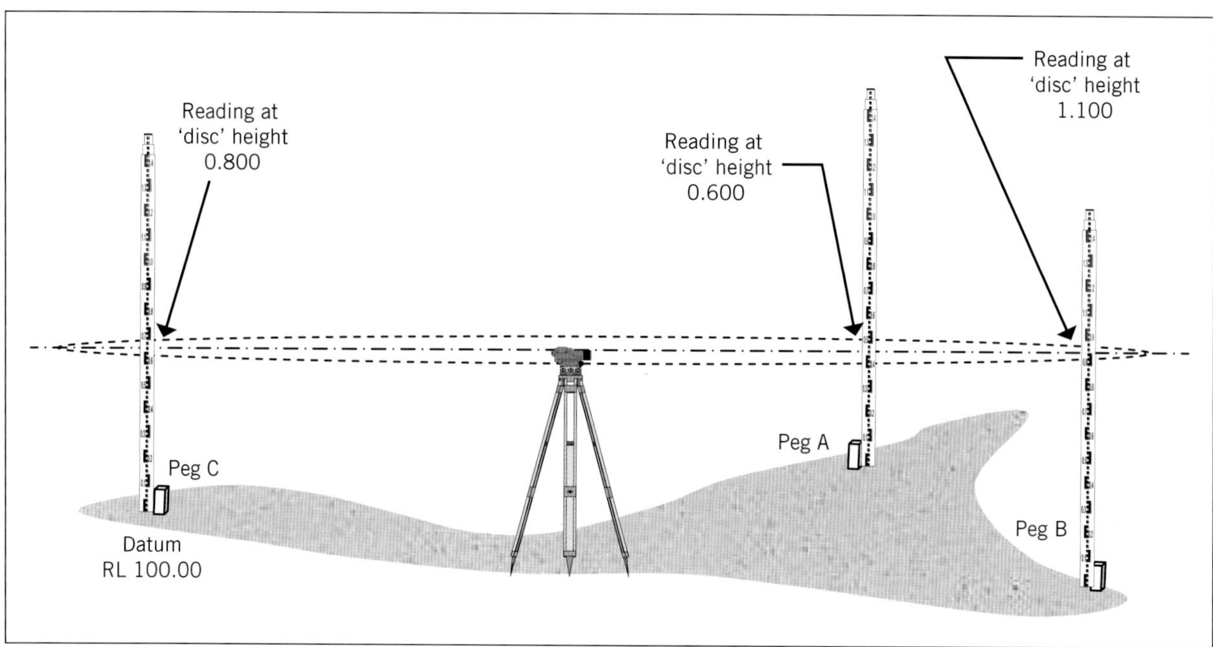

FIGURE 10.25 Determining the difference in height or elevation at various locations

This is referred to as the 'height of instrument' or 'line of collimation' method (discussed more fully later in the chapter – see p. 261). Though simple, it clearly demonstrates the how and why behind more complex approaches, helping to explain what is happening in the following procedures.

It is not the preferred approach for doing multiple sightings. In such cases you should use the *rise and fall* method of 'booking' readings.

The rise and fall approach

This is a system for booking (recording) the readings taken during a survey. The booking sheet looks something like that shown in Figure 10.26.

Readings are written down or 'booked' in the order that they are taken. The term 'rise and fall' refers to the relationship between one reading and another. One peg or staff position is seen as either higher or lower than the previous peg, or that to be read next. Figure 10.27 shows how this works. Peg B records a rise from A, while C records a fall from B.

One of the key benefits of this system over the height of collimation approach is that it deliberately causes cumulative error if you do not book, or derive your RLs from the booked values, correctly. A system of checks is then applied that identifies if an error has been made; allowing you to find and fix it before using the field data in application.

Surveying terminology

Before looking closely at the rise and fall approach, there are a few terms that need to be understood. These are station, backsight, foresight, intermediate sight, rise and fall; and are depicted in Figures 10.28 and 10.29.

- Station: This is the location of the instrument. To 'change' station is to pick up the instrument and relocate it elsewhere. The new location becomes another station.
- Backsight: This is the first sighting of any survey (Figure 10.28). It is also the first sight taken when you change position of the instrument – known as changing station – and begin taking readings again. It is called a 'back' sight because you are looking back to a reference point of known elevation or RL, usually your datum or the last sighting you

Backsight	Intermediate Sight	Foresight	Rise	Fall	Reduced Level	Notes

FIGURE 10.26 Typical rise and fall booking sheet

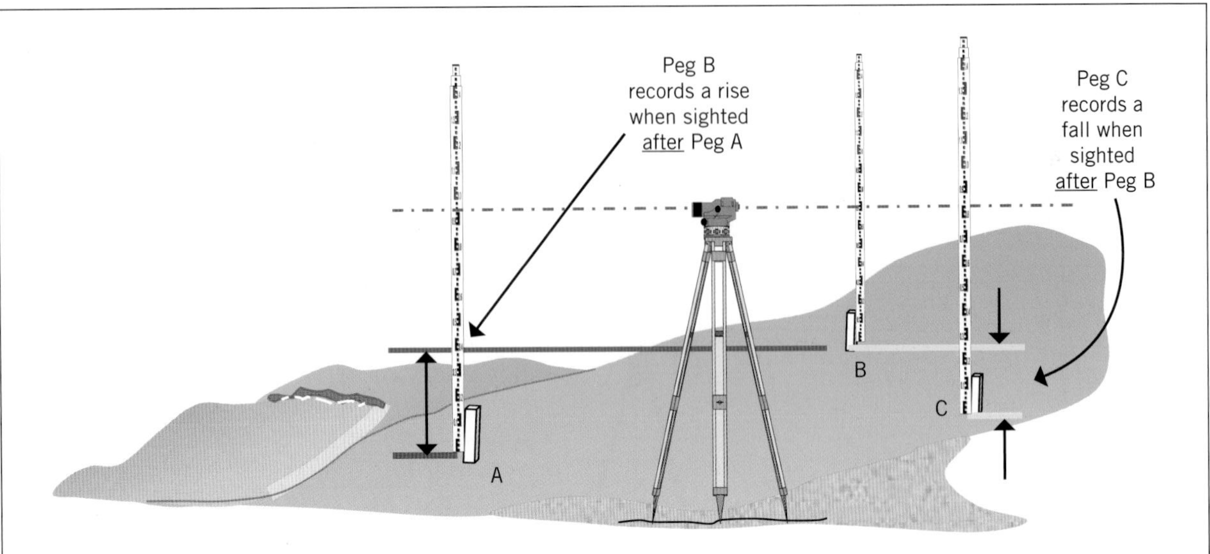

FIGURE 10.27 Basic principle of the rise and fall method

took before moving the instrument. Think of it as looking back to that which has gone before, and is known.

- Foresight: This is the last sighting taken before moving your instrument (changing station) or ending the survey (Figure 10.28). You are looking for(e)ward to the future.
- Intermediate sight: These are all the sightings taken in between the backsight and the foresight (Figure 10.29).
- Rise: An increase in height between a reading and the one taken immediately before it.
- Fall: A decrease in height between a reading and the one taken immediately before it.

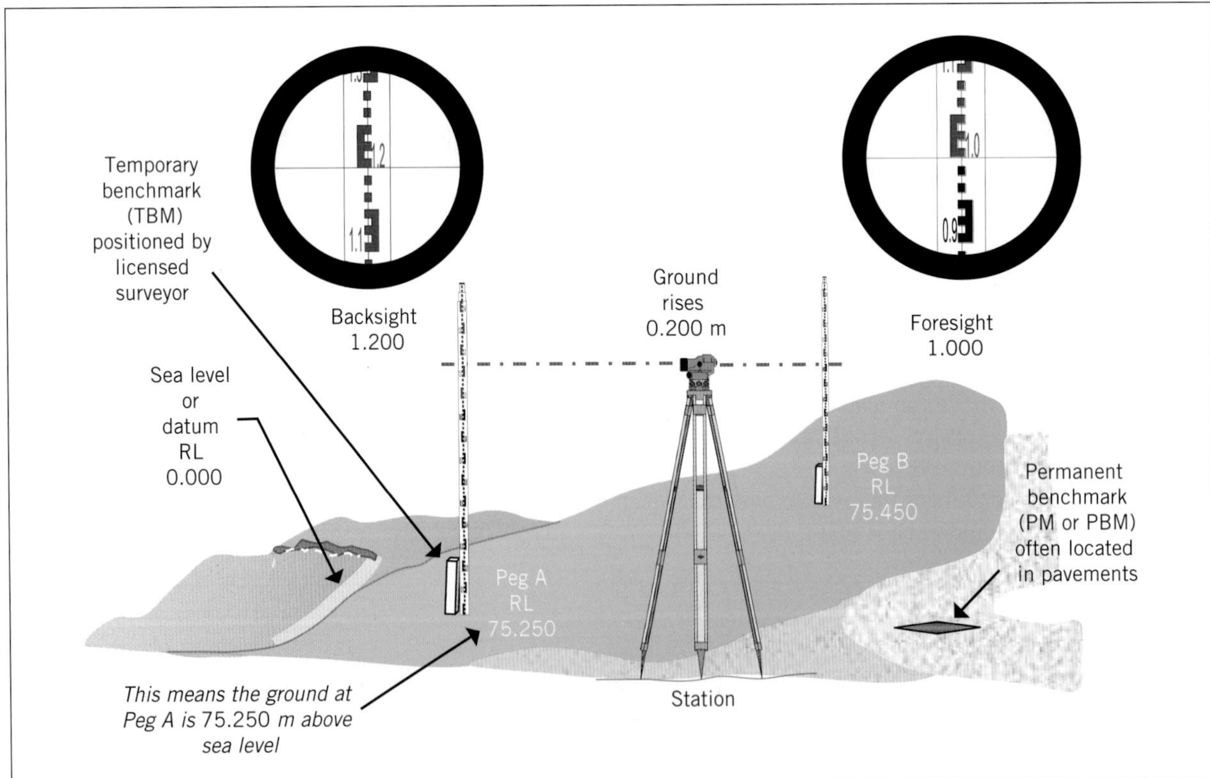

FIGURE 10.28 Descriptions of key surveying terms and elements

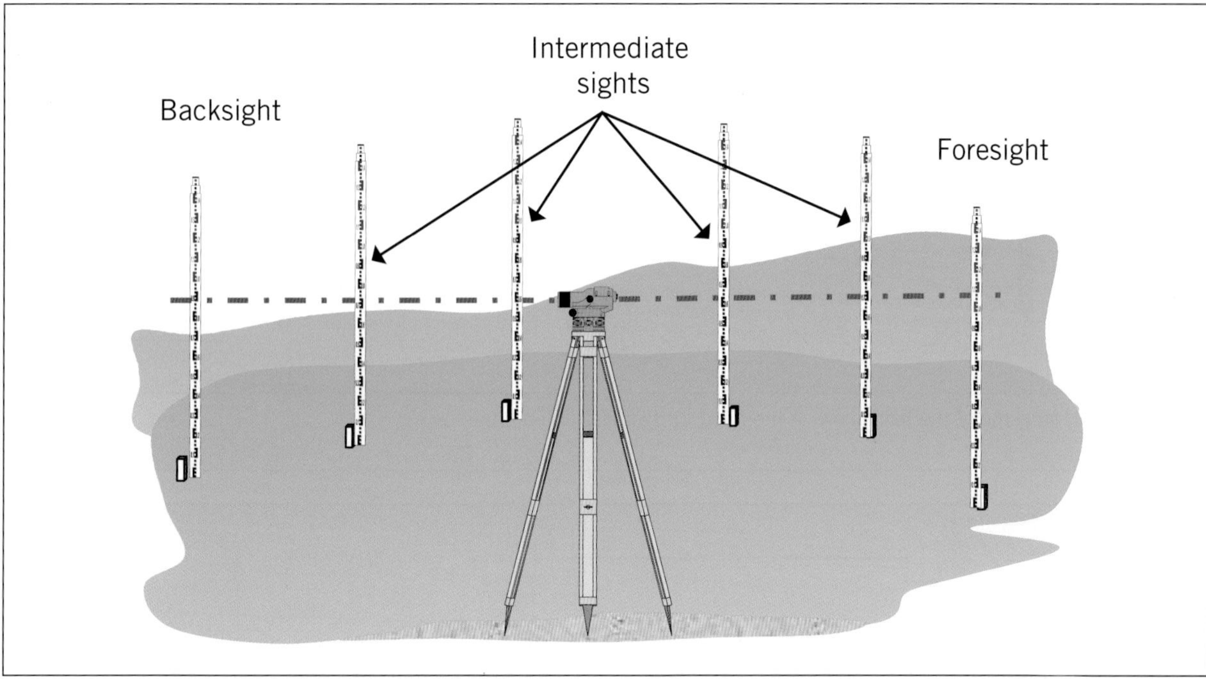

FIGURE 10.29 Intermediate sightings

Booking a basic site survey

Figure 10.30 shows a rectangular strip footing for brick work on a sloping block. As Figure 10.31 shows, the land is sloping too much for the instrument to sight all points from the one position: the line of sight is either running into the ground, or being too high to read the TBM. This means that two station positions will be required.

Figure 10.32 shows the instrument set up at Station 1 so that a sighting can be taken 'back' to the temporary benchmark (TBM). This first reading is a backsight, and is 'booked' in the backsight column as shown in Figure 10.33. Note that included in this entry is the known reduced level or RL for the TBM and that the name of the peg (TBM) is entered in the Notes column.

The next sighting (Figure 10.34) is a foresight to peg NW – the last sighting taken from this station before the instrument is moved.

This information is then booked as shown in Figure 10.35. Note that this ***change point*** (***CP***) – the point at which the level is removed to a new location – is listed in the 'Notes' column beside the name of the peg: in this case 'NW'. The next sighting will look 'back' to this peg.

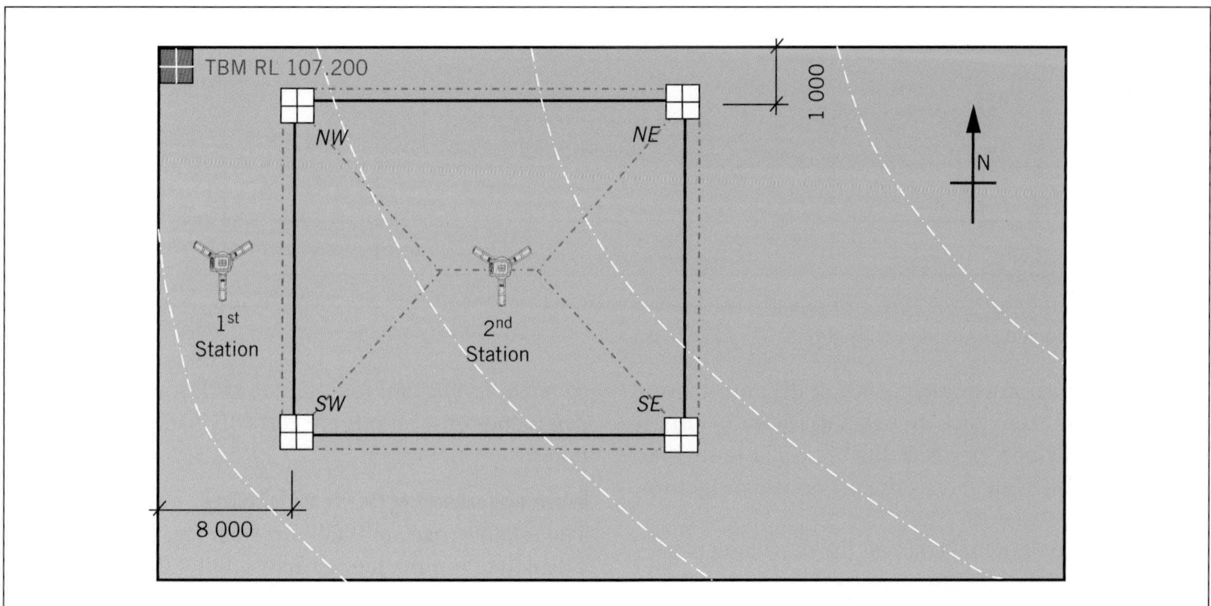

FIGURE 10.30 Plan of a rectangular strip footing pegged for survey and location of instrument stations

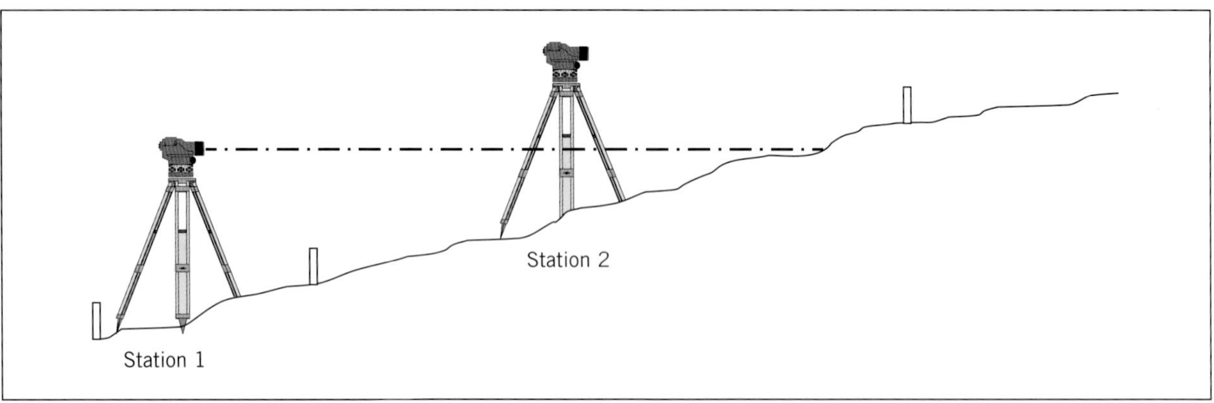

FIGURE 10.31 The slope of the land showing the need for two instrument stations

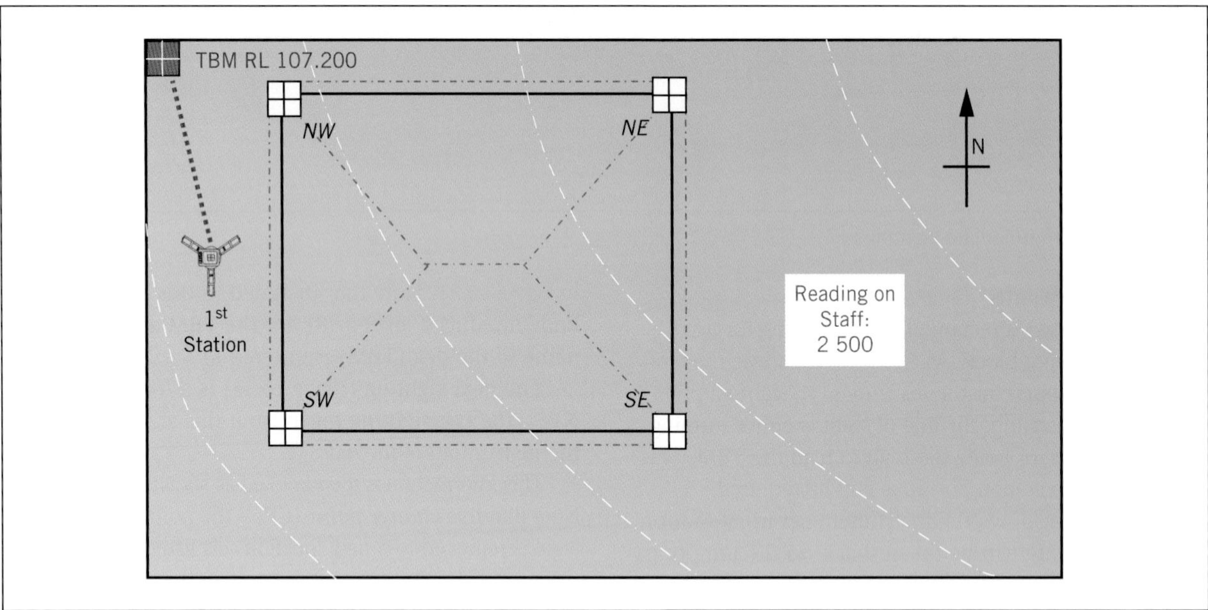

FIGURE 10.32 1st station and backsight to TBM

Backsight	Intermediate Sight	Foresight	Rise	Fall	Reduced Level	Notes
2.500					107.200	TBM

FIGURE 10.33 Booking of initial backsight

With the instrument established at the 2nd station, a backsight is taken 'back' to peg NW (Figure 10.36). As there is only one 'peg NW' the backsight is always recorded on the same line as the previous foresight to the same peg (Figure 10.37).

Sightings are now taken to all the remaining pegs. These will be 'intermediate sights' except for the last one: which will be a foresight (see Figure 10.39).

It is important that sightings are recorded *as they are taken, in the order that they are taken, and in the correct columns and rows* (Figure 10.38). Also, that all peg names or locations are correctly identified in the notes.

Determining and booking the rise and fall values

The relative 'rise' and 'fall' for each location is now calculated by subtracting each sighting from the one preceding it. This is done in the order that they were taken, and within the readings taken from a single station (see green loops in Figure 10.40). If the result

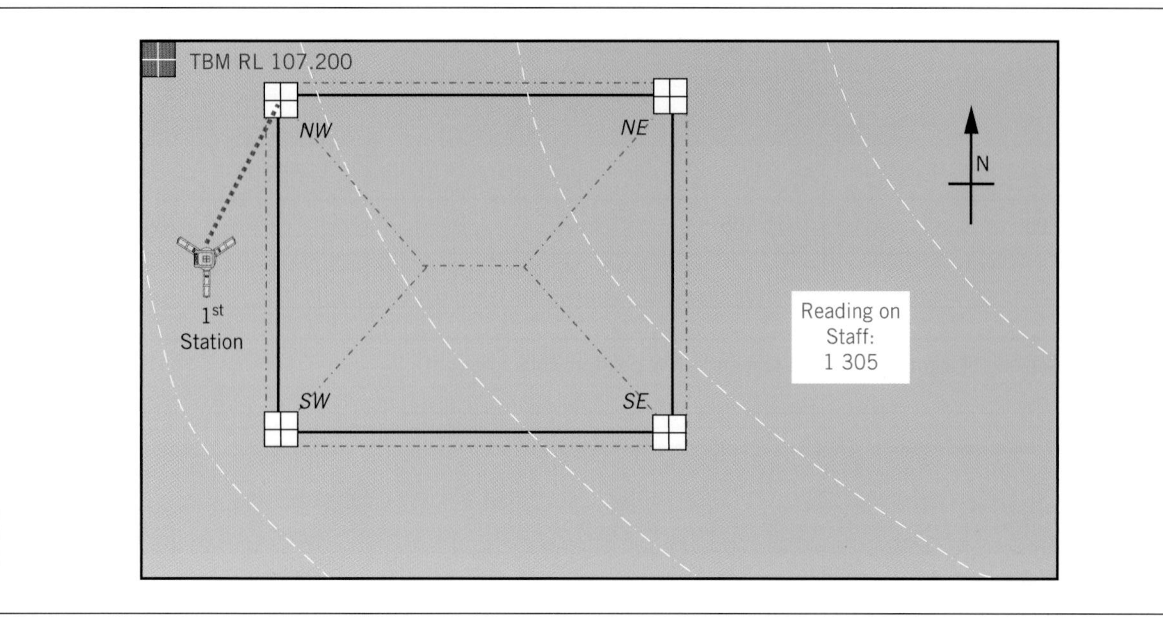

FIGURE 10.34 1st station foresight. Once taken and booked, the instrument may be moved.

Backsight	Intermediate Sight	Foresight	Rise	Fall	Reduced Level	Notes
2.500					107.200	TBM
		1 305				NW - CP

FIGURE 10.35 Booking of 1st station foresight

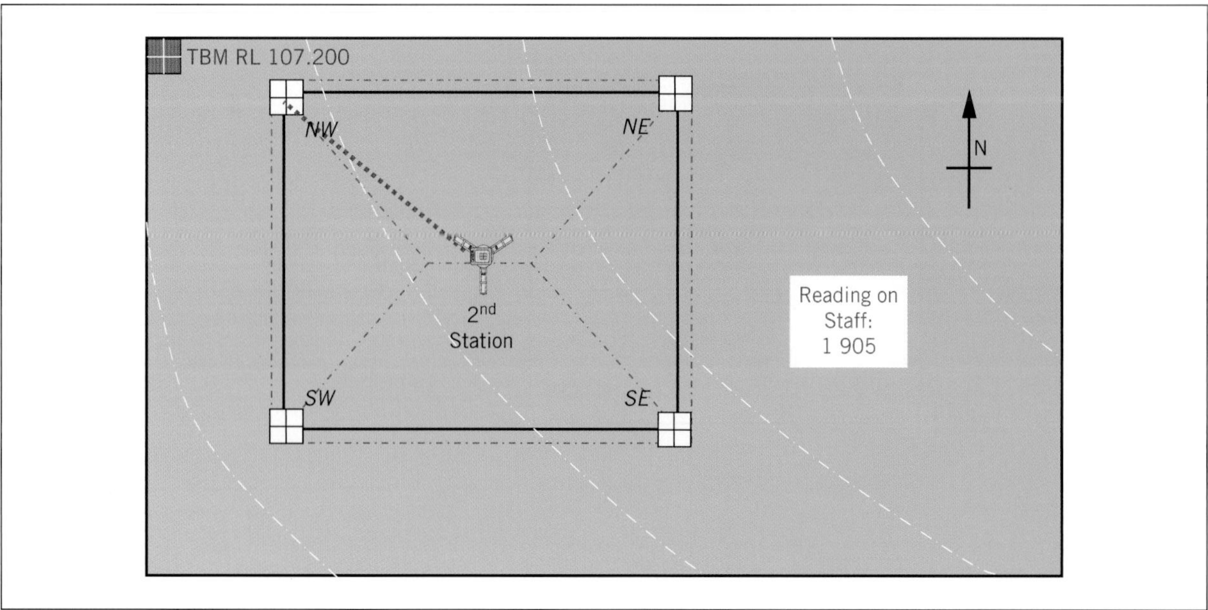

FIGURE 10.36 Backsight from 2nd station to NW peg (previous foresight)

Backsight	Intermediate Sight	Foresight	Rise	Fall	Reduced Level	Notes
2.500					107.200	TBM
1905		1 305				NW - CP

FIGURE 10.37 Booking of sighting from 2nd station to NW peg (previous foresight)

Backsight	Intermediate Sight	Foresight	Rise	Fall	Reduced Level	Notes
2.500					107.200	TBM
1.905		1.305				NW - CP
	0.910					NE
	1.255					SE
		2.365				SW

FIGURE 10.38 Intermediate sights and notes completed

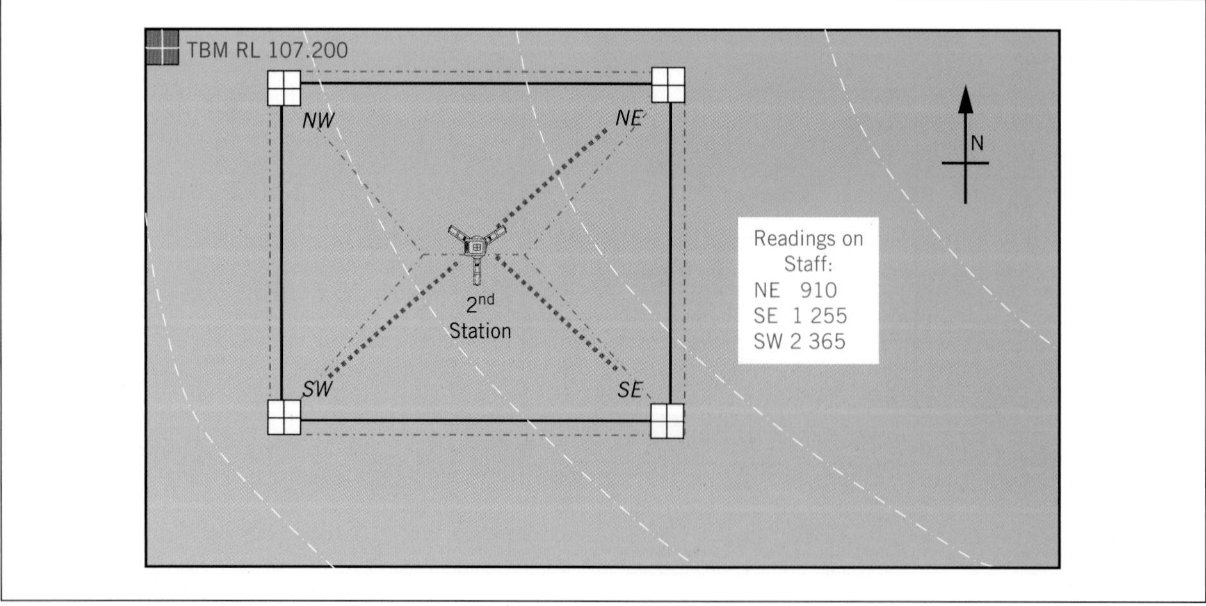

FIGURE 10.39 Taking the intermediate sights (NE and SE) and foresight (SW)

Backsight	Intermediate Sight	Foresight	Rise	Fall	Reduced Level	Notes
2.500					107.200	TBM
1.905		1.305	1.195			NW - CP
	0.910		1.040			NE
	1.255			0.345		SE
		2.365		1.110		SW

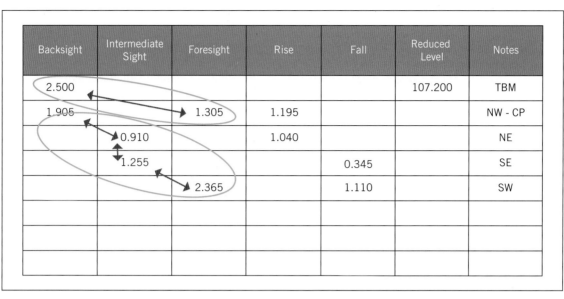

FIGURE 10.40 Entering the rise and fall developed from each sighting. Green loops cluster the sightings from a single instrument station.

is 'positive' it is recorded as a rise. If 'negative', it is recorded in the fall column.

For example:

Peg NW

2.500 – 1.305 = 1.195 (rise)

Peg NE

1.905 – 0.910 = 1.040 (rise)

Peg SE

0.910 – 1.255 = – 0.345 (fall)

Peg SW

1.255 – 2.365 = – 1.110 (fall)

Determining and booking the reduced level values

The reduced levels (RLs) for each location are found by adding the 'fall', or subtracting the 'rise', to the RL of the peg preceding it. In this example:

Peg NW

107.200 + 1.195 (rise) = 108.395

Peg NE

108.395 + 1.040 (rise) = 109.435

Peg SE

109.435 – 0.345 (fall) = 109.090

Peg SW

109.090 – 1.110 (fall) = 107.980

These values are entered into the Reduced Level column as shown in Figure 10.41.

Backsight	Intermediate Sight	Foresight	Rise	Fall	Reduced Level	Notes
2.500					107.200	TBM
1.905		1.305	1.195		108.395	NW - CP
	0.910		1.040		109.435	NE
	1.255			0.345	109.090	SE
		2.365		1.110	107.980	SW

FIGURE 10.41 Entering the calculated reduced levels (RLs)

Checking the booking

As stated previously, the rise and fall method of booking and calculating RLs deliberately builds in the potential for what may be called 'cumulative error'. This is because each RL is calculated from the one prior to it. Such being the case, error in any one RL calculation will flow on, and be reflected in each of the following RLs and the final RL. This allows for some simple check procedures to be undertaken to identify if an error has occurred. The steps following are graphically shown in Figure 10.42.

Step one:

- Add all the backsights together.
- Add all the foresights together.
- Subtract the foresight total from the backsight total.

Step two:

- Add all the rises together.
- Add all the falls together.
- Subtract the falls total from the rises total.

Step three:

- Subtract the initial RL from the last RL.
- Check that totals found from each of these steps are equal.

It is important that these figures are found to be equal. If they are not, then there is a booking or calculation error which must be found and corrected. Errors can occur in one of two ways:

- errors in booking
- errors in calculation.

Booking errors usually occur when you have:

- placed a sighting in the wrong row, such as not keeping the backsight and foresight readings in the same row at a change point
- placed a sighting in the wrong column; e.g. a foresight in the intermediate column
- placed a calculated rise in the fall column, or fall in the rise column.

Calculation errors can result from simple input mistakes to writing down the wrong number.

To find these errors you will need to go back over booking sheets carefully row by row and then each of the rise and fall and RL calculations until the mistake is identified, and then recheck.

Another error that can occur may develop through instrument error. This is most likely to show up when taking sightings of backsights and foresights over varying – close then long – distances. The importance of checking instruments for error is therefore a critical aspect of any surveying or levelling project. The how and why of these checks are explained in the following section.

Testing equipment for error

No levelling instrument should be taken for granted, be it optical or laser. Instrument error is always a possibility as the equipment is frequently knocked and shuffled around in utes, trailers and on site while in use. Even a small knock can put an instrument out of calibration, leading to significant errors that can be time consuming to fix. To counter this possibility, you should test all levelling equipment prior to use.

Backsight	Intermediate Sight	Foresight	Rise	Fall	Reduced Level	Notes
2.500					***107.200***	TBM
1.905		1.305	1.195	+	108.395	NW - CP
+	0.910	+	1.040		109.435	NE
	1.255		+	0.345	109.090	SE
		2.365		1.110	***107.980***	SW
4.450	**—**	**3.670**	**2.235**	**1.455**	**—**	
	= 0.780			**= 0.780**	**= 0.780**	
	Step one			Step two	Step two	

FIGURE 10.42 Booking checks

Simple spirit levels

Small spirit levels are often incorporated into even the most advanced digital and laser levelling instruments. This may be in the form of a simple blister or bubble level as found on dumpy or automatic levels, or it may be a short, straight spirit level used for ensuring absolute accuracy in the set-up of theodolites and total stations.

Testing these levels is relatively easy and quick using the procedure outlined below:

1 Mount the instrument on the tripod or set it up on a bench.
2 Level the instrument such that the spirit level, whichever type, is showing the bubble in the exact centre as shown below

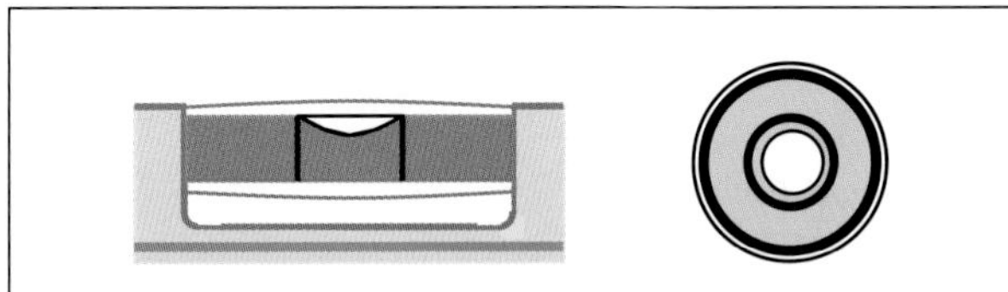

3 Now rotate the instrument 180°. If the bubble continues to show in the middle as before, then the spirit level is accurate. If, however, it sits a little to the left or the right as depicted below, then the spirit level is in error and needs to be adjusted.

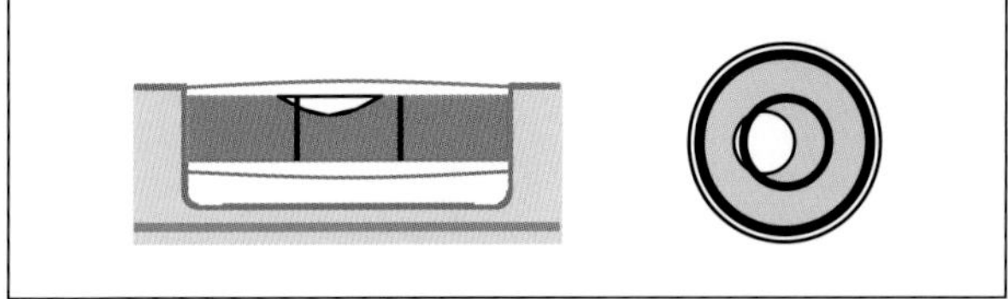

Adjustment of individual levels is not complex, but is still best performed by a professional who can then test the instrument for other irregularities.

Testing the optics

Running a field test on the optics of dumpy or automatic levels, theodolites and total stations is also not complex, but does take more time. For dumpy and automatic levels, the technique is known as the two-peg test.

HOW TO

THE TWO-PEG TEST

Step 1 (Figure 10.43)

- Establish the instrument on reasonably flat ground giving clear sighting for 10 to 15 m in each direction if possible.
- Locate the first peg so that a staff can be easily read, but far enough away to make any error in the instrument noticeable.
- Locate the second peg exactly the same distance from the centre of the instrument on the opposite side.

Where possible make the two pegs and the instrument in line with each other.

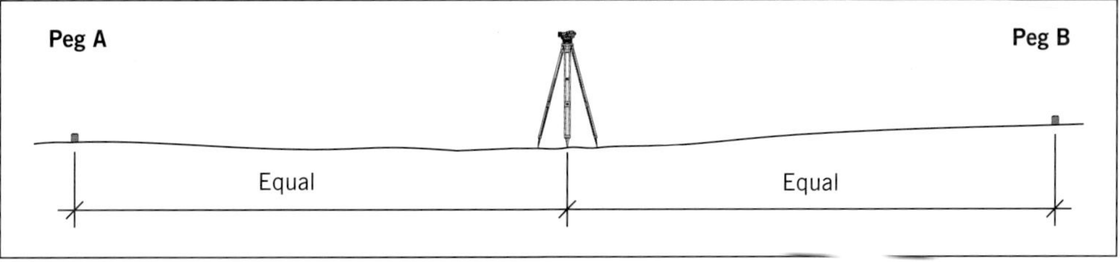

FIGURE 10.43 Setting up for a two-peg test

Step 2 (Figure 10.44)

- Take a reading at Peg A and book as shown below in the 1st station column.
- Take a reading at Peg B and book in the row below A.

For this example, these readings shall be 1.200 m and 1.000 m, respectively.

- Calculate and book the difference between these two readings; i.e. 0.200 m.

	1st station	2nd station
Peg A	1.200	
Peg B	1.000	
Difference	0.200	

FIGURE 10.44 Establishing the difference in height between the two pegs

>>

>>

What you have done:

In carrying out these two steps correctly, these heights above the peg are level with each other; i.e. a level line of known height over each peg has been identified. As Figure 10.45 shows, if this level line passes through the optical centre of the level, the line of collimation, then the instrument is without error; however, we cannot identify this just yet.

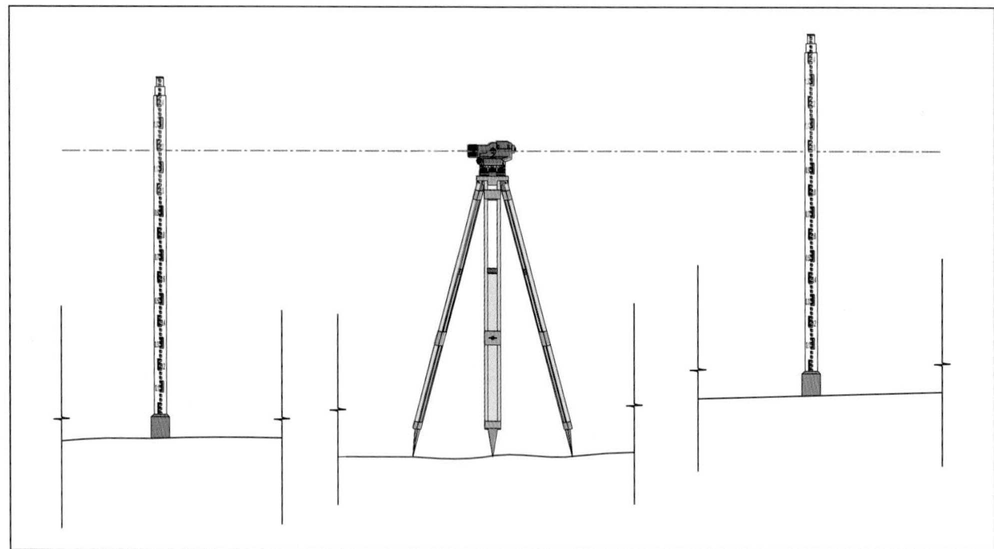

FIGURE 10.45 A correctly calibrated instrument establishing a level line height

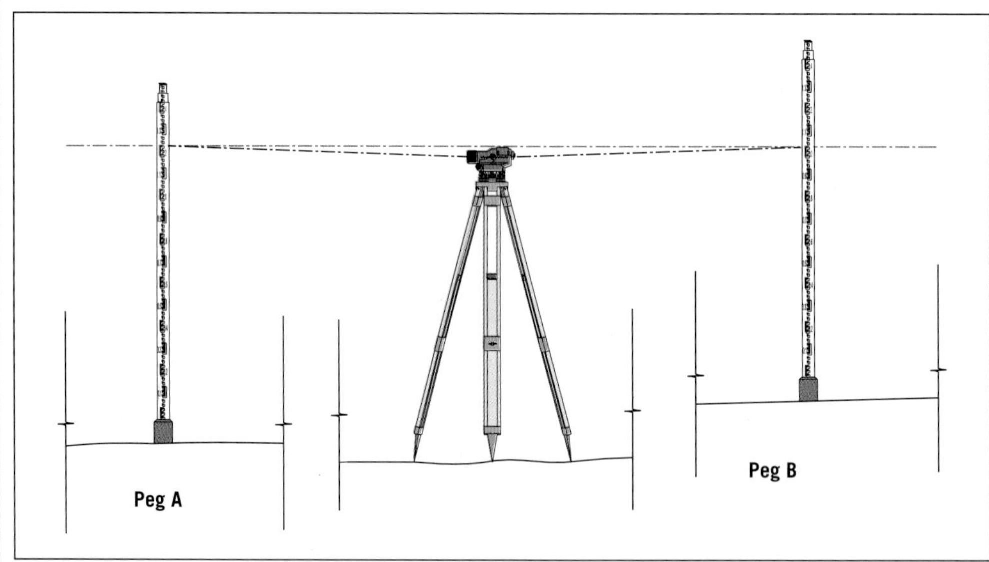

FIGURE 10.46 Level line height established with upward pointing instrument

If the instrument is out of calibration, then it may be reading slightly 'uphill' (Figure 10.46), or 'downhill' (Figure 10.47). In such cases a level line has still been established, except that this line is actually above the centre of the instrument, or below it, as shown.

>>

>>

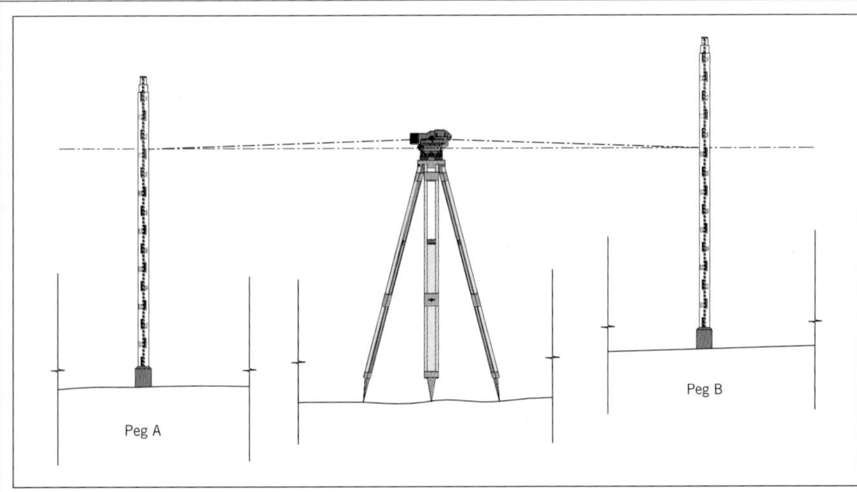

FIGURE 10.47 Level line height established with an instrument pointing down

Step 3	Having identified and booked the heights over the pegs, the instrument is moved to a 2nd station approximately 5 to 10 m beyond, but in line with, pegs A and B (see Figure 10.48).

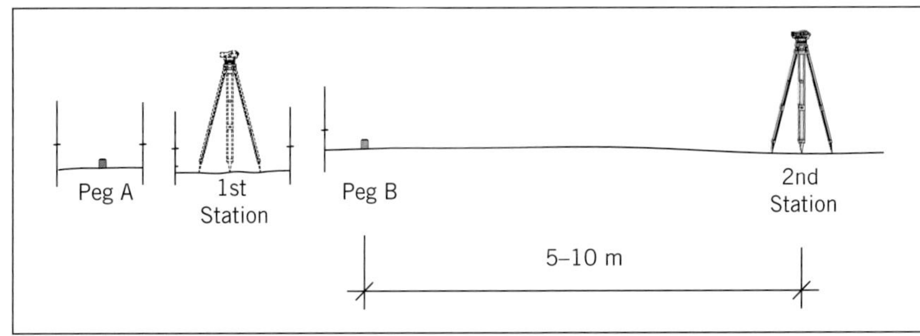

FIGURE 10.48 Locating the 2nd station

Step 4	From this 2nd station, take readings back over the two pegs and book as shown in Figures 10.49 and 10.50. Again, calculate the difference between the two sightings. This result is then compared to the difference between the previous two sightings. If there is a difference of greater than three (3) millimetres over the distances described in this example, then the instrument is suspect and in need of professional servicing.

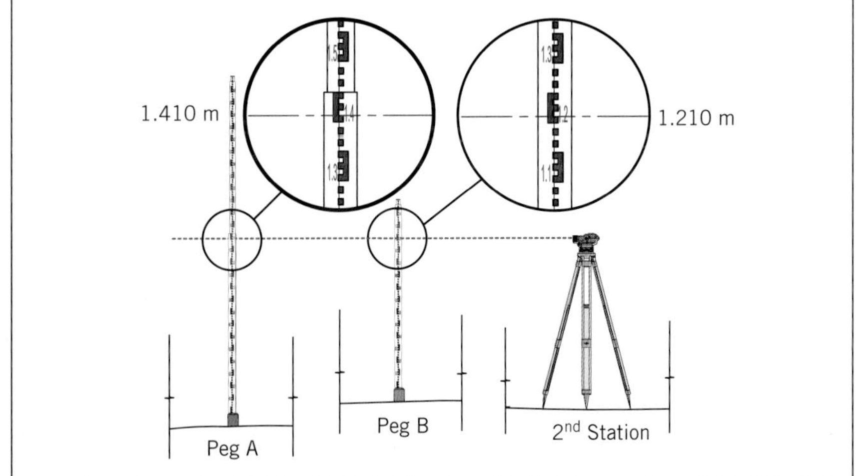

FIGURE 10.49 Taking readings from the 2nd station

>>

>>

	1st Station	2nd Station
Peg A	1.200	1.410
Peg B	1.000	1.210
Difference	0.200	0.200

Nil Error – Instrument correctly calibrated

FIGURE 10.50 The difference between 1st and 2nd station readings

An explanation

It's important to understand not just how to undertake this test, but how it works, and in so doing knowing what you are testing for to be sure that you have made your conclusions correctly. Study Figures 10.51 to 10.53. In Figure 10.51 the instrument is sighting uphill; i.e. it is out of calibration. The result is that the readings get progressively further above the horizontal plane the further away they are taken from the instrument. This led to the higher difference in height at peg A than at peg B.

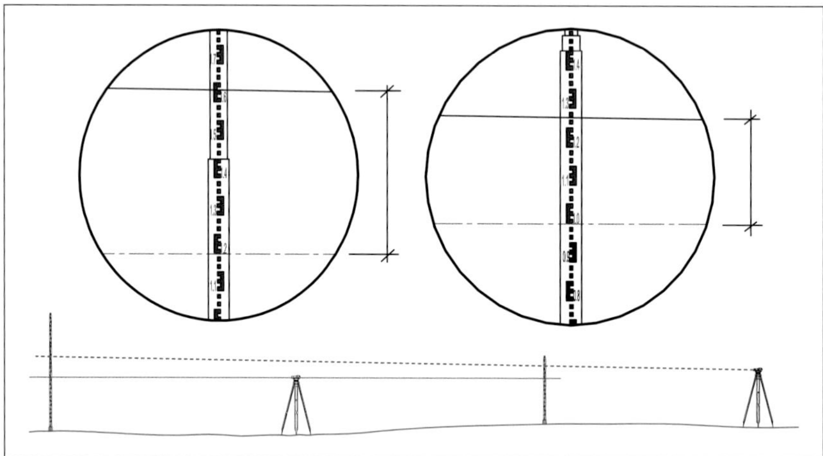

FIGURE 10.51 Level sighting up: difference in heights unequal

Figure 10.52 shows the opposite circumstance, the instrument sights downhill and so the reading shows a lesser difference in height at peg A than peg B.

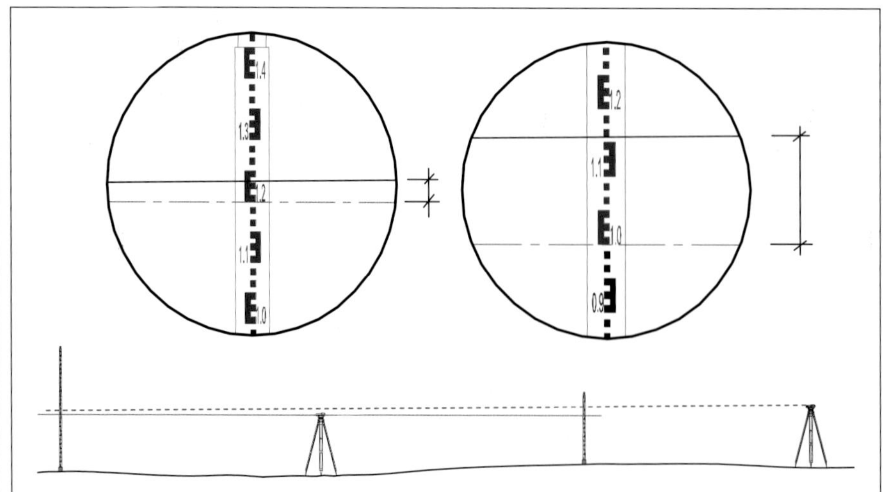

FIGURE 10.52 Level sighting down: difference in heights unequal

>>

>>

Figure 10.53 then shows an instrument correctly calibrated. In this instance the difference in the height at each peg is equal. That is, both lines are level and running parallel with each other, and the instrument is confirmed to be correctly calibrated.

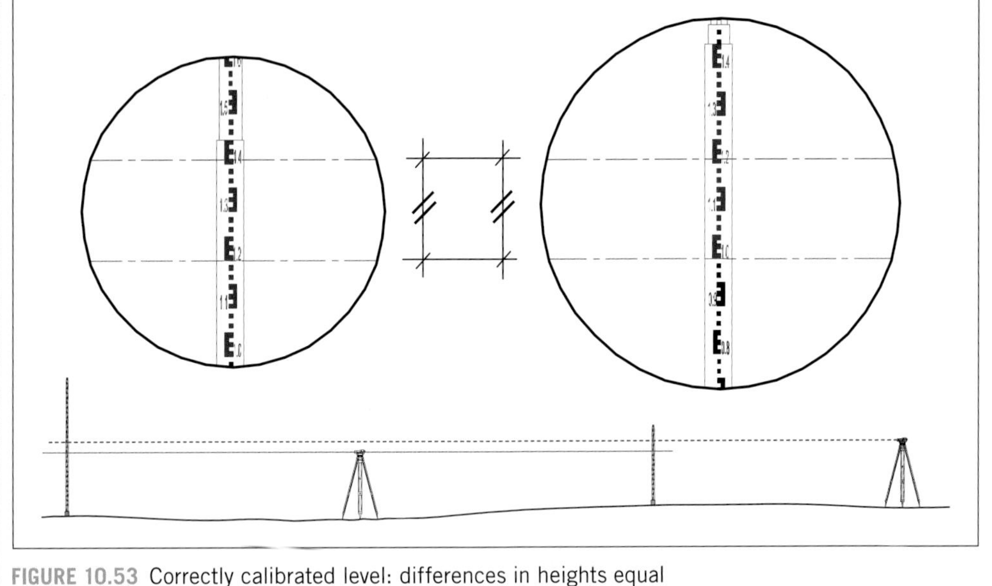

FIGURE 10.53 Correctly calibrated level: differences in heights equal

Testing theodolites

Despite the greater complexity of the instrument, testing the optics and angular calibration on a theodolite is actually a fairly simple process. We will begin by looking at what is known as collimation in altitude – collimation being the line sight through the cross-hairs (reticule) and the centre line of the optical train (all the lenses in line).

HOW TO

COLLIMATION IN ALTITUDE (VERTICAL PLANE)

Step 1	• Establish the instrument on reasonably flat ground giving clear sighting to a vertical surface or post approximately 20 m away. • Place the theodolite in face left; i.e. the vertical disc or large flat side of the instrument is on the left (see Figure 10.54). • Set the telescope up so that it is reading 90 degrees in the vertical scale; i.e. the telescope is reading level according to the scale or readout. Lock this axis. • Ensure you have removed parallax error in the eyepiece of the telescope (see parallax error, p. 237) • Sight to the wall or post and have an assistant mark where the horizontal line of the cross-hair appears on the wall. Check that the mark is exact.
Step 2	• Rotate the theodolite so that it is now face right; i.e. the vertical disc or large flat side is on the right. • Flip the telescope over 180 degrees and set it up so that it is reading 270 degrees in the vertical scale; i.e. the telescope is again reading level according to the scale or readout. Lock this axis. • Sight to the wall or post and again mark where the horizontal line or the cross-hair appears to be. • Check the vertical distance between these marks. Over a distance of 20 m, there should be less than 6 mm of error between these two marks.

>>

>>

COLLIMATION IN AZIMUTH (HORIZONTAL PLANE)

Step 1	• Establish the instrument on reasonably flat ground exactly midway between two timber profiles (two pegs in the ground approximately 1 m apart and a horizontal member screwed to them) or fences or walls approximately 30–40 m apart. That is, you will be sighting over a distance of 15–20 m, but most importantly, exactly the same distance in either direction. • Fix a nail, or mark, clearly on to just one of these target surfaces. • Place the theodolite in face left (see Figure 10.54). • Sight the nail or mark accurately by locking the horizontal scale and using the fine adjustment. • Set the horizontal scale to zero. • Rotate the instrument through 180 degrees so that it is now pointing at the other target and lock the axis. • Ensure you have removed parallax error in the eyepiece of the telescope (see parallax error, p. 237). • Sight to the wall or profile and have an assistant mark where the vertical line of the cross-hair appears on the target. Check that the mark is exact.
Step 2	• Rotate the theodolite and flip the telescope so that it is now face right (i.e. the vertical disc or large flat side is on the right) and pointing at the original nail or mark. • Lock the horizontal axis and accurately align the vertical cross-hair with the mark using the fine adjustment screw. • Again, set the horizontal scale to zero. • Rotate the theodolite 180 degrees so again you are looking at the second target and lock the horizontal axis. • Have an assistant mark where the vertical line of the cross-hair appears on the target. • Check the horizontal distance between these marks. Over a distance of 20 m, there should be less than 6 mm of error between these two marks. 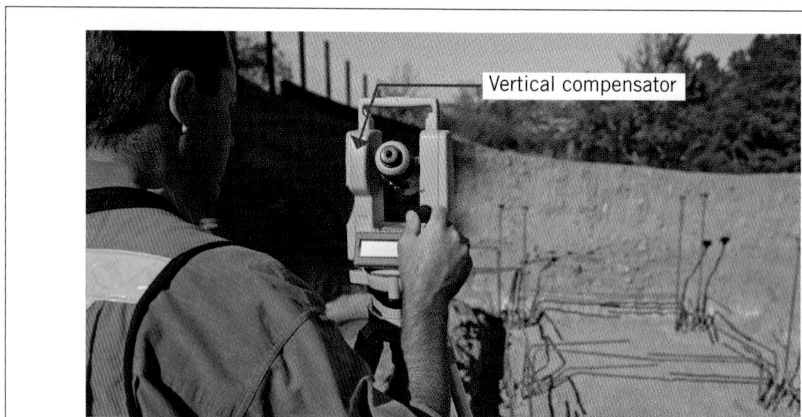 Source: Position Partners Pty Ltd FIGURE 10.54 Theodolite in face left. Note screen is facing the operator and broader, rounded, element of the frame – known as the vertical circle or vertical compensator – is on the operator's left-hand side. If the instrument is shown to be in error by more than the amounts described above (6 mm over 20 m) then it will need to be recalibrated and serviced by a professional.

Testing rotating laser levels

Field testing rotating laser levels can also be done; however, as they put out a circular plane or disc of light, this disc may be out of level in one or both planes. How this is tested is outlined in the section below.

Rotating laser levels: the four-peg test

Effectively this is the two-peg test performed twice: Once 'north–south' and again 'east–west'. A 'plan' view of the layout of pegs and stations is shown in Figure 10.55.

The approach to booking and the calculations required for this test are shown in Figure 10.56. When performing the test ensure the following factors are adhered to:

- The instrument is centred in the circle of pegs (+/- 100 mm).
- Pegs must be directly opposite each other and their axes are opposed at approximately 90 degrees.
- The instrument head must be aligned in the opposite direction when set up outside the circle to how it sat when in the centre.

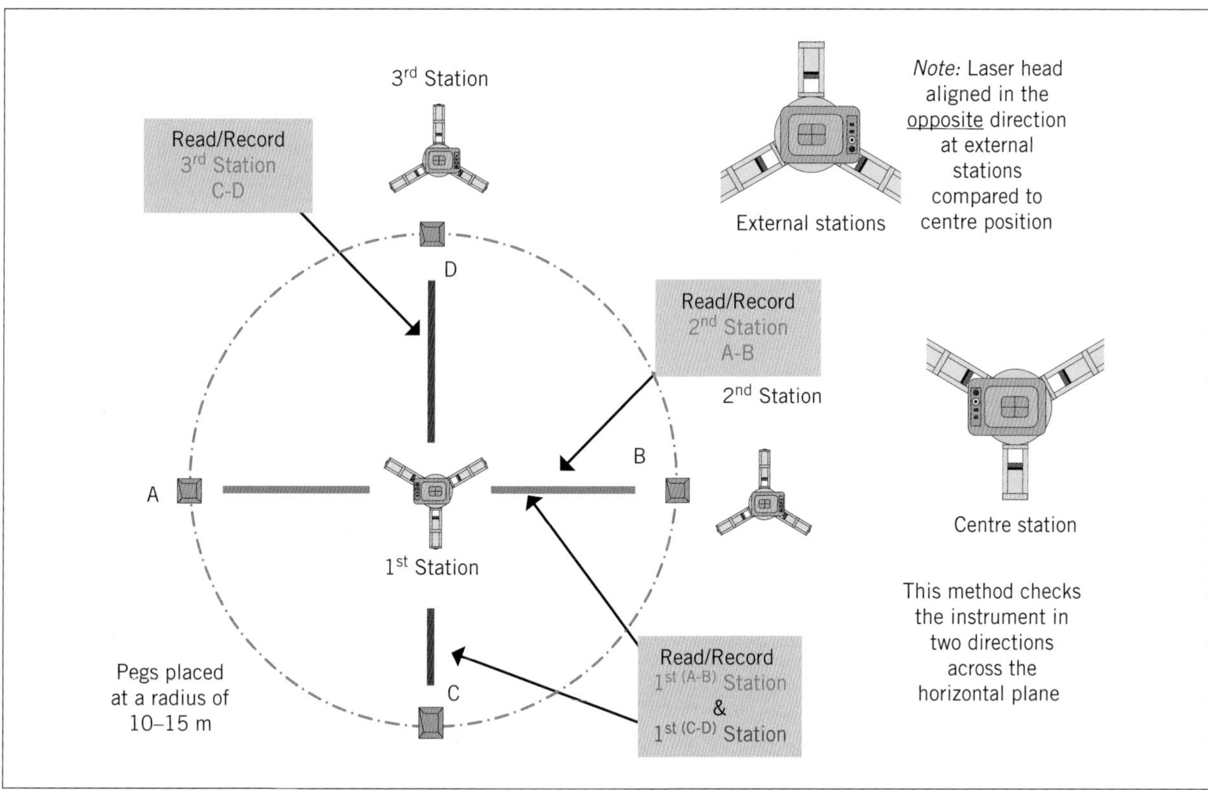

FIGURE 10.55 Four-peg test arrangement

Note that you are looking for a discrepancy across each axis only, not between axes. In the example table in Figure 10.56 the result is nil error and so the tool is correctly calibrated. If the error is greater than 3 mm (or the stated manufacturer's tolerances – see the specifications section of the operator's guide) across the distances described in this example, then the equipment needs to be professionally serviced.

Closed loop levelling

Frequently referred to as a closed traverse or closed level traverse, closed loop levelling is defined by the first backsight and last foresight both being taken to the same point – generally, the site datum or temporary benchmark. A closed traverse is the only true check of the accuracy of a survey. This is because the last reduced level, as calculated at the end of the traverse, should be identical to that given for the datum or benchmark from which you began – they being RLs of the same point.

There are two methods of booking a closed traverse: the rise and fall method discussed previously, and the height of collimation approach. Both these methods are outlined below.

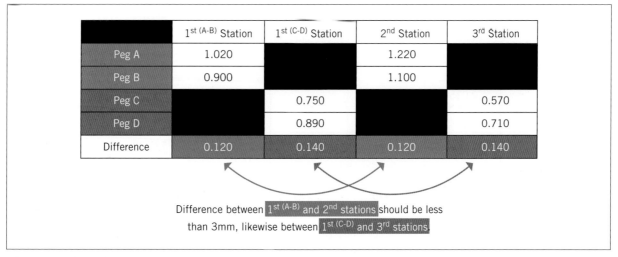

	1st (A-B) Station	1st (C-D) Station	2nd Station	3rd Station
Peg A	1.020		1.220	
Peg B	0.900		1.100	
Peg C		0.750		0.570
Peg D		0.890		0.710
Difference	0.120	0.140	0.120	0.140

Difference between 1st (A-B) and 2nd stations should be less than 3mm, likewise between 1st (C-D) and 3rd stations.

FIGURE 10.56 Laser four-peg test – booking the results and carrying out checks

The rise and fall approach

As mentioned earlier, this is always the preferred method of booking surveys due to it forcing the exposure of errors that may then be checked for easily. The advantage of a closed loop survey is that the last entry is to the same peg from which you began, and so an additional and obvious check becomes available. That is, that the RL determined by the traverse for that peg is the same, or within the allowable error, as that given at the outset. This is demonstrated in Figure 10.57.

Allowable error

Often referred to as 'allowable misclose', this error is distance dependent. In the example offered in Figure 10.57, a total of 10 sightings were undertaken over nine pegs and a distance approximating 500 m. It is this distance and the class or type of levelling undertaken, not the number of sightings or pegs, that determines the allowable error for this survey. The allowance is derived from the industry agreed formula shown below:

$r = n\sqrt{k}$

where:

k is the distance traversed in kilometres

n is the value ascribed to the class of levelling being undertaken

r is the allowable error in millimetres.

In cadastral surveys – surveys associated with properties and boundaries – the value of 'n' may be 12 or 18 depending upon the level of association with the Australian survey control network. The lower the number, the higher the accuracy required. The classes of survey work most associated with low-rise domestic and commercial works are:

- LC – Cadastral control surveys or 3rd Order Levelling where 'n' = 12
- LD – Cadastral and other surveys or Levelling where 'n' = 18
- LE – Approximate and lower order surveys or Levelling where 'n' = 36.

As a general rule, you should be working towards at least the Class LD; i.e. where 'n' equals 18.

In the example offered below, using Class LD, the allowable misclose is found by:

$r = 18\sqrt{0.5}$

r = 13 mm

That is, upon return to the starting point you may be in error by plus or minus 13 mm and still be within the tolerances accepted by both national and state or territory standards. Most practitioners, however, will frequently work to much tighter standards, arguing for plus or minus not more the 5 mm over such a distance: this being one-half of one graduation on a standard levelling staff (see Figure 10.12).

Backsight	Intermediate Sight	Foresight	Rise	Fall	Reduced Level	Notes
2.550					**57.300**	**TBM**
	1.995		0.555		57.855	A
	1.785		0.210		58.065	B
2.795		1.630	0.155		58.220	C - CP
	1.835		0.960		59.180	D
1.825		1.590	0.245		59.425	E - CP
1.240		1.705	0.120		59.545	F - CP
	2.035			0.795	58.750	G
1.260		2.510		0.475	58.275	H - CP
		2.235		0.975	**57.300**	**TBM**
9.670 ←		→ ***9.670***	***2.245*** ←	→ ***2.245***		
	0.000			**0.000**	**0.000**	

FIGURE 10.57 Closed traverse booking sheet showing checks. The first RL, when subtracted from the last RL, should equal zero or be within the allowable tolerances or 'misclose'.

Height of collimation approach

The basic principle behind the height of collimation approach to levelling has been outlined earlier in the chapter. Unlike the rise and fall method, this system relies solely upon the line of collimation for each station as the means of determining the reduced level. There is, therefore, no direct correlation between each sighting such as the rise or fall relationship in the previous method.

Backsight	Intermediate Sight	Foresight	Collimation Height	Reduced Level	Remarks
2.550			59.850	**57.300**	**TBM**
	1.995			57.855	A
	1.785			58.065	B
2.795		1.630	61.015	58.220	C - CP
	1.835			59.180	D
1.825		1.590	61.250	59.425	E - CP
1.240		1.705	60.785	59.545	F - CP
	2.035			58.750	G
1.260		2.510	59.535	58.275	H - CP
		2.235		**57.300**	**TBM**
9.670 ◄	──────	► **9.670**			
	Diff = 0.000			**Diff = 0.000**	

FIGURE 10.58 Height of collimation booking example

Determining the collimation height

Using the same figures from the rise and fall booking example previously outlined, the booking sheet in Figure 10.58 shows the different columns used in the height of collimation approach. In this example, you can see that there is a different height of collimation for each station at which the level is established – identified by the change point (CP) note in the Remarks column.

To determine the height of collimation the reduced level of the first backsight must be known. In this case the marker is a temporary benchmark or TBM with a surveyed height of 57.300 m. The backsight to this mark read as 2.550 m: i.e. the height of collimation – the line of sight of your instrument – is passing 2.550 m above the marker. The height of collimation is therefore:

Collimation height = 57.300 + 2.550
= 59.850 m

This figure is then inserted into the collimation height column. All future RLs are developed from this height until such time as a change of station occurs. The RL calculations are demonstrated below:

RL = Collimation height – the sighting taken (intermediate or foresight)

e.g. RL Peg A = 59.850 – 1.995
= 57.855

RL Peg B = 59.850 – 1.785
= 58.065

RL Peg C = 59.850 – 1.630
= 58.220

Peg C represents a change point (CP) and so a new collimation height must be determined. This is done in the same manner as previously: only this time using the newly established RL for peg C:

Collimation height = 58.220 + 2.795

= 61.015 m

This figure is now entered into the collimation height column in the row aligning with peg C and the calculations for the RLs to follow are completed as previously shown. This pattern is then repeated for all future change points.

Checking height of collimation bookings

Only very basic checks may be done upon a height of collimation booking sheet. When the traverse is open – i.e. it does not return to the original peg – there is only the one check that may be made, this is:

Sum of all backsight – Sum of all foresights = First RL – Last RL

In the example given in Figure 10.59, the traverse finishes at peg H.

The backsights are tallied to equal 8.41 and the foresights to 7.435, giving a difference between the two of 0.975. The first RL (the TBM) has an RL of 57.300, the difference between this and the last RL of 58.275 (Peg H) is also 0.975. Unfortunately, this does not guarantee that all the intermediate RLs are correct as these are individually developed from the height of collimation.

E.g. the RL for Peg G is developed from:

Peg G RL = Collimation height 60.785 – Intermediate sight 2.035

= 58.750

However, while inputting the value 2.035 you could mistakenly type 2.350, or 60.785 could become 60.875. The resultant RL would be greatly in error in either case and go undetected as the final RL for that station does not require this figure for its calculation: the final RL (Peg H) requiring only the collimation height and the last foresight.

I.e.

Peg H RL = Collimation height 60.785 – Foresight 2.510

= 58.275

It is for this reason that errors in any intermediate RLs are not detectable using the height of collimation approach. And this is why it is not considered an appropriate booking method for anything more than the most basic levelling activities.

Backsight	Intermediate Sight	Foresight	Collimation Height	Reduced Level	Remarks
2.550			59.850	**57.300**	**TBM**
	1.995			57.855	A
	1.785			58.065	B
2.795		1.630	61.015	58.220	C - CP
	1.835			59.180	D
1.825		1.590	61.250	59.425	E - CP
1.240		1.705	60.785	59.545	F - CP
	2.035			58.750	G
		2.510		**58.275**	H
8.410		**7.435**			
	Diff = 0.975			**Diff = 0.975**	

0 Difference = OK

FIGURE 10.59 Checking height of collimation booking of open traverse

LEARNING TASK 10.1 RECORDING INFORMATION FOR USE AT SITE

Using the information in the readings taken list and the sketch below, fill out the blank level book and carry out the required checks to ensure you have properly documented the rises and falls from this work scenario.

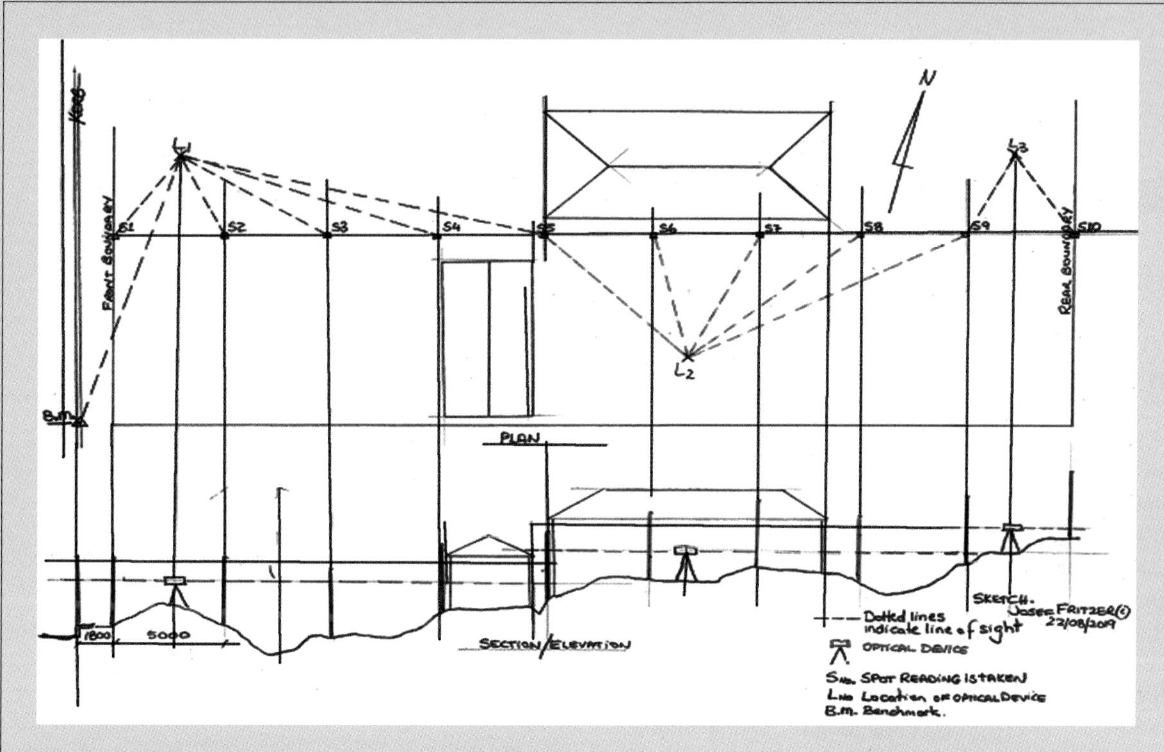

The readings taken were:

- BM is nominated to be at an RL of 14.000
- L1: BM = 1980
- L1: S1 = 1730
- L1: S2 = 1545
- L1: S3 = 2165
- L1: S4 = 1390
- L1: S5 = 970
- L2: S5 = 1860
- L2: S6 = 1255
- L2: S7 – 725
- L2: S8 = 1570
- L2: S9 = 440
- L3: S9 = 1360
- L3: S10 = 320

Record your field readings information in the booking sheet below, making sure that all the checks are properly shown.

B/S	I/S	F/S	Rise	Fall	RL converted to metres (m)	Notes

>>

B/S	I/S	F/S	Rise	Fall	RL converted to metres (m)	Notes

CHECKS =	CHECKS =	CHECKS =

Setting out: techniques and calculations

The setting out of a building relies on a sound understanding not only of the basics of levelling, but also the capacity to perform a range of calculations confidently. This section begins with an outline of the most common calculations common to set out practices before following on with the techniques and procedures of setting out regular and irregular buildings.

Common levelling calculations

The calculations required for setting out and levelling are not onerous, and most will be familiar with them to some degree. It is only their application in the construction context that you may find a bit different. Aside from the use of standard formulas for the determination of areas and volumes, these calculations are based either on trigonometry or Pythagorean principles in the main. We will therefore begin with some simple examples of the applied use of Pythagoras' theorem before following with the basics of trigonometry.

Pythagoras' theorem

This formula applies only to right-angled triangles and yet is used extensively in construction, allowing us to determine the lengths of a multitude of angled components such as rafters, skewed end wall plates and the like. In set out and levelling it is used primarily for determining right angles (90° corners) off a given base line.

The basic formula is generally expressed as:

$$A^2 = B^2 + C^2 \text{ or } A = \sqrt{B^2 + C^2}$$

The easiest proof of the formula is with the 3:4:5 right-angled triangle as described in Figure 10.60.

How this is used to 'prove' the theorem is shown in Figure 10.61 that follows. Here it can be seen that to 'square' a number (multiplying the number by itself) effectively finds the area of a square with sides of that number. That is, $4^2 = 16$, the area of a 4 by 4 square.

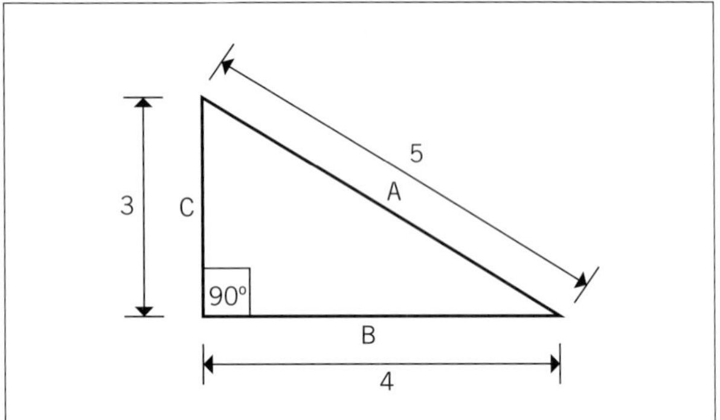

FIGURE 10.60 The 3:4:5 triangle always gives a 90° base angle

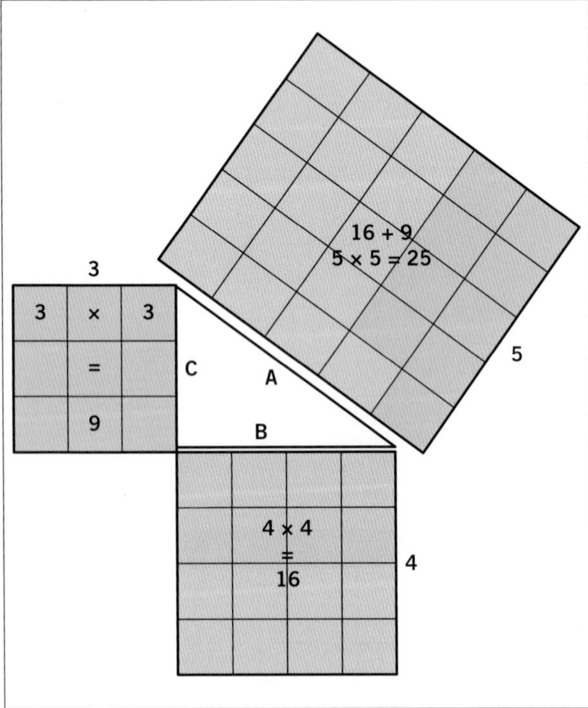

FIGURE 10.61 Proving Pythagoras' theorem

What Pythagoras' theorem shows is that if you take the areas formed by turning the two shorter sides into squares and add them together, you have the area of the longer side squared. Conversely, if you can take the area of one of the shorter sides away from the area of the long side, this will find you the area of the second of the smaller sides.

To take the square root of a number is effectively to ask, 'I have the area of a square, what is the length of the sides?'

As Figure 10.61 shows, the area of side 'A' is 25, the square root of 25 is 5: i.e. 5 × 5 = 25. So, the length of 'A' is 5. Which, of course we already knew, and so the principle of the theorem is proven. Mathematically it looks like this:

$$A = \sqrt{4^2 + 3^2}$$
$$A = \sqrt{16 + 9}$$
$$A = \sqrt{25}$$
$$A = 5$$

Application

In setting out, we can use the 3:4:5 triangle to generate a set out line at 90° to any base line. For most buildings, however, the triangle produced by using 3 m, 4 m and 5 m is too small for the accuracy we desire. To produce a larger triangle, we need only factor up these sides equally to whatever size we require; i.e. multiply each side by the same amount. Figure 10.62 offers some examples.

Note: You do not have to use the base unit of '1' when seeking an appropriately sized 3:4:5 triangle. That is, you could use multiples of 200 mm, 300 mm, 500 mm or indeed any length that works for the context at hand. Using 300 mm, for example, means your triangle has sides of 900 mm (3), 1200 mm (4), and 1500 mm (5).

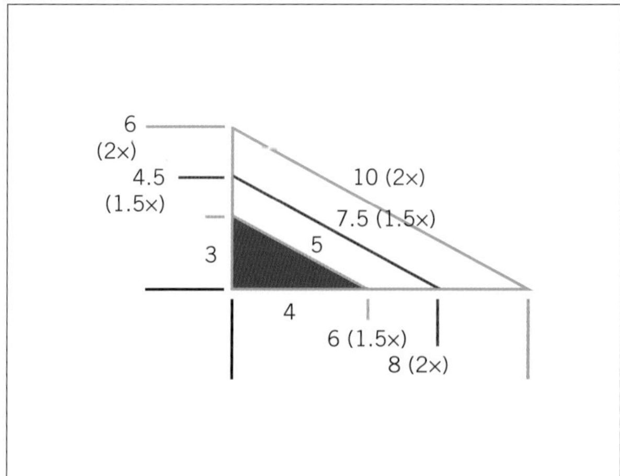

FIGURE 10.62 Increasing the size of a 3:4:5 triangle using ratios or multipliers

We can then use the desired size triangle to set out our 90° corner as shown in Figure 10.63.

The set out shown in Figure 10.63 can be done using a single tape measure and two or more people, or two tapes and a single person. See the 'How' to box.

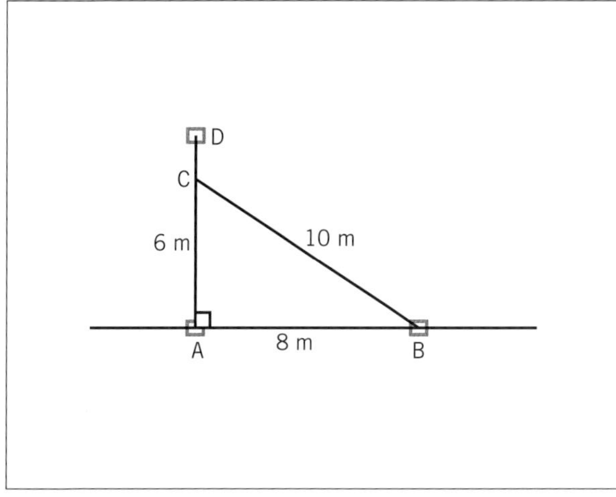

FIGURE 10.63 Using the 3:4:5 principle to develop a 90° corner

HOW TO

DEVELOP A 90° CORNER

The procedure:

1. Establish peg A with a nail driven in directly under the base line and exactly positioning the corner of the building.
2. Locate peg B with a nail driven in at 8 m along, and directly under, the base line.
3. If two tapes are available, hook one over the nail at peg A and the other likewise at peg B.
4. Pull the tapes out so that they cross each other at C at 6 m and 10 m as shown above.
5. Establish a peg at D just beyond the tape crossing point C.
6. Install a string line at Peg A and pull through over D.
7. Check the accuracy of the line to the crossed tapes and adjust the string line at D accordingly.

Common trigonometric calculations

The basic trigonometry formulas also apply only to 90° triangles. However, they work with the simple ratios found if you divide one side into another and the angle for that specific ratio of sides. The common descriptions of the right-angled triangle for trigonometry are shown in Figure 10.64.

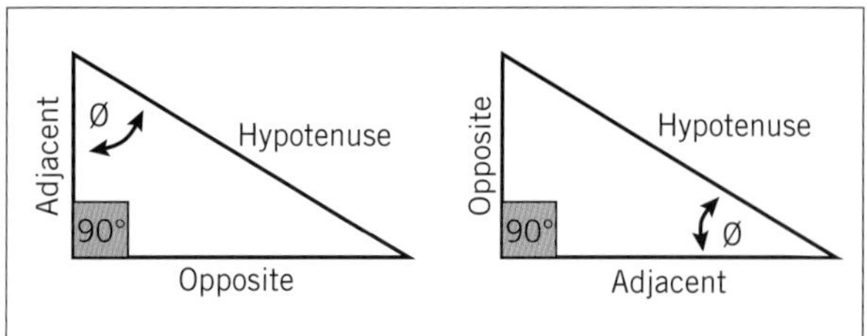

FIGURE 10.64 Naming the sides of a right-angled triangle based upon the angle under consideration

The ratios were first put in tables, to be looked up manually; now we use a calculator to look up the table for us. The three possible ratios, developed and named many centuries ago, are as follows:

Sine Ø = opposite/hypotenuse

Cosine Ø = adjacent/hypotenuse

Tangent Ø = opposite/adjacent

where Ø is the angle being considered or is known.

What these ratios tell us is that for any given angle, the number found by dividing the two specified sides together will be the same for any triangle no matter how large or small it is – as long as the angle remains constant. For example, if we have a right-angled triangle such as that shown in Figure 10.65, where the known angle is 30° the ratios are as given below:

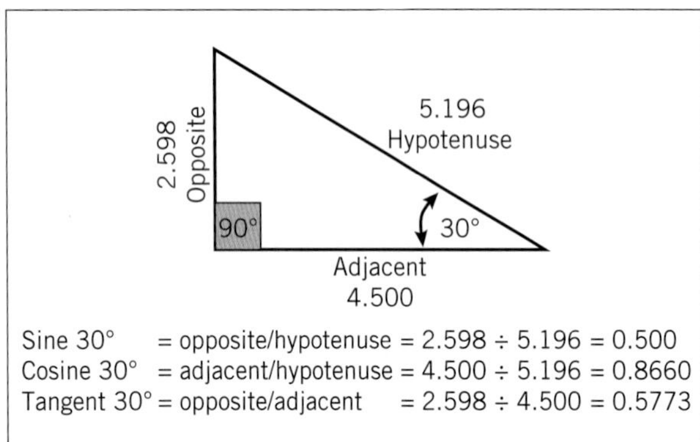

2.598
Opposite
5.196
Hypotenuse
90°
30°
Adjacent
4.500
Sine 30° = opposite/hypotenuse = 2.598 ÷ 5.196 = 0.500
Cosine 30° = adjacent/hypotenuse = 4.500 ÷ 5.196 = 0.8660
Tangent 30° = opposite/adjacent = 2.598 ÷ 4.500 = 0.5773

FIGURE 10.65 Trigonometric ratios for 30° right-angled triangle

And as Figure 10.66 shows, if we were to triple the size of the triangle, but remain with 30°, those ratios would remain the same. Indeed, they will do so no matter how large or small we make it.

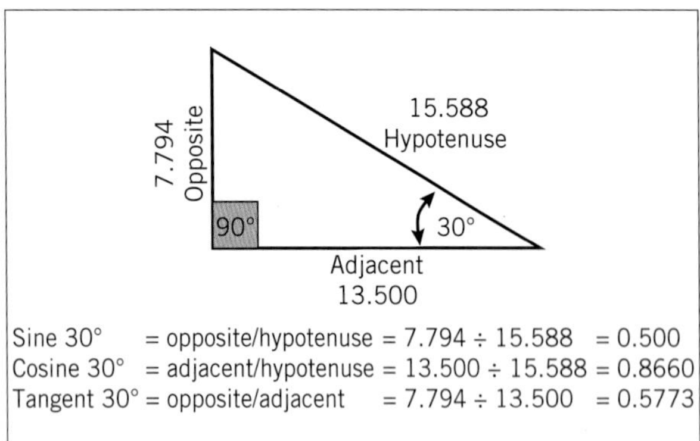

FIGURE 10.66 Trigonometric ratios for a larger 30° right-angled triangle

It is these ratios that are tabulated and called up when you press the Sin (Sine), Cos (Cosine), or Tan (Tangent) buttons on your calculator. For example, type in 30 in your calculator and press the Sin button and the answer will be 0.5, enter 30 and press Cos and you will get 0.866, etc.

All the other angles

The Sin, Cos and Tan buttons on your calculator allow you to access the ratios for any angle in degrees and/or fraction of degrees. Type in 20, for example, and hit Sin and you will get 0.342; this is the ratio you would find if you divided the opposite and hypotenuse sides of a 20° triangle. Type in 56.25 (56° 15')*, press Tan and you will get 1.497; again, this is the ratio you will find if you divide the opposite side to the angle by the adjacent side of a right-angled triangle with that as one of its two other angles.

Application

The value of trigonometry to construction generally, and surveying or setting out specifically, cannot be understated. Given any angle, and one side of a right-angled triangle, we can find any of the other two sides. This is because we have only three variables to consider in our equation: one known side, one unknown side and the ratio of those two sides for the known angle. Alternatively, we can find the angle if we know the lengths of two sides.

For example: given a sloping site, we need to find the horizontal or level distance between two points. As Figure 10.67 shows, we can determine the angle of the slope by using a theodolite; the length down the slope we can measure. The horizontal distance we can now calculate.

To solve this problem we need to look at what values we have, and how they relate to the trigonometry ratio tables available to us through our calculator.

From Figure 10.68 we can determine that our information, and 'X' our unknown side, relate to the trigonometric ratio known as cosine; i.e. the adjacent side divided by the hypotenuse. So, mathematically our solution looks like this:

Cos Ø = adjacent/hypotenuse

Cos 12° = X ÷ 21.540 m

0.978* = X ÷ 21.540 m

X = 21.540 m × 0.978

X = 21.066 m

*Note 1: 0.978

Calculators vary on input flow. Some require typing 12 into the calculator and pressing the Cos button; others require that you press the Cos button first, and then enter the angle (12 in this case).

Be sure also that your calculator is functioning in degrees, not radian or gradient mode.

Note 2: Always perform a quick logic check that you are multiplying or dividing correctly. In this case, the side we are finding must be shorter than the hypotenuse as by definition the hypotenuse is the longest side of the triangle. But it will not be much shorter as the angle is not very steep.

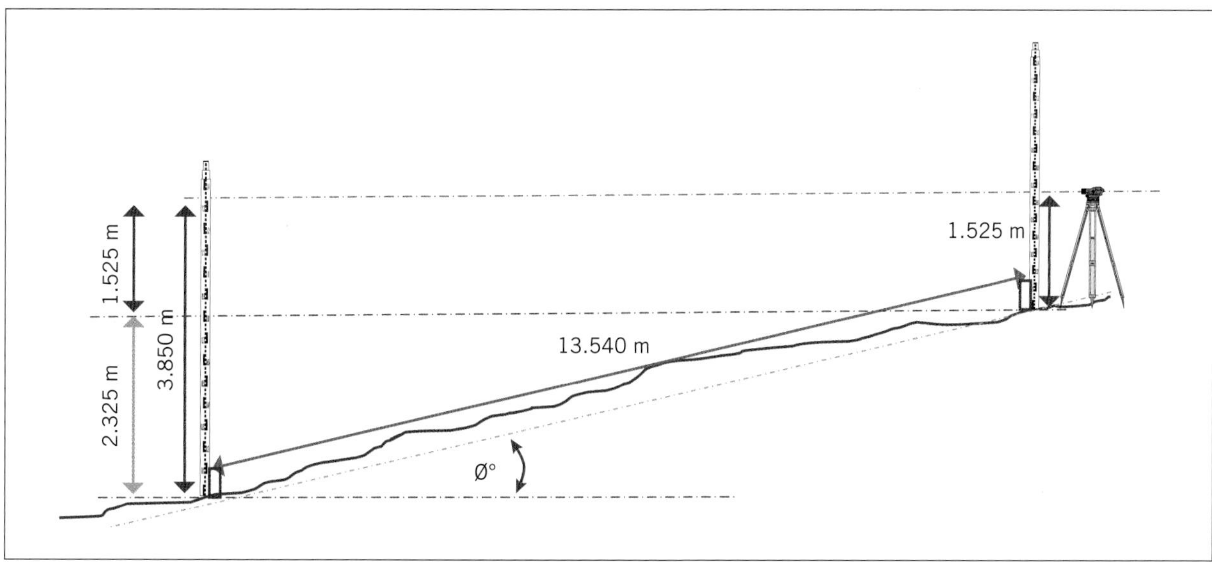

FIGURE 10.67 Determining the horizontal distance from measured slope distance and angle

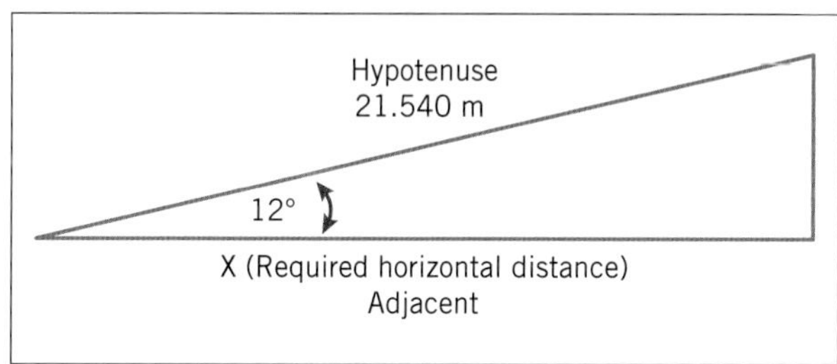

FIGURE 10.68 Sketching the information as a trigonometric problem

Finding an angle

On other occasions in setting out a structure, or plotting the location of a feature on a block, we may have a range of measurements but need to know the value of an angle. For example, we may not have a theodolite, but need to identify the pitch of a sloping block for determining compliance of access with the maximum angle allowed for a sloping path.

From Figure 10.69 we can identify the lengths of two sides of our right-angled triangle. This is enough to determine the angle we seek as is shown in Figure 10.70 and the associated calculations. In this case, the relationship of the known sides to the trigonometric ratios is opposite the hypotenuse and hence we use the sine ratios.

Sin Ø = opposite/hypotenuse

Sin Ø = 2.325 m ÷ 13.540 m

Sin Ø = 0.1717

Ø = Sin^{-1} 0.1717*

Ø = 9.887°

*Note: Sin^{-1}

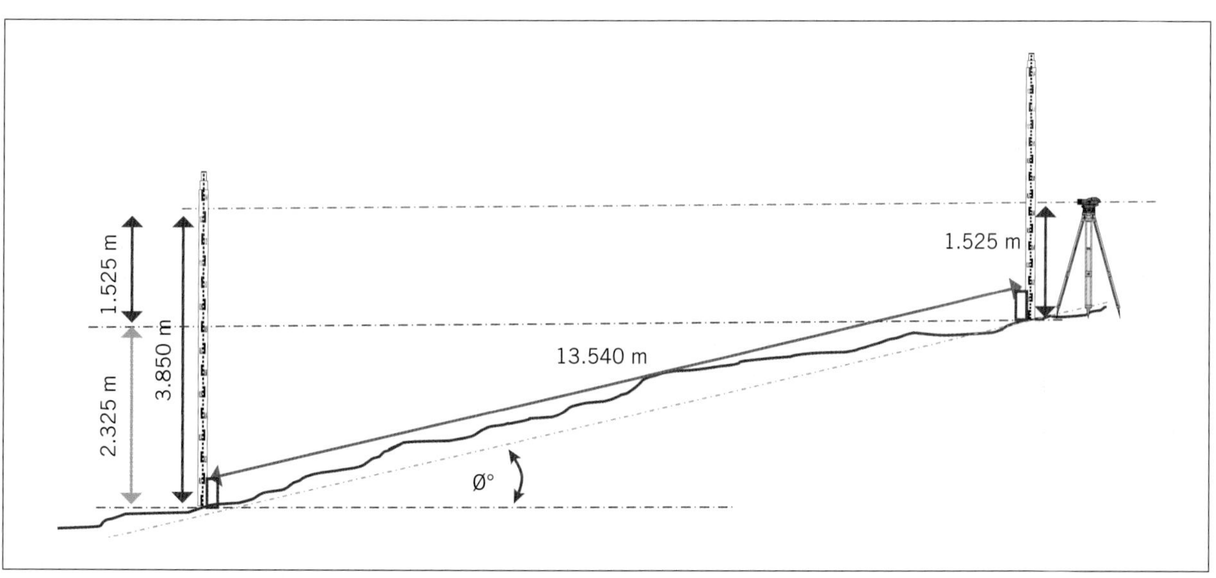

FIGURE 10.69 Gathering the information to find the pitch of a sloping site

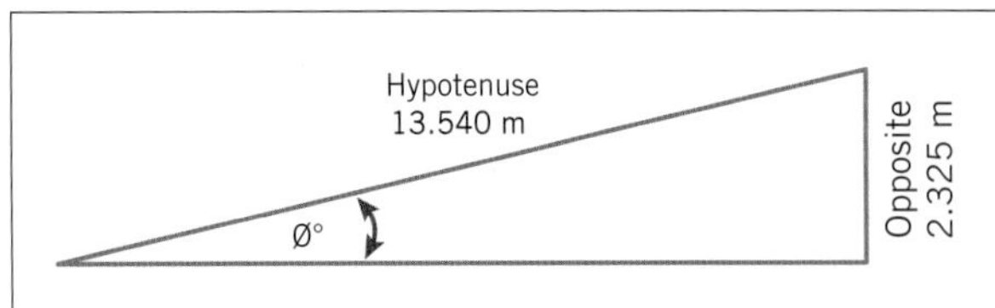

FIGURE 10.70 Sketching the information as a trigonometric problem

This is known as the inverse of Sin (or Cos or Tan). It is effectively saying, 'I have the sine ratio value for two sides of a triangle, please find me the angle in the tables that this relates to.'

With the calculator, enter the number and then press the 'inverse' button and then the Sin button.

Other trigonometric formulas and calculations

On some occasions you will find that your surveying for a set out or projected excavation work requires you to work around existing structures or other impediments such as trees, pits, or large water masses. In such cases, knowledge of the following formulas is of extreme advantage.

The sine law (law of sines)

The sine law or 'law of sines' is an extension of Pythagoras' theorem and allows us to find angles and lengths of non-right-angled triangles when information is not sufficient to do so by other, simpler means (Figure 10.71). The law looks like this:

$$\frac{\text{Sin A}}{a} = \frac{\text{Sin B}}{b} = \frac{\text{Sin C}}{c}$$

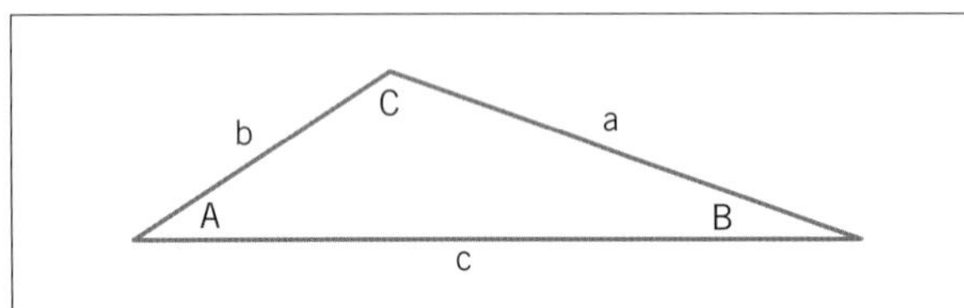

FIGURE 10.71 Standard notation of a non-right-angled triangle

In application it works as exampled in Figure 10.72 and the associated calculations.

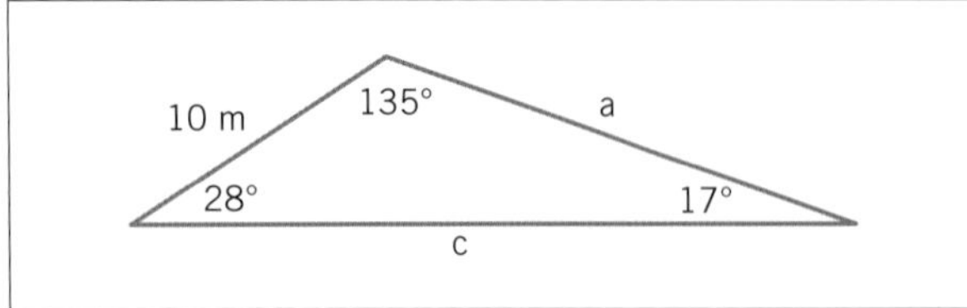

FIGURE 10.72 Standard notation of a non-right-angled triangle

$$\frac{\text{Sin A}}{a} = \frac{\text{Sin B}}{b}$$

Note: you only need two of the ratios to begin solving the problem.

$$\frac{\text{Sin } 28^\circ}{a} = \frac{\text{Sin } 17^\circ}{10}$$

$$\frac{0.4695}{a} = \frac{0.2924}{10}$$

$$\frac{0.4695}{a} = 0.0292$$

$$a = \frac{0.4695}{0.0292}$$

$$a = 16.078 \text{ m}$$

With this formula you can continue to find for side 'c' by first determining angle 'C'. 'C' is found by acknowledging the rule that the sum (addition) of all angles in a triangle is 180°. Hence 'C' must be:

$B = 180 - (17 + 28)$

$B = 135^\circ$

The cosine law (law of cosines)

This law is framed around the same standardised triangle shown in Figure 10.71 and used for the sine law. The formula is somewhat different however.

$c^2 = a^2 + b^2 - 2ab \text{ Cos C}^\circ$

Using the information from Figure 10.72, the solution to 'c' can be found as follows:

$c^2 = 16.078^2 + 10^2 - 2 \times 16.078 \times 10 \times \text{Cos } 135^\circ$

$c^2 = 258.502 + 100 - 321.56 \times -0.7071$

$c^2 = 358.502 - -227.377$

$c^2 = 585.879$*

$c = \sqrt{585.879}$

$c = 24.205$ m

*Note: 358.502 – 'Negative' 227.377

That is, 227.377 is added, not subtracted.

Result is 585.879

With these two laws you can solve for triangles of any size or angle provided you know either:

- two angles and any one side
- two sides and any one angle.

Converting degrees, minutes and seconds to decimals

Angles are generally provided in degrees. They may also include fractions of degrees, and in such cases this fraction will often be in the form of 'minutes' and 'seconds'. To input these into a calculator it may be necessary to convert the angle into a decimal format (some calculators have a function that can do this for you). There are a number of online tools for doing this conversion, but it is at times necessary to do this yourself.

An angle in degrees, minutes and seconds looks as below:

36° 24' 45"

where ' symbolises minutes, and " indicates seconds.

There are 60 minutes to a degree, and 60 seconds to the minute, but more importantly, 3600 seconds to the degree. It is this latter figure that will be used to determines the 'seconds' element of the decimal solution as it needs to reflect its part of degree, not a minute.

To convert an angle in this form to digital you do the following:

36° 24' 45"

=> 36.0 + (24 ÷ 60) + (45 ÷ 3600)

=> 36.0 + 0.4 + 0.0125

=> 36.4125°

For an online converter, the Australian Antarctic Division has a range of conversion tools at: https://data.aad.gov.au/aadc/calc/

The first two tools deal with converting degrees to decimal and, conversely, decimal degrees to minutes and seconds.

Standard set out procedures

The purpose of this section is to develop an understanding of the basic set out procedures for low-rise domestic and commercial structures. In each case the base principles are the same; however, with larger commercial structures, particularly those greater than two stories, other lines and control points may need to be established.

In both commercial and domestic contexts, the building site, block, or allotment must be accurately surveyed and the boundaries correctly aligned. This is known as cadastral surveying and must be conducted by a licensed surveyor, or a survey technician supervised by a licensed surveyor.

The cadastral survey

This survey will establish all boundary lines through the installation of approved survey pegs at boundary corners. The regulated size of these pegs varies from state to state to some degree, but as a minimum they will be 300 mm long, not less than 50 mm square (NSW tends to favour 75 × 50), be white painted on their exposed upper surface and be driven deep into the ground – Victorian regulations stipulating, for example, that the top be within 20 mm of being flush with the ground.

Because the pegs are driven so deep into the ground, they are sometimes hard to spot in thick vegetation, may get accidentally covered with earth, or may be driven over by vehicles and lost, damaged or displaced. When boundary pegs cannot be found or clearly identified the only solution is to have them re-established by a registered surveyor.

Without these clearly identified boundaries you cannot begin the task of setting out.

The set out

Identifying the boundaries of an allotment, however, is not sufficient to begin the set out. There remain a few preliminaries you must undertake to establish that you are on the *correct* building block. You do this by checking:

- the lot number is correct – these numbers can sometimes be switched or lost
- the location of the lot in relation to street corners and other blocks is correct
- the dimensions and shape as shown on the plan reflect the block you are observing
- all other identifying features on the site/location plan match those that you see in the surrounds; e.g. trees, bend in road, towers, major buildings, towers, water courses, etc.

With the block confirmed, we can begin with a standard set out procedure applicable to all building types. Figures 10.79 to 10.87 show the procedure as a set of logical steps that, while not the only way to set out, is one of the more streamlined and hence quicker approaches.

However, before outlining this approach it is appropriate to introduce a few basic principles and structures required to ensure your set out not only runs smoothly, but also remains sound and repeatedly usable over the requisite period of construction.

Profiles

In order to establish string lines to 'frame' your building and allow you to excavate footings, plot services and the like, you need to create some sort of system for fixing these lines to. This is known as profiling – the establishment of profiles. There are numerous ways by which profiles may be constructed; however, they generally fall into one of three categories:

- Saddles (Figure 10.73) – are constructed with the 'head' fixed to the cut tops of the stakes. These are seldom used as the process of cutting, or driving in, all the stakes to the one level is time consuming, and the action of nailing on top can put them out of level anyway.
- Hurdles (Figure 10.73) – are the more common approach on most construction sites. It is also common today to find the timber stakes replaced with steel star pickets, allowing for greater stability and reuse.
- Continuous (Figure 10.73) – are used when there are a lot of internal walls or other features of the building that are close together. It is not unknown for continuous profiles to be lengths of timber fixed directly to a handy boundary fence as the most convenient location.

For profiles to be used effectively and accurately, the heads – that to which the sting lines shall be fixed – should be all at the one level. This ensures measurements are taken horizontally rather than down a slope. For preference they should also run parallel with the exterior walls of the proposed building and be in line with one another. There are ways to work

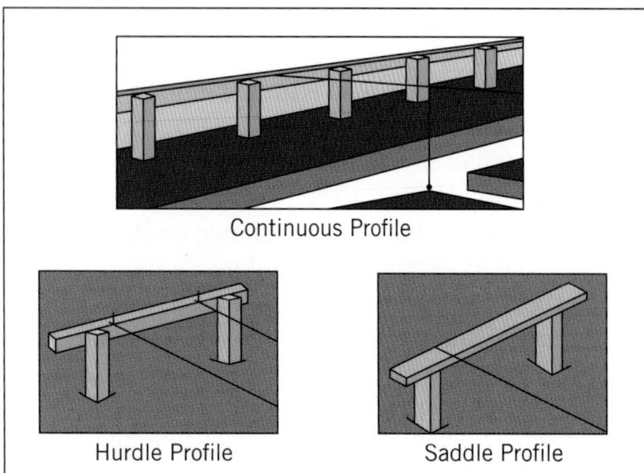

FIGURE 10.73 Profiles

around these criteria, which at times must be utilised due to extreme slopes or obstructions, but they tend to take more time to achieve and increase the risk of introducing error.

Tall or high profiles, or any profile that shows a tendency to move easily, must be braced.

The closer a profile is to the proposed structure, the less chance there is for error or movement in the string lines. However, profiles should be installed with regard to the excavation techniques that will be deployed; manual excavation needing less room than mechanical. In most cases, leaving a space of 1.5 m to 2.0 m between the profile and the excavated edge is sufficient to avoid accidental damage; however, if a greater distance is available then consider using it.

Setting out a 90° corner

Previously you were shown the use of the 3:4:5 triangle as a means of setting out a 90° corner. While it most certainly works, it is not overly efficient as a system and should only be used if there is not room to do something else. More typically the isometric triangle is used in such situations. This method is demonstrated in Figure 10.74 and involves the following steps, as shown in the 'How to' box.

Alternative means of setting out a 90° corner include:

- an optical square – hand-held units are available (Figure 10.75) but only tripod-mounted instruments are considered accurate enough for set out purposes
- a theodolite (see Figure 10.18 – not a common instrument on a domestic construction site).

Both the above instruments would be set upon a tripod and positioned directly over peg A by either optical plummet or a plumb bob suspended centrally under the instrument. Other means of setting out a 90° corner include:

- a folding building square – also rarely seen on a contemporary site but easy to make on site if required (see Figure 10.77)
- a fixed building square – simply three lengths of timber nailed together using the 3:4:5 principle to ensure square – also easy to make on site.

While all the above can and are used by various practitioners on site, in most cases it is far quicker and, with experience, just as accurate (excepting the

HOW TO

ISOMETRIC TRIANGLE METHOD FOR SETTING OUT A 90° CORNER

1. Establish peg A with a nail driven in directly under the base line and exactly positioning the corner of the building.
2. Locate pegs B and C equally distanced from peg A with a nail driven in at 8 m along, and directly under, the base line. This distance should be 'logical' to the size of the structure being set out; i.e. the longer the length of the building, the further pegs B and C should be apart to improve accuracy.
3. If two tapes are available, hook one over the nail at peg B and the other likewise at peg C.
4. Choose a length equal to or greater than the distance between pegs B and C and pull the tapes out so that they cross each other at D, as shown in Figure 10.74.
5. Establish a peg at E just beyond the tape crossing point D.
6. Install a string line at Peg A and pull through over peg E.
7. Check the accuracy of the line to the crossed tapes and adjust the string line at E accordingly.

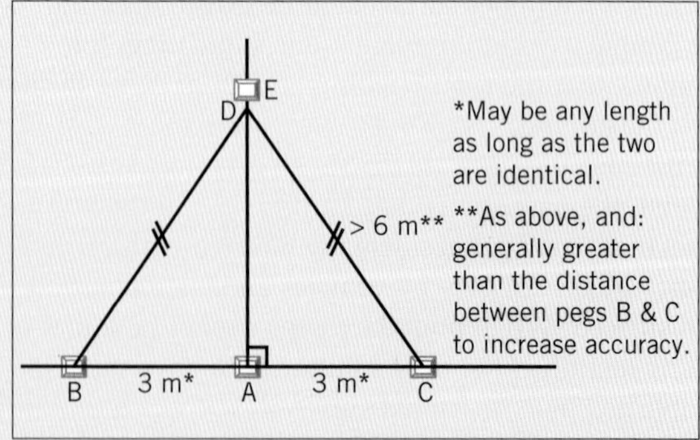

FIGURE 10.74 Isometric triangle used for setting out a 90° corner

FIGURE 10.75 Optical square

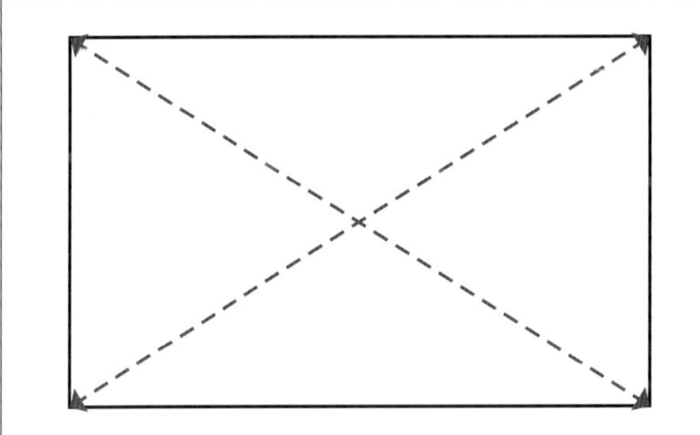

FIGURE 10.76 Checking for square by measuring diagonals for equal

theodolite approach), to have one person pull a string out from peg A as far is required for that wall length, stand over the top and look! If the strings seem to make a 90° degree angle then put a peg in the ground, pull the string tight and set up a profile.

The reason for adopting this seemingly rather vague approach is that ultimately you will be checking for square using diagonals (Figure 10.76 and 'checking for square'); you are seldom, if ever, reliant upon the initial squaring of a single corner.

Checking for square

Once the first square or rectangle of a building is established it must be checked to determine if the corners are exactly at 90°. We call this checking for 'square'. We do this by measuring the two diagonals and ensuring they are the same length. If they are not, then we must make the appropriate adjustments.

Three things may lead to unequal diagonals:

- the lengths of the sides have not been correctly measured and are out of parallel
- the initial 90° corner was not set out accurately
- a combination of the above.

Before adjusting anything, study Figure 10.78:

- *Identify the critical line (side)*: i.e. the line or side that cannot be moved because it represents the correct distance from a boundary or another building. Check that this is parallel and accurate first. In this case it is the **orange line over pegs C, F**, as this is the measured offset from the road frontage.
- *Identify the critical corner (peg)*: i.e. the peg that represents the corner of the building that is a set distance from another boundary or building. In this case **peg C**.

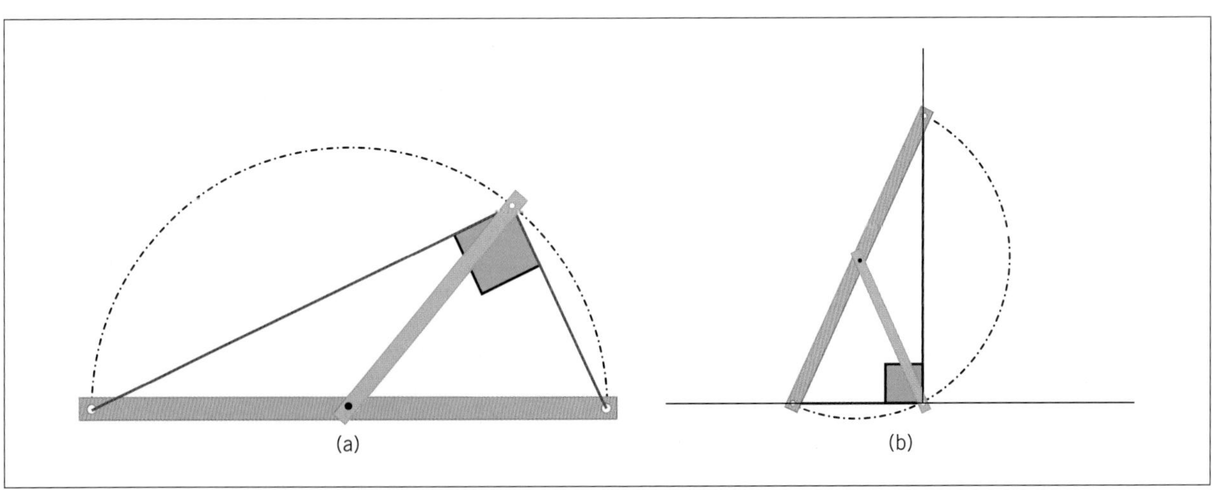

FIGURE 10.77a Principle

FIGURE 10.77b Application

FIGURE 10.77: The folding building square. The blue arm is fixed by a bolt at centre of the yellow arm and allowed to swing freely. A hole is drilled towards the end of the blue arm, and equally through the ends of the yellow. Any triangle formed within a semicircle, for which the hypotenuse is the diameter, is a 90° triangle (Figure 10.77a).

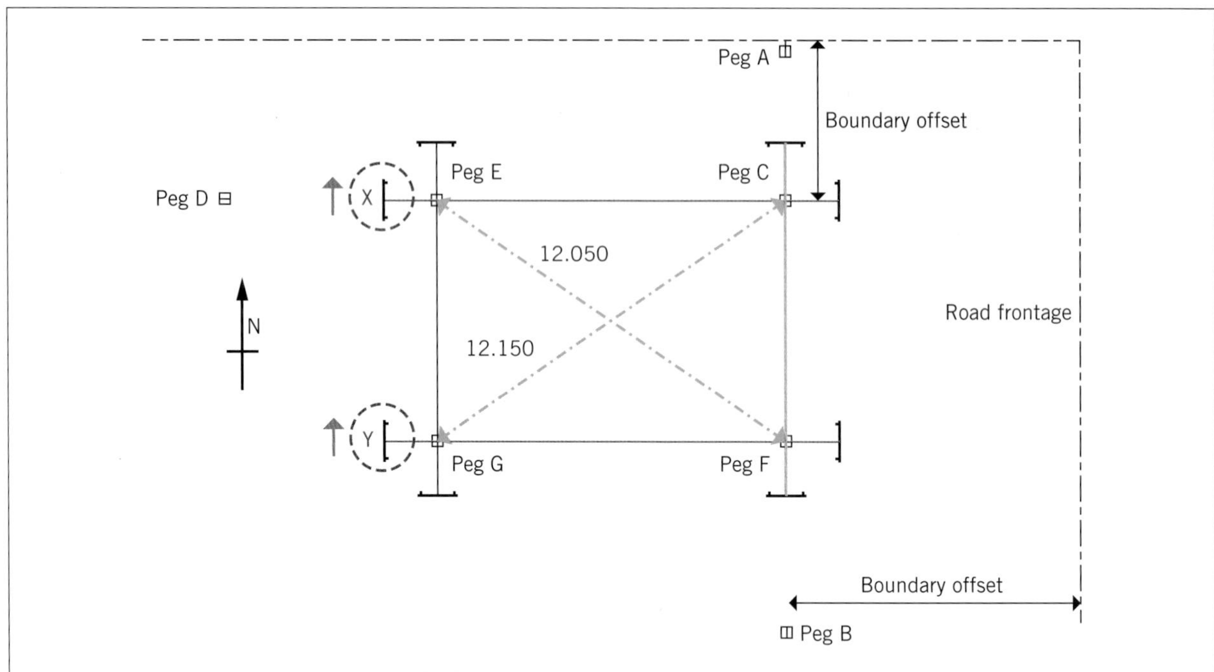

FIGURE 10.78 Squaring a rectangular set out

Now check all side measurements for error in length. This should be measured profile to profile, or profile to string, but not string to string, to ensure accuracy.

Ensure adjustments are made at the non-critical profiles – in this case those around Pegs E and G.

In this case:

- If length C–F is wrong, only red line over peg F may be moved.
- If length C–E is wrong, only black line over peg E can be moved.
- If length E–G is wrong, only red line over peg G should be moved.
- If length F–G is wrong, only black line over peg G can be moved.

Now re-check the diagonals. If the diagonals still show discernible error (greater than 3 mm):

- Move the **length** that is parallel to the critical line in the direction that will shorten the long diagonal, and so lengthen the other. In this case **the lines at X and Y will be moved north by 100 mm**.

Note:

- you must move both points that create this length equally in the one direction
- you should move them by the full amount of the difference found in the diagonal measurements.

Once this adjustment has been made, recheck the diagonals and repeat the procedure if any further error remains.

The set out procedure

Having confirmed the block, and determined the appropriate boundary lines, you will generally need to run a string line along at least one boundary line, frequently two, to give you something to measure from.

When setting out, where possible you should set out the largest rectangular or square shape first. This is then checked for square by means of the diagonals. Only after this do you define all the other external and internal walls and features.

HOW TO

SET OUT AN L-SHAPED BUILDING

1 Establish the first offset line by measuring the exact distances required from the boundary line (as shown on the plan) and locate pegs A and B (Figure 10.79).

Notes:

a Peg A and peg B are located such that the line extends well past the likely location of any profiles that will need to be installed later.

b Check the established line for correct distance from the boundary at both ends before proceeding further.

>>

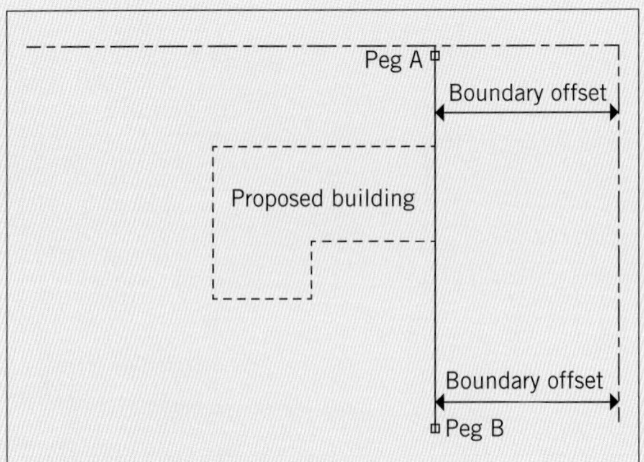

FIGURE 10.79 Boundary offset

c Depending on the slope or nature of the block, it is possible to install profiles straight away, instead of pegs. This saves having to transfer the string line up to the profiles later. However, it is not always easy to gain an accurate measurement using this process, hence it is not the preferred approach.

2 Locate peg C (Figure 10.80) such that it provides a corner of the building that is the correct distance from a second boundary line. Then double-check your offsets for correct and accurate measurement.

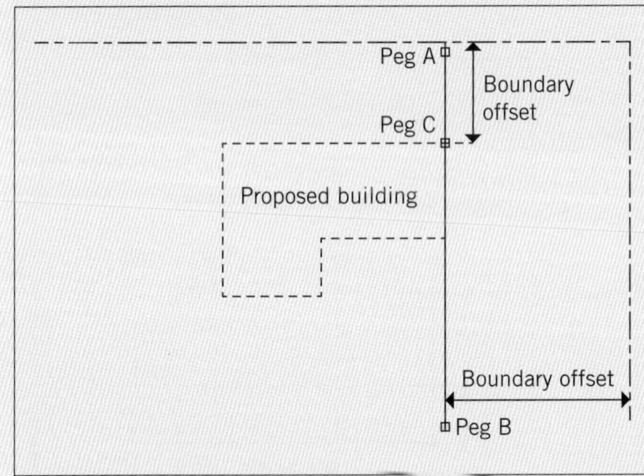

FIGURE 10.80 Second offset peg

3 Using your preferred method of developing a 90° line, locate peg D and run a string line accordingly. As with pegs A and B, be sure to locate peg D past the likely location of any profiles (Figure 10.81).

4 The approximate positions of pegs E and F may now be established by measuring the building length and width along the two existing string lines from Peg C. Peg G is found by measuring the respective lengths out from pegs E and F once in place (Figure 10.82).

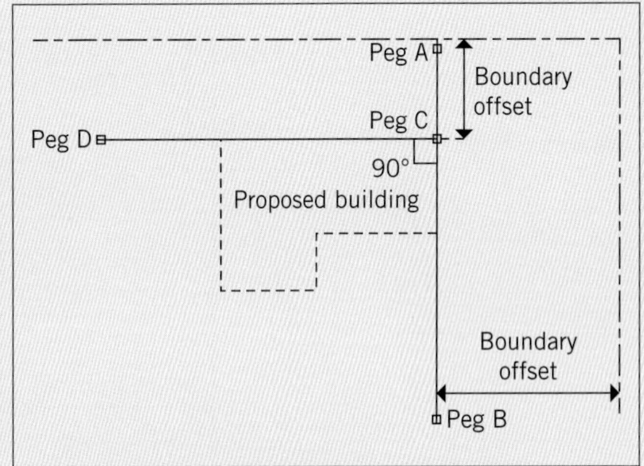

FIGURE 10.81 Establishing 90° line from boundary offset

Note: These pegs are approximate only as you may have to measure up or down a slope. If this is the case, hold the tape measure as level as possible and locate the peg by eye, spirit level or plumb bob: either way the position remains approximate.

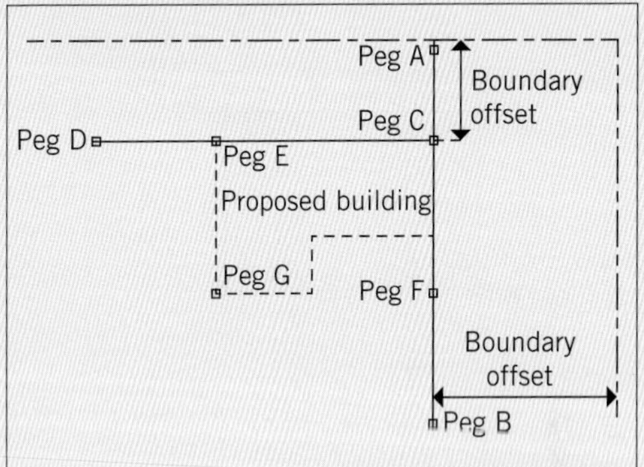

FIGURE 10.82 Approximate position of additional corner pegs

Notes:

a Locate profiles as far as is practicable from any proposed excavation work; i.e. a minimum of 1.5 m if mechanical excavation is required.

b Wherever possible, all profiles should be placed in line, parallel with the proposed building, and with the heads or cross-pieces at the same level. This ensures measurements are applied horizontally rather than at a slope.

5 Having located all four corners of the base rectangle, you can now install the first eight (8) profiles as shown (Figure 10.83).

6 Lines A–B and C–D must now be transferred to their associated profiles. This may be done by using a level, plumb bob or oil bath reflector.

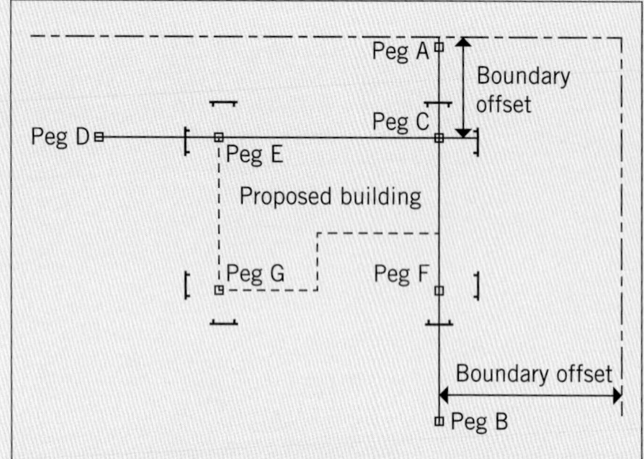

FIGURE 10.83 Establishing profiles

Once done, direct measure the length and width of the building to the other profiles as required. Then establish the string lines to provide you with a visible external shape of the proposed building (Figure 10.84).

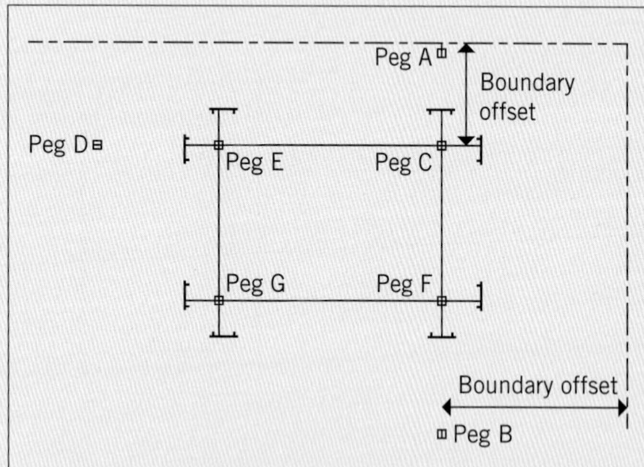

FIGURE 10.84 Transferring lines to profiles

7 Check for square by measuring the diagonals of the rectangle formed by the string lines (Figure 10.85). If not, check your length and width measurements at the profiles and, if necessary, adjust accordingly (see previous section on 'checking for square').
8 Once square, locate any additional profiles as per the plan. Be sure they are in line with those profiles already established and level (Figure 10.86).
9 On these additional profiles, measure and mark the other walls and locate string lines accordingly (Figure 10.87).

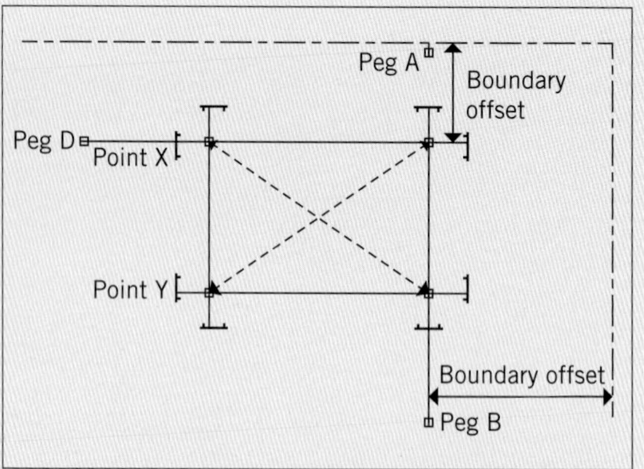

FIGURE 10.85 Checking for square using the diagonals

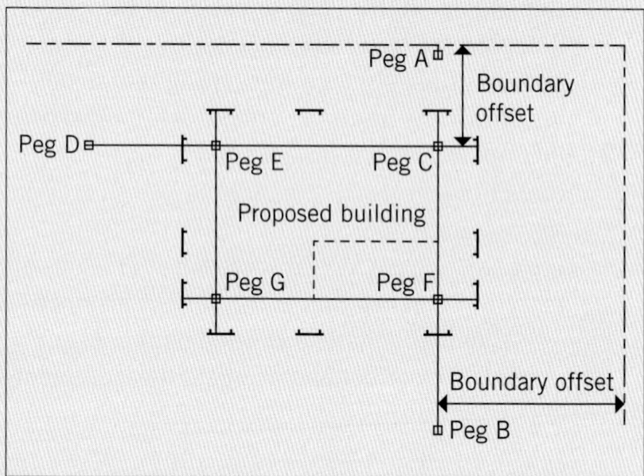

FIGURE 10.86 Locating additional profiles

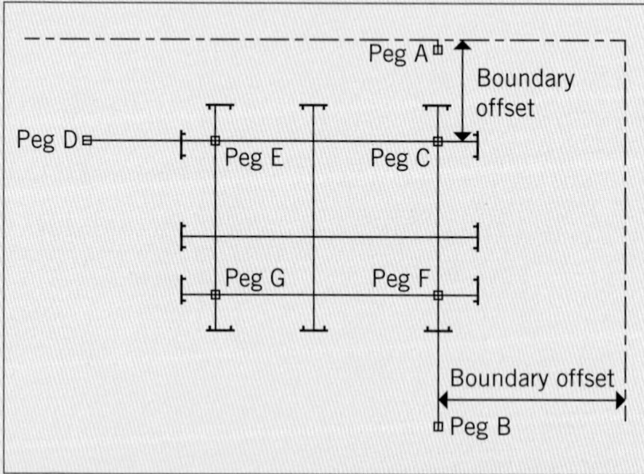

FIGURE 10.87 Location of additional wall lines on profiles

10 Check all dimensions and then mark out all additional information required on the profiles, depending upon the type of building being constructed.

Information contained on profiles

The position and degree of information to be marked on a profile will depend upon the type of construction used on the building. Examples of four common construction techniques, and the information each requires on the profile, are offered in Figure 10.88, which demonstrates typical applications.

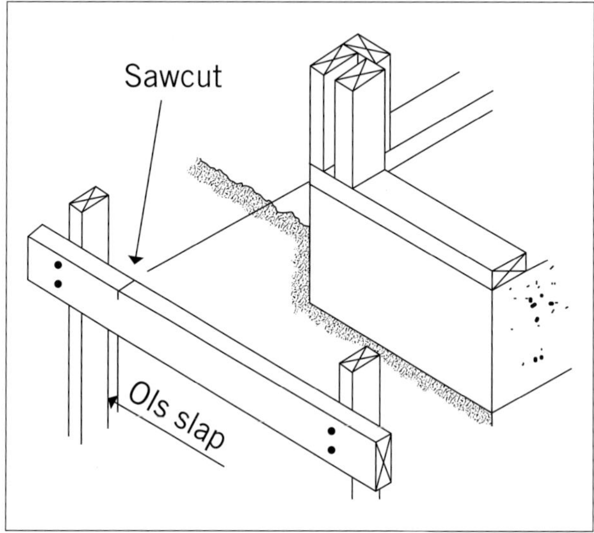

FIGURE 10.88 Typical application of strings and information for a concrete slab, timber-framed and clad building

Slab on ground: brick veneer

A brick veneer structure built on a slab generally requires a rebate to be formed around the perimeter. This rebate allows the bricks to sit lower than the top of the slab, reducing the visible edge of the building. With brick veneer it is the outside face of the brickwork that represents the overall size of the building.

The information required on a profile for this structure is shown in Figure 10.89.

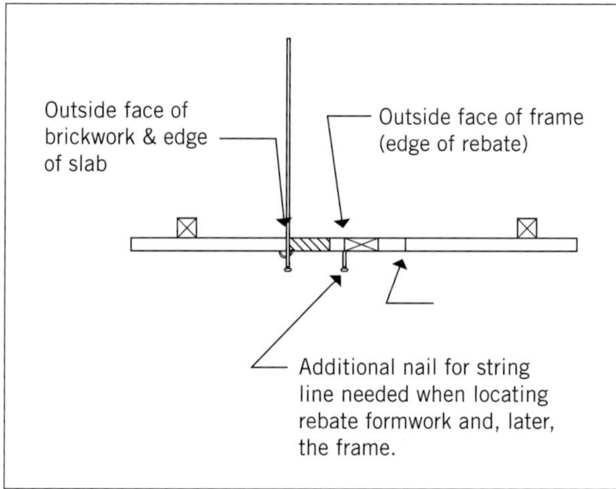

FIGURE 10.89 Transferring lines to profiles

Slab on ground: cavity brick

As with brick veneer, the overall dimensions of a solid brick building are measured to the outside face of the external brickwork. Once again, there is generally a rebate to be formed which must be shown, and additionally the inside face of the internal skin of brick (Figure 10.90).

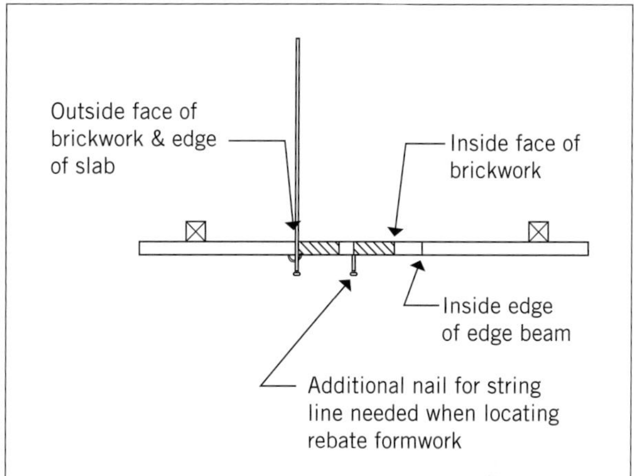

FIGURE 10.90 Profile information for slab on ground: cavity brick

Strip footing: brick veneer

A brick veneer building on a continuous strip footing will require much the same information as shown previously. The only difference is that with a slab, the outside edge of the building aligns with edge of the footing: with strip footings, the width of the strip footing is balanced beneath the wall (Figure 10.91).

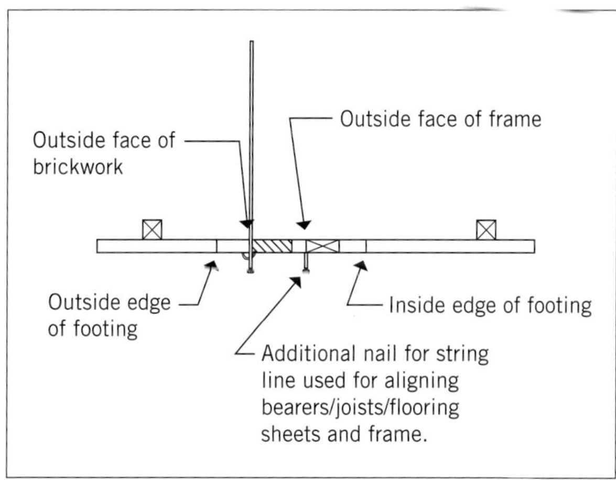

FIGURE 10.91 Profile information for strip footing: brick veneer

Strip footing: Cavity brick

Cavity brick construction likewise will require the width of the footing to be shown, and the following:

- outside face of the external skin of brickwork
- inside face of internal skin of brickwork Figure 10.92).

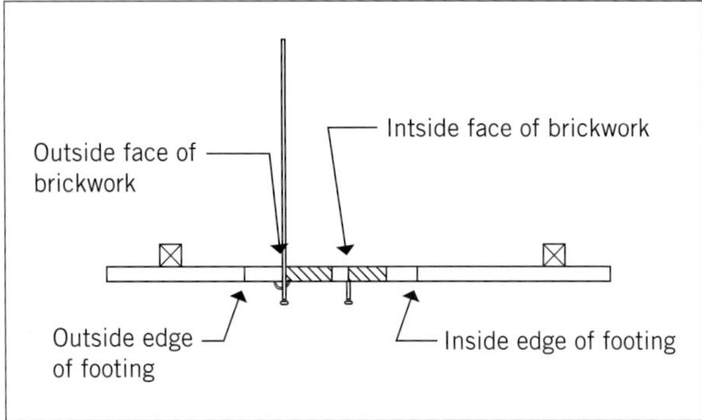

FIGURE 10.92 Strip footing under cavity brick

Dealing with error

Working in the field is to work with the risk of error from a variety of sources. Many of these errors can be avoided by way of sound work practices, careful use and maintenance of instruments and systematic checking of work undertaken. You also need to be honest with yourself and know when to return to the field when the data doesn't make sense.

Types of error

Most errors fall into one of three categories:

- gross errors – generally human error or mistakes
- systematic or constant errors – generally equipment based
- random errors.

Each of these forms of error need to be understood before they can be tackled with any chance of, if not eliminating them, at least reducing their influence upon the project.

Gross errors

There are numerous reasons why human errors may occur when in the field. This can include inexperience with, or lack of knowledge of, the equipment and or the task; lack of care, or rushing to complete to a deadline; simple tiredness; or poor team work. In addition, we sometimes make assumptions about a particular feature, or the state of a piece of equipment, and fail to check before proceeding.

Remaining vigilant, working in teams or pairs, constantly checking your and your partner's actions, and being aware of where common mistakes tend to occur is your best defence. The common areas where errors tend to creep in are:

- misreading the staff – this may occur in two ways:
  - Reading against a stadia line rather than the central horizontal cross-hair. Solve this by beginning your observation at the edge of the field of view – only the central cross-hair goes to the edge – then trace back to the staff and make your reading.
  - Misreading the markings on the staff itself, always read it twice at least. Also, assuming the wrong metre marking when it is not in view – ask the person holding the staff to read it for you.
- misplacing the decimal point on a reading
- calling out or hearing and recording a rearrangement of the figures; e.g. 2.120 as 2.012
- booking figures in the wrong columns
- making calculation errors when determining rise, fall, or RL values
- using an instrument that has not been levelled
- not eliminating parallax error
- using the wrong initial values – coordinates or benchmark RL
- using the incorrect settings on an instrument – radians on a calculator, or continuing to use a grade level setting on a laser level
- reading a staff that is leaning towards or away from the instrument.

Systematic or constant errors

These are perhaps the most serious form of error as they will perpetuate and accumulate throughout the project. They can also go unnoticed for quite some time before something shows the readings, or calculations based on those readings, to be outside logic or the expected range.

As a rule, these errors derive from defects in the instrumentation, or the choice of incorrect settings within the instrument. Examples include:

- line of sight or collimation error within the instrument; i.e. it sights up or down fractionally
- wear on joints of the staff causing a constant mismeasurement
- software issues in more advanced theodolite, total stations, or electronic measuring devices.

Non-equipment-based systematic errors also occur, though generally these simply exacerbate or accentuate an equipment error. The most common ones include:

- significantly unequal distance back and foresights; this will accentuate the errors from an instrument that is not sighting level
- poor booking systems, such as not putting readings to the same peg or point of interest on the same line.

Random errors

Random errors are perhaps the hardest to spot and solve for. Luckily, they tend to be small and self-eliminating, though not always. Common errors of this kind are:

- wind, humidity and high temperatures – these can cause atmospheric conditions (such as heat shimmer) that can deflect both lasers and line of sight observations
- expansion or contraction of metal staffs due to extremely high or low temperatures
- expansion and contraction of metal tape measures in high and low temperatures
- random software or instrumentation errors due to excessive solar exposure (heat)
- random electronic instrumentation errors – particularly those reliant upon global navigation satellite systems (GNSS), which may be affected

by solar activity such as sun spots and other electromagnetic influences.

Reducing the potential for error

The complete elimination of error is not possible when dealing with measurement or where humans are involved. However, its influence upon the outcome can be reduced by following the advice offered in the examples above. In addition, you must understand that the concepts of repeatability and acceptable error.

- **Repeatability**: when a measurement is taken, you, or someone else, should be able to repeat what you have done and get the same result. The reality is their result may be fractionally different due to a range of conditions (such as some of those mentioned previously), not least the accuracy limitations of the equipment used. This is where acceptable error comes in.
- **Acceptable error**: This is the amount of error considered acceptable in industry for the type of work being undertaken. Examples have been offered previously in this chapter, such as the error acceptable in a theodolite or laser level, or that for a closed loop survey based upon distance.

Maintaining equipment so that it remains within these tolerances, repeating sightings were necessary as a check, using closed loop surveys, ensuring back and foresights are of similar distances – these all help in reducing the potential for error.

Cut and fill calculations

The expression 'cut and fill' as used in the construction industry refers to the removal of high ground and the filling of low ground to level a site ready for set out and construction. This is an excavation procedure that, while having a range of issues regarding appropriate compaction and footing support over the filled areas, is common to many sloping sites where a concrete floor slab is involved.

Estimating the amount of earth to be removed, and hence available for fill and compaction, is a critical part of planning and costing prior to works commencing. In this section we will explore two methods by which these calculations may be performed:

- on continuously even sloping ground
- on land that falls in more than one direction or unevenly.

The first task in both cases is to determine the area of ground to be cleared, as this informs you on such issues as on-site storage of topsoil, or quantity of contaminated earth for removal.

Total site area

Site clearing is the first step on any construction site, and involves the stripping and storage of topsoil for spreading later back on to exposed areas. As with the excavation itself, knowing how much is to be stored or removed is an important part of planning and costing.

Figure 10.93 provides a plan view of a typical cut and fill excavation.

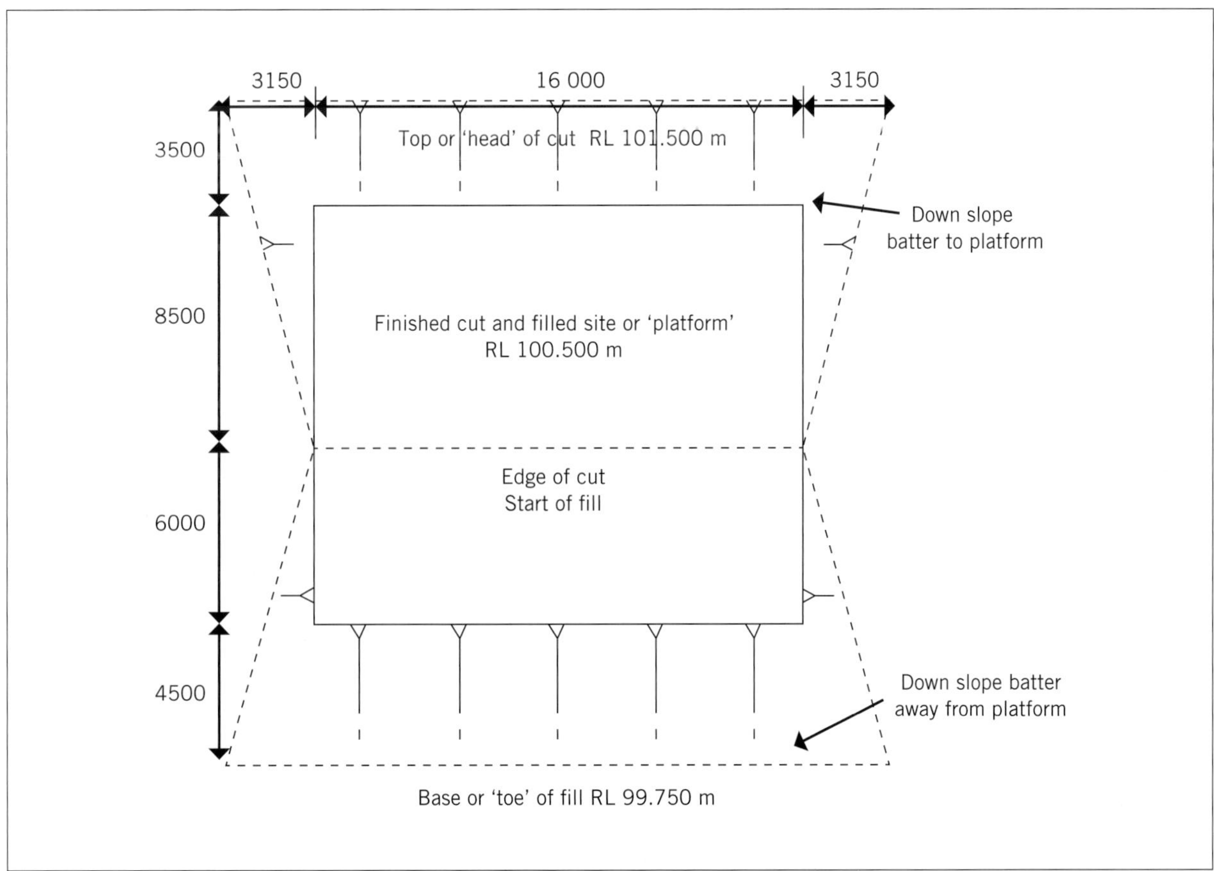

FIGURE 10.93 Plan view of cut and fill site

Calculation of area to be scraped and volume of topsoil to be removed

The typical area comprising a cut and fill site is depicted in Figure 10.93 and can be considered as two trapezoid shapes butted together. The formula for a trapezoid is shown here:

$$\frac{(A+C)}{2}\times D$$

where:

A is the long side of the trapezoid

C is the short side

D is the depth or height of the trapezoid

In this case:

For the first trapezoid: A = 22.3 m, C = 16.0 m and D = 12.0 m

For the second trapezoid: A = 22.3 m, C = 16.0 m and D = 10.5 m

The area of the first trapezoid is therefore:

$$\frac{(22.3+16.0)}{2}\times 12 = 229.8\ m^2$$

For the second trapezoid:

$$\frac{(22.3+16.0)}{2}\times 10.5 = 201.075\ m^2$$

Total area to be cleared = **430.875 m^2**

Turning this into a volume depends upon a site inspection to determine the depth of soil to be removed. This may be as little as 50 mm or as much as 300 mm.

Assuming a scraping of 100 mm, the volume of topsoil to be removed would equal:

$430.875 \times 0.100 = 43.0875\ m^3$

In turn, this figure would need to be adjusted by what is known as the 'bulking factor' for that particular soil type. This gives us the volume of 'spoil' (the soil expands when excavated) to be stored on site or transported elsewhere.

A typical bulking factor for topsoil is 1.3.

The final volume of soil under consideration is therefore:

$1.3 \times 43.0875\ m^3 =$ **56.01 m^3**

Cut and fill volume: continuously even sloping ground

Using the same cut and fill project as depicted in Figure 10.93, both the 'cut' excavation, and the fill volume, may be viewed as being made up of three components as depicted in Figure 10.94. The centre section (A) is a triangular prism, while the outer elements (B & C) are triangular-based pyramids. The volumes of the two outer elements are identical, so calculations for the centre (A) and either of the smaller elements (B or C) multiplied by 2 is all that is required.

Cut volume

The depth of the cut is found by subtracting the RL of finished site platform (100.500) from the RL for the top or 'head' of the cut (101.500); i.e. 1.000 m.

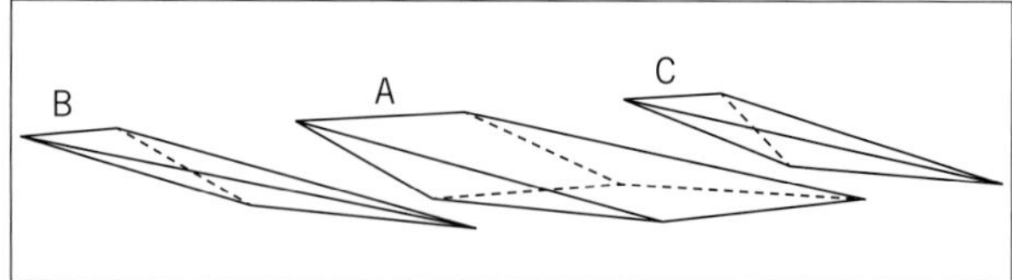

FIGURE 10.94 Elements of the 'cut' portion of a cut and fill site

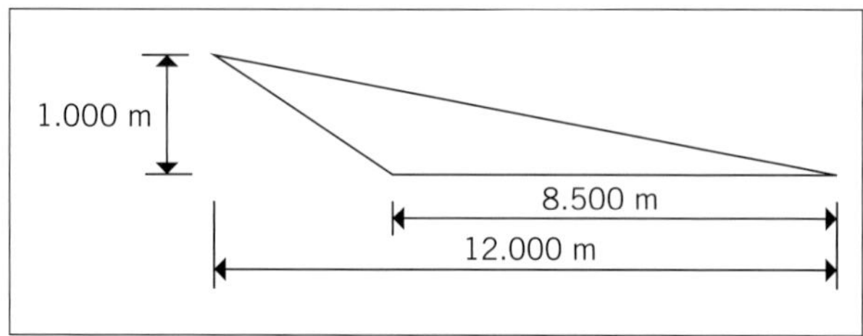

FIGURE 10.95 Cross-section of the 'cut' of a cut and fill site depicted in Figure 10.93

The calculation of sectional area is then as for any triangle:

i.e. ½ base × height

In this case:

Sectional area = 0.5 × 8.5 × 1.0

= 4.25 m^2

Volume of the centre section (A) is then:

Sectional area × width

In this case:

Volume of A = 4.25 × 16.0

= 68.0 m^3

Volume for the side elements is as for a pyramid; i.e.:

[Sectional (or 'base') area × width (or 'height')] ÷ 3

In this case:

Volume of B (or C) = (4.25 × 3.150) ÷ 3

= 4.4625 m^3

The total volume of the cut is then:

Volume of A + (2 × Volume of B (or C))

In this case:

Total volume = 68 + (2 × 4.4625)

= 76.925 m^3

As outlined previously, there is a difference between the volume to be excavated, and the volume of 'spoil' once removed. The bulking factor for this exercise will be taken as for clay which is 22% or a multiplier of 1.22. So, the spoil that must be handled is:

Total spoil = 76.925 × 1.22

= 93.8485 m^3

Fill volume

The same geometrical and mathematical approach may be taken to determine the fill volume. That is, find the volume of A and 2 × the volume of one of the side elements (B or C – see Figures 10.96 and 10.97).

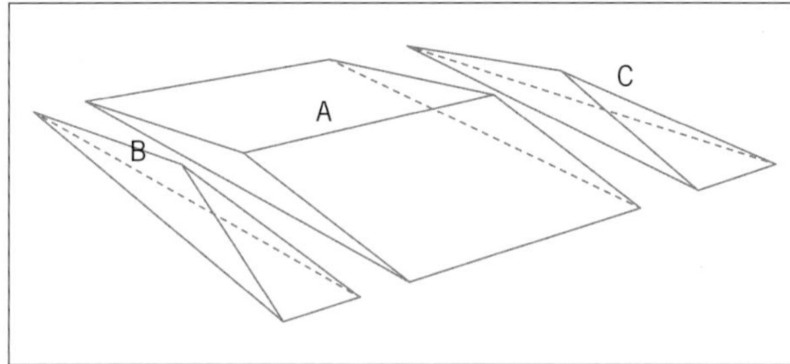

FIGURE 10.96 Elements of the 'fill' portion of a cut and fill site

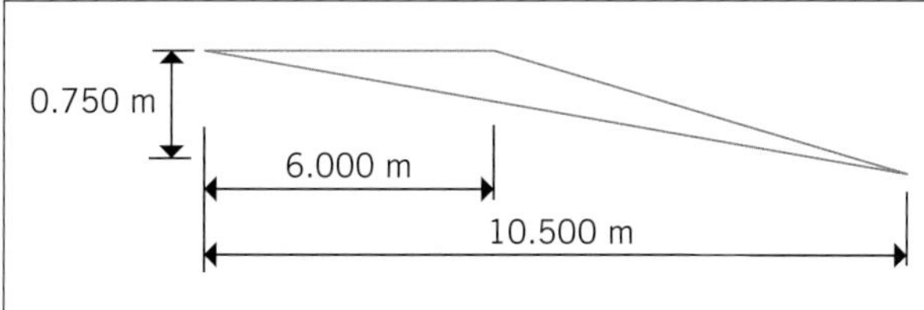

FIGURE 10.97 Cross-section of 'fill' of cut and fill site depicted in Figure 10.93

The previous mathematics may be simplified into a single formula that looks like this:

= [Area of section A × width] + (2 × [Area of section B × width ÷ 3])

The height, however, is different to that of the cut. But it is found in a similar manner:

RL of the finished site platform (100.500) minus the RL of the base or 'toe' of the fill (99.750);

i.e. 0.750 m

The fill volume may then be calculated as follows:

Fill volume = {[(6.00 × 0.75) ÷ 2] × 16} + 2({[(6.00 × 0.75) ÷ 2] × 3.150} ÷ 3)

= 36 + 4.725

= 40.725 m^3

As fill, the bulking factor needs to be applied to determine the volume required in the uncompacted state. Assuming clay once more (1.22), this gives:

Uncompacted volume = 1.22 × 40.725

= 49.685 m^3

From the above calculations, it may be determined that there is sufficient spoil from the 'cut' element to create the filled area of the project. It also informs us that there will be approximately 40 m^3 left over that may have to be removed from the site for disposal elsewhere.

Note: Using controlled compaction, the cut or extracted volume may be reduced by 10–15% when placed as fill. That is, 100m^3 compacted down with a vibrating roller or the like produces only 75–90 m^3 of actual fill depending upon soil type and moisture content.

Excavations on uneven ground or complex slopes

The above approach to the calculation of excavation or fill volumes is simple but not overly realistic given that a construction site is seldom a perfect slope only in one direction. Nor is it always the case that the platform is going to be sited perfectly as suggested by the example offered.

When the ground is more complex you will need to adopt an alternative strategy, of which there are several. Most of these alternatives rely upon some understanding of topography and the development of grid systems used to map uneven terrain. These alternative systems will be described in the next section 'Topography'. There is one method, however, that does not work on grids and is one of the most common approaches used – this is described below.

Excavation and fill volumes using sectional areas

This approach may be undertaken in several ways. The simplest is the calculation of the two end areas, adding them together and finding their average by dividing by two. This average area is then multiplied by the horizontal distance separating them.

Figures 10.98 and 10.99 offer an example of fill showing two end areas separated by the distance 'L'. The end areas are trapezoid, so the formula for their calculation is simply:

End area = ½ (A + B) × height

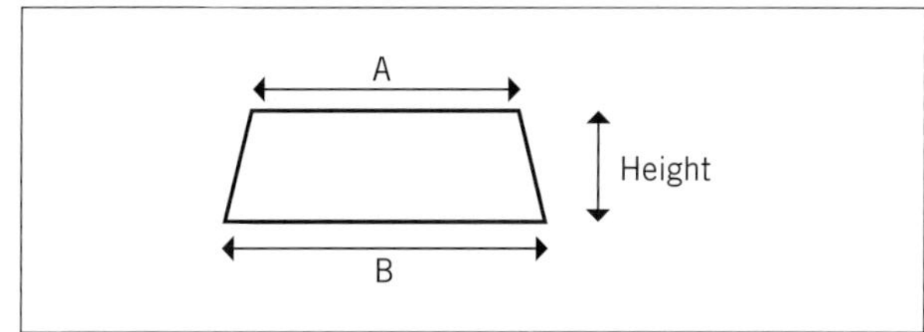

Figure 10.98 Trapezoid area calculation

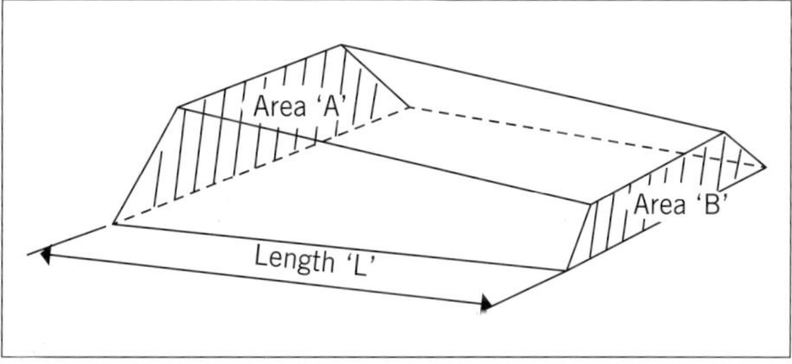

Figure 10.99 Volume of fill estimated using end sections

These two areas are then averaged by adding them together and dividing by two. The result is then multiplied by the distance 'L' to find the approximate volume of fill required. This same approach may be used for excavation volumes as well.

E.g. approximate volume = [(Area A + Area B) ÷ 2] × L

Where the excavation or fill tapers off completely; i.e. there is only one end area, the result is a pyramid and the volume found by:

Approximate volume = (Area A × L) ÷ 3

This system works for construction sites that are fairly regular in surface prior to excavation. When the surface becomes noticeably irregular then multiple cross-sections are used. The site is 'broken up' into a series of shorter lengths covering the excavated zone as shown Figure 10.100.

The area of each cross-section is then calculated, added together to find an average, then multiplied as before by the overall length of the excavation. That is,

Approximate volume = [(Area A + Area B + Area C + Area D + Area E) ÷ the number of sections*] × L

*In this case 5 as there were five sectional areas developed.

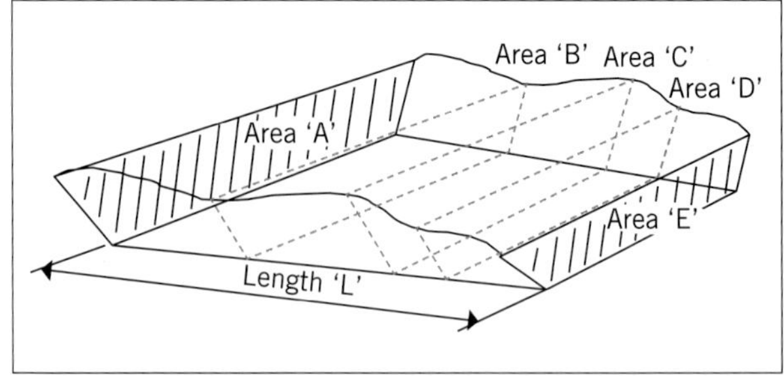

FIGURE 10.100 Estimating excavation volumes using multiple cross-sectional areas

LEARNING TASK 10.2 USING BASIC CALCULATIONS TO CARRY OUT WORK

1 You have measured along the ground for a distance of 17.5 m on a sloping site. Needed is a horizontal measurement of 16.350 m.
Draw the appropriate sketch and calculate the min. height of any long pegs that you might need, allowing for the pegs to go at least 200 mm into the ground and allow at least 150 mm above the horizontal line to fix off and brace the peg.
What minimum height should the pegs be?

2 Someone has told you that the ground you will be working on looks to be sloping at around 25°. You know that you will need to do a 'cut and fill' by digging into the ground at least 1700 mm at the high point for a retaining wall. Calculate how far horizontally from the anticipated cut point that the fill will need to be started from.

3 You have been asked to calculate the volume of spoil to be removed from an excavation that has the following overall dimensions (see diagram).
How many m^3 need to be removed before allowing for any 'bulking' of the material?

Note: The trench sides will be sloped at an angle of 10° off the vertical, with the ends being plumb.

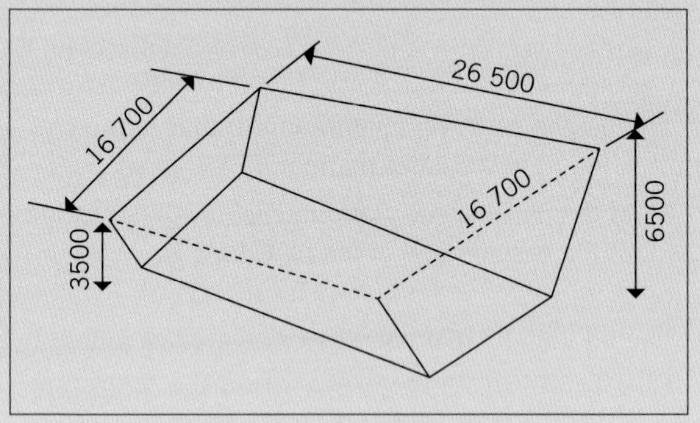

Topography

With regards to construction, topography is the study of surface features of the landscape; in particular, their description and location in relation to each other. Generally, such descriptions are satisfied by means of maps and 'legends' – tables of figures, symbols and colour coding – that allow the reader to interpret the physical shape of hills, valleys, rivers, roads, towns and the like.

Contour lines

One of the major features of a topographic map that allows this interpretation is the contour line. This line (Figure 10.101), or more correctly, these lines, follow a continuous height above sea level or elevation and are evenly separated vertically by an amount relative to the scale and detail desired for a specific map.

Characteristics

A contour line, or cluster of contour lines, have certain typical characteristics; they:

- are continuous lines that do not stop at any point unless a perfectly vertical plane is met, such as a cliff, which in nature is very rare
- can never cross or meet each other except in the situation just described
- can never split into two or more lines
- indicate the same height or elevation above the datum for the entire length of that contour.

Generally, the vertical spacing between contours should be the same for the entire site. However, this may change if a particular area requires greater detail. In such cases, a sub-plan should be produced for that area.

This allows us to visualise the shape of the ground, including steepness of slopes, depressions, hills, valleys and the like. For example:

- contour lines that are close together show steeper slopes than those far apart – the ground dropping elevation faster over the same horizontal distance
- the steepest slope at any given contour is at 90° to the line at that point
- a number of contours all evenly spaced denotes a steady slope
- contours that are variably spaced means changes in the slope.

Figure 10.101 shows how contours may be used to depict two small hills with varying slopes.

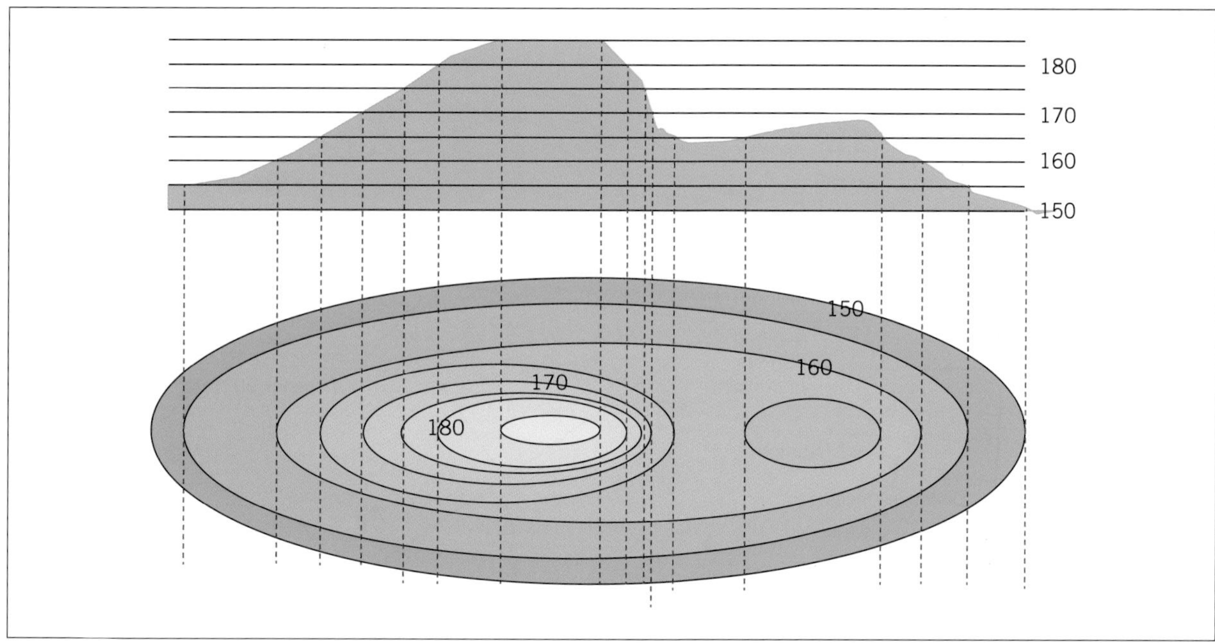

FIGURE 10.101 Contour lines representing two small hills

Note: Without the elevations given for some of the contours, you would not be able to determine if the plan view represented hills or depressions. Not every contour needs to have its elevation given. Generally, only the 'primary' elevations are shown – in this case every 10 m even though contours are at 5 m intervals.

Mapping

The mapping of contour lines for major topographic map makers such as Geoscience Australia is undertaken today by a combination of airborne Light Detection and Ranging (LiDAR) systems, and global navigation satellite systems (GNSS – see p. 241). This is linked with a massive digital database drawn from decades of previous land-based and aerial surveys.

Currently, mapping contour lines on a construction site using any of the advanced techniques mentioned above is not economical. Instead there are a range of fairly simple, though time consuming, approaches available to us. These fall into one of two categories:

- direct methods
- indirect methods.

Your choice of method will generally depend upon one of five factors:

- time available
- size of site being surveyed
- steepness of terrain
- plan scale or level of detail required
- purpose of contouring.

Mapping contour lines for construction: direct methods

These will only be discussed in brief as they are generally the most time consuming.

The direct method plots the contour on to the landscape of the site – literally painting the line as a series of dots or dashes on the ground using spray marking paint or steel pins/pegs with coloured cloth to make them more visible. This is done by setting up a level and taking a reading that represents the height of the first contour you wish to plot. The staff is then walked around the slope, finding and marking locations of the same height reading at a distance apart suitable to the required accuracy.

Once the contour is marked on the ground, the position of the contour must be surveyed – generally by measuring from a known point, or a series of known points (such as survey instrument stations) and angle from north (measured bearings), or by plotting straight chain lines and measured offsets (see Figure 10.102).

Mapping contour lines for construction: indirect methods

There are two indirect methods typically used in the plotting of contours:

- the casting of cross-section lines
- the development of a grid.

The casting of cross-section lines is both reasonably simple and appropriately accurate for most work. In short, lines known as 'traverse' lines are cast along the face of a slope so that they are both straight and reasonably level (Figure 10.103). Cross-section lines are then cast down the slope at roughly 90° to these traverse lines. Points are then plotted down these sectional lines such that a fair representation of slope can be plotted (see Figure 10.103). The number of sectional lines and the number of points down them is dependent upon the desired accuracy.

The alternative, the development of a grid system, is by far the most common system in use as it is only marginally slower in the field, and much quicker and easier to plot as a drawing either on paper or by CAD. This is discussed fully in the following section.

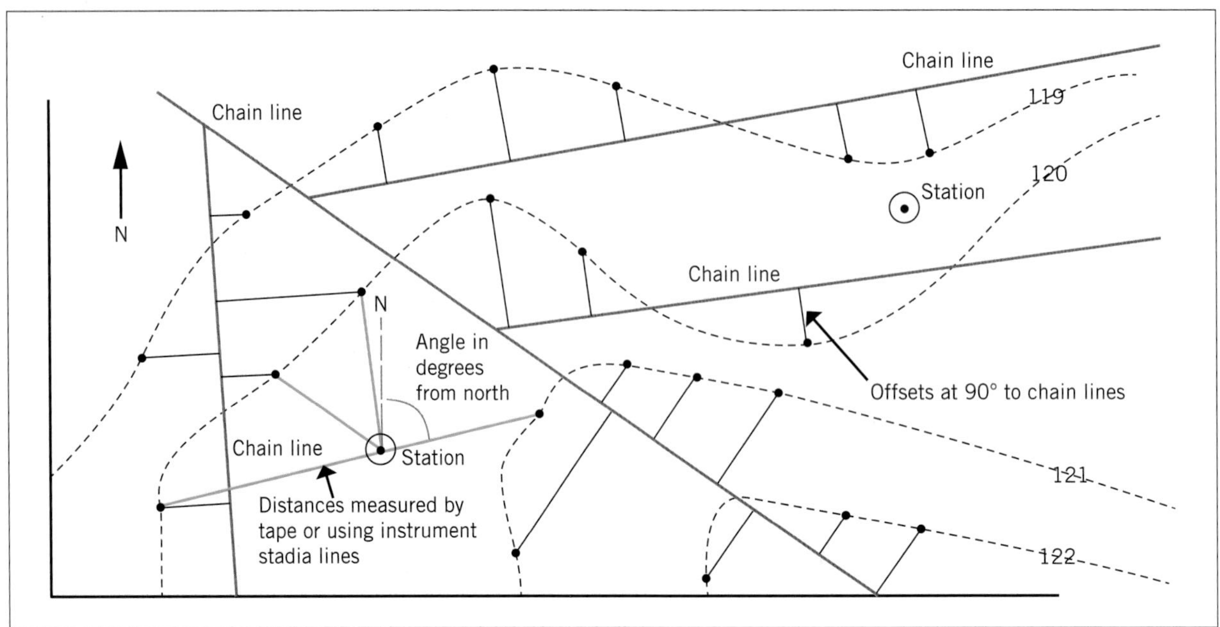

FIGURE 10.102 Direct plotting of contour lines: chain lines and offsets; measured bearings

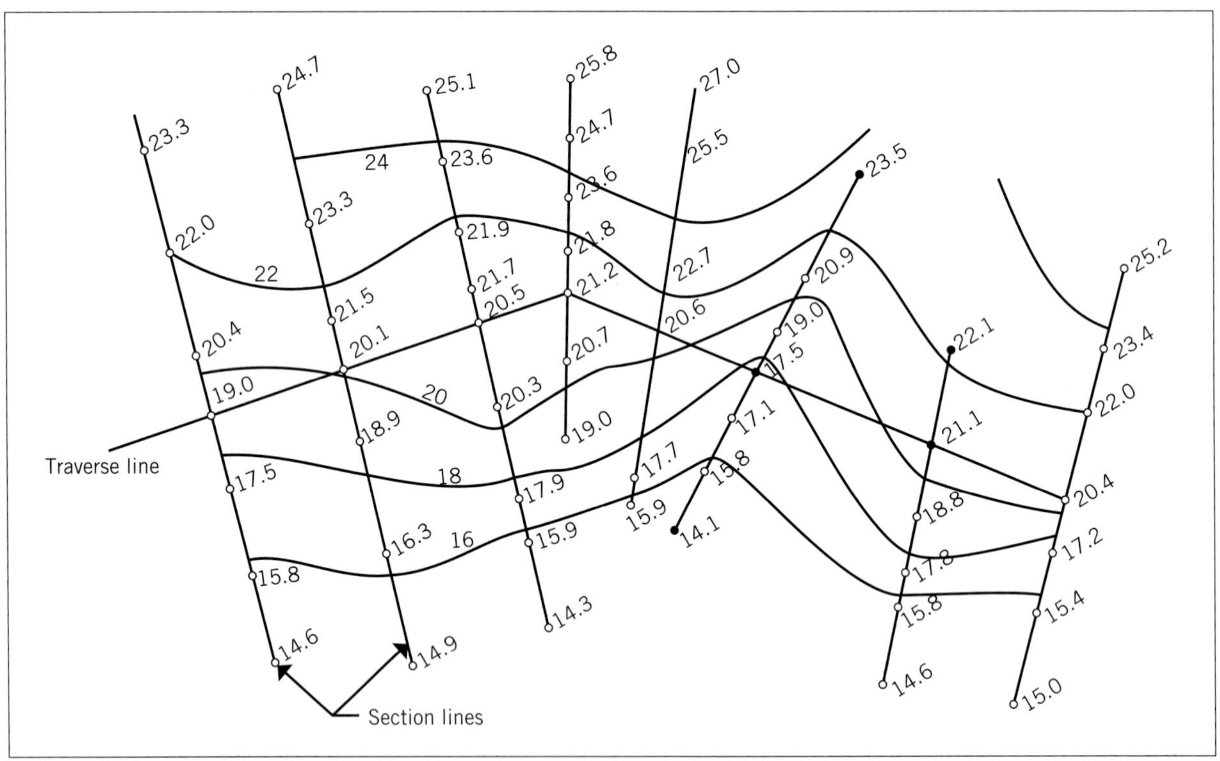

FIGURE 10.103 Indirect plotting of contour lines: traverse and cross-section lines

The basic grid

The use of a grid for surveying contours is common practice when the area is not overly extensive or of complex terrain. The grid may be established either as squares or rectangles but must be developed such that the lengths of the sides are accurate in plan (horizontally), not down the slope. Once established, this grid allows for reduced levels to be surveyed at known points. These may then be easily plotted onto a drawing and the contours then developed, as will be discussed in the following sections.

Establishing the grid

The important characteristic of the grid is that all measurements are taken horizontally so that when viewed in plan all the lengths are correctly proportioned. The following 'How to' procedure explains how this should be done (see also Figure 10.104).

HOW TO

ESTABLISH A GRID

1. Determine the size of the grid squares based upon:
   a. required scale and detail
   b. the available time for the survey
   c. complexity of the ground being surveyed
   d. the vertical interval between contours.
2. Establish a base line along one side of the survey area.
3. Place pegs or ranging poles at the chosen intervals along this line, e.g. 5 m, 10 m, or the like.
4. Establish a second line at 90° to the base line. This line may demark a side of the survey area, or, when the site is not rectangular, any of the mid lines that will pass all the way through the survey area to the other side.
5. Establish two more lines at or as close to the survey boundary as possible. These will be parallel to the first lines, thus making a closed rectangle.
6. Place pegs or ranging poles at the chosen intervals along these survey boundary lines.
7. From here it is a matter of placing pegs along each of the lines running between the ranging poles/pegs so that a grid is produced, as shown in Figure 10.104.

This grid is then drawn or sketched in the field and a means of identifying each of the survey points (where lines cross or abut), may be easily identified. In Figure 10.104 this is achieved by using an alphanumerical system; i.e. one boundary line is labelled A, B, … F; the other 1, 2, … 5. This allows the survey points to be easily identified such as B3 or E5, etc.

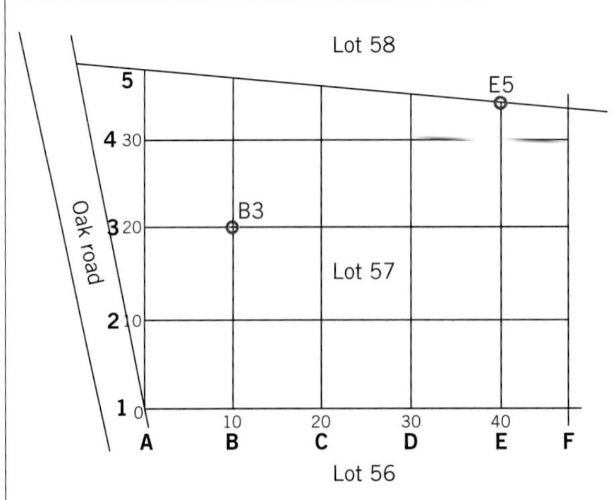

FIGURE 10.104 The grid

Taking levels

With the grid in place, the reduced levels (RLs) of each grid reference point may now be taken. This is best done using the closed loop rise and fall system of booking (see earlier in the chapter). The closed loop system is preferred as it ensures that you have found any errors prior to leaving the field.

The RLs for each grid reference are then plotted onto the grid in preparation for the development of the contour lines.

Preparing contour plans

The development of contour lines is undertaken using interpolation. With regards to graphical or numerical data, this is a means of determining a specific point between two other points. This may be completed either:

- mathematically
  or
- graphically.

We will deal with the graphical approaches first as these are the quickest and easiest paths to understand. This will also help you interpret the mathematical solutions.

Graphically

There are multiple ways of graphically finding the location of the contour lines. We will look at the two most common approaches:

- using a ruler and scaling
- using a radial interpolation graph.

In the example worked below, the vertical separation between is taken to be 1.0 m.

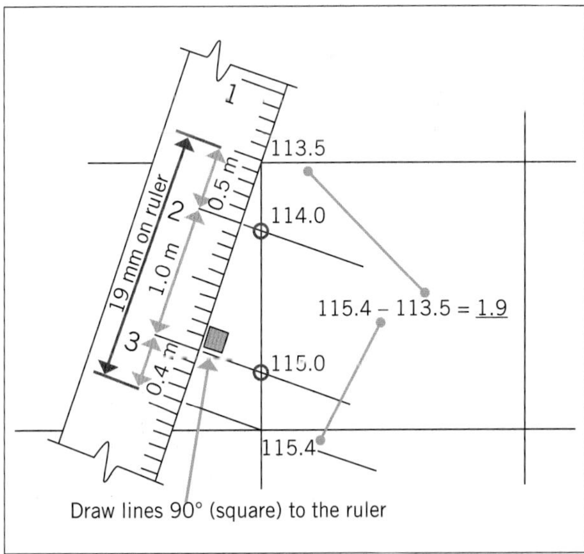

FIGURE 10.105 Interpolation using a ruler

Using a ruler

Figure 10.105 shows how to use a ruler and an appropriate unit of measure to locate the required contour line(s). The task may be broken down into the steps outlined in the following 'How to' box.

HOW TO

LOCATE CONTOUR LINES WITH A RULER

- Identify the number and height of contours that are to be found between the two grid points.
  - In this instance the grid references are 113.5 and 115.4.
  - With a vertical separation of 1.0 m, there are two contours to be plotted between these two grid points: 114 m and 115 m.
- Determine the total height difference between the grid points.
  - In this instance: 115.4 – 113.5 = 1.9 m.
- Choose a suitable scale or measurement unit that allows a breakup of the distance between the two drawn grid points into 19 parts.
  - In the example shown in Figure 10.105, this is a 1-mm graduation, with the ruler held at an angle such that 'zero' sits on the grid point with the lowest RL and the 19th graduation aligns with the second grid point when a line is drawn square off the ruler's edge as shown.
- Arrange the ruler in a way the logically suggests the contour location(s) relevant to the lowest grid point RL
  - In the example, the ruler is arranged such that 5 mm clearly sits between the first grid point and the first contour. That is, 0.5 m brings you from the RL of the first grid point (113.5 m) to the 114 m contour.
- If there is a second contour within the one grid line, this contour will be 10 mm away from the first.
  - In the example, this is shown as being at the 30-mm mark on the ruler as the first contour was at the 20-mm mark.

Provided the lines you draw to identify the location of the contours are drawn at, or very close to, 90° off the face of the ruler then this system provides a very accurate means of identifying the contour location.

Using a radial interpolation graph

Figure 10.106 shows the basic radial graph that may be used to divide up a grid line into the required number of parts to find the contour line(s). This is drawn upon a piece of thin perspex or other clear plastic. Depending upon the vertical distance between any two grid points, the 20 divisions shown may indicate 1, 2, or 3 units as required. The secondary line is used to help you keep the graph parallel to the grid lines.

Using the grid from the previous example, Figure 10.107 shows how the radial graph simplifies the division of a grid line.

Note that the radial graph must be applied so that the base or mid line is parallel to the grid line. In the example

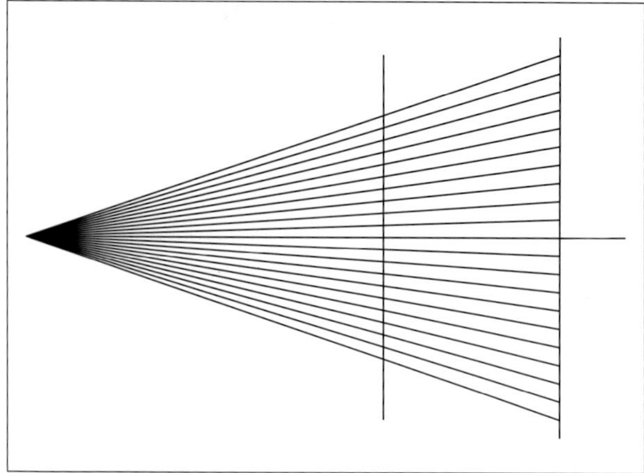

FIGURE 10.106 The radial graph

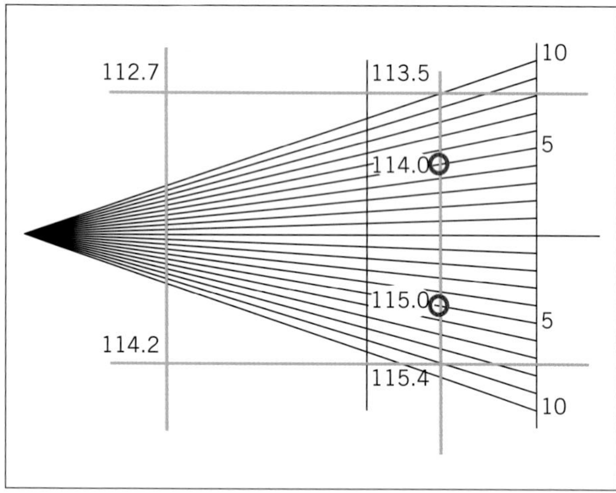

FIGURE 10.107 The radial graph used to divide a grid line into 19 parts and plot contours

(Figure 10.107), the grid line is divided neatly into 19 equal parts. It is then a simple count of five from the lower RL towards the higher to find the first contour (114 m), 10 more and the second contour (115 m) is found.

Mathematically

From the previous examples we know that there is a 1.9 m difference in vertical height between the 113.5 and 115.4 grid points. Removing the decimal point allows us to think of 19 equal-sized parts between these two grid points, where each part is 100 mm in height. We make this choice based upon what is a logical number of parts we can divide the grid line into without getting lost in large numbers.

Figure 10.108 shows that the hypotenuse of a triangle allows us to easily proportion the triangle's base graphically based upon this division of its height. If we can do it graphically, we can do it mathematically (Figure 10.109).

Drawing the contours

Using any one of the approaches described above, the location of each contour's intersection with a grid line is plotted. The contour lines are then drawn as clean

EXAMPLE 10.1

If we accept that the grid squares are drawn at 50 mm on a side, the base of our triangle then becomes 50 mm long. The 'rise' or height becomes 19. We can see from Figure 10.108 that the 19 parts of the base line are each larger than those of the 'rise'. Their size is found by:

50 mm ÷ 19 = **2.632 mm**

The 114.0 m contour strikes the grid line 5 parts from the 113.5 grid point towards the next grid point (115.4).

So:

5 × 2.632 = **13.2 mm**

The 115.0 m contour strikes the grid line 15 parts from the 113.5 grid point towards the next grid point (115.4).

So:

5 × 2.632 = **39.5 mm**

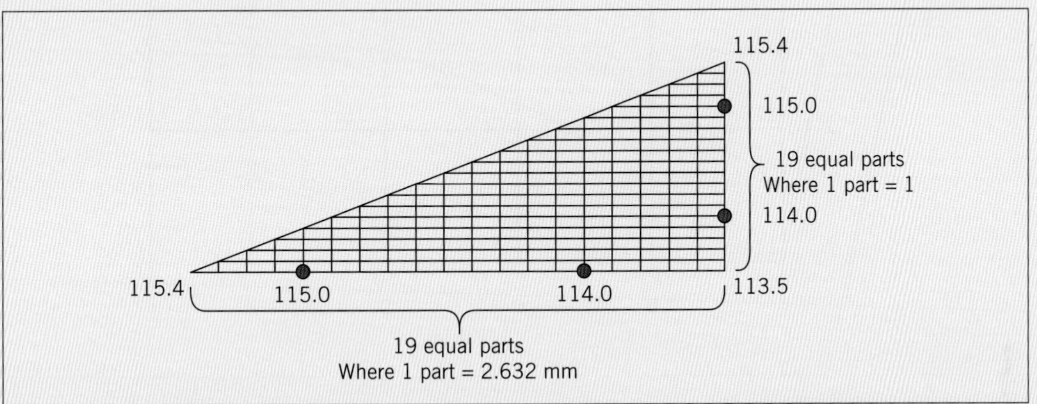

FIGURE 10.108 Graphical representation of proportioning the grid line based upon the vertical separation of two grid points

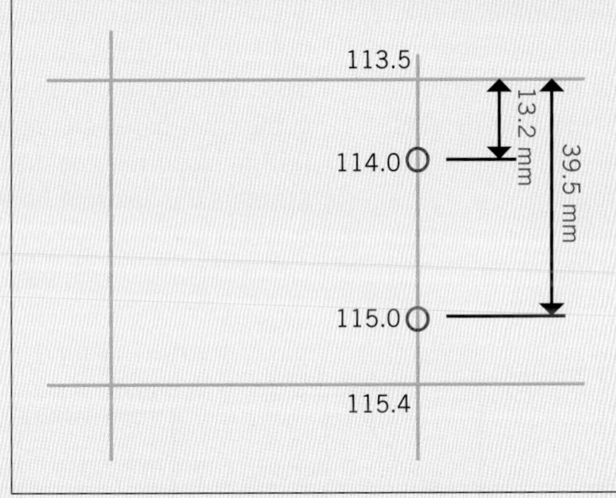

FIGURE 10.109 Mathematically plotting the contour intersection with the grid line

curves passing through each of these identified points. In so doing, ensure that:

- each contour is complete in its length – entering and exiting the site logically
- no contour crosses another
- contours do not 'split' into two
- contours to not intersect in any other way with each other.

Figure 10.110 gives you an example of the how this will look. Take the time to look at where each contour passes through a grid line and understand why the 'dots' are where they are.

Determining volumes

The calculation of excavation and fill volumes through the use of end sections was discussed previously in the second section of this chapter. In this section, we look at calculating these volumes using spot heights and an understanding of topography and contours.

Volumes using spot heights

The use of spot heights to calculate excavation volumes is most useful in three instances:

- when large-scale earthworks are to be undertaken
- when a contour plan is already available for quotation purposes
- when the ground is undulating significantly over the excavation area.

The spot height approach uses a system effectively identical to that of contouring; i.e. the establishment

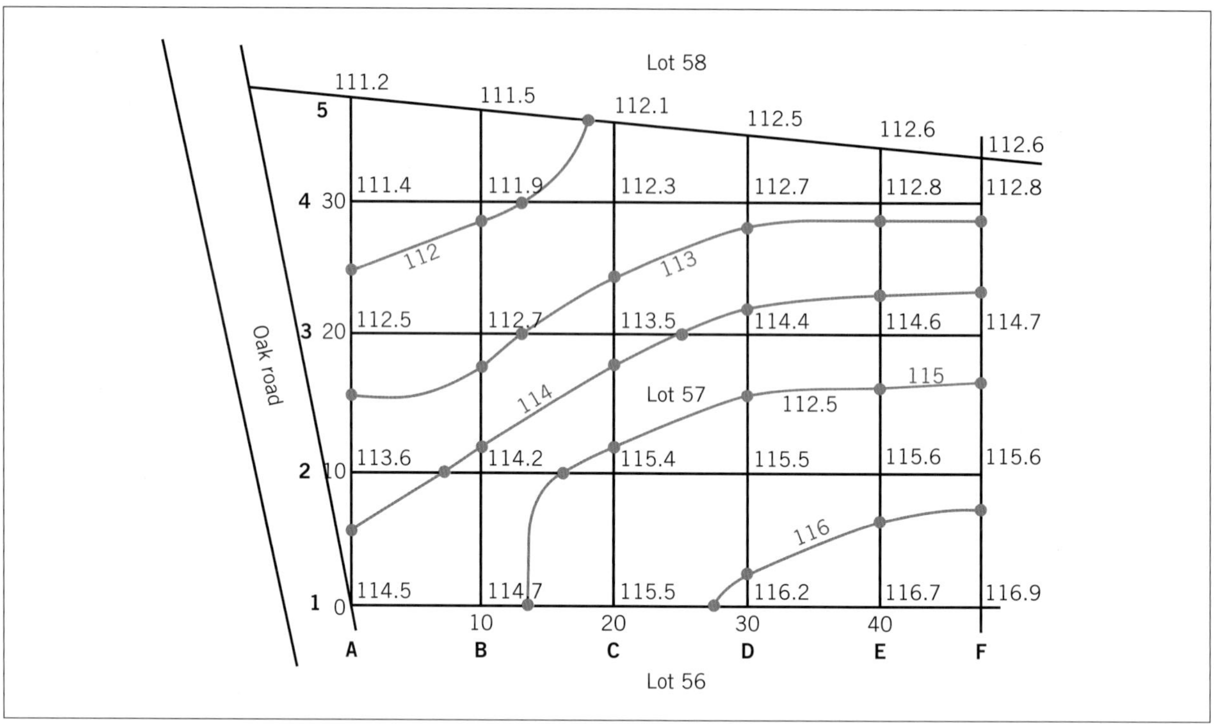

FIGURE 10.110 Contours plotted onto the grid

of a grid on the site, and then determining the reduced levels (RLs) for each grid position ***in relation to the platform level***. In so doing, spot heights are then known for each corner of the square or rectangular blocks demarking the excavation. These four heights are then averaged as shown in Figure 10.113, and the volume for each block or prism is found.

Interpolation is used when the edge of the excavation crosses a grid line, much as a contour line might. Figure 10.111 shows the proposed excavation area of a basic rectangular cut. In this instance the platform level of the cut is based on the top left or north-east corner (assuming north to be the top of the page). This is therefore the first spot height that needs to be found.

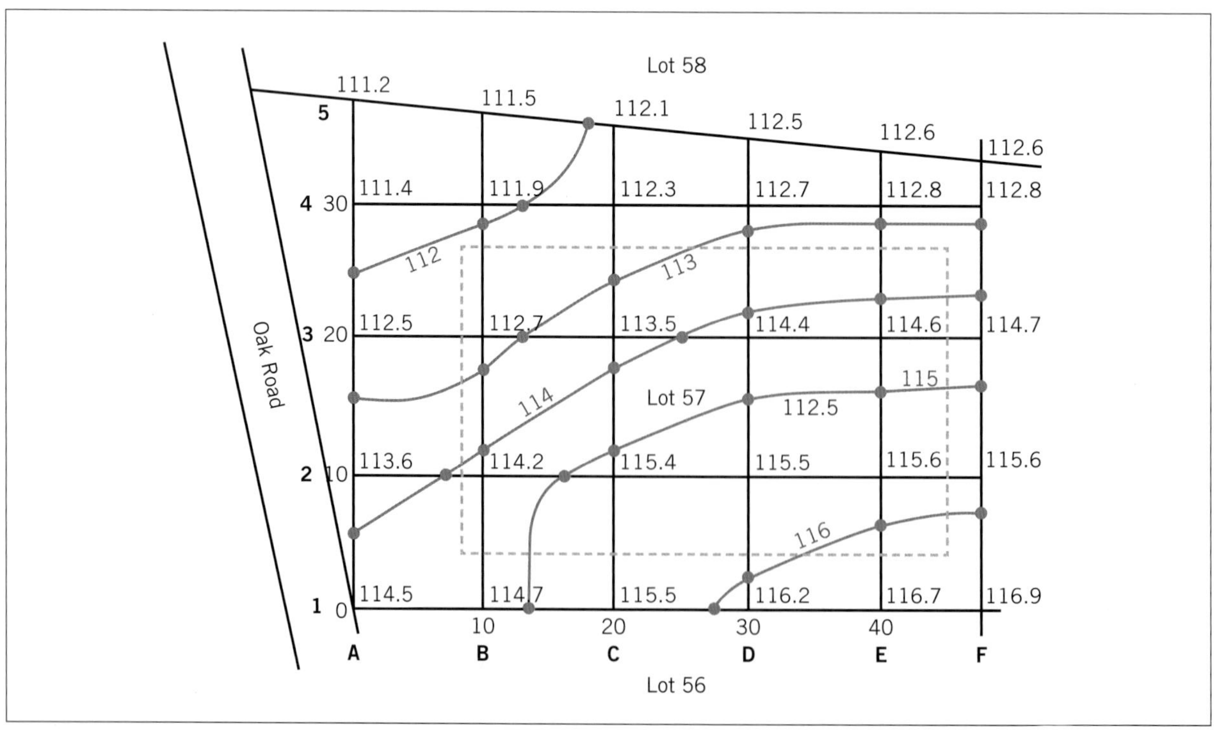

FIGURE 10.111 Excavation plotted onto the grid

The detail image in Figure 10.112 shows one way to find the spot height of an excavation corner using an existent contour map. In this instance, a red line is drawn from the 12-m contour through the corner of the proposed excavation to a logically selected grid point. The short black line shows that the distance from this contour to the excavation corner is about one-tenth of the length of the red line. The height difference between the 12-m contour and the grid point (12.7 m) is 0.7 m. So, the excavation corner spot height is 12.07 m, found by:

$12 + (0.7 \div 10) = 12.07$ m

This is done for all corners and areas where the excavation crosses the grid lines. In this manner, all sectors of the excavation have a square or rectangle for which all corners have a known height relative to the proposed platform.

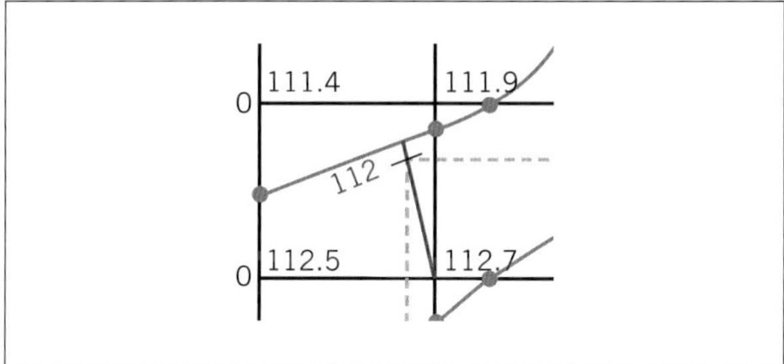

FIGURE 10.112 Interpolation of the excavation corner spot height

Frequently, no contour map is available and these spot heights will be established on site. Doing so is the same, however, as developing a contour map; i.e.:

- the establishment of a grid of appropriate size
- conducting a survey so as to identify and record the RLs of each grid point
- relate these RLs to the proposed platform height.

Calculating the volume

The volume of each block or prism of the excavation may now be calculated. This is done by averaging the heights of each prism, as shown in Figure 10.113.

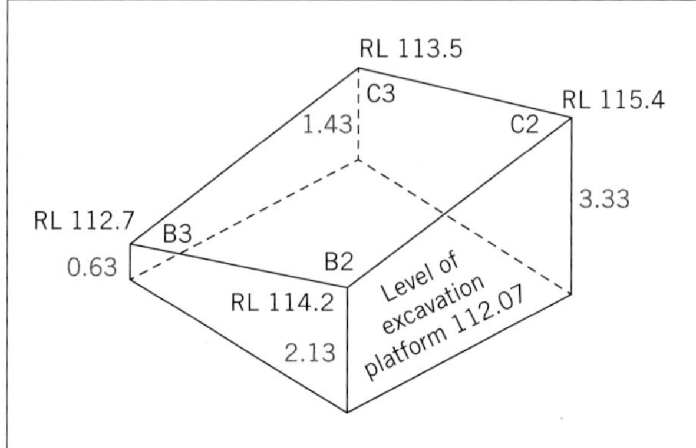

FIGURE 10.113 Heights of the corners of the excavation prisms

The heights above the excavation platform (shown in red) are found by subtracting the platform height from the known RL for the chosen grid point.

For example:

Prism corner height = spot height – platform height

Prism corner height = 115.4 – 112.07

= **3.33 m**

This is done for all RLs on the grid, and so all corners of the prisms, as shown in Figure 10.113.

Note: When spot heights are collected in the field it is possible to set the datum or zero point at the known excavation platform height. The heights surveyed above this point may then be taken without any need to subtract RLs.

Allowing that the grid was based upon a 10.0 m square, the volume of the prism shown in Figure 10.113 may be calculated as follows:

Volume of prism = Average height of the prism × prism base area

Where average height of prism is found by adding all the corner heights and dividing by 4; e.g.:

Average height = (0.63 + 1.43 + 2.13 + 3.33) ÷ 4

= 1.88 m

Volume of prism = 1.88 × 10.0 × 10.0

= 188.0 m^3

The calculation is then conducted for each prism in the excavation area and the results added together to give a total excavation volume that is highly accurate. It is, however, a rather long procedure. This may be simplified by means of a spreadsheet and the recognition that some heights are used up to four times depending upon where in the excavation the prism is located. This is demonstrated in Figure 10.114.

For this example, Figure 10.113 is taken as the total excavation area and the excavation platform RL as 111.5. Finding the total volume may be expedited by the use of a table or spreadsheet system as shown in Table 10.1. This table allows us to find the average depth of the excavation for the whole site, rather than the sectors. This may then be multiplied by the

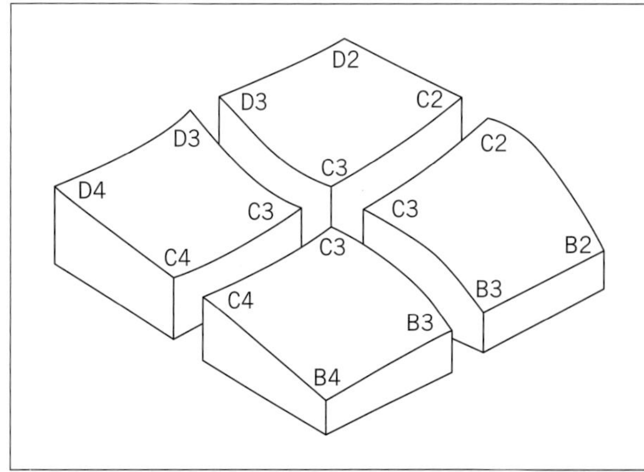

FIGURE 10.114 Spot height duplication in excavation prisms

TABLE 10.1 Spreadsheet to determine total collective excavation depth

Grid point	RL	Ex' platform RL	Spot height (ex' depth)	Number of uses	Collective excavation depth
B2	114.2	**111.5**	2.7	1	2.7
B3	112.7	"	1.2	2	2.4
B4	111.9	"	0.4	1	0.4
C2	115.4	"	3.9	2	7.8
C3	113.5	"	2	4	8
C4	112.3	"	0.8	2	1.6
D2	115.5	"	4	1	4
D3	114.4	"	2.9	2	5.8
D4	112.7	"	1.2	1	1.2
			Totals	**16**	**33.9**

area of the whole site and so the total volume of the excavation is quickly found.

Average depth of excavation = total collective excavation depth ÷ total spot height uses

= 33.9 m ÷ 16

= 2.119 m

Total area of site (10 m grid) = 20 m x 20 m

= 400 m^2

Total volume of excavation = 847.6 m^3

This figure would then be multiplied by the soil bulking factor appropriate to the site.

Longitudinal sections: pipelines and drainage

Drainage and other pipe works, such as sewers and waste lines, will require some sort of vertical section of the ground to be undertaken. This is partly for the purposes of construction, partly for the purpose of plotting the completed or 'as constructed' works. These generally fall into one of two categories:

- cross-sections
- longitudinal sections.

Longitudinal sections particularly show the following information:

- the profile of the existing surface
- the profile of the proposed works

LEARNING TASK 10.3 ESTIMATING BULK EXCAVATION FOR EARTHWORKS

Using the RL readings in the table, carry out the required calculations to determine the amount of spoil that needs to be removed from the area (shown dotted) that requires bulk excavation to be carried out. The required RL at the bottom of the excavation needs to be level at all points to achieve an AHD 34.760.

The readings taken at the surface are based on a benchmark (BM) of AHD 39.470 (RL 10.000)

Using the table below, showing the readings recorded at the site, calculate the anticipated amount of spoil to be removed.

The shaded area has been measured as 30.0 m × 31.5 m. Do not allow for any bulking allowance at this stage.

Hint: Set up an Excel spreadsheet and create formulas to speed up the calculations process, you will then just need to enter the correct data – your spreadsheet will become a template for you to use in the future.

The nominated grid-based RL readings, taken at the GL surface where excavation is to take place, are based on a benchmark of AHD 39.470 and the recorded readings are given in the table based on a nominated datum of RL 10.00.

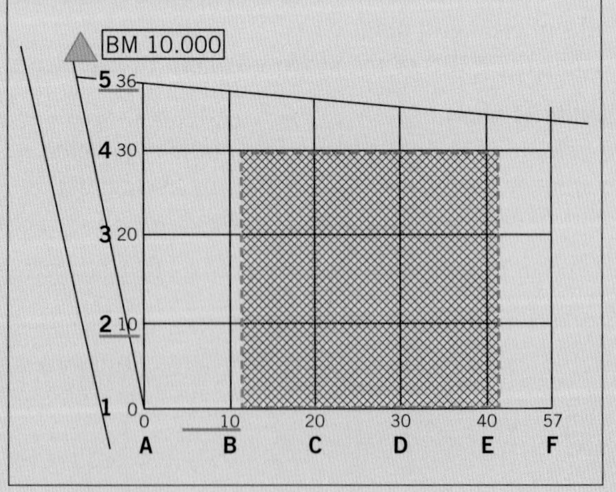

B1	10.450	C1	11.500	D1	11.900	E1	13.650
B2	10.570	C2	11.800	D2	12.300	E2	14.400
B3	10.750	C3	11.180	D3	12.650	E3	13.700
B4	10.900	C4	11.250	D4	11.800	E4	13.400

- the distances between points along the proposed works
- the gradient of the proposed works (pipeline) by way of invert levels
- the location of cross-sections.

Cross-sections will also show the existing surface and the location of piping, but are more useful in the determination of batter levels.

Drawing longitudinal sections

The surveying and drafting of longitudinal sections generally differ from normal sections in having different vertical and horizontal scales. The purpose of this difference is twofold:

- to accentuate the expression of the fall or 'grade' of the piping or base line of the trench
- to allow the overall length of the survey to fit onto the one drawing.

To develop the sections, the heights and horizontal distances are generally provided by way of *invert levels* and *chainages*. It is the invert levels that tell us the *grade* at which a pipe or drain rises or falls.

Invert levels

Longitudinal sections will generally include what are known as invert levels. An invert level is the reduced level (RL) of the bottom inside level of a pipe, sometimes referred to as the 'floor' level of a pipe. Most councils and water authorities will require these RLs to relate to the Australian Height Datum (AHD).

This is important levelling and installation information in pipe or drain laying to ensure adequate flow. What is adequate flow depends upon what is being drained; adequate being sufficient flow to ensure debris does not collect and foul or block the pipe or drain over time.

Chainages

The term chain line has been used once previously to describe straight lines in contour surveying from which offsets have been cast. Historically a 'chain' referenced an actual chain of specific length (22 yards) made up of steel links. While that tool is no longer used, nor the length it referred to, the concept of 'chaining', 'chain lines', and 'chainages', remains part of linear surveying terminology.

Chainage refers to a distance, chainages to multiple distances, generally of the same nominated length for that particular survey. For example, a chainage for one survey could be 10 m, for another 15 m.

In longitudinal sections, each chainage refers to the distance between invert levels and/or the location of cross-sections.

Grade

The grade at which a pipe or drain rises or falls refers to the angle of the pipe to the horizontal. A grade of 1 in 40 means that the pipe rises (or falls) 1 metre over a distance of 40 metres. Grade may be expressed in a number of ways including ratio, percentage, decimal, total fall, or an angle. Some of these expressions and their calculation will be explored more fully in the section following.

Drawing

The development of a longitudinal section begins with the creation of a table of data that includes the chainages (distance between survey points on the centre line of the drain) and, for each chainage:

- reduced levels of the surface
- invert levels of drain
- depth of trench or required fill of the excavation allowing for bedding material.

Above this table will be drawn a scaled diagram that shows this data graphically as demonstrated in Figure 10.115. Note the variation between the vertical and horizontal scales used to ensure a clear indication of the grade being depicted.

Once developed, the longitudinal section may be used to calculate excavation or earthwork volumes. It also acts as a guide to excavation contractors with regards to equipment choices and extracted volumes or required fill; in the two latter cases, this also allows for the estimation of any off-site disposal issues including transportation.

Note: With regards to drainage pits, the invert level indicates the inside surface of the bottom of the concrete pit. Where pipes enter the pit, the invert level is taken as where the inside bottom of the pipe would strike the centre line of the pit (see Figure 10.116).

Calculating and expressing grades

The grade of a pipe or drain may be expressed in a number of forms, the most common being as a ratio of rise to run as exampled in the previous section. Using that previous example of a 1 in 40 metre grade, we can explore how it may be expressed in three other commonly used manners:

- angle
- percentage
- total fall or rise.

Having explored these three, we will return to the ratio form to demonstrate how to develop invert levels for a chosen grade.

Grade as an angle

The grade at which a pipe or drain rises or falls refers to the angle of the pipe to the horizontal. A pipe that rises 1 metre over every 40 metres of horizontal run may be expressed as this angle and found through basic trigonometry, in particular the tangent function. The basic formula is:

$$\text{Angle of rise or fall} = \tan^{-1}(\text{rise} \div \text{run})$$

Using the 1 in 40 example:

$$\begin{aligned}\text{Angle of rise or fall} &= \tan^{-1}(1 \div 40)\\ &= \tan^{-1} 0.025\\ &= 1.43^\circ\end{aligned}$$

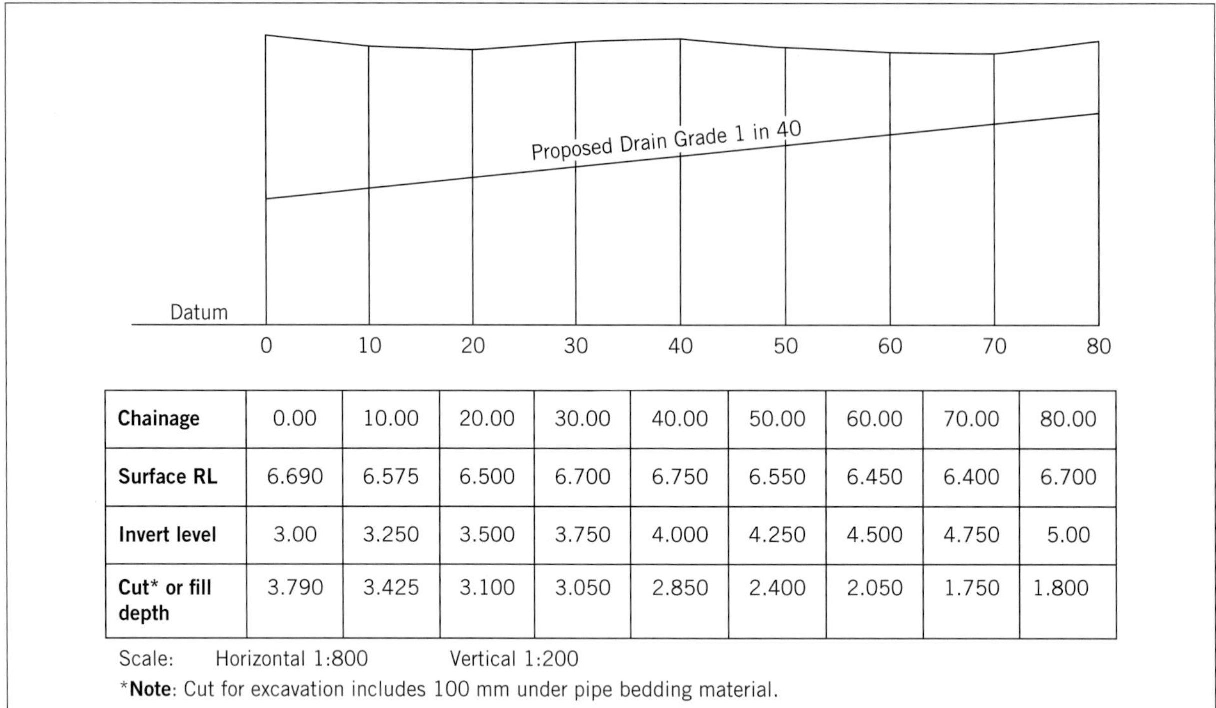

Chainage	0.00	10.00	20.00	30.00	40.00	50.00	60.00	70.00	80.00
Surface RL	6.690	6.575	6.500	6.700	6.750	6.550	6.450	6.400	6.700
Invert level	3.00	3.250	3.500	3.750	4.000	4.250	4.500	4.750	5.00
Cut* or fill depth	3.790	3.425	3.100	3.050	2.850	2.400	2.050	1.750	1.800

Scale: Horizontal 1:800 Vertical 1:200

*Note: Cut for excavation includes 100 mm under pipe bedding material.

FIGURE 10.115 The longitudinal section

Grade as a percentage

The grade of a pipe or drain may also be viewed as the fall or rise being a percentage of a specified distance. Taking our 1 in 40 example, we can express this as 1 metre being a percentage of 40 metres; i.e.:

Rise/fall as percentage of run = 100 × (rise or fall ÷ run)

Using the 1 in 40 example:

Rise/fall as percentage of run = 100 × (1 ÷ 40)

= 100 × 0.025

= 2.5%

Note: Grade may also be expressed simply as a decimal (sometimes referred to as a decimal percentage); e.g. 0.025.

Grade as total rise or fall

A further way of expressing rise or fall is the total rise or fall that a pipe undergoes over its total run, or to a specific point in that run where the grade may change, or where a pit is installed, or simply where the pipe ends. This is done using the invert levels of the two nominated end points.

Total fall = invert level high end point – invert level low end point

Using the high and low end point invert levels from the longitudinal section shown in Figure 10.115:

Total fall = 5.000 – 3.000

= 2.000 m

Total fall may then be compared to the total horizontal distance run of the pipe to ensure adequate fall has been achieved. This is usually based on one of the other expressions such as angle or ratio.

E.g. achieved ratio = total fall in total run

= 2 in 80

Or 1 in 40, which is the required fall in the example offered by Figure 10.115.

Levels and clearances for grades

Figure 10.115 offered a completed table and diagram for a typical longitudinal section. In that example the figures for the invert levels and excavation depths were provided without any explanation of how they were derived. As determining those levels is an important skill, we will discuss them here.

Calculating required invert levels for a given grade

Using the Figure 10.115 example, the calculation of the required invert level at each chainage is as follows:

Determine the invert levels at 10-metre chainages over 80 metres of gravity drain line.

- Initial (low end) invert is 3.0 m (relative to Australian Height Datum).
- Grade is 1 in 40.
- Minimum bedding material depth is 100 mm.

The rise over any given chainage is found by:

Rise = Grade × length of chainage

e.g.:

Rise = (1 ÷ 40) × 10 m

= 0.025 × 10 m

= 0.250 m

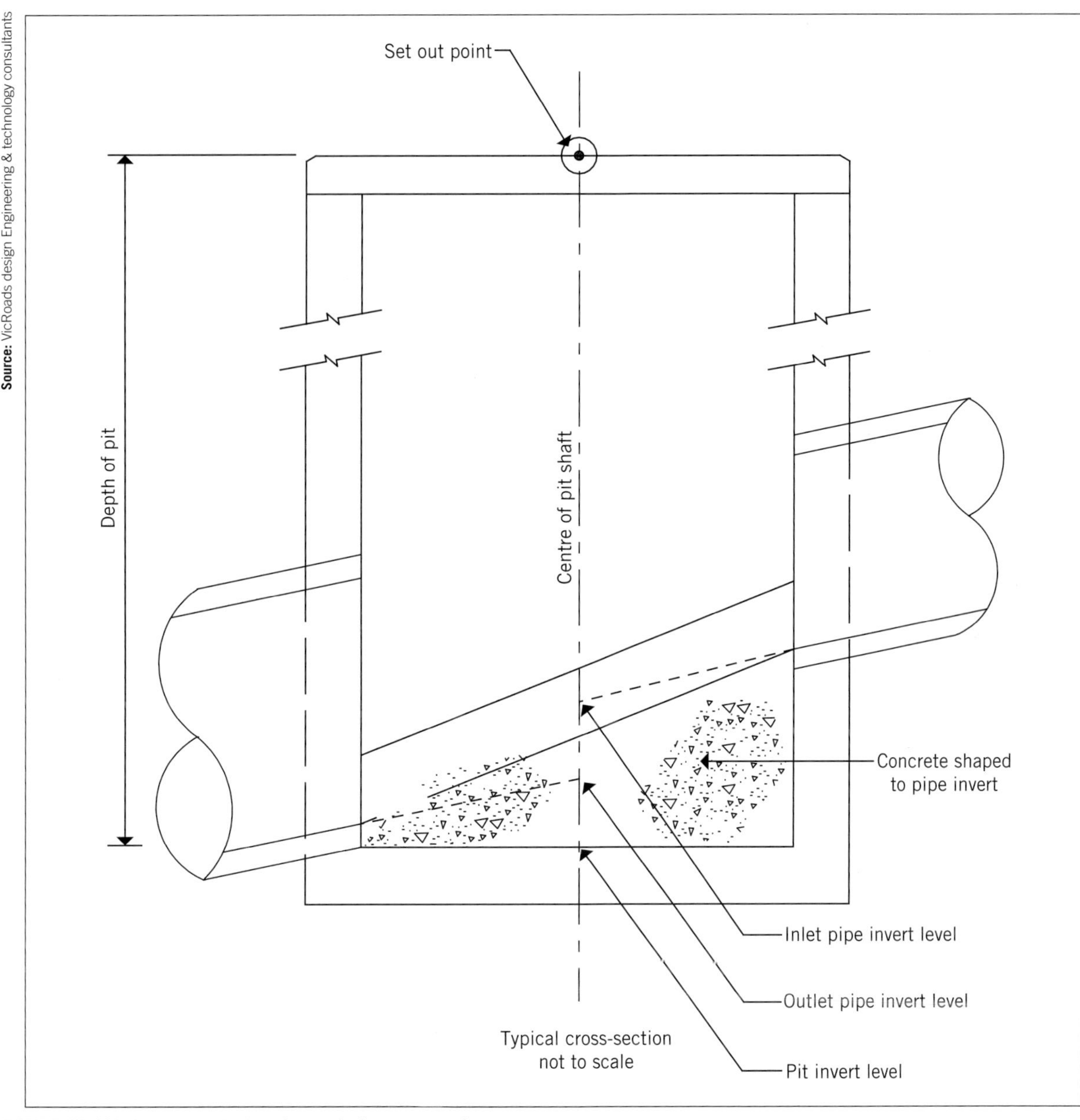

FIGURE 10.116 Invert levels for pits and pipe connections

The invert levels are then found cumulatively by:

Invert level at 'x' chainage = Previous invert level + rise over chainage

e.g.:

Invert level at 0.00 m (start) = 3.000 m

Invert level at 10.00 m = 3.000 + 0.250 = 3.250

Invert level at 20.00 m = 3.250 + 0.250 = 3.500

Invert level at 30.00 m = 3.500 + 0.250 = 3.750

And on ...

Calculating required excavation depths as reduced levels

These calculations are based upon the developed invert levels for a given grade and the reduced level for the surface at each chainage. In addition, they must include any requirement for bedding material that goes under the pipe or drain.

The calculation is as follows:

Excavation depth = surface RL – (invert level – depth of bedding material)

This calculation is then done at each chainage point.

EXAMPLE 10.2

Using the Figure 10.115 example:

Excavation depth at 0.00 chainage = 6.690 – (3.000 – 0.100) = 3.790

Excavation depth at 10.00 chainage = 6.575 – (3.250 – 0.100) = 3.425

Excavation depth at 20.00 chainage = 6.500 – (3.500 – 0.100) = 3.100

And on ...

Application of invert levels in the field: tolerances

In the field, absolute precision is not generally possible, and some level of tolerance must therefore be allowed for. This level of tolerance may be stated in any working drawings or associated specifications, or in a technical specification pertaining to any works undertaken in the jurisdiction of a given water or sewage authority. An example specification may be found for most states and territories online, such as STD-SPE-C-004 from ICON Water in Victoria; the source for the tolerances shown in Example 10.3.

Tolerances are generally specified by the type of infrastructure or piping being installed. The vertical tolerance (the only one we are considering here) will differ for a pressurised pipe vs a gravity flow pipe. A gas pipe, for example, may have an allowance of plus or minus 50 mm, whereas a gravity sewer or drain is likely to be more specific with a range of additional limitations.

EXAMPLE 10.3

The ICON Water STD-SPE-C-004 vertical location tolerances for gravity flow pipelines are as follows:

- 10 mm higher and 25 mm lower as applied to pipe invert level provided that:
- the grades specified in Table 7.2 are complied with and no localised low points exist.

The referenced Table 7.2 then provides further restrictions based upon the drain's grade stated as a percentage. For pipes with a grade less than or equal to 1% the pipe may run up to 10% steeper, but 0% (i.e. zero tolerance) flatter. For those with a grade greater than 1%, again there is zero tolerance in being flatter, but up to 15% steeper is allowed.

The purpose of these tolerances is clear. Gravity-fed pipes and drains may easily become blocked if the engineered grade is not provided, or if there are areas where the pipe flattens out and so allowing for debris to build up or adhere to the pipe or drain walls.

Calculating batter levels

The sides of excavations are prone to collapse and so designers will generally specify a 'batter', the angle for the excavation wall, as a preventive measure. As Figure 10.117 shows, the batter is specified as a ratio, much like grades, only with the horizontal first, and vertical as 1. The angle formed by these ratios is dependent upon a range of factors including:

- the type of soil involved
- the location of the trench to other buildings and infrastructure
- closeness of traffic or heavy machinery to the trench edge
- the depth of the excavation.

Batters are to some degree an expression of the angle of repose of a particular soil type. However, it is a mistake to impose such an angle too directly. Instead

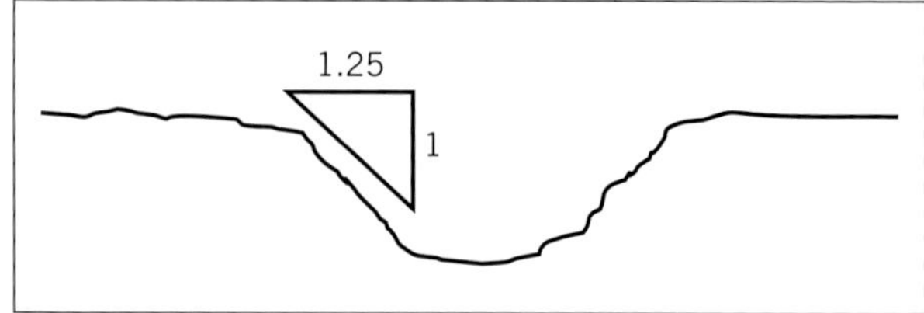

FIGURE 10.117 Batter ratio of 1.25 to 1: horizontal first, vertical always 1

most authorities, be they WHS, road and transport, or council will require a minimum batter of 2 (horizontal) to 1 (vertical). The Australian model code of practice for earthworks, for example, limits all batters to not more than 1 to 1 or 45°.

Batters also apply to fill when producing an embankment of some kind. Figure 10.118 offers a typical example of a path running from a cutting to a filled embankment. Sectional views are also offered (Figure 10.119) which demonstrate the shift in appropriate batter between cut material and compacted fill.

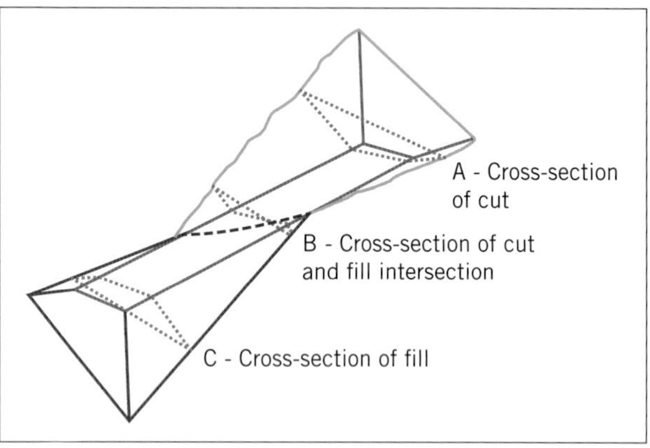

FIGURE 10.118 Depiction of path going from cut to fill and associated cross-sections

Calculating batter slope measurements

A range of measurements are required for the setting out and construction or excavation of earthworks and their associated battered sides. In combination with the engineer and/or designer of the earthworks, the following information may be found and/or developed:

- cross-sectional areas
- field data, including levels and measurements. This includes:
  - ground levels
  - formation levels (from longitudinal section)
  - formation width(s)
  - batter slopes
  - material type for cut and/or fill.

This data is used to determine the points at which the battered slope meets the natural ground line and so informs the development of the cross-sections. In most cases today, the cross-sections are developed using CAD programs which reduce the requirement for specific calculations. However, there are times when it is useful

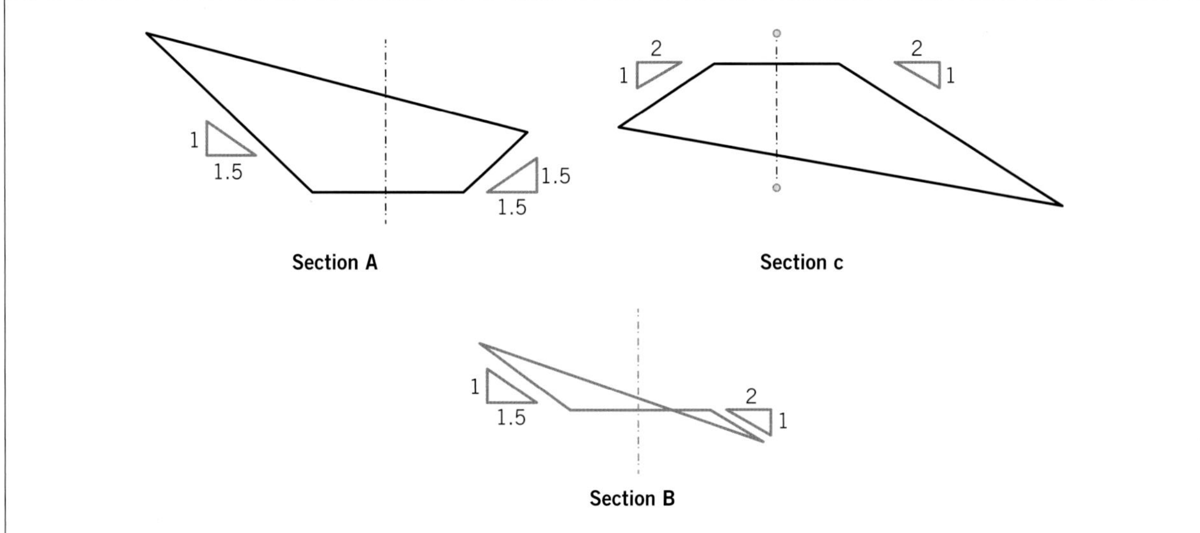

FIGURE 10.119 Cross-sections associated with Figure 10.118, showing batters

to know the basic steps for a manual calculation to be undertaken. The two examples below show how this might be conducted for both a simple excavation and a slightly more complex shape.

Simple excavation

Figure 10.120 depicts a simple uniform excavation. Batters are equal and the upper and lower surfaces are level and parallel.

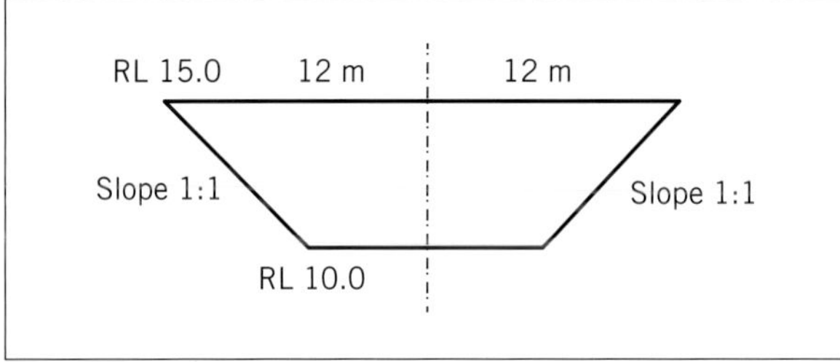

FIGURE 10.120 Cross-sections of basic excavation

The calculation for this is as follows:

EXAMPLE 10.4

Determine the height or depth of the excavation

Excavation depth = difference in RLs

e.g.: excavation depth = 15.0 – 10.0

= 5.0 m

Identify or calculate the horizontal distance of batters

With a batter slope of 1:1 or 45° we know that in this case the horizontal distance of the batter equals the vertical height.

E.g. horizontal run of batter = 5 m.

Calculate the width of the excavation base

Given that the width at the top of the excavation is 12 m either side of the centre line, the overall width is 24 m.

To find the width at the base we subtract the two batter horizontal runs, i.e.:

Base width = top width – horizontal batter run of both sides

= 24 – (5 + 5)

= 14 m

Calculate area of section

In this case we have a simple trapezoid. The formula for which is:

$$\frac{(A+C)}{2}\times D$$

where:

A is the long side of the trapezoid (top in this case)

C is the short side (base)

D is the depth or height of the trapezoid

In this case:

A = 24 m, C = 14 m and D = 5 m

The area of the cross-section is therefore:

$$\frac{(24+14)}{2}\times 5 = 95\ m^2$$

More complex shapes

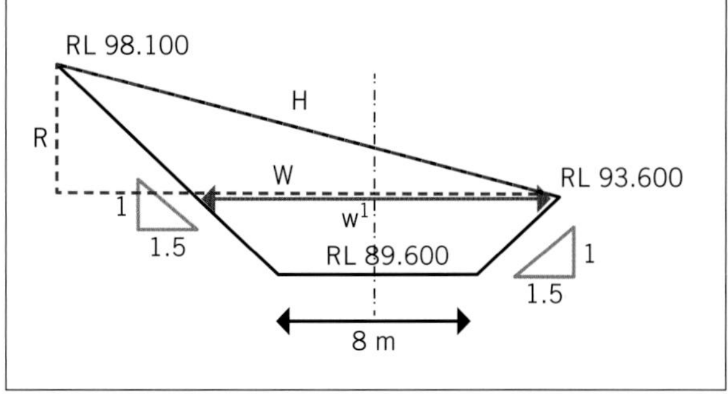

FIGURE 10.121 Cross-sections of a more complex excavation

EXAMPLE 10.5

Determine the height or depth of the excavation

In this instance (see Figure 10.121) we have two depths to find: the left- and right-hand batter vertical heights. The principle remains the same.

Excavation depth = difference in RLs

e.g.: excavation depth left = 98.1 – 89.6

= 8.5 m

Excavation depth right = 93.6 – 89.6

= 4 m

Identify or calculate the horizontal distance of batters

The batter slopes in this case are both 1.5 to 1. This allows us to calculate the horizontal run for each by multiplying the height by ratio factor of 1.5.

E.g. horizontal run of left batter = 1.5×8.5

= 12.75 m

Horizontal run of right batter = 1.5×4.0

= 6.0 m

Calculate the loping width of the excavation opening or top

The information we have gathered now allows us to give dimensions to the dotted red triangle superimposed on the excavation in Figure 10.121. The height of the triangle (R) is the difference between the right and left RLs.

I.e.: R = 98.1 – 93.6

= 4.5 m

The base width of the triangle (W) is equal to the width of the excavation base and the two previously calculated batter runs.

i.e.: W = 8.00 + 12.75 + 6.00

= 26.75 m

Using Pythagoras' theorem, we can find the hypotenuse of this triangle or sloping width of the excavation opening (H).

I.e.: $H = \sqrt{(R^2 + W^2)}$

$= \sqrt{(4.5^2 + 26.75^2)}$

$= \sqrt{735.8125}$

= 27.125 m

Calculate area of section

In this case we have two areas to find. The first a simple trapezoid as before, the other a basic triangle. For each we must first find the length of the top of the trapezoid (w^1). This is found by adding the horizontal run of the right batter to the base width twice.

$w^1 = 8.00 + 6.00 + 6.00$

= 20 m

We have now the dimensions of a simple trapezoid. In this case:

A = 20 m, C = 8 m and D = 4 m

The area for the trapezoid section is therefore:

$$\frac{(20+8)}{2} \times 4 = 56 \text{ m}^2$$

The second area, the triangle, is found by the normal triangle formula of ½ base × height. In this case:

Area = $0.5 \times 8 \times 4.5$

= 18 m²

The total area of the cross-section is then:

Total area = 56 + 18

= **74 m²**

LEARNING TASK 10.4 LONG-RUN TRENCH EXCAVATIONS: PUTTING THE THEORY INTO PRACTICE

1 Using the drawing and correctly booked RLs from the first Learning Task (p. 263), you now need to excavate for a trench that will be 0.8 m wide to various grade depths for a stormwater pipe that will be going through the existing property as an easement.

The original readings (see the first Learning Task in this chapter) taken were:

- BM is nominated to be at an RL of 14.000
- the following readings were taken:

L1: BM = 1980	L2: S5 = 1860	L3: S9 = 1360
L1: S1 = 1730	L2: S6 = 1255	L3: S10 = 320
L1: S2 = 1545	L2: S7 = 725	
L1: S3 = 2165	L2: S8 = 1570	
L1: S4 = 1390	L2: S9 = 440	
L1: S5 = 970		

Easement pipe trench excavation (the required work)

Specification

The BM. is now given as AHD 25.700, with anticipated grades depths at the stations where readings have previously been taken being:

- Invert of pipe depth at S1 is to be AHD 24.300 so as to allow connection to the street stormwater system in the future
- S1 to S5 rising grade of 1:40
- S5 to S10 rising grade of 1:10.

Respond to these tasks

a Convert the readings taken based on a datum given as 14.000 and convert to the AHD now stated as 25.700.

b Using a table similar to the one shown, determine the relative depths at each station that the excavation will need to finish at:

>>

>>

Trench width (m) =													
Grid distances – (m) =													
	S1	S2	S3	S4	S5	S6	S7	S8	S9	S10			
Chainage	0	5.00	10.00	15.00	20.00	25.00	30.00	35.00	40.00	45.00			
Surface RL – based on initial readings													
Straight line invert level (depth) at S1													
Grade required	S1 to S5 1:40					S5–S10 now 1:10							
Distance between points	Base	5.0	5.0	5.0	5.0	5.0	5.0	5.0	5.0	5.0			
Changed depth between points													
New invert level to incl. grade change											Total length	Trench width	Anticipated Vol: m^3 Excavation
Changed cut/fill depth at invert to incl. rising grade													
											Averaged depth at each station		

c Using the original sketch provided and the information in your table, draw a 'straight line' diagram of the levels at ground level, where readings S1 to S10 are taken from.

d Draw a straight line on the diagram to represent the invert level, based on S1 – without taking into account the nominated grades between points.

e On the same diagram, draw a 'straight line' diagram of the pipe invert level, allowing for the nominated grades in the specification, where readings S1 to S10 are taken from.

f Will any depth of the proposed pipe excavation 'invert' depth line be less than 600 below the existing ground line?

g What depth will the excavation need to be, in terms of AHD, if starting the trench at S10?

h What anticipated volume of spoil (without bulking) will need to be excavated from the 800-mm-wide trench based on the straight line heights between reading locations (S1 to S10) along the easement line?

2 Provide a reason as to why any part of the 300-mm-diameter pipe to be used in the easement trench might be above the existing ground lines that are currently shown in the sectional drawing given.

SUMMARY

Accurate set out and surveying is at the core of most contemporarily built structures. The purpose of this chapter is to provide you with the basic knowledge of how to undertake these two core activities on low-rise domestic and commercial buildings. These skills, and the knowledge developed that underpin them, will aid not only in on-site activities, but also in the essential quantity and estimation practices leading to successful tendering.

To reiterate the key areas covered:

- surveying equipment, their selection, use and maintenance
- set out and measuring techniques
- excavation volume calculations
- contour levelling and volume estimation
- development of longitudinal sections
- excavation grades and batters for pipelines and drainage applications.

With this knowledge, and an adequate period of on-site application and experience, you will soon find the setting out of most structures a rewarding, rather than daunting, task.

REFERENCES AND FURTHER READING

Intergovernmental Committee on Surveying and Mapping (ICSM) and Permanent Committee on Geodesy (PCG) (2014). Standard for the Australian Survey Control Network: Special Publication 1, v2.1, Canberra, available at: **https://www.icsm.gov.au/publications/standard-australian-survey-control-network-special-publication-1-sp1** (accessed 17 October 2019).

Laws, A. (2016), *Site Establishment, Formwork and Framing* (3rd edn), Cengage Learning Australia, Melbourne, Vic.

11

SIMPLE BUILDING SKETCHES AND DRAWINGS

Chapter overview

This chapter describes the process for capturing design concepts and details from architectural and engineering documentation in order to communicate that information to other members of the construction team. Builders, experienced trades people, project managers and estimators need to be able to interpret information contained within construction plans and provide simplified sketches and drawings to estimators, subordinate trades people and contractors for the purpose of facilitating the construction process. These simplified sketches and drawings can have a range of uses, such as estimating, explanation of construction details and demonstrating critical dimensions required for setting out and constructing components of the building.

Elements

This chapter provides knowledge and skill development materials on how to:

1. prepare to make sketches and drawings
2. create simple sketches and drawings
3. notate and process drawings.

To gain the most from this chapter, you must have access to the National Construction Code, Volumes One and Two. These codes are available for free download, after free registration from the following internet site: https://ncc.abcb.gov.au/ncc-online/NCC.

Download the following documents:

- NCC 2019, Volume One (Building Code of Australia Class 2 to Class 9 Buildings)
- NCC 2019, Volume Two (Building Code of Australia Class 1 and Class 10 Buildings)

The Australian standards that relate to this chapter are:

- AS 2870 Residential slabs and footings
- AS 3740 Waterproofing of domestic wet areas
- Australian standards set AS 1100.301–2008 (includes Amdt 1–2011) Technical drawing – Architectural drawing

Other references

Chapter 5 Plans and specifications

Introduction

Communication is a critical component of the building and construction process, and the preparation of sketches and drawings is a vital part of that communication skillset. Construction plans (the full set of drawings and specifications from which a structure is constructed) are increasingly becoming more complex and detailed; senior trades people and building supervisors need to be able to interpret the information contained within the construction plans and re-present it in a more simplified form for use by other construction team members.

There are a range of ways by which this may be achieved, from simple roughly proportioned sketches to neat hand-drawn scaled diagrams accurately dimensioned and notated. In each case, however, they will frequently be drawn on site, and often in front of the target audience. Such being the case, this is an important skill area that should be developed through practice over time; particularly being able to visualise and then impart images to other team members in a two- and three-dimensional (2D/3D) format.

To help you begin the journey, this chapter will workshop you through a case study example to demonstrate the process of preparing simple sketches and drawings, from starting to gather the required information, to the development of a final hand-drawn, yet scaled, drawing.

Preparing to make sketches and drawings

In order to complete the sketches and drawings that are required to communicate specific construction tasks, it is necessary to identify and collect the relevant information from the construction plans. For a sound introduction to plans and plan reading you should review Chapter 5 of this book, which outlines the various aspects of a construction plan set.

As that chapter states, a critical part of the communication process is the confirmation that proposed building details comply with relevant building regulations and standards. At times you may find that creating a basic sketch is the means by which this confirmation is obtained, or at least clarified.

In addition, the format, context and the relative experience of the intended audience also need to be carefully considered when preparing sketches and drawings.

Identify and select relevant information from construction plans

The construction process has become steadily more complex over the past decades, involving large teams of trades people and contractors to complete even standard volume-built homes. Builders, experienced trades people and project managers are increasingly required to break projects down into smaller trade-specific tasks for delegation to junior trades people and contractors.

Interpretation of the information contained within the construction plans is the critical first step to breaking down the project into specific tasks. The required information may be spread across multiple locations, such as: the National Construction Code (NCC), specific Australian standards, architectural drawings and specifications, engineering drawings, soil test reports and the like. Sometimes only one or two of these documents will be required to produce the drawing you require. On other occasions, many will be required, with the information amalgamated into one simplified sketch/drawing to be provided to an estimator, tradesperson, or contractor.

Figure 11.1 provides a typical floor plan for a contemporary Australian home. The companion drawing, Figure 11.2, offers the engineering specification for the concrete slab of the same home. These are the drawings from which our example sketch will be developed. In addition, there is Figure 11.3; this provides the timber flooring specification relevant to the element we need to describe.

The case study

There is a Learning Task at the end of this section that revolves around the figures mentioned above. This activity effectively outlines the case study that you will workshop throughout the remainder of this chapter. The focus of this case study is the step, or 'set', down required between the concrete level in the ensuite and the floor of the remainder of the house (see Figure 11.4).

This setdown in the slab is required because although the floor levels are shown to be the same – both show finished floor levels (FFLs) of 20.00 m – they have different floor finishes. This is identified in two ways:

From the floor plan (Figure 11.1):

- FF01 for the main floor
- FF02 for the ensuite floor.

This tells us that there is a difference between the two floor finishes.

The second piece of information is in the specifications (Figure 11.3). This tells us that:

- the main areas of the house have solid timber flooring (tallowood)
- the ensuite, bathroom and laundry are tiled.

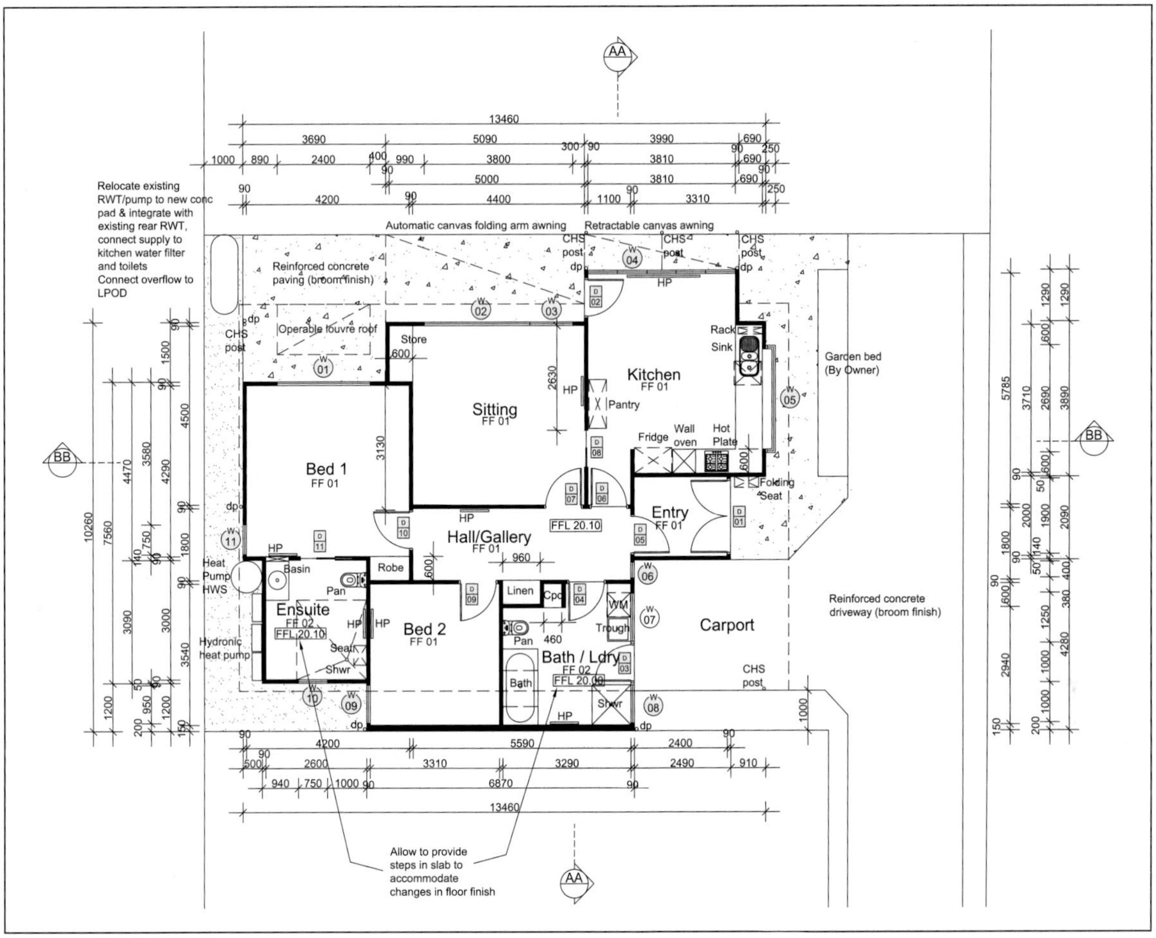

Source: P.J. Yttrup & Associates Pty. Ltd

FIGURE 11.1 Example architectural floor plan

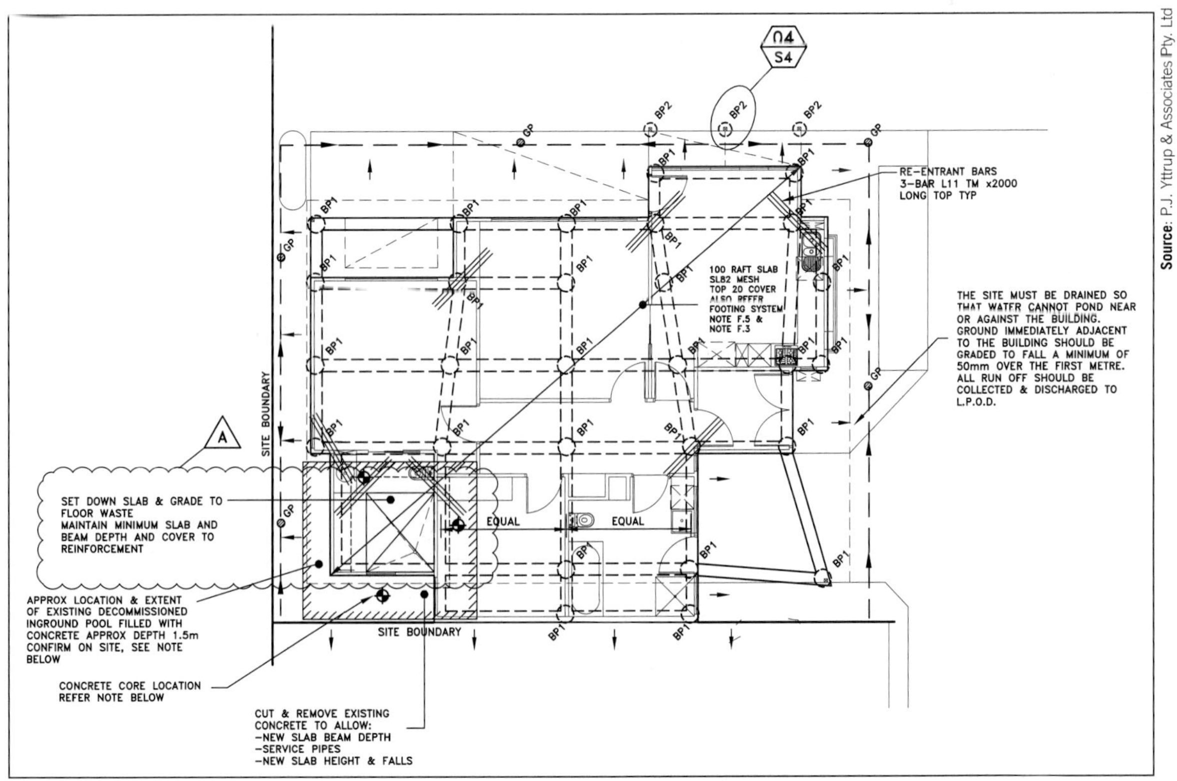

Source: P.J. Yttrup & Associates Pty. Ltd

FIGURE 11.2 Example structural engineering drawing

Source: P.J. Yttrup & Associates Pty. Ltd

EXTERNAL CLADDIIG		SUPPLIER	LOCATION	
EC01	Material: Colorbond Custom Orb Colour: Pale Eucalypt	Supplier: Roof Plumber	Roof	
EC02	Material: Wealhertex Ecogroove 150 Colour: White	Supplier: Bunnings Torquay Ph: (03) 5261 1800	Refer Architectural dwgs	
EC03	Material: Colorbond Custom Orb Colour: Evening Haze	Supplier: Roof Plumber	Wall cladding	
FLOOR FINISHES				
FF01	Type: Tallwood overlay Hard wood floor Size: TBC Finish: Refer Applied Finishes	Supplier: Timberzoo	Hall, Bed 2, Sitting, Entry, Kitchen	
FF02	Type: Floor Tile Hawkesbury sandstone 600×300	Supplier: Elegance Tiles Ph: Ph: 5241 2271	Ensuite, Bath/Ldry	
JOINERY		**SUPPLIER**	**LOCATION**	
AS = Adjustable Shelves, CD = Cupboard Fronts, DR = Drawer Fred, FP = Filled Panel, JN - Joinery Carcass, OS = Open Fixed Shelves				
CD01	Material: Re-use existing cubpboard door	Supplier: Owner	Linen	
CD02	Material: Redcote Colour:TBC Finish: Paint	Builder	Robe, Broom	

FIGURE 11.3 Example architectural specification excerpt

What the specification doesn't tell us is how thick each of these flooring elements are. The complexity is that timber overlay flooring on concrete slabs can be completed in a number of different ways: direct fixing, glued, or batten fixed. In either case there is going to be a difference in thickness between the timber system and tiles; hence the requirement for a setdown (Figure 11.4).

In addition to the above, when reviewing the engineer's drawing of the slab, it is apparent that there is a stiffening beam passing under the location of the setdown. Potentially, a review of the geotechnical engineer's report may also bring further important information to light which could alter the slab design.

This information too, must be included in the sketch to be generated.

Source: P.J. Yttrup & Associates Pty. Ltd

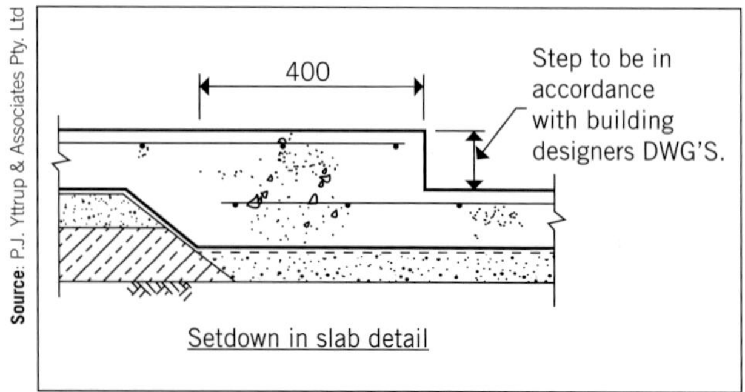

FIGURE 11.4 Structural engineering drawings construction detail

Filling the information gaps: identifying those who know

Identifying those who can give you clear and accurate answers to questions arising from the specifications or plan set is usually not particularly difficult. Sometimes it simply takes a call to the architect or designer, at other times a supplier or tradesperson may provide the information. In this case, the best way to determine exactly what the variation in height between the two floor finishing materials, is to talk to the contractors concerned; i.e. the tile layer and the floor fixer.

For the purposes of our case study example, the following information shall be said to have been obtained from these contractors:

FF01 Flooring system: solid timber on battens

- flooring is 19 mm thick
- battens 70 × 35 mm laid on the flat (i.e. 35 mm thick)

FF02 Flooring system: tiles direct fixed

- tile and glue combination: 20 mm thick

This is the information that will need to be provided to the concreter by way of a sketch so that they may create the setdown as required. Review the Learning Task below before reading on to the next section.

LEARNING TASK 11.1 SIMPLE SKETCH

Prepare a simple sketch, based on the engineering drawing detail (Figure 11.4), describing the required concrete slab construction detail, which can be used to communicate to the concreter so they can calculate the required materials and prepare the on-site set-out. The sketch should describe the height of the required setdown to accommodate the proposed timber overlay flooring based on Figures 11.1 and 11.2.

Confirming construction plan information is compliant

Confirmation that construction plan information complies with building and construction regulations and codes is a fundamental component of the building construction communication process. This process requires an implicit understanding of the appropriate regulations and the ability to analyse the information contained within construction plans.

Identifying the relevant codes and standards

Chapter 1 of this book offers a clear introduction to the standards and codes applicable to construction in Australia. In addition, that chapter provides advice on how to navigate the National Construction Code (NCC) for the purposes of identifying specific requirements and/or particular codes relevant to each aspect of the construction. If you have not done so already, you are advised to review that chapter before proceeding further.

The case study example used here is predominantly about the correct formation of a concrete slab: specifically, a step in that slab. The typical reference standard for domestic slabs of this nature is AS 2870. Figure 11.4 shows that step or setdown by way of a sectional elevation. This sectional view also provides the designer's proposed location for the reinforcement. The reinforcement lap indicated in this diagram can be confirmed by reference to AS 2870, which details the accepted lapping requirements, as shown in Figure 11.5.

As noted when first outlining the case study, the step in the slab occurs over a stiffening beam not shown in the engineer's sectional detail offered in Figure 11.4. The dimensions of this beam need to be confirmed from the same standard (AS 2870) or the NCC, Volume Two, despite being provided in another detail found among the engineer's specifications (see Figure 11.14). In addition, upon checking the geotechnical report it is noted that concrete piles are required to be provided to a depth of 1.5 m below the stiffening beam in order to reach appropriate foundation bearing pressure. The dimensions of these pilings will also need to be checked according to AS 2870 and AS 2159.

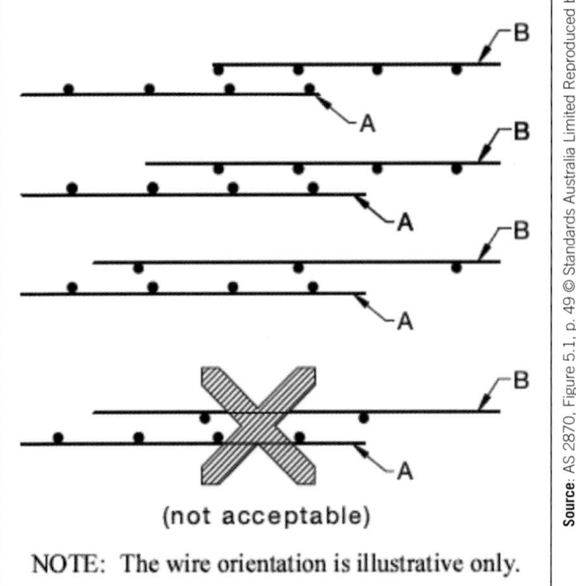

FIGURE 11.5 AS 2870 excerpt for reinforcement lapping

Other sections of AS 2870 that should be checked include:

- the minimum concreted cover to the reinforcement for indoor locations
- slab thickness
- required compressive strength of the concrete.

Our case study is a fairly simple exercise and so in this instance no other standards are of particular import. In more complex situations you may need to look at multiple standards and/or multiple parts of the NCC to confirm that all is as required.

Considering purpose, presentation and the intended audience

The primary purpose of preparing simple building sketches and drawings is to synthesise the information contained within construction plans and convey that information to the trades people and contractors working on a particular project.

The purpose

In larger building companies, builders, experienced trades people and project managers may also be required to prepare simple sketches and drawings for the purposes of estimating quantities of materials and labour required for a specific task. Variations to building contracts will also require builders to provide costings for proposed changes to the building design; this process may require the production of simple building sketches and drawings in order to accurately describe the revised task to estimators, trades people and contractors.

In the case study example, the purpose is to clarify the depth of the setdown in the concrete slab so that formwork may be established accurately for the concrete pour. Likewise, to clarify the correct positioning of the steel mesh reinforcement. The sketch will then be used for the instruction of possibly three distinct parties – the concreter, the formworker(s) and the steel fixer.

The audience

Who the audience is can at times be critical to the manner in which the information is to be imparted. This is in part due to that audience's skill level and/or experience with the reading of plans and sketches, and in part due to their knowledge and experience in the area of work to which the sketches relate. In some instances, the audience will need further guidance on the reading of the original plan set from which the sketch derives. On other occasions, they may need further sketches before they grasp the full implications of what it is you are trying to impart.

In other instances again, it could be that you are sketching for a draftsperson or even the architect in an endeavour to get a specific point across. In such cases, you may well want to produce a drawing of a higher standard so as to not have the sketch dismissed without the consideration it requires. It is important to understand drawing, even sketching, as a language – one that can influence people simply through how it is expressed or presented. And so, like a language where words can have less meaning for the audience than the expression, drawing can influence positively or negatively through its presentation.

The presentation

Despite being simply a sketch, there are a number of ways by which you may wish, or be required, to present it to ensure the information is being understood. As stated previously, your choice may also be influenced by the intended audience.

The simplest form of sketch is a two-dimensional (2D) diagram offering a plan, elevation, or sectional representation of the area or object(s) concerned (see Chapter 5 for an explanation of these terms). Such diagrams can be quickly hand drawn without instruments, and need not be to scale, but rather 'in proportion': in proportion meaning that lengths and heights 'look' right, but are not exactly measured. No scale is assigned to these drawings.

The next form is also not to scale but attempts a three-dimensional (3D) perspective of the area or object. These can also be drawn freehand (without instruments) and frequently are used to provide the viewer with a clearer perception of particular details. This form of sketch usually borrows from the isometric or oblique styles (see Figures 11.6 and 11.7).

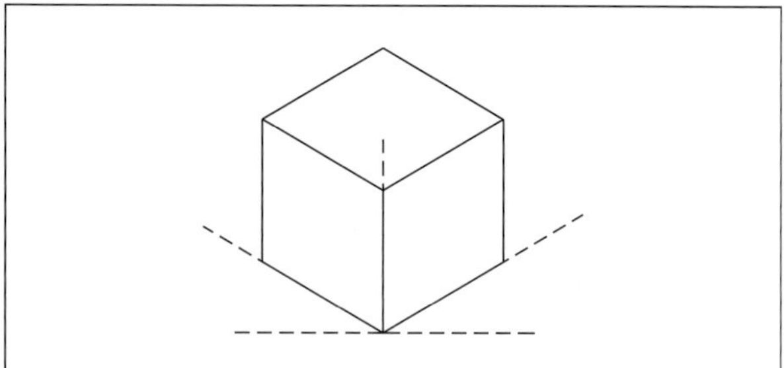

FIGURE 11.6 Isometric

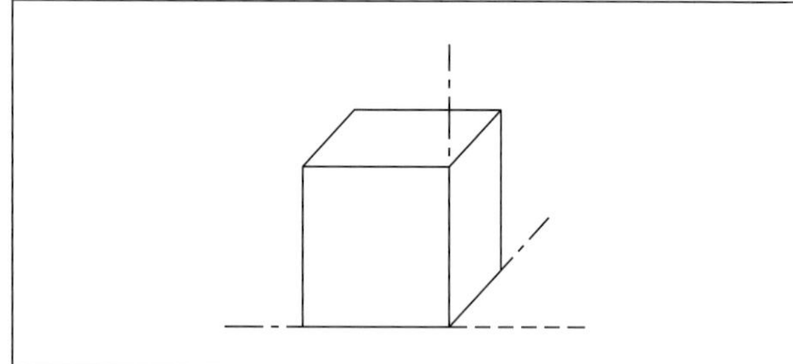

FIGURE 11.7 Oblique

Either type of sketch, 3D or 2D, despite not being to scale, can have dimensions added to give the important information required.

The next 'level' of sketching is where the drawing is given a defined scale and all lengths are drawn accordingly. As with the main plan set, there are 'typical' or common scales that you should use. These include 1:2, 1:5, 1:10, 1:20, 1:50, 1:100, and so on. The scale you choose is dependent upon the size of the object or area of the building you need to draw.

Scaling of three-dimensional sketches is also possible and sometimes preferred, though it is frequently unnecessary as you are simply trying to impart a general representation rather than an exact image.

Scaled sketches will almost always have dimensioning detail provided, and may also be given a title, along with other information identifying the drawing, who drew it and when.

Determining format and inspecting equipment

The development of an easily replicated process for preparing sketches is a valuable management tool for building professionals, especially when sketches need to be prepared on site where facilities and conditions are often limited and rudimentary.

Pre-printed, pro-forma site instructions which include scaled grids can be used. Figure 11.8 shows such a sheet, where each box may represent any scale of your choosing: e.g. 500 mm at 1:100 scale, 250 mm at 1:50 scale or 50 mm at 1:10 scale, and

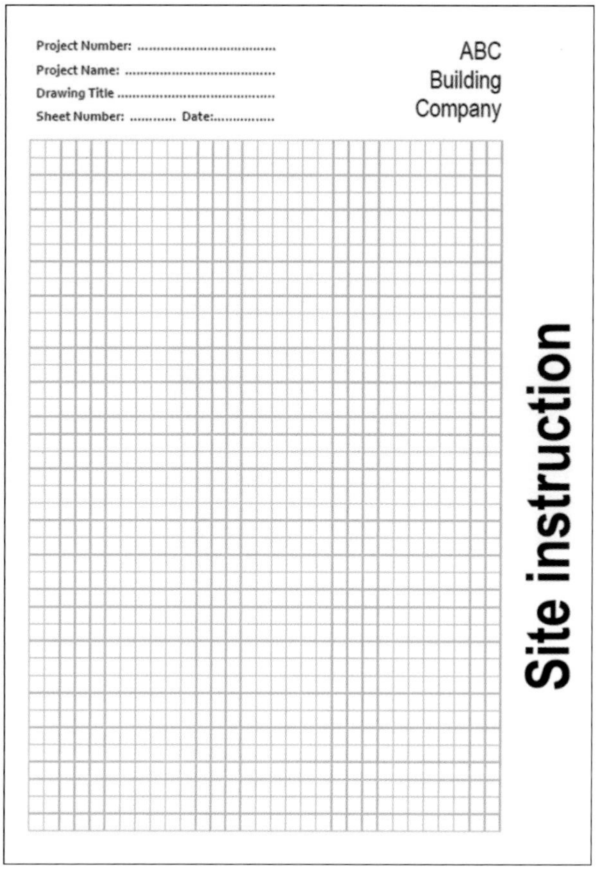

Project Number: ....
Project Name: ....
Drawing Title ....
Sheet Number: .... Date:...............

ABC
Building
Company

Site instruction

FIGURE 11.8 Example of pre-printed/scaled site instruction

so on. This helps to facilitate the rapid scaled production of simple sketches and drawings without instrumentation.

Another form of drawing, not mentioned earlier, is the making of a direct copy of a portion of the construction plans. This is usually done through the use of tracing paper to produce quick 'overlay' sketches. These may then be reworked to give the required information before being scanned and printed when such equipment is available in the site office.

Ideally, sketches should be produced on an A4 sheet size format so they are readily replicated using this same on-site office equipment.

Digital sketching applications

With the advent of digital technology there are also many phone and tablet-based digital applications (apps) that can be used for on-site sketching, estimating and communication (Figure 11.9).

These digital apps often originate from international sources and so may be using imperial measurement systems as their default setting. In such cases the settings will need to be adjusted to suit the Australian system of metric measurement.

A range of products are available, some for free, some needing to be purchased, some simple while others are quite powerful, such as Google Sketchup (Figure 11.10). This program can be used on site via a laptop or tablet to produce simple 3D drawings remarkably quickly once you have become proficient with the system.

In addition, digital laser measuring devices have also found their role in the building industry as their accuracy has increased and their price has rapidly decreased, making them more accessible. The value in using such equipment in measuring long or numerous distances is that in some instances where compatibility is stated, the measurements found can be automatically input into the drawing app using Bluetooth or wireless connections. In choosing to adopt such instruments it is important to ensure they are frequently checked for accuracy by either the supplier, the manufacturer or by standard measuring devices such as tape measures.

Creating simple sketches and drawings

Once you have identified the task, what you need to draw and for whom, the methodology, the form of drawing and the system you intend to use, it is possible to commence the preparation of the proposed sketches. The basic information needed to prepare the sketches is likely to come from various sources including architectural and engineering documentation, and occasionally site measurements. This will then be compiled into the required sketches using recognised drawing conventions at an appropriate scale and format.

Establishing and recording the information required

As discussed above, the required information may be located in different parts of the construction plans. These may include architectural drawings and specification, structural engineering drawings and the like. In addition, depending upon your role in the construction project, you may also need to source information from builders, experienced trades people, project managers or estimators, depending upon the complexity of the sketch, or the elements requiring to be drawn.

Compiling the information

In the example case study being discussed, it has been recognised that a sketch needs to be provided to the concreter so that they may accurately set-out the required stepdown from the ensuite to the remaining

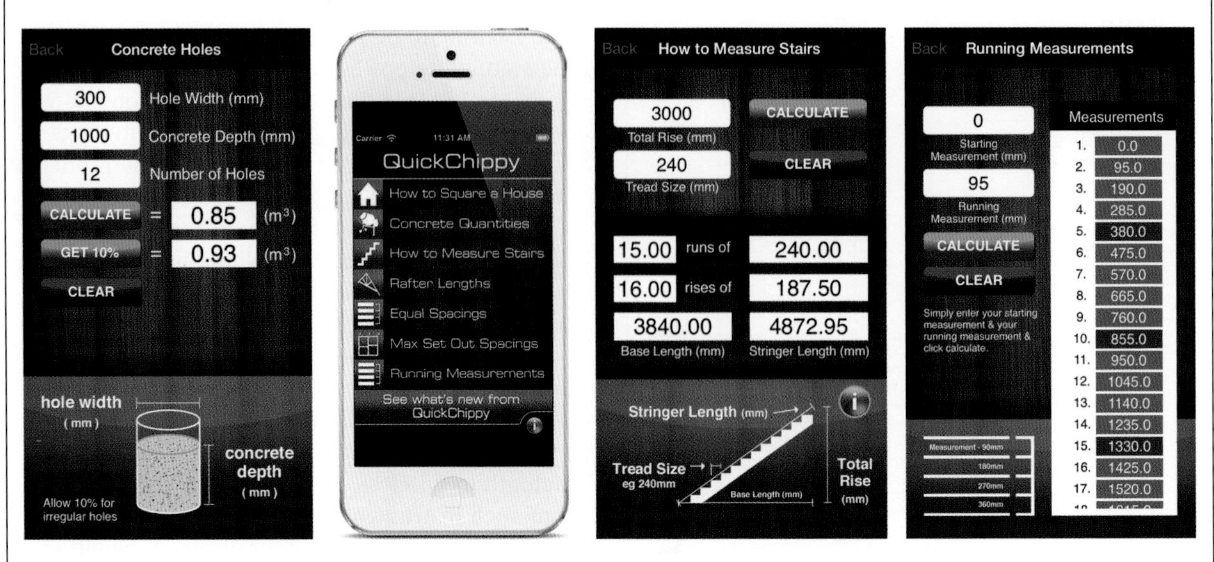

Source: Quick Chippy

FIGURE 11.9 Example of phone and tablet-based digital applications

FIGURE 11.10 Example of Sketchup 3D construction detail

floor area. This is important as the proposed design is for a partially disabled client whose brief is for there to be no steps or changes in floor levels throughout the house. The following information is required to be collated into the sketch for the concreter:

1 finished floor levels (refer to Figure 11.1 architectural drawings – floor plan)
2 dimensional set out of ensuite floor plan (architectural drawings – floor plan)
3 flooring specification – (architectural specification – P56)
4 stepdown (structural engineering drawings – construction detail)
5 internal rib beam (structural engineering drawings – rib beam detail).

The finished floor levels (FFL 20.00) are detailed on the floor plan of the architectural drawing set (Figure 11.11) and reflect the requirement to provide a flat floor surface throughout the proposed dwelling. The dimensional set out of the ensuite floor plan is likewise detailed in this same floor plan.

The flooring specification is included in the architectural specification and is also cross-referenced on the respective parts of the floor plan (Figures 11.12 and 11.11 respectively).

The proposed construction detail is included in the structural engineering drawings and indicates the thickening required under the proposed setdown and also the reinforcement overlapping requirements (Figure 11.13).

As has been identified earlier in the chapter, the height of the proposed setdown for the case study has not been specified by the structural engineer in recognition of the multitude of systems that may be used in the laying of both timber flooring and tiling.

The setdown detail provided is what is known as a 'standard detail'. In so being, it also fails to depict the location of the stepdown directly above an internal rib beam. This means that the construction requirements

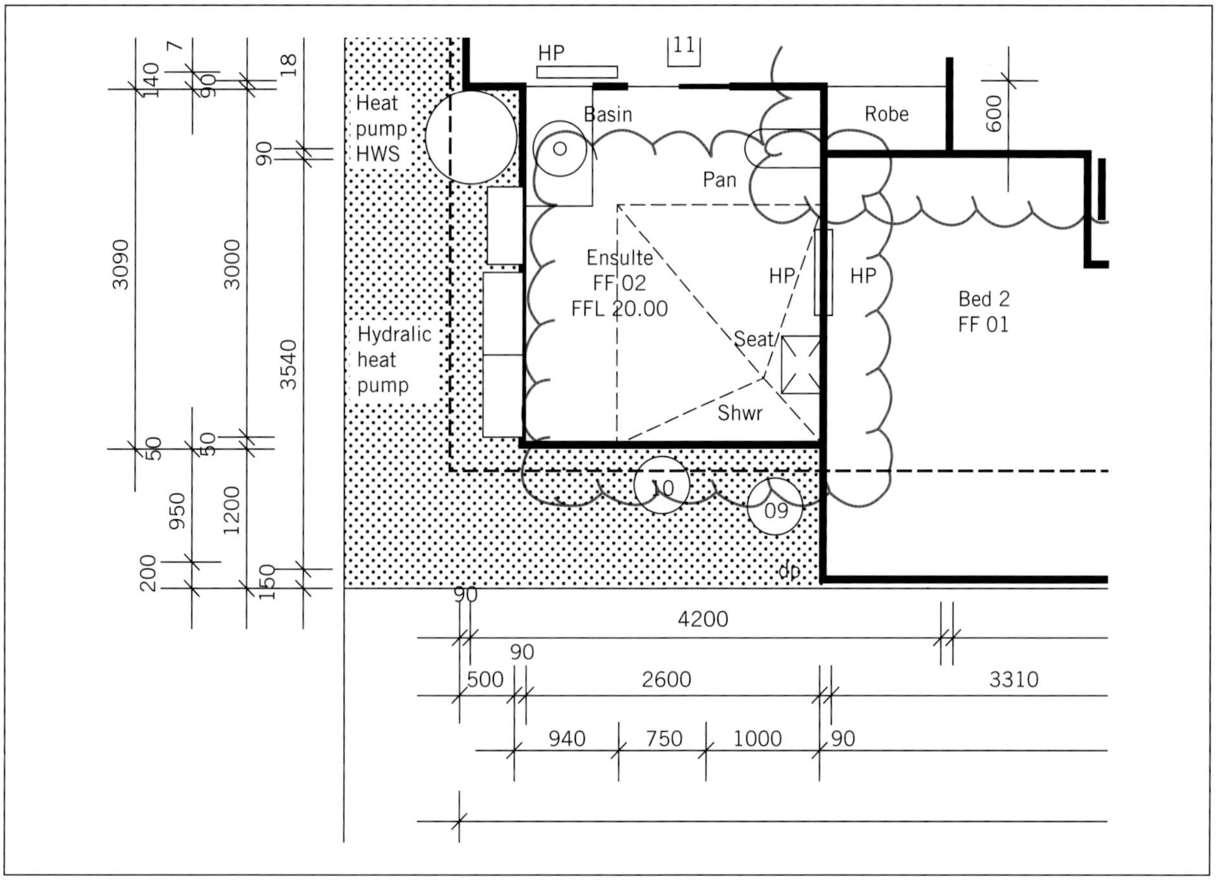

FIGURE 11.11 Architectural drawings excerpt showing FFL, dimensions and flooring notations

FLOOR FINISHES

FF01	Type :Tallowood overlay Hard wood floor Size: TBC Finish : Refer Applied Finishes
FF02	Type: Floor Tile Hawkesbury sandstone 600x300

Source: excerpts from engineering drawings by P.J. Yttrup & Associates

FIGURE 11.12 Architectural specification flooring

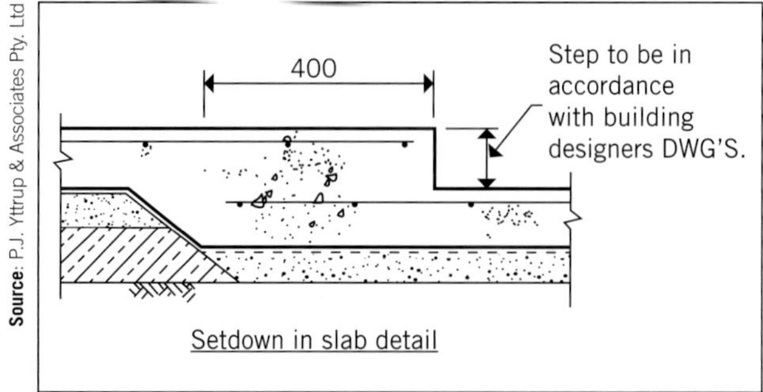

Source: P.J. Yttrup & Associates Pty. Ltd

FIGURE 11.13 Structural engineering setdown detail

for that element will also need to be combined into the sketch for the concreter (see Figures 11.2 and 11.14).

In addition, the geotechnical site investigations for the project dictated that mass concrete piles will need to be excavated to a depth of 1500 mm below natural ground. The dimensions of these additional foundation improvements are indicated on the structural engineering drawings (see Figures 11.2 and 11.14)

It is a combination of all of these details and the information gained about the size of the setdown that must be subsumed into the one sketch as clearly as possible.

LEARNING TASK 11.2 SIMPLE SKETCH: STUD FRAME

Working from the case study detailed in the above architectural and engineering drawings and specifications (Figures 11.1–11.14), prepare a simple sketch describing the required stud frame, lining and fixing details at the stepdown in the slab. Ensure you describe the following:

1 Stud and bottom plate sizing (include relevant Australian standard reference).
2 Required fixing/tie down to the concrete slab (include relevant Australian standard reference) based on a wind speed of 41 m/s.
3 Propose a 'proprietary' fixing mechanism to the slab that will meet the AS requirements for tie down (e.g. Ramset or similar).
4 Include detail of internal lining specification including wet area waterproofing requirements (include relevant Australian standard reference).

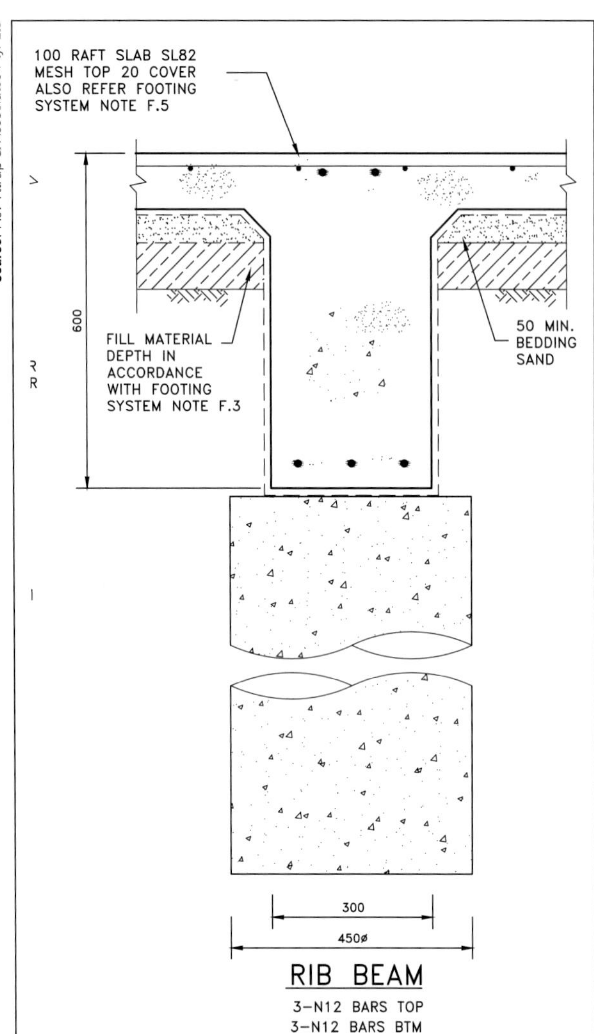

FIGURE 11.14 Structural engineering drawings rib beam detail

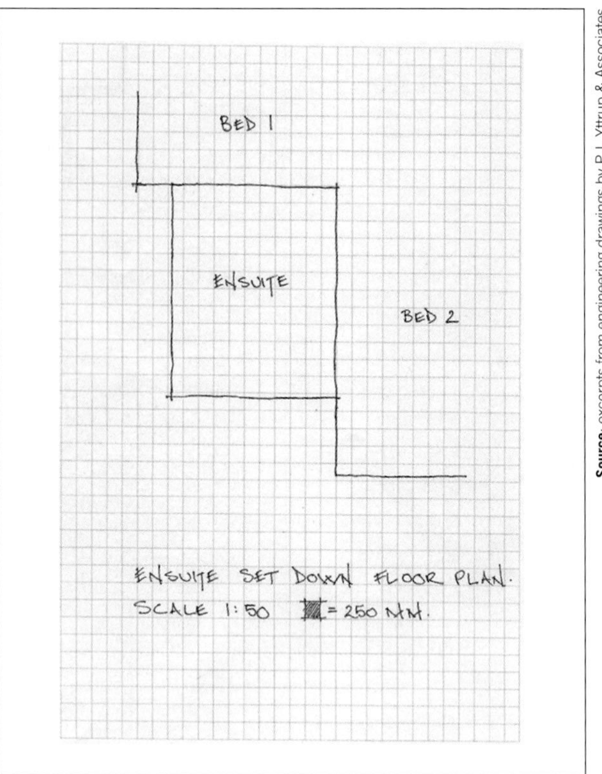

FIGURE 11.15 Ensuite setdown preliminary floor plan

Two- and three-dimensional drawing and sketches

With the required information collected, it is possible to begin the compilation of sketches which will describe the task to be completed. Before commencing, however, the scale of proposed sketches should be carefully considered in order to allow the best depiction of the information.

Selecting scales and beginning the drawing

Scales for this type of sketch can range widely depending upon what is being shown. Typical scales are:

- floor plans and elevations are typically presented at 1:100 or 1:50 scale
- sections are typically presented at 1:50
- construction details are typically detailed at 1:10 or 1:5 scale.

In this case, to fit the page and still provide the detail required, a scale of 1:50 is appropriate.

Utilising the pro-forma 'site instruction' template (Figure 11.8), the outline of the floor plan sketch should be plotted. This is done by carefully adhering to the scaled dimensions described on the architectural floor plan (Figure 11.11).

Figure 11.15 is the beginning of the floor plan sketch based on information compiled for the case study.

Three-dimensional oblique drawings (refer Chapter 5 of this book for more detailed description of three-dimensional drawing techniques) can also be useful tools to describe construction details; particularly for those not well versed in the reading of construction details. Figure 11.16 shows the proposed setdown required from the ensuite to accommodate the different floor finishes in order to achieve the client's 'brief' requirement of a flat floor surface throughout the dwelling. You will note that the sketch, though neatly drawn and proportioned, is not drawn to scale. By using both techniques, the concreter or formworker has a clearer picture in their mind of just how the step is to occur.

Standard drawing conventions

Standard drawing conventions are primarily detailed in AS 1100.301. This standard outlines the methods for depicting different construction elements as well as prescribing the use of specific graphic symbols, hatching and colour (see the section on 'Common symbols, abbreviations and terminology' in Chapter 5).

The basic conventions

Depending upon the details being described in the sketch, several of these symbols and drawing conventions may need to come into play, such as depictions of floor levels, sectional view locations and dimensioning practices. The more common of these are outlined below.

Detail references

Detail references are used on plans, elevations and sections to refer the reader to other larger-scale construction details that describe particular construction elements more fully. The top number in the circle refers to the number of the detail and the bottom number in the circle refers to the sheet number where the construction detail is located within the plan set (see Figure 11.17).

One thing that is not shown in the detail referencing system in Figure 11.17 is the direction in which the viewer is looking when studying the identified sectional detail. Generally, it is considered best practice to put an arrow, or some other identifiable mark, giving the direction of view. Typical examples for construction plan sets may be found in Chapter 5. For sketches, however, even simpler indicators may be used, as shown in Figure 11.18.

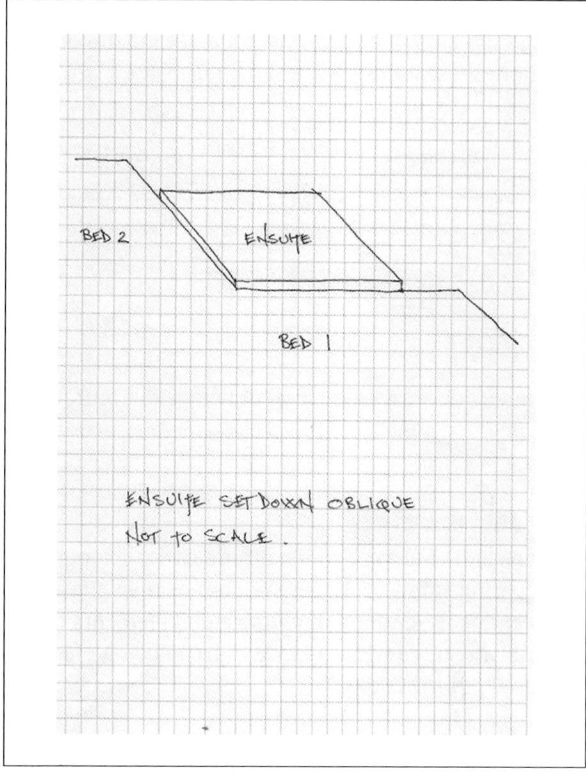

FIGURE 11.16 Ensuite setdown preliminary oblique drawing

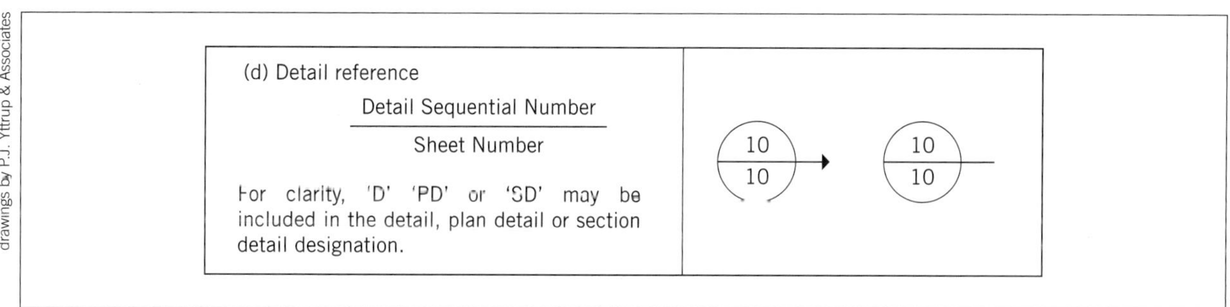

Source: excerpts from engineering drawings by P.J. Yttrup & Associates

FIGURE 11.17 Detail reference identifiers

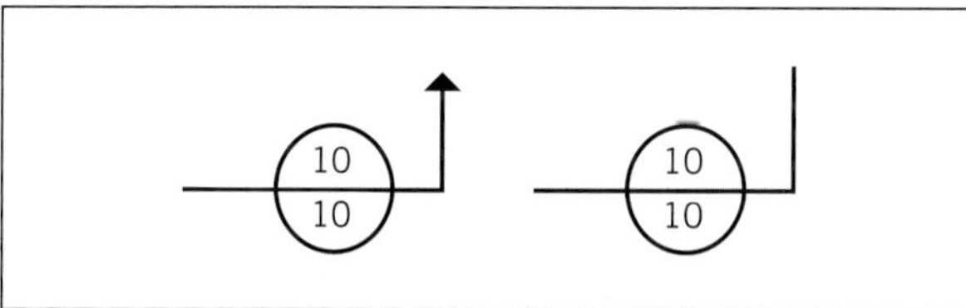

FIGURE 11.18 Direction of view indicators for sketching. Basic arrow or just a line running in the direction of view.

Expression of levels

It is standard practice to express levels in increments of 5 mm or 0.005 m. The level should then be described in a box, frequently with the letters RL (reduced level) placed in front to ensure the figure is not confused with some other factor such as a length (see Figure 11.19). When a level for some existing element is to be changed, then the original figure is shown outside the box and directly above, as also depicted in Figure 11.19.

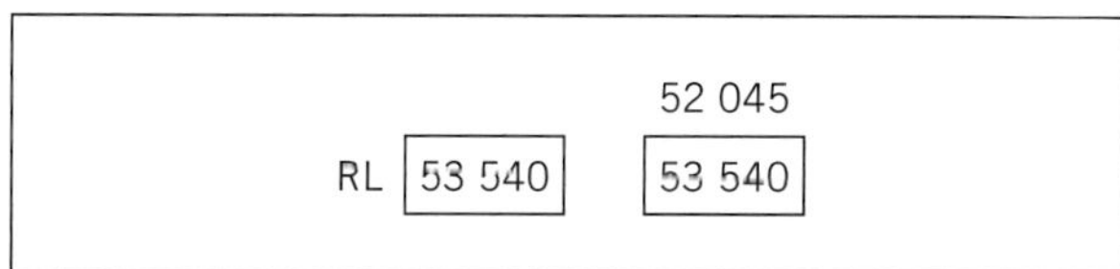

FIGURE 11.19 Depiction of reduced levels

These levels may be finished floor levels (FFLs) or the reduced levels (RLs) of particular points around the structure, or the block of land on which the building is to be sited. They may be used in plans, elevations and/or sections and details to indicate the relative levels at which the building elements need to be constructed.

Dimensioning

Dimensions are generally depicted in one of two ways: by arrow-headed lines running between two clearly demarked points; or by what are known as 'architectural ticks', small lines running at an angle through the beginning and end points of the dimension line. Both these methods are shown in Figure 11.20.

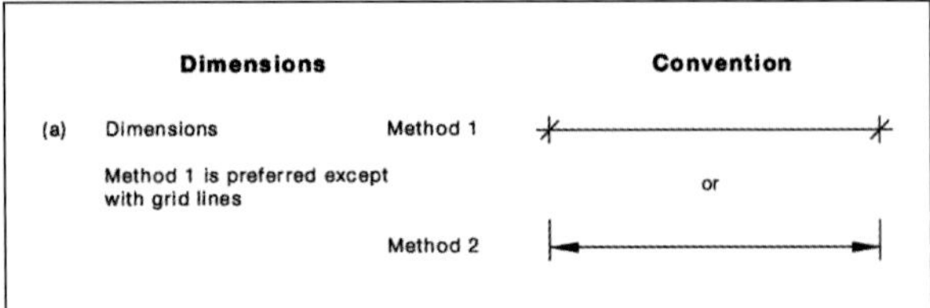

FIGURE 11.20 Dimensioning conventions

Dimensions are used on all drawings to accurately describe the geometry of the proposed construction works.

Sectional drawings

Sections and construction details are critical elements of preparing simple building sketches as they allow the accurate description of construction elements and materials required to be used in order to complete the proposed construction.

Expanding the case study

In the example case study scenario, the architectural and engineering drawings have been shown to contain various elements that need to be combined into a series of sketches used to describe the task to the concreter, formworker and steel fixing subcontractors. This information will be coupled with the information provided earlier by the floor tiler and timber floor installer. This includes:

1. finished/reduced floor levels (Figure 11.1 architectural drawings – floor plan)
2. flooring materials (architectural specification – P56 and contractor data)
3. stepdown and slab thickening and lapping of reinforcement (structural engineering drawings – construction detail)
4. internal stiffening beam and concrete pier location (structural engineering drawings – rib beam detail).

As there is an internal wall at this point as well, it is worth putting that element into the drawing also. Figure 11.21 shows this wall detail in place.

Notating and processing drawings

Once the preliminary sketches are completed, additional layers of information need to be added to the drawings using standard drawing conventions such as title blocks, dimensions, descriptive notes, symbols and hatching to adequately describe the proposed construction techniques and the materials to be used.

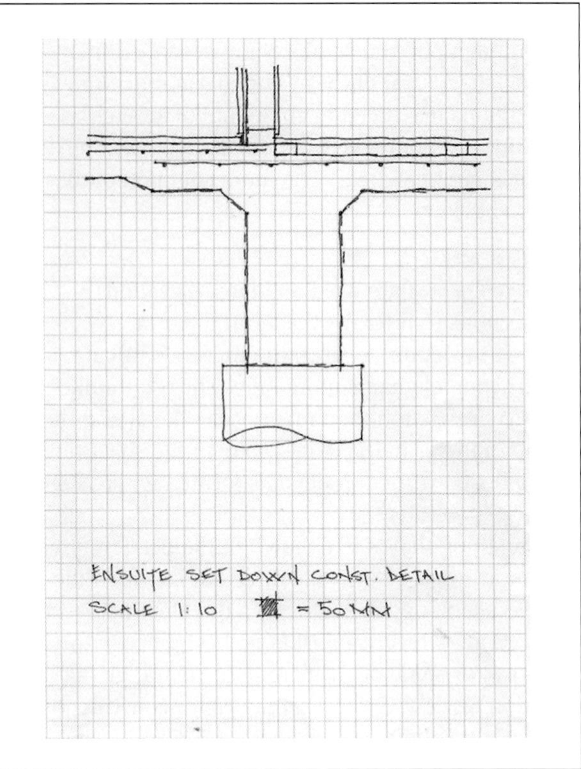

FIGURE 11.21 Ensuite setdown preliminary construction detail sketch

Applying essential information to the drawing

The example case study sketches now include dimensions, symbols, abbreviations, notes and hatching to form the final set of sketches that accurately describe the required task.

The sketches now include the following graphical elements:

1. scale, which varies for each sketch in the set
2. dimensions that have been transferred from the architectural drawings
3. material hatching, which is based on conventions in AS 1100
4. detail references, which are described in AS 1100 and assist to provide cross-referencing between the sketches
5. notations, which assist to describe the various elements contained within the sketches
6. construction abbreviations, which are also described in AS 1100 (e.g. RL = reduced level).

In Figure 11.22, take particular note of how the location of the sectional view (shown in Figure 11.24) is identified, and in which direction you are looking when viewing that sectional detail. Pay strict attention to the identified scale: in this drawing (Figure 11.22) it is 1:50, so each small square represents 250 mm or 0.250 m.

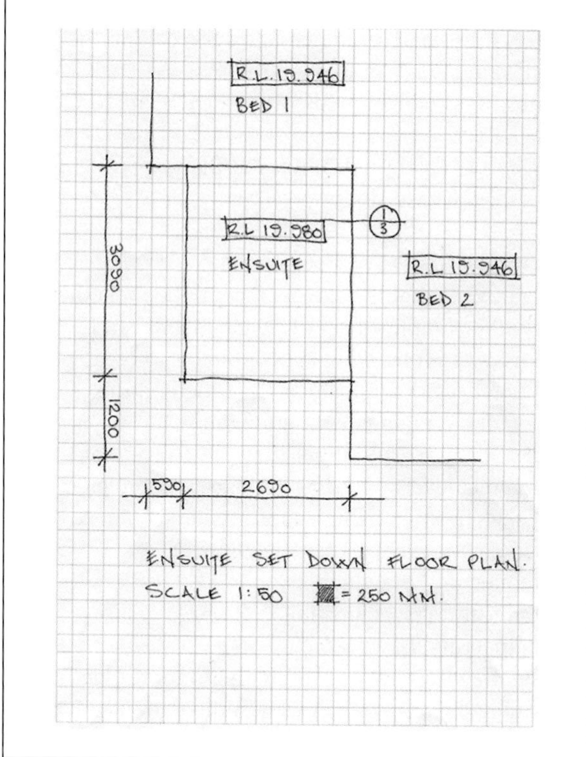

FIGURE 11.22 Ensuite setdown floor plan

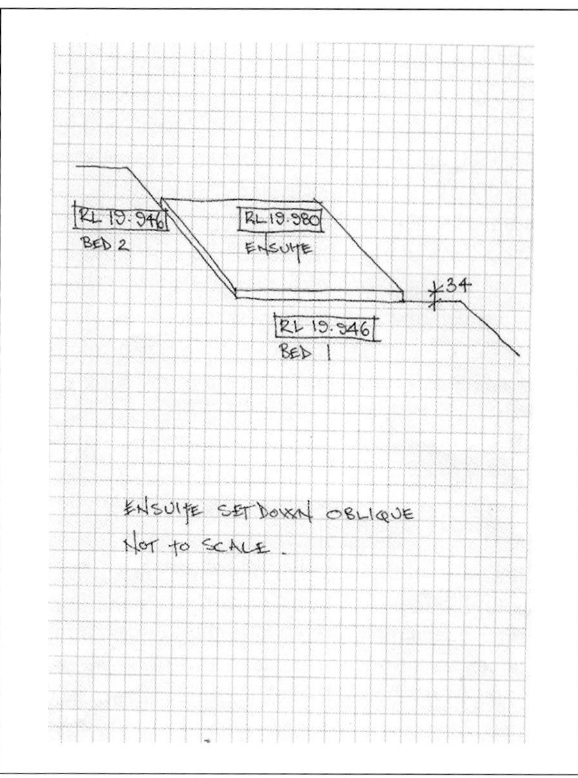

FIGURE 11.23 Ensuite setdown oblique drawing

You can also see in this diagram that the RLs for each floor have been identified, based upon the information gained from the flooring contractors. This is repeated in each of the other sketches as well:

- ensuite concrete surface RL is 20.00 m minus 20 mm of glue and tiles, so 19.980 m
- main floor area then becomes 19.980 m minus 34 mm of flooring and battens, so 19.946.

The value of the 3D sketch (Figure 11.23) is evident here in that it shows how the floors relate to each other with regards to the 34-mm step very clearly. And it does so despite not being to scale.

The final sketch, Figure 11.24, then provides all the detail that the contractors need to accurately form up the slab. This sketch will also prove valuable to the excavator in pointing out the necessity for accuracy in their pile boring. As may be seen, the detailed sectional drawing, despite being simply a sketch, is loaded with information, is clear and, most importantly, neat and easily read. Note again the scale; in this instance it is 1:10, so the sides of each small square represent 50 mm in length.

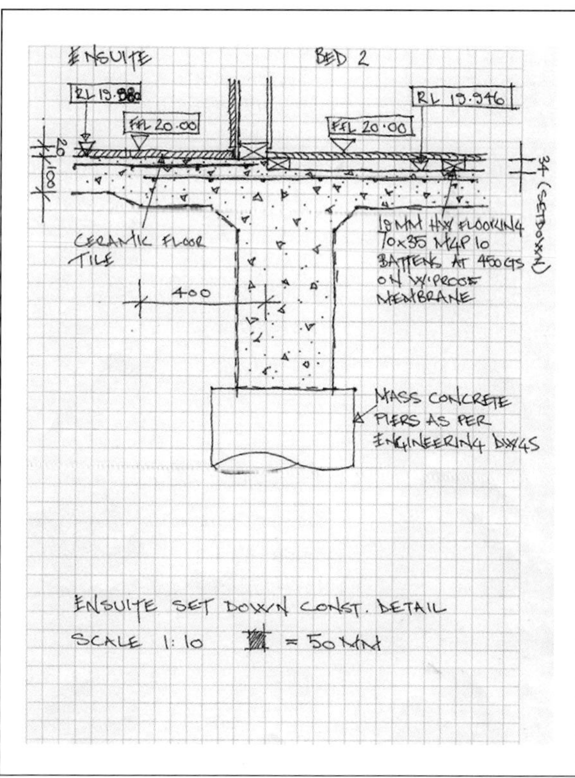

FIGURE 11.24 Ensuite setdown construction detail

LEARNING TASK 11.3 SIMPLE OBLIQUE DRAWING

Prepare a simple oblique drawing sketch of the construction detail described in Learning Task 11.2.

The title block

Title blocks perform a vital function on drawing sheets as they describe the project details and the project context for the specific sketches. Though not always required when sketching, in the production of a series of sketches, such as described in this case study, they can be just as important as when used on more formal construction drawings. Particularly if they form part of a 'variation' to the contract (see Chapters 3 and 5); as these must be fully described and signed off by the client.

The title block should include the project reference/number (e.g. A 17002) so the sketch or drawing can be correctly filed with other project information, and also the client's name and/or address.

Sheet numbers are indicated as a function of the total number of sheets in the drawing set (e.g. sheet number 1 of 3), as this allows the recipient to know that there are multiple sketches in the drawing set that all need to be reviewed in order to become familiar with the proposed construction process/method being described.

Scale can be shown in title blocks or it can also be shown on the title of respective drawings, which can allow for multiple scales to be used on the same drawing sheet. Utilising the pro-forma site instruction template allows for the scale to changed based on the type of sketch required. In our case study example, the floor plan is completed at 1:50 and the section detail is completed at 1:10.

The final element that needs to be included in the title block is the date, as this helps to position the sketch sequentially in relation to other sketches that may be produced while the project is under construction.

Figure 11.25 shows examples of the respective title blocks for the case study.

Completed drawings

Now all the elements have been combined together to form the completed set of sketches that can be utilised by the concreting subcontractor to complete the concrete setdown task, as described on the architectural and engineering drawings.

The title block and other drawing conventions have all been included in order to produce a completed set of sketches that includes information from the following construction plan elements:

1 architectural drawings
2 architectural specification
3 engineering drawings.

The drawings are now of an appropriate level of detail and clarity to be given to the various subcontractors with confidence that the concrete step setdown will be constructed as described.

One cautionary note should be made at this point. Communication is a two-way interaction. You have provided one direction of the communication only. In passing over the sketches, you must make certain

Project Number: A 17002
Project Name: GARDENER
Drawing Title ENSUITE SETDOWN FLOOR PLAN
Sheet Number: 1 OF 3 Date: 09/10/19
ABC Building Company

Project Number: A 17002
Project Name: GARDENER
Drawing Title ENSUITE SETDOWN OBLIQUE
Sheet Number: 2 OF 3 Date: 09/10/19
ABC Building Company

Project Number: A 17002
Project Name: GARDENER
Drawing Title ENSUITE SETDOWN CONST. DETAIL
Sheet Number: 3 OF 3 Date: 09/10/19
ABC Building Company

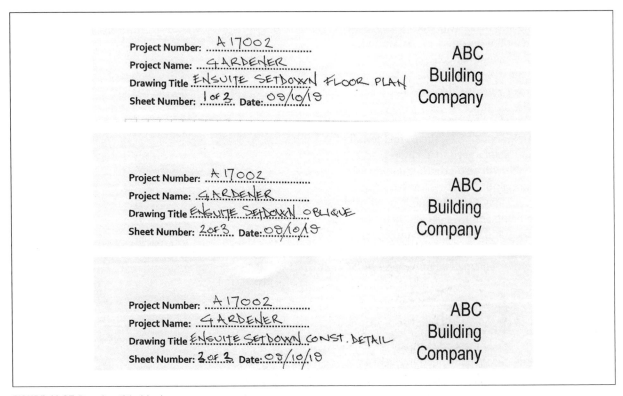

FIGURE 11.25 Drawing title blocks

that the contractors receiving them can feed back to you their understanding of what you have drawn. Do not be surprised if they ask for further clarity of a particular point – instead, welcome that feedback as a sign that they are studying the information intently and with purpose.

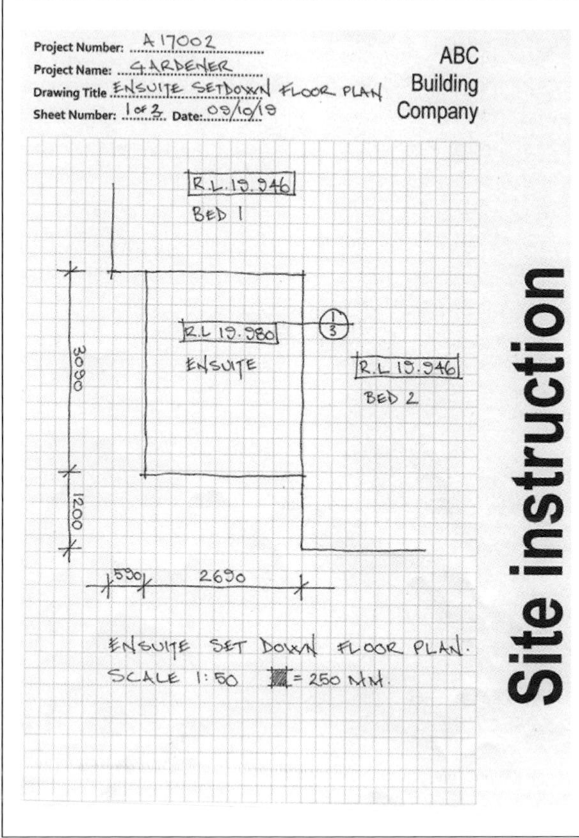

FIGURE 11.26 Ensuite setdown floor plan final sketch

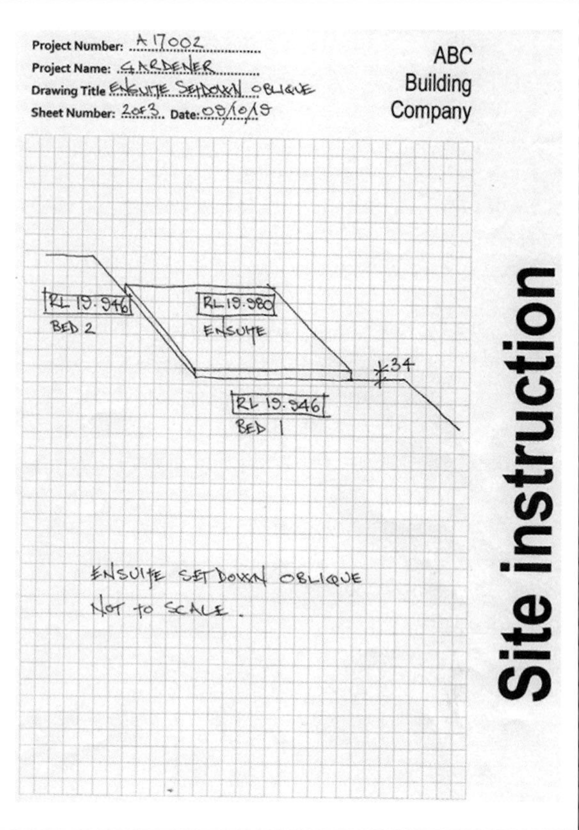

FIGURE 11.27 Ensuite setdown oblique final sketch

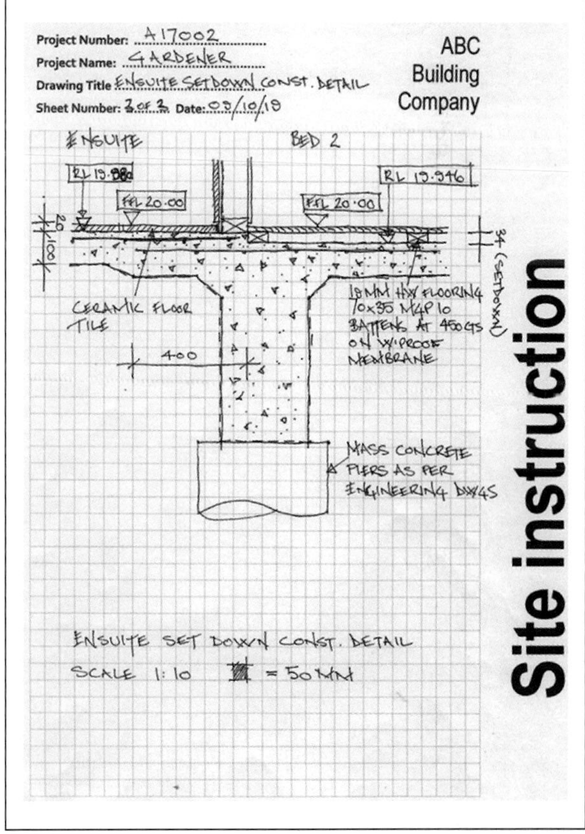

FIGURE 11.28 Ensuite setdown construction detail final sketch

SUMMARY

This chapter has described the process for how to prepare simple building sketches and drawings by the use of a case study scenario, which relies on extracting information from a range of sources within the construction plan set. Along with the information provided in Chapter 5 (Plans and specifications), you should now be familiar with the process for completing simple construction sketches that can then be used for describing specific construction tasks to other members of the project team.

The sketching process described utilises a simple graphic tool – the 'site instruction' graph template. This makes your sketching easier to produce, but it is not unknown for sketches to be produced on blank paper, or with digital applications using iPads and the like. They are also done, in very informal fashion yet quite commonly, on pieces of cardboard, ply, plasterboard, convenient walls, or small blocks of wood!

Drawing of this type also relies on developing familiarity and knowledge of drawing conventions contained within AS 1100. Similarly, some knowledge of relevant Australian standards and/or the Building Code of Australia and the NCC is required in order to ensure that proposed construction methods meet the regulatory requirements.

Most importantly, never forget that basic principle of communication – feedback. When you sketch, you are drawing a picture to inform others. You need to know that they understand what you have drawn and so always seek feedback to confirm that they can indeed interpret what you have offered. And never be offended if they need more sketches to clarify – that in itself is proof of their intent to understand and carry out your instructions to the betterment of the project as a whole.

PART 4

STRUCTURES

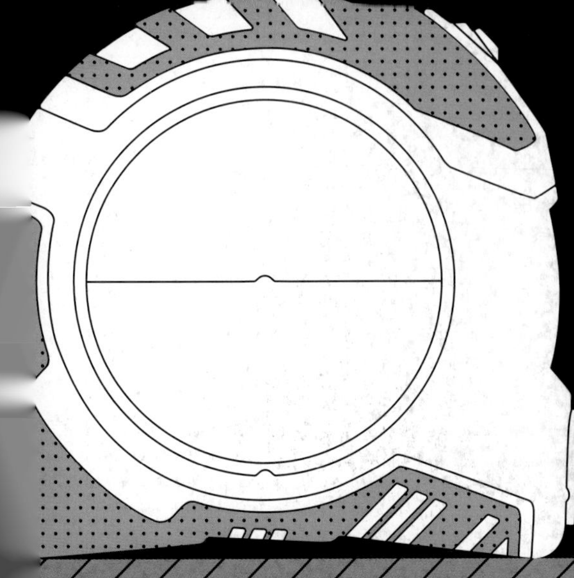

12 STRUCTURAL PRINCIPLES

Chapter overview

This chapter introduces you to the main concepts underlying structural principles, and how these principles inform planning for the erection and/or deconstruction of low-rise residential and commercial constructions. Though limited to two-storey structures, the chapter ranges over all classes of buildings as defined by the National Construction Code (NCC) with the limitation of Type C buildings for Classes 2–9 only. Likewise, this chapter also underpins Chapter 13, which applies this knowledge to the various elements of a building; i.e. footings, floors, walls, roofs and the like.

The chapter begins with an overview of the main structural principles involved, including forces and loads, and the determination of reaction forces found within structural systems common to most contemporary buildings. This is followed by standardised approaches to bracing structures against lateral loads such as wind, before moving to the basic principles behind truss components used in floors and roofs.

The final sections expand upon the key structural properties of building materials, and how these properties apply to some of the main elements of a building, such as slabs, columns, beams and retaining walls. The chapter concludes with the application of this knowledge of structural principles to safe and timely demolition of a structure.

Elements

This chapter provides knowledge and skill development materials on:

1 structural principles related to forces
2 structural principles related to loads
3 moments and force system solutions
4 bracing systems
5 truss systems – floor and roof components
6 the influences of material properties on structural performance
7 the structural characteristics of common building elements
8 structural principles applied to demolition.

Introduction

The safe construction of a building has always required an appreciation of at least some basic structural principles – the triangle for bracing, the capacity of a material to withstand a load, the value of good foundations. The same may be said when building a structure that we hope may have some level of longevity. Today, our understanding of materials, our supporting technologies and the depth of knowledge surrounding the many aspects of our built environment has increased many-fold. Hence, we can create structures that would have seemed impossible only a century ago.

Parallel to this has been an increase in codes, regulations and legislation. These have the purpose of ensuring that those who practice construction are knowledgeable and experienced, and build according to approved practices or better. These issues, along with the Australian standard units of measure, are dealt with in a brief introductory section at the beginning of this book. You are advised to acquaint yourself with that material prior to delving too deep into this chapter.

Structural principles: the basics

This is an extensive section covering the main structural principles as they apply to the erection or demolition of a structure. It begins with an introduction to forces and loads before outlining the basics of how to solve force systems that may be found when these loads are applied to some of the more common elements of buildings. The principles of bracing against such forces is then followed by how these concepts are applied in trusses.

Forces 1: Defining and describing forces – vectors

Force in engineering terms is the result of an interaction between two bodies or objects. The resultant interaction is a push or pull action that will attempt to change the position, attitude (angle or tendency for rotation), or shape of one or both of the objects.

Simple ('contact') forces occur when objects actually touch each other. For example, friction between two surfaces, tension on a steel cable between a mass and the cable's fixing point, the pressure applied by a heavy object on a surface.

Complex ('action at a distance') forces are harder to see, but are a daily occurrence. Gravity is the prime example. This is the action (technically not a 'force' by itself) between two masses that tries to bring them together. Magnetic and electric forces similarly act at a distance though they may repel or attract.

The magnitude (size/amount) of a force is measured in newtons (N), though in construction this is more generally kilonewtons (kN).

Any individual force is defined by four characteristics: magnitude, direction, sense of direction, and position or point of action on an object (see Figure 12.1).

Vectors and vector diagrams

The first three characteristics also define what we call a vector. This is how we graphically depict a force.

Vector diagrams use drawings of arrows to represent the size or magnitude of a force. These arrows are often drawn to scale, with resultant forces measured directly off the diagram. However, they can also be a nominal sketch from which the resultant force is then mathematically calculated using trigonometry. In either case they allow the determination of the magnitude, direction and sense of direction of a force resulting from the interaction of several forces on the one object.

Figure 12.2 offers a basic construction example: in this case, a combination of forces deriving from wind acting against the roof of a house. Note that the mass of the roofing materials is also given as a force, being described as 8 kN rather than 800 kg. The '***resultant***' force is 14.2 kN.

There are various online tools for calculating vectors. One very simple and useful tool may be found at https://www.mathsisfun.com/algebra/vector-calculator.html. It is worth accessing this site and exploring the input of a range of forces and the

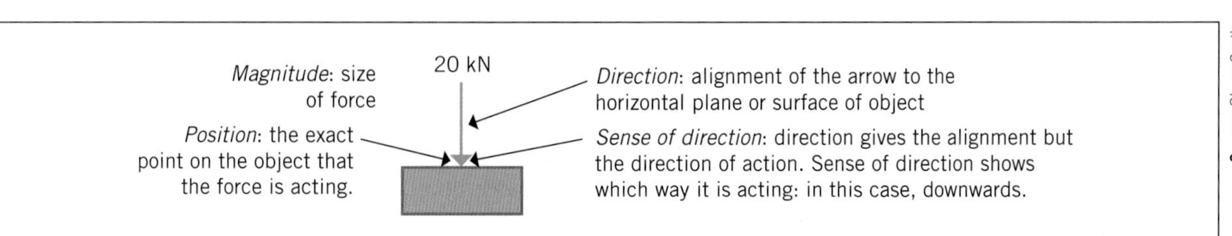

FIGURE 12.1 Characteristics of a force

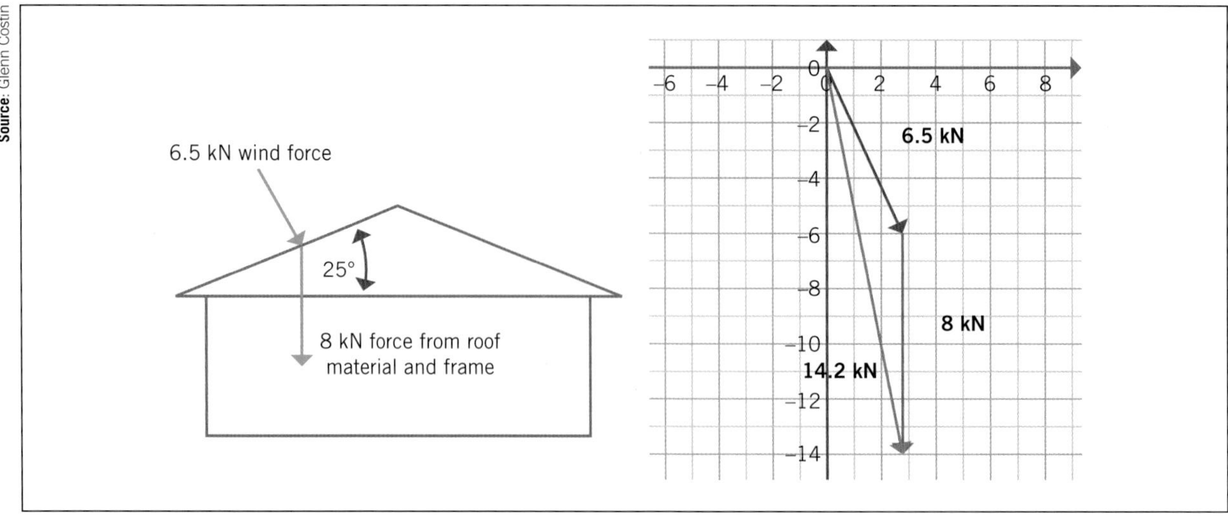

FIGURE 12.2 Forces acting on a house

derived outcomes. You can also use this site to test your mathematical skills in calculating vector solutions using trigonometry.

Note: Vectors are always arranged 'head-to-tail'. The order in which you choose to do this makes no difference to the outcome (Figure 12.3). The ***resultant*** vector will always be correct in magnitude, direction and sense of direction. This is demonstrated below using different arrays of the same four vectors.

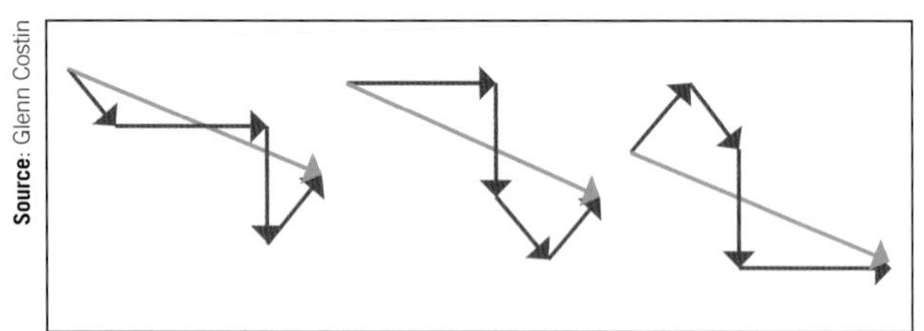

FIGURE 12.3 Vector addition: 'head-to-tail' order makes no difference to the result

Forces 2: Newton's laws

Sir Isaac Newton (1642–1727) developed a range of laws, calculations and theories concerning the Earth's gravitational pull. This pull means objects near the Earth will accelerate towards its centre at a rate of 9.81 m/s^2. It is this acceleration that gives force.

This figure is generally rounded up to 10 for both convenience and error allowance when applied to the construction sector. This means that for ease of calculation we can use the following rounding:

- A 1-kg object will produce a downwards force of 10 Newtons (N) or 0.01 kN.
- A 10-kg object produces 100 N or 0.1 kN.
- A 100-kg object produces 1000 N or 1kN.
- And 1000 kg (1 tonne) produces 10 kN.

Newton's three laws of motion form the foundation of structural engineering as well as many other fields of science, such as physics, in which bodies and forces interact.

Newton's first law of motion

Every object remains at rest or in uniform motion in a straight line unless acted upon by other forces or agencies.

Also known as the law of inertia (after Galileo who developed a similar concept), when applied to construction this effectively means that a building's components will stay still unless acted upon by some external force.

In reality, any structure on Earth is always being acted upon by that agency known as gravity; for which Newton has a solution in his third law, while helping us to calculate that solution using his second law.

Newton's second law of motion

This law states that:

Force is the product of mass times acceleration; i.e.
F = M × A

There are two ways in which this law is of considerable use to construction:

- **One**: This tells us that when a mass (a building component) is acted upon by a force or an agency such as gravity, it will accelerate. What is more, it will continue to accelerate unless we create some level of resistance – an opposing force – to stop it.

  That is, Gravity will cause our component to try and accelerate towards the ground – fall – unless we do something to hold it in place.
- **Two**: This means that given a known rate of acceleration (gravity at 9.81 or 10 m/s^2), and the mass of the object (in kg) we can calculate the force generated and, just as importantly, how much resistance we need to make the component stay in place.

Newton's third law of motion

For every action there is an equal and opposite reaction.

This law tells us that for components of a building to remain stationary, despite the action of gravity and other applied loads trying to make it fall, there must be an equal and opposite reaction holding it up. This means that if a component and its loads produce a downwards force of 15 kN, then the supporting structure, footing, components, or the like, must be producing an upwards force of 15 kN.

This upward, resistant or supportive force is sometimes referred to as either a 'normal' or 'reaction' force. It's called 'normal' when this resistant force is acting at 90°, perpendicular or 'normal' to the surface upon which the object is acting. 'Reaction' may be used in all cases where one object is making contact with another.

Forces 3: Statics and equilibrium

Newton's first and third laws form the basis of what we know of in engineering as 'statics' or equilibrium.

The concept of statics or equilibrium derives from Newton's first and third laws in application. Statics requires that for a building, or its component parts, to remain stationary – static – then the sum of all the forces acting upon it – wind, gravity, people, etc. – must equal zero. This must be so irrespective of the direction from which those forces are applied; up, down, sideways, or at an angle.

Note that this does not necessarily mean not moving: it requires only that the structure or its parts are not constantly accelerating. Things may vibrate in situ, deflect a little, or even stretch a little, but they must otherwise stay in place and be essentially 'static'. This is best explained by way of what is known as a free-body diagram. We used the basic principles of these diagrams when calculating vectors previously.

The first diagram (Figure 12.4) shows a body or building component that is not moving, that is static, but being acted upon by a combination of three forces. For the body to remain static, those forces must cancel each other out. Figure 12.5 shows that this is indeed the case. When we align the three vectors up head-to-tail they come back to the same point; i.e. no movement is occurring.

You can use the following data to confirm this using either the online vector calculation tool, or work it out mathematically using trigonometry:

- Vector 'a': 3.41 kN at 161°
- Vector 'b': 9.42 kN at 70°
- Vector 'c': 9.97 kN at 270°

Note: These angles are based on standard CAD practice of 3 o'clock as being zero, and angles sweeping anti-clockwise.

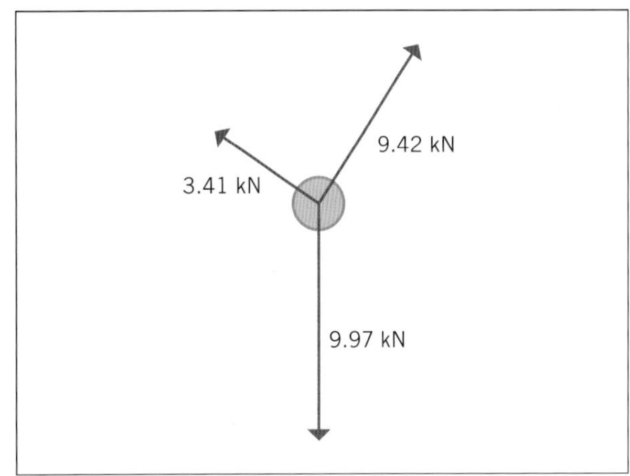

FIGURE 12.4 Loads acting on a static component

Source: Glenn Costin

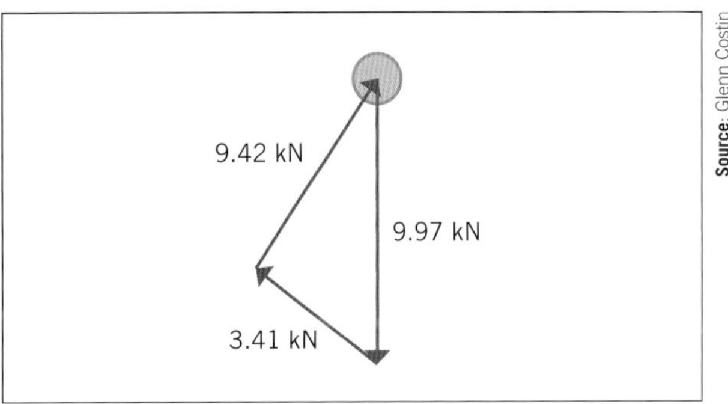

FIGURE 12.5 Vector solution showing summing to zero

Source: Glenn Costin

The basic equations of statics

There are three basic equations governing the static systems. They each must hold true if the system is to remain stationary. These are:

$\sum Fv = 0$	The sum of all vertical forces (Fv) in the system must be zero.
$\sum Fh = 0$	The sum of all horizontal forces (Fh) in the system must be zero.
$\sum Fm = 0$	The sum of all turning forces – also called moments (M) – in the system must be zero.

Note: the symbol $\sum$ means to add everything up or 'sum' them up.

If any one of these are not summing to zero, then the system is not in equilibrium and the object being acted upon is moving. In construction, that means failure.

At times, it is preferred to sum up some of the forces being applied to a particular location. This is called a ***resultant*** force, and was exampled earlier in the vector diagram showing the forces of wind and material loads applying to the roof of a building (Figure 12.2).

The ***reaction*** force is that which must be applied to make the system stable.

Concurrent forces

These are forces or point loads acting '***concurrently***' on an object. This means that they are acting at the same time and upon the same point, but possibly from different directions. For an object to remain static the sum of these forces must also equal zero.

Note: If they do not act upon the same point then the object will rotate; that is, not be static.

Stress and strain on materials

Stress and strain are concepts bound to each other by the effects exhibited on a material when a particular force is applied. This is generally depicted as a stretching or ***tensile*** force, but this is only one of its forms.

Stress is the force, strain is the material's response to that force. Up to a point, when stress is increased, so does strain proportionally. Enough stress, and a material will deform, and ultimately it will break. This can be plotted on a graph for each material type.

Stress

Materials under load undergo stress. The level of force applied over a given area dictates the amount of stress a component is withstanding. Hence:

Stress = Force ÷ Area or Stress = F/A

Where:

Force is in newtons (N)

Area is in m^2

The product, stress, comes out as N/m^2 which is equal to 1 pascal (Pa)

As 1 Pa is too small to be of much relevance to construction, we normally use kilopascals (kPa) or megapascals (MPa). Concrete, for example, has a nominated compressive strength stated in MPa; i.e. 25 MPa or 30 MPa for most domestic footings.

For ease of conversion remember that 1 kPa = 1 kN/m^2 and 1 MPa = 1 MN/m^2.

As mentioned above, there are many ways stress may be applied to a material:

- Tensile stress is when forces try to stretch the material – place it in tension. This is one of two forms of 'axial' stress – those that tend to change the length of the 'axis', the line running down the centre, of a material.
- Compressive stress is when forces try to squash the material – to compress it. This is the second form of 'axial' stress – as it too tends to change the length of a material.
- Shear stress occurs when a force acting perpendicular to one surface is being counteracted by another equal and opposite force acting on the opposite side. Think of a pair of scissors or 'shears': that is the action they apply to the material they cut. In construction, think of a bolt or screw holding a metal plate and the metal plate is pulled sideways, or you are trying to 'shear' the head off the bolt with a sideways blow with hammer and cold chisel. (See Figure 12.6.)
- Bending stress is a combination of compressive, tensile and shear stress acting within the one piece of material. Figure 12.7 shows bending stress acting in a beam under load. Shear stress is occurring between the (nominal) parallel laminations or fibres running the length of the beam.
- Torsional stress occurs when a material is twisted or rotated while one end remains relatively stationary. (See Figure 12.8.)
- Bearing stress occurs at the point where one material supports the load of another, such as a brick column supporting a lintel near its edge, or that part of the steel plate that the side of a bolt makes contact with, as shown in Figure 12.9.

Strain

Strain is the response demonstrated by a material undergoing stress. Strain is effectively a measure of the effect that stress has upon a material or component. Strain is calculated by finding the ratio between how far the material has stretched or 'deformed' relative to its original length. This deformation is referred to as a material's '***extension***'. As a calculation it is as follows:

Strain = Extension ÷ Length

or

Strain = $\Delta L \div L$

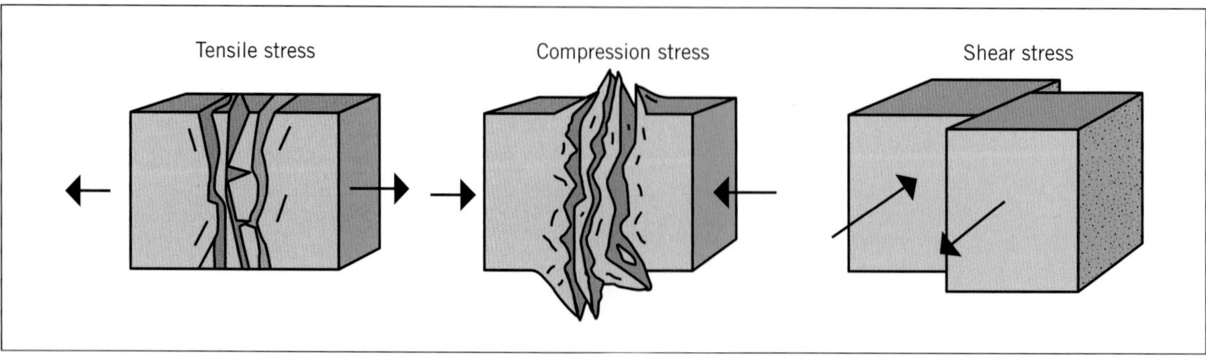

FIGURE 12.6 Examples of stress upon materials

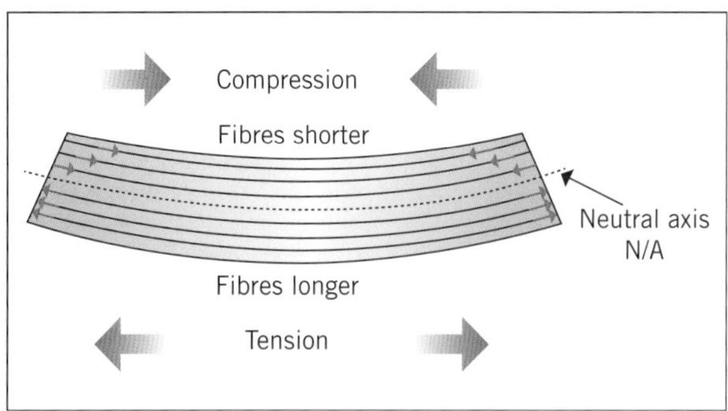

FIGURE 12.7 Bending stresses on a beam under load

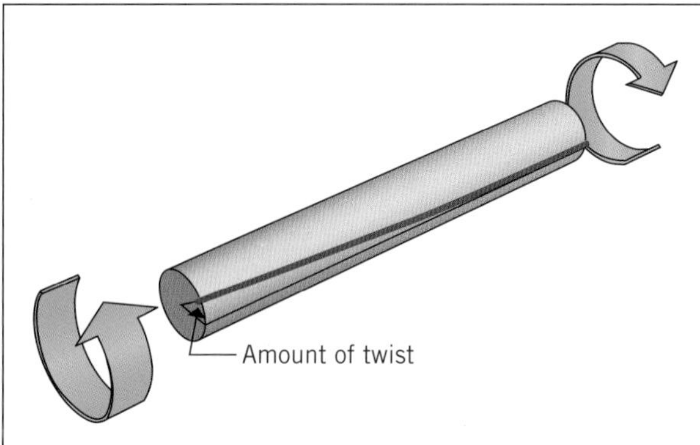

FIGURE 12.8 Torsional stress

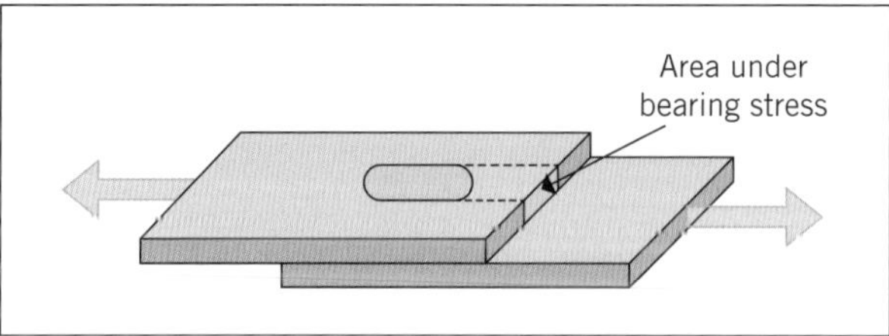

FIGURE 12.9 Bearing stress produced by a bolt from two plates under tension

Where: ΔL* or extension is the difference between the original length of the material, and the length after stress has been applied (after it has stretched) measured in metres (m). For example, if a material was originally 1.500 m in length, and then measures 1.525 m after stressing, then the extension (ΔL) is 0.025 m.

L or length is the original length of the material measured in metres (m).

Because both lengths (ΔL and L) are measured in metres (m) there is no specific unit for strain.

* **Note**: The symbol Δ is called delta.

The relationship between stress and strain

When a material is stressed, the amount of strain it incurs is directly proportional to that stress – up to a point. Hooke's Law is the name given to this direct relationship between stress and strain: on a graph this shows as a straight line, as depicted in Figure 12.10. The angle of the line is particular to the material being studied.

While a material is operating within Hooke's Law it is said to be exhibiting 'elastic' behaviour. This means that when the stress is released, it will return to its original length with no detrimental effect.

Within this zone it may also be given a value or 'constant' (something that will always be the same for that particular material). This constant is known as Young's modulus of elasticity or E.

A Young's modulus value is particular to a given material. It is a property of that material and so applies no matter how large or small a piece of that material you have. As a formula it looks like:

E = stress ÷ strain

Looking at the graph in Figure 12.10, Young's modulus (E) is effectively the gradient of the line on the graph; i.e. stress ÷ strain.

Stress, strain, yield and fracture

Young's modulus only holds up for a certain amount of stress or applied force. After this a material is said to yield. Once a material, for example mild steel, has 'yielded', it will not return to its former length. It has stretched to the point of, if not no return, at best only partial return.

At this point, in some materials such as steel, the amount of stress required to induce significantly more strain actually reduces for a time. Then a new point of resistance is found and again the stress must increase until the material's ultimate strength is found, the material becomes significantly elongated (necking) and finally fractures. Figure 12.11 shows how this looks when plotted on a graph.

A highly informative and visual example of stress or 'tensile' testing of materials is available online at the following site: https://www.youtube.com/watch?v=D8U4G5kcpcM

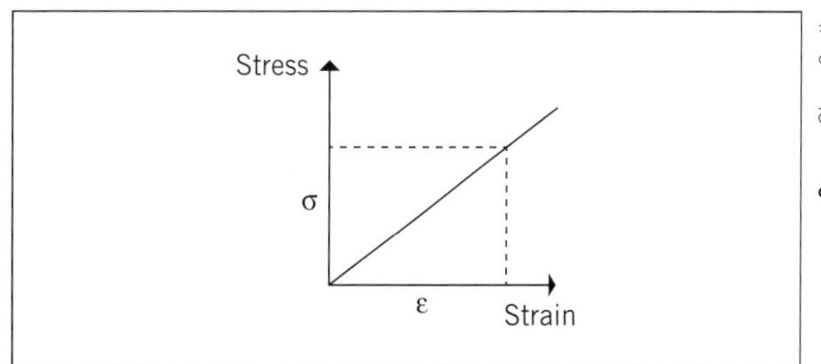

Source: Glenn Costin

FIGURE 12.10 Basic stress strain graph

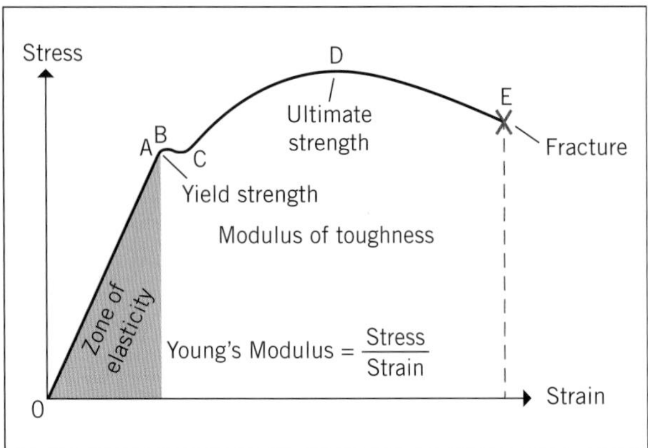

FIGURE 12.11 Simplified stress strain graph for mild steel

LEARNING TASK 12.1

IDENTIFYING HOW FORCES ACT ON STRUCTURAL MEMBERS

Stress and strain

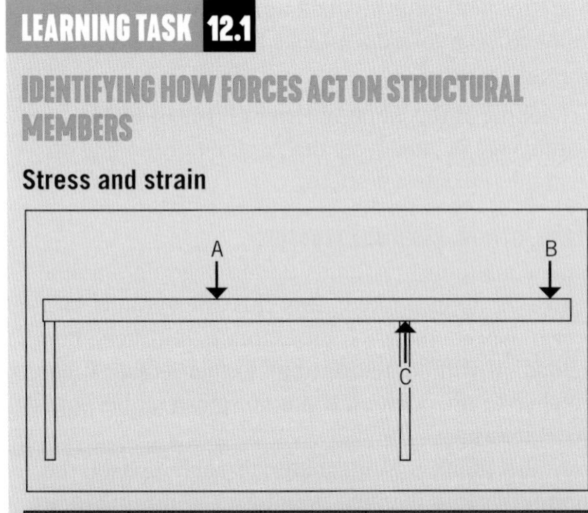

Force	Name of forces acting on slab	Way force will affect the slab
A		
B		
C		

Using the information shown on the suspended floor slab shown above, identify the name of the 'effect' of the forces acting on the slab at A, B and C. Describe the way those forces will affect the suspended slab member.

Structural principles: loads

A force acting on a body or surface induces what is known as a load. However, when we speak of loads in construction, we are referring to a force of a particular type and nature, not just its size and direction. The range of loads acting on a building is broad and, surprisingly, it took us a long time to realise that some of these were up, not just down. Figure 12.12 depicts just some of the loads that have an effect upon a domestic home.

Architects, engineers and builders all have a role in resolving these loads to ensure that the resultant building is structurally sound and stable.

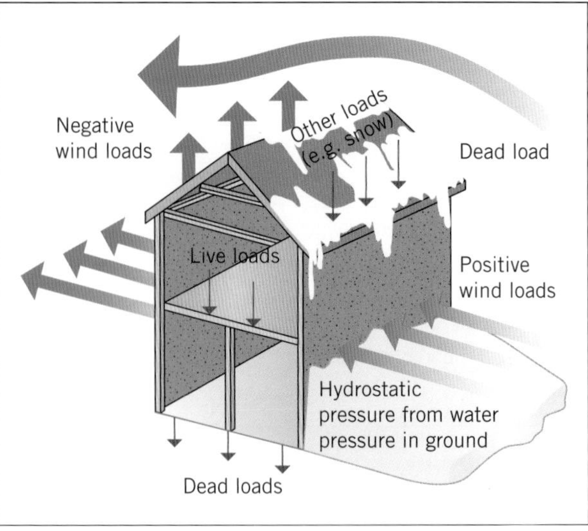

FIGURE 12.12 Examples of loads on a typical two-storey construction

Loads 1: Live and dead loads

The purpose of this section is to identify and define the loads common to most domestic structures. The section begins with ***dead loads*** and ***live loads*** before addressing ***wind loads***. A range of other common loads important to structural design are then discussed followed by the identification of ***load paths***. The section closes with a range of basic load calculations based upon Newton's laws and the concept of statics and equilibrium discussed earlier.

You should note from the outset that when dealing with loads on structures you should refer to the following standards:

- AS/NZS 1170:2002 Structural design actions and its parts:
  - General principles (1170.0:2002)
  - Permanent, imposed and other actions (1170.1:2002)
  - Wind actions (1170.2:2011)
  - Snow and ice actions (1170.3:2003)
  - Earthquake actions in Australia (AS only 1170.4:2007).

Dead loads

Dead or permanent action loads result from the mass of the component parts of a structure. As the phrase 'permanent action' suggests, these are forces considered to act continuously upon a structure with insignificant variation over time. The more common description 'dead load' is given because these loads are also said to be not moving. They are consistent and generally permanent loads that tend to be cumulative as they are assessed vertically down through the structure.

In general, dead loads are considered to act only in the vertical plane. In certain instances, however, they can have a lateral element. This may be exampled by earth pushing against a retaining wall, or uncoupled rafters pushing outwards on their supporting walls.

Dead loads may be calculated for a single component, a cluster of components such as a wall or flooring system, or for the structure as a whole. When assessing loads on a component or system, you must be confident that it can carry the combination of:

- its own weight (self-weight)
- the combined weight of anything else it supports
- any of the other load types (live, wind, etc.) that may be imposed upon it.

Self-weight is a characteristic of a given material and is generally provided as a force in N or kN, hence the term 'weight'; i.e. the effect of gravity is already factored in. It is usually expressed as unit weight and offered as a volume in m^3, or per lineal metre of a specific sectional size.

On occasions unit mass will be provided instead of weight; i.e. the effect of gravity has not been included. This will be offered in kg. In such cases you must multiply the kg by 10 to convert to newtons (N). Steel Universal Beams, for example, have their mass per metre given in their designation. For example, 150 UB 14.0; the 14.0 is the unit mass in kg/m. This must be converted to N or preferably kN. Multiply by 10 to convert to N, divide that result by 1000 to convert to kN, or simply divide the original mass by 100; i.e.: the unit weight of a 150 Steel Universal Beam is 0.14 kN.

Other materials with supplied units to be wary of:

- bricks – frequently offered in kg/m^2 based upon 110 mm of thickness (1 brick)
- glass – kg/m^2 based upon a given thickness (e.g. 4 mm, 6 mm, 8 mm, etc.)
- most sheet materials.

The dead load of a component is found by multiplying its volume or length by its unit weight as is relevant.

Table 12.1 offers the unit weight of some of the common materials.

TABLE 12.1 Unit weights of common materials

Material	Unit weight kN/m^3
Aluminium	24
Brick	18 ~ 22 Product variable
Reinforced concrete	24
Glass	25.5
Mild steel	77
Timber	3.5 ~ 12 Highly species variable

Example use of unit weight

Drawing upon the above units we could calculate the load of the suspended reinforced concrete slab described below as follows:

Concrete slab dimensions: 4.8 m × 4.5 m × 0.1 m

Unit weight: 24 kN/m^3 (from table above)

Dead load = volume (m^3) × unit weight (kN/m^3)

Slab dead load = (4.8 m × 4.5 m × 0.1 m) × 24 kN/m^3
= 2.16 m^3 × 24 kN/m^3
= 51.84 kN

Note: The m^3 of volume and '/m^3' part of kN/m^3 (sometimes writing as kNm^{-3}) of unit weight cancel each other out, leaving only kN.

Alternative method:

This alternative calculates the dead load for a 1.0-m length of the component, then multiplies by the actual length. In this case we will use a 250 mm × 110 mm reinforced concrete lintel. For example:

Dead load/m = (1.0 m × 0.25 m × 0.11 m) × 24 kN/m^3
= 0.0275 m^3 × 24 kN/m^3
= 0.66 kN for every lineal metre of this size lintel

Dead load for a 2.5-m lintel = 2.5 m × 0.66 kN/m
= 1.65 kN

Dead load for a 3.0-m lintel = 3.0 m × 0.66 kN/m
= 1.98 kN

Note: Even though our calculation came out at 0.66 kN, we know that this is the unit weight for every lineal metre or 'per metre': So, we can once again write it as kN/m.

As the example shows, the advantage of this method applies when a project requires several components of the same material and sectional size, but of differing lengths and acting on separate parts of the structure.

Live loads

Live or imposed action loads are all those items and materials within a building that may move, or may be moved. AS/NZS 1170.0 defines an 'imposed action' simply as a set of forces … 'resulting from the intended use or occupancy of the structure…'. One way to visualise the difference between dead and live loads is to think of dead loads as anything that is of the building, or is fixed to it, while live loads are just about anything that's in or on the building that is not fixed to it.

Live loads can be highly significant, particularly with regards to decks and verandahs, where large groups of people can congregate, dance to music or simply stand in tight groupings.

Examples of live loads include people, furniture, vehicles, computers and TVs, books – anything that you can imagine that may be brought into, moved within, or taken out of a building.

Live load ratings

Table 3.1 of AS/NZS 1170.1 and Section 3 of AS 1684.2 provide the floor loads that should be taken into account when designing a structure. Those loads applicable to a residential building are shown in Table 12.2.

TABLE 12.2 Live loads on various construction elements

Specific use	Live load kPa or kN/m² (uniformly distributed action)
General floor areas; deck areas less than 1.0 m above ground	1.5
Decks over 1.0 m above ground	3.0*
Garages for vehicles under 2500 kg	2.5
Garages for vehicles over 2500 kg but less than 10,000 kg	5.0

Source: Derived from Table 3.1 of AS/NZS 1170.1, and Section 3 of AS 1684.2

*Note: AS/NZS 1170.1 calls for only 2.0 kPa for decks over 1.0 m above the ground, whereas AS 1684.2 and AS 1684.3 call for 3.0 kPa. Despite the NCC specifically calling up AS/NZS 1170.1 (rather than AS 1684), it is considered best practice that when there is disparity between codes or standards you should work with the higher loading.

Loads 2: Wind

This is possibly the most critical load for contemporary housing. Over the years our materials have become lighter, jointing systems simpler – butt and nail as against mortice and tenon of studs to plates, for example – and, through changing climate, incidences of high winds more prevalent. In the past we tended to focus on holding up; today much more attention is being applied to holding down. (See Figure 12.13.)

In brief, wind loads are becoming increasingly significant due to changing:

- climate
- building design
- building techniques
- building materials.

The effects of climate

Climate change has brought higher wind speeds and shifting weather patterns to all regions of Australia. Extreme weather events, such as cyclones in the north and tornadoes and other funnel winds in the south, are becoming more common. The structural design of the contemporary Australian home has had to change accordingly to cater for the increased loads, particularly in uplift and lateral pressures. Though this section focuses upon wind loads, you should be aware that changes in climate also effect salinity, bushfire and flooding issues; all of which have an influence upon housing design and construction. For further information on the effects of climate change on architecture and communities more generally, download: https://www.be.unsw.edu.au/sites/default/files/upload/pdf/cityfutures/cfupdate/Neuman_Posters.pdf
or
https://acumen.architecture.com.au/environment/place/climate/climate-change-adaptation-for-building-designers/

Source: Fairfax Syndication/Angela Wylie

FIGURE 12.13 Wind damage to house due to hold-down failure

Changing house design

Aside from Australian having, on average, perhaps the largest homes in the world, home designs today differ from earlier eras in having larger rooms and windows. That is, bigger areas, but less walls. Less solid wall area means less area for bracing against the lateral loads or the racking forces of wind. Likewise, there are fewer locations for holding these larger roofs down against the uplift due to wind.

Construction techniques

Early Australian timber-framed homes used mortice and tenon joints between plates and studs. Today, almost all joints are butted and nailed together. It's a faster technique, but one that offers only what is known as a 'pinned' joint as against the 'fixed' joint of the former style. Fixed joints offer better racking (bracing), tension (uplift), and torsional (twisting) resistance than do pinned joints. These nominal advantages are no longer built into the contemporary timber house frame.

Further, most homes today use trusses for the roof structure. These are lighter, span greater distances and transfer roof loads to the outside walls. Light alloy window frames, which rely upon the wall frame for rigidity, have replaced the solid timber frames of previous eras. The loads needing to be supported by the outside wall are therefore much greater, and wind loads only compound the issues.

Contemporary materials

Timber-framed brick veneer is still the main construction technique in Australia; however, many lighter construction materials are gaining favour among architects, builders and homeowners. The most common of these include:

- light steel framing and trusses
- polystyrene cladding
- engineered timbers
- rendered hard sheet cladding

- ply claddings
- autoclaved aerated concrete
- lightweight steel cladding
- plastics.

These materials, and their associated construction systems and methods have made homes easier and faster to assemble while tending to put less load upon footings (see Figure 12.14). Combined, they can significantly reduce costs. However, many of these materials have little or no inherent bracing capacity. Logically, they also offer little resistance by way of mass. This means that wind loads will have a greater effect upon the structure unless other means of bracing are designed in.

Source: Shutterstock.com/Image Supply

FIGURE 12.14 Modern home design using contemporary materials

Structural design responses to wind loads

An Australian standard, AS 4055 Wind loads for housing, has been developed around these issues and it is this standard that informs much of AS 1684 Residential timber-framed construction. In brief, our designs today include much greater attention to bracing capacities: designing in 'units' of bracing that are carefully engineered and located based upon expected loads, defined by the region in which the home is being built. These bracing units, also based upon lightweight materials, include tensioned steel, ply and compressed fibre boards that may be quickly installed rather than cut in.

Bracing will be discussed in detail later in the section 'Bracing systems'.

Loads 3: Other important loads

While the above are the most commonly regarded or primary loads, there are many more that a builder and designer or engineer must cater for. Often referred to as 'secondary' loads, they are just as important to the structural design, particularly on a regional basis. The more significant of these are addressed under the headings that follow.

Seismic

These are loads brought about by movement of the Earth's surface. Commonly referred to as earthquakes or tremors, these can happen in any part of the world. In Australia they are reasonably rare and mostly light as the country resides on the one tectonic plate. However, even the one plate can have what are known as hot spots, places where currents in the Earth's mantle tend to rise or push against the plate. Hence, quakes causing significant devastation such as Newcastle in 1989 and Ellalong in 1994 do occur. These loads are addressed in AS 1170.4 Structural design actions – Part 4 Earthquake actions in Australia.

Geoscience Australia has developed an interactive Earthquake Hazard Map of Australia which may be found at: http://www.ga.gov.au/interactive-maps/#/theme/hazards/map/earthquakehazards

This map is shown below (Figure 12.15). Note the large hot spot around Tennant Creek in the Northern Territory: the areas of yellow trending to orange and red are the areas where earth movement is known to occur. The darker the colour, the higher the risk. It is worthwhile accessing this map to view the likely actions that may influence structural design in your region.

Snow

These are applicable to alpine and sub-alpine areas. They are also defined in the AS 1170 suite of standards. Snow loads are deemed a form of live load in that they are not consistent; the load shifts over short periods of time as the snow melts. Design considerations include steeper roof pitches which decreases the depth to which the snow can form, and changes the magnitude of the downward force.

Dynamic

These are complex loads that can be short lived, or of long, even continuous, duration. They may be sudden, random, or vibrating in nature. Wind on a tall flue rising out of a roof is dynamic in that it is sporadic and less likely to blow it over as rattle it loose. Music is a dynamic load in that it can cause damage through strong, repeated vibration of components of the building, as can dancing to that beat.

Shrinkage and expansion

All building materials expand and contract to some degree. Timber and brick are particularly renowned for such actions due to moisture gain and loss. Metal likewise expands and contracts notably with changes in temperature. Timber will seldom change much in length, but it may shrink or expand as much as 10% in cross-section depending upon the species. Brick and steel may expand and contract in all dimensions. This expansion and contraction over time can cause

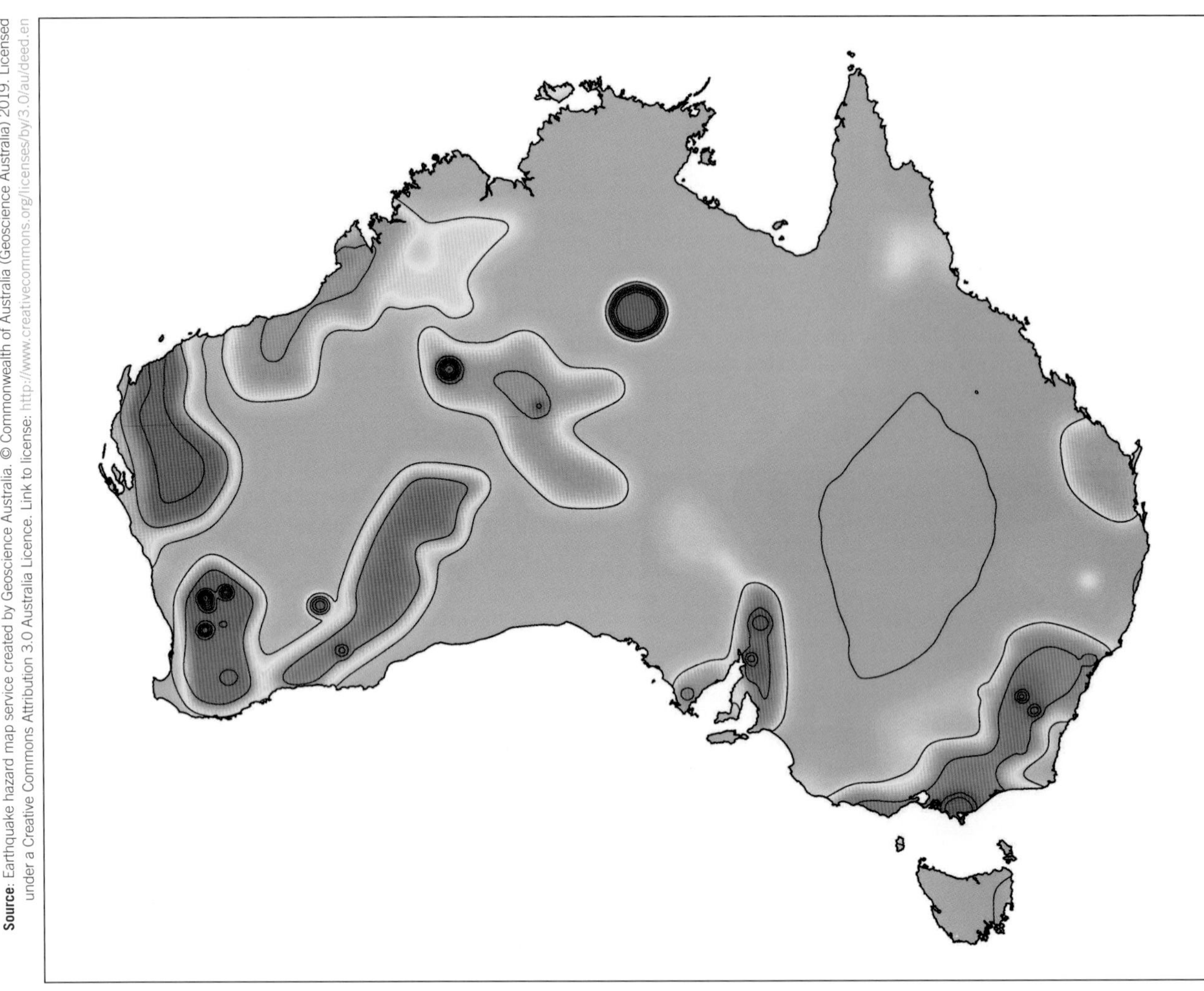

FIGURE 12.15 Earthquake risk map of Australia

significant damage to a structure if not allowed for in the design or component assembly.

Thermal

As noted with metals, changes in temperature will also cause materials to expand and contract, doing so more rapidly than moisture uptake or loss. Heat and extreme cold – Antarctica is the driest continent on Earth – may also cause increased moisture loss from materials such as timber and brick. Thermal loads can also change the strength of some materials, generally lowering them, though others may become more brittle.

Settlement

Footings may settle into the foundation material over time. If this occurs inconsistently around the structure then disproportional loads may occur over the length and width of the structure. This effectively leads to differential support leading to significant damage: cracking of walls and ceilings; doors and windows jamming; brickwork and rendered surfaces cracking. Dealing appropriately with settlement loads is more than just a matter of good footing design: it may also require articulation of materials unlikely to be able to flex sufficiently such as brick, plaster walls, rendered panels and the like.

Hydrostatic pressure

Water build up or trapped behind a surface will produce a load known as hydrostatic pressure. This is a common and significant load consideration for basement and retaining walls. This is why water proofing is applied wherever practicable on the 'positive' side of such walls; i.e. the side the water will be coming from. When waterproofing is on the inside of the wall it is the action of hydrostatic pressure that pushes material off the face of the wall. If water cannot penetrate through the wall it will tend to push it over. So, the wall must be able to contend with both dry earth and the much heavier wet earth, as well as water that is attempting to flow through it.

Lateral earth pressure

This load is also of significant relevance to basements and retaining walls. Earth is a dead load acting vertically down; however, earth has a natural 'angle of repose' that varies from one type of soil to another. The more 'fluid' the earth, such as sand or rounded gravels, the lower the angle of repose and hence the greater the pressure applied to the wall. Lateral earth pressures can be increased further by loads such as neighbouring structures or traffic that acts close to a wall.

Loads 4: Tracking load paths

Identifying the path that loads will take as they pass down through a structure is critical to sound building practice. Just as critical is the ability to identify and calculate these loads. It is from this knowledge that footings are designed. What is often not recognised is that footings are like the meat between the sandwich: their design being based upon the loads supported, and the load carrying capacity and nature (tendency to move) of the foundation material the footing rests upon. The tracking of loads generally begins at the top of a structure, though there are times when you will need to track them laterally (sideways). Such situations are discussed below. The loads are then traced downwards through the structural members. In so doing, the self-weight of each of the components must be taken into account, as well as any imposed actions (live loads) where applicable. Figure 12.16 shows the typical load paths for a domestic house.

Concentrated loads

A frequently missed factor in load path analysis is concentrated loads. These may occur beside openings when loads that would otherwise be carried by the window or door are instead transferred to the opening studs by way of lintels or other beams. Not only must the wall carry a greater load at that location: such loads are then transferred to the footing, making for a greater load at that point than at others.

Likewise, girder trusses, hanging, strutting, and combination beams will all bring a concentration of loads to a wall. Such concentrated loads must be accounted for by a corresponding increase in strength of the supporting members (studs or columns) below. Again, this load must ultimately find its way to the footing to be dispersed to the foundation material.

Lateral load paths

There will be times when you will need to track load paths laterally. Wind loads and the bracing required to prevent racking are classic examples. Hydrostatic and earth loads on retaining walls likewise must be accounted for latterly. The images in Figure 12.17 provide examples of the load path that can occur in such cases. Note that they can pass upwards, and travel through truss or ceiling components before the load is transferred to the ground. What is not shown is that they can also pass into internal bracing walls and then to the footing and foundations.

Misjudging load paths can mean that footings are of an inappropriate dimension, leading to differential settlement across the building.

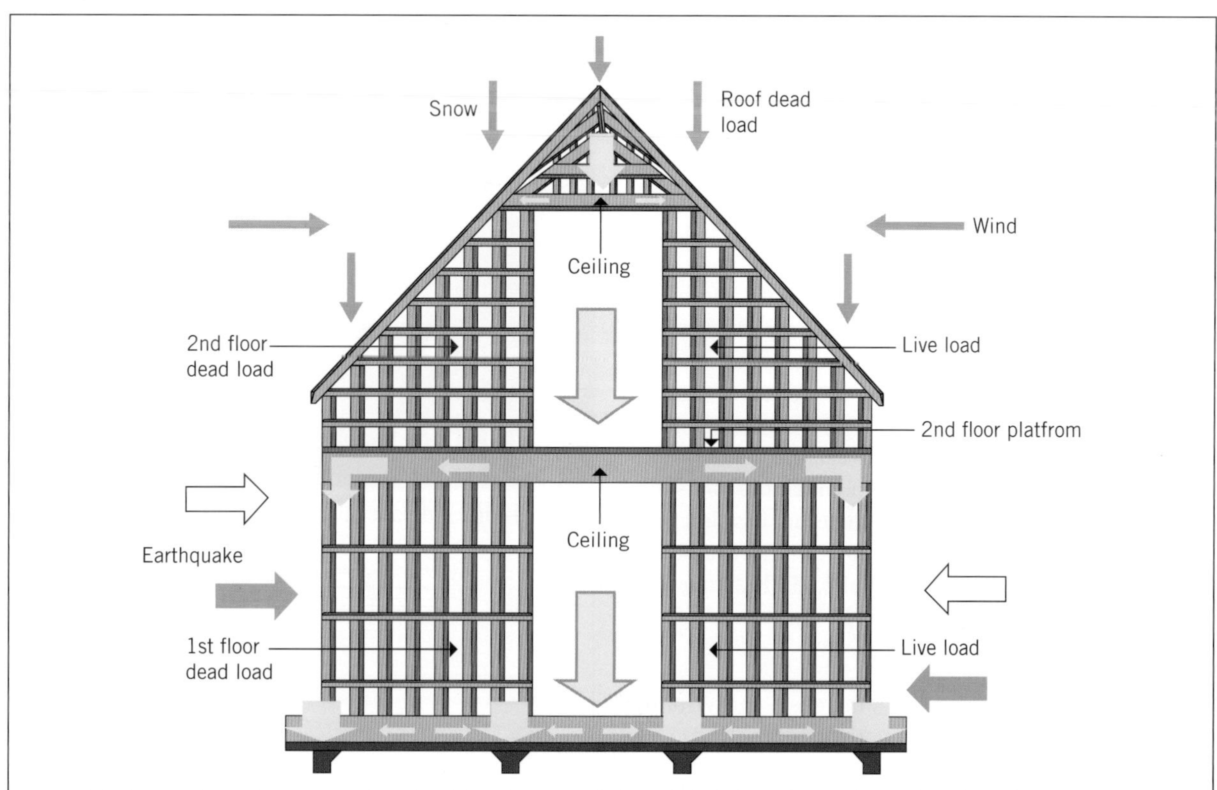

FIGURE 12.16 Typical load paths on a domestic house

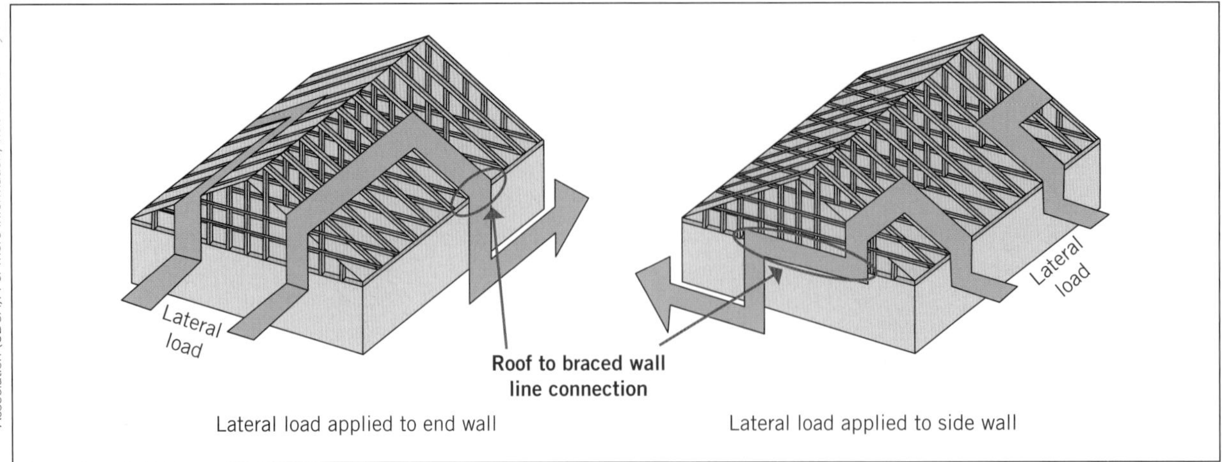

FIGURE 12.17 Lateral load paths

Loads 5: Load characteristics

The type of load, and the path that a load may take, is only part of the story. We must also take into account the way loads are imposed upon a structure. These are the load characteristics and they may be categorised in three ways: point loads, uniformly distributed loads and uniformly varying loads.

Point loads

Point loads are those deemed to be acting upon a specific point. The area of action is regarded as too small to not warrant comment, or it is sized so that a penetration by shear action may be assessed. For example, a lintel picking up the concentrated load of a girder truss is deemed to be receiving a point load at the location that the truss sits. In Figure 12.18, the roof load from a ridge beam delivers a point to the lintel, which in turn distributes this to either side of the radial headed window. These two opening stud clusters (three studs nailed together) then take the concentrated loads down to the floor and on to the footing. That is, the load path has been tracked via a series of point or concentrated loads. In this image you should also note the steel strapping applied to counteract uplifting wind loads.

Point loads in calculation diagrams, another form of the free-body diagrams shown previously, are shown by an arrow. Figure 12.19 shows an example of this type of diagram and the manner in which the loads are offered. The arrow head denotes the exact location of the action on the supporting member: in this example the red arrow shows a point load of 2.45 kN acting 2.9 m from support B.

FIGURE 12.18 Point loads (note strapping to prevent uplift)

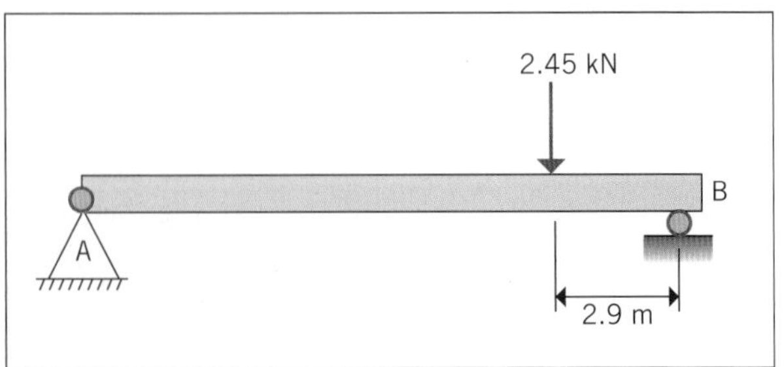

FIGURE 12.19 Free-body diagram showing loads on a beam

Uniformly distributed loads

In the next diagram, Figure 12.20, there is a cluster of smaller red arrows linked by a horizontal line and spread over a width of 4.2 m. Above this a force of 1.1 kN/m is indicated. This is one of the various ways by which a uniformly distributed load or UDL may be

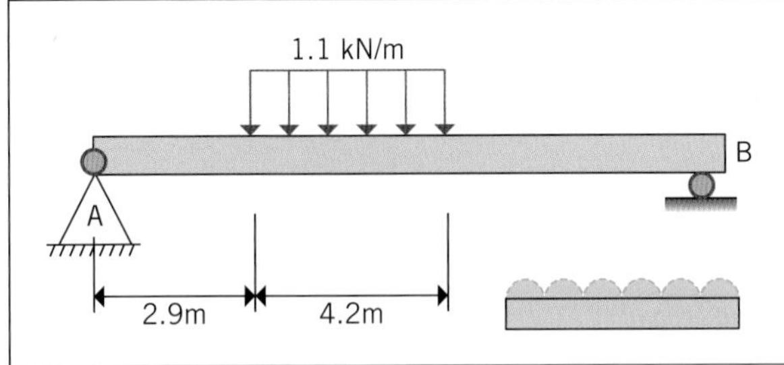

FIGURE 12.20 Uniformly distributed load (UDL) on a beam. An alternative depiction of UDL is shown at bottom right.

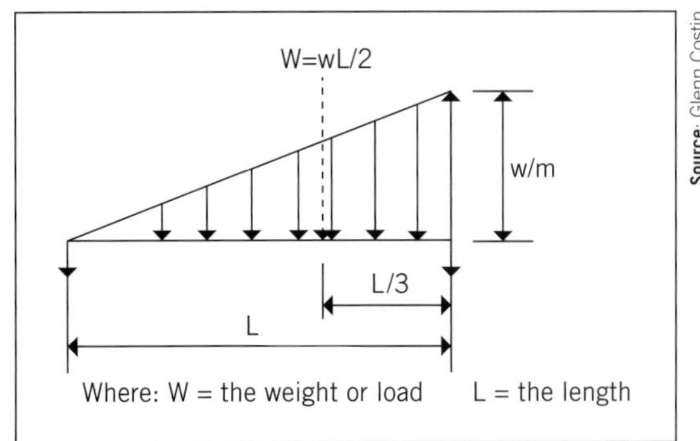

Source: Glenn Costin

FIGURE 12.21 Uniformly varying load and considered point of action

depicted; a common alternative is offered in the same figure. The load is always offered per unit of length (generally metres), but the load itself may be given as a mass (kg) rather than a weight or force (N or kN). In such cases you must account for gravity and multiply by 10.

UDLs loads are deemed to be evenly distributed over a given area or length. Imagine a person standing on thin ice. Standing, they have two-point loads, their feet, delivering the whole of their weight to a very small area. Like this, the ice will probably break. If the person were to lie down upon the ice sheet, however, their weight would be evenly distributed so the effective weight per square metre is radically reduced and the ice less likely to fail.

UDLs in construction are very common loads: suspended concrete slabs resting on a beam; roof or floor loads on beams; water in swimming pools; the load of a footing upon the foundation – these being the more obvious examples. We may also use the principle of the UDL in place of small point loads. This is exampled in how we deal with a series of floor or ceiling joists on a beam.

Uniformly varying loads

Like UDLs, uniformly varying loads (UVLs) distribute a load across a surface or support member. The difference is that UVLs have a load that is greater at one end than the other: the load tapers off as it is distributed along the beam. In a load diagram it might look like Figure 12.21.

Knowing how to interpret these diagrams is fundamental to being able to solve basic structural load calculations. In reality, you will seldom have to do such calculations: they are in the realm of the engineer. However, knowing the basics allows you to visualise the loads being applied in practice. You are therefore more informed of the risks and challenges faced in day-to-day construction, even if your role is confined to timber wall framing, concrete formwork, or scaffolding. It also increases your capacity to have constructive input into the building design process, and to better interpret engineering responses.

LEARNING TASK 12.2

CREATING SIMPLIFIED DIAGRAMS EXPLAINING LOAD PATHS

Types of loads (AS 1170) and load paths diagrams

Using the 'sectional' diagram shown below, demonstrate your understanding of the elements that would be shown on a 'force diagram' where it is anticipated there will be:

- a balustrade made from glass along the outside edge of the cantilevered part of the slab
- an internal 400-mm-long masonry wall approximately one-third in from the left-hand side of the slab
- office furniture and cabinets will be evenly spaced around floor areas
- columns that will be placed above the existing so as to create an additional storey above the current suspended slab.

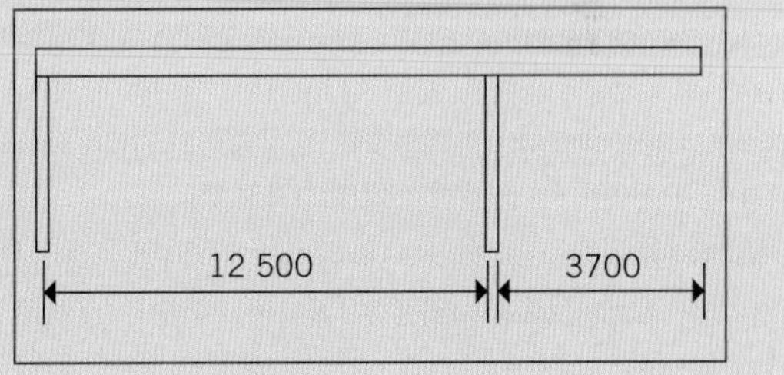

Moments and force system solutions

The diagrams of the beams above (Figures 12.19 and 12.20) show point and UDL loads acting away from supports. When a force is applied at a distance to a support or point of fixture, it tends to make the beam try to rotate around that support or point.

This is exampled when you push on a door. In pushing on the door, the door rotates away from the force applied; pivoting on the hinge. If the door was replaced with a sheet of ply fixed on one edge without a hinge then it would still try to rotate, if pushed, but would resist. If you continued to push, one of four things would ultimately happen: you would give up;

the ply would break; the fixing at the edge of the ply would fail; or the material that the ply is fixed to would begin to rotate instead of the ply.

The measure of this tendency to try and resist the turning effect is called a 'moment'.

When calculating a moment, you are applying the principles of statics. You are effectively asking, 'What is needed to resist the applied force so that the object does not rotate?' The direction of the moment you find is therefore opposite to the direction of the applied force – it is someone pushing back on the other side of the door with the same strength you are applying.

The value of a moment is found by multiplying the amount of force applied by the distance that this force is applied from the support.

I.e. $M = F \times d$

where M = Moment (N.m or kN.m is the resultant unit)

F = Force (N or kN)

d = distance (m)

Figure 12.19, for example, shows a force of 2.45 kN applied 2.9 m away from support B: so the moment is:

$M = 2.45 \text{ kN} \times 2.9 \text{ m}$

$M = 7.105 \text{ kN.m}$

Moment solutions

Solving for moments is done using free-body diagrams similar to those already shown. The diagram in Figure 12.22 offers perhaps the most basic example, a simply supported 8-m-long beam with a single centrally located point load.

This is a fairly straightforward problem for which the reaction forces, or loads supported, at A and B are found by simple logic. With a load of 39 kN in the middle, the reactions must be 19.5 kN at each support for the 'system' to sum to zero. That is:

$\Sigma Fv = 0$

In this case: 19.5 (up) + 19.5 (up) – 39.0 (down) = 0

This is the first step towards the static state. We know the forces equal out so the beam is not moving up or down.

There are no horizontal forces so:

$\Sigma Fh = 0$

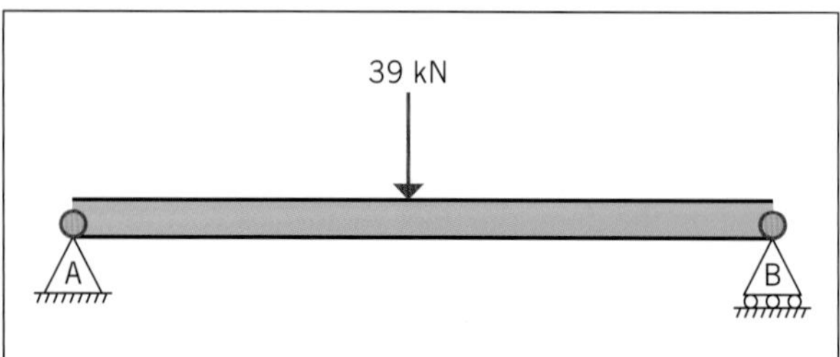

FIGURE 12.22 Free-body diagram for simply supported beam and centred point load

This leaves us with the last equation, that of moments:

$\Sigma M = 0$

When solving for moments we have the same sort of diagram, but done a bit differently; and we need a bit more information. Using the beam shown previously, we will add the extra information, and then prove our reaction forces first, as moments, before finding the moments for the load. Remembering that moments are a rotational response to a load, we talk of solving 'around' or 'for' each support; doing so by effectively removing one support and assuming the beam will 'hinge' around the other as if on a pivot.

Figure 12.23 now shows us our beam pivoting around support 'A'. We have to ensure this beam stays still: stays 'static'. In this diagram we replaced the support at 'B' with an arrow indicating the amount of force we would need to use to hold that end up. It shows us that this reaction at R_b, the ***moment***, is an upwards ***anti-clockwise*** rotational force acting in opposition to the load – a downwards ***clockwise*** rotational force.

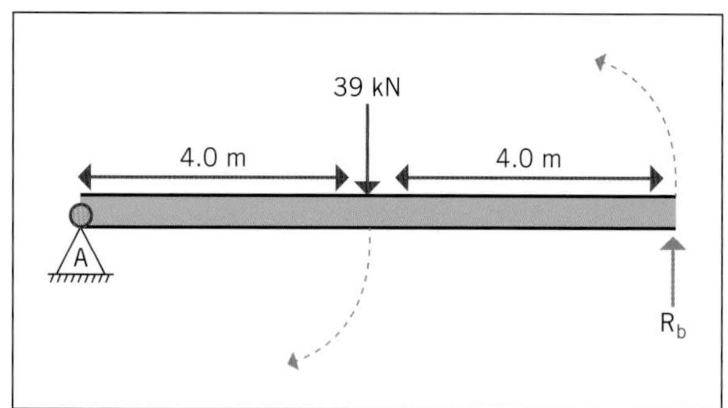

FIGURE 12.23 Moment diagram for simple beam and point load: solving around 'A'

And here we will step back a little and remind you of levers, a principle most tradespeople use on a daily basis. In the two images in Figure 12.24, two people are trying to hold up similar loads. Experience tells you that the person at right is making lighter work of it; i.e. uses less force to hold up the load.

Looking back at our diagram in Figure 12.23 the same principle applies, clearly less force will be needed at R_b than would be required if the support was located directly under the load; i.e. it will be less than 39 kN.

The same principle applies in our next diagram, Figure 12.25. In this case we have replaced support B and removed 'A'. Only in this case we see the reverse. Reaction at R_a is trying to go ***clockwise***, while the load is trying to spin the beam ***anti-clockwise***.

Because we have distance from the axis (support point) we can calculate the moment in each case. This is done in the following manner.

- **Step 1** Determine which reaction you will solve for (R_a or R_b)

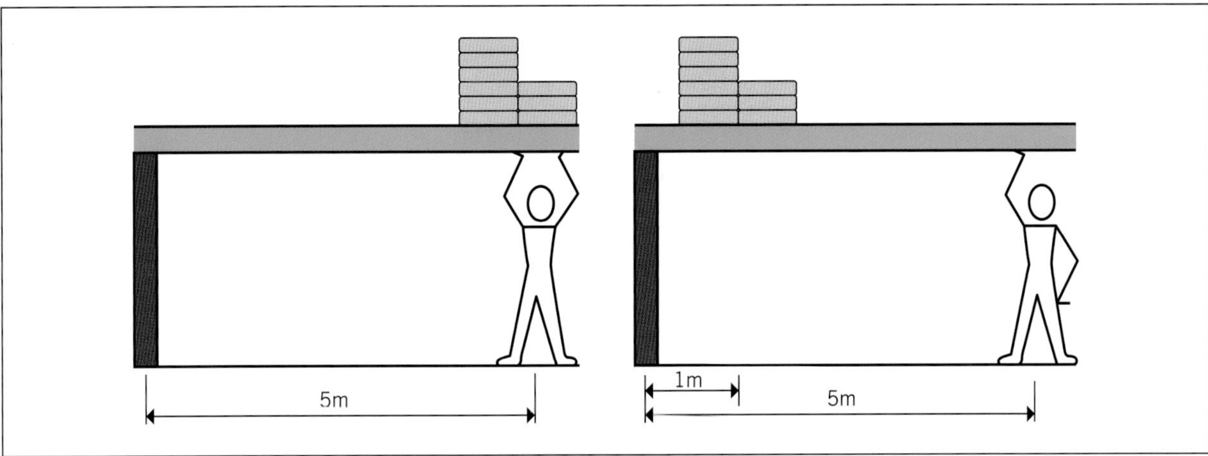

FIGURE 12.24 Principle of levers. Person at right needs less effort to hold up the load.

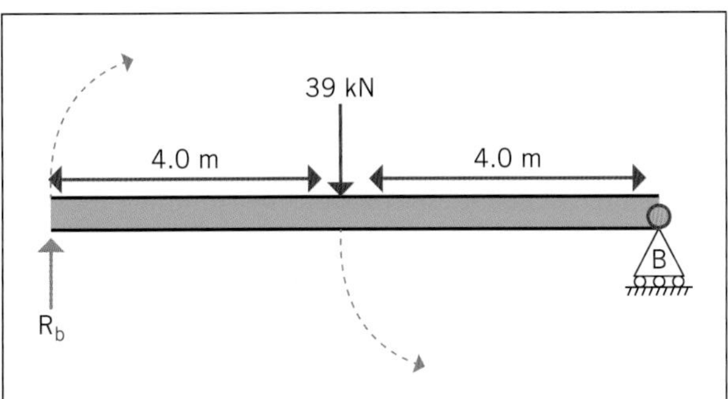

FIGURE 12.25 Moment diagram for simple beam and point load: solving around 'B'

- **Step 2** Identify the direction of rotation (clockwise or anti-clockwise) of both the reaction and the loads applied.
- **Step 3** Identify the distance the load is applied from the axis point; i.e. support A or B. Hint: if you are looking for reaction R_a, then your axis is B.

There are various ways that the calculations can be performed. Some use positive and negative values for clockwise or anti-clockwise. Others use a balanced equation approach. We will use the balanced equation as it is more visual in presentation and so the concepts easier to interpret.

As shown below, we begin with a heading: 'solving for …' followed by our two sides of the equation headed up as clockwise on the left and anti-clockwise on the right. For the beam to stay still, the forces we calculate on each side must equal each other.

Solving for moments around A (Figure 12.23)

Clockwise	=	**Anti-clockwise**
39 kN × 4 m	=	R_b × 8 m
156 kN.m	=	R_b × 8 m
156 kN.m ÷ 8 m	=	R_b
19.5 kN	=	R_b

This is the amount of force our logic told us earlier; i.e. we have 'proved' our reaction at B.

Solving for moments around B (Figure 12.25)

Clockwise	=	**Anti-clockwise**
R_a × 8 m	=	39 kN × 4 m
R_a × 8 m	=	156 kN.m
R_a	=	156 kN.m ÷ 8 m
R_a	=	19.5 kN

Again, 'proving' our earlier belief for the force at support A.

You now check to see if the sum of the reaction forces equals the sum of applied forces. This is based on the fact that our reaction forces at the supports must equal our moment forces at the supports or:

$\Sigma F = \Sigma M$

That is:

Total loads	=	**Total reactions**
39 kN	=	$R_a + R_b$
39 kN	=	19.5 kN + 19.5 kN
39 kN	=	39 kN

If the supporting reactions had come out less than the total combined force then the beam would collapse. Conversely, if the reaction forces had summed as being greater than the applied force, then our beam is floating away …!

This is the key point. You are working towards a static state. That is, your building, or a given structural component, is being designed to remain stationary, neither rising nor falling.

More complex force systems

In summing up forces, we make a distinction between up and down, left and right; and, for moments, clockwise and anti-clockwise. Up is conventionally given a positive value and down a negative or minus value. Likewise, a force acting towards the left is negative, while a positive force acts toward the right.

With regards to moments: clockwise is positive whereas anti-clockwise is negative. You might see different conventions used but whatever convention you choose, you will arrive at the correct answer as long as you are consistent.

Such conventions are useful in cases where the total forces or loads on a system under consideration do not sum to zero. For example, the uplift calculations on roof sheets, will result in – (minus) value. This means the chosen hold down, screws, bolts, straps, etc., will need to be equal to or greater than this found value: we have to add a 'plus' to counter the 'negative'.

In the following example, continue to use the balanced equation of 'clockwise must equal anti-clockwise'.

Example 1: Single offset point load

Figure 12.26 shows an offset point load on our 8-m beam. Solving the moments for this uses the same principles as previously; i.e. removing a support and solve moments around the remaining one. Remember, to find a moment, you must multiply the load by the distance it is applied from the support.

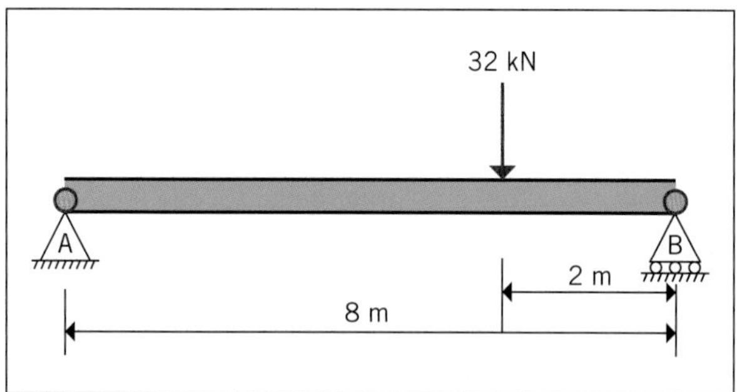

FIGURE 12.26 Single offset point load

Solving for moments around A

Clockwise	=	**Anti-clockwise**
32 kN × 6 m	=	R_b × 8 m
192 kN.m	=	R_b × 8 m
192 kN.m ÷ 8 m	=	R_b
24 kN	=	R_b

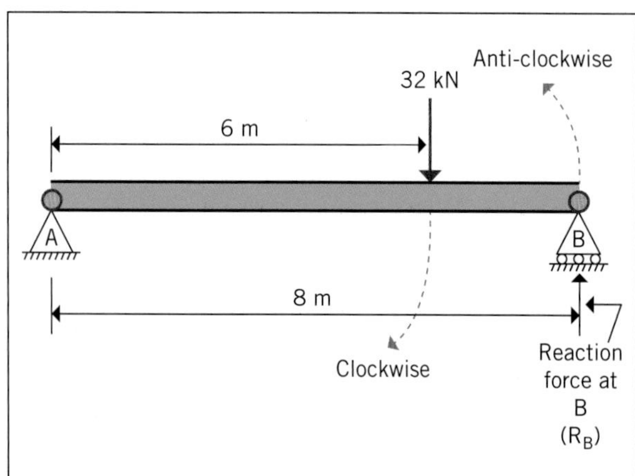

FIGURE 12.27 Calculating for reaction force at support 'B'

Solving for moments around B

Clockwise	=	**Anti-clockwise**
R_a × 8 m	=	32 kN × 2 m
R_a × 8 m	=	64 kN.m
R_a	=	64 kN.m ÷ 8 m
R_a	=	8 kN

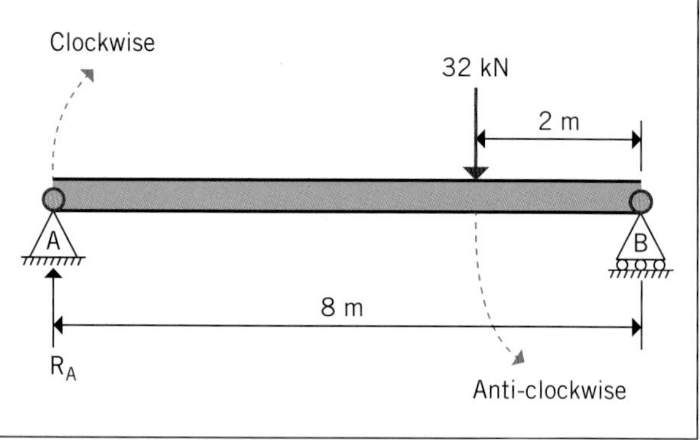

FIGURE 12.28 Calculating for reaction force at support 'A'

We then check these solutions by summing the forces for equilibrium; i.e. $\Sigma F = \Sigma M$

Total loads	=	Total reactions
32 kN	=	$R_a + R_b$
32 kN	=	24 kN + 8 kN
32 kN	=	32 kN

See **Appendix 4** for more examples, where we gradually add more complex loads, moving from single and multiple offset point loads, through universally distributed loads, combinations and finally cantilevered loads. In each case we shall continue to use the balanced equation approach.

This concludes the section on forces and force systems. By understanding the principles behind these calculations, you are significantly better positioned to engage with other construction professionals in making considered structural choices.

LEARNING TASK 12.3

PRACTICAL USE OF MOMENTS IN FORCE SYSTEMS

You have been asked to move a steel beam 3.6 m in length, weighing approximately 34.5 kg/m.

- How much weight (force) will you need to 'notionally' lift the beam and keep it held above the ground while moving it, if only one end of the beam was being lifted and you were then going to rotate the beam through 180°?

Bracing systems

Supporting vertical loads in a structure is a relatively simple exercise; however, catering for lateral loads can be more complex. Holding against these loads is known as bracing, and this is a critical part of any structure. This section introduces you to the basic principles behind bracing, and the common strategies adopted by builders, designers and engineers.

As discussed previously when looking at loads and forces, wind can act to push a wall over, as can lateral earth and seismic forces. There are three main approaches to solving for the thrust of these forces:

- shear walls
- triangulation
- fixed joint systems.

Each of these have been used in some form or another for many centuries, and the principles have remained much the same. As the shear wall is the most basic, it will be discussed first.

Shear walls

The easiest way to hold up a wall is to use another wall. This principle you most likely applied when building a house of cards as a child. The first card (wall) is held up by another card placed at right angles to it. Pushing against the first card only makes the system stronger – up to a point …

The second card is your shear wall. It is rigid in being a solid sheet that when pushed from the end must buckle, fold, or slide before the other wall can collapse. In construction the role of this card may be played by concrete panels, timber frames sheathed in ply, or even a solid brick wall. Located, much as the playing card, at 90° to the wall it intends to support, it can provide significant resistance to racking forces. Figure 12.29 shows the location of a shear wall and how it acts to resist lateral loads stemming from forces being applied to the wall running at 90° to it.

Penetrations in shear walls reduce their capacity to withstand racking forces and so must be avoided. When penetrations are unavoidable, reengineering may be required depending upon the size of the penetration and the forces involved.

Shear walls are the basis of most contemporary bracing strategies for domestic homes today.

Triangulation

Triangulation is another very old principle and also one you will have used many times to prop up something that is leaning over with an angled stick. Triangles are an extremely strong shape, being only as weak as the components or jointing systems by which they are made. Even with pinned joints – connections that allow the coupled members to rotate or move easily at the joint – a triangle cannot change shape without one of the members, or the joints, failing.

As a bracing technique it is relatively simple: squares and rectangles are locked from racking by something fixed diagonally from corner to corner. In so doing, a triangle is formed and the shape cannot change without twisting to the side or the brace or joints breaking.

It is also important to understand that the materials and fixing methods used to form this brace have a strong influence upon the capacity bracing achieved. Timber for example, is good in compression; however, timber joints are not particularly good under tension. Conversely, steel strap and rod is very good in tension, and joints for holding this tension easily achieved: but such materials are very poor in compression. This is exampled in Figure 12.30, where only one steel brace is used: lateral forces coming from the right are easily resisted whereas those coming from the left cause failure.

Solid timber bracing functions the reverse of this, acting more like the prop you may have used to hold something up discussed earlier. This is demonstrated in Figure 12.31, where it is the joints that are seen to have failed.

To overcome these issues, both timber and steel bracing is best applied in pairs – or in the case of steel strap, as a crossed pair.

Triangulation was once the main principle behind the bracing of timber structures in this country. As available wall area has reduced through changes in home design, the ability to deploy multiple triangles has diminished. Despite being replaced in many cases

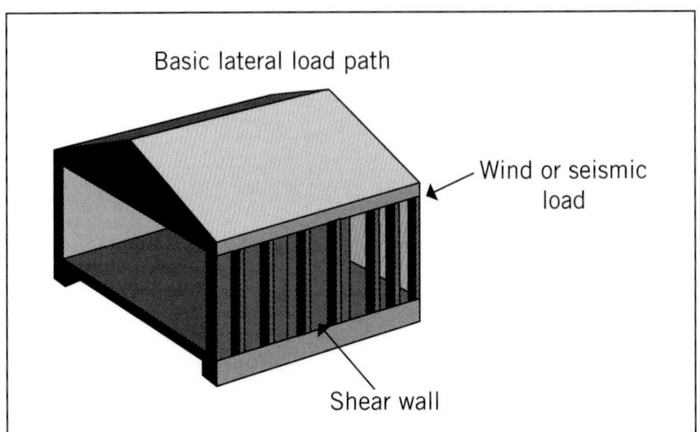

FIGURE 12.29 Basic lateral load path

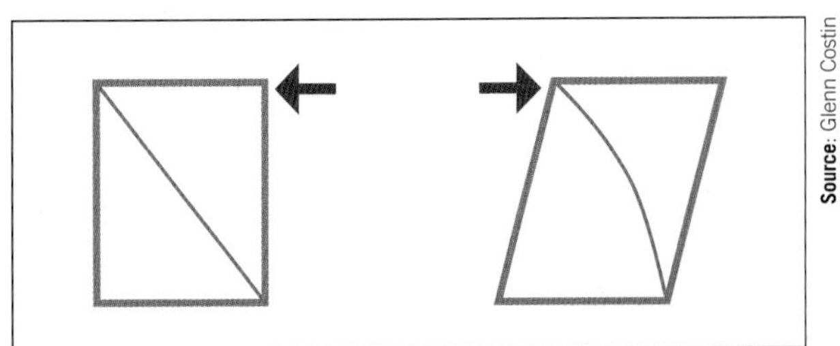

Source: Glenn Costin

FIGURE 12.30 Lateral forces on single steel strap bracing

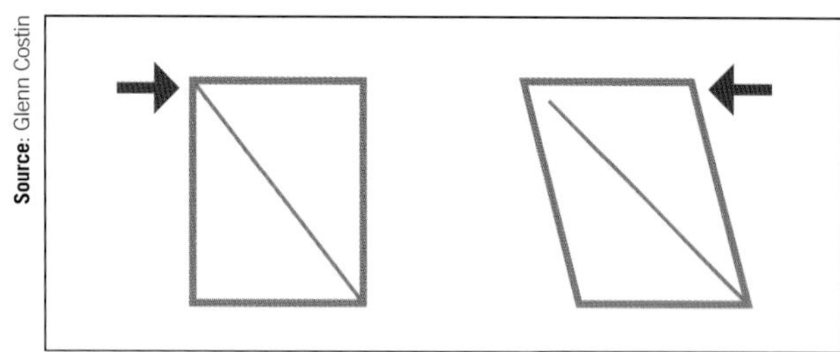

FIGURE 12.31 Lateral forces on single solid timber bracing

by shear wall systems, it is still commonly used where wall length allows.

Fixed joint systems

Fixed or ridged joints have historical beginnings in timber through the use of pinned mortice and tenon joints. Large frames could be made which could resist a fair degree of lateral load. Today, this form of bracing allows for the construction of portal frames in timber or steel. The ridged joint system – in timber now usually achieved through flat steel 'fish' plates – limits the amount of movement at the joint, transferring the load instead to the column and base. In steel portal frames, the joints may also use fish plates, direct bolting, or be fully welded. Provided the column is of sufficient integrity at or near the joint to withstand this load, the frame is capable of withstanding significant lateral forces. (See Figure 12.32.)

Steel and timber portal frames have an extensive history in the construction of sheds, factories and warehouses. However, they are also appearing in domestic home design as a means of quickly establishing the outer shell, prior to creating spaces and rooms with non-loadbearing walls.

Floors and ceilings as bracing

Upper floors, ceilings and roof surfaces can all act as bracing within a structure, while many require bracing to be structurally stable. Suitably braced, using any one of the methods offered above (though most are usually crossed steel strap or rod bracing), these surfaces provide significant lateral restraint.

Ceilings or upper floors, for example, when braced and tied appropriately to braced walls, turn the structure into a ridged box. For failure to occur, one or more theses surfaces must buckle, and the joints must tear. Even pitched roof surfaces, once made rigid through bracing, afford similar characteristics.

Application with regards to codes and regulations

The bracing of a structure is carefully regulated in Australia, particularly in the light of changing climatic conditions. The NCC and various Australian standards must be adhered to and so bracing diagrams form part of the plan set of most contemporary homes. These will dictate the type and load-carrying capacity of each bracing 'unit', such units then being defined by AS 1684 with regards to timber structures. This shall be addressed more fully in the next section of this chapter, where discussion will focus upon the requirements of materials, fixings and the load capacities for each method adopted.

LEARNING TASK 12.4

SOLVING BRACING PROBLEMS

Access AS 1684.2 Bracing section and respond to the following questions.

1 What section shows the bracing requirements for residential timber buildings?
2 Is temporary bracing required during the construction process for residential buildings?
3 If the answer is yes, what clause in AS 1684.2 requires this and what amount of temporary bracing is required?
4 a How are racking forces calculated from AS1684.2?
  b Provide the relevant clause number.
5 You have calculated that you must resist 44.6 kPa based on an assessment of the calculated pressures on the southern wall and roof surfaces according to the method provided in AS 1684.2. Nominate the wall/s north, south, east or west where the bracing will need to be placed, according to AS 1684.2, so as to effectively resist the pressures on the southern building surfaces subject to wind forces.

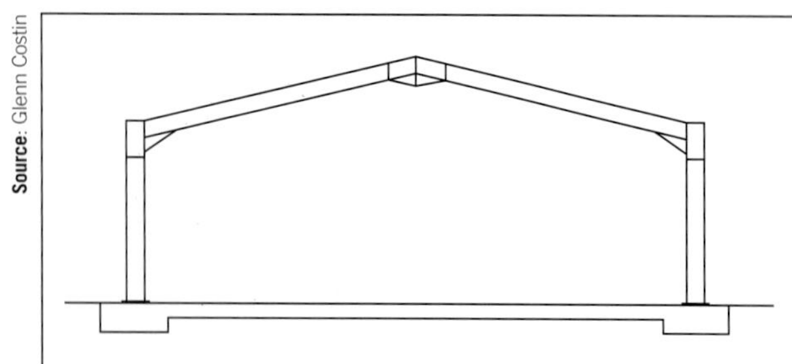

FIGURE 12.32 Typical portal frame

Truss systems: floor and roof components

Trusses are frames that make use of the structural characteristics of triangles to bridge large spans and carry both live and dead loads. Most contemporary homes in Australia are constructed using either timber or steel trusses that displace roof loads to the outside walls; thereby leaving the internal space free of loadbearing walls and allowing greater adaptability of design.

Whole books could be, and have been, written about trusses, most of which is best obtained from the manufacturers themselves – one of which you will be directed to at the end of this section. The purpose of this part is but to offer some insights, to 'nutshell' this plethora of information so that the basics are clear.

Basic principles

The purpose of a truss is to efficiently transfer loads outwards to its extremities where such loads and forces can be received by walls, posts, or beams, thus leaving a clear safe span underneath. Trusses are made up of a number of lightweight components. These components interact with each other such that when loaded downward or laterally each component is in *either* compression or tension (see Figure 12.33). This makes trusses more efficient at supporting loads over wider spans than simple beams, which experience tension and compression at the same time.

From an ideal engineering perspective, all the components of a truss would have pinned joints. These types of joints impose no bending moments on the attached members, thereby eliminating the need for such calculations in the design. The reality, however, is that most timber trusses use some form of fixed jointing which requires more complex computations.

When designing trusses, it is important to identify ties – components under tension, as against struts – and components under compression. In typical loading conditions the common fink truss components may be identified as shown in Figure 12.33.

This basic principle of tie and strut applies to all of the truss types to be described below; even in parallel chord trusses used in floor, or floor and ceiling designs. What must be understood, however, is that with severe uplift, all these components may reverse their roles.

Common types of trusses

The roof truss system has been a part of construction for many centuries and you may find both simple and complex forms the world over. The modern lightweight timber or steel truss is, however, a relatively modern invention arising in the 1950s. In Australia today, trussed roofs have come to dominate the residential construction sector. There are a range of truss types in use and these are depicted in Figure 12.34. Of them all, the Fink truss is currently the most commonly used form in Australian housing.

Aside from roof trusses, there are trusses used to support floor, and floor and ceiling loads in place of deep heavy beams. These are known as parallel chord trusses and may be made from steel, timber or, more commonly, timber and steel in combination (see Figure 12.35). Parallel chord trusses can also be used for roofing when a vaulted or cathedral ceiling is desired.

Jointing systems

Although this section's focus has been primarily upon proprietary truss systems, architectural or bespoke trusses – those made specifically to a client's design requirements and usually highly visible – are still required.

Proprietary systems tend to use some form of gang nail, or multi-nail plate that is machine pressed over the joint (Figure 12.36a). Although not visually appealing, this is acceptable as they are normally hidden within the ceiling space. Architectural trusses, on the other hand, tend to be exposed, particularly those of large-section timber. These are frequently jointed using steel fish plates that are through bolted. These plates may be exposed on the face or cut into the ends of the components to offer a cleaner look and

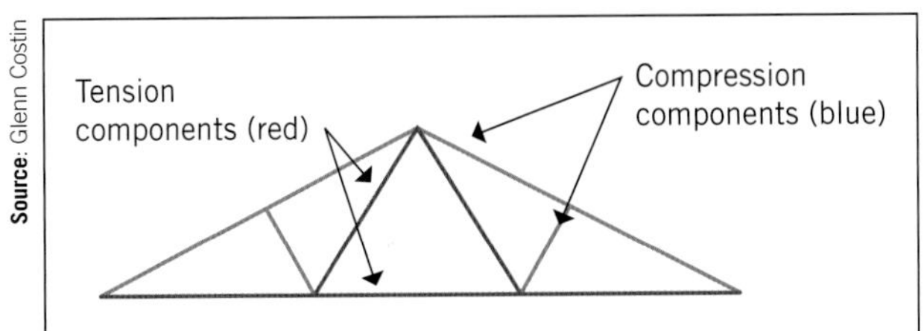

Source: Glenn Costin

FIGURE 12.33 Fink truss showing tension and compression components

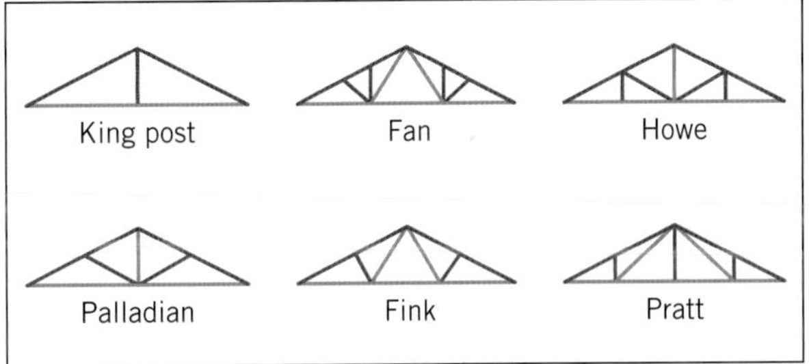

FIGURE 12.34 Common truss types used in Australia

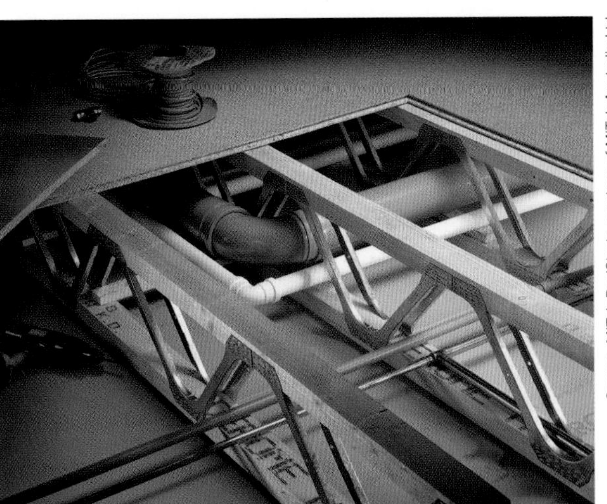

Source: MiTek PosiStruts courtesy of MiTek Australia Ltd

FIGURE 12.35 Parallel chord truss

show off the timber and the joint in a more striking contrast (Figures 12.36b and 12.36c).

Older style timber jointing is also used in these type of bespoke trusses (Figure 12.36d); however, their use is limited due to the time and skills required in their manufacture. When complete though, their effect is striking; and their strength, though seldom if ever tested by engineering, has been tested by time in buildings standing many centuries after their assembly using these processes (for example, the Harmondsworth Great Barn – 1427).

LEARNING TASK 12.5

CONSIDERATIONS: SHOULD I ALLOW TRADES TO MODIFY OR CUT STRUCTURAL MEMBERS?

Truss systems: jointing

Consult AS 4440 and the MiTek website.

1 Go online and do a search for truss manufacturer installation instructions similar to that shown in the text. Using the information you have accessed, what parts of a truss are:
   a tension members (use 'T' to indicate on the diagram)
   b compression members (use 'C' to indicate on the diagram).

   Note: All lines shown should have a 'T' or 'C' notation.

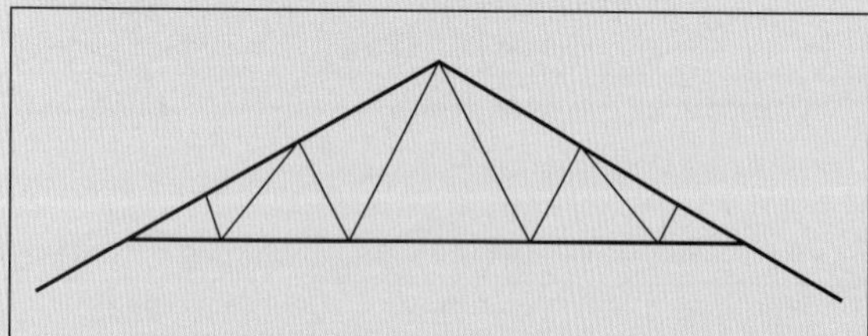

2 Which truss members should never be cut on site, the tension or compression members? State your reasoning why.

Source: iStock.com/Lakeview_Images

FIGURE 12.36(a) Gang-nail plates

FIGURE 12.36(b) Internal fish plates

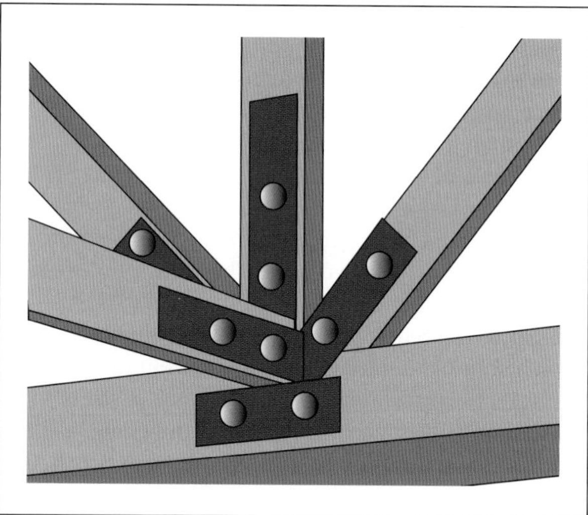

FIGURE 12.36(c) External fish plates

Source: Alamy Stock Photo/Alex Ramsay

FIGURE 12.36(d) Timber joints

Material properties: influence upon structural performance

Structural performance of a building component under load, or its suitability in any given circumstance, is highly dependent upon three main factors:

- the material(s) the component is made from – a component's mechanical properties
- the sectional shape or profile of the component – its sectional properties
- the way that the component is aligned in relation to that profile – an extension of the sectional properties.

We will begin by briefly discussing the mechanical properties before dealing with the influence a profile and its alignment can make to a component's effective strength.

Mechanical properties

Aside from flammability – a chemical property – the many mechanical properties of a material are highly influential in determining the appropriateness or otherwise of a component for a given situation. Each of these properties are discussed in Table 12.3.

Note that in reading the table, many properties can be viewed as pairs of opposites; such as elasticity and plasticity. You should also understand that a single material may be identified by multiple properties, not just the one. Further, that materials may have varying levels of a property; e.g. that in suggesting a type of steel is highly ductile does not mean that it can be stretched out like a piece of gold. Rather, that it is perhaps more ductile than other materials, or other steels, being considered for a particular use.

Sectional properties

Knowing the mechanical properties of a material provides us with some important insights into how that material will behave under given loads. However how we shape that material, such as a steel beam, can also make a significant difference to its performance. The shape of concern is known as its ***section*** or

TABLE 12.3 Mechanical properties of materials

Property	Definition	Test or example
Strength	Describes the resistance the material can continue to offer to a load just prior to either permanently deforming or fracturing.	May be measured in many ways including tension, compression and deflection; concrete strength, for example, is rated on its capacity to withstand a given pressure via a compression test.
Hardness	Resistance to abrasion or indentation.	A common arbitrary test is the material's capacity to withstand a blow from a pointed object – the basic test for determining if steel is suitable for fixing with a powder-actuated tool.
Toughness	The ongoing resistance to fracture by a material after repeated bending or twisting.	There are several tests, such as noting the frequency with which the material may be twisted prior to fracture, or how many bends it can withstand under given conditions. We often declare a material to be 'tough', such as a piece of plastic, having given up trying to break it by constant bending, and resort to the chisel or knife.
Brittleness	A brittle material is one that is easily snapped or fractured without stretching.	Many materials may become brittle with repeated straining or bending, commonly exampled when we bend a thin piece of metal back and forth to break it.
Elasticity	The ability of a material to flex or deform under load and then spring back to its original shape or length.	Elastic (rubber) bands and pencil erasers example this behaviour, but this same attribute may be said (to a lesser degree) of some metals, plastics and even timber.
Plasticity	The opposite of elasticity, plasticity allows a material to be easily deformed or reshaped under load. Unlike elasticity, however, it will not spring back to shape, but remains in the new form.	Many fillers exhibit this nature prior to hardening.
Ductility	Ductile materials can be stretched or bent significantly without breaking.	Gold is highly ductile in that a single ounce (about 31 grams) can be stretched without breaking to approximately 80 km. In being stretched or otherwise reshaped, the material's other properties such as strength, hardness, toughness, etc. are not changed.
Malleability	This is the capacity for a material to be reshaped. In being reshaped, the material's other properties are not changed.	In this regard it is similar to ductility. However, a material may be highly malleable, yet actually have fairly low ductility because stretching it snaps it easily. Conversely, like gold, it may be both malleable and ductile.

cross-sectional shape; and each shape has properties that differ depending upon how they are aligned with respect to the loads applied.

Take careful note: These properties have nothing to do with the strength of the material itself. Sectional properties are based solely on the shape of the section, not the material they are made of. Sectional properties explain why one shape may be more efficient at a supporting load than another, despite both components being made of the same material; or that they are both of the same material and shape but are aligned to the load differently.

With regards to construction, the main sectional properties that you need to be aware of are:

- centre of area (or centroid)
- second moment of area (I)
- section modulus (Z)
- radius of gyration (r).

The importance of each of these to your understanding of a component's response to loads is offered below, along with the basic calculations by which each is determined. These calculations are generally undertaken for both alignments or *axes of symmetry*; i.e. the x–x axis and y–y axis as shown in Figure 12.37.

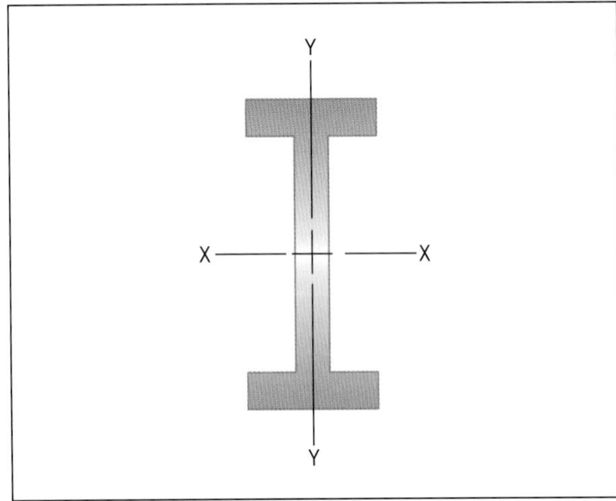

FIGURE 12.37 The two axes of a component regarded in calculating sectional properties

Centroid

Also referred to as the centre of area, the centroid is effectively a shape's centre of gravity. Pick up something from a point below the centre of gravity and it is likely to rotate. The centroid is identified for each of the sectional shapes shown in Figure 12.38. In each case, you will most likely have been able to identify the location intuitively, logically. On the more complex shapes, however, finding the exact point can require some calculation.

Knowing the location of a component's centroid allows us to either align it such that it is inherently unlikely to topple over or, alternatively, stabilise or brace the component when structural requirements dictate the arrangement. Typical examples you are likely to be familiar with are tall, thin, hanging beams or deep beam joists.

The centroid is used in determining other properties, such as the section modulus.

Second moment of area (I)

Effectively, the second moment of area (I) is a statement of a shape's ability to resist bending when a load is applied to one axis or the other. It is a measure of the shape's resistance to the load, not the material's resistance. This tells us which way to lay a component so as to take best advantage of both the material's and the shape's capacities. Figure 12.39 highlights this choice in a well-known building scenario.

Often referred to as the ***moment of inertia*** (though technically there is a difference) the second moment of area has a direct relationship to the area of a section and how that is displaced around the centroid. The calculation of the second moment of area (I) is taken for both axes. The higher the value found, the less bending will take place.

Remember, this is a measurement of the shape's capacity, not the material's. Hence, a shape with a high

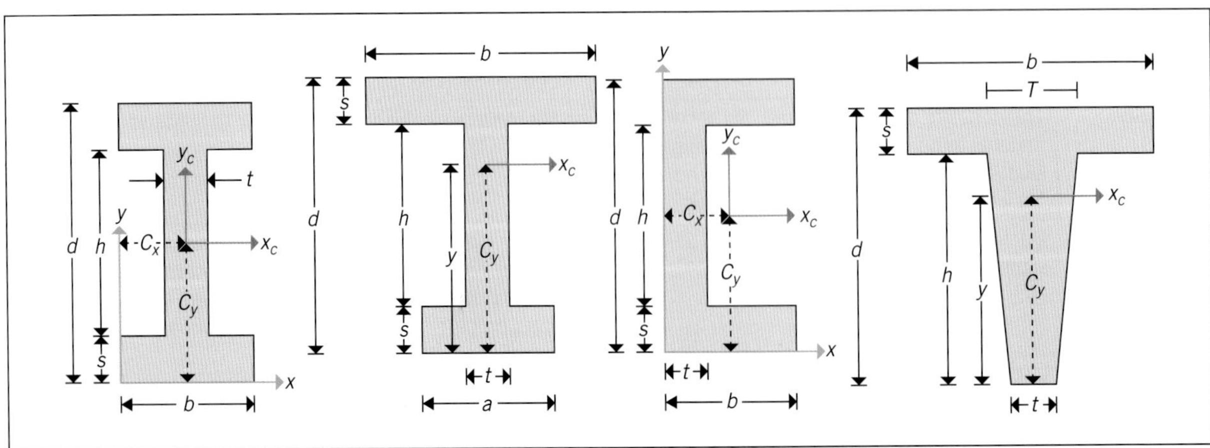

FIGURE 12.38 Identifying the location of the centroid on various sectional shapes

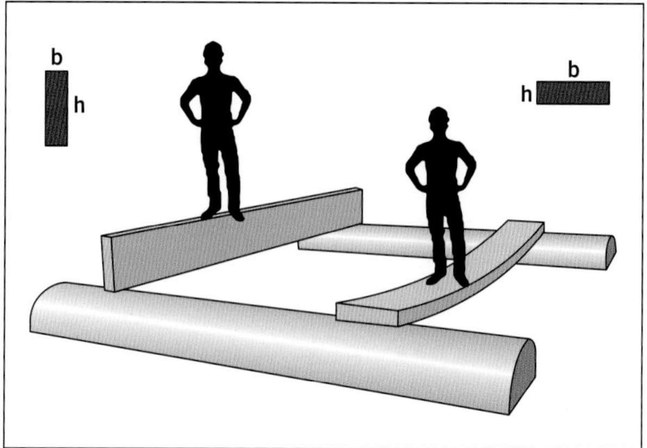

FIGURE 12.39 Beam alignment and bending

'I' value made from radiata pine is unlikely to be as strong as shape with a smaller 'I' value but made of steel.

The formulas for determining the second moment of area for basic shapes are given below:

- square $I_{xx} = I_{yy} = bd^3 \div 12$
- rectangle $I_{xx} = bd^3 \div 12$ and $I_{yy} = bd^3 \div 12$
- circle $I_{xx} = I_{yy} = \pi D^4 \div 64$
- Where b = breadth; d = depth; D = diameter

The result in each case will have the unit of mm^4.

As millimetres are used in these calculations, very large numbers will result. If need be, revise the section on scientific and engineering notation when following the example calculations that follow. Typically, the result should be given as a value × 10^6 mm^4. For example, the calculated result may be:

84,547,365 mm^4

This becomes:

84.55 × 10^6 mm^4 (note the rounding of .547 to .55)

Comparing the 'I' for different beam orientations

A value for 'I' is required about the x–x axis for a simple rectangular beam, as depicted in Figure 12.40. The expected load is being applied vertically down on the beam.

Orientation A has the beam aligned 'on edge' or vertically. In orientation B, the same beam is arranged horizontally. Logic and experience tell us the answer, the calculations should therefore support our belief that in orientation B the beam will bend more.

Calculating I

Being a rectangle, we will use: $I_{xx} = bd^3 \div 12$

Orientation A: $I_{xx} = 75 \text{ mm} \times (225 \text{ mm})^3 \div 12$

$I_{xx} = 71\ 191\ 406.25 \text{ mm}^4$

$I_{xx} = \mathbf{71.19 \times 106\ mm^4}$

Orientation B: $I_{xx} = 225 \text{ mm} \times (75 \text{ mm})^3 \div 12$

$I_{xx} = 7\ 910\ 156.25 \text{ mm}^4$

$I_{xx} = \mathbf{7.91 \times 106\ mm^4}$

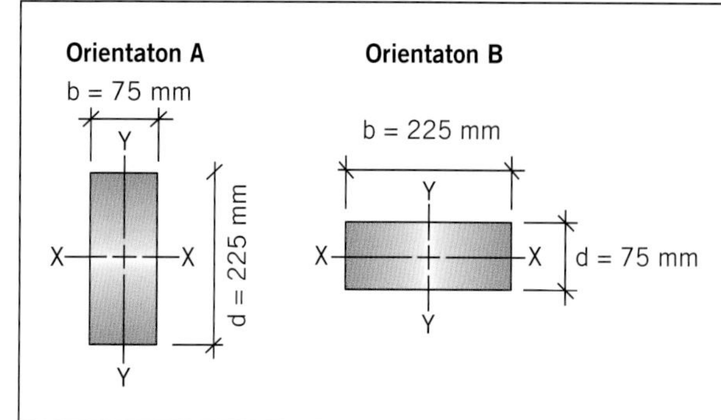

FIGURE 12.40 Beam alignment and relation to axes

The result backs our experience in this case, clearly showing that the beam's shape performs significantly better in orientation A than laid flat as in orientation B.

Section modulus (Z)

A very important property for beam designers, the section modulus (Z), offers a direct measure of a beam's strength. Note again that statement of 'strength' is a property of sectional shape, not material: hence, when comparing beams of the same material, the higher the value found for the section modulus, the stronger a beam may be said to be. It does not relate between materials.

It is the calculation of this property that led to the development of the 'I' beam. The section modulus tells us that the majority of the work in a beam takes place in the upper and lower fibres. This allows material to be withdrawn from the centre section, leaving only that which is sufficient to hold these upper and lower fibres apart as shown in Figure 12.41.

In principle, the further the structural material – known as the flange – is from the centroid, the more structurally efficient the beam is. In practice, however, if these flanges are too far from the centroid, the beam becomes overly slender and becomes prone to buckling.

The basic formula for section modulus is:

$Z = I \div Y$

where: Y is the distance from the centroid

and I is the second moment of area.

As Figure 12.42 shows, for symmetrical sections the value of Z is the same above or below the centroid.

In asymmetrical sections, as depicted in Figure 12.43, there is a Y_1 and Y_2 value. This is because distance from the centroid is different. In such cases, two calculations for Z are required: Z_{max} and Z_{min}.

The calculation of the value of Z for symmetrical shapes such as a rectangle thus becomes:

$Z_{xx} = I_{xx} \div y$

where: $I_{xx} = bd^3 \div 12$

and $y = ½$ depth or $d \div 2$

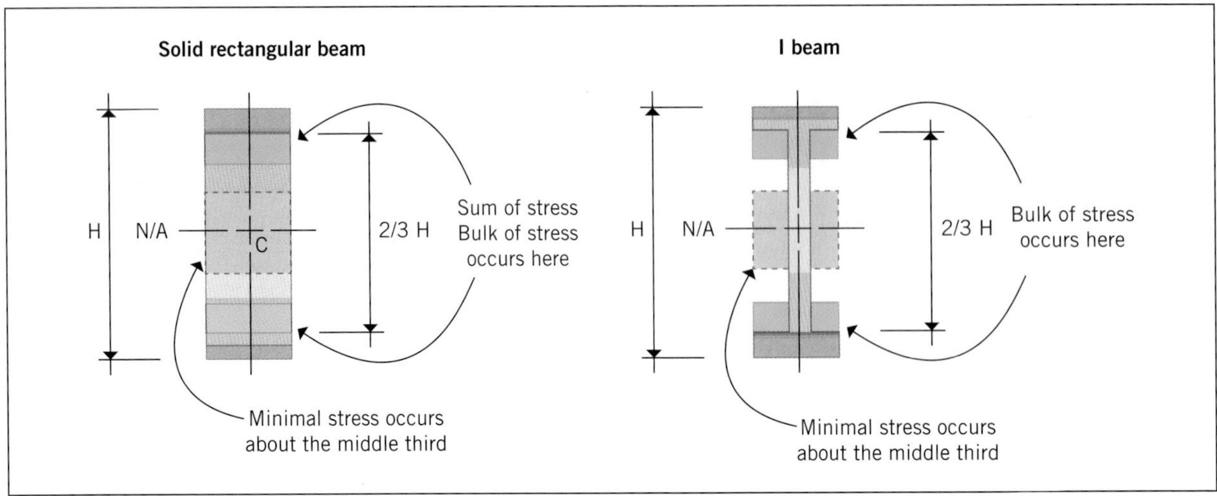

FIGURE 12.41 The mechanics of an 'I' beam

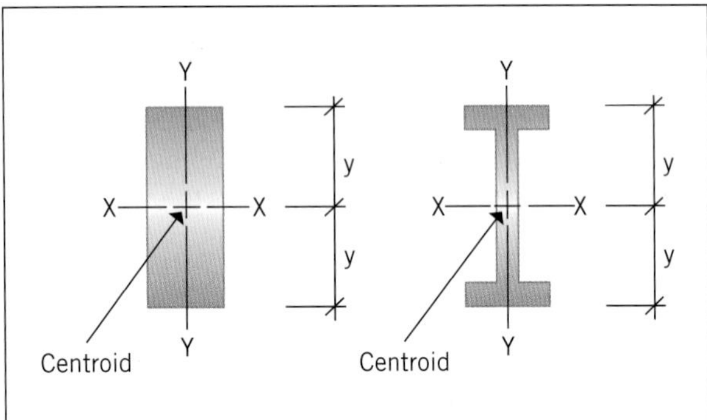

FIGURE 12.42 Symmetrical sections showing the centroid as equally distanced from both the top and bottom fibres of the beam

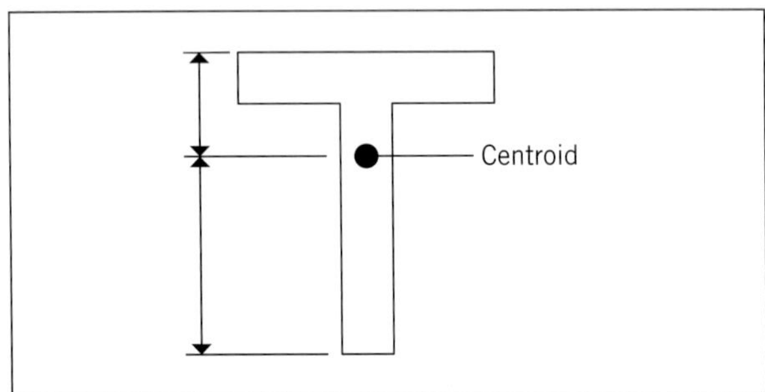

FIGURE 12.43 Extreme fibres at unequal distance from centroid

Put together the formula for Z becomes:

$Z = bd^2 \div 6$

The resultant unit is mm³.

The value found for Z is usually expressed as a value × 10³ mm³.

Comparing Z for differing beam orientations

Using the same beam as when calculating the second moment of area (I), we can again use experience to tell us the expected result.

Orientation A: $Z = 75\text{ mm} \times (225\text{ mm})^2 \div 6$

$Z = 623{,}812.5\text{ mm}^3$

$Z = \mathbf{623.81 \times 103\text{ mm}^3}$

Orientation B: $Z = 225\text{ mm} \times (75\text{ mm})^2 \div 6$

$Z = 210{,}937.5\text{ mm}^3$

$Z = \mathbf{210.94 \times 103\text{ mm}^3}$

Again, the calculations confirm our experience that in orientation A the beam will effectively be stronger or stiffer than in orientation B.

Radius of gyration (r)

With the second moment of area we found the best orientation of a beam to resist bending when loaded in span. The radius of gyration (r) finds which way a beam is likely to buckle when loaded on end.

The easiest way to visualise this property is to stand a ruler upright on a desk and push down on its end as shown in Figure 12.45. No matter how you apply the pressure, the ruler will buckle outwards on its flat face.

As with the other sectional properties using simple shapes, this is somewhat self-evident. However, when a beam or column is of a complex sectional shape it will need to be calculated.

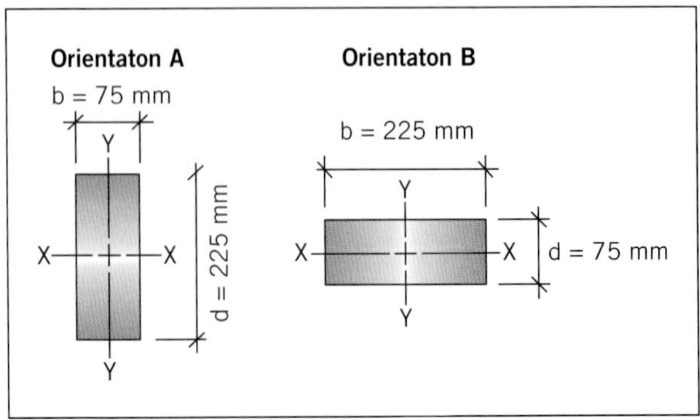

FIGURE 12.44 Beam orientation and section modulus

FIGURE 12.45 Flexible ruler under load depicting the deflection of a column under load

The basic formula for the radius of gyration is:

$$r = \sqrt{\frac{I}{A}}$$

where r = radius of gyration

I = second moment of area

A = area

The resultant unit will be millimetres (mm).

Calculating the radius of gyration

In using the same 75 mm × 225 mm beam as previously for our example, we can use the I_{xx} and I_{yy} values already found (note that I_{yy} for this beam was effectively calculated as I_{xx} for orientation B):

i.e.: $I_{xx} = 71.19 \times 10^6 \text{ mm}^4$

$I_{yy} = 7.91 \times 10^6 \text{ mm}^4$

The sectional area for this beam is:

$$A = 225 \text{ mm} \times 75 \text{ mm}$$

$$A = 16{,}875 \text{ mm}^2$$

The radius of gyration for each axis is therefore:

$$r_{xx} = \sqrt{17.19 \times 10^6 \text{mm}^4 \div 16{,}875 \text{mm}^2}$$

$\mathbf{r_{xx} = 64.95 \text{ mm}}$

and

$$r_{yy} = \sqrt{7.91 \times 10^6 \text{mm}^4 \div 16{,}875 \text{mm}^2}$$

$\mathbf{r_{yy} = 21.65 \text{ mm}}$

Once more the calculations follow our experience in suggesting that the beam will buckle about the y–y axis based upon the significantly lower *r* value.

So why do the calculations?

While each of these calculations (for Z, I and *r*) using a basic beam tell us no more than we knew already it is tempting to say they are unnecessary. The reality is that with light-weight construction becoming more common, component sectional shapes are becoming more complex, particularly extruded aluminium and plastics. It is only through such calculations that engineers can determine the structural properties of these new forms.

Influence upon component structural performance

As we work through the following sections of this chapter the influence of these properties upon design will become more evident. The main influences should, however, be clear: the choice of material, the shape of the material and the manner of a component's orientation all have an influence upon the structural performance of that component. Steel reinforcement for concrete, for example, is often specified with a requirement for low ductility, but not always. It is dependent upon the context and the engineering required (such as pre- or post-tensioning – see Chapter 13).

Some cautions

These calculations work on the premise of homogenous materials such as metals and plastics. There are some complexities when working with non-homogenous materials such as timber. In the main, these sectional properties will hold true, but natural products like timber, with its inherent grain and knots, can act differently than calculated when the difference between the two axes is minor. Likewise, engineered timbers such as ply, LVL and cross laminated timber (CLT) may vary due to glues and the direction of laminations, allowing greater flexibility in one direction or another.

LEARNING TASK 12.6

SHOW YOUR UNDERSTANDING OF TECHNICAL TERMS

Simplify the terms to show your understanding of what the following section properties are:

Technical term	Simplified language explaining the technical term
Centroid	
Second moment of area (I)	
Section modulus (Z)	
Radius of gyration (*r*)	

Structural characteristics of common building elements

Understanding the basic structural performance characteristics of slabs, floors, beams, columns and retaining walls is critical for your role of builder or even a tradesperson within the construction sector. These components are the basis of most contemporary domestic housing in Australia and failure of any one is to cause significant failure to the structure as a whole. Beginning with beams, we will look at each of these components in light of their general applications, key characteristics and the actions they most commonly must resist or support.

Beams

There are three essential attributes that all builders should know about beams:

- the common arrangements or 'types' of beams; i.e. the manner of their use
- the general characteristics of beams such as bending, shear, influence of shape
- the principles by which beams support the loads applied to them.

Without this knowledge, planning your construction so that it will perform as expected, or otherwise be suitable for the use intended, can be problematic. With regards to the arrangement of beams, or types of beams, there are three typical forms: simple, continuous, or cantilever. These arrangements are discussed below.

Simple beams

Simple beams or 'simply supported' beams are those that are supported at each end only. When loaded the beam will tend to bend in the middle between these two supports. This is the sort of beam used in most of our calculations of moments and force systems (see Figure 12.46).

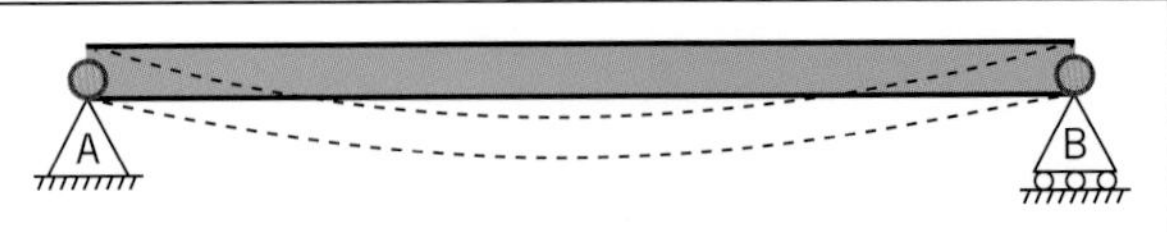

FIGURE 12.46 Simple beams

Continuous beams

Continuously supported beams are ones that are supported at three or more points along their length. Beams in this arrangement deflect less than a simple beam over the same span. This is because a load down on one span tends to cause the section of beam over the neighbouring span to rise; effectively increasing its capacity to support a load. Generally, a beam in continuous span is 20% more efficient than a simple span, thereby enabling it to span longer distances. This is reflected in the span tables of all beams, be they timber, engineered timber, or steel (Figure 12.47).

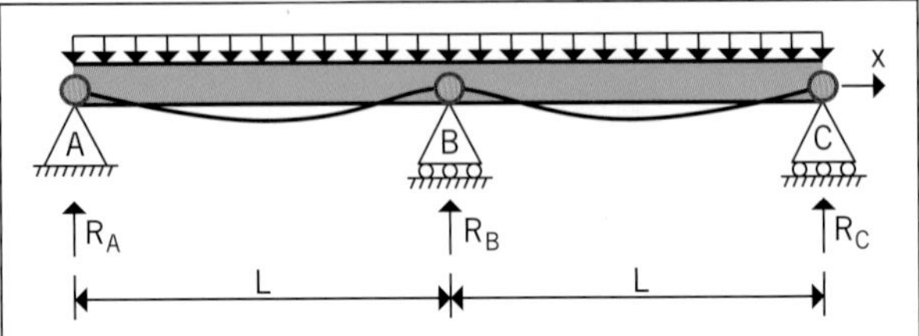

FIGURE 12.47 Continuous beams

Cantilever beams

Cantilever beams are those that overhang or travel past their support at either end or both ends.

The cantilever section of the beam is supported to a degree by the back span in the same manner as a continuously supported beam; i.e. the load on the back span tends to cause the cantilever to flex upwards, increasing its load-carrying capacity. However, the cantilever cannot support the same loads as the back span – the section of beam between the supports. A typical allowance for the amount of cantilever in an LVL beam, for example, is ¼ of the backspan: a 6.0-m beam could have a cantilever section approaching 1.2 m long assuming 4.8 m of allowable backspan. With other materials, cantilever can in some instances be capable of extending to 1/3 of the backspan (see Figure 12.48).

Load characteristics

Beams under load exhibit complex internal behaviour as the various forces act on the upper and lower fibres of the material. Three forces are in play as the load increases, these being:

- tensile
- compressive
- shear.

Figure 12.49 shows where in the beam each of these forces are in operation. For the neutral axis to remain its original length, the upper fibres must become compressed, while the lower fibres are stretched. Due the different directions the fibres are acting, the fibres also undergo horizontal shear as each layer attempts to act differently to its neighbour (see Figure 12.50).

Shear forces in a beam under vertical load actually take two forms: horizontal shear, as already depicted,

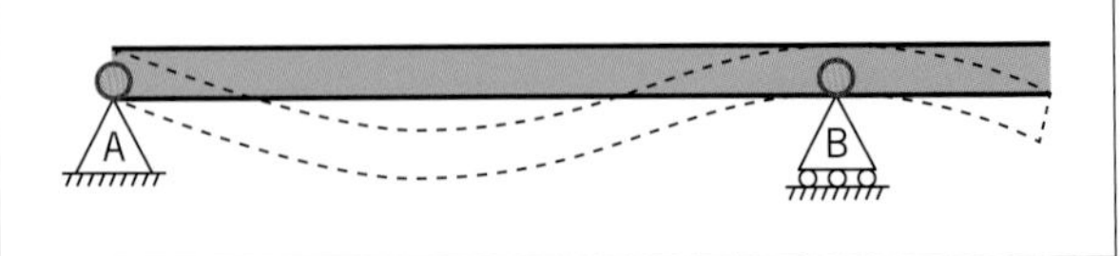

FIGURE 12.48 Cantilever beams

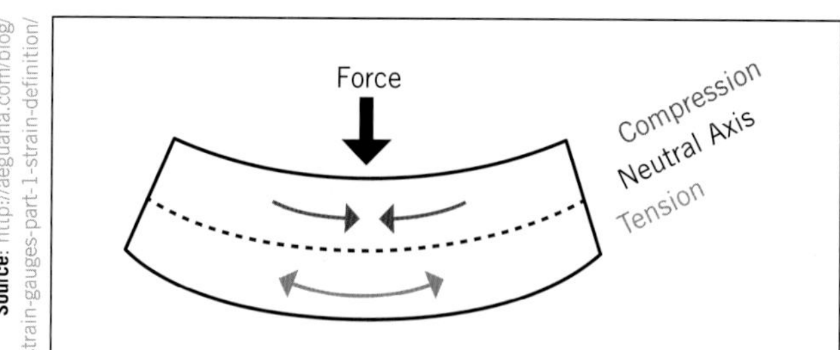

Source: http://aeguana.com/blog/strain-gauges-part-1-strain-definition/

FIGURE 12.49 Tension and compression in a beam under load

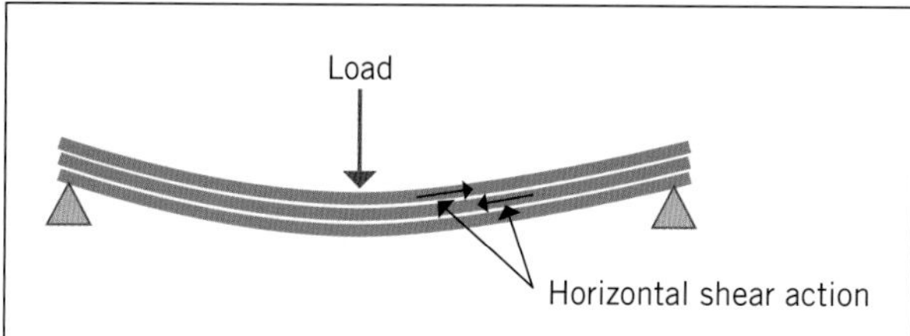

FIGURE 12.50 Horizontal shear acting on beam fibres

and vertical shear. Vertical shear forces tend to act to cut through the beam, much as a pair of scissors or shears will cut through paper. In a beam, the down force in shear is provided by the load, the up force is the reaction supplied by the supports. The magnitude of the shear action therefore varies from the mid span of the beam to the points of support. The maximum vertical shear force generally taking place at close proximity to those points of support (see Figure 12.51).

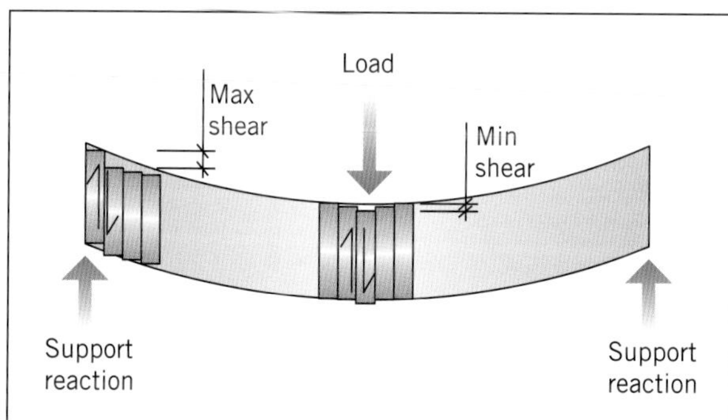

FIGURE 12.51 Vertical shear action

While there are a number of calculations that can be made of these forces, they are outside the scope of this unit. What is important is knowing where in a beam the tensile and compressive forces are acting. This becomes critical in concrete beam construction, a common task for most builders. Likewise, it is critical in determining the location of checkouts and penetrations in any beam.

Penetrations

The drilling of holes or cutting of notches in beams can significantly reduce a beam's performance under load. Poorly located, such penetrations may cause a beam to fail. The following points are your main considerations when a penetration is required:

- maximum shear is close to the supports
- maximum bending is at mid span
- maximum tension is in the lower third of the beam
- maximum compression is in the upper third of the beam.

This mean notching or drilling holes in these areas is to be avoided. The previous information on bending, compression and tension help you to understand the mechanics of why this is the case. For example, making penetrations in the top flange of a steel beam reduces the beam's capacity in the following ways:

- reduces the material available to withstand compression, thereby increasing the tendency to buckle
- changes the location of the neutral axis as more of the upper portion of the beam takes up the compression load; i.e. the neutral axis is lowered.

AS 1684.2 (p. 42) and AS 1684.3 (p. 39) Residential timber-framed construction provide detailed information and diagrams on penetrations and notching applicable to solid timber sections.

The diagram below, Figure 12.52, applies to all other materials including engineered timber and steel 'I' beams and rectangular or square hollow steel sections (RHS and SHS).

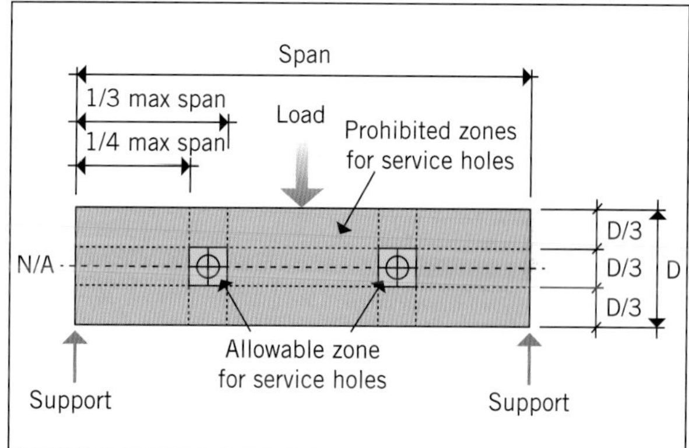

FIGURE 12.52 Acceptable locations for penetrations

Types of penetrations

When penetrations cannot be avoided, wherever possible these should be circular, oval, or round-cornered squares and rectangles. Sharp corners tend to focus stresses at the very point of the corner, leading to tears, fractures and delamination of the beam.

Deflection

Deflection is the amount a beam bends under a given load. For reasons of safety and serviceability, beams are offered what are referred to as deflection limits. AS/NZS 1170.0 Structural design actions: General principles, provides guidelines for these serviceable limits in table C1 of Appendix C. These apply to all materials (e.g. steel or timber beams) and are offered for various situations. You should review these whenever either specifying or installing a beam where excessive deflection may interrupt some other part of the structure, such as pushing down upon a ceiling lining. Limits are provided with regards to visible deflection and sensory deflection (bounce and vibration).

The most common beam deflection limits you should keep in mind are:

- general roof and floor beams: span ÷ 300
- beams as lintels in walls: span ÷ 240 but not exceeding 12 mm maximum.

Every beam deflects under load, even if only marginally. Due to the properties of both the beam's section and material, some beams can bend quite significantly without risk of failure. Excessive deflection, however, can be off-putting visually and experientially. Too much 'bounce' also feeds dynamic loads back to the rest of the structure, which may lead to the cracking or failure of fixings or other elements.

Influence upon deflection

Aside from the aforementioned beam properties such as the material it is made of or its sectional shape, there are two other factors (aside from load) that can influence beam deflection:

- end fixity
- camber.

End fixity is the manner in which the ends of the beam are secured. As Figure 12.53 shows, if the ends of a beam are rigidly fixed, it will tend to deflect less than if one or both ends are free or loose. The capacity to make such fixtures is of course based upon the expected expansion and contraction of the beam: steel, for example, can seldom be provided with rigid fixings due to excessive movement that may rupture the support.

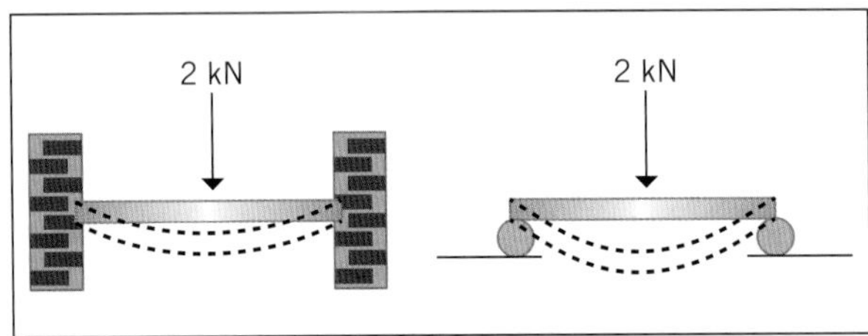

FIGURE 12.53 Rigid and free end fixtures: effect on deflection

Camber is a slight curve or bend that a designer may deliberately incorporate into a beam. Generally, this camber will be an upward curve based upon the known loads being applied vertically downwards. This deliberate curve is designed to counter the amount of deflection likely to be incurred under load. Correctly applied, a cambered beam will become straight under load. This is most commonly seen in domestic construction in the bottom chord of roof trusses (Figure 12.54).

Columns

Columns are vertical supports within a structure receiving mostly compressive forces. Over the centuries they have taken various forms and names depending upon location and so may be named struts, pillars, posts, or stanchions as well as 'columns'.

Despite the clear difference in loading, columns may be made of exactly the same materials, and be of the same sectional shape, as beams. The main difference is that with columns, the most ideal form is square or circular; with the most efficient being circular hollow and square sections (in that order), as shown in Figure 12.55. This is based upon the section property *radius of gyration (r)* discussed earlier that shows a member with equal 'r' values on both axes is less likely to buckle.

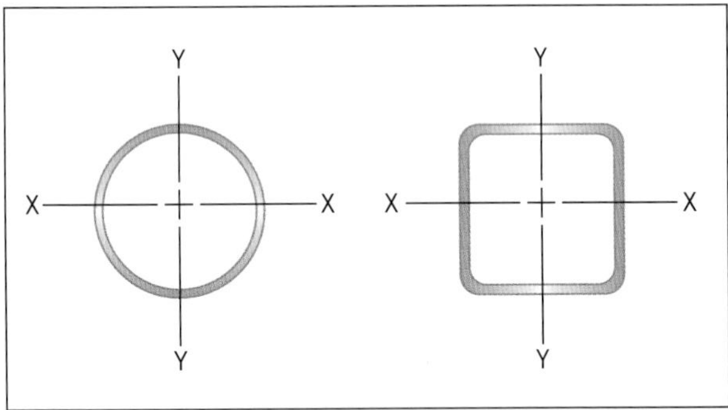

FIGURE 12.55 Typical location of radius of gyration for symmetrical columns

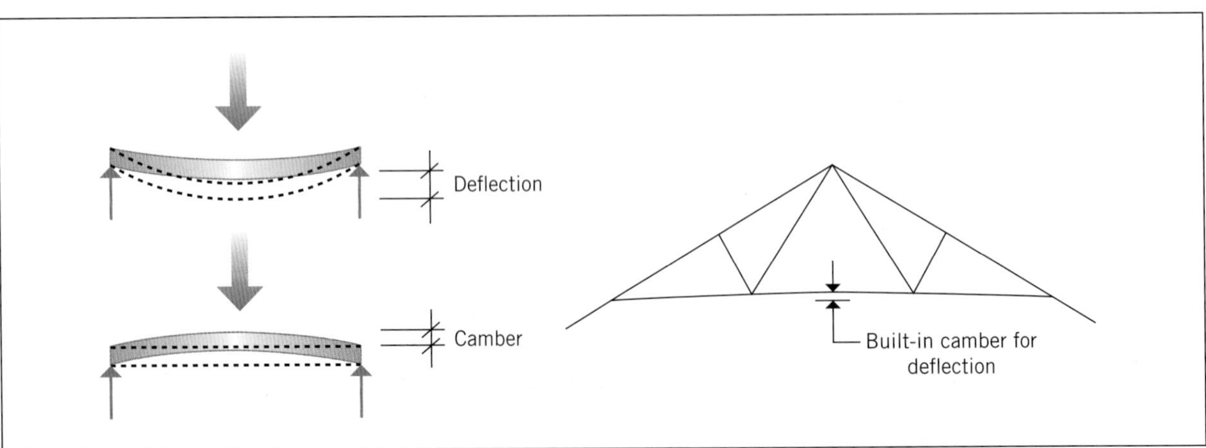

FIGURE 12.54 Camber in beams and bottom chord of truss

Loading of columns: axial and non-axial loads

Columns may be required to support loads that are placed directly on top and in the centre, or they may be attached to the side or on top, but off-centre. When loads are centred, they are known as axial loads as the loads are applied directly over the beam's central axis or centre of gravity. When loaded to one side of this centre line, they are referred to as non-axial loads. These two loads are demonstrated in Figures 12.56 and 12.57.

Source: Glenn Costin

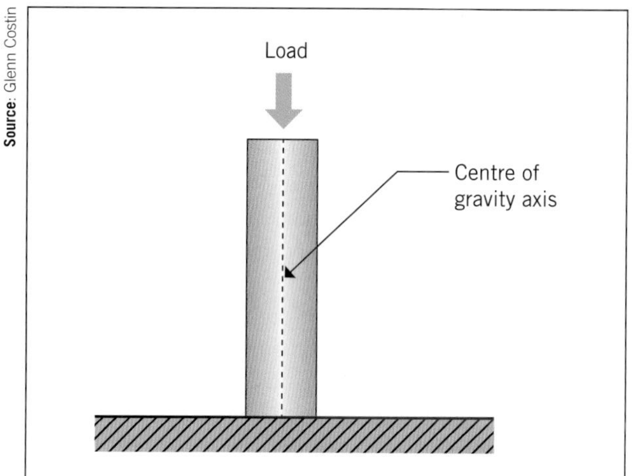

FIGURE 12.56 Axial load

Source: Glenn Costin

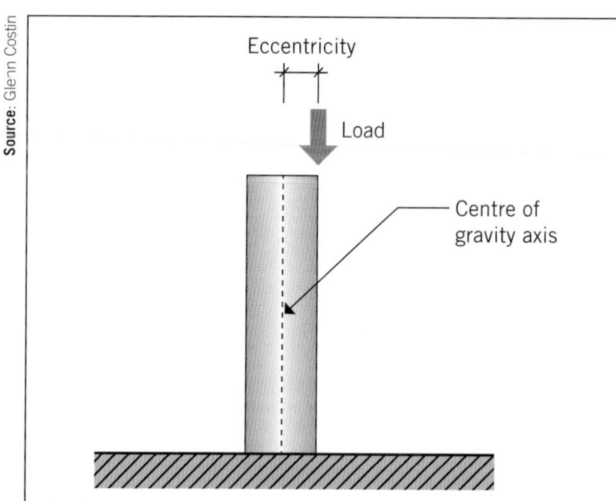

FIGURE 12.57 Non-axial load

Axial loading allows a column to take up the compressive forces equally throughout its structure. Axially loaded columns are sometimes referred to as 'concentric columns'.

Non-axially loading of a column results in bending and compressive stress being applied to the column unequally. Figure 12.57 shows the load applied to the right of the column's centre of gravity or central axis. This distance is referred to as its 'eccentricity' and is used in calculating the bending moments within the beam. For this reason, these columns are frequently referred to as eccentric columns.

Buckling and crushing

Buckling is most likely to occur with non-axially loaded columns. Technically, its occurrence is observed when the compressed member deflects at 90° to the direction of loading. As columns are vertical, this deflection is horizontal (see Figure 12.58). Buckling is more likely to occur in tall columns, rather than in short ones of the same sectional dimensions. As mentioned earlier, it is also more likely to occur in non-symmetrical columns.

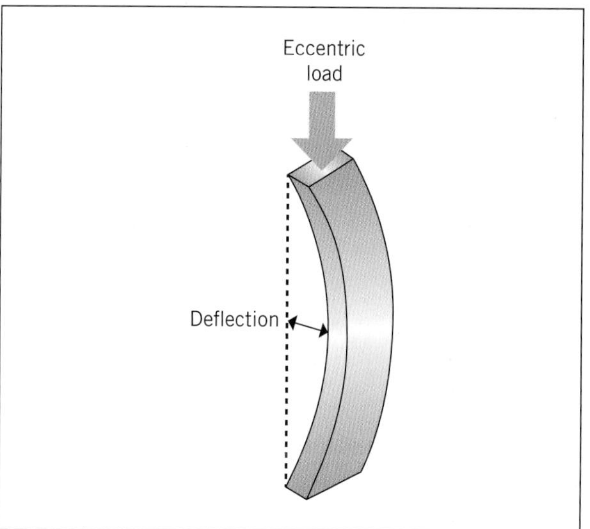

FIGURE 12.58 Non-symmetrical column buckling under eccentric loading

Due to the greater likelihood of buckling, non-symmetrical columns often require lateral (sideways) support. The radius of gyration calculations dealt with previously provide us with the most likely direction of deflection, and hence the appropriate location of these supports.

However, a column may still fail under extreme loads through crushing. Crushing is where it is the buckling and delimitation of the external fibres of the column that lead to failure. Generally, this happens only to severely overloaded short or squat columns, such as that shown in Figure 12.59.

The slenderness ratio

The load that a column may support depends to some measure upon what is known as the ***slenderness ratio***, coupled with the material from which it is made. There are three elements to the slenderness ratio:

- effective length of column
- sectional size
- end fixity.

Effective length is a derivative of a column's height and ***end fixity***. How rigidly a column is fixed at either end influences its susceptibility to buckling. In so doing, the effective length of the column is open to buckling changes. There are four ways by which a column may be coupled to its surroundings, which are described in Figure 12.60. Also within these

Source: Shutterstock.com/john vlahidis

Source: Shutterstock.com/john vlahidis

FIGURE 12.59 Columns failing under crushing load

figures is the description of effective length and how this relates to the end fixity offered. The dotted line represents the static centreline of the column, the curved line represents the part of the beam prone to buckling.

The formula for slenderness is:

Slenderness = effective length ÷ least radius of gyration

The lower the slenderness ratio the more resistant a column is to bucking.

It must be understood that the slenderness ratio is not a statement of a column's strength, merely a statement of its likelihood of buckling. Also, as with sectional properties, the slenderness ratio can only compare columns of the same material; i.e. the resultant slenderness ratio of a timber column cannot be compared to that of a concrete or steel column.

Concrete slabs

As one of the most common flooring systems used in the contemporary Australian house, steel-reinforced concrete slabs are a critical part of a builder's knowledge base. Concrete slabs used in residential construction may be categorised as one of the following:

- slab on ground and stiffened raft slabs
- suspended slabs
- precast, pre-stressed and post-stressed slabs.

The principles of reinforced concrete slabs are similar to those of concrete beams and columns – deriving their structural capacity by making strategic use of the strengths of the two components:

- concrete, which has high resistance to compression but is poor in tension
  and
- steel rod, which has high resistance to tension but buckles easily in compression.

Together, these two materials form a remarkably stable and strong composite product – provided it is appropriately designed. Appropriate design requires that the concrete is appropriately specified, and the steel is likewise appropriately specified, but also

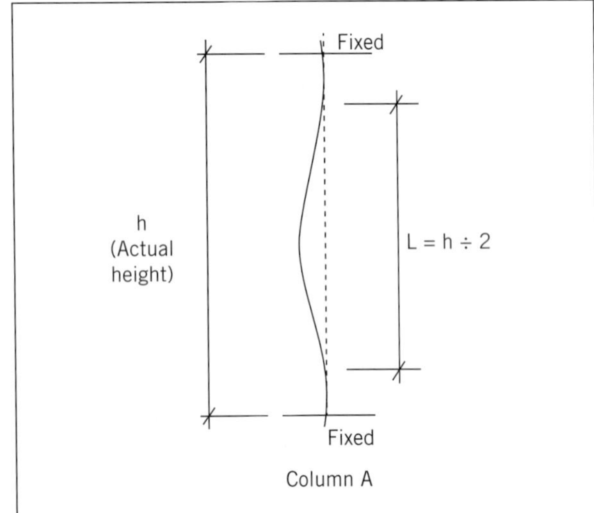

FIGURE 12.60(a) Column fixed top and bottom

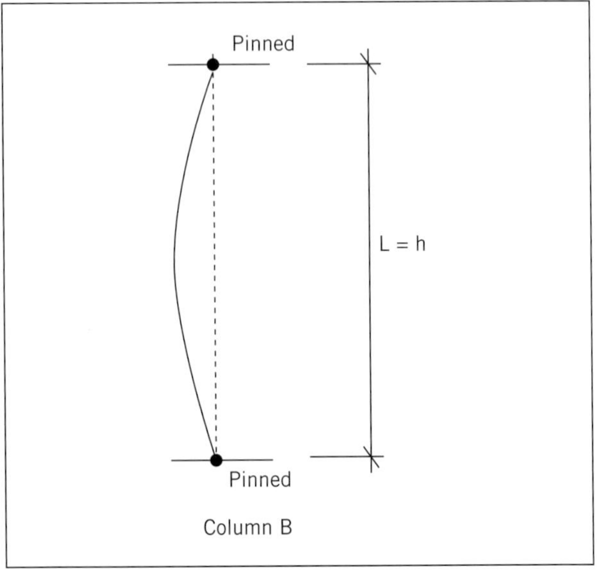

FIGURE 12.60(b) Column pinned top and bottom

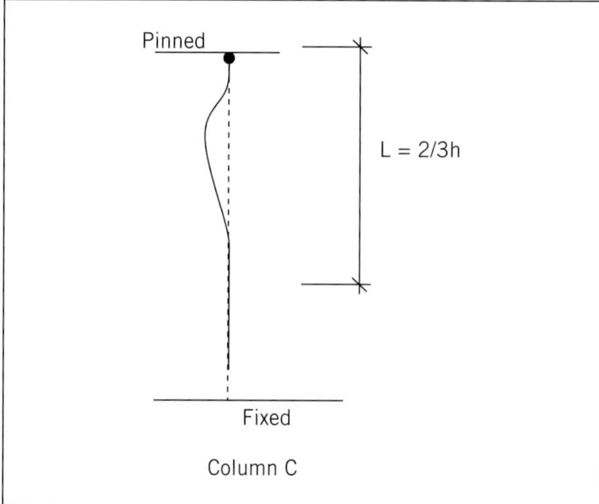

FIGURE 12.60(c) Column fixed at base, pinned at top

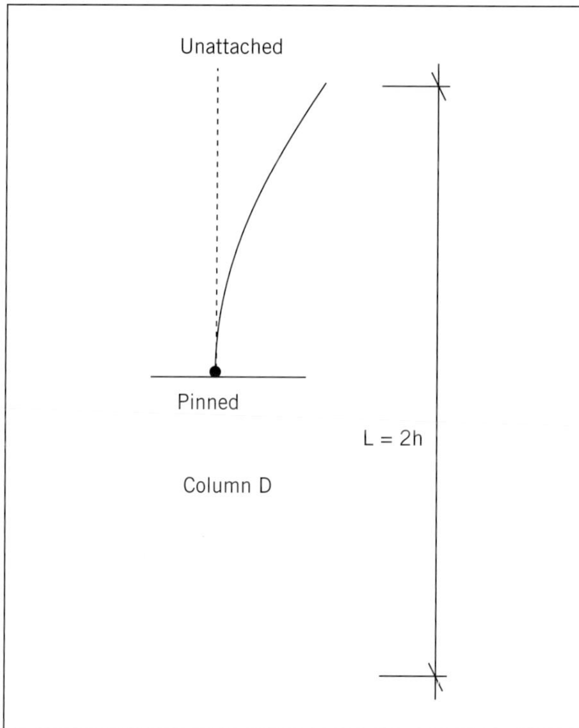

FIGURE 12.60(d) Column pinned at base only

correctly located to counteract the expected tensile forces within the slab.

To identify where the tension within a slab is likely to occur, you should review the section on beams, as the same principles apply. The key is in determining from which direction, and at which point, the significant loads are being applied. From this you may determine the direction of deflection. Figure 12.61 is an example of the steel location in a typical house slab based upon the principles of bending in beams.

What follows is a brief description of each of the slab types and their key structural considerations.

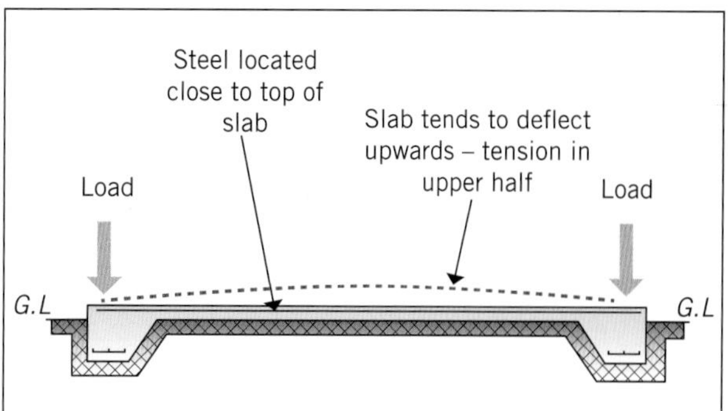

FIGURE 12.61 Typical domestic house slab with loads from roof trusses applied to external walls

Source: Glenn Costin

Slab on ground and stiffened raft slabs

These are the most common concrete slabs of Australian homes. After excavation, these types of slabs are poured directly into forms on the ground, allowing the sand-buffered foundation material beneath the slab to act as the main support to the slab itself. House slabs of this type are typically little more than 100 mm in thickness.

However, where there are expected loads such as outside walls carrying roof loads from trusses, internal load-bearing walls, or point-loaded columns, there will need to be thickening to varying degrees. This thickening is generally in the form of an attached beam, edge beam, or thickening beam as shown in Figure 12.62. AS 2870 defines a slab on ground as a slab with edge beam only (no internal beams). Once internal beams are required it is referred to as a stiffened raft slab.

The stiffened raft slab supplants the basic slab on ground which has limited application due to soil instability and load-carrying capacity. Stiffened raft slabs use a grid work of internal beams under the slab to improve resistance to these torsional forces – the twisting cause by upheaval or subsidence. Aside from improving load-carrying capacity, the internal beams stop the spreading of the external beams: of particular import for slabs created on reactive soils (Figure 12.63).

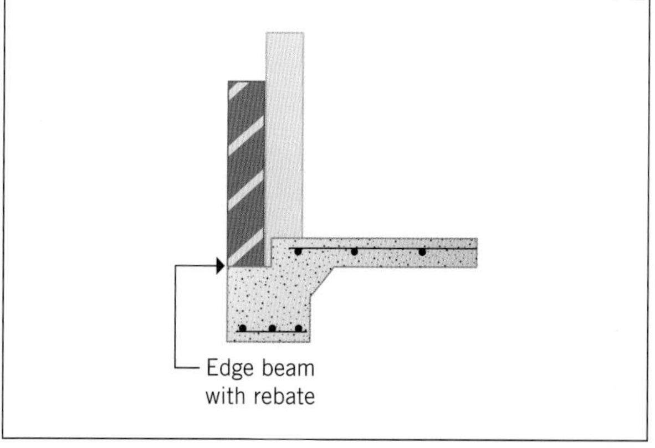

FIGURE 12.62 Typical edge beam configuration of slab on ground

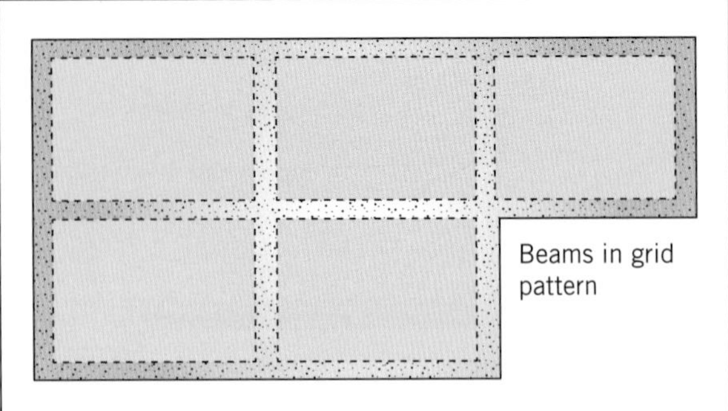

FIGURE 12.63 Grid layout of internal beams to stiffened raft slab

Concrete slabs on ground (stiffened or otherwise) are designed around three main influences:

- dead loads of the building
- load-bearing capacity of the foundation material
- expected movement of the foundation material.

The last two of these are obtained from soil reports which offer a site classification for the building's location. This testing is done by soil or geotechnical engineers who determine allowable bearing pressure (ABP) of the soil at various depths and the soil or clay type. The NCC requires that the minimum ABP for residential housing must be not less than 50 kPa.

The site classification and the expected loads, based upon construction type – e.g. brick veneer, timber clad, solid masonry, or the like – allows the development of basic slab designs based upon AS 2870 Residential slabs and footings and/or the NCC/BCA Vol. Two, Part 3.2. These two documents underpin all concrete slab designs in Australia and must be adhered to in all their respective parts.

Suspended slabs

Unlike slab on ground, a suspended slab is reliant upon its own internal structure to remain static and not deflect excessively under load. This form of slab is suspended in the air by supports at least at two edges, or around all four edges. Suspended slabs supported at two ends only are referred to as one-way slabs as the download forces will cause the slab to deflect much as if it were a simple beam; i.e. in one direction only (see Figure 12.64). The main reinforcement in these slabs needs to run from support to support.

When supported on all four edges, suspended slabs are known as two-way slabs in that the deflection can cause a 'dishing' deflection to occur; i.e. in two directions. Reinforcement for these slabs will need to run in both directions, as shown in Figure 12.65.

The tension forces introduced to slabs of this form act in the lower half of the slab. This requires that the main reinforcement rods be located in the lower third of the slab.

Note: The reinforcement in these depictions show only the location of the main reinforcement rods. Depending upon loads, and other support members, reinforcement is most likely to be required in other locations if only to control shrinkage and cracking.

Pre-stressed slabs

Pre-stressed concrete components are usually the realm of major construction works, such a bridges and large commercial structures. However, with a slow, but increasing, interest in prefabrication, such techniques are finding their way into the domestic sector as wall, floor and even footing elements.

Pre-stressed concrete has internal tension introduced to the material prior to the concrete being put into service. This pre-stressing is achieved by one of two paths: pre-tensioning and post-tensioning.

Pre-tensioning is achieved by placing the steel reinforcement in tension prior the pouring of the concrete. Once the concrete has reached a predetermined strength the tension on the steel is slowly released, thereby transferring the tension to the concrete.

Post-tensioning is achieved by running the main reinforcement through sleeves or ducts to the outside of the formwork; *no* tension is put on the cables at this time. Once again, the concrete is poured and allowed to reach a predetermined strength. Only at this point is tension slowly put onto the cables through the use of

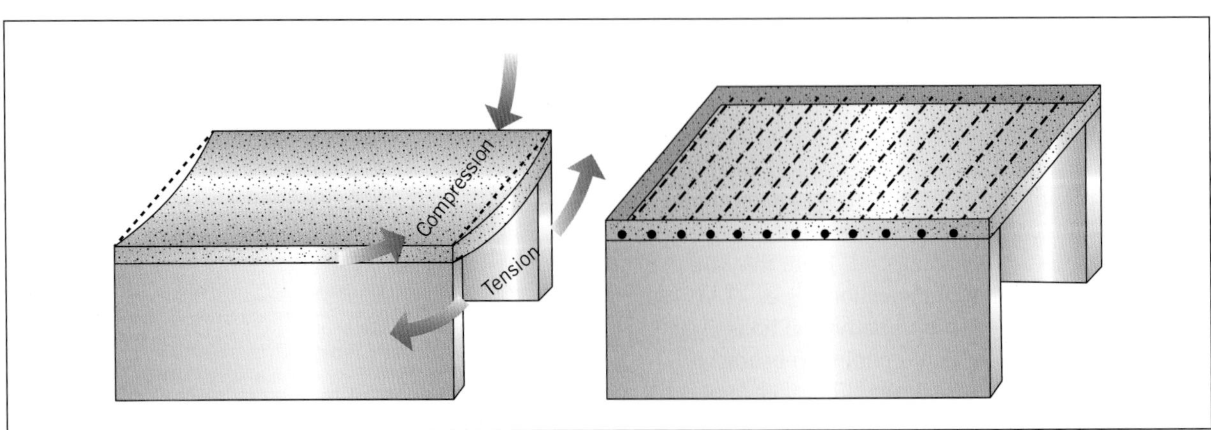

FIGURE 12.64 One-way suspended slab

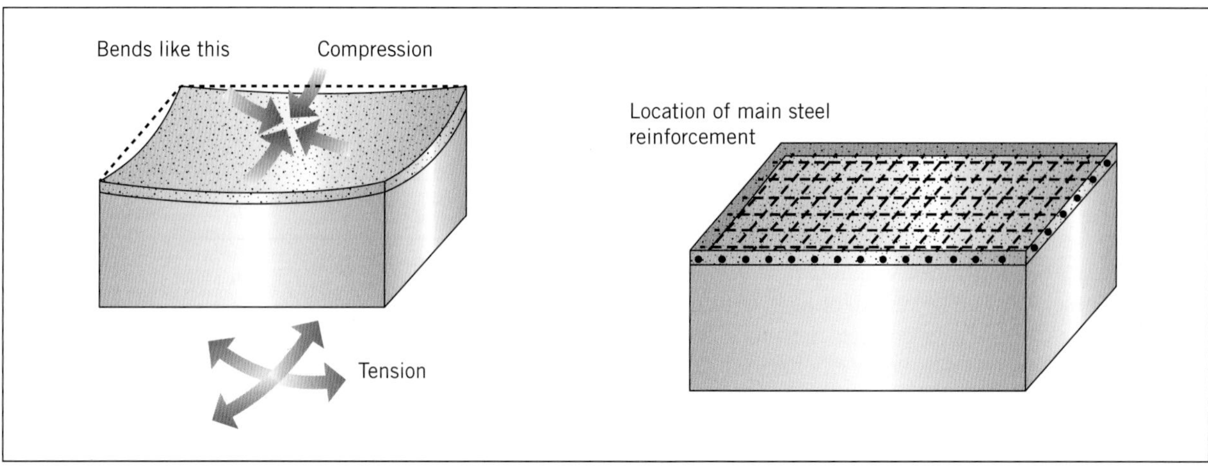
Bends like this
Compression
Tension
Location of main steel reinforcement

FIGURE 12.65 Two-way suspended slab and reinforcement arrangement

hydraulic rams. This tension is maintained by lock pins or by the rods being attached to large treaded rod and nuts. The sleeves or ducts are usually back-filled with grout by pump and suction methods.

These techniques allow components to be made off and on site, as well as increasing the accuracy with which the strength of components can be achieved. Because of this accuracy, less concrete and steel may be required in a component, moving projects to greater sustainability and energy efficiency.

Failure risks

While the potential benefits of these systems are significant, there are risks. In highly saline soils, for example, the locking systems holding post-tensioned reinforcement rods can corrode and fail. Because the slab is totally reliant upon the tension on these rods, it becomes susceptible to cracking and upheaval. Once this has occurred the only path is demolition.

Retaining walls

Sloping sites often require retaining walls to hold back earth to maintain a site cut and/or produce level lawns, gardens or entertainment areas. Retaining walls must contend with complex forces and so, excepting some proprietary systems, walls over 1.0 metre in height must be designed by a registered engineer. Figure 12.66 describes these forces.

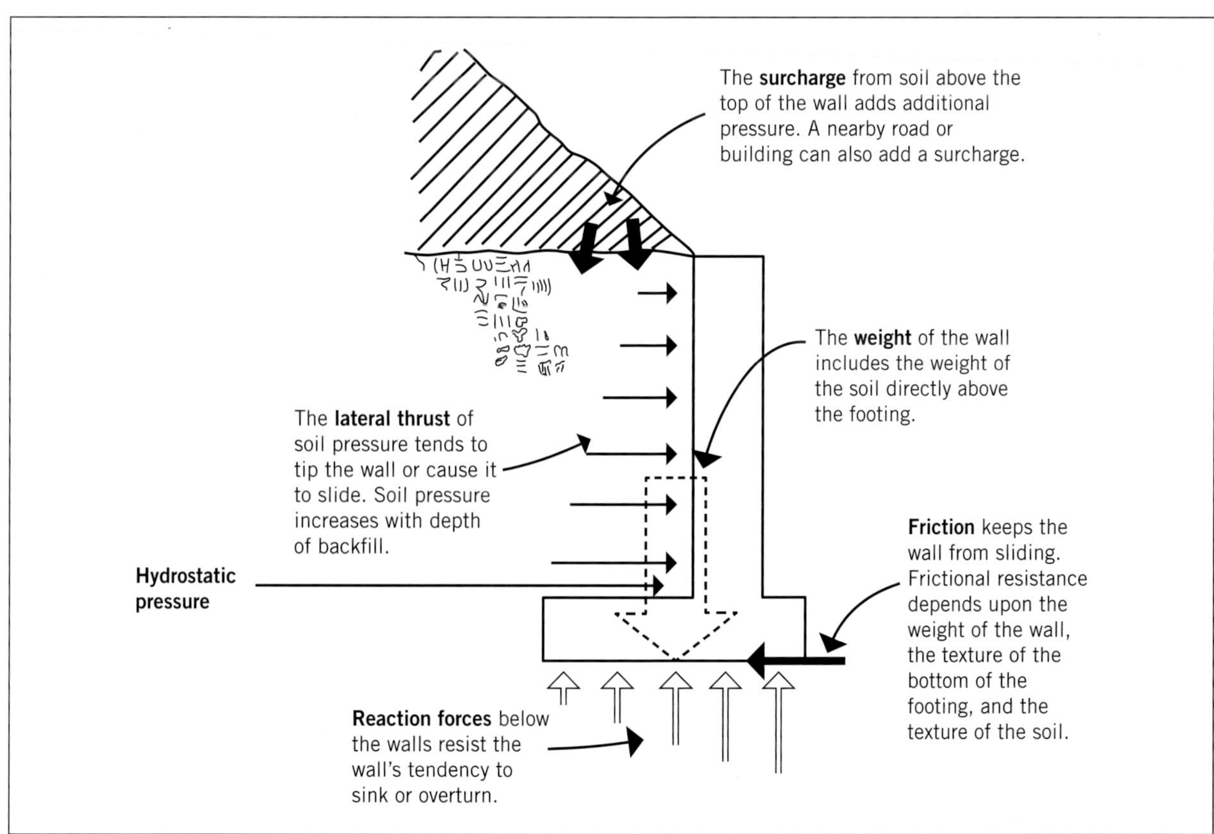
The **surcharge** from soil above the top of the wall adds additional pressure. A nearby road or building can also add a surcharge.
The **weight** of the wall includes the weight of the soil directly above the footing.
The **lateral thrust** of soil pressure tends to tip the wall or cause it to slide. Soil pressure increases with depth of backfill.
Friction keeps the wall from sliding. Frictional resistance depends upon the weight of the wall, the texture of the bottom of the footing, and the texture of the soil.
Hydrostatic pressure
Reaction forces below the walls resist the wall's tendency to sink or overturn.

FIGURE 12.66 Forces acting on a retaining wall

Basic principles

The forces shown in Figure 12.66 act to overturn or topple the wall, push or slide the wall away horizontally, or simply have it sink under its own weight. A retaining wall must therefore be designed such that it may counter all of these actions. Of the many approaches used, most retaining walls fit one of three basic forms (Figure 12.67).

- Gravity walls rely upon their weight to hold back the earth. These walls can be vertical or near vertical and may be made of large concrete or stone blocks, or poured in-situ concrete. Some proprietary systems also work on this method, using hollow blocks that are then back-filled with concrete and stiffened with reinforcement rods tied into the footings (Figure 12.67a).
- Inclined walls are those that lean back against the earth, relying, much as do gravity walls, on their weight to hold the earth back. The wall 'falls' against the earth, thereby restraining the material. Many proprietary systems use this approach, using stepped interlocking blocks to create the specified angle which may be only a few degrees off vertical (Figure 12.67b).
- Cantilever walls (Figure 12.67c) are generally of engineered reinforced concrete. The cantilever may extend forward or backward from the wall depending upon the foundation material or the available space for excavation. Reinforcement rods are tied into the footings and are met with horizontal bars as they proceed vertically through the wall (see Figure 12.68).

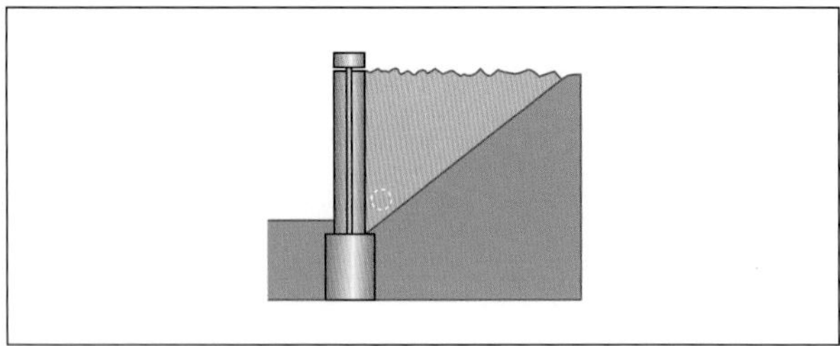

FIGURE 12.67(a) Gravity walls

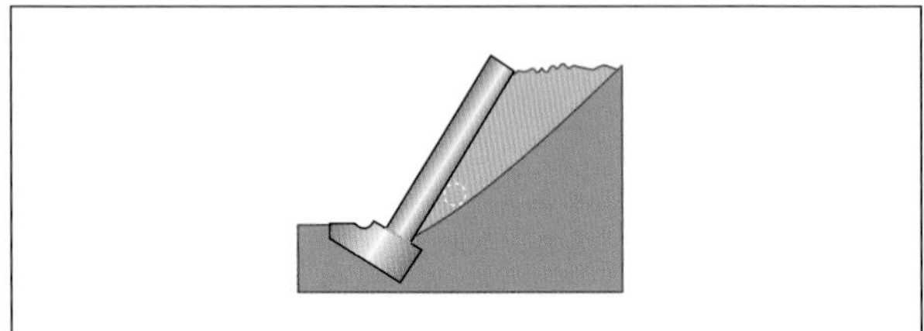

FIGURE 12.67(b) Inclined walls

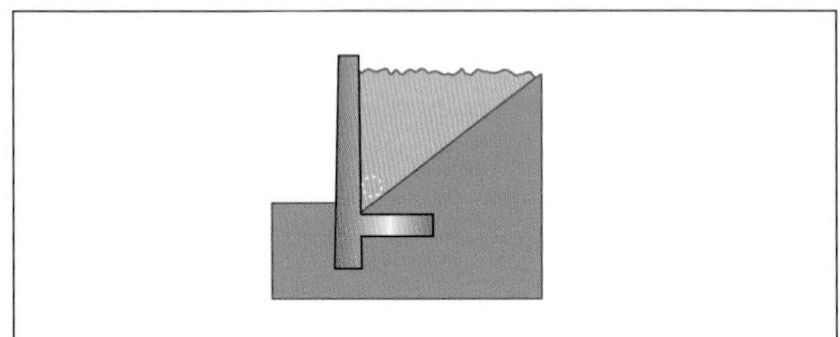

FIGURE 12.67(c) Cantilever walls

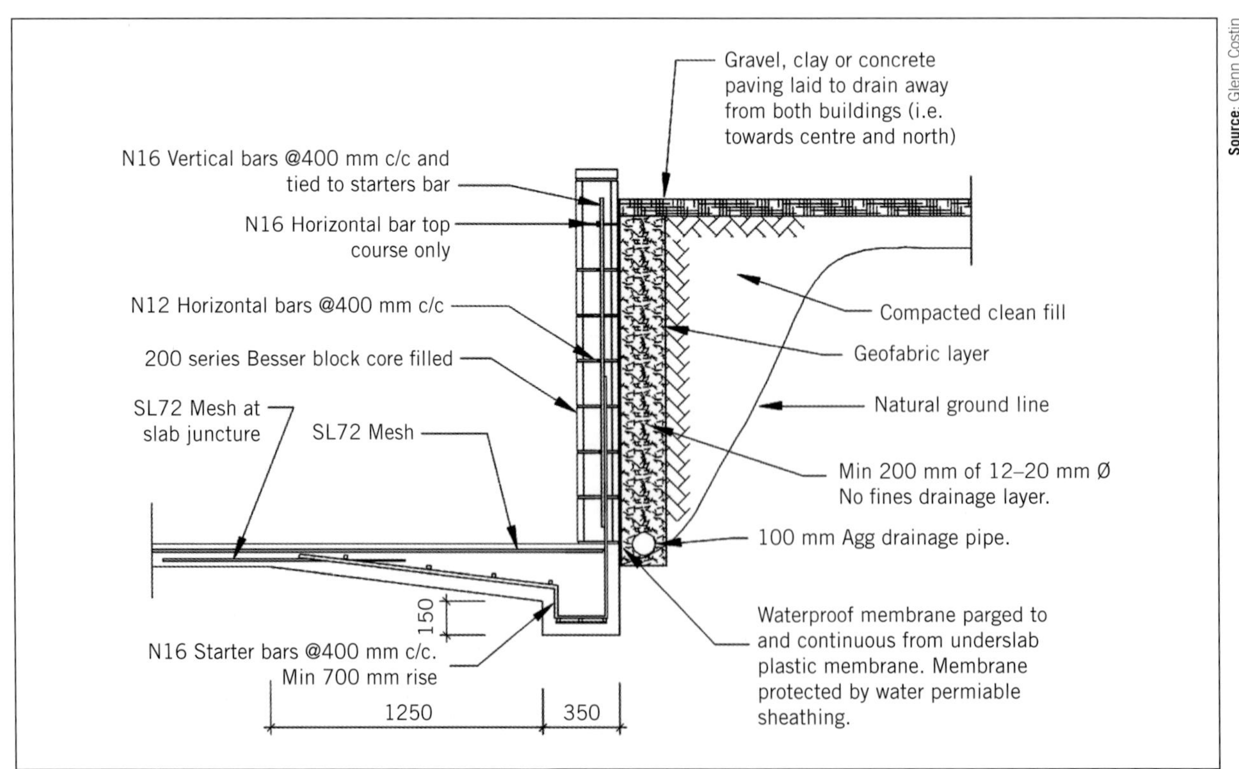

FIGURE 12.68 Cantilever retaining wall tied into a garage slab footing

Hydrostatic pressure and drainage

As Figure 12.68 demonstrates, retaining walls may form part of a house, basement, or garage, where the building has been set back into the earth. This diagram also shows the drainage systems that need to be in place to draw water away from the back of the wall. This water applies a force known as hydrostatic pressure. Through the build-up of water, significant lateral pressure can form behind a wall, either toppling it, fracturing it, or simply penetrating it. In the latter case, this may be an engineered desirable, achieved through weep holes when the wall is an exterior feature. When the wall is part of a house or garage, however, then this must be avoided using approaches such as those shown.

Application and design

As the builder, you need to be able to make at least a basic assessment of a design with regards to the context in which the retaining wall is to be constructed. The actual engineering calculations are not your concern, but the overall design is. Looking at Figure 12.68 you can see that it at least has the features you would expect for the situation. Key features in this example are:

- the lower footing is stepped 150 mm offering some physical resistance to outward drift aside from the slab mass itself
- the wall has both vertical, tied-in reinforcement bars bent to form, as well as horizontal bars
- the wall is core filled
- there is a waterproof membrane protected by sheathing
- a drainage layer exists which is sheathed in geofabric
- there is additional mesh at the point where the cantilever meets the extension of the main slab. This mesh is in the lower one-third of the slab to counter any downward deflection caused by forward rotation of the wall and associated cantilever.

However, questions remain. The questions you may need to ask in this example include:

- What grade of geofabric is required? (It is unstated in the diagram.)
- Is one drainage pipe sufficient? (Is there a spring or other high-water issues behind the wall?)
- What may be used as water permeable sheathing to protect the membrane?
- Where can the water drain to?

With these questions answered, you could safely construct the wall, paying strict attention to the waterproofing knowing that to return and repair will be a costly exercise.

LEARNING TASK 12.7

EXPLAIN TECHNICAL STRUCTURAL TERMS AND ACTIONS IN SIMPLE LANGUAGE

Provide simple explanations of the features of the 'typical behaviour' (negative features) that a structural member might experience for the following terms.

Technical term	Negative feature – simply explained
cantilevers	
tensile	
shear	
compressive	
deflection	
penetrations	
buckling	
crushing	
columns – non-axial loading	
beams – single span	
beams – continuous span	
one-way slabs	
two-way slabs	
slenderness ratio	
retaining walls	
hydrostatic pressure	

Structural principles in demolition

Whenever renovation or extensions are being considered, demolition becomes integral to your planning. Demolition refers to the complete or partial dismantling of a building or structure. Such activities should be undertaken in a planned and controlled process, using as a basic principle that what went up last, comes down first.

In recent years the term ***deconstruction*** has been adopted in order to refocus the industry towards the reuse and recycling of building components and materials wherever possible. This is addressed fully in Chapter 15 Minimising waste.

Controlled demolition, particularly deconstruction, relies heavily upon your knowledge of structures. This is the key to you and your co-workers remaining safe throughout this process – an issue that applies equally to all the other stakeholders in a demolition process, such as neighbours, pedestrians and passing motorists.

Due to these safety concerns, councils impose stringent conditions ensuring that demolition is undertaken in accordance with all legislative and planning requirements, environmental standards and safe working procedures.

Permission for demolition

As is common, state and territory legislation varies with regards to demolition requirements. Generally, demolition is regarded a 'development', requiring one or more permits from the local council. This may be in the form of a development application or a dedicated demolition permit. In addition, and depending upon the scale of the demolition, a demolition licence may be required.

There are five steps to ensuring a streamlined demolition project:

1. Check whether you need council approval and comply with any requirements.
2. Check if a permit is needed when access or work crosses council footpaths or roadways.
3. Ensure your WHS and environmental controls are in place.
4. Ensure all insurances are in place.
5. Check whether you need approval from service authorities, such as power, water, telecommunications or the like.

Irrespective of state or territory, any application for a permit to demolish or remove a building will need to include the following information:

- a description of the building
- a site plan showing the building in relation to boundaries, other buildings on the site, adjacent sites, streets, footpaths and crossings
- structural computations demonstrating the structural adequacy of the building when partial demolition is proposed
- details of perimeter protection such as hoardings, barricades and protective awnings
- a written demolition procedure
- evidence that the demolisher has the necessary knowledge, experience, equipment and facilities to properly conduct the demolition operations.

Some states and territories also require that where a structure is over 15 m in height, there are chemicals, tower cranes, pre-stressed concrete or explosives involved, a site supervisor will be required at all times.

Planning for demolition

The key element to planning for demolition is being able to recognise the potential hazards and their associated risks. Your knowledge of structural principles thus becomes critical to the safety of you and your work team. It is only through this knowledge that you can avoid the greatest risk in any demolition or deconstruction project, which is uncontrolled or unplanned collapse. Any unplanned collapse of elements, even small timber framing components, can lead to injury, lost time, fines and, in the worst-case scenario, death.

The basics of your planning revolve around the elimination of collapse where possible, controlled and timed collapse when necessary, and good communications. Planning should cover the following key areas:

- structure
- services
- site
- environment
- recovery: recycling or reuse.

As this unit deals with the structural aspect of demolition, it is on this that we shall focus.

Structural considerations in demolition

As stated previously, the principle concern of the builder is uncontrolled collapse. Knowing how this may occur can help you plan against it. The lead causes of collapse are generally due to:

- incorrect demolition sequence and ignorance of stability issues
- overloading of floors by machines or debris
- unexpected voids and hidden basements
- instability in high winds
- over-stressed elements caused by altered load paths
- risk-taking individuals.

Each of these are avoidable through an understanding of the structural principles addressed so far and applying them logically to the structure under consideration. Altered load paths, for example, is a key factor requiring careful deliberation before the work commences. This is, at least in part, the purpose of a pre-demolition investigation or survey.

Pre-demolition investigation

A pre-demolition survey or investigation is a careful study of the structure as it now stands, along with any associated plans that may remain in existence. The survey seeks to determine a range of critical aspects including:

- key dimensions, including distances to neighbouring structures
- existence of services – gas, electricity, water and the like
- materials and construction methods involved (e.g. steel, wood, glass, concrete, bricks)
- presence and location of hazardous materials (e.g. asbestos, lead)
- equipment and materials storage areas
- areas where public protection may be required
- what currently maintains structure stability
- areas where reinforcement or stabilisation is required to prevent early collapse
- alterations to the building from the original plans.

Consideration will need to be given to what services may be required during the various stages of demolition and which services require disconnecting before work can commence.

The work plan

Having identified the main structural considerations, a work plan must be created. Before writing out the plan, you must first identify and assess the hazards associated with the demolition process and develop detailed procedures designed to eliminate or control the risk arising from them. Once this is complete, the approach to the task can be more fully developed and the risk analysis revised as new hazards come to light in the planning.

Once complete, the work plan may be documented beginning with a brief description of the type of building involved, including its overall height above ground level, and the distance from the structure to each site boundary. This is then followed by descriptions of actions and key information based upon the site survey and the risk analysis, such as:

- outline of structural support systems and principal materials involved
- description of the demolition methods and sequence proposed
- methods proposed for handling and disposing of demolished and hazardous materials
- maintenance of access and egress to the work site
- timeframe
- description of proposed hoardings, scaffolding, fencing, or any overhead sidewalk protection
- key safety requirements
- first aid and emergency access provisions
- worker amenities.

Of greatest import is that demolition should be sequential, meaning that it follows a series of well-thought out, pre-planned steps designed to ensure safe deconstruction. Generally, this means removing building elements in the reverse order of their assembly.

The work plan should also include points at which secondary inspections of the structure can be made. For example, a secondary structural inspection might be conducted after internal linings and/or non-structural external cladding have been removed; this may show evidence of termite damage, or that the structure is otherwise less sound than originally thought. Once a concrete slab is exposed, checks should be made to ensure no pre-stressing was involved in its creation, noting that even domestic house floor slabs can have this feature.

LEARNING TASK 12.8

PLANNING FOR DECONSTRUCTION

You have been asked to demolish an existing two-storey house. The house is a 1980s slab-on-ground building with cavity brick construction and a suspended concrete slab first-floor (180-mm deep) supported on internal cement-rendered masonry walls, conventional hardwood timber roof covered in concrete glazed tiles. The roof pitch is 30°. The house takes up most of the block of land with side access at boundary walls 1.1 m both sides. The boxed eaves are 450-mm wide with metal gutters.

- Provide your client with a logical deconstruction schedule as to how you would approach this job, based on the requirements of AS 2601 Demolition of structures.

SUMMARY

The purpose of this chapter was to provide you with the grounding necessary to identify and interpret the loads and forces acting upon the various elements of a structure, be it a small domestic home, or a much larger commercial structure. In addition, the chapter offered explanations of those systems traditionally used to counter or stabilise structures under such loads and forces. It is notable that principles are much the same, no matter the scale of the project under consideration; the variance being in scale or magnitude rather than anything fundamental. What can change, however, is the approach to dealing with these greater loads, through more complex solutions or with materials less common to the domestic or residential area.

The chapter to come, Chapter 13 Apply structural principles to low-rise construction, goes on to apply these principles to both commercial and residential structures. It therefore covers the likes of footings, floors, walls and roofs, and the various interpretations of these elements.

REFERENCES AND FURTHER READING

Weblinks

https://www.be.unsw.edu.au/sites/default/files/upload/pdf/cityfutures/cfupdate/Neuman_Posters.pdf

https://acumen.architecture.com.au/environment/place/climate/climate-change-adaptation-for-building-designers/

Geoscience Australia earthquake map: **http://www.ga.gov.au/interactive-maps/#/theme/hazards/map/earthquakehazards**

APPLYING STRUCTURAL PRINCIPLES TO LOW-RISE CONSTRUCTIONS

13

Chapter overview

The previous chapter, Chapter 12 Structural principles, provided the knowledge underpinning both construction and deconstruction works. This knowledge allows for safe work practices, security of the structure during the work and, when a new building, stability over time. In this chapter, we will explore the application of these principles to the more common elements of both commercial and residential buildings defined by the National Construction Code (NCC) as Classes 1 through to 10. Though limited to two-storey structures of Type C only, this captures a broad extent of our built environment, including: domestic houses, hostels and motels, shops, warehouses and even low-rise medical centres. In so doing, the chapter analyses those elements not explored previously in Chapter 12.

Elements

This chapter provides knowledge and skill development materials on how to:

1. analyse and plan for the structural integrity of Class 1 and Class 10 buildings
2. plan, coordinate and manage the laying of footings
3. plan, coordinate and manage the laying of floor systems
4. plan, coordinate and manage the building of structural and non-structural wall systems
5. plan, coordinate and manage the building of roof systems
6. plan, coordinate and manage the external wall cladding of structures.

In addition, the Australian standards of relevance to this chapter are.

- AS 1684 Residential timber-framed construction
- As 2601 The demolition of structures
- AS 2870 Residential slabs and footings
- AS 3600 Concrete structures
- AS 3959 Construction of buildings in bushfire-prone areas
- AS 4055 Wind loads for housing
- AS/NZS 1170.2 Structural design actions – wind actions.

Introduction

As stated in the previous chapter, there has been a significant rise in regulations and legislation surrounding and supporting the construction industry. This shift has been brought about to some extent by changes to the way we live, coupled with how we design our homes and other low-rise buildings such as shopping centres and the like. In addition, this shift responds to the increased potential for profit taking that the building industry is perceived to hold; i.e. it limits the less scrupulous from creating buildings that are not fit for purpose or, more concerning, dangerous to both inhabitants and surrounding structures.

This chapter therefore starts with a brief overview of these regulations, codes and standards. It is imperative that you understand these documents as they form the basis from which the principles covered previously may be appropriately applied.

The National Construction Code

Though covered in some depth in Chapter 1, the NCC must now be approached in application. To complete this current chapter, you will need to access the following:

- **NCC 2019, Volume One, Class 2 through to Class 9 Buildings**
- **NCC 2019, Volume Two, Class 1 and Class 10 Buildings**

These are required to interpret much of this chapter's content. See the introductory section at the front of the book on the best means of access.

REFER TO AS 1684

AS 1684

This standard underpins most timber-framed construction, which remains at the forefront of the nation's domestic building practices. Being able to interpret this standard fully is an essential part of what it is to be a builder in any Australian state or territory.

AS 1684 is a four-part standard as shown in Table 13.1.

Given the importance of this standard to your studies, and your future building practice, you are advised to either purchase the version most relevant to you, or access it via one of the paths laid out at the beginning of this book.

Choosing the relevant standard

The relevant standard depends upon the location of the project being considered. Generally, your choice will be between AS 1684.2 (non-cyclonic) and AS 1684.3 (cyclonic) based upon the map provided in Figure 13.1 and the information below.

- Region A or B: AS1684.2 is appropriate.
- Region C or D: AS1684.3 will be required.

If you remain uncertain, the local council, other regional builders, or local architects or engineers will usually be able to steer you to the correct choice. It is suggested that you do not purchase AS 1684.4 (Simplified) as this will not adequately cover the areas of study for this chapter.

Low-rise construction: a definition

The concept of low-rise construction is used very broadly throughout Australia with no definitive national description. For the purposes of this chapter, the construction limitations of AS 1684 and AS 4055 have been adopted, extended to include all NCC building classes. That is:

- buildings classified by the NCC as Class 1 through to Class 10
- not more than two storeys in height
- not greater than 8.5 m at the highest point, with a maximum of 6.0 m to underside of eaves
- not greater than 16.0 m in width
- length not greater than five times its width
- a roof pitch not exceeding 35°
- Type C buildings only.

TABLE 13.1 The four parts of AS 1684

AS 1684.1 – 1999 Residential timber-framed construction – Design criteria	Contains criteria by which the span tables and underpinning engineering of the other volumes has been developed.
AS 1684.2 – 2010 Residential timber-framed construction – Non-cyclonic areas	These two volumes are developed such that their relevance is specific to non-cyclonic or cyclonic regions as identified in the NCC.
AS 1684.3 – 2010 Residential timber-framed construction – Cyclonic areas	
AS 1684.4 – 2010 Residential timber-framed construction – Simplified – Non-cyclonic areas	A simplified text for non-cyclonic regions whereby the most commonly used span table are printed within the volume itself. The commentary on loads and fixings is also much reduced.

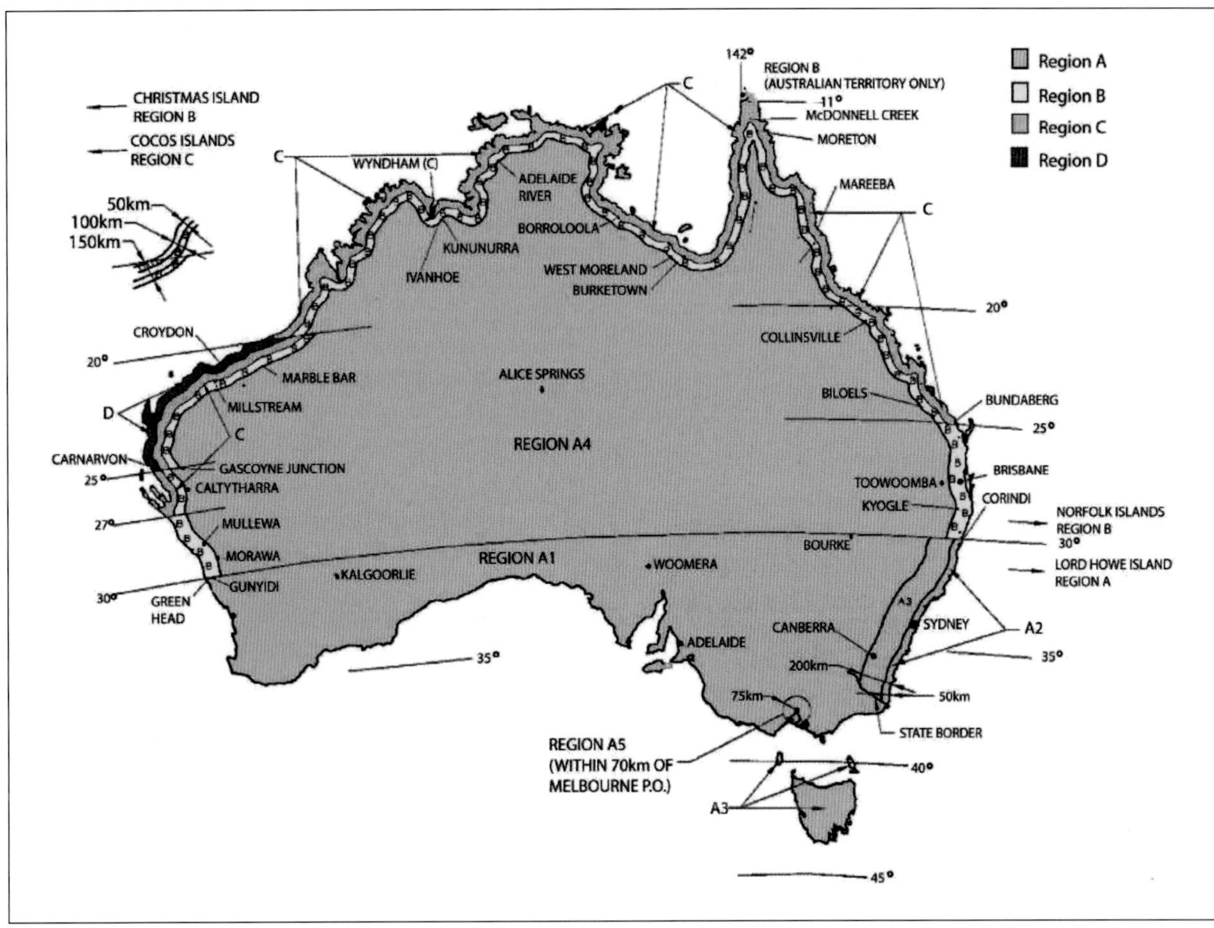

FIGURE 13.1 Wind regions derived from AS/NZS 1170.2

There are three 'types' of buildings defined under the NCC: Types A, B and C. Aside from Classes 1 and 10, this chapter only deals with Class 2 to Class 9 buildings that fall into the Type C classification. Building type is framed around levels of required fire resistance, which in turn is based upon maximum floor areas of a fire compartment or atrium, as shown in Table 13.2 below.

- An atrium is defined as a space within a building that connects two or more storeys, enclosed at the top by a floor or roof (excluding stairwells).
- A fire compartment means either the total space of the building when there are no fire separation walls internal to the structure, or a part of the building that is bounded by fire separation walls. That is, there may be multiple fire compartments within the one structure.

Type C is the least fire-resistant category; the requirements for this building type are found on pages 97 to 99 of NCC 2019, Volume One.

Planning for structural integrity

Chapter 12 provided you with the structural principles, the theory, that goes towards the erection of a stable, durable building. With this topic you begin the journey of applying that theory.

Such a journey begins with the analysis of the building documentation: the plans, specifications and other relevant documents. The purpose of this

TABLE 13.2 Maximum size of fire compartments or atria

Classification	Type A construction	Type B construction	Type C construction
5, 9b or 9c	Max *floor area*—8 000 m² Max volume—48 000 m³	Max *floor area*—5 500 m² Max volume—33 000 m³	Max *floor area*—3 000 m² Max volume—18 000 m³
6, 7, 8 or 9a (execept for *patient care areas*)	Max *floor area*—5 000 m² Max volume—30 000 m³	Max *floor area*—3 500 m² Max volume—21 000 m³	Max *floor area*—2 000 m² Max volume—12 000 m³

analysis is in part to ensure that the proposed building conforms to the NCC and associated Australian standards. It is also your first step towards planning the construction.

This initial path can be daunting at times, even for the most experienced builder. New technologies, materials and techniques are becoming available at an ever-increasing rate. Likewise, Australian standards and building codes are under constant review and revision. You should not feel alone in this process, and so the first section of this part details the professionals from whom you may choose to seek advice.

Consulting the professionals

While it's expected that you will have a sound grasp of structural principles, the reading of plans and the relevant building codes and standards, it is understood that you cannot know it all. Not only is it understood, but it is preferred that you seek advice whenever you are uncertain or stepping into new areas of construction. The following outlines just some of those to whom you may turn, and with what areas of advice they may assist you.

Architects, designers and engineers

Each of these are tertiary-qualified professionals, usually registered with an organisation or institution after satisfying strict licensing requirements. As other chapters cover most of these professionals in some manner, the focus here will be on what they can offer by way of assistance in structural matters.

Architects

Architects employed on both domestic and commercial constructions may offer concept design, design development and some of the more rudimentary engineering principles. It is through the architect's office that all the construction drawings are created, though they may not draw them themselves.

Issues can sometimes arise in a design where spans or spaces may appear, or may actually be, inadequate for appropriate beams, columns or other structurally significant components to be installed. This may also apply to matters of bracing. It is to the architect that you should initially turn for clarity on these matters. Invariably, if the concern shows itself to be valid, an engineer will need to be engaged to solve the issue. Different materials may be required or the design subtly changed.

Note: Never substitute materials or beam sizes yourself. The architect must be informed first as changes in the space occupied by the 'shell' or fabric of a building (floor thickness, wall thickness, etc.) changes room spaces and external aesthetics; i.e. there is always a knock-on effect.

Designers and draftspeople

Architects will frequently engage draftspeople for the drawing (drafting) of specific plan details, engineers' drawings and the like. The registered plan that you will build from is almost always drawn up by a draftsperson.

Draftspeople may also work independently or with a designer, or may be the designer. In either case the structural issues you would otherwise have referred to an architect should be discussed with whichever of these professionals were responsible for the design. Again, no changes should be made without this initial consultation.

Engineers

There are a range of engineers offering very specific advice on standards, codes and the structural implications of a design as it pertains to their field. These include:

- geotechnical engineers – site classification and other foundation conditions surrounding footing design
- services and mechanical engineers – fire control systems, mechanical access and the like
- hydraulic engineers – boilers, pumps, storage and transportation of fluids
- electrical engineers
- structural engineers – design of structural components including concrete slabs and footings.

Engineers also may:

- prepare reports on how to protect neighbouring properties during excavation and/or demolition
- design retaining walls, columns, beams and loadbearing walls
- prepare reports on the structural adequacy of existing buildings and demolition projects.

It is the engineers that design the structure of a building and so while you may initially turn to the designer or architect, it is invariably an engineer who will be called in to solve the problem. In most cases, the registered building surveyor (council or private) will require an engineer's approval for any structural changes prior to a structural variation being accepted.

Building inspectors and surveyors

A building surveyor ensures building works comply with relevant Australian standards and the NCC. They may be involved in obtaining approvals required under the planning Acts, building regulations, by-laws and other relevant legislation. Most importantly, they can undertake structural surveys as part of renovation or demolition works.

Building inspectors may be directly employed by councils or they may be independent, licensed as or work under the authority of the relevant building surveyor. This is a much misunderstood and under-appreciated role, as the experience and skills of these individuals can be extensive and their advice on structural matters is generally very well founded.

Building inspectors are engaged to inspect new domestic constructions at four specific stages of the works:

- foundation and footings
- steel reinforcement
- frame
- final.

It is during the first three inspections that commentary provided by an inspector must be respected and acted upon with regards to structural matters. Often these may be minor, or appear minor; on other occasions the matter may be of some significance and you should take their concerns in the manner they are offered – in the best interest of the building's structural integrity.

In the past, building surveyors were only engaged on very large projects; however, as regulations and codes have tightened, their role is becoming more relevant to all construction, including domestic housing. Aside from engaging them during a demolition or renovation project, or for the mandatory inspections listed above, you may turn to these professionals via the architect or designer when you have concerns that may require structural changes.

Land surveyors

Generally, land surveyors are associated with on-site measurements of land, allotments, building sites and existing buildings. However, from a structural perspective they may assist a builder with:

- setting out the building line, marking the location of columns, beams and even the external perimeter of a building
- checking the vertical alignment of a building during construction
- checking the exact positioning of structurally important features or members both laterally and in height.

While surveyors have little training in structural principles directly, through precision services such as those listed above, they can ensure that your actions are as per the engineer's and designer's specifications.

Analysing project documentation

Project documentation and its analysis has been extensively covered in the following chapters:

- Chapter 1 Building codes and standards
- Chapter 5 Plans and specifications.

This section will therefore be brief, focusing only upon where in the documentation information about structural adequacy is most likely to be found, and in what form it may take. In addition, your attention is drawn to those parts of the specifications covering structural matters.

Identifying and collating the relevant documents

Plans and specifications should be obtained and reviewed at the earliest moment possible in the planning stages. It is then that your knowledge of structure must be brought to bear so that you can pick up any areas that may seem vague or inadequate. It is at this early stage that new, alternative, or innovative construction methods or materials may be identified. As such times you may need to seek further advice from manufacturers, suppliers, engineers and the like – this is covered in more detail in the second section of this chapter.

Likewise, this is the time to review other key areas of compliance specific to location and context, such as: bushfire attack levels, high wind, earthquake and alpine environments. The second section of this chapter also covers these issues in more detail.

What this means is that just the plans and specifications are not sufficient. You may need documents from a range of professionals, other tradespeople, suppliers and manufacturers. Documents covering the bushfire rating of materials or products may need to be matched with engineer's reports on structural components such as footings or foundation material. Each will need then to be analysed for their implications for the structural integrity of the building as a whole.

Analysing the documents

Plans and specifications must be read in their entirety, even in regards to key structural matters where you will often find yourself focused upon specific sectional drawings or details. This constant stepping back and reflecting on the project as a whole allows you to see how a small detail can influence the structure overall.

For example, the partial drawing (Figure 13.2) taken from the footing and subfloor layout of a major domestic demolition and extension project shows a requirement for continuous strip footings to tie into the existing home's strip footing (circled areas). A review of the rest of the drawings shows that there are no details of this junction. An inspection of the elevations, shown in Figure 13.3 (our stepping-back exercise), shows that an existing column is to be retained and adopted as part of the new structure.

This is structurally complex as these new and old footings are likely to articulate, hinge, to some degree. The drawings and specifications would need to be very carefully analysed for any information on how this was supposed to be accounted for. If nothing is found, you would need to discuss the issue with the designer

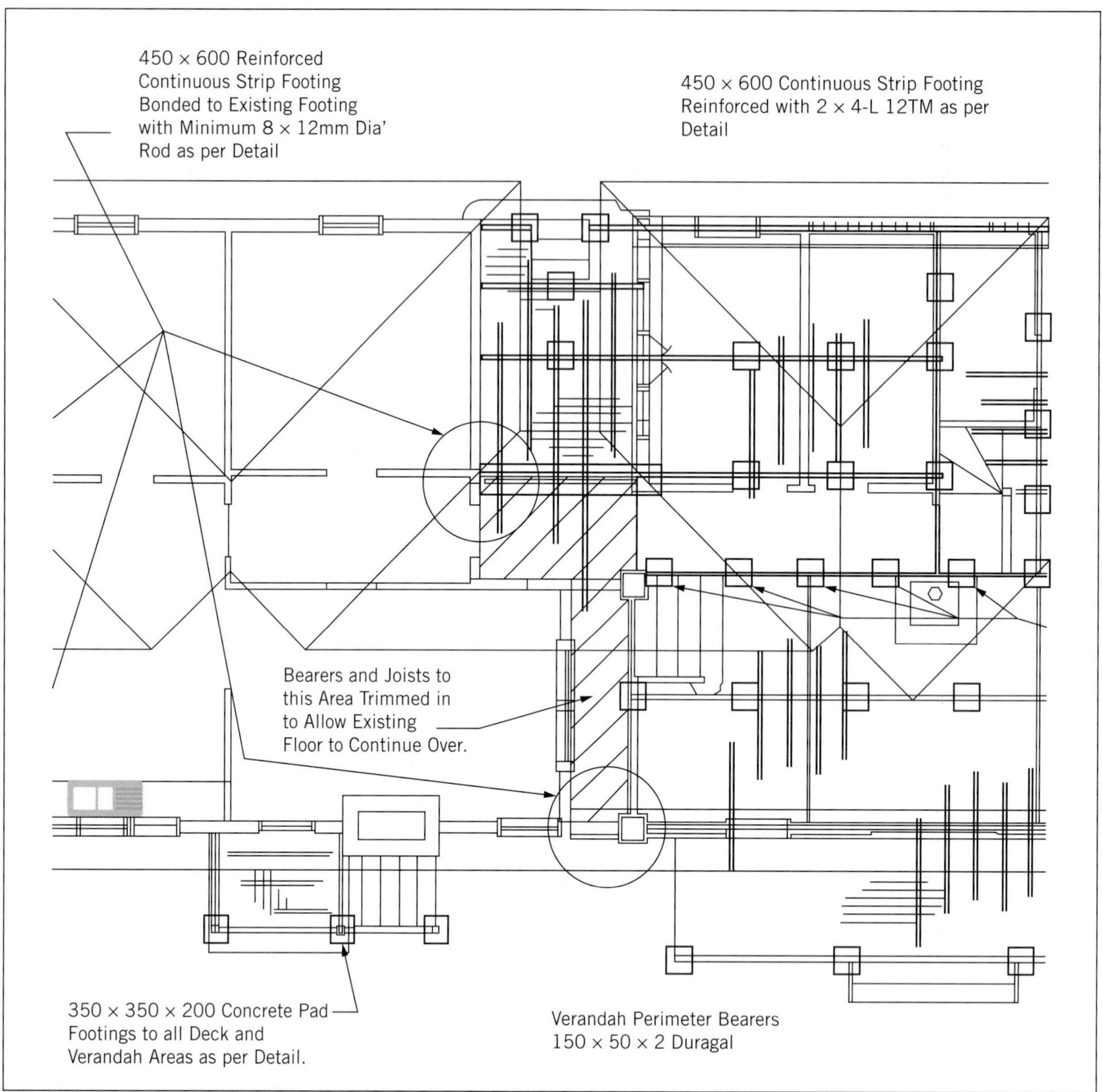

FIGURE 13.2 Partial footing layout

and possibly with those involved in the cladding (the project uses rendered Hebel panels). The discussion may end up including the manufacture and/or supplier of the cladding as well.

From Figure 13.2 alone, there are many things of structural importance that need to be assessed and not taken for granted. The spacing of bearers, for example, and therefore the location of pad footings must be checked in accordance with the manufacturer's specifications and the specifications for the subflooring components proposed.

Even the size, depth and reinforcement proposed in the drawing needs to be checked for appropriateness based on the foundation material and expected loads. This will generally require a review of the geotechnical engineer's report; a document that sits outside of the plan and specification set, though the key information may be included in these documents. It is wise to locate this report, if existent, just in case any misinterpretation has occurred.

And this is the key point at this stage – you need to be sure you have all the documentation on hand, and that this documentation can actually provide you with the structurally significant information you require. Know that the plans are seldom, if ever, enough.

Project documentation and NCC compliance

Having identified and collected the relevant project documentation it needs to be analysed for compliance with the NCC. In this section we will focus upon the requirements for bushfire, high wind, earthquake and alpine environments.

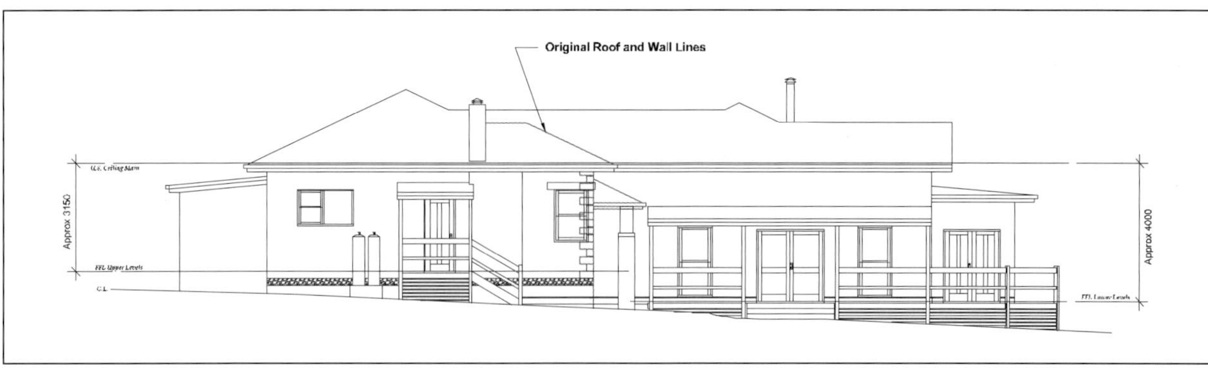

Source: Glenn Costin

FIGURE 13.3 Elevation showing included column and area where new and old meet

The National Construction Code (NCC)

The NCC has been extensively covered in Chapter 1. You are directed to that chapter for information about this code in general, how it interacts with the Australian standards, and how it should be read and interpreted.

The key areas of the two volumes of this code that are of import to this section are:

NCC 2019 Volume One

- Section B – Structure
- Section C – Fire resistance
- Section G – Ancillary provisions. Particularly:
  - Part G4 Construction in alpine areas
  - Part G5 Construction in bushfire prone areas

NCC 2019, Volume Two

- Performance provision:
  - Part 2.3 Fire Safety
  - Part 2.7.4 Buildings in alpine areas
  - Part 2.7.5 buildings in bushfire areas:
- Acceptable construction
  - Part 3.0 Structural provisions
- Additional construction requirements:
  - Part 3.10.2 Earthquake areas
  - Part 3.10.4 Construction in alpine areas
  - Part 3.10.5 Construction in bushfire areas

Note: High wind is a pervasive element within many parts of the code including glazing and cladding. This is an issue addressed later in the section.

Relevant Australian standards

Many Australian standards are called up by the NCC as '***acceptable construction manuals***'. The import of these were discussed in Chapter 1 Building codes and standards and you are referred back to that chapter for detailed information pertaining to their use. Under the relevant headings, the key standards are listed and their application discussed.

Bushfire zones

Bushfire-prone areas are those which have been designated in legislation as being subject, or likely to be subject, to bushfires. With regards to the requirements of Volume One of the NCC, consideration is required only with regards to building Classes 2 or 3. In such cases, reference is made to clauses G5.0, G5.1 and G5.2 of the same volume. NCC Volume Two, however, calls up two external standards within Part 3.10.5.0 as acceptable construction manuals for construction of homes in these regions:

- AS 3959 Construction of buildings in bushfire-prone areas
- NASH Standard – Steel Framed Construction in Bushfire Areas

REFER TO AS 3959

AS 3959 Construction of buildings in bushfire-prone areas
NASH Standard – Steel Framed Construction in Bushfire Areas

These standards are extensive and their many provisions must be read carefully. You are directed to the standards themselves for further guidance; however, the main points with which you should familiarise yourself are as follows.

- **Bushfire attack level (BAL)** – This is a measure of a building's potential exposure to the extremes of bushfire including ember attack, radiant heat and direct flame. The building site is assessed by a competent person with regard to the worst-case direction of approach. This assessment includes slope, type of vegetation and volume of vegetation. The BAL rating given is a statement of the expected heat flux in kW/m^2. The rating then becomes the basis for all construction provisions for that building. (See Figure 13.4.)
- **BAL Rating Provisions** – The standard is then broken up into divisions that describe each element of the structure with regards to required bushfire resistance. Timber can still be used in most cases; however, the emphasis is upon bushfire-resistant timbers which, in the main, are dense Australian

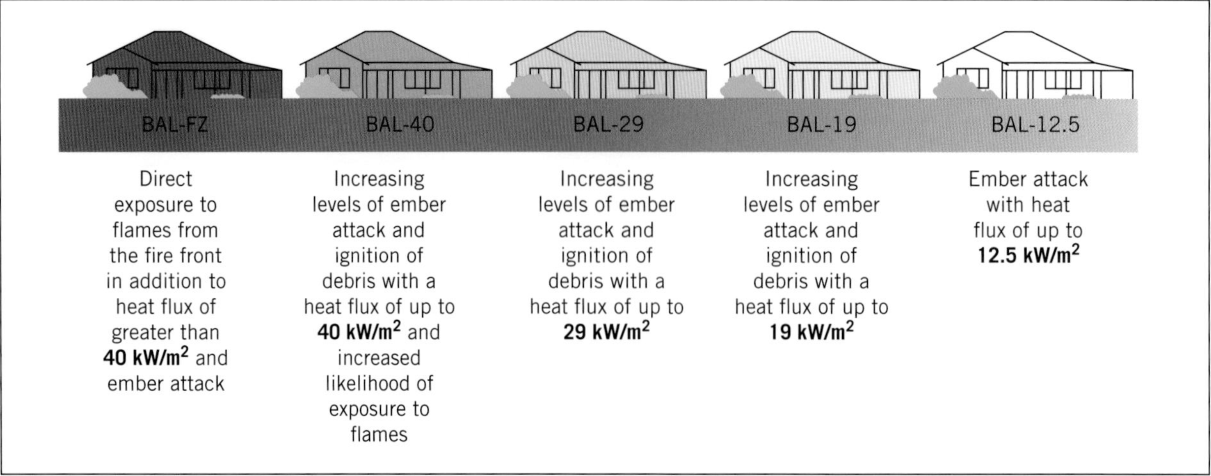
BAL-FZ
BAL-40
BAL-29
BAL-19
BAL-12.5
Direct exposure to flames from the fire front in addition to heat flux of greater than 40 kW/m2 and ember attack
Increasing levels of ember attack and ignition of debris with a heat flux of up to 40 kW/m2 and increased likelihood of exposure to flames
Increasing levels of ember attack and ignition of debris with a heat flux of up to 29 kW/m2
Increasing levels of ember attack and ignition of debris with a heat flux of up to 19 kW/m2
Ember attack with heat flux of up to 12.5 kW/m2

FIGURE 13.4 BAL ratings showing proximity to the flame zone

hardwoods such as red gum, ironbark and blackbutt. Other factors that are defined by these provisions include maximum gaps in decking and other spaces based upon ember attack, window types, glass thickness, metal screens to windows and the like. Restrictions tighten as the BAL rating increases.

- **State and territory limitations** – There are some significant restrictions on the use of these standards by various states and territories so you must confirm, either from within the NCC or from local councils or registered building surveyors, their applicability on a given project. For example, NSW will not accept the 'Section 9' of BAL FZ provision in AS 3959 or the whole of the BAL FZ provision of the NASH standard.

Alpine areas

Alpine areas are defined by the NCC as those parts of NSW, ACT and Victoria that are 1200 m or more above the Australian Height Datum (AHD), or 900 m above the AHD (Tasmania) for Tasmania (see Figure 13.5).

There are no standards that the NCC specially calls up as acceptable construction manuals for alpine areas in either volume. Indirectly, AS/NZS 1170.3 Structural design actions Part 3: Snow and ice actions is offered for the determination of individual actions and must be read in conjunction with AS/NZS 1170.0. This is specifically engineer's territory, but from a structural perspective they are your most important sources of information.

Beyond these two standards there are a range of standards governing energy efficiency; these are covered in Chapter 14 Thermal efficiency and sustainability, and you are directed there for the specifics. In reviewing that chapter, take particular note on the issues of condensation and the implications of such to structural members. These incidental standards aside, construction in alpine areas is governed alone by Part 3.10.4 of Volume Two of the NCC, and Part G4 of Volume One. The emphasis of these Parts is not upon structure, however, but upon the restricted access and egress that can come with sub-zero temperatures.

From a structural perspective your considerations differ. It is therefore to engineers, designers and material manufacturers that you will need to turn for specifications on allowable snow loads and thermal stresses. The key point to note is that these stresses should not be underestimated. Snow and ice can change the shape of a building and therefore change its profile to high winds. Dead loads increase, and as water freezes it expands, which can crack stone, brickwork or split adhered render coatings from sheeting. Experienced material suppliers, subcontractors and other builders are therefore another key source of information in this area.

Earthquake zones

As the NCC states, most buildings in Australia are not required to be specifically designed to withstand earthquakes. However, acceptable design manuals offered in Part 3.10.2 of Volume Two and Schedule 4 of Volume One include the Australian standard set AS/NZS 1170, and specifically AS 1170.4 Structural design actions Part 4: Earthquake actions in Australia. This standard provides maps with Earthquake Hazard Factors (Z) marked upon them. These maps are drawn

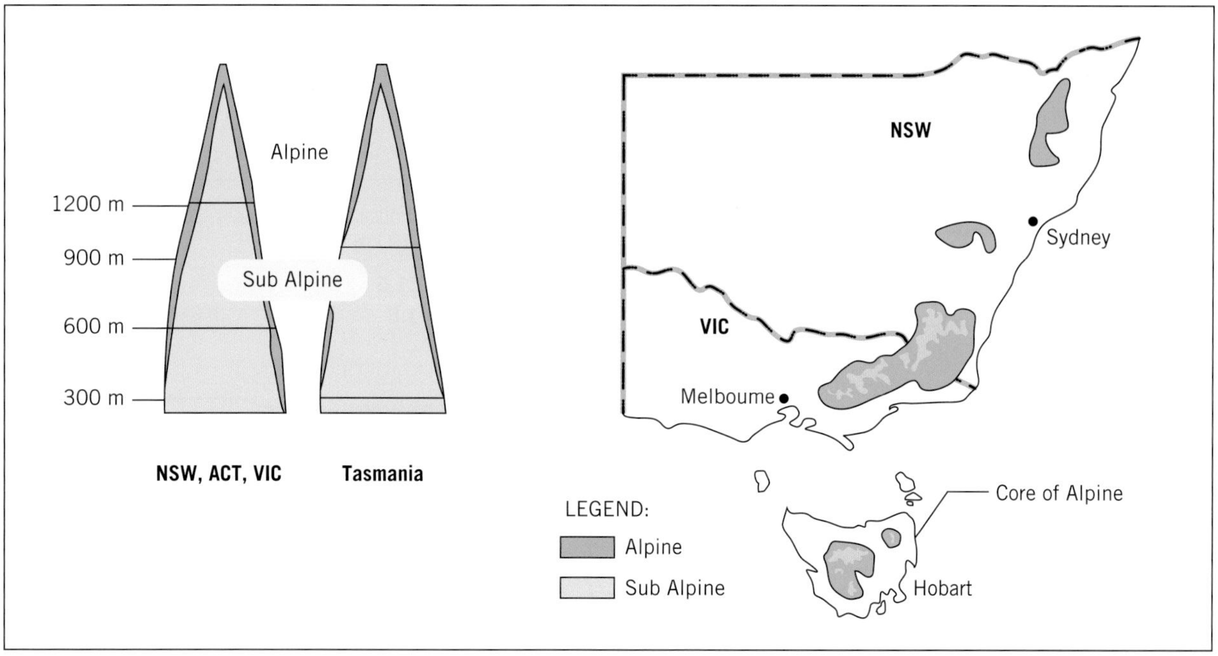

FIGURE 13.5 NCC-declared alpine areas in Australia

much like the contour maps with which you will be familiar. Figure 13.6 shows an updated map of this type from Geoscience Australia. The values given here are the Z values offered in AS 1170.4.

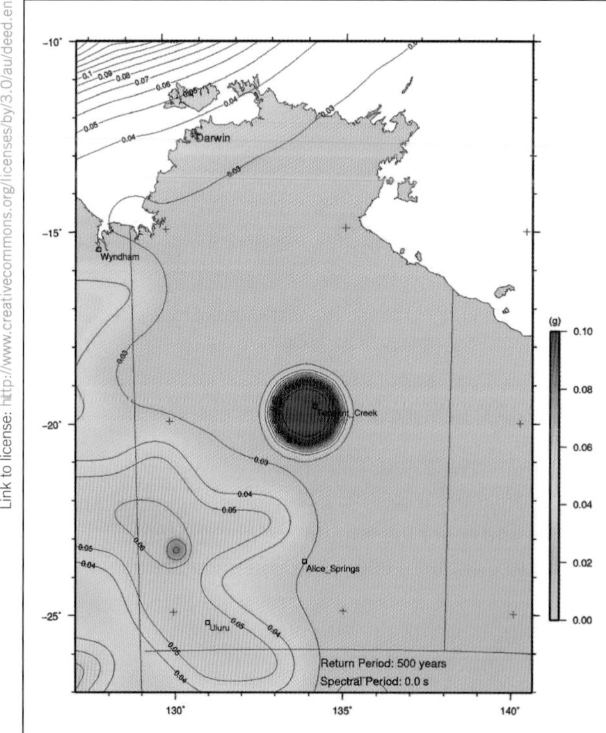

FIGURE 13.6 Earthquake hazard map for Northern Territory. Note the 'hot spot' at Tennant Creek.

This Hazard Factor (Z) is an acceleration coefficient. This is used to inform engineers of the expected severity of earthquake and the associated earth movement. Engineers and designers must accommodate this additional load into a building's bracing design. Generally, existing bracing based upon wind loads is adequate. In severe regions, however, further bracing may be required, coupled with improved footing. This is the sort of thing that you must look for if you know that you are in an earthquake-prone area.

High wind areas

The NCC defines **high wind areas** as regions subject to wind speeds *greater* than those classified as N3 or C1, which equate to 50 m/s or 180 km/h. Part 3.0 of Volume Two of the NCC holds the main provisions for housing in high wind areas. This section also calls up two Australian standards as acceptable design manuals:

- AS 4055 Wind loads for housing
- AS 1170.2 Structural design actions – Wind actions

REFER TO AS 4055, AS 1170.2

AS 4055 Wind loads for housing
AS 1170.2 Structural design actions – Wind actions

Based upon the definition of high wind areas within the NCC, it is important to understand that a building can be in a defined high wind category irrespective of its state or territory location; i.e. it *does not* have to be in the north of Australia and subject to cyclonic wind forces. It is based upon the wind category as defined for a specific construction site. Notably, Volume One of the NCC makes no definitive statement upon high wind areas other than offering the definition offered at the beginning of this section.

Wind classification determination

There are 10 wind classifications identified within these standards and referenced by the NCC. These range from N1–N6 and C1–C4. 'N' stands for non-cyclonic and 'C' stands for cyclonic. To determine a site's classification, you must first identify the region in which the structure is set, and then apply parameters such as terrain, topography, shielding and the like.

Wind region

The wind region applicable to a construction is defined by the likely winds for an area and is identified by a full map of Australia in both the NCC (Figure 3.10.1.4) and AS/NZS 1170.2. This map has been replicated previously as Figure 13.1. The regions are classed as A, B, C, or D, the latter two being cyclonic zonings. It is by this map that you identify which version of AS 1684 applies to your region.

Once this is determined, other factors must be identified before a wind classification can be finalised. These are: terrain category, topographic class and shielding.

Terrain category

The terrain category is based upon how rough the surface is within 500 m of the site. It allows for how this area is 'likely' to be five years after the home is complete. AS/NZS 1170.2 identifies six terrain categories; however, AS 4055 provides only five, arguing that the wind forces for the sixth, TC4.0, being based on central city landscape, are difficult to calculate.

The terrain categories are shown in Figure 13.7.

Example: The site plan and photo in Figure 13.8 shows the location of a proposed new home. Based on this information a terrain category of 2.0 is appropriate.

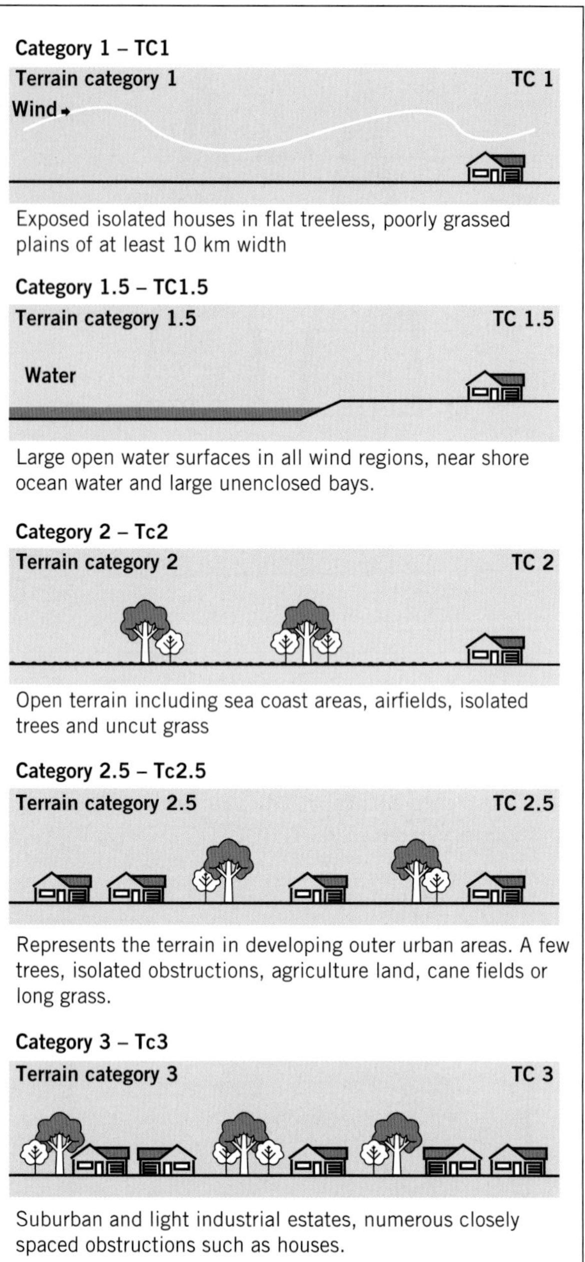

Category 1 – TC1
Terrain category 1
TC 1
Wind
Exposed isolated houses in flat treeless, poorly grassed plains of at least 10 km width
Category 1.5 – TC1.5
Terrain category 1.5
TC 1.5
Water
Large open water surfaces in all wind regions, near shore ocean water and large unenclosed bays.
Category 2 – Tc2
Terrain category 2
TC 2
Open terrain including sea coast areas, airfields, isolated trees and uncut grass
Category 2.5 – Tc2.5
Terrain category 2.5
TC 2.5
Represents the terrain in developing outer urban areas. A few trees, isolated obstructions, agriculture land, cane fields or long grass.
Category 3 – Tc3
Terrain category 3
TC 3
Suburban and light industrial estates, numerous closely spaced obstructions such as houses.
Source: Copyright BlueScope Steel Limited 3 May, 2017

FIGURE 13.7 Determining the terrain category

Topographic class

This defines the effect of wind on the structure based upon its location on a hill, valley, ridge, or the like, and the slope of the hill or escarpment in question. There are six topographic classes ranging from T0 to T5. Slope is variously described in degrees or as a ratio, with location on the hill defined as being either in the lower, mid, or upper thirds. The top third is split again into three parts (see Figure 13.9).

Shielding

A shielding class is derived on the basis of obstructions of a similar size to the building under consideration. Again, it is based on what is likely to be there in five years' time. Trees may be considered obstructions only in wind Regions A and B. In cyclonic Regions C and D, trees and other vegetation are not considered as shielding. The three classes of shielding are shown in Figure 13.10.

Example: Using the site plan and photo in Figure 13.8 once more, we can see that the house is to be located on the top of a low hill – based upon the contours on the site plan – and will have no shielding worth considering. The topographic class will therefore be T1, and the shielding class NS (no shielding).

This information is now inserted into Table 2.2 of AS 4055, as shown in Figure 13.11.

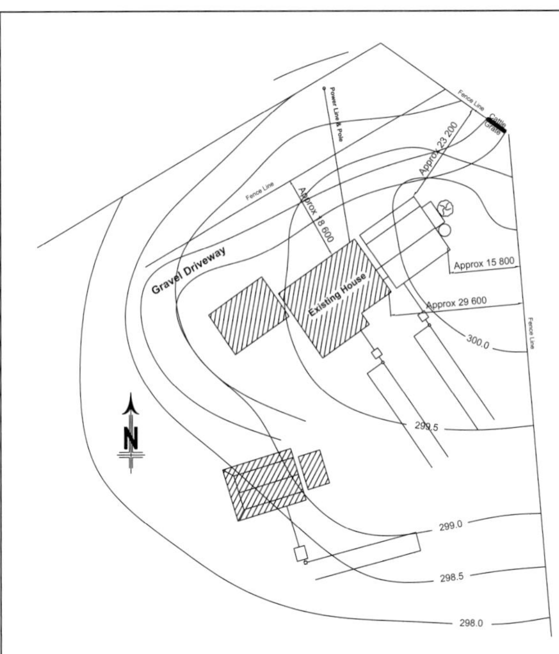

Source: Shutterstock.com/Ton Anurak

FIGURE 13.8 Site plan and photograph of terrain for proposed house

Based upon this determination, the house in question is *not* in a high wind area, but clearly homes in the area could be if located on the higher neighbouring hills.

In many cases, the determination of this wind classification will be undertaken by the relevant building surveyor or council, but frequently this may occur in consultation with you as the builder. If you are doing the determination yourself, you will have to demonstrate how the determination was made.

Application

Having made the determination, the wind classification is used to determine all bracing, structural components, hold-down fixings and even window glass thickness. Provided the classification is not greater than N3 or C1, these elements may be determined by the relevant acceptable design manuals such as AS 1684.2, AS 1684.3 or the NASH standard for steel framing. Glazing elements and fixing requirements for cladding are found in the NCC, Volume Two. Each of these are discussed under the appropriate headings later in this chapter.

Where the wind classification is greater than N3 or C1, the relevant sections of the above standards may still be used, coupled with: AS 2047 and AS 1288 for glazed assemblies; and AS 3700 and AS 4773 for masonry. In Wind Areas C and D (cyclonic), greater attention must be paid to AS/NZS 1170.2.

New and emerging technologies

Because the NCC is a performance standard, new and emerging building technologies are allowed for by way of the Performance Requirements laid out in Part 2 of Volume Two and similarly in each part of Volume One. Solutions other than those within the deemed-to-satisfy provisions (Part 3 of Volume Two) may be developed and assessed using one or more of the applicable verification methods identified in the NCC.

Known as Performance Solutions, the process for demonstrating and providing evidence of compliance was fully discussed in Chapter 1. You are directed to this chapter for detailed information. Alternatively, you may download the ABCB guidance document from https://www.abcb.gov.au/Resources/Publications/Education-Training/Development-of-Performance-Solutions

In brief, compliance may be demonstrated by one or more of the following:

- Evidence of Suitability – i.e. evidence, as per the General Requirements, is supplied
- Verification Methods – tests, calculations, inspections, or the like as deemed appropriate within the NCC or by an appropriate authority (as defined by the NCC)
- Comparison with the deemed-to-satisfy (DTS) provisions – the proposed solution is compared with existing deemed-to-satisfy examples offered within the NCC
- Expert judgement – a qualified and experienced person judges that a particular approach complies with the Performance Requirements.

A Verification Method is defined by the NCC as a '... test, inspection, calculation or other method that determines whether a Performance Solution complies with the relevant Performance Requirements' (NCC, Vol. Two, p. 519).

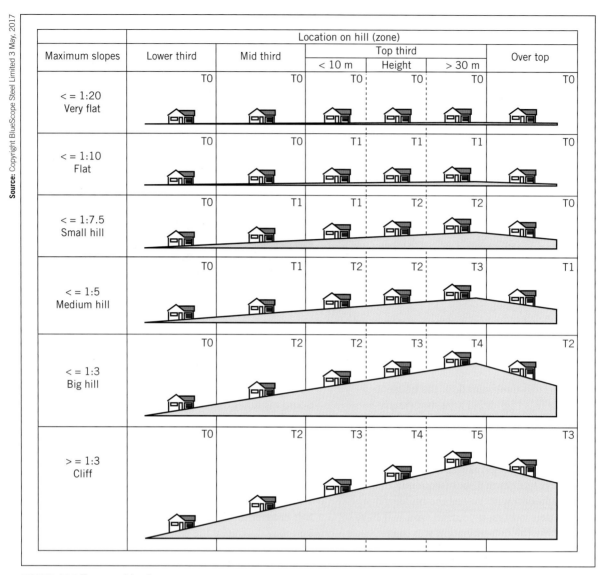

Maximum slopes	Location on hill (zone)					
	Lower third	Mid third	Top third			Over top
			< 10 m	Height	> 30 m	
< = 1:20 Very flat	T0	T0	T0	T0	T0	T0
< = 1:10 Flat	T0	T0	T1	T1	T1	T0
< = 1:7.5 Small hill	T0	T1	T1	T2	T2	T0
< = 1:5 Medium hill	T0	T1	T2	T2	T3	T1
< = 1:3 Big hill	T0	T2	T2	T3	T4	T2
> = 1:3 Cliff	T0	T2	T3	T4	T5	T3

FIGURE 13.9 Topographic class

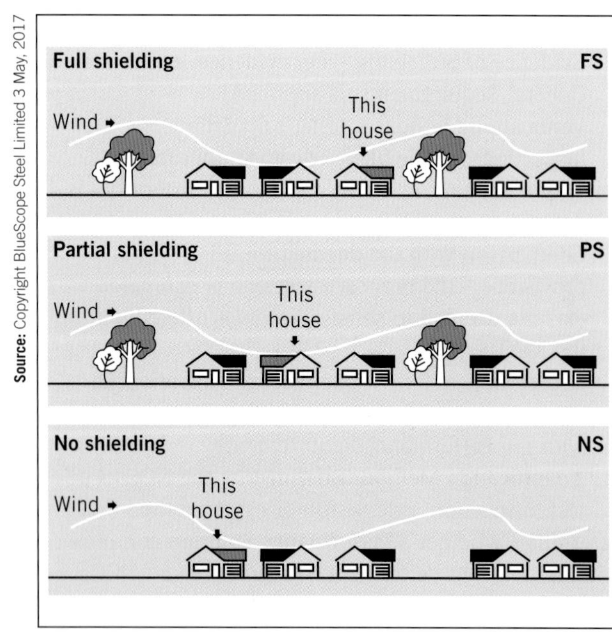

FIGURE 13.10 Explanation of shielding

From a structural perspective, compliance will need to address the following key areas:

- structural stability
- suitability of materials
- fire resistance
- thermal and energy efficiency.

Structural stability

From a structural perspective, this is obviously the key area of concern: that the technique, method, or component being proposed can hold the intended loads. Evidence of this will need to come from the following sources:

- engineers – documentary evidence of calculations, testing procedures, comparisons and judgement
- manufacturers and developers – proof of independent testing and external verification
- examples – evidence of use of the approach, or similar approaches in other similar contexts including comparisons with the DTS provisions.

TABLE 2.2 WIND CLASSFICATION FROM WIND REGION AND SITE CONDITIONS

Wind region	TC	Topographic class												
		T0			T1			T2			T3		T4	T5
		FS	PS	NS	FS	PS	NS	FS	PS	NS	PS	NS	NS	NS
A	3	N1	N1	N1	N1	N2	N2	N2	N2	N2	N3	N3	N3	N4
	2.5	N1	N1	N2	N1	N2	N2	N2	N3	N3	N3	N3	N4	N4
	2	N1	N2	N2	N2	N2	N3	N2	N3	N3	N3	N3	N4	N4
	1.5	N2	N2	N2	N2	N3	N3	N3	N3	N3	N3	N4	N4	N5
	1	N2	N3	N3	N2	N3	N3	N3	N3	N4	N4	N4	N4	N5

FIGURE 13.11 AS 4055, Table 2.2

Most importantly, this material must show evidence of performance over time: again, either by time-simulation testing, or through comparison with like materials and methods with a viable history.

Suitability of materials

This is the acknowledgement that a method or system may be made up of many parts, or that a single part, while perfectly adequate from a structural perspective, may fail in context. Here you should reflect upon the earlier section on the structural characteristics of materials, such as one that is overly brittle, for example. Other factors of context may include insect attack, or failure in moist conditions. Hence the materials of a system may need to show the capacity to resist any one or more of the following:

- wind, snow, rain, heat, UV radiation and other climatic conditions
- repeated and extensive thermal changes (broad temperature range)
- termites and/or borers
- seismic activity
- fire (including bushfire)
- fungal attack
- moisture (condensation, rising damp)
- saline conditions (salt)
- chemicals and/or polluted atmospheres.

This is not an exclusive list, but it demonstrates that a material may be structurally adequate, yet fail, sometimes quite early, due to other circumstances that have not been considered.

Resistance to fire

Though mentioned above, it is treated as a separate issue by the NCC through specific requirements for structural and non-structural systems in a building. As such, substantial independent testing must be undertaken and demonstrated. This is particularly the case when a material is completely new or at least new to the field of construction.

Depending upon the role and location of a system or structural approach, a required fire-resistance level or FRL will be required. Discussed in full in Chapter 1, an FRL is a statement of timed resistance which may, for example, read 60/60/60. Derived from AS 1530.4, these figures in this example would represent:

- structural adequacy (60 minutes before failure)
- integrity (will prevent fire from penetrating to the other side for 60 minutes)
- insulation (will not allow the other side to become excessively hot for 60 minutes).

In addition to the FRL there are Flame and Smoke indexes that must be considered. Though non-structural, these are integral to the system and identify the likelihood of materials within the system producing further flames, smoke, or fumes.

Testing of such systems is often undertaken by the likes of the CSIRO; however, other independent private organisations are also available.

The role of the site inspection

Reading the material is one thing, seeing the land is another. Context is often critical to the ability of a structure to maintain its integrity over time. The only way to assess this context and validate the assumptions made in your analysis of the documentation is to undertake a site inspection. From this inspection you should produce a condition report.

The condition report

The **condition report** reflects your observations of a range of factors based upon the information in the documentation you hold, the type of project involved and the site and its neighbouring areas. Condition reports, also known as site inspection reports, can be extensive, and are frequently based on checklists looking at things like access, existing services, obstructions and the like.

From a structural perspective, however, your main areas of concern will be:

- neighbouring properties – possible undermining of existing footings and thereby impacting upon the structural integrity of another building
- access of large structural components
- evidence of backfill, mine shafts, dam or swamp reclamation, creeks or the like
- mineral water springs

LEARNING TASK 13.1 WHO SHOULD I CONSULT WHEN I AM REQUIRED TO WRITE A REPORT?

1 After reading the text and doing some additional research, write the 'technical name' given to the specialist person you would best consult or use for the following design or building and construction work.

1	Person who would design a stormwater detention tank for commercial premises	
2	Person who would confirm that the building is the correct height above a known flood zone corridor	
3	Person responsible to ensure that the NCC requirements have been met	
4	Person responsible for ensuring that the bracing requirements for a large residential timber-framed building have been met	
5	Person responsible for detailing the existing foundation profile that will be built on	
6	The person who would be used to draw up plans and provide specifications based on architectural design sketches	
7	The person who would ensure that a 2.4-m retaining wall would not fall, move, or tip over after construction	
8	Person responsible for determining the amount of work and materials that would be needed	
9	Person who would determine the type and sizing of a ventilation, air conditioning and heating system (HVAC) for a building	
10	The location and method of getting power to and around a medium-rise industrial building	

2 a How would you be able to identify where or if there are any bushfire-prone areas in your local area?

b Based on your response and using a 50-km radius from where you currently live, investigate and provide evidence to your trainer about the identified local bushfire-prone areas that may exist, which affect the way residential building and construction will be carried out in your local area.

3 Create a site condition report based on a property next to a proposed or existing building site; or access an existing site condition report and determine the main items the report has as headings. Document the headings, summarise the information contained and show your trainer from where you obtained the information.

- subsidence
- large subsoil boulders
- earth retainment (slope of land, retaining walls and the like)
- water in flows and drainage.

Where a soil engineer's report has been provided, you should track their reported findings with what it is you can see on the ground to ensure that they have not inadvertently missed some change in soil or foundation type within the zone of your construction.

When a demolition or partial demolition is involved you should look back at Chapter 12 for advice on how to carry out a demolition report alongside the condition report just described. A good example of a domestic site condition report template may be found at: https://www.fairtrading.nsw.gov.au/__data/assets/pdf_file/0009/367875/Residential_site_condition_report.pdf

The footings

From the previous chapters you will be aware of the basic footing types, designs and associated materials including reinforcement. You will also have an appreciation of the codes and standards applicable to their construction more generally, these being: AS 2870, AS 3600, and AS 1684. The purpose of this section is to apply that knowledge to the following:

- checking of proposed domestic house footings for compliance
- planning, coordinating and managing the laying of footings.

In each case the focus is upon structural integrity and so the knowledge you have been developing so far in this chapter is used to inform each of your judgements. What follows is therefore a basic description of the various footing types you may encounter, matched with where in the standards and codes you must look to ensure compliance.

Assessing structural integrity

The assessment of structural integrity is determined through an evaluation of the proposed footings' compliance with Volume Two of the NCC. Which part of the code is applicable is dictated by what type of footing is under review. The type of footing, in turn, is determined by the desired flooring system and, to a degree, the proposed cladding or wall system. The required footing, in terms of strength, size, shape and depth, is determined by the building structure as a whole, coupled with the classification of the site with regards to foundation material and wind.

The determination of wind categories was discussed earlier, so this section will begin with the site classification, which has only been briefly referred to previously.

Site classifications

The site classification should be listed both on the footing plans and within the specifications – as would the wind classification. If a soil report has been conducted, this will also state the classification. For building Classes 1 and 10, the site classifications are provided in Part 3.2.4.1 of Vol. Two of the NCC; further detail is also provided in AS 2870. In brief, a site may be classified as in Table 13.3.

Sites classed E or P must be fully engineered and are not within the scope of the DTS provisions of Volume Two of the NCC. In checking footings of this type, you look for a match between the listed site classifications on all documents and if concerned with any apparent discrepancy or inadequacy in the design you must check with the engineer and designer.

For building classes 2 through to 9, it is possible to use the soil classification system in Table 13.3 provided the structure does not exceed the limitations expressed in Volume Two, Part 3.2.1; i.e. not more than 30 m long or 8 m in height and other provisions as listed in AS 2870. In all other cases, footings must be engineered on the basis of a geotechnical engineer's report. This may involve multiple on-site test bores to ascertain specific load-bearing capacities.

Having confirmed the site classification (and wind classification) you can now assess the offered footing sizes and dimensions in relation to the structure and type of footing proposed.

Footing types

The structural principles behind concrete slab footings have been discussed in some detail in Chapter 12. When not required to be fully engineered for commercial structures, the required construction details are to be found in Part 3.2, of Volume Two of the NCC, with further commentary available in Part 3.1 of AS 2870.

However, stiffened raft slabs, or slabs with an integrated edge beam, are only one form of the many footing systems available. The other common forms are described below.

Note: With regards to the NCC, care must be taken when following the construction practices outlined as their application has very specific limitations, just some of which are:

- slabs may not be longer that 30 m
- the building is a maximum of two storeys
- the building is not sited in an alpine area.

The many other limitations may be found in Clause 3.2.1 Application, of Volume Two of the NCC.

- Pad or blob footings: these are round or square mass pour concrete, generally without steel reinforcement. They are sized to carry the loads of stumps or piers as part of the subfloor system of bearer and joist construction, or for verandah posts, columns and other such point loads. However, in determining pad footings, consideration must also be given to uplift forces as well as bracing capacities for columns and posts.

  The specific dimensions and design capacities of pad footings may be found in three documents:
  - Part 3.7 of AS 1684.2 or 3.6.6 AS 1684.3 for timber-framed construction loads; bracing capacities are within Part 8.3.5 on the basis of no uplift. With uplift, AS 2870 is the reference.
  - Clause 3.2.5.6 of Volume Two of the NCC
  - Appendix E of AS 2870.
- Waffle pod slabs: A system that borrows from the stiffened raft slab design, but uses polystyrene foam void forms (boxes) laid on the ground to create

TABLE 13.3 Soil classification

Class	Foundation	
A	Stable sand and rock, negligible movement from moisture change	
S	Slightly reactive clay, very little movement from moisture change	
M	M	Moderately reactive clay/silt, moderate movement with shallow moisture change
	M–D	Moderately reactive clay/silt, moderate movement with deep moisture change
H	H	Highly reactive clay, high movement with shallow moisture change
	H–D	Highly reactive clay, high movement with deep moisture change
E	E	Extremely reactive clay, extreme movement with shallow moisture change
	E–D	Extremely reactive clay, extreme movement with deep moisture change
P	Problem sites – soft or loose soils, high reactive soils, subsidence and the like	

Source: NCC 2019 Building Code of Australia - Volume Two; AS 2870

internal stiffening beams. By this means excavation is reduced. The design has limited application; i.e. mostly flat sites, and many engineers are now calling for the edge beams to be excavated into the ground, or edge strip footings to be laid into the ground first. The design specifics of these slabs are found in Volume Two of the NCC and AS 2870.

- **Strip footings**: Generally, a steel-reinforced strip of concrete that runs around the perimeter of a building to support the external walls and what is known as a dwarf wall. This use of strip footings is therefore common to stump and bearer construction. However, strip footings are also used in a number of other ways:
  - supporting internal load bearing walls
  - footings to brick or concrete fencing
  - footing slabs.

  Strip footings may run level, or they may be stepped to follow sloping ground. These steps are governed by both Volume Two of the NCC and AS 2870.
- **Footing slabs**: In this design the slab edge sits upon a strip footing, to which when constructed in Class A soils it needs not be connected. In other classes it must be tied in by way of R10 reinforcement rod at a minimum of 600-mm centres. See Volume Two of the NCC and AS 2870.
- **Pier** or **pile** and **edge beam footings**: These are governed by AS 2159 and are of multiple types and forms. AS 2159 makes no distinction between piers and piles. Piles, in this standard, cover all pier and pile types in that they may be bored and filled, drilled, screwed, jacked, vibrated, or driven including:
  - screw piles
  - bored piles
  - driven piles.
- **Deep beam footings**: These are an extension of the edge beam of a stiffened raft slab, with AS 2870 showing them as between 750 and 1500 mm from finished floor level to under the beam. You will find details of these footings in Part 3.5 of AS 2870.

Reinforcement

The purpose and location of reinforcement has been discussed earlier in this chapter. With regards to the specifics for any given footing, these are dealt with alongside the dimensions of the footings in Part 3.2.3.2 of Volume Two of the NCC and in AS 2870.

Importance of wind classifications

The details of determining a wind classification have been covered earlier. The import to the footing needs, however, a brief mention. All footings are based upon expected or assumed loads, up, down and also laterally – wind loads being of concern to all three.

Uplift loads you should now be familiar with, so the concept of needing hold down by way of the footing may be taken as a given. AS 1684 treats most uplift on footings as negligible for wind classifications N3 or less. N4 and above and you are directed to AS 2870. Bracing (lateral loads) and down pressures, however, are factored into all the footing designs. Hence, footings are framed around a given wind classification. Loads on a footing from the bracing of lateral forces can increase downward pressures on a footing as well as toppling. These are all accounted for in designs offered in the standards.

Your role in checking compliance is to ensure that the footings are matched to the wind classification, and when it is evident that a classification (such as N4 or C2) is outside the available designs, that the footings have been signed off by a qualified engineer.

What's on top: type of structure

The type of structure and its intended use clearly has an influence upon the footing design as it is this that dictates both the dead and live loads expected. For homes on slab footings (of whichever type), the dimensions of the slab are governed by Volume Two of the NCC and, where this volume directs, by AS 2870 and AS 2159; each framing the footing base on wind and soil classification as well as the type of structure involved. That is, timber or steel-framed brick veneer, clad timber or steel frame, solid masonry, articulated masonry, as well as overall house dimensions. These designs are also limited around the number of storeys involved.

For stump and bearer subfloors with strip and pad footings, again the structure influences loads but AS 1684 handles them somewhat differently to slabs. With slabs, except for particular point loads, the whole of the slab tends to the one design. With pad and strip footings, the use of the space – i.e. the expected live loads – above can change the pad dimensions, as well as point loads (increase dead loads). These must all be calculated together to come up with a pad size. The path for these calculations may be found in Part 3.5 of AS 1684.2.

Modelling a calculation

Decks and verandahs are common fare for most domestic builders, and the design of the footings is often the role of the builder when part of a small extension or addition. This being the case, it makes such footings appropriate for an example procedure in carrying out the necessary calculations (see Example 13.1). Figure 13.12 offers the information for a typical deck.

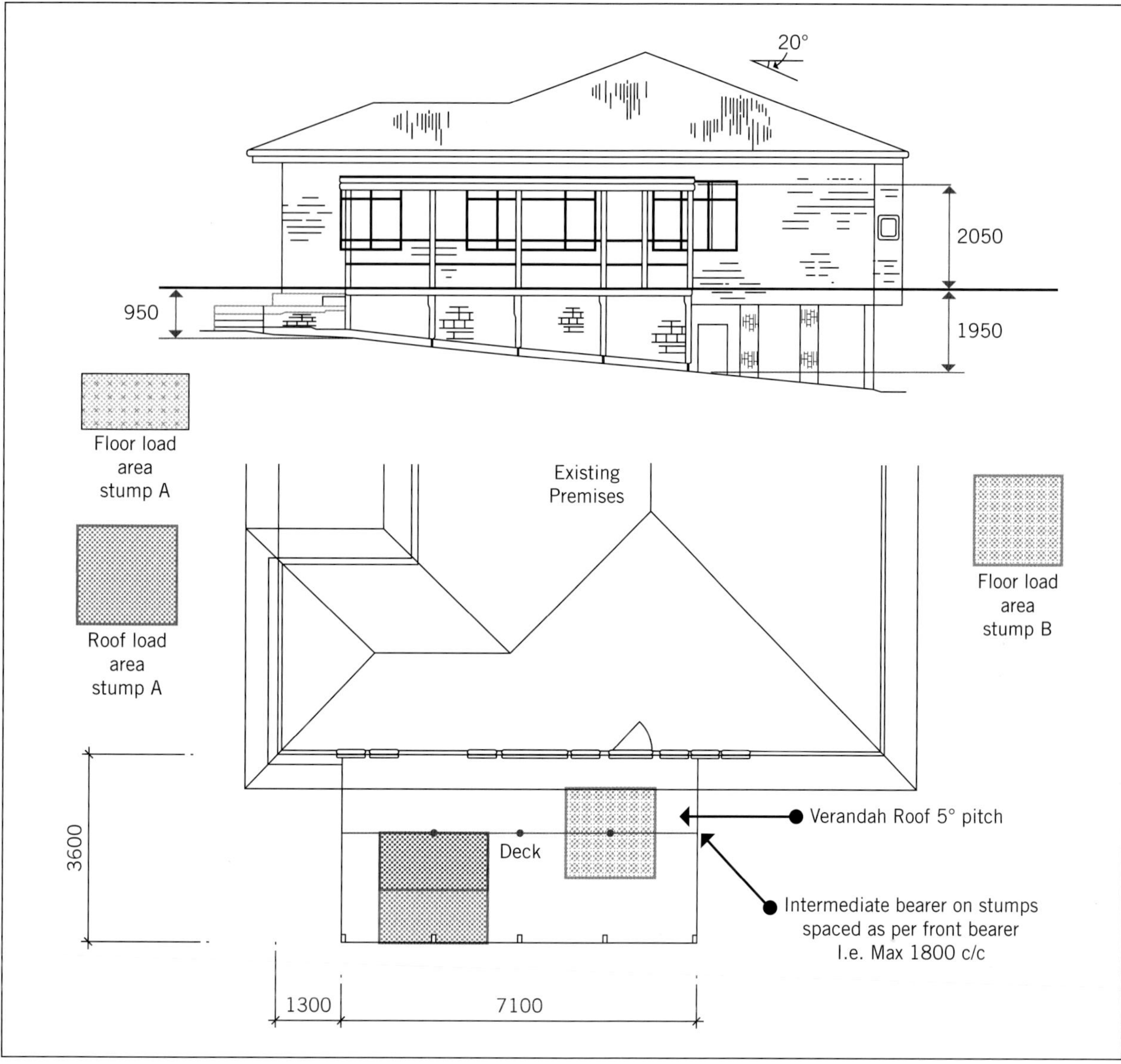

FIGURE 13.12 Deck elevation and plan

EXAMPLE 13.1

DECK AND VERANDAH FOOTINGS

Site criteria

In this example we will use the following criteria:

- Wind category N2
- Site (soil) classification S
- Proposed pad size on plan is 250 mm Ø × 100 mm deep
- Loads determined from AS 168.2, pp. 35–7.

Footings on a structure of this type would normally all be bored to the one depth and size. To determine the required size, the footing carrying the largest load must be found.

To make this determination the following must be found:

- Length of rafter
- Stump A:
  - Area of roof load:
    - Clause 2.6.4 Roof load width (RLW)
    - Clause 3.6.4.2 (C) Permanent (dead) loads
  - Area of floor load:
    - Clause 2.6.2 Floor load width (FLW)
    - Clause 3.6.4.2 (a) Permanent (dead) loads and Table 3.1
  - Live load: Clause 3.6.4
- Stump B:
  - Area of roof load
  - Area of floor load
  - Live load

>>

>> **Solution**

Rafter length 3.6 ÷ Cos 5° = 3.613

Stump A

Roof load = 1.775 × 1.806 × 0.4^ = 1.278 kN
Floor load = 0.900 × 1.800 × 0.3* = 0.486 kN
Live load = 0.900 × 1.800 × 3.0# = 4.860 kN
^AS 1684 p. 36, Clause 3.6.4.2 (C) Roof loads
*AS 1684 p. 35, Table 3.1 (lightest floor available is used for decks)
#AS 1684 p. 36, clause 3.6.4.3 (a) For decks greater than 1.0m above the ground ...
Total load = 1.278 + 0.486 + (0.5* × 4.860) = **4.194 kN**

From AS 1684, Table 3.2 (p. 37):
min. 225 × 225 × 150
or
250 dia. × 150 deep.

From AS 1684, Table 3.3 (p. 37):
Type 1 min 230 × 230 × 100
or
250 dia. × 100 deep.

Stump B

Roof load = nil load = 0.0 kN
Floor load = 1775 × 1800 × 0.3 = 0.959 kN
Live load = 1775 × 1800 × 3.0 = 9.585 kN
Total load = 0.0+ 0.959 + (0.5 × 9.585) = **5.752 kN**

From AS 1684, Table 3.2 (p. 37):
min 275 × 275 × 150
or
300 dia. × 150 deep.

From AS 1684 Table 3.3 (p. 37):
Type 2 min 300 × 300 × 150
or
350 dia. × 150 deep.

Bearer span check – Table 3.5 (p. 39): max 3.2m OK. (actual 1.775)

Conclusion: The original pad size offered is *inadequate* as it was based on the assumption that the combined roof, deck and live loads would be greater than a stump-carrying deck and live loads only. This is a common error when it has not been identified that the floor area of a single stump in the middle of a deck is greater than that at the edge of the structure.

Note: Even though AS 1684 section 3.6.4.3 (a) concedes that live load may be reduced to 1.5 kN/m^2 for decks over 40 m^2, this is not considered best practice (Duragal systems ignore it, for example). Larger decks just mean larger gatherings, and groups still cluster in small groups. It is best to stay with 3.0 kN/m^2 for all decks.

Set out

This performance criteria of the unit requires that footings, upon being confirmed for compliance, are set out in accordance with the building's plan. The information for this action is covered in Chapter 10. Only a few points are therefore required here.

The plans: identifying relevant information

Your main task at this juncture is to identify the information required for the set out to proceed. Several tasks will need to be completed for slab on ground in particular with regards to services. This is because in excavating for the footing, excavation for the services such as plumbing, sewer, electrical, gas and the like will need to take place. In many instances this will require services to rise up through the slab.

Footings

The footing dimensions are critical, particularly with regards to depth in reactive soils such as M–D and E–D (see soil classifications in Table 13.3). This also has implications for plumbing services where fall to main sewer is an issue; i.e. the services may have to pass under or penetrate this deepened footing.

Where stepped strip footings are required, the location of the steps must be considered in relation to any masonry walls above, and ensuring appropriate required minimum depth below the finished surface.

Other key pieces of information from a structural perspective are:

- required slab thickness
- location and depth of steps in slabs
- finished floor level
- relation of slab to zones of cut and fill
- depth of piles and resistance pressures required for screw piles.

Reinforcement

The location of reinforcement in the footing must be exactly as specified by the engineer or the codes and standards being applied. Structurally, this is one of the most important undertakings in constructing the footing, as Example 13.1 demonstrated. The details of reinforcement location with regards to the edges and surfaces of a slab in line with either the engineer's requirements or those of the applicable standards is equally important from a structural perspective. Failure to adhere to these strictures can lead to the reinforcement rusting and the structural integrity coming into question.

Services

The location of service penetrations in the slab must be plotted exactly as changes are difficult and expensive and repairs to the moisture barrier are complex. The dimensions for these locations must be checked with the subcontracting plumber to ensure fall will be adequate and any articulation of services that must penetrate a footing can be safely affected.

A failure to correctly locate and protect services the first time can lead to structural failure when the repaired penetrations allow either moisture or termites to rise into the house frame (steel or timber). Likewise, if penetrations through footings (waste to sewer, or water in feeds) are not appropriately protected for movement, the structure can suffer undermining or upheaval as moisture is allowed to form under the slab or strip footing.

Formwork and falsework: accuracy and integrity

What is often referred to collectively as 'formwork' is actually made up of two components: formwork, which provides the shape, the 'form' of the finished concrete; and falsework, which supports the formwork and ensures it is capable of handling the loads. This section discusses the structural implications of both.

Formwork

There are three basic principles governing formwork: these are quality, safety and economy. Of these this section is most concerned with quality. Safety is also of concern, but is covered under 'Falsework' below.

Quality is based upon four further factors:

- accuracy
- finish
- strength and rigidity
- tightness of joints.

The focus of this section is on all but the finish. Accuracy is critical as errors can lead to wall plates and other load-bearing elements to be inadequately supported; i.e. the plate can end up sitting only partially on the slab edge, or in the worst instances, a corner may hang clear of the edge of a slab altogether. The same may be said of masonry on strip footings.

Strength and rigidity are also essential as the loads on formwork during a pour, and in finishing, can be substantially greater than the dead load of the finished concrete itself. In the main this is through live loads, which are significant. In addition, the placement of concrete by pump, barrow, or shovel has high-impact loads which again can cause structural failure. This is addressed further under 'Falsework' below.

Tightness of joints is also a structural concern as significant leakage can lead to exposed reinforcement which can then rust and fail over time.

Falsework

The structural importance of falsework must be reflected upon from two perspectives – one being the quality and accuracy of the finished product to the required specifications; the other being the capacity of the formwork as a whole to carry the concrete and placement loads involved. It is the second of these that also directly influences safety.

When forming up for footings, even formwork of relatively low height must be adequately supported and braced against placement-induced live loads. These will tend to push outwards on formwork, and so the falsework must be strategically located to ensure failure cannot occur: limiting deflection and stopping collapse. Taller formwork, for deep-exposed edge beams or large steps in footings, carry significant loads and so present very real safety risks to those involved in the pour.

Your role in creating the formwork and supporting falsework is therefore one where your knowledge of structural principles is essential, taking into account appropriate materials and load path tracking particularly. Failure of falsework means failure of formwork and so all the issues of accuracy, tightness of joints and rigidity are at risk.

Laying and checking

The pour is almost always a time of frantic activity and even the best and most experienced concreters will tell you it is a time of stress for them. This is because once the concrete is in, it's too late for anything else to change what will result. From a structural perspective there are a few things you should keep in mind.

Before pouring

Prior to the concrete arriving you have an opportunity to make some final checks to ensure the product will be to specification upon completion.

Final check measures

These are generally simple and straightforward. You are looking for the following:

- overall dimensions are within tolerances for the type of work being undertaken
- thickness of floor slab is equal to or slightly greater than that specified – but not less than
- width of beams, edge or otherwise, are to specification or slightly greater
- services are correctly located and protected from movement
- falsework is structurally sound
- finished floor levels are to specification
- steps in footings are appropriate to the masonry where required.

Reinforcement, membranes and services

The reinforcement also requires check measurements to ensure that the distance to slab edge, membrane and finished surface is as specified. At this time, you would also check its location in relation to the slab as a whole, and that the types of reinforcement used are as specified.

Membranes need to be checked to ensure no penetrations have occurred during placement of the reinforcement, or where such has occurred, they have been patched. All tapping should be complete and service penetrations sealed. Termite prevention barriers should also be in place, such as collars to services. All these have long-term structural implications as moisture and insect attack can destroy the structural integrity of components in remarkably short periods of time.

Cleanliness and access

Concrete must be able to reach the various parts of the form. Dirt and debris can inhibit this access and cause voids, which, if minor are not necessarily of great concern; but when it prevents steel from being properly encased or causes larger voids, then structural issues can arise.

Access for completing a pour is not necessarily of great concern structurally unless the obstruction causes the formwork to be damaged in carrying out the pour. To ensure this does not occur it is therefore best practice to check with those conducting the pour that access is appropriate. This can involve temporary bridges over the formwork edges to ensure no downward loading during placement by barrows, for example, will be exerted.

Pouring and finishing

From a structural outlook this has been addressed in the main in previous sections, particularly with regards to placement loads. Strict monitoring of formwork during placement should be maintained throughout the pour to identify any area of potential failure. Additional to this, there are three points that need to be addressed.

Slump and test sampling

The strength of concrete is listed in the documentation on the footings and the concrete itself must comply with AS 3600. These strictures must be adhered to. Frequently, however, the concrete seems too stiff when conducting the laying. Despite this, the invitation by the driver to 'give it a drink' (add water) to improve 'slump' (the fluid nature of concrete) must be declined. The adding of water serves to reduce concrete strength, whereas slump should be controlled by the producer by means of coarse and fine aggregate and cement ratios. The strength then is governed by the cement to water ratio.

Slump may be tested on site if necessary and, if further deliveries are planned, a request for greater slump made: but not at the expense of strength.

Strength testing is, unfortunately, always post-pour, post-curing; i.e. the footing is in place and reached full strength before any test can be made to see if it is of the specified strength. Test sampling is seldom done on domestic sites; however, test sampling by the producer on random batches is common practice and, in some locations, all batches. Samples are made from the mixed concrete before delivery. This is usually in the form of a 150-mm Ø × 300-mm cylinder which is held for 28 days. This then undergoes a compression test to ascertain the maximum pressure it can withstand in mega pascals (MPa). The usual strength for domestic slabs is 20 to 25 MPa.

Vibrating

This is the most common means of ensuring the concrete reaches the extremities of the formwork. In domestic formwork, it is usually completed using a vibrating rod that this thrust into the concrete. The structural concern is that with excessive vibration, segregation can occur with the coarse aggregate settling to the bottom and the fine to the top. This effects the finished structural capacity and can make the surface of the concrete weak.

Screeding and trowelling

These have limited to no structural implications other than ensuring that there are not surface air pockets and the surface is appropriately hard – as against the structure being appropriately strong. Poor trowelling in particular can lead to a very soft surface, whereas a well-finished steel trowel effect can make a major difference to surface hardness.

After pouring and finishing

After the pour and surface finishing is complete there are still factors that can influence the final integrity of footings with regards to its structural capacity being to specification.

Formwork removal

Practices vary on the removal of formwork; however, the longer it can be retained, the stronger the concrete is likely to be. This is because it helps to retain the moisture within the slab, preventing drying and promoting curing.

In instances where a footing involves deep beams or other elements that have tall formwork and falsework in place you must take heed of AS 3600 Clause 17.6.2, or of the engineer's determinations for when removal may take place. This is generally 24 hours for forms up to 600 mm in height, and 48 hours for those up to 3.0 m.

Curing

This is spelled out in some detail in AS 3600 and AS 2870 based upon exposure classifications and required target strengths. Exposure classifications are in part based upon soil types (sulphates and saline conditions) and in part on exposure to different air qualities (coastal fringes, for example). In the main these resolve to three days curing for strengths to 25 MPa, and seven days for other strengths. Longer curing time continues to improve concrete surface hardness particularly.

Test results

When test results are available, they should be accessed and reviewed with regards to the specified strengths required. In the unlikely event that a test fails, advice from a structural engineer must be obtained for the implications to the load-bearing capacity of the footings. This may involve increased spread of load to reduce point loading, underpinning and the like. All such design changes must be handled through the architect or designer and new load paths considered.

Damp-proofing and building out termites

Damp-proofing and termite prevention all have a role in long-term structural integrity. These must be planned for and installed as per the plans and specifications, and in accordance with the relevant codes and standards. These are briefly discussed below.

Vapour barrier or damp-proof membrane?

In all states and territories other than NSW and South Australia, a vapour barrier manufactured in accordance with AS 2870, Clause 5.3.3.3 must be installed continuously under a domestic concrete slab, including all Class 1 and Class 10 buildings. This barrier must be of 0.2 mm medium impact polyethylene film.

In NSW and South Australia, a damp-proof membrane is used. This is also of 0.2 mm polyethylene film except that it is a denser and tougher product that is of high impact grade and branded accordingly.

In all cases, 200-mm laps are required at all joints, which must be tape sealed. The barrier or membrane must continue up the edge of the slab to ground level. All penetrations must be tape sealed as with any tears or punctures.

Termite barriers

Termite barriers are governed by Part 3.1.3 of Volume Two of the NCC and AS 3660.1. There are a range of approaches available and these will be specified in the documentation, either on the footing plan or in the specification, usually both.

Chemical treatments

Where a chemical barrier has been described, there are again a range of options including full under-slab spray, in built pumped rticulation systems and perimeter-edge spraying only. Chemicals used in the treatment system must be registered with the appropriate authority for use in this manner. A durable notice is required to be affixed in a prominent but secure location such as a meter box stating the life expectancy of the spray and frequency of inspections required.

Once the spray has been applied it must not be disturbed or penetrated by formwork and falsework. For this reason, perimeter-edge spaying is usually done after all works are completed and the home ready for handover.

Physical barrier 1: termite shields

Termite shields are usually used in stump or pier and bearer construction, but sheet shielding can be used in a variety of ways including the edges of raft slabs. Termite shields sit on top of the pier, stump or dwarf wall and though they do not prohibit termites from reaching the bearers or joists, they require the termites to build larger, more visible galleries.

When installing or inspecting the installation of these systems you are seeking compliance with Section 5 of AS 3660.1. In the main this means continuous galvanised or other approved sheet metal barriers with joints as per Clause 5.3.11 of AS 3660.1, which includes triple folding, brazed welding or termite-resistant adhesives.

Physical barrier 2: stainless steel mesh

Stainless steel mesh may be used in a variety of ways that cover all penetrations and perimeter access points. The standards for this mesh are laid out in Section 5 of AS 3660.1 but the application is best viewed at https://www.termimesh.com.au/specifications. In checking the installation of this barrier, you are looking particularly for tears or cuts to the material caused by dropped tools or sharp bricks down the wall cavity of full masonry and brick veneer homes.

Where used as collars to service penetrations (waste pipes and the like), ensure that the collars are tightly bound to the pipes and at the correct height to ensure appropriate penetration into the concrete slab. Another location where this material may be used is at junctions between slabs. In this case you are looking for a fold that will allow for movement and that this fold is protected from concrete slurry binding the folds together.

Physical barrier 3: crushed and graded stone

Covered by AS 3660.1 Section 6, this approach must be handled by licenced installers. It may be used for all construction types including penetrations and slab edges. Under stump and bearer construction it must be laid at specified depths and be contained at the perimeter. Joints between slabs may also be protected in this manner; however, the method is not always easily achieved.

In assessing the application of this system, you are looking for disturbance of the material, incomplete filling at penetrations, or lack of compliance with the regulated depth requirements. These requirements may be viewed in Section 6 of AS 3660.1 or at http://granitgard.com.au/

Hybrid solutions

Vapour barriers with chemical treatments impregnated into them are available as replacements for normal membranes. Some are laminations of membrane and a thin foam layer that holds the chemical, others have the chemical bound into a single layer of plastic. Inspection is as for normal membranes, though product specifications may vary with application; see: https://kordon.net/ or http://fmcaustralasia.com.au/our-products/homeguard/sheeting/

Other preventions

Naturally termite-resistant timbers, such as white cypress pine, ironbark, tallowood, turpentine and the like may be used in the construction of susceptible structures; a full list is available in AS 3660.1.

Inspection requires confirmation of species and suitability of application (in ground or above ground).

Likewise, treated timbers, including some LVL, hardwood and radiata pine framing, and structural particleboard flooring products, have been treated for prevention of termite attack. In some cases, this is with a light organic solvent-based termicide, in others it may be with copper chrome arsenate (CCA), which turns the timber a distinctive green colour.

With most framing material treatments, the preventative is only a surface coating. However, cut-end treatment is not always required so inspection will be dependent upon a full review of the manufacturer's specifications.

There are also paintable termite proof compounds – similar in application to waterproof compounds. Inspection of application requires a review of the material specifications, but generally you are looking for fibrous strips over joints and the filling of holes and gaps prior to painting.

LEARNING TASK 13.2

WHAT CAN I DO TO SUPPORT MY BUILDING?

1 Access AS 2870 and determine the locations that reinforcement might be placed in a stepped footing that will be supporting a timber-framed house on a sloping site.
Initially draw a simple (scaled) sketch that details the slope of the site of around 1:8 over a 15.0-m length from one end of the building to the other.

2 Provide a checklist of items to check for a slab on ground for a single-storey 'L' shaped house separately identifying the following stages. The house will have a number of bathrooms and two (2) small terraces, approx. 8–10m^2, accessed through sliding doors to the outside with the garage attached to the house under the main roof.
Your checklist should cover:
- set out
- formwork and falsework
- reinforcement
- laying and pouring the concrete slab.

Flooring systems

Flooring systems is rather a broad topic that can, and does, cover anything from concrete slab floors and suspended concrete slabs to stump and bearer systems. The purpose of this section is not to explain how each of these is constructed, though the basic principles are outlined, but rather to focus upon the key elements of structural performance. These elements are then discussed with a view to checking the proposed system's compliance with the relevant codes and standards.

Structural integrity and compliance

Concrete slabs have been covered extensively in previous sections, though not in light of compliance of a specific design. As the flooring is integral to the overall slab footing design, it is appropriate that this analysis is presented in this section.

Bearer and joint configurations will also be reviewed for compliance in this section, and to a lesser extent, truss or deep beam joist systems.

Concrete slabs

As has been outlined previously, slab designs are governed by Part 3.2 of Volume Two of the NCC, AS 2870, and AS 3600. In Example 13.2 that follows, the focus shall be on the first two codes. The example will be considering the main slab only.

The National Construction Code and AS 2870

Figure 13.17 shows the footing plan for a proposed dwelling that has been fully described in the figures immediately preceding it (see Figures 13.13, 13.14 and 13.15). To check this design for compliance, the following information must be obtained:

TABLE 13.4 Compliance information for footings

Site classification	M–D
Wind classification	N3
Site location (state or territory)	NSW
Construction type	Clad frame
Roof cladding	Sheet
Overall dimensions of building	As per Figure 13.13 (14.0 m × 10.5 m)
All dimensions of the proposed slab and footing	As per Figure 13.17
Proposed reinforcement to footing	As per Figure 13.17
Proposed concrete strength	As per Figure 13.17 (25 MPa)

The initial investigation takes place with Part 3.2 of Volume Two of the NCC. Your procedures are as follows.

EXAMPLE 13.2

SLAB DESIGN

Read any ***explanatory information*** and check for state variations – in this case NSW requires a damp-proofing membrane. (See Figures 13.13 to 13.17 for plans and elevations.)

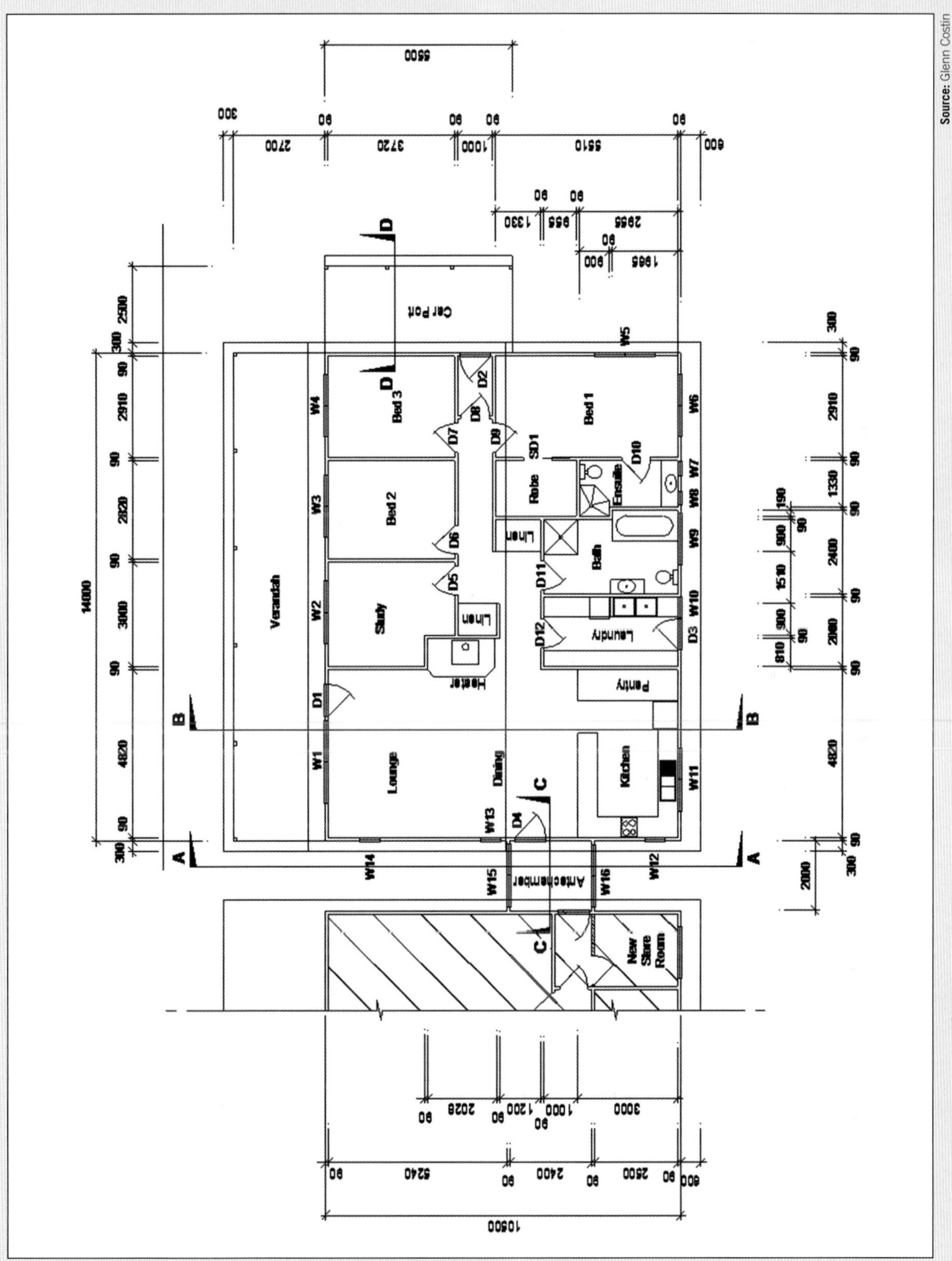

Figure 13.13 Floor plan

>> Check ***Application*** (Clause 3.2.1) to ensure the slab is within the scope of Part 3.2. In this case: M–D, single-storey, slab length is under 30 m and does not contain construction joints, it is not in an alpine area, there are no masonry walls, nor arches, the site is considered 'normal'.

- ***Clause 3.2.2.6*** is replaced in NSW with that requiring a damp-proofing membrane.
- ***Part 3.2.3*** has no applicable state variations, concrete strength is above spec.
- ***Part 3.2.5, Clause 3.2.5.2 – Footings and slabs to extensions of existing buildings***. This clause could be used to specify the footing beam dimensions after an inspection of the existing premises. However, given the size and effective separation of the structure, best practice on this occasion would be to ignore the clause and design as if new.
- ***Clause 3.2.5.3 Shrinkage control (a)***: This applies to tiled floor areas greater than 16 m^2. The kitchen itself is less than this; however, there is a clear walk zone to the antechamber that runs between the kitchen and lounge. Floor coverings are unspecified. A check with the designer is required. Response is tiles. This clause therefore applies.
- ***Clause 3.2.5.3 Shrinkage control (b)***: Applies.

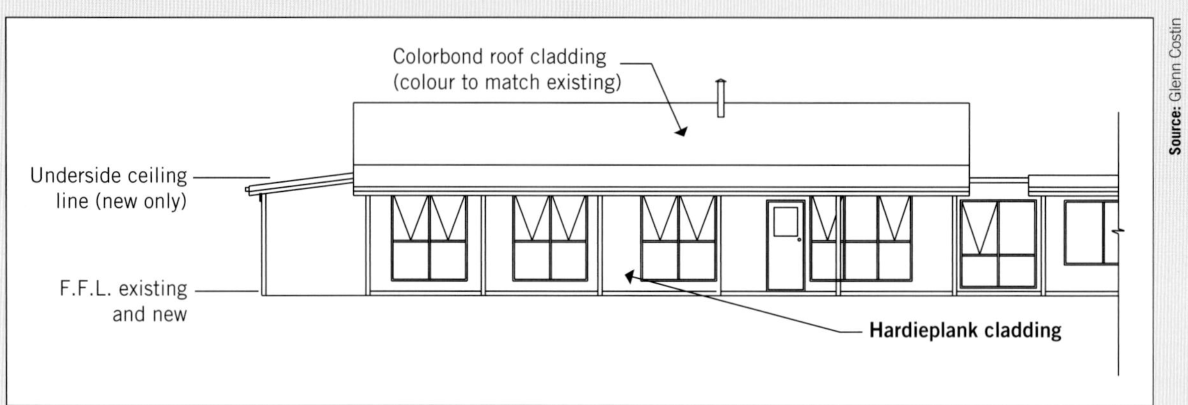

FIGURE 13.14 East and west elevations

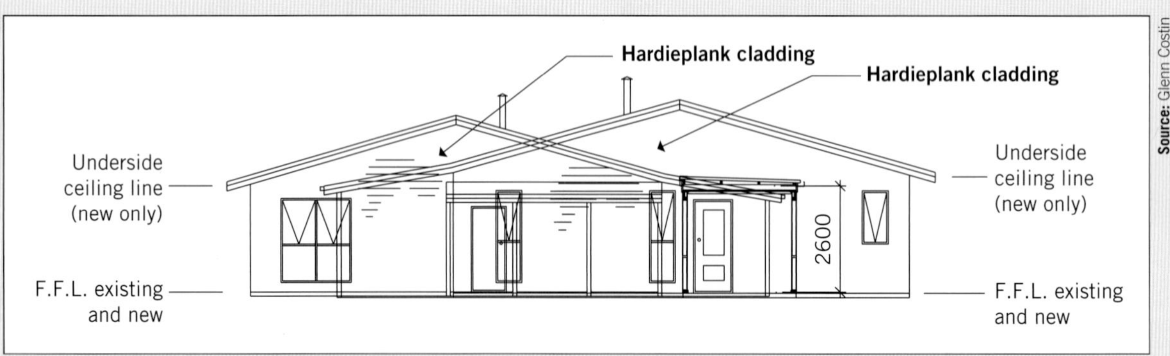

FIGURE 13.15 North and south elevations

- ***Figure 3.2.5.3 (b)***: This is the table from which the required design is defined. The requirements for Class M–D are as follows:
  - depth 400 mm
  - bottom reinforcement is 3-L11TM
  - max spacing is 5.50 m*. This is based on ***Note 2*** of Figure 3.2.5.3 (b) which allows for 10% increase when beam spacing in the opposite direction is 20% less than specified, which in this case they are.
  - Slab mesh SL72.
- ***Note 1 of figure 3.2.5.3 (b)*** states that the internal and external beams must be arranged as per clauses 5.3.8 and 5.3.9 of AS 2870. These require:
  - beam continuity:
    - internal beams are to be continuous from edge to edge of slab
    - re-entrant corners require the external beam to continue as an internal beam
    - Figure 5.4 does not apply as side lengths are greater than 1.5 m.
  - The distance to the first internal beam from any edge beam is not more than 4.0 m.

>>

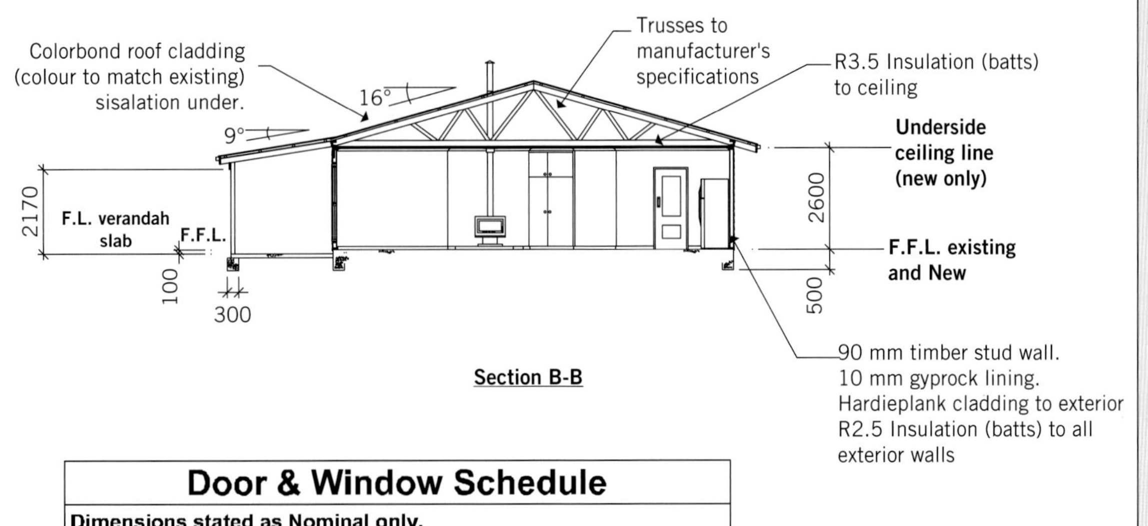

Door & Window Schedule	
Dimensions stated as Nominal only. **Glazing Design Wind Speed: 41m/s** **Double glazing to all windows excepting W9** **All doors treated with weather seals all round**	
W1	2100 x 2400 Aluminium Awning
W2, 3, 4, 5, 6, 15, 16	2100 x 1800 Aluminium Awning
W7, 8	1200 X 600 Aluminium Obscure
W9	1200 x 1500 Glazing Bricks
W10	900 x 600 Aluminium Awning
W11	1200 x 1800 Aluminium Awning
W12	1200 x 600 Aluminium Awning
W13, 14	2100 x 600 Aluminium Awning
D1, 2, 3	2040 x 820 External Solid Core
D4 - 12	2040 x 820 Internal Flush Panel
SD1	2040 x 820 Internal Cavity Slider

Source: Glenn Costin

FIGURE 13.16 Section B–B

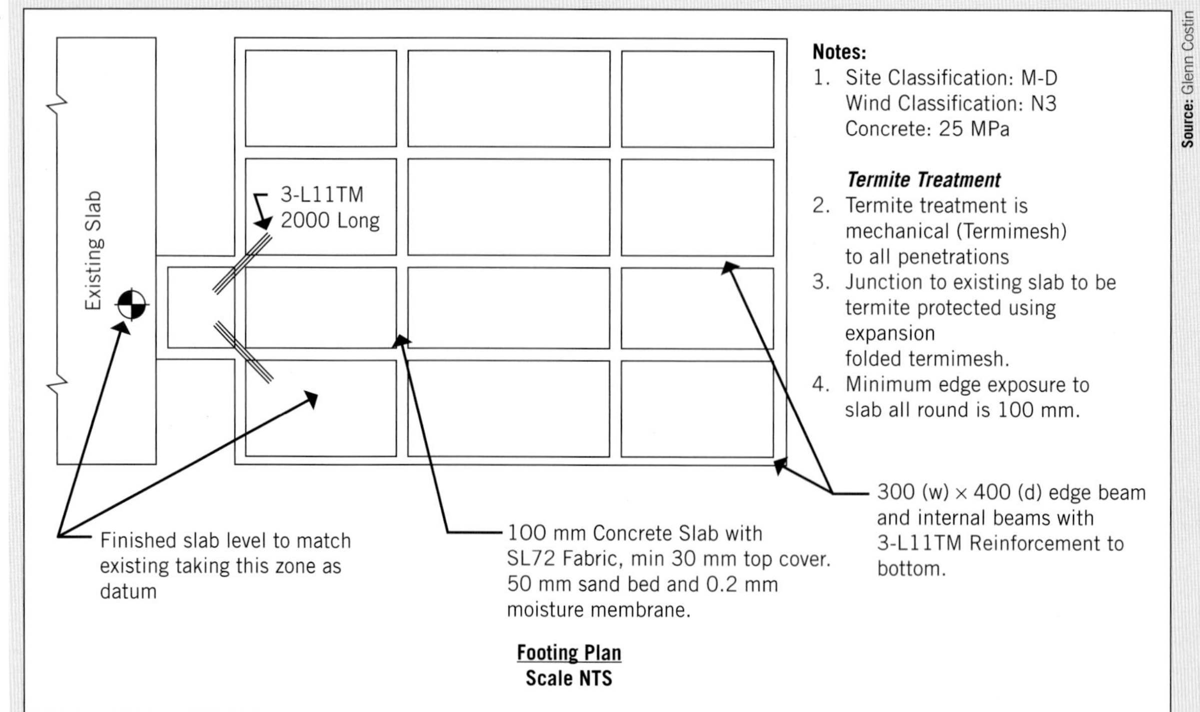
Existing Slab
3-L11TM
2000 Long
Notes:
1. Site Classification: M-D
Wind Classification: N3
Concrete: 25 MPa
Termite Treatment
2. Termite treatment is mechanical (Termimesh) to all penetrations
3. Junction to existing slab to be termite protected using expansion folded termimesh.
4. Minimum edge exposure to slab all round is 100 mm.
300 (w) × 400 (d) edge beam and internal beams with 3-L11TM Reinforcement to bottom.
Finished slab level to match existing taking this zone as datum
100 mm Concrete Slab with SL72 Fabric, min 30 mm top cover. 50 mm sand bed and 0.2 mm moisture membrane.
Footing Plan
Scale NTS
Source: Glenn Costin

FIGURE 13.17 Footing plan

Summary conclusions

Based on the above information, the proposed slab ***complies*** in all parameters *excepting* the issue of tiled flooring in the kitchen/walkway area. Accordingly, a discussion would be had with the designer, client and also the subcontracting tiler reflecting on the options listed in this clause. Best practice would be to increase the mesh as required using either Clause 3.2.5.3 (a) (i) (A) or (B). The latter, using a second layer of SL72, is likely the most cost-effective and practical solution.

Suspended slabs and stripping times

The design of suspended slabs is based upon Section 3 of AS 2870. Again, you must note the limitations of this section, which includes limits on its application to slabs to the upper floor being spelled out in Clause 3.9. The general design offered for suspended slabs in AS 2870 Clause 3.8 is only valid for site classifications Class A and Class S. Outside of these site classifications an engineer must be engaged.

Stripping times for suspended slabs is always extended to ensure the concrete has reached the required strength. In addition, props may be required after stripping to ensure support until full strength is reached – this is particularly necessary when further works will continue on or around the slab before the end of 28 days.

Standard stripping times may be found in Tables 17.6.2.4 and 17.2.6.5 of AS 3600, which is backed by AS 3610. These times are temperature-dependent for the period between casting and proposed stripping, and are as shown in Table 13.5.

TABLE 13.5 Stripping times

Average temperature	Time before stripping of supports
Greater than 20	12 days
20 or less but more than 12	18 days
12 or less but greater than 5	24 days

Note: These times are for floors not supporting loads. AS 3600 aside, stripping of formwork and supports to suspended slabs should not be undertaken without engineering advice.

Source: AS 3600, Table 17.6.2.5

Bearer and joist systems

Determining compliance of these systems is based almost exclusively on AS 1684 when dealing with timber systems, the NASH standard for steel, or one of the many proprietary manuals for steel, LVL or other engineered timber or composite products, most of which may be obtained from the internet. In Example 13.3, we will be using AS 1684.2 and the associated tables.

The example offered is that of a simple deck as defined in Figure 13.12, previously used to check pad footings for compliance.

EXAMPLE 13.3

The offered specifications for this deck are as follows.

Two bearers run the length of the deck (a long way), supported on posts at max. 1800-mm centres as shown in the plan. A third bearer is affixed to the existing house external wall running parallel with those on posts. Joists run continuously from house to external bearer, supported on bearer at midpoint in the deck.

- Bearer: 2/90 × 35 F17 Seasoned Hwd
- Joist: 90 × 35 @ 450 mm centres. F17 Seasoned Hwd
- Wind classification: N3

This information determines our table selection and the specific columns within those tables. In this example, the tables required will be from:

- ***AS 1684.2 N3 Supplement 8: Timber framing span tables – Wind classification N3 – Seasoned hardwood – Stress Grade F17***

  Specifically:
- Table 49: Deck Bearers – more than 1 m off the ground
- Table 50: Deck Joists – more than 1 m off the ground.

Evaluation of compliance

From this information, both bearers and joist are of continuous span. Both are F17 Hwd and the wind classification is N3.

Your procedure is as follows:

- Check application and limitations of the standard in Clause 1.4.
- Identify applicable load widths as per Clause 2.6.2 Floor load width (FLW).

Bearer

Clause 2.6.2 gives the middle bearer a FLW of 1800 mm and the spacing of the post gives a continuous span of 1775 mm. The middle bearer is used as this has the greater FLW.

Inserting this information into Table 49 we find the following (Figure 13.18):

TABLE 49 DECK BEARERS — more than 1 m off the ground

Size D×B (mm)	1200		2400		4800		1200		2400		4800	
	Maximum Bearer Span (mm)											
	Span	Cantilever	Span	Cantilever	Span	Cantilever	Span	Cantilever	Span	Cantilever	Span	Cantilever
	Single Span						Continuous Span					
2/90×35	1900	500	1500	400	1100	300	2200	600	1600	400	1100	300
2/90×45	2100	600	1700	500	1200	300	2500	700	1800	500	1200	300
2/120×35	2600	700	2000	600	1500	400	2900	800	2100	600	1500	400
2/120×45	2800	800	2200	600	1700	500	3300	900	2400	700	1700	500
2/140×35	3000	900	2400	700	1700	500	3400	1000	2400	700	1700	500

FIGURE 13.18 AS 1684.2, Table 49

Finding: Bearers are 2/90 × 45.

Interpolation is allowed in these tables; using this approach, the following would be found:
(2400 + 1200) ÷ 2 = 1800; i.e. the required FLW.

Therefore
(2200 + 1600) ÷ 2 = 1900 as allowable maximum span.

Finding: i.e. 2/90 × 35 is permissible.

Which to use comes down to experience. Most builders would take the 90 × 45 as this will reduce spring, stiffen the deck and give the client a greater sense of security and satisfaction with the end result. However, the proposed size is compliant.

Joists

The joists are seen to have a continuous span of 1800 mm and a spacing of 450 mm. Input this into Table 50 and we find (Figure 13.19):

>>

>> **TABLE 50 DECK JOISTS — more than 1 m off the ground**

Size D×B (mm)	Joist Spacing (mm)											
	300		450		600		300		450		600	
	Maximum Floor Joist Span (mm)											
	Span	Cantilever	Span	Cantilever	Span	Cantilever	Span	Cantilever	Span	Cantilever	Span	Cantilever
	Single Span						Continuous Span					
70×35	1300	300	1200	300	1200	300	1600	400	1600	400	1400	400
70×45	1500	400	1400	400	1400	400	1800	500	1700	500	1600	400
90×35	2000	600	1900	500	1800	500	2300	600	2200	600	2000	550
90×45	2300	600	2100	600	2000	600	2600	750	2400	700	2300	600
120×35	2800	800	2600	700	2600	750	3100	900	3000	800	2700	750

FIGURE 13.19 AS 1684.2, Table 50

Finding: Joists as 90 × 35 are compliant.
Note: Interpolation is permitted in this table also, however, the next size down, at 70 × 45, is limited to 1700-mm span so a reduced size is not possible.

Truss and deep beam systems

A deep beam joist, or simply deep joist, is defined in AS 1684 as having a depth equal to or greater than four times its breadth or thickness. Deep joists are a common option for flooring where a reduced number of bearers are desired. Deep joists can be either solid timber, LVL, or cross-laminated timber, or any number of other engineered options, such as timber 'I' beams.

Parallel trusses are also used in these settings, which may allow for even greater spans and lighter dead loads overall.

Lateral support of deep joists: blocking

Clause 4.2.2.3 of AS 1684 requires that deep joists be supported against lateral loads or toppling by means of either:

- a continuous end joist
- solid blocking of joist ends at not more 1.8 m centres and repeated at 1.8 m centres down the joist length (Figure 13.20)
- herringbone strutting applied as above.

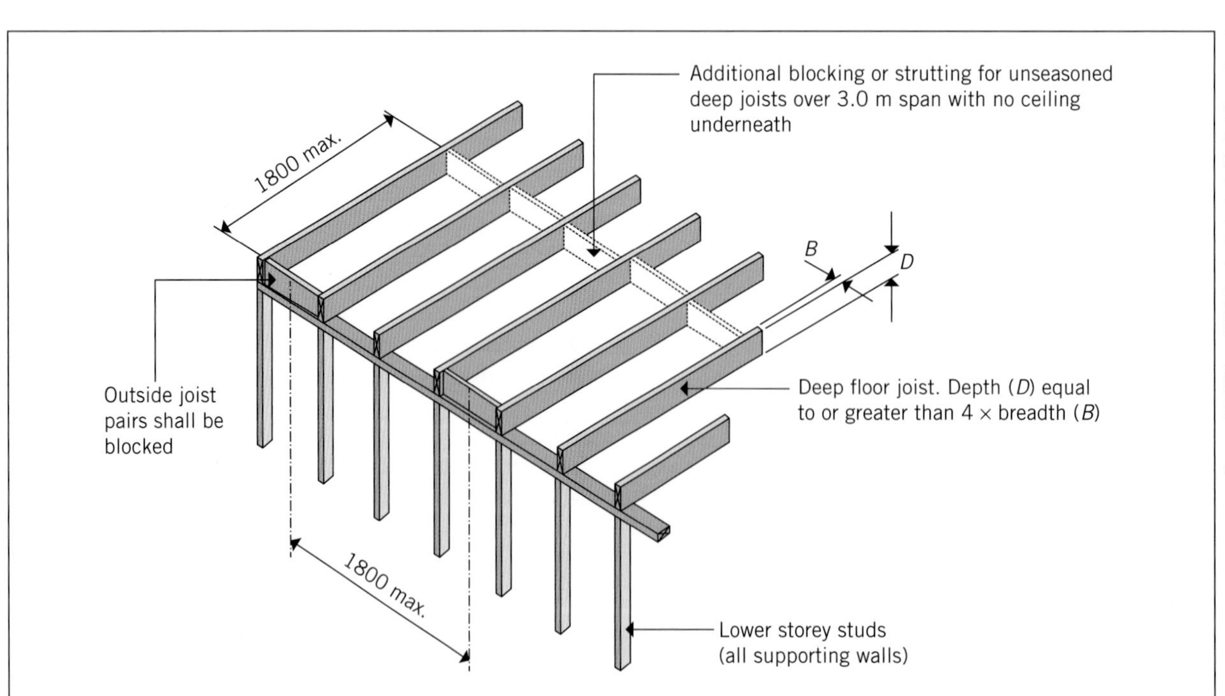

FIGURE 13.20 Blocking of deep joists

EXAMPLE 13.4

Looking at the deck in our previous example, a deep joist system may be proposed whereby the middle bearer and associate posts are eliminated.

Framed as an alternative solution to that proposed by the designer, you would need to redo your calculations along the following lines:

- bearer span: 1755 mm continuous (as per previous example)
- joist span: 3600 mm single span @ 450 centres.

Bearer

Clause 2.6.2 gives the bearer a FLW of 1800 mm as before.

All inputs are therefore as before.

Table 49 finding is as per previously, i.e.: ***2/90* × *45 (preferred)***

Joists

On this occasion or data is very different, inputting to Table 50 (Figure 13.21):

TABLE 50 DECK JOISTS — more than 1 m off the ground

Size D×B (mm)	Joist Spacing (mm)											
	300		450		600		300		450		600	
	Maximum Floor Joist Span (mm)											
	Span	Cantilever	Span	Cantilever	Span	Cantilever	Span	Cantilever	Span	Cantilever	Span	Cantilever
	Single Span						Continuous Span					
70×35	1300	300	1200	300	1200	300	1600	400	1500	400	1400	400
70×45	1500	400	1400	400	1400	400	1800	500	1700	500	1600	400
90×35	2000	600	1900	500	1800	500	2300	600	2200	600	2000	550
90×45	2300	600	2100	600	2000	600	2600	750	2400	700	2300	600
120×35	2800	800	2600	700	2600	750	3100	900	3000	800	2700	750
120×35	3000	900	2900	800	2800	800	3400	1000	3200	900	3100	800
140×35	3200	900	3100	900	3000	850	3700	1100	3500	950	3200	850
140×45	3500	1000	3400	1000	3200	950	4000	1150	3800	1050	3600	950
170×35	3900	1100	3700	1100	3500	1050	4400	1300	4200	1150	3800	1050
170×45	4300	1200	4000	1200	3700	1100	4800	1400	4600	1250	4300	1150
190×35	4400	1300	4100	1200	3800	1100	5000	1450	4700	1250	4300	1150
190×45	4800	1400	4400	1300	4100	1200	5400	1550	5100	1400	4900	1250
240×35	5400	1600	4900	1400	4600	1300	6300	1800	6000	1600	5400	1450
240×45	5800	1700	5200	1500	4800	1400	6800	2000	6500	1750	6100	1600
290×45	6700	2000	6000	1800	5600	1600	7200	2100	7200	2100	7100	1900

FIGURE 13.21 Determining joists for a deck that is more than 1 m off the ground

Finding: Deep joists at 170 × 35 will be compliant.

A discussion would need to follow this finding. This would include new pricing and whether this joist might still offer too much spring, and if to go deeper again (190 × 35) or wider (170 × 45) which, in not being a defined deep joist, reduces the need for blocking.

Manufacturers' requirements

When using a proprietary deep joist approach the likes of the NCC Volume Two and AS 1684 must be read in conjunction with the manufacturer's specifications. These are usually available online and are highly detailed. Example span tables along with descriptions of lateral supports and the like may be found at: http://www.dindas.com.au/; https://www.hyne.com.au/; http://www.crosslamtimber.com.au/; http://www.pryda.com.au/architects-builders-designers-engineers/product-information/floor-rafter-trusses/

Subfloor ventilation and access

Sunlight and fresh air flow are the best means of preventing build-up of moisture under the floor of a home. In so doing the risks of various structurally damaging issues can be significantly reduced. This includes termite infestation and fungal attack.

Part 3.4.1 of Volume Two of the NCC deals directly with this requirement, specifying minimum ventilation openings in mm²/m of external wall based upon climate zones. There are four climate zones based on average expected relative humidity; these are identified on the map of Australia provided in Figure 3.4.1. It is essential that these figures be adhered to or exceeded.

In assessing compliance, you are looking also to ensure there are no 'dead zones' where air pockets may stay stagnant, hence the requirements for vents not more than 600 mm from corners.

Access to subfloors is also required for the purposes of inspection. These sections therefore also give minimum bearer to ground clearances depicted in Figures 3.4.3, diagrams a and b.

Flooring systems: construction and compliance

Ensuring that an adopted flooring system has been laid correctly is an important part of ensuring both compliance and structural integrity. Concrete slab floors have been adequately handled in preceding sections, including their laying to specification. This section will therefore focus only upon timber flooring systems, specifically bearer and joist installations.

Bearer and joint installation: specifying and jointing for integrity

Checking the specifications of bearers and joins was handled in the main through the previous section. There are, however, other factors that apply to ensure compliance and to ensure that the various loads and forces are appropriately catered for.

Wind categories and fixing strategies

How bearers are tied to stumps or piers, and joists, in their turn, to the bearers must be given careful consideration based upon the wind category for the site. In addition, there a range of clauses within AS 1684 of import to the installation of floor systems components, such as:

- Bearing areas – Clause 1.10 specifies the minimum bearing of components on their supports.
- Stress grades – Clause 1.11 requires that you identify and confirm the grades of timber being used.
- Steel grades and corrosion protection of connectors and fixings – Clause 1.15.

There are numerous other clauses that must be read; however, seven are of particular importance:

- Vertical lamination nailing patterns – when a component is listed in the tables as being two or more sections of the same-sized timber vertically laminated to make the one: as exampled in the bearers to the verandah deck exampled earlier (2/90 × 35). This is covered in Clause 2.3.
- Permissible penetrations and notches of bearers and joist: Clause 4.1.6 is very specific on this matter.
- Load paths and offsets – Clause 1.7 and Section 4.
- Openings in floors: Clause 4.3.2.5 specifies increase sizes of components in these instances.
- Joint groups – Clause 9.6.5; Table 9.15; Figure 9.6 and Table G1 of Appendix G.
- Uplift load width (ULW) – Clause 9.6.2; Figure 9.5; Tables 9.5 – 9.14.
- Nominal fixings.

AS 1684.2 and 1684.3 both provide extensive commentary on fixing requirements based upon uplift derived from wind classifications in Section 9. The flow diagram of Clause 9.3 (Figure 9.4) outlines the required approach.

Within the approach the concept of nominal fixings is raised. Clause 9.4 explains that nominal fixings are to be used when no specific fixings or tie-down is required. Tables 9.2 and 9.3 are offered as a quick means of determining if specific fixings will be required or not at a given part of the building. Clause 9.5 then explains what the nominal or minimum fixings are to be through Table 9.4; such as 2/75 × 3.05 mm Ø nails holding floor joist to bearer.

Specific tie-down fixings are then addressed in Clause 9.6 and the associated tables and figures.

EXAMPLE 13.5

For this example, we will use the same set of drawings used for the assessment of a house slab; i.e. Figures 13.13 to 13.17. Only in this case, the floor slab is replaced by a stump and bearer approach. We will be considering the bearer to stump and joist to bearer connections only. Our path to compliance for fixings and tie down is then as follows:

- wind classification: N3
- roof type: sheet
- structure: clad frame – bearers 90 × 63 LVL; joists 90 × 45 LVL; concrete stumps.

From Table 9.2 (Figure 13.22), we can see that specific fixings are required. Incidentally, Table 9.3 – Shear – also calls for specific connections for bearers to stumps.

Using ***Figure 9.5*** of AS 1684 as a guide, the uplift load width (ULW) is determined from Section B–B (Figure 13.16) and the floor plan (Figure 13.13) of our plan set. This shows that the trusses clear span the width of the building, and our situation matches Figure 9.5(c) of AS 1684.

As the greatest ULW will be on the wall supporting the verandah and main house roof, we will find for this bearer and joist set only; i.e. half the building width plus half the verandah width.

>>

>> **TABLE 9.2 UPLIFT**

Connection	Wind classification							
	N1		N2		N3		N4	
	Sheet roof	Tile roof	Sheet roof	Tile roof	Sheet roof	Tile roof	Sheet roof	Tile roof
Roof battens to rafters/trusses — within 1200 mm of edges — general area	 S S	 S S	 S S	 S S	 S S	 S S	 S S	 S S
Single- or upper-storey rafters/trusses to wall frames, floor frame or slab	S	N	S	N	S	S	S	S
Single- or upper-storey floor frame to supports	N	N	N	N	S	S	S	S

FIGURE 13.22 Uplift

- Building width: 10.500 m
- Verandah width: 2.700 m
- Stump spacing: 1.800 m (based on typical spacings for this form of construction)
- Joist spacing: 0.450 m (based on typical spacings for this form of construction).

The ULW is therefore:

ULW = (10.5 ÷ 2) + (2700 ÷ 2)

ULW = 6.6 m.

We now have two options: use Table 9.5 of AS 1684 or the appropriate table selected from Tables 9.6 to 9.14. Both paths are explained below (Figures 13.23 and 13.24).

Using Table 9.5

TABLE 9.5 NET UPLIFT PRESSURE, kPa

Connection/tie-down position	Wind classification							
	N1		N2		N3		N4	
	Sheet	Tile	Sheet	Tile	Sheet	Tile	Sheet	Tile
Frame of slab	0.12	—	0.42	—	1.01	0.01	1.82	1.42
Single- or upper-storey floor frame to supports	—	—	—	—	1.01	0.61	1.82	1.42

FIGURE 13.23 Net uplift pressure, kPa

Net uplift force = Net uplift pressure × ULW × spacing

In this case:

Net uplift force = 1.01 kPa × 6.6 m × 1.8 m

Net uplift force = 12 kPa.m^2 or 12 kN (see units of force and pressure at introduction).

Using Table 9.9

The result is 13 kN, similar but slightly greater than that found by the other approach. (See Figure 13.23.)

Interpolation is allowed so:

6.600 – 6.000

7.500 – 6.000 × (13 – 10) + 10 = 11.2 kN

I.e. similar but fractionally lower than the first approach.

TABLE 9.9 NET UPLIFT FORCE—BEARERS—SINGLE STOREY OR UPPER STOREY—TO COLUMNS, STUMPS, PIERS, OR MASONRY SUPPORTS

Wind uplift load width (*ULW*) mm	Fixing spacing mm	Wind classification							
		N1		N2		N3		N4	
		Tile roof	Sheet roof	Tile roof	Sheet roof	Tile roof	Sheet roof	Tile roof	Sheet roof
	1800	N	N	N	N	1.6	2.7	3.8	4.9
6000	1800	N	N	N	N	6.5	10	15	19
	2400	N	N	N	N	8.7	14	20	26
	3000	N	N	N	N	10	18	25	32
	3600	N	N	N	N	13	21	30	39
	4200	N	N	N	N	15	25	35	45
7500	1800	N	N	N	N	8.2	13	19	24
	2400	N	N	N	N	10	18	25	32
	3000	N	N	N	N	13	22	31	40

FIGURE 13.24 Net uplift force – bearers

>>

>> **Which approach to use?**

Logically, the simpler the better, hence using Table 9.5 gets an acceptable figure quickly with only minor calculations. Using Table 9.9 is quick provided you read the table correctly. However, if your ULW falls in between those offered on the table then you may find that getting the required tie-down fixing is harder as it must match the higher specified load. Interpolation, though, is allowed and you should know how to do it, although it adds unnecessary complexity.

We will therefore use Table 9.5 to find the net uplift force of the joists. The same factor of 1.01 used on bearers is used for the joists.

Hence:

Net uplift force = 1.01 kPa × 6.6 m × 0.45 m

Net uplift force = 3 kPa/m^2 or 3 kN

Selecting tie-down connection

Before selecting the tie-down approach, the joint group of the two components must be identified. Both bearers and joists have been identified as being LVL, which is seasoned *pinus radiata*. Following Clause 9.6.5 and Table 9.15 we find the joint group to be JD4 (which matches the manufacturer's specifications).

As Figure 9.6 in AS 1684 explains, the connection required is selected on the basis of the material with the weakest joint group classification. In this case both are the same, and the bearer is connected to a concrete stump.

Selection is then based on matching the required net uplift with the specified material joint group and an appropriate connection in Tables 9.16 and 9.17. This is shown in Figure 13.25.

Bearers

Position of tie-down connection		Uplift capacity, kN					
		Unseasoned timber			Seasoned timber		
Bearers to stumps, posts, piers		J2	J3	J4	JD4	JD5	JD6
Bolt as per table; 500 mm for M10 and M12, 600 mm for M16; Bolt tied to footing	Bolts						
	M10	18	18	18	15	12	9
	M12	27	27	26	20	16	12
	M16	50	50	46	35	28	21

FIGURE 13.25 Uplift capacity and bolts

As shown, the stump must have a significant connection bolt cast into it to hold the force. If these are not available, consideration would be given to steel stumps with adjustable heads. These have a stated manufacturer's tie-down capacity of 25 kN. Though more expensive, costs can be saved through the simplicity of installation.

Joists

UPLIFT CAPACITY OF FLOOR JOIST TIE-DOWN CONNECTIONS

Position of tie-down connection		Uplift capacity, kN					
		Unseasoned timber			Seasoned timber		
Floor joists to bearers or top plates		J2	J3	J4	JD4	JD5	JD6
Framing anchors as per table, 4/2.8 Ø nails in each leg	No. of framing anchors						
	1	4.9	3.5	2.5	3.5	2.9	2.2
	2	8.3	5.9	4.2	5.9	4.9	3.7
	3	12	8.4	5.9	8.4	6.9	5.2
	4	15	11	7.7	11	8.9	6.8

FIGURE 13.26 Uplift capacity of floor joist tie-down connections

As may be seen (Figure 13.26), the simplest approach will be through the use of a single framing anchor fixed as shown (image shows two, but the table allows for single anchors and up to four).

Floor laying, live and dead load implications

Contemporary construction practice for timber tongue and groove (T&G) floors favours the installation of particleboard or plywood sheet flooring first, then the T&G boards are laid over this surface once the building is fully enclosed. Often referred to as platform flooring, this makes for the safer and quicker assembly of frames and roof trusses, provides extra insulation, and also reduces expansion and contraction of the T&G boards that would otherwise occur though changes in moisture and humidity under the floor.

With this approach, T&G floor boards are non-structural and so only the sheet flooring, its selection and installation, is of import here.

Selection

This is based upon the spacing of the joists and the likely live loads applied, as this dictates the distance the sheet flooring must span. Point loads and other dead loads must not be supported by the floor at all, but rather by direct transfer to subfloor members.

There are a number of different flooring sheet options, some that have insulation layers included which raise their R-value (see Chapter 14). From a structural perspective, the issue is span and thickness of the sheeting used.

The selection of plywood sheeting is based upon manufacturer's specifications and Table 5.3 of AS 1684. In many instances, plywood designed specifically for flooring can span greater distances than the equivalent thickness particleboard flooring. In addition, with correct sealing (following EWPAA recommendations: www.ewp.asn.au) these sheets can be used in wet areas. An example manufacturer's handbook may be found at: http://www.ozbuildmaterials.com.au/pdf/boral%20plywood%20handbook.pdf

With regards to particleboard flooring, Table 5.3 of AS 1684 requires that you access AS 1860.1 for selection based on spans and thickness. However, only AS 1860.2 offers a suggestion on this aspect which is:

- 19-mm particleboard flooring to span 450 mm
- 22-mm particleboard flooring to span 600 mm.

Otherwise you must deal with the manufacturer's specifications, an example of which may be found at: http://www.bigrivergroup.com.au/wp-content/uploads/2016/05/CHH-STRUCTAflor-Installation-Manual.pdf

Installation

Installation of particleboard flooring is guided by AS 1860.2. This standard is indirectly called up by the NCC through AS 1684 as a normative standard; i.e. one that must be followed. The laying and fixing requirements are expressed in Clause 5.5.4 of AS 1684.

The laying of plywood flooring has no associated Australian standard, but is governed by clause 5.5.3 of AS 1684.

Installation in wet areas

All the standards mentioned above, and the various manufacturer's guides, allow for the use of particleboard flooring in wet areas provided it is appropriately sealed and protected by a waterproof membrane. Experienced builders will not take this path as history has shown that despite these measures, particleboard flooring can 'sweat' under the membrane, causing the sheeting to expand. So, even though the waterproof membrane may not fail, the particleboard still takes up moisture leading to failure of the system as whole.

In such instances, it is preferred practice to install one of the alternative fibre cement sheet options in the shower and immediate surrounding area in place of either plywood or particleboard flooring.

Commercial variances

The previous sections, though applicable to the commercial sector, have a greater focus upon residential or domestic structures. In the commercial sector, new materials and techniques are tending to be used, particularly for two-storey buildings where floors may need to span much greater distances, such as in car parks, shopping centres, or large open-plan office spaces. Although such structures require specific skilled engineering, as the builder or site supervisor it is necessary for you to have a sound appreciation of these systems and materials.

Pre-stressed concrete

Pre-stressing of concrete is now a common and well-understood system for creating concrete elements that can span significantly greater distances than that previously possible with standard reinforced concrete. With normal reinforced concrete, the steel provides the tensile strength that concrete cannot achieve. The steel, however, is simply laid in place and concrete formed over and around it. Steel will give a little under tension, how much being load-dependent. When spanning large distances, part of that load is the concrete and steel itself. The floor slab or beam then can deflect downwards under such loads as the steel stretches.

Pre-stressing puts tension on the steel before the concrete is fully set. That tension transfers to the concrete, attempting to compress it before service or applied loads are introduced. The result is a slab or beam capable of spanning greater distances, sometimes with an induced upwards deflecting camber. The term pre-stressing includes two systems: post-tensioning and pre-tensioning.

Post-tensioning

This system is conducted on site with in-situ poured concrete floor slabs and/or beams. Steel reinforcement cables or 'tendons' are laid strategically through the slab. They often look as if they have been 'draped', with the cable laying low in the centre of the span, and rising higher at points of support such as columns or walls (see Figure 13.30, on p. 387). Tension is applied by way of hydraulic strand stressing jacks after the concrete has cured for four to seven days at the engineer's determination (Figure 13.27 below).

There are two approaches to post-tensioning: bonded and unbonded. Both may use single or multi-strand cables; however, there is a significant difference between the two systems.

Unbonded post-tensioning

In this system, the tendons are housed in smooth, greased sleeves which allows them to be tensioned without undue resistance caused by friction. Once the tension is applied, the cables are held from slipping back by mechanical wedging (see Figures 13.27, 13.28 and 13.29). This system has the advantage of being able to have the tension released prior to remedial or demolition works. With care, the cables may also be cut and re-tensioned to create penetrations in the slab for renovations or alterations.

The downside of this system is that the tendons are totally reliant upon the wedges to maintain tension in the slab. Should the wedges fail, then the slab itself may fail under load as the steel is simply drifting inside the greased sleeves.

Bonded post-tensioning

The cables in this system pass through corrugated sleeves without grease. Tensioning is by the same mechanism of hydraulic jacks and wedges again to hold the tension in the tendons. This is where the similarity ends. The sleeves surrounding the cables have tubes rising at strategic points along their lengths to enable them to be backfilled with grout. The grout is both pumped and vacuum-drawn into the tendons which then sets and bonds cables to the sleeves. In addition, the sleeves are rough on the external faces to ensure they are in turn bonded to the slab (see Figure 13.31).

The obvious advantage of this system is that it is no longer reliant upon the wedges as the sole means of maintaining tension. The downside is that re-tensioning is not possible during any renovation or remedial works. It is also more time consuming and more costly, and requires a greater level of skill to install. It remains, however, the superior system from a long-term structural perspective.

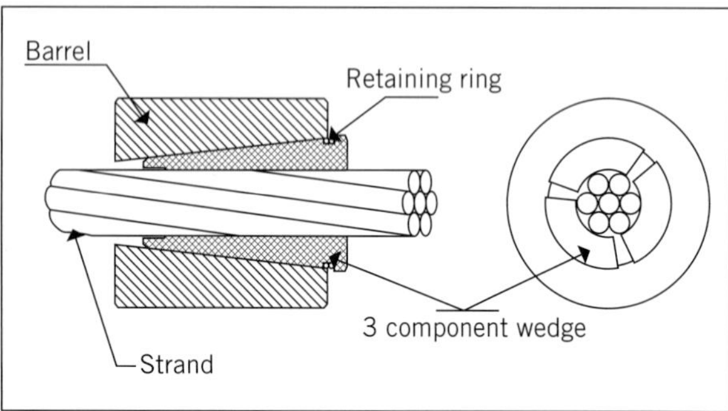

FIGURE 13.27 Wedge system for holding tension on post-tensioning tendons

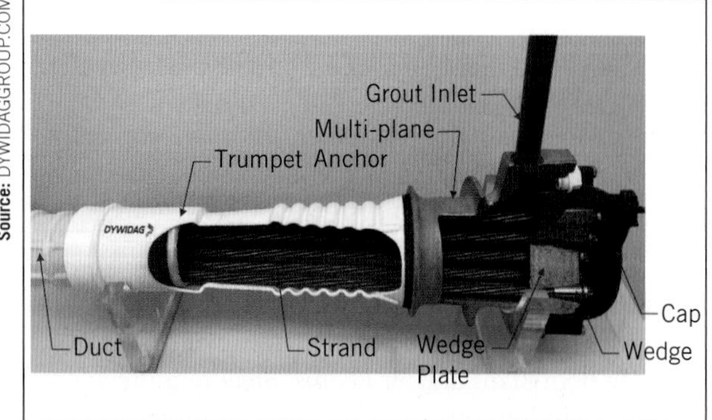

FIGURE 13.28 Unbonded multi-strand post-tensioning tendon showing end anchor and wedge detail

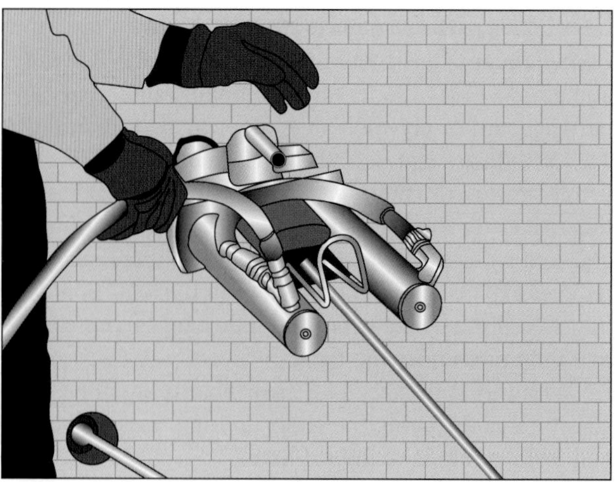

FIGURE 13.29 Hydraulic tensioning jack applied to a single-strand unbonded tendon

Pre-tensioning

Pre-tensioning is generally conducted off site in the manufacture of precast elements such as floor and wall panels, shear walls, beams and bridge components (see Figure 13.32). Tendons or cables are laid through the off-site formwork mould and tensioned before the concrete is poured. In this instance, there are no sleeves around the cables as the tension is already on the tendon and the intent is for them to bond strongly to the concrete. Once poured, the concrete is allowed to cure as required by the engineer, generally four to seven days. At this time the cables are released and the tension is transferred directly to the concrete component (Figure 13.33).

Pre-tensioning is a superior system to either bonded or unbonded post-tensioning as the tension is directly applied and directly bonded to the concrete surrounding the tendons. There is no reliance upon wedges or secondary bonding such as grout. However, it is not at all easy to produce on site.

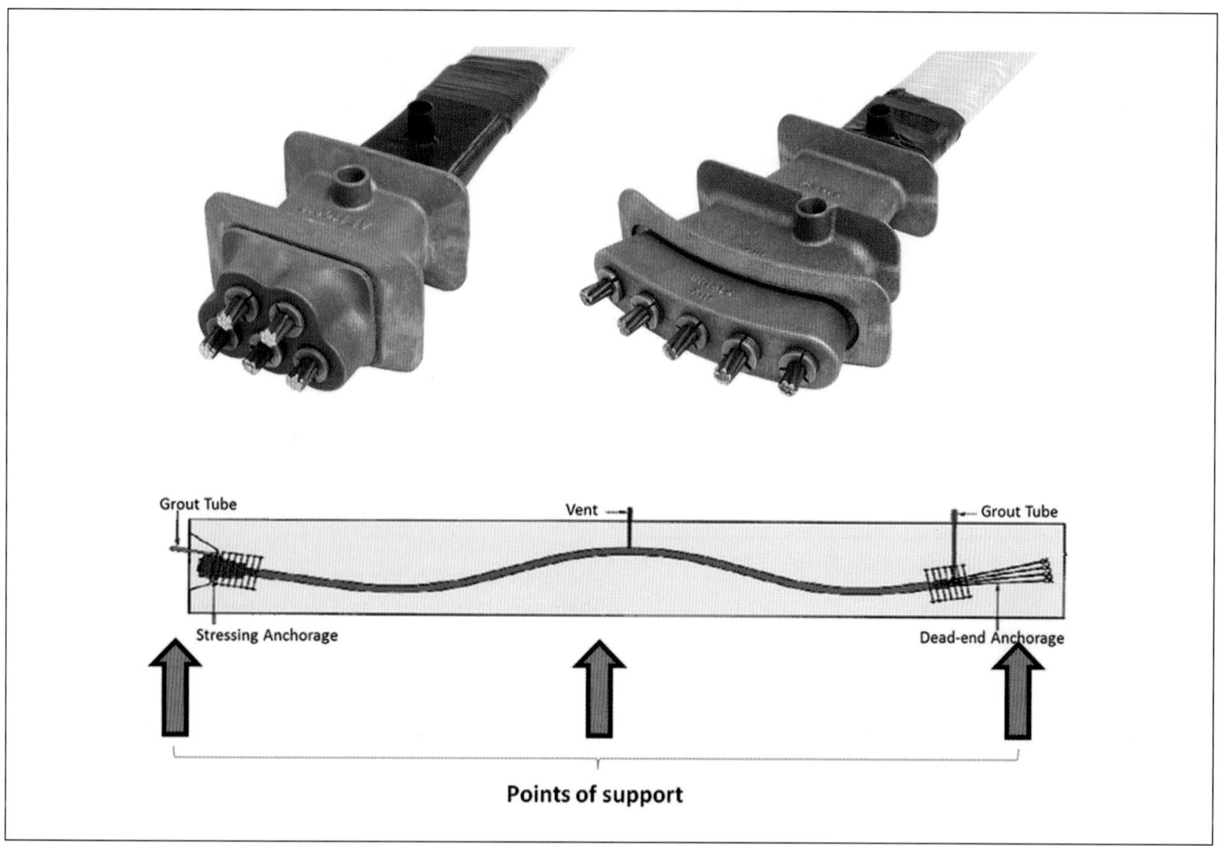

Source: Photo by Ultracon

FIGURE 13.30 Bonded post-tensioning cables strategically 'draped' through the slab

Source: Shutterstock.com/Aisyaqilumaranas

FIGURE 13.31 Bonded tendons strategically 'draped' ready for the concrete pour and post-tensioning

Structural timber floors

Cross-laminated timber or CLT has found a distinct place in the creation of major structural elements such as floors and walls. Lending itself to wide panels and long spans, the material is frequently used to manufacture off-site components and is not uncommon in multi-storey, high-rise developments such as 25 King Street, Brisbane (Figure 13.34). As with most commercial work, an engineer will need to be engaged to determine span capacities, loads and the like. In the case of 25 King St, the columns and beams are solid timber Glulam.

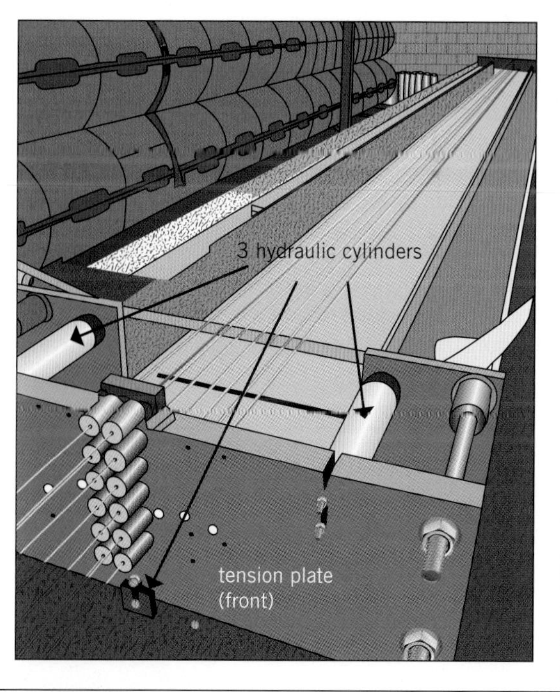

FIGURE 13.32 Pre-tensioning system for casting of a beam

Composite floors

Composite flooring elements are those that use more than one material to create the component (ignoring reinforced concrete). Again, not a complete reserve of the commercial sector, composite floors do tend to

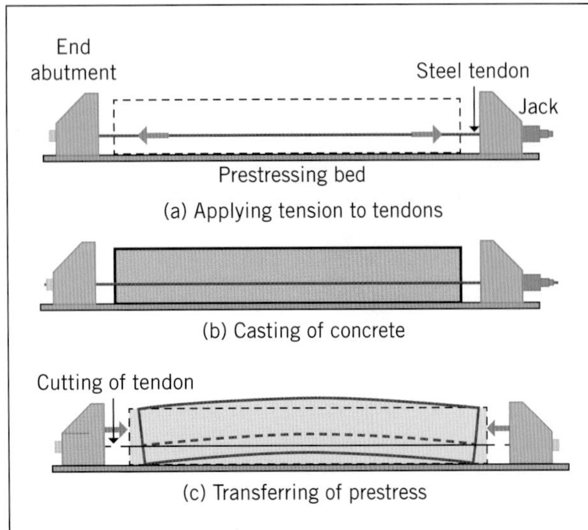

FIGURE 13.33 Stages of pre-tensioning

Source: AJ Moller Photography

FIGURE 13.34 Cross-laminated timber (CLT)

appear more in this sector than the standard domestic home. The advantage of these systems is reduced weight, increased spans and thinner floor zones – the vertical distance between the finished floor level and the underside of the ceiling.

Steel and concrete

The most common composites are steel and concrete, where the steel – such as Bondek – is placed as a tray over steel, concrete, or timber beams. This tray is effectively acting as sacrificial formwork, bonded to the concrete though its shape and/or shear studs (Figure 13.35). Shear studs particularly act to prevent slippage between the face of the concrete and the face of the steel form; i.e. they prevent a shear-like movement between the two unlike materials. Concrete is then poured in situ over the top to provide the floor. As the steel is under the concrete it is placed in tension, while the concrete plays its role by resisting compression.

Timber and concrete

Timber is now being used to replace steel as sacrificial formwork, as well as playing a key role in the structural capacity of the floor (much as steel does). Timber is being used in a range of forms in this hybrid role, the most common being CLT and parallel strand lumber (PSL) or longitudinal strand lumber (LSL). As with steel, the sheets are laid and affixed to the beams prior to concrete being poured over the top to form the floor. The bond between the concrete and the timber is either through fine aluminium mesh fixed vertically into the face of the sheets, or by screws acting much as do shear studs. Depending upon the context, hard insulation may be placed between the concrete and the timber to reduce heat loss or to reduce noise transfer. The timber elements may also be arranged such that they form a composite beam, as shown in Figure 13.36.

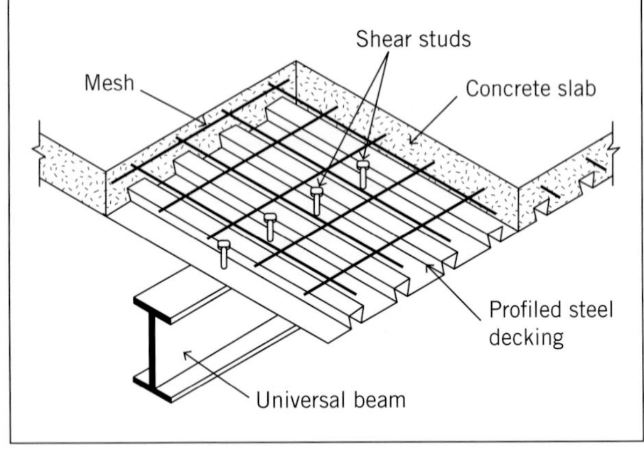

FIGURE 13.35 Metal composite floor slab with shear studs

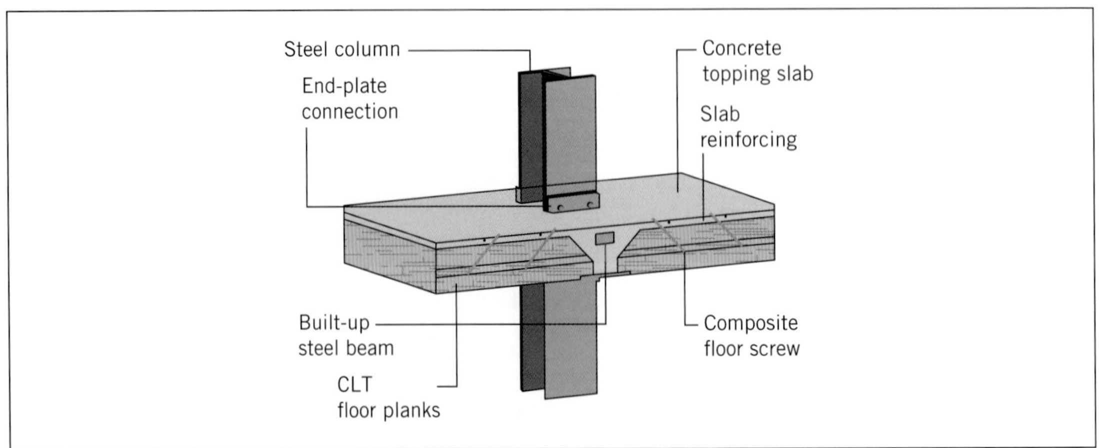

FIGURE 13.36 CLT and concrete composite flooring

LEARNING TASK 13.3

CALCULATING MEMBER SIZES FOR A TYPICAL TIMBER-FRAMED FLOOR

Floor systems

Figures 13.13 to 13.15 show a floor plan and elevations for a slab on ground single-storey house.

- The client is now looking at other options and a timber-framed floor is being investigated.
- Demonstrate your ability to draw a simple diagram showing the main loads that will be placed on the floor frame. Using the 'worst-case scenario' loads, provide a fully detailed summary of compliance with AS 1684.2. This should include a proposal for the size of floor framing members such as bearers and joists based upon the span tables. The following are the proposed specifications.

Specifications

- Colorbond-type metal roof materials – shape and pitch will stay the same.
- Floor framing: MGP10, H3 treated pine.
- Wind category – N2.
- Minimum pier or stump span of 2.4 m, bearers will be continuous over at least three supports.
- Utilise the overhang (cantilever) provisions from AS 1684.2 to have as few piers/stumps as possible.
- Client advises that they will be using 22-mm particle board floor sheets.
- Ensure economical use of materials while but complying with AS 1684.2 sizing options.

Wall systems

This section is focused primarily upon contemporary timber-framed wall systems as described by AS 1684.2 or AS 1684.3. However, the principles apply equally to steel wall framing, which is governed by the equivalent NASH standard. There are other walling systems available to the domestic construction sector; however, they are less commonly used and are not the required focus of this element of the competency unit. These include:

- tilt-up reinforced concrete walls
- poured in situ concrete walls
- full masonry construction
- load-bearing autoclaved aerated concrete
- load-bearing concrete block
- load-bearing mud brick and like alternatives.

Where some of these materials have relevance to load-bearing timber-framed walls they will be discussed with regards to their basic construction principles and performance as materials within a total walling system.

Construction principles and material performance

The basic construction principles of walls from a structural perspective have been adequately handled in previous sections of this chapter. This section will therefore focus on the more technical aspects and required performance of key materials.

Relevance of wind classification

The significance of wind classification to the structural performance of a building has been repeatedly shown in past sections. It is of even greater relevance to walls as these are the surfaces that take the brunt of the forces involved, face on and laterally in support of other walls, through uplift, and through vertical down or dead loads.

The bracing required to support walls against these actions will be covered shortly; however, the wind classification is also required, as with previous elements of the building, for the appropriate selection of framing components and the tie down or fixing of those components, including when planning for cladding.

Identifying the construction system

This is the first step in planning for the construction of the walls of a building. There are multiple systems available, many which do not require an internal load-bearing frame. These solid wall structures are usually self-bracing, at least laterally, and so rely only upon similar walls being built at appropriate spaces perpendicular in plan to themselves. This is reflected in concrete tilt-up panels, solid masonry, mud brick and concrete block, among others. Each of these have their own Australian standard or handbook to which you may refer.

Additional to these are post and beam systems whereby the roof is constructed on top of columns or posts first. The walls are then non-structural and act as in fill only.

The focus of this section however, is ***timber- or steel-framed loadbearing wall systems***.

Both these systems are based upon standardised framing practices using identified structural components. This section, while making some commentary with regards to steel systems and the applicable NASH standard, will focus its examples upon timber, and the associated Australian standard AS 1684.2. The practices in this standard are closely aligned with AS 1684.3 (cyclonic) and so the principles can be used in either text. Both these and the NASH standards are called up in the relevant part of Volume Two of the NCC.

REFER TO AS 1684.2

Contemporary framing materials

The components of the contemporary wall frame are generally of timber or steel and their selection governed by the standards previously listed. The required qualities of these materials are also expressed in these standards.

Timber

Timber framing is addressed through Part 3.4.3 and 3.10.1 of Volume Two of the NCC. These parts effectively direct attention to AS 1684; excepting Queensland, where a specific note on materials selection is given in the state and territory variations.

- Queensland requires that any timber species used in Class 1 or Class 10 buildings must be selected from the nominated schedules offered. These may be found at: http://era.daf.qld.gov.au/id/eprint/3623/. The required schedules provide information on a range of timber properties including stress grade, durability class, moisture content and insect resistance.

For other parts of the country, the durability of the timber components is declared in Part 1.8 of AS 1684 which in turn directs you to Appendix B. Other factors called up in relation to the required characteristics of the timbers are found in Appendices E and G, which cover allowable moisture content and general properties of specific species including the important joint group.

The joint group is required, as exampled previously, for the determination of the tie-down capacities of fixings between components. It is also the basis of many assumptions in the engineering of bracing options offered in Clause 8.3.6. The importance of selecting framing materials to which bracing is fixed, based on joint groups, is outlined in Clause 8.3.6.3.

Tangential shrinkage for species is also provided for in this table, which should be considered where movement is of particular concern.

A strength group is also offered, though this is less well known or used. Its use is in identifying an appropriate stress grade through AS 2878 for less well-known timber species.

Engineered timber products (ETPs)

With a shortage of large-section timbers for common domestic construction purposes, engineered timber products have found a firm and growing place in the industry. ETPs are produced by binding or gluing veneers, strands, short sections of timber, or whole boards together to make long beams, broad sheets or large-section components. Produced to precise engineering specifications these products are highly stable and durable. The common ETPs used in wall framing are:

- laminated veneered lumber (LVL): all framing components including studs and plates
- glue laminated timber (Glulam): beams and lintels
- cross-laminated timber (CLT): beams and lintels
- lumber glue laminated (LGL): beams and lintels
- plywood (Ply): wind bracing
- finger jointed timber: all framing components
- 'I' Beams: beams and lintels.

Stress grade

Stress grades are the better-known criteria by which framing timber is selected, being the basis of most of the structural 'hold-up' engineering. There are two main grading systems employed in Australia:

- F-grades as defined in AS 1720.1 – F-grades are based upon the basic engineering properties generic to a timber species. The actual grading is initially based on the testing of a small clear sample, after which individual lengths may be determined visually.
- MGP grades, also defined in AS 1720.1 – MGP grades refer to machine graded pine and the use of rapid light-deflection machine-grading systems at the point of milling. This allows for fast yet accurate evaluation of each individual piece of timber and reduces (but does not eliminate) the need for visual inspection.

Some species will have more than one stress grade, while others may vary only in being seasoned or unseasoned. Radiata pine (*pinus radiata*), may have one of three stress grades attributed to it: MGP 10, 12, or 15. The latter is seldom available, though strongest, as it is often prone to splitting when nailed.

Treatments

Timber is frequently not used in its natural state. Timber preservative treatments are common and vary from full penetrative approaches such as CCA (copper chrome arsenate) to surface coatings of LOSP (light organic solvent preservatives). Each is rated for exposure to certain in- or above-ground conditions with ratings ranging from H1–H6 with subgroups in between; H6 being the highest level of resistance.

Making the selection

When making your selection of timber species you should ensure you have addressed the following:

- stress grade matches or exceeds the specifications
- species is compatible with the requirements of durability both of design and location (e.g. Qld)
- species joint group matches the requirements of tie-down and bracing fixing requirements
- durability matches or exceeds the requirements of site location
- termite and/or borer resistance is as specified
- fungal resistance.

Steel framing

Steel framing is addressed in Part 3.4.2 of Volume Two of the NCC. Clause 3.4.2.1 of this part stipulating that the framing material have a yield stress not less than 250 MPa (for yield stress, see the section 'Stress, strain, yield and fracture' and Figure 12.11 in Chapter 12).

There are also stipulations regarding corrosion protection listed in this Part – particularly in coastal fringe areas and areas of heavy industry.

Due to the electrical conductivity of the material, the whole of a steel frame must be correctly earthed by a licenced electrician and all wiring penetrations in components must be insulated. Likewise, due to actions of galvanic corrosion, all brass and copper fittings must be prevented from making contact with the frame.

A careful reading of the both the NASH standard and this Part of Volume Two of the NCC are important when planning the construction; particularly when cladding is to be direct-fixed to the steel frame stud work. This is because there are no voids within which to run services.

In such cases, the steel framing system selected should have pre-punched holes, or a means of bringing services through the ceiling space or suspended under the floor will need to be planned in.

Other steel components

Steel is also used as connectors, fixings, and as lintels and beams in timber. Requirements for selection of structural steel components is found in Clause 3.4.4.2 of the NCC, Volume Two. AS 1684 also gives very specific advice concerning bolts, screws, straps and framing anchors in Parts 9.2 and 9.5; this includes specifying flat-head nails to connectors – clouts must not be used as the heads may pull through. Connectors and fixings must be appropriately protected against corrosion relevant to the exposure and materials concerned; i.e. galvanised, stainless steel, or coatings that protect against specific timber preservatives.

Also highlighted in AS 1684 – Appendix E – is the differential movement between steel and timber components; particularly unseasoned timber. Allowances must be made for shrinkage of the timber, and expansion of the steel.

Timber members

This section deals specifically with the selection of timber framing members as required by AS 1684. AS 1684.2 will be referenced throughout, acknowledging that the approach taken is very similar to that required by AS 1684.3 (cyclonic).

Excepting some of the more major home building groups, it is usual for the builder to have to determine all components for a specific structure. On occasions, decisions may need to be made on site. The process for making these determinations must therefore be clearly understood.

The section is framed in the main by a worked example based on the plan set in Figures 13.13 to 13.17.

Wind classification

As stated in other sections, no determinations of components, spans, or connections can be made without first determining the wind classification of a site. As with floor-framing components, the section of the supplementary tables is based on this classification, and the characteristics of the timber being used.

Timber characteristics

The various characteristics of timber components have been described earlier in this section, as were the key factors in making a determination of the suitability of a given timber or its engineered equivalent to a particular situation; i.e. stress grade, species, and seasoned or unseasoned.

This information is then used to determine the required sectional size of a component based on the various loads and forces in play.

Traditional timber framing: the components

The common components of contemporary timber house framing are described Section 6 of AS 1684.2. For simplicity, the arrangement of these components is described in Figure 13.37.

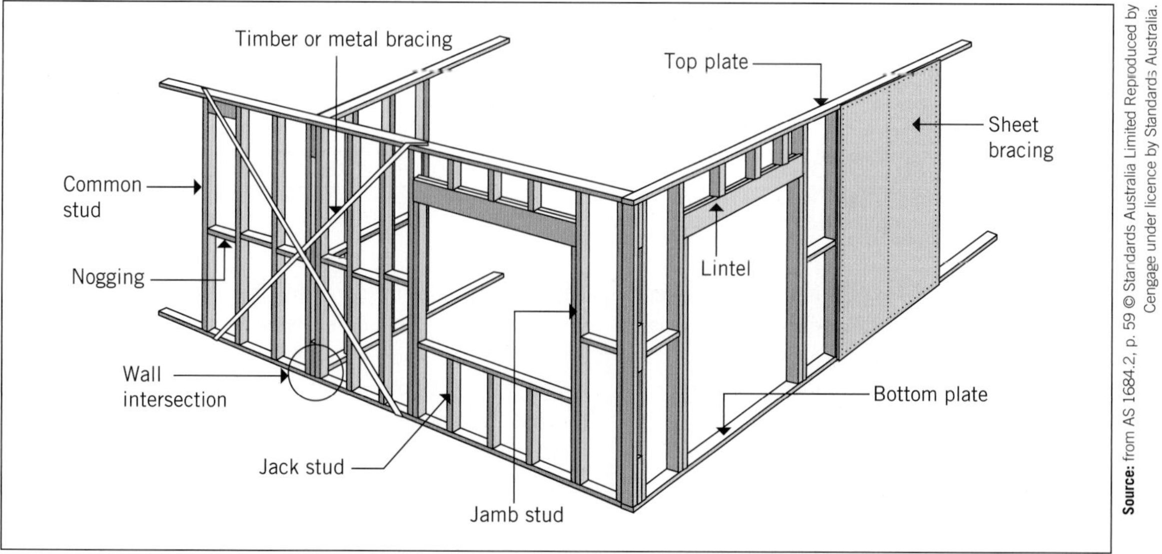

Source: from AS 1684.2, p. 59 © Standards Australia Limited Reproduced by Cengage under licence by Standards Australia.

FIGURE 13.37 Typical timber-frame construction

EXAMPLE 13.6

DETERMINING COMPONENT SECTIONAL SIZES

This example uses the plan set found in Figures 13.13 to 13.17 and is based on the north wall of the lounge only. The key specifications used in the example are:

- wind category: N3
- timber; species and stress grade: MGP12 *pinus radiata*
- wall frame: to be 90 mm thick to allow adequate expansion of insulation
- lintels: LVL F17
- wall cladding: timber weatherboards direct fixed
- roof cladding: sheet
- roof framing: timber trusses at 900 mm centres
- roof pitch: 16°
- stud and floor joist spacings: 450 mm centres
- double top plates will be used (sometimes called a pitching plate)
- internal walls non-load-bearing, excepting bracing
- window W1: 2100 (h) × 2400 (w) aluminium awning
- door D1: 2040 × 820.

Bracing shall be considered in the next section.

Note on LVL: although manufactured from *pinus radiata*, the use of F17 Seasoned Hardwood span tables for selection is permitted.

Wall plates

Bottom plates:

AS 1684.2: Clause 6.3.3 and Figure 6.18

The sectional size of bottom plates for single-storey (or upper-storey) walls is determined from Span Table 14 of whichever wind category, timber species and stress grade is applicable.

Required information: Floor joist spacing, stud spacing, rafter/truss spacing, roof load width (RLW).

RLW determination [from Clause 2.6.4, Figure 2.16 (c)]: Top chord length including ½ verandah:

RLW = RLW of main roof + ½ verandah rafter length

RLW_M = ½ span ÷ cosine of roof pitch in degrees

RLW_M = (10.5 ÷ 2) ÷ cos 16°

RLW_M = 5.462 m

½ verandah rafter length = (span ÷ cos 9°) ÷ 2
= 1.367 m

RLW = 5.462 + 1.367
= **6.829 m**

Sectional size

AS 1684.2 N3 Supplement 5 Seasoned softwood MGP12, Table 14 (Figure 13.38)

TABLE 14 BOTTOM PLATES – SUPPORTING SINGLE OR UPPER STOREY

Size DxB (mm)	Joost Spacing (mm)	Rafter / Truss Spacing (mm)							
		450	600	900	1200	450	600	900	1200
	Maximum Roof Load Width (mm)								
		Sheet Roof				Tile Roof			
	----	----	----	----	----	----	----	----	-----
35×90	300	7500	7500	7500	5700	7500	6300	3900	3000
	450	7500	7500	5000	2900	6800	4800	2600	NS
	600	7500	7500	4000	2400	4900	3700	2100	NS
45×90	300	7500	7500	7500	7500	7500	7500	7300	5500
	450	7500	7500	7500	6500	7500	7500	4600	3700
	600	7500	7500	7500	5100	7500	6300	3800	2600
2/35×90	300	7500	7500	7500	7500	7500	7500	7500	7500
	450	7500	7500	7500	7500	7500	7500	6400	5000
	600	7500	7500	7500	6900	7500	7500	4700	3200

FIGURE 13.38 Bottom plates supporting single or upper storey

Clause 6.3.3 allows for 90 × 35-mm bottom plates if studs are supported by blocking under the plate when not sitting directly on top of a joist. The additional labour of such actions means **90** × **45** will be a more economical option.

Top plates

AS 1684.2: Clause 6.3.4 and Figure 6.19

The sectional size of top plates for single-storey (or upper-storey) walls is determined from Span Tables 15 and 16 of whichever wind category, timber species and stress grade is applicable.

Required information: stud spacing, rafter/truss spacing, roof load width (RLW), tie-down spacing.

RLW (from above) = **6.829 m**

>>

>> **Tie-down spacing** (see the fourth section of this chapter for commentary):

From Table 9.5 Net Uplift Pressure = 1.33
ULW (from floor frame example) = 6.6 m
Net uplift force at each stud = 1.33 × 6.6 × 0.450* (*stud spacing)
= **3.950 kN**
Net uplift force at each second stud = 1.33 × 6.6 × 0.900*
= **7.900 kN**

Based on this level of uplift force, either a typical U or L proprietary metal stud tie will be required at each stud, or metal strap fixed as per Table 9.19 (d). The later allows for up to 8.4 kN and some specific proprietary long-legged U brackets will allow up to 10.5 kN (e.g. Pryda SB103).

I.e. tie-down spacing may be 450 mm or 900 mm.

Assume 900 mm (every second stud).

Tie-down locations

The locations of tie-down connections are provided in Table 9.21 (i). This is acceptable for all locations except the sides of openings. For the sides of openings, see Table 9.20.

Sectional size

AS 1684.2 N3 Supplement 5 Seasoned softwood MGP 12, Table 15 (Figure 13.39)

Two plates are the preferred contemporary practice as this eliminates the need for blocking out under trusses when they do not land on a stud (Clause 6.2.2.3 and Figure 6.8). Your choice is either 45-mm or 35-mm thickness based on Table 15. Each offers identical maximum RLWs. Cost and ease of assembly suggests:

Top plate and ribbon plate combination: **2/90 x 35**

Studs and noggins

AS 1684.2: Clause 6.3.2 and Figures 6.12 and 6.14

Common studs

The sectional size of common studs and noggins* for single-storey (or upper-storey) walls is determined from Span Tables 7 and 8 (notched) of whichever wind category, timber species, and stress grade is applicable.

***Note**: Noggins are required at max. 1350 centres but do not need to be stress graded. See Clause 6.2.1.5.

Required information: stud spacing, stud height, rafter/truss spacing, roof load width (RLW), notched or not-notched.

RLW (from above) = **6.829 m**

Always plan for ***notched studs*** as electrical wiring, plumbing lines and the like may be required or installed in any location.

Stud height = wall height – combined plate thickness* (*assumes no notching to plates)

Wall height = underside ceiling to finished floor level + plaster thickness + plaster batten + floor covering thickness

Wall height will be assumed as 2600 + 45 mm (10 mm plaster, 15 mm batten, 20 mm floor)

Stud height = 2645 – (35 + 35 + 45)
= 2530 mm

Sectional size

AS 1684.2 N3 Supplement 5 Seasoned softwood MGP 12, Table 8 (Figure 13.40)

Common studs may be **90** × **35**.

Jamb studs

AS 1684.2: Clause 6.3.2.3 and Figure 6.14 (a)

The sectional size of jamb studs for single-storey (or upper-storey) walls is determined from Span Tables 11# of whichever wind category, timber species and stress grade is applicable.

#Unless carrying additional concentrated loads; see Clause 6.3.2.3.

Required information: stud spacing, stud height, roof load width (RLW), opening size.

RLW (from above) = **6.829 m**
Stud height (from above) = **2530 mm**
Opening size = **2400 mm***

*Actual frame opening sizes must be gained from the window manufacturers.

Sectional size

AS 1684.2 N3 Supplement 5 Seasoned softwood MGP 12, Table 11 (Figure 13.41)

Jamb studs may be **2/90** × **35 mm**.

Lintel

AS 1684.2: Clause 6.3.6 and Figure 6.20 (a)

The sectional size of lintels for single-storey (or upper-storey) walls is determined from Span Tables 17 and 18 of whichever wind category, timber species and stress grade is applicable.

Required information: Rafter or truss spacing, roof load width (RLW), opening size (lintel span)

RLW (from above) = **6.829 m**
Truss spacing = **900 mm**
Lintel span = **2400 mm***

*Actual frame opening sizes must be gained from the window manufacturers.

Sectional size

AS 1684.2 N3 Supplement 5 Seasoned softwood MGP 12, Table 17 (Figure 13.42)

Multiple options are frequently available when making lintel selections. In this case the first choice might be the ***190*** × ***45*** on the basis that it meets the requirements – just. However, it is also the deepest lintel, and if head room is limited, then the shallower lintels may be required. The immediate alternatives both provide a stiffer beam with room for error (2.6-m and 2.8-m spans). The ***2/170*** × ***45*** may be preferred over the 35-mm option as if fills out the wall thickness (90 mm) and eliminates packing for cladding or lining.

>>

TABLE 15 TOP PLATES – SHEET ROOF – SINGLE OR UPPER STOREY

		Rafter / Truss Spacing (mm)														
		600					900					1200				
Tie-down Spacing(mm)		0	600	900	1200	1800	0	600	900	1200	1800	0	600	900	1200	1800
Size DxB (mm)	Stud Spacing (mm)	Maximum Roof Load Width (mm)														
35×90	300	7500	7500	7400	5200	2200	7500	6300	5100	4400	2900	5600	4800	3500	2900	2200
	450	7500	7500	7100	5000	2100	5200	5200	4700	4100	2700	2900	2900	2900	2600	2000
	600	7500	7500	6900	4800	1900	3800	3800	3800	3800	2600	2400	2400	2400	2300	1800
45×90	300	7500	7500	7500	7400	2900	7500	7500	7400	5200	3700	7500	6500	5100	4400	2900
	450	7500	7500	7500	7300	2800	7500	7500	7200	5000	3500	6600	6100	4900	4200	2800
	600	7500	7500	7500	7100	2700	7500	7500	7500	4800	3400	5300	5300	4600	4000	2600
2/35×90	300	7500	7500	7500	7500	3700	7500	7500	7500	7400	4400	7500	7500	6600	5100	3700
	450	7500	7500	7500	7500	3600	7500	7500	7500	7200	4300	7500	7000	6300	4900	3500
	600	7500	7500	7500	7500	3400	7500	7500	7500	7000	4100	7500	7000	6100	4700	3300
2/45×90	300	7500	7500	7500	7500	5900	7500	7500	7500	7500	6700	7500	7500	7500	7400	5200
	450	7500	7500	7500	7500	5800	7500	7500	7500	7500	6500	7500	7500	7500	7200	5000

FIGURE 13.39 Top plates – sheet roof

TABLE 8 WALL STUDS Notched 20 mm – single or upper storey

Size D×B (mm)	Stud Height (mm)	Rafter/Truss Spacing (mm)							
		450	600	900	1200	450	600	900	1200
		Maximum Roof Load Width (mm)							
		Sheet Roof				Tile Roof			
		Stud Spacing 450 mm							
90×35	2400	7500	7500	7500	7500	7500	7500	7500	7500
	2700	7500	7500	7500	7500	7500	7500	7500	7100
	3000	7500	7500	7500	6600	7500	7500	5700	4300
	3600	4200	3300	2100	1500	2700	2100	NS	NS
90×45	2700	7500	7500	7500	7500	7500	7500	7500	7500
	3000	7500	7500	7500	7500	7500	7500	7500	6500
	3600	7500	6900	4800	3500	6100	4700	3100	2300

FIGURE 13.40 Wall studs

TABLE 11 JAMB STUDS – Single or upper storey

Size D×B (mm)	Stud Height (mm)	Roof Load Width (mm)									
		1500	3000	4500	6000	7500	1500	3000	4500	6000	7500
		Maximum Width of Opening (mm)									
		Sheet Roof					Tile Roof				
90×35	2400	3000	2500	2100	1800	1600	2800	2200	1700	1500	1300
	2700	2300	1800	1600	1400	1200	2100	1600	1300	1000	NS
	3000	1600	1300	1200	1000	NS	1400	1200	NS	NS	NS
90×45	2400	4100	3400	2800	2500	2200	3800	2900	2400	2000	1700
	2700	3100	2500	2100	1800	1600	2800	2200	1800	1500	1300
	3000	2200	1900	1600	1400	1300	2100	1600	1400	1100	NS
2/90×35	2400	4500	4500	4500	4400	3800	4500	4500	4300	3600	3100
	2700	4500	4500	3900	3400	2900	4500	4100	3300	2700	2400
	3000	4000	3700	3100	2600	2300	4000	3200	2600	2100	1800
	3600	1600	1600	1600	1500	1300	1600	1600	1500	1300	1200
2/90×45	2400	4500	4500	4500	4500	4500	4500	4500	4500	4500	4100
	2700	4500	4500	4500	4400	3900	4500	4500	4400	3600	3200
	3000	4500	4500	4100	3500	3100	4500	4300	3500	2900	2500
	3600	2200	2200	2200	2100	2000	2200	2200	2100	1800	1500

FIGURE 13.41 Jamb studs

TABLE 17 LINTELS – SHEET ROOF – SINGLE OR UPPER STOREY LOADBEARING WALLS

	Roof Load Width (mm)									
	1500		3000		4500		6000		7500	
Rafter/Truss Spacing(mm)	600	1200	600	1200	600	1200	600	1200	600	1200
Size DxB (mm)	Maximum Lintel Span (mm)									
2/140×45	3500	3500	3000	3000	2700	2700	2400	2500	2200	2200
170×35	3300	3300	2700	2800	2400	2400	2200	2100	2000	2000
2/170×35	3900	3800	3300	3300	3000	3000	2700	2700	2500	2600
170×45	3500	3500	2900	2900	2600	2700	2300	2300	2200	2100
2/170×45	4100	4100	3500	3400	3100	3100	2900	2900	2700	2800
190×35	3600	3500	3000	3000	2700	2700	2400	2400	2200	2200
2/190×35	4200	4200	3600	3500	3200	3200	3000	3000	2800	2800
190×45	3800	3700	3200	3200	2900	2900	2600	2700	2400	2400
2/190×45	4400	4400	3800	3700	3400	3400	3200	3200	3000	3000

FIGURE 13.42 Lintels

Bracing

Bracing plans are generally part of the construction plan set for domestic buildings, but not always. Likewise, builders can be involved with clients, designers, architects and draftspeople in determining the best bracing layouts. To ensure compliance, the builder must know how the required bracing is calculated, as well as the capacities of the various bracing options available.

Bracing is intimately bound with tie-down capacities and techniques, which have been dealt with by calculation in the previous sections. Part of compliance checking, however, is ensuring that these systems are applied as required. Likewise, that the component parts of the wall are assembled appropriately with regards to AS 1684 and Volume Two of the NCC.

Calculating bracing requirements

The bracing requirements for a domestic structure, single, split level or two storey, are covered in Section 8 of AS 1684.2. The bracing required is found by multiplying the lateral wind pressure per m^2 by the area of the building's wall and roof elevation for any given wind direction. This gives the racking force in kN which must then be countered by sufficient wall bracing in walls running perpendicular to that being impacted upon by the wind.

The explanation for making a determination of the bracing required will be covered in Example 13.7. The example is based upon the plan set found in Figures 13.13 to 13.17, earlier in this chapter.

EXAMPLE 13.7

Only one wind direction shall be discussed: wind from the south striking the ***southern elevation***. The key specifications used are as follows:

- wind category: N3
- timber; species and stress grade: MGP12 *pinus radiata*
- roof cladding: sheet
- roof pitch: 16°

Steps and associated clauses

TABLE 13.6 **Wind specifications and clauses**

1	Determine the wind classification	Clauses 1.4.2 & 1.5 & AS 4055 / AS 1170.2
2	Determine the wind pressure	Clause 8.3.2 & Tables 8.1 to 8.5
3	Determine the area of elevation	Clause 8.3.3 and Figure 8.2(A) or (B)
4	Calculate the racking force	Clause 8.3.4
5	Design the bracing systems i subfloors ii walls	 Clause 8.3.5 Clause 8.3.6, Tables 8.18 and 8.19
6	Check even distribution and spacing	Clause 8.3.6.6 and 8.3.6.7
7	Determine connection of bracing to roof/ceilings (at walls) and floors	Clause 8.3.6.9 and 8.3.6.10

Wind pressure

AS 1684.2: Clause 8.3.2 and Table 8.2. Note if calculating for the gable end of the building, **Table 8.1** would need to be used. This table gives a single pressure value for each wind classification.

Wind pressures are given in Tables 8.1–8.5 and selection is based upon: wind classification, roof pitch and building width.

Required information: Wind classification, roof pitch, width of building, type of elevation, wall height, roof height.

The southern elevation shows a gabled roof surface pitching away from the viewer at 16°. This replicates the image in Table 8.2.

- Wind classification = N3
- Roof pitch = 16°
- Width of building = 10.500 m (see Figure 13.43)

Interpolation is allowed, and so an alternate (lower) figure of 0.79 kPa is possible; however, for ease of determination, and surety of capacity, it is often better to stick with the higher value.

Area of elevation

AS 1684.2: Clause 8.3.2 and Figure 8.2 (A).

As the image below shows (Figure 13.44), only the elevation of the new structure is of concern in this calculation.

From the plan set offered previously (Figures 13.13 to 13.17), the following will be needed to make the calculation:

- roof pitch = 16°
- wall height = 2.600 m
- building length = 14.000 m
- building width = 10.500 m
- gable eave width = 0.300 m

roof height => Tan 16° × ½ building width
= Tan 16° × (10.5 ÷ 2)
= 1.505 m. Allow 200 mm for top chord, battens and roof cladding
= 1.705 m

>>

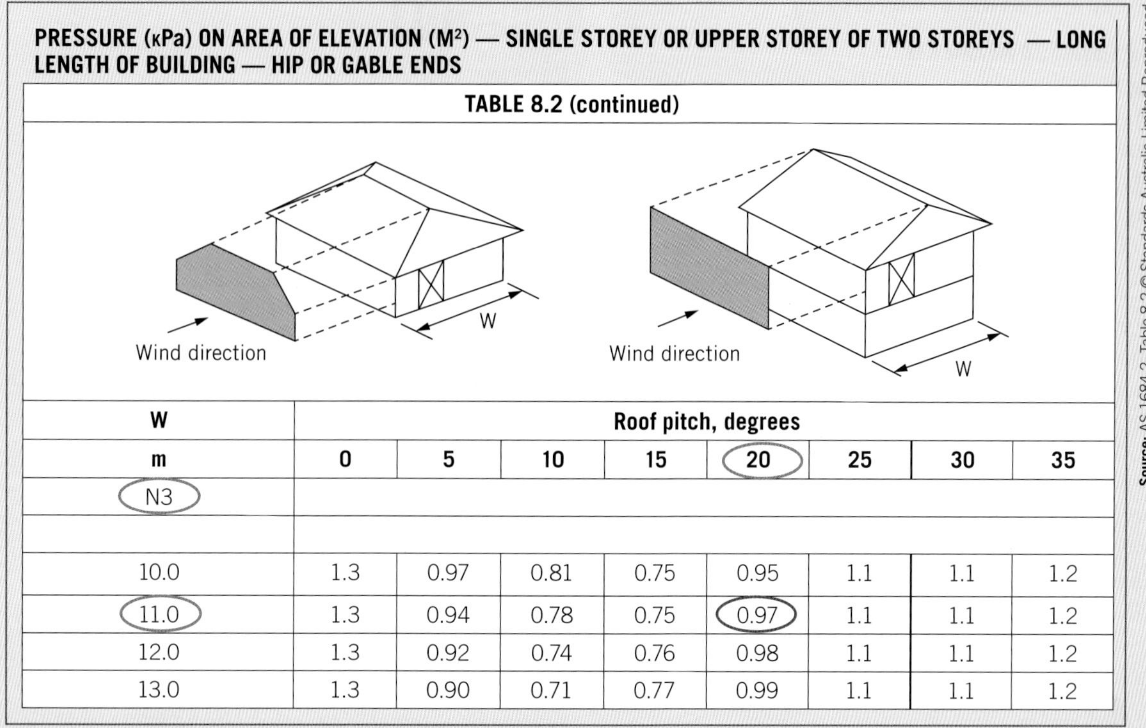

PRESSURE (kPa) ON AREA OF ELEVATION (M²) — SINGLE STOREY OR UPPER STOREY OF TWO STOREYS — LONG LENGTH OF BUILDING — HIP OR GABLE ENDS

TABLE 8.2 (continued)

W	Roof pitch, degrees							
m	0	5	10	15	20	25	30	35
N3								
10.0	1.3	0.97	0.81	0.75	0.95	1.1	1.1	1.2
11.0	1.3	0.94	0.78	0.75	0.97	1.1	1.1	1.2
12.0	1.3	0.92	0.74	0.76	0.98	1.1	1.1	1.2
13.0	1.3	0.90	0.71	0.77	0.99	1.1	1.1	1.2

FIGURE 13.43 Pressure on area of elevation

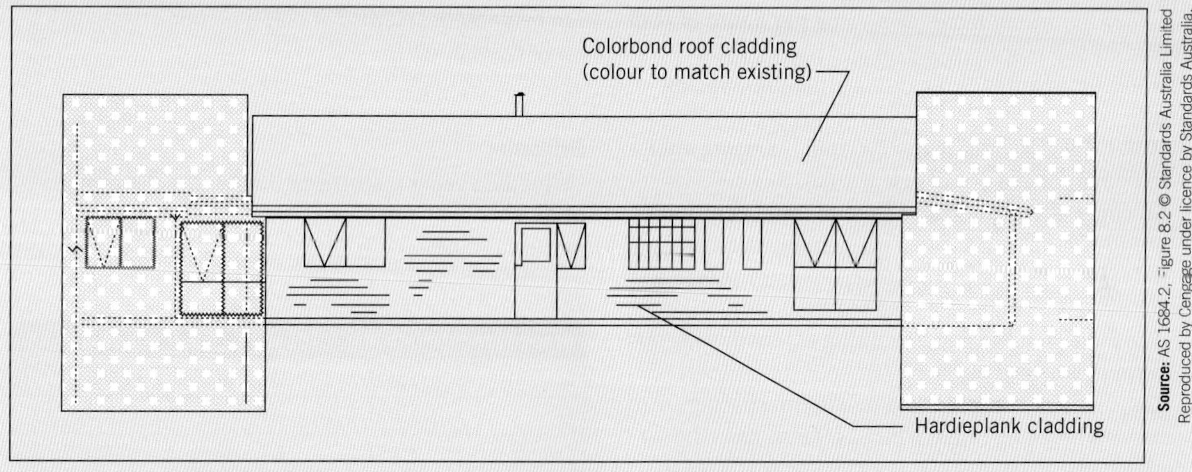

FIGURE 13.44 Area of elevation

Area of elevation

Elevation area of roof surface + ½ Wall area*

***Note**: Clause 8.3.2 and Figure 8.2 (A) show that only the upper half of the wall area is required.

Area of elevation = (14.6 x 1.705) + (14 x 2.6 x 0.5)
= 16.706 + 13.65
= **43.173 m²**

Racking force

AS 1684.2: Clause 8.3.4

The racking force is found by:

Total racking force = area of elevation (m²) x lateral wind pressure (kPa)
= 43.173 m² x 0.97 kPa
= 41.80 kPa.m²

i.e. = **41.80 kN** (see Australian units of measure in the introduction to this book).

Design and distribution of wall bracing

AS 1684.2: Clause 8.3.6 and Table 8.18

Having found the racking force, bracing units must be installed in the walls perpendicular to that which is receiving the force. The design and arrangement of these bracing units must comply with Clause 8.3.6.3. Bracing is considered in two forms:

- nominal wall bracing (Clause 8.3.6.2)
- structural wall bracing (Clause 8.3.6.3).

Nominal wall bracing

Nominal wall bracing is the resistance to racking provided by wall linings such as plasterboard, fibre cement sheets and the like. Clause 8.3.6.2 states that up to 50% of the total calculated racking force may deemed to be resisted by nominal bracing. However, it also seriously qualifies this statement.

Nominal bracing is to be ignored if it is not evenly distributed throughout the building. Given that this is seldom if ever the case, it is best practice to always ignore nominal bracing altogether. That is, give it no resistance value.

Structural wall bracing

Structural wall bracing is designed as units of bracing as defined by Table 8.18. These bracing units provide resistance to racking forces as a value per metre of wall covered by the brace: i.e. kN/m.

Notes of importance regarding structural bracing include:

- Wall heights greater than 2700 mm – bracing values to be reduced as per Clause 8.3.6.4.
- Bracing is based on being fixed to framing of joint group JD4 or better, otherwise bracing value is reduced as per Clause 8.3.6.3.
- Clause 8.3.6.6 stipulates the required location and distribution of bracing, favouring external walls and corners where possible.
- Clause 8.3.6.7 specifies the maximum spacing between bracing units, which is 9.0 m for wind classifications under N3. For classifications N3 or greater, Tables 8.20 and 8.21 apply.
- Clause 8.3.6.9 outlines the required top fixing of bracing walls through Table 8.22.
- Clause 8.3.6.10 outlines the required bottom fixing for bracing walls and also through Tables 8.23 and 8.24.

Choosing a bracing unit

Choosing a particular bracing unit type is often based on the wall space available. Typically, sheet bracing will be used on external walls where space is available, and cross-braced metal strap or metal angle brace for internal walls. The sum of the bracing units' resistance capacities must equal or exceed the total calculated racking force.

EXAMPLE 13.8

BRACING UNIT

- Maximum spacing (from Table 8.20): 9.0 m
- Wall height: 2600 mm (standard – no reduction required)
- Framing is JD4 (standard – no reduction required)
- Preferred locations: corners where available

Bracing types preferred are:

- Type D: tensioned metal straps with stud straps – 3.0 kN/m (Table 8.18, AS 1684.2, p. 144)
  Length between 1.8 and 2.4 m. Use 1.8 m for calculating capacity.
  Capacity = 3.0 kN/m × 1.8 m
  = 5.4 kN
- Type H: plywood Method B – 6.0 kN/m (Table 8.18, AS 1684.2, p. 146)
  Length = 0.9 m
  Capacity = 6.0 kN/m × 0.9 m
  = 5.4 kN

Based on these units, space for 8 units is required.

Note: Frame is to be timber clad; use of plywood bracing on external walls would require either reduction of the framing width (90 mm – thickness of plywood) at the location of the brace, or the full wall will need to be packed. Likewise, if only a portion of an internal wall is sheathed.

Distribution

Using the sketch plan (Figure 13.45), the distribution of bracing is planned using the following table of properties (Table 13.7).

TABLE 13.7 Bracing table of properties

Brace type	Length mm	Location	Restraint offered kN
D	1800	a	5.4
D	1800	b	5.4
D	1800	c	5.4
D	1800	d	5.4
H (B)	1800	e	10.8
D	1800	f	5.4
D	1800	g	5.4
Total restraint			43.2 kN
Total required restraint			41.80
			Compliant

Note: This is the minimum requirement. Best practice may include bracing of other walls to ensure rigidity of particular internal walls. The '1' in D1a (b, c, d ...) denotes wind direction 1; in bracing layouts there will always be two (or more) wind directions specified.

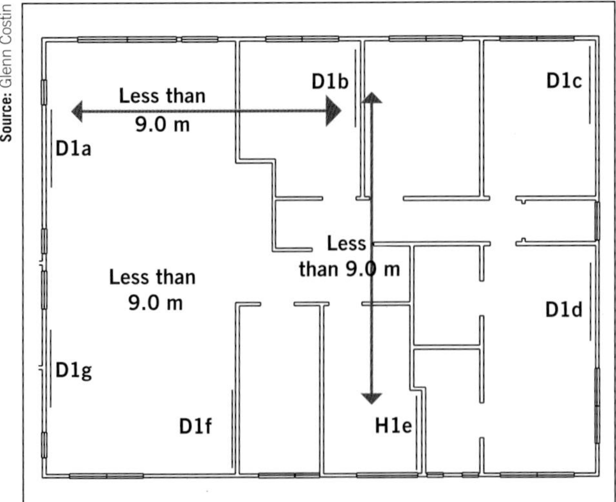

FIGURE 13.45 Bracing sketch plan

Connections

Bracing units are designed to resist the racking force applied laterally to the frame to which they are fixed. In this manner they support walls running at 90° to them. However, the forces pushing laterally on the bracing unit will try to either topple the frame by lifting either end, or slide the frame sideways by shearing the fixings (Figure 13.46).

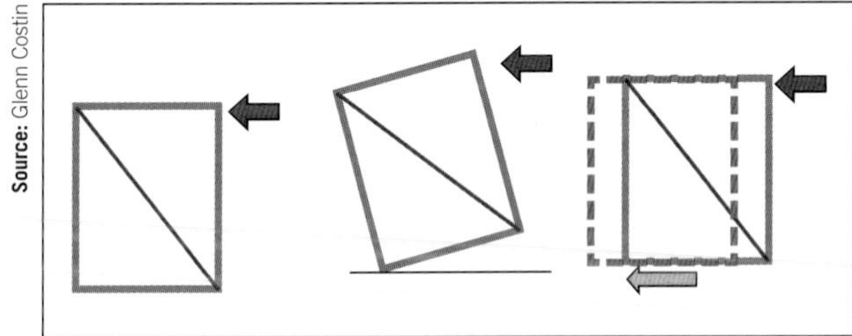

FIGURE 13.46 Bracing unit being toppled or pushed in shear

Clause 8.3.6.10 addresses this issue, requiring that bottom fixings for bracing units be determined by one of three paths.

- Nominal: for bracing systems with a capacity up to 3.4 kN/m.
- Table 8.18: as per description of bracing system.
- Tables 8.23 and 8.24: for when no specific fixing is offered in the Table 8.18 description.

For nominal fixings, see AS 1684.2, Table 9.4 (Clause 9.5).

Note: These fixing requirements may be exceeded by the identified uplift forces found in Table 9.2; likewise, by the shear forces identified in Table 9.3. In such cases, specific fixings designed to match these greater loads must be used – see 'Tie-down spacing' under 'Top plates', page 393.

Tie-down locations

The locations of tie-down connections are provided in Table 9.21 (i). This is acceptable for all locations except the sides of openings. For the sides of openings, see Table 9.20.

Application to example

The example being worked through began with the selection of wall frame components. In so doing the specific tie-down spacings were identified as being required based on Table 9.2. This influences the required hold down of bracing walls.

Shear loads must also be checked using Table 9.3; these are identified as nominal only.

There are two types of braces given in the example.

- Type D braces: rated at 3.0 kN/m
  - Internal walls, nominal only; i.e. one 75-mm masonry nail, screw, or bolt at not more than 1200 mm centres*.

    *From Table 9.4. This is not likely to be adequate to prevent movement of the wall plate over time, or offer sufficient rigidity of the plate to internal fixing out (skirting and the like). A minimum of 900 mm is recommended.
  - External walls, specific (from Table 9.2). Based on tie-down spacings and forces calculated for top plates, this is: 900 mm centres; load capacity 7.9 kN#.

    #Table 9.18 provides three fixing options for plates to concrete slabs: driven nail (inadequate tie-down); cast in bolt (easily strong enough, but slow); proprietary anchors – for which manufacturer's specifications must be sought.

    The latter is the most common option; in this case a 10-mm Ø AnkaScrew may be used, which offers 9.8 kN tie-down (see http://www.ramset.com.au/Product/Detail/80/AnkaScrew-Screw-In-Anchors).
- Type H brace: rated at 6.0 kN/m
  - Specific bracing required as per Table 8.18 (h); one 13 kN capacity connector# at each end of the sheathed area of wall, and intermediate fixings at max 1200 mm centres.
  - The only H brace unit in the planned bracing approach is internal and 1800 mm long. The above fixings satisfy, and would also satisfy the identified uplift requirements for an external wall.

    # As per previous commentary on Table 9.8. In this case a 12-mm Ø AnkaScrew may be used which offers 14.4 kN tie-down.

Quality control

Quality control and management processes are fully explored in Chapter 6 On-site supervision and Chapter 8 Tender documentation. With regards to on-site framing practices there are three types of framing that may be met with:

- factory-assembled (prefabricated) frames that need standing and corner joining only

- factory pre-cut components that must be assembled
- full on-site cut and assembled frames.

In discussing these approaches to framing, quality control will be focused upon structural integrity rather than aesthetics, material wastage, or economics.

Prefabricated frames

Factory-assembled frames require inspection upon delivery more than anything else. You are looking for quality of transport, then appropriate offloading procedures first. After this, the quality of the frame is inspected. Key points to look for are:

- the jointing and connecting systems used
  - do they match the required fixings identified (nominal or specific) for up-lift loads?
  - have they remained secure during transport (open joints, failed connectors)?
  - have they been installed correctly (connector pins not correctly pressed home)?
  - nails protruding from frame?
  - framing split through poor nailing/fixing practices?
- materials used
  - split, bowed, or warped timber
  - excessive knots, or knots in key load areas
  - incorrect size of components
  - lintel stress grades and sizes
- bracing strategies
  - notches for metal angle brace cut too deep (beyond 20 mm)
  - strap bracing end fixture
  - sheet bracing thickness and stress grades
  - sheet bracing fixings
- framing layout
  - correct array of components
  - noggins at correct heights
  - jamb studs straight and appropriately clustered to carry loads
  - point load stud arrays correctly located
- dimensions (though not strictly structural, unplanned on-site changes can reduce integrity)
  - overall dimensions correct
  - opening dimensions correct
  - openings correctly located
  - wall junctions correctly located.

Once the walls are stood, this inspection should be repeated prior to having a frame inspection take place.

Factory-cut frames

Factories will sometimes supply cut components only, which are then site-assembled. This process can speed up manufacture while increasing accuracy and reducing waste. The checks listed above regarding material quality apply equally here. An initial delivery inspection should be made; however, as the materials will arrive in strapped packs, the inspection of individual pieces tends to take place upon assembly. As assembly progresses, all the other checks listed above and in the next section flow through as each wall is made ready for standing.

Site-cut frames

Everything listed for prefabricated frames effectively apply to frames made on site. The key difference is that when site cutting, the responsibility for quality of materials and finish and the appropriate arrangement of the components lays firmly in the hands of the builder.

The design of timber frame walls is covered extensively by AS 1684; however, within that standard there is significant room for individual approaches. Hence, it is seldom that any two builders will have exactly the same framing practices. This means structural quality control is a critical aspect of any on-site framing approach.

The previous sections have outlined how to select each component, and how they should be fixed together or the frame connected to floor frame or concrete slab. However, there remains a few points with regards to the arrangement of these components, and the standards that make for quality framing practice that need to be covered.

Corner and wall intersection arrays

Clause 6.2.1.3 and Figure 6.3 of AS 1684 stipulate the accepted arrangement of studs and wall junctions to ensure stability. From a structural integrity perspective, the quality issues in these arrays are:

- poor block fixture
- splitting of stud ends
- poor stud selection – bowed, split, excessive knots
- incomplete arrays – blocks missing, studs missing
- incorrect nail procedure or incorrect fixing selection
- bottom plate connection to slab, flooring or subfloor framing.

Studs and wall plates

Clause 6.2.1.1 allows for the straightening of studs by way of crippling. This process involves cutting halfway through a stud's depth and cleating either side. Nailing patterns and minimum cleat lengths are all clearly stipulated. No more than 20% of the studs in a given wall may be straightened by this method.

Studs may also be planed straight provided that the minimum design depth of the stud is maintained. In the worked example given previously with regards to stud sizes, if you were to check Table 8 of the supplements carefully you will note that a 70 × 45-mm stud will suffice if not for the stipulation to have a 90-mm thick wall for insulation purposes, but 70 × 35 will not. The studs chosen were 90 × 35. Clause 1.13 of AS 1684.2 does not allow any reduction of seasoned timber when deriving sizes from the span tables. This means that these studs cannot be planed straight.

Notching of studs and wall plates is governed by Clause 6.2.1.4, Figure 6.4 and Table 6.1. This information should be reviewed prior to making any cuts or holes, and heed taken when making inspections. Of particular importance is the need to monitor and inspect the work of other trades with regards to holes and notches for services. It is not unknown for key load-bearing studs to be excessively drilled to install wiring or plumbing lines.

From a structural perspective, the straightness of wall plates is to some degrees incidental, though it can have an influence upon the appropriate arrangement of trusses or rafters. Of greater concern is that stud walls be plumb or vertical, unless the structure as a whole is engineered specifically to be otherwise. Each state and territory produces some version of a *Guide to Standards and Tolerances*. Most are very similar. The accepted standard Australia-wide is that ***frames must not be more than 4 mm out of plumb (off vertical) in any 2-m length***.

This same standard applies to wall plates in both horizontal planes. That is, the wall cannot bow out or in by more than 4 mm over a 2-m length; and a plate cannot bow or buckle up or down to the same degree.

Noggins must also be in place and well fixed. This is a structural imperative that is often overlooked. The placement of noggins should also be checked after other trades have been through as they are sometimes relocated for use as supports for fixtures.

Connectors and fixings

These have been covered to a degree previously in the identification of bracing and top plates. Where specific fixings have been identified through Clause 9.4 and Tables 9.2 and 9.3, these must be confirmed. Where nominal fixings are allowed, these must match the requirements of Table 9.4 of the same clause.

The role of temporary bracing

Temporary bracing plays a significant role in the support of a house prior to it being fully enclosed. Wind forces on frame skeletons can be seriously underjudged, as can construction loads due to worker's movements and placement loads such as truss installation. Temporary bracing is covered by Clause 8.2 and requires that it be equivalent to at least 60% of the permanent bracing required.

Sarking and other membranes

Often referred to as building paper, wall wrap, reflective insulation, or by the trade name sisalation, sarking is used to control the passage of moisture through a walling system. There are two common forms applied to exterior walls:

- vapour-permeable sarking
- vapour-resistant or non-permeable sarking.

Incorrect application and selection of sarking can lead to the rapid degradation of timber wall frames so it is of structural significance.

Selection

Selection is based upon a number of factors that are discussed in Chapter 14 Thermal efficiency and sustainability. With regards to walls, the two main criteria are:

- climate
- cladding.

Climate

When dealing with cavity wall construction (brick veneer or the like) consideration should be given to the regional climate in determining the appropriate wall sarking. In hotter climates, a non-permeable or vapour resistant sarking is generally promoted. For colder climates a highly vapour-permeable sarking should be used to prevent structurally damaging condensation build-up.

Cladding

Volume Two of the NCC stipulates that, irrespective of climate, vapour-permeable sarking shall be used between the timber cladding and the frame. For metal cladding and other types of claddings, the type and application of sarking must be guided by the manufacturer's installation instructions.

Installation

Installation must be as per the manufacturer's instructions but always so that water flows are away from the frame.

LEARNING TASK 13.4

SEQUENCING THE ERECTION OF A STEEL-FRAMED BUILDING

Wall systems

Access a steel-framed building website and provide a step-by-step method (in point form) of the correct method as recommended by the publication you accessed, in the erection and bracing of a steel-framed, residential low-rise building.

Roofing systems

There are a range of roofing systems possible for residential structures, but the most common today is the trussed roof, which is outlined in Chapter 12 under 'Common types of trusses'. From a planning and coordination perspective these are also by far the simplest. The other common systems include:

- pitched or 'stick' roofs
- cathedral ceilings
- architectural exposed trusses
- proprietary truss systems.

Of these, the first two are the most demanding from a structural compliance perspective. The others, including standard trusses, will be externally engineered and hence your role will be one of confirming adequate documentation and installation practices.

Structural integrity of components

Evaluating the structural integrity of components depends upon the type of roofing system employed. With proprietary or architectural truss systems, assessment will be based upon the available documentation and visual inspection upon delivery. Hand- or site-cut roofs are dealt with through AS 1684, much as you would a wall frame.

REFER TO AS 1720

The trussed roof

The components and design of roof trusses are governed by AS 1720.1 Timber structures Part 1 Design methods and AS 1720.5 Timber structures Nailplated timber roof trusses. The design of roof trusses delivered to a site are based on these documents and the information that the builder supplies.

Architectural trusses are generally designed by engineers. The documentation must be signed off accordingly and the truss plan provided must be highly explicit with regards to locations and all members that will support or be supported by these frames. With architectural systems there are generally fewer trusses, and they are usually exposed as part of a raked ceiling system. The jointing approach is dependent upon the aesthetic requirements of the designer and client, so inspection for compliance is based on ensuring that the delivered components match the designed jointing systems specified by the engineer.

Initial inspection

All aspects of a truss are crucial to the structural integrity of the system as a whole. Upon taking delivery of the trusses they must be checked as would prefabricated wall frames or any other structural component. This includes quality of transport, and appropriate offload handling procedures. After this the trusses should be inspected, looking particularly at:

- the jointing and connecting systems used
  - correct installation, plates correctly pressed home
  - transportation has not opened joints or otherwise caused connector failure
- materials used
  - split, bowed, cracked warped components, excessive knots, or knots in key load areas
- truss design
  - correct array of components, particularly at cantilevers
  - specified point load features handled as per advice provided to manufacturer
- dimensions
  - overall dimensions correct as no site changes are allowed.

The truss plan

After or during the initial quality inspection, the trusses should be checked against the truss plan provided. This plan will indicate the number of each truss type and their location on the home. There are a range of basic truss types that you should be able to readily identify: Figure 13.47 describes these and their specific location.

From this plan you want to identify specific locations of key load-bearing components such as girder trusses. These deliver point loads to the walls and you need to ensure that the designed location is in alignment with the point load (concentrated load) stud arrays. Where this is not the case you should reconfirm the truss layout and, if correct, install new concentrated load studs as required.

REFER TO AS 1684

The pitched (hand-cut) roof

This section deals specifically with the selection of timber roof framing components as required by AS 1684. AS 1684.2 will be referenced throughout, acknowledging that the approach taken is very similar to that required by AS 1684.3 (cyclonic). Like those before it, this section is framed around a worked example based on the plan set in Figures 13.13 to 13.17.

A pitched roof is almost always site cut, and as such the determination of components falls upon the builder's skills and knowledge based upon the aforementioned standards. This is very much the case with regards to hip and valley construction. With cathedral roofs, however, it is common for the main rafters and ridge beams to be specified; though often the builder has been involved in making that specification. Either way, these components need to be checked for compliance.

The cathedral roof examples a major distinction in roofing that must be identified when either designing or checking a pitched roof for compliance: that of coupled and non-coupled roofs. Failure to recognise the structural difference between these two forms can lead to complete failure of a building.

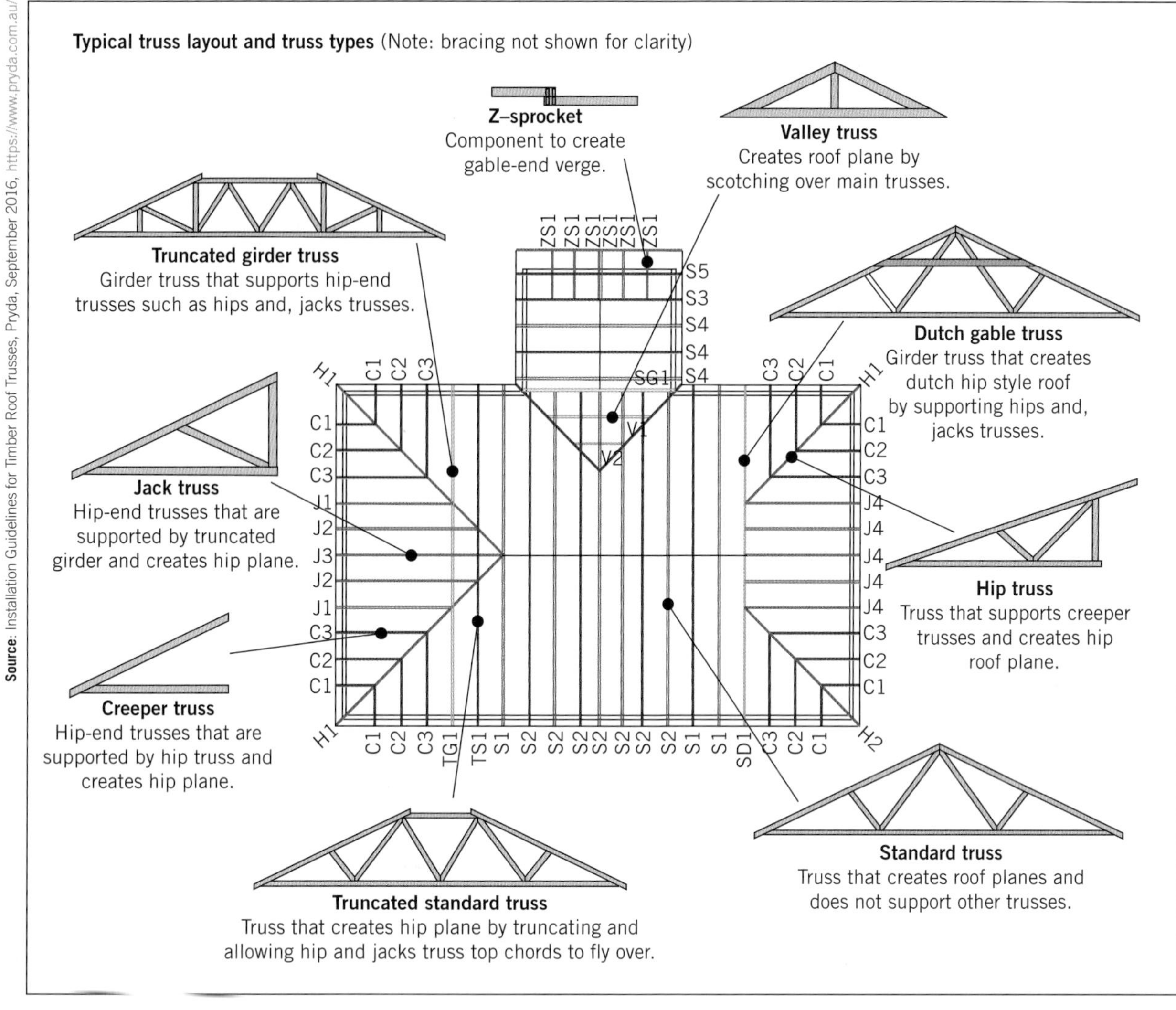

Typical truss layout and truss types (Note: bracing not shown for clarity)
Z–sprocket
Component to create gable-end verge.
Valley truss
Creates roof plane by scotching over main trusses.
Truncated girder truss
Girder truss that supports hip-end trusses such as hips and, jacks trusses.
Dutch gable truss
Girder truss that creates dutch hip style roof by supporting hips and, jacks trusses.
Jack truss
Hip-end trusses that are supported by truncated girder and creates hip plane.
Hip truss
Truss that supports creeper trusses and creates hip roof plane.
Creeper truss
Hip-end trusses that are supported by hip truss and creates hip plane.
Standard truss
Truss that creates roof planes and does not support other trusses.
Truncated standard truss
Truss that creates hip plane by truncating and allowing hip and jacks truss top chords to fly over.
Source: Installation Guidelines for Timber Roof Trusses, Pryda, September 2016, https://www.pryda.com.au/

FIGURE 13.47 Truss types for common roof configurations

Coupled roofs

Coupled roofs are defined by 7.1.2.2 and Figure 7.1 of AS 1684.2. Figure 13.48 shows the basic configuration of this roof form.

The defining element of the coupled roof is the ceiling joist. This triangulates the structure, tying the walls and rafters together and preventing the walls from spreading under roof load. Without the ceiling joist the structure will collapse. Like a truss bottom chord, this component is under tension. Though a coupled roof resembles a truss to some degree, the difference is in that components are generally lighter, and loads are transferred to internal, as well as external, walls. The ridge of a coupled roof effectively carries little to no gravity (downward) loads; acting mainly to maintain rafter separation, though ridge struts can be implemented on occasions.

One of the key issues with designing roofs of this type is the need to design for the ceiling and the roof loads at the same time as they are frequently interdependent upon each other. As Section 7 of AS 1684.2 shows, strutting beams can be acting also as hanging beams, or counter beams. In such cases, both ceiling and roof loads must be allowed for in beam selection.

Non-coupled roofs

Non-coupled roofs are defined by 7.1.2.3 of AS 1684.2. Figure 13.49 shows the key difference in configuration.

In a non-coupled roof, the rafters are attached to and supported by a ridge beam, and if required or desired, an intermediate beam. The ridge and intermediate beams are then supported at either end by walls or posts. On occasions, intermediate posts may be required on larger spans. These beams must be designed to carry their own weight and the proportion of roof that they carry as a clear span from support to support. Rafters are designed to clear span from wall to ridge or intermediate beam.

Coupled roof components

There are numerous components in a coupled roof, and many variations of this roof form; of which the gabled roof – the type in the plan set used for examples

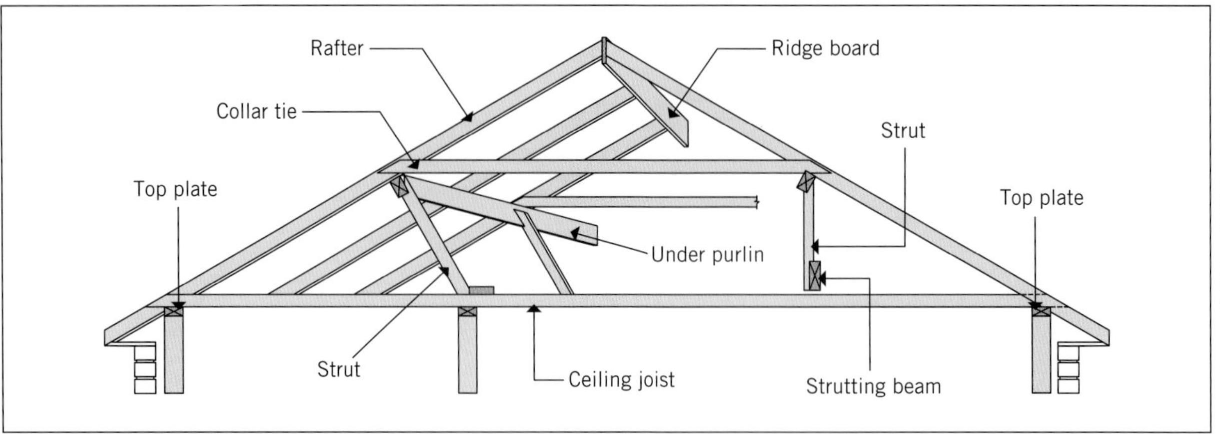

Source: AS 1684, p. 138 © Standards Australia Limited
Reproduced by Cengage under licence by Standards Australia.

FIGURE 13.48 Coupled roof

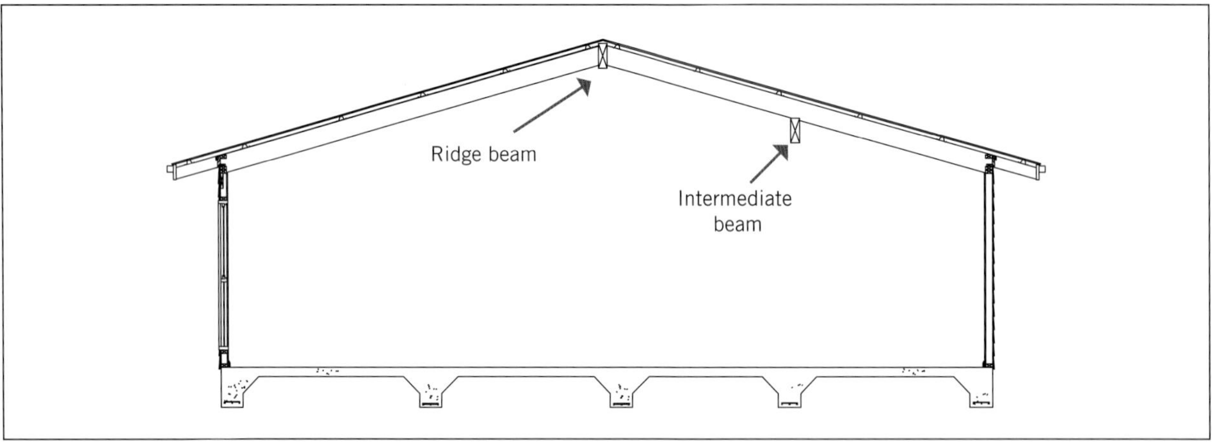

FIGURE 13.49 Non-coupled roof

so far – is the most basic. These components, their descriptions, arrangement within the roof and ceiling structure, and their connections are outlined in Section 7 of AS 1684.2.

The sizes for each of these components are defined by Clause 7.3 (Member sizes) which directs you to the appropriate tables within the document or the supplementary span tables. Table 7.6 provides the sectional sizes for less critical components. Figures 7.19 through to 7.30 explain the various load calculations required to make span table selections. One example of these calculations is offered below.

EXAMPLE 13.9

COMBINED HANGING/STRUTTING BEAM

AS 1684.2: Clause 7.3.9 and Figure 7.22

For this example, a pitched gable roof is proposed to replace the truss option previously discussed for the plan set in this chapter (Figures 13.13 to 13.17). The roof and ceiling area being discussed is that over bedroom 1.

A combined hanging/strutting beam is required to support the ceiling joists, which cannot span the length of the room; likewise, the purlin supporting rafters running up to the ridge. Ridge struts will be assumed so only one side of the roof is required to be considered as load. This component has compounding loads from roof and ceiling. It is a common beam in homes seeking larger room sizes. (See Figure 13.50.)

Information required:

- proposed hanging/strutting beam material – LVL F17, joint group JD4
- rafter length (total span from plate to ridge)
- hanging/strutting beam span – room width
- total underpurlin span
- ceiling load width (CLW) – see Clause 2.6.3 and Figure 2.12
- roof pitch: 16°
- ½ span of building: 5.250

>>

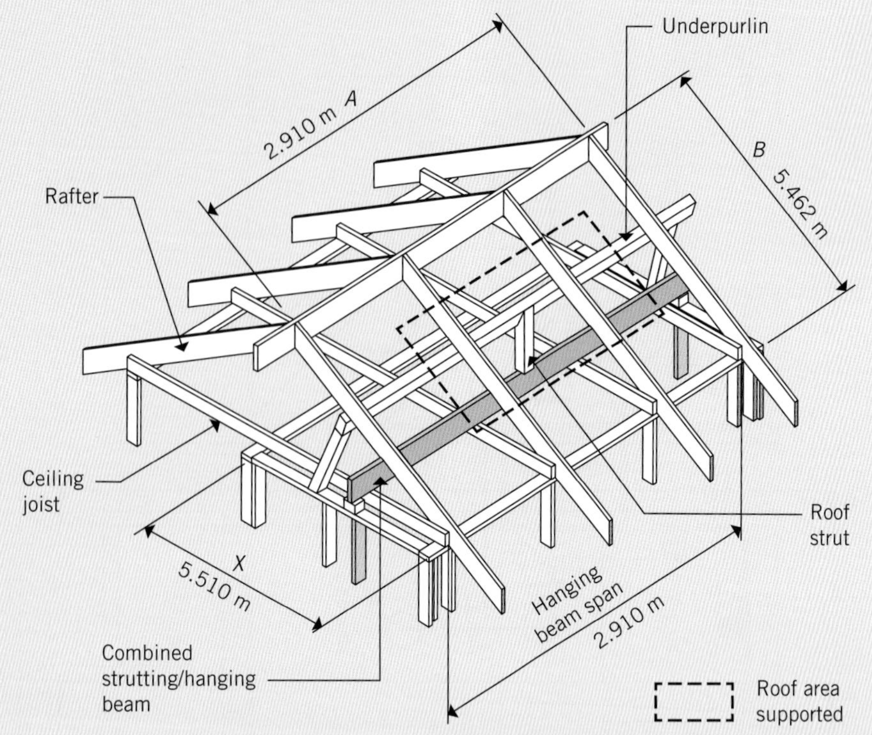

Source: AS 1684.2, p. 104 © Standards Australia Limited Reproduced by Cengage under licence by Standards Australia.

FIGURE 13.50 Identifying the roof loads on a strutting beam

Rafter length (span) = 5.250 ÷ cos 16°
= 5.462 m

Roof area supported = ½ underpurlin span × ½ rafter span
= (½ × 2.910 m) × (½ × 5.462 m)
= **3.974 m²**

Ceiling load width = ½ × hanging beam span
= ½ × 2.910 m
= **1.455 m**

Hanging/strutting beam determined from AS 1684.2 N3 Supplement 8 Seasoned hardwood Stress Grade F17, Table 25 (Figure 13.51).

Required counter/strutting beam span: **2.910 m**

Required counter/strutting beam = **2/140** × **35**

Note (vi) of this table requires multiple members such as required here (2/140 × 35) to be nailed together as per Clause 2.3.

TABLE 25 STRUTTING/HANGING BEAMS Supporting roof and ceiling loads

	Ceiling Load Width (mm)											
	1800						3600					
Roof Area Supported (m²)	2	4	6	8	10	12	2	4	6	8	10	12
Size D×B (mm)	Maximum Beam Span (mm)											
	Sheet Roof											
2/120×35	3000	2600	2200	2000	1800	1700	2500	2200	2000	1900	1700	1600
2/120×45	3300	2800	2500	2200	2100	1900	2700	2500	2200	2100	1900	1800
2/140×35	3600	3100	2700	2500	2300	2100	3000	2700	2500	2300	2100	2000
2/140×45	3800	3400	3000	2800	2500	2400	3200	3000	2700	2500	2400	2200
2/170×35	4200	3800	3500	3200	3000	2800	3600	3400	3100	2900	2700	2600
2/170×45	4400	4100	3800	3500	3300	3100	3900	3600	3400	3200	3000	2900
2/190×35	4500	4200	3900	3700	3400	3200	4000	3700	3600	3300	3200	3000

Source: AS 1684.2, Table 25 © Standards Australia Limited Reproduced by Cengage under licence by Standards Australia.

FIGURE 13.51 Strutting/hanging beams

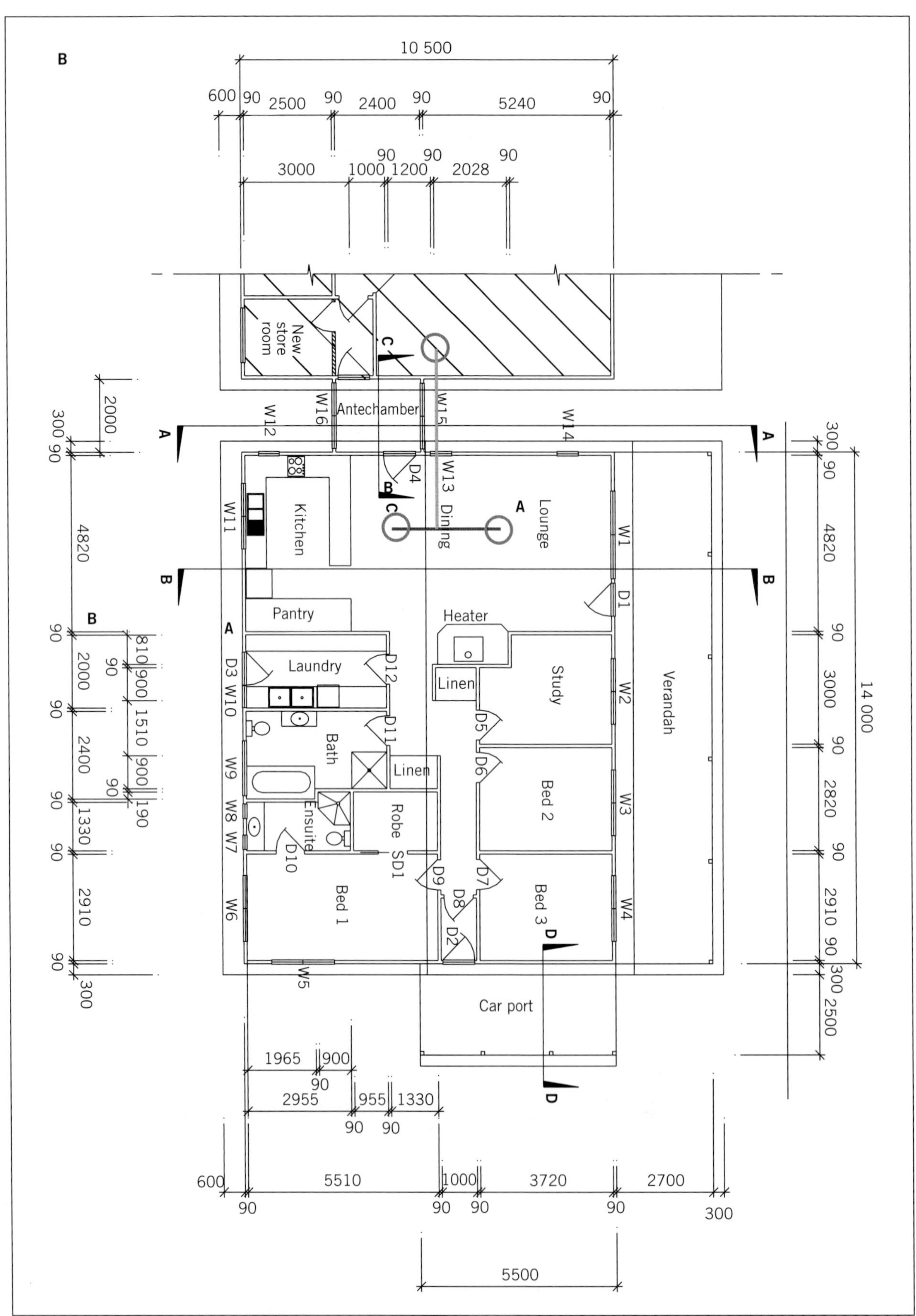

FIGURE 13.52 Ridge beam and strutting beam locations

Other components

All other components must be confirmed as compliant in the same manner; i.e.:

- identify the wind classification
- identify the component as it relates to AS 1684.2 definitions
- identify the material the component is made from (timber species and stress grade)
- identify the clause concerned with the determination of this component
- follow the clause requirements and identify the appropriate supplement table
- from the table, determine the required sectional size.

In making the determination, be sure to read all notes associated with the tables and the relevant clauses.

Non-coupled roof components

Components for non-coupled roofs are found in a similar manner. The loads and spans, however, are substantially different. In the example to follow – again based on the plan set in Figures 13.13 to 13.17 – the trussed roof over the kitchen lounge area is to be replaced by a cathedral (raking) ceiling. This means a large ridge beam must be installed, and large section rafters brought up to meet it, as shown in Figure 13.52.

In making the change, two options for supporting the ridge beam (green) at the 'heater' end are possible:

- A proprietary girder truss is installed over the study/laundry walls that will clear span over the heater and pick up the end of the ridge beam.
- Install a strutting beam spanning the heater and hall space (red). A strutting beam must be specified instead of a lintel, it must be capable of supporting the ridge beam as a point load from a vertical (king) post siting off centre of the beam span.

For this example, we are focused on the ridge beam only, so either may be assumed. The determination of the strutting beam is discussed briefly, however, at the end of this section.

EXAMPLE 13.10

RIDGE BEAM

AS 1684.2 Clause 7.3.14 and Figure 7.29

This example has been chosen as it exceeds the capacities of the tables provided in the supplements to AS 1684.2, and so offers the opportunity to demonstrate that alternative tables can be accessed for engineered timber products such as LVL.

Figure 13.53 shows the location of the beam and the information required to determine its size. This is a single or 'simple' span beam.

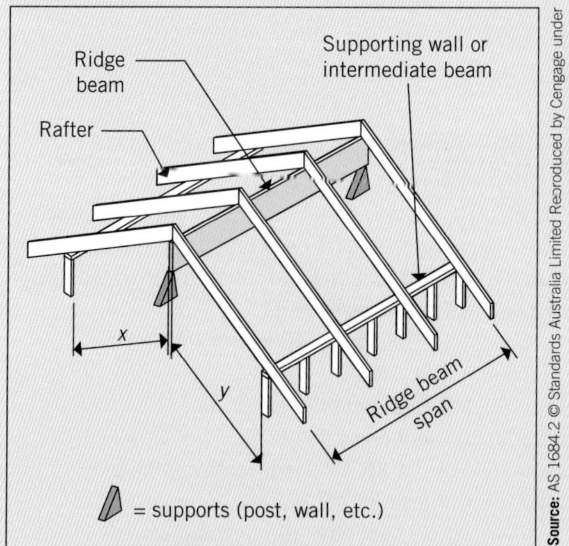

FIGURE 13.53 Ridge beam in an non-coupled roof

Information required:

Proposed material for ridge beam: LVL F17 joint group JD4

Span of ridge beam = **4.820 m**

Roof load width (RLW) = (rafter length + rafter length) ÷ 2
= rafter length
= **5.462 m** (from previous, p. 405)

Roof and ceiling mass = 30 kg/m^2* or 40 kg/m^2#

*This mass is derived from AS 1684.2, Appendix A, Table A1.1. Interpolation of the tables is required.

#This mass is derived from the LVL manufacturer's preferred roof and ceiling mass.

Determination

An initial review of Span Table 30 (single spans) in AS 1684.2 N3 Supplement 8 Seasoned hardwood Stress grad F17 shows that for these dimensions and a 30 kg/m^2 roof/ceiling mass, the table is at its limit. A beam can be identified – 2/290 × 45 showing an interpolated maximum span of 5.050 m. This beam leaves little room for error; it also requires the purchase of two beams and coupling them together.

Figure 13.54 is drawn from the Carter Holt Harvey website, one of the manufacturers of LVL beams.

>>

>>

hySPAN® solutions range

Roof Beams

Ridge, Intermediate, Eave and Bressumer Beams

Table 16

hySPAN SECTION D x B (mm)	SHEET ROOF AND CEILING							
	ROOF LOAD WIDTH 'RLW' (m)							
	2.4	2.7	3.0	3.6	4.2	4.8	5.4	6.0
	MAXIMUM SINGLE SPAN (m)							
300 x 45	5.3	5.1	4.9	4.6	4.3	4.1	3.9	3.8
300 x 63	5.9	5.6	5.4	5.1	4.8	4.6	4.4	4.2
300 x 75	6.2	5.9	5.7	5.4	5.1	4.8	4.6	4.4
360 x 63	7.0	6.7	6.5	6.1	5.8	5.5	5.2	5.0
400 x 45	6.8	6.4	6.1	5.5	5.1	4.7	4.4	4.2
400 x 63	7.7	7.4	7.2	6.7	6.4	6.1	5.8	5.6
400 x 75	8.1	7.8	7.6	7.1	6.7	6.4	6.1	5.9
450 x 63	8.6	8.3	8.0	7.6	7.2	6.8	6.5	-

Source: Carter Holt Harvey Woodproducts Australia Pty Limited

FIGURE 13.54 A manufacturer's roof beam span selection

Ridge beam = **400 × 63 LVL F17**

The beam identified here is significantly deeper (by 110 mm) but narrower (63 vs 2 × 45). It gives greater structural 'head room' (space for error or unexpected increased loads). It also is not reliant upon correct nailing procedure to couple two beams together.

Strutting beam

AS 1684.2 Clause 7.3.11 and Figure 7.24

The strutting beam is determined in a similar manner by selecting the appropriate table and determining the required span of the strutting beam and area of roof carried. In this case the roof area is effectively half of the ridge beam span (2.410 m) plus half a rafter spacing (0.45 ÷ 2) multiplied by the RLW developed for the ridge; i.e. the length of the rafter (5.462 m). Supported area therefore equals 14.392 m^2.

Assuming the beam to be LVL F17, Span Table 27 of AS 1684.2 N3 Supplement 8 Seasoned hardwood Stress grade F17 may be used; or a manufacturer's table as in the case of the ridge beam. This load (40 kg/m^2) and load area is outside the capacity of AS 1684.2 Supplement [see Note i)], so from the manufacturer's span tables using the span of 3.318 (from plan in Figure 13.52) a 240 × 63 beam is appropriate.

Implications of design changes

Making changes to the roof structure as proposed in these examples has implications for the rest of the structure, particularly footings. In the cathedral roof (ridge and strutting beam) example, point loads are now at either end of the strutting beam, and at the antechamber end of the ridge (see blue circles 'A' and 'B' on plan in Figure 13.52).

These point loads need to be calculated based on the roof loads found for the beams themselves and the footings increased accordingly. For a slab footing, this may mean an additional tie beam running under the heater hallway area aligned with the strutting beam.

For stump and bearer construction, pad footing beneath the circled end of the strutting beam will be required.

EXAMPLE 13.11

To calculate the additional roof loads applied to each pad, a bending moments calculation is required. The load on the strut is found by roof area supported × roof and ceiling load mass/m^2; i.e.:

Load = 14.392 m^2 × (40 kg/m^2 ÷ 100*)

Load = 5.76 kN

*See Australian standard units of measure in the introductory section of this book. Alternatively, the same load of 0.4 kN for sheet roofs can be drawn from Clause 3.6.4.2 (c) of AS 1684.2.

Figure 13.55 describes the forces involved: where 'A' is the northern end of the strutting beam.

>>

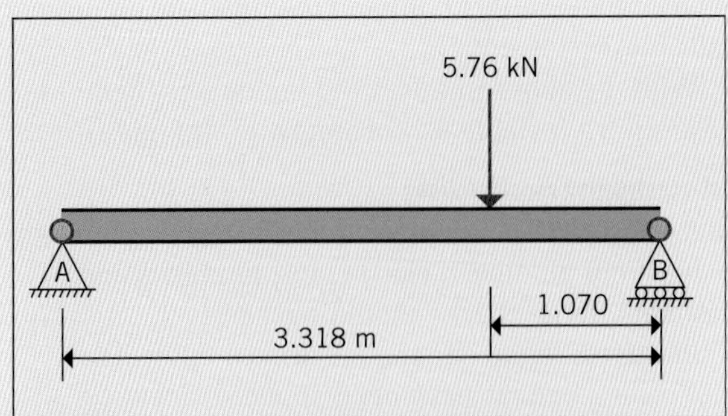

FIGURE 13.55 Strutting beam and load bending diagram

Using the calculation approach covered in 'Moment solutions' in Chapter 12 (see also Appendix 4)' the resultant forces are:

A = 1.86 kN
B = 3.90 kN

In calculating the pad footing sizes, these loads must be added to the dead and live floor loads developed following Clause 3.6.4 of AS 1684.2. Using a typical floor area supported by stumps of 1.8 m × 1.8 m (3.24 m^2) and following the total vertical gravity calculation in Clause 3.6.5 of AS 1684.2, the result is:

Floor load = 3.24 m^2 × 0.60 kN/m^2 (from Table 3.1 of AS 1684.2)
***Floor load* = 1.944 kN**
Wall load = 3.24 m^2 × 0.4 kN/m^2 (from Clause 3.6.4.2 (b) of AS 1684.2)
***Wall load* = 1.296 kN**
Live load = 3.24 m^2 × 1.5 kN/m^2 (from Clause 3.6.4.3 (a) of AS 1684.2)
***Live load* = 4.86 kN**
For pad footing B (largest load):
total load = total dead loads + 0.5 × live load
total load = (1.944 + 1.296 + 3.90) + (0.5 × 4.86)
***Total load* = 9.57 kN**
From Table 3.2 of AS 1684.2 the required footing size is 400 mm ∅.

Erecting and checking roof trusses

The installation of prefabricated roof trusses is outlined in AS 4440 Installation of nailplated timber trusses. Although this standard is not called up as an *acceptable construction manual* within Volume Two of the NCC, AS 1684 is; and this standard does call up AS 4440 in Clause 7.2.21.

In addition, all installation guides produced by truss manufacturers reference AS 4440 as their guide to best practice. In most cases, truss assembly will be directed by one of these manufacturer's installation guides. A good example guide may be found at: http://www.mitek.com.au/Products/Structural/Roof-Trusses/. In reading this guide, you must make note of the wind categories applicable in each area of the document, and other limitations such as joint group considerations for fixings and maximum jack truss spans (Table 2, for example).

This being the case, it is not the purpose of this section to replicate all that is in these guides, nor in the standard. This section will therefore focus solely on the key structural issues with which you should be concerned. Appropriate work health and safety (WHS) practices will be assumed.

Planning assembly

Frequently, when taking delivery of trusses, the component parts will be offloaded directly onto the top plates of the walls. This must be planned for with additional temporary bracing to ensure such loads may be supported. If offloading to ground level, trusses must be supported on gluts or otherwise off the ground and flat to avoid distortion or rupturing of joints.

Planning for assembly is then based on the following key points:

- checking that supporting structure aligns with the concentrated or point loads in the truss plan
- ensuring non-load bearing walls are suitably below the transit of the truss bottom chord
- checking of truss design documentation to ensure specific service loads required have been allowed for, such as: roof space hot water header tanks, on-roof solar hot water tanks and panels, photovoltaic cell (solar panels) and the like
- identifying specific trusses, particularly those designed for key loads such as those listed above, girder trusses, truncated trusses and the like (see Figure 13.46)
- setting out the top plates for the truss locations
- identifying the appropriate connectors and fixings (no clouts to be used)
- establishing a lifting approach that ensures trusses are loaded to the top plates in a manner that does not cause them structural strain, undue bending, or stress of joints.

Once trusses are confirmed to be accurately manufactured and an appropriate lifting and assembly strategy is in place, trusses should be installed following the manufacturer's guidelines.

Compliant assembly

From a structural perspective, the main issues with truss assembly is in the temporary bracing, fixings, alignment, final bracing and clearances between truss bottom chords and internal wall top plates.

- **Alignment** issues tend to be that of plumb. Trusses must be installed as close to the vertical plane as possible unless specifically designed otherwise. Compliant practice is stated as being:

  The lesser of 50 mm out of plumb measured from the apex to the bottom edge of the bottom chord, or this height divided by 50.

For example, a 3.0-m-high truss may be out of plumb by only 50 mm, not 60 mm. While technically acceptable, it is still not good practice, and certainly not appropriate for gable end or gambrel (Dutch gable) trusses.

In addition, the bottom chord of a truss must not be bowed by more than its length divided by 200 or a maximum of 50 mm.

- **Temporary bracing** must be installed as per the manufacturer's instruction or AS 4440. In the main this requires top chord ties at each panel point, and solid props to ground ties at gable ends. Bottom cord ties are should be placed at every 3000 mm. Angled braces affixed to the webs may also be used. Excepting bottom cord ties, all other temporary bracing should be removed after completed assembly and final bracing.
- **Fixings** must be to AS 4440 and the manufacturer's specific instructions for each connection point. These are many and variable but must be complied with completely. Note particularly that clouts must not be used in place of identified flat head nails or proprietary nails specifically designed for the purpose of affixing truss and frame connectors. See the guide listed previously for connection requirements for each component.
- **Final bracing** must be completed precisely as the manufacturer specifies in the documentation. A general guide is provided in the example installation guide provided above. Particular attention must be given to the following:
  - speed brace overlaps
  - brace end fixings
  - tension and straightness of speed brace
  - blocking out of trusses at brace ends (heels).
- **Truss bottom chord and internal wall top plate clearance** is an essential part of truss engineering design. The bottom chord of the truss must be able to flex under load either upwards or downwards without making contact with any obstruction. The bottom chord is designed for tension only. Interruption of the bottom chord through contact to internal walls introduces a point load which may fracture or crack the chord, leading to truss failure. That internal walls are designed to be non-load-bearing is incidental to this issue.

Wall wind-bracing considerations

After installation, consideration must now be made to correct tie-in with wall bracing units. There are specific clauses in AS 1684.2 (and 1684.3) regarding appropriate connection procedures to internal bracing walls by way of the bottom chords. These are found in Clause 8.3.6.9 and Table 8.22.

The important point of these systems is that the bottom chord is prevented from lateral movement, but *not* up or down movement. The required clearance between the bottom chord and the top plate of internal walls must be maintained.

With the above in mind, the preferred systems are those shown in Table 8.22, Figures (e), (g) and (j).

Post-construction considerations

After construction and other trades have been through, such as roof plumbers, electricians, plumbers and plasterers, some checks should be made.

Plasterers' work should be checked immediately after ceiling battens have been applied or before sheeting is installed. You are looking to ensure that ceiling battens have not been chocked up by wedging between the internal non-loadbearing walls and the bottom chord of a truss. This is unfortunately a common practice and goes counter to truss design, whereby the bottom chord must remain clear of obstruction. The same must apply to all trades, so piping, wiring and the like must also remain clear.

Hand-cut roofing systems and quality control

To ensure the structural integrity of hand-cut roofing systems there are a few basic considerations to keep in mind. The first is material choices; both the timber and the fixings and connectors used. The second is the installation of all required components and that such installation satisfies the tie-down requirements.

The basics: materials and components

Section 7 of AS 1684, Roof framing, provides the necessary guidance in this area. Component selection has been covered in Part 6.1; the critical observations to be made at this point is in ensuring that all the necessary components are available and that the material choices are sound. In this case, we are looking for materials that are not only of the appropriate stress grades, but also in sound condition; i.e. unsplit, without fractures, knots, or other compounding issues that may make them unsuitable for the structural role that they must play.

Clause 7.3.17 and Table 7.6 provide for the miscellaneous roof framing members for which no span tables exist. These are still structurally important components, including the likes of ridge boards, valley boards, collar ties and roof struts. Components must match the requirements of this table and again be of sound materials.

Size tolerances as per Clause 1.13 must be maintained for all components.

Connectors and fixings

Fixings and tie down is covered by Section 9 of AS 1684.2 and has been covered in the specifics in earlier sections of this chapter. Ensuring quality control over fixings from a structural perspective requires that all fixings comply with the standard whether specific or nominal.

Nominal fixings are outlined in Table 9.4 of AS 1684.2 including the minimum length and diameter of nails for each application. These fixings must comply with Clause 1.15 and through this be protected from corrosion in accordance with the Australian standards listed.

Specific tie-down connections are identified through Table 9.2 and defined by Table 9.16. Connections are joint group specific.

In checking fixings and connectors, attention must be made to ensuring no splitting or cracking of the components has occurred and that skew nailing particularly has been correctly carried out.

Assembly: monitoring compliance

Compliant assembly is achieved through ensuring that all components have been fitted and correctly spaced according to the planned for design, span tables and roof type. Unless a particularly different style of roof design is being undertaken, ridges should be parallel with wall plates; rafters should be 90° to wall plates and parallel with each other. Hips and valleys should bisect corners and all roof surfaces should be in the one plane without twist or wind.

Key supporting components such as strutting beams must be positioned such that they do not interrupt ceiling surfaces – providing a 25-mm gap at mid span – and when required by the span tables or their relevant AS 1684 clause, laterally braced. Although only 'deep joists' are specifically defined in this manner (see Clause 4.2.2.3), when the depth of a beam exceeds four times the breadth, lateral constraint should be considered when not otherwise supported by another member.

In checking fixings and connectors, attention must be made to ensuring no splitting or cracking of the components has occurred and that skew nailing particularly has been correctly carried out.

The quality of cuts and joints is not particularly a structural consideration unless no direct contact between surfaces is made. However, tight joints mean less potential movement and hence overall greater rigidity.

Bracing

The bracing of pitched roofs is governed by Clause 8.3.7 of AS 1684.2. Bracing requirements are determined in relation to:

- roof type – hip (all forms) or gable (including cathedral types)
- pitch angle
- wind classification
- gable width.

Clause 8.3.7.1 considers hipped roofs to be self-bracing and therefore not requiring of additional specific bracing. Good practice suggests that for very large hipped roofs, additional strap bracing should be installed. Gable roof bracing is specified in Table 8.25 while the required connections are offered in Table 8.26. Dutch gable or gambrel roofs should be treated as gables.

Clause 8.3.7.1 (b) (ii) allows for metal bracing to be installed according to 'engineering principles'. This allows for you to follow the bracing applications of gabled trussed roof systems as the principles are identical. Alternatively, an engineer must be engaged.

Clause 8.3.7.2 is applicable for wind classifications up to N2 only.

Sarking and roof cladding

Both the sarking and the cladding of roofs have potential structural implications. Cladding also has a direct structural implication. These elements must therefore be planned for and installed appropriately to ensure compliance with the relevant codes, standards and manufacturer's guidelines.

Sarking

Reflective insulation on roof is used as a first level of insulation, and as a barrier from condensation that may form on the underside of the metal roofs, and as a vapour control membrane to prevent or reduce condensation more generally within the roof structure. It is these latter issues that are important for ongoing structural integrity.

As with wall sarking there are effectively two forms available:

- vapour-permeable sarking
- vapour-resistant or non-permeable sarking.

Incorrect application and selection of sarking can lead to the rapid degradation of timber roof frames and connectors.

Selection

Selection is based upon a number of factors that are discussed in Chapter 14. With regards to a roof's ongoing structural integrity, the two main criteria are similar to that for walls:

- climate
- type of roof cladding.

Sarking is also a mandatory requirement in all bushfire zones rated BAL 12.5 upwards as a means of defence against ember attack. However, as this is not a structural matter it is not discussed further here.

Climate

Sarking can help control water vapour from entering or being trapped in the roof space from either direction – from outside the home in; or inside the home upwards. In hot humid climates, the water vapour is generally travelling inwards due to it being cooler and less humid inside than out (when air conditioned). In colder climates, it is the reverse as the air is generally hotter and more humid inside than outside.

- In hot or warm climates with mild winters, a non-permeable sarking is advised.

- In locations with warm summers but also cold winters the two options are available dependent upon the bulk insulation used: non-permeable when insulation is not more than R3.5; permeable when using higher levels of insulation in the ceiling.
- Locations with very cold winters and mild summers should use permeable sarking for all insulation levels.

Cladding

There are potentially many roof cladding options available for the domestic home including slate, timber shingles and shakes, and even thatched straw. The bulk of modern homes, however, will be clad either in steel sheet or some form of concrete or terracotta tile. With regards to these latter two, the choice of sarking aligns with the climate issues outlined above, but also with the form that the sarking may take or even if it is required at all.

Sarking for tiled roofs is governed by Clause 3.5.1.2 and Table 3.5.1.1 of Volume Two of the NCC. According to this document, whether sarking is required for this cladding is dependent upon the pitch of the roof and the rafter length. Specific installation requirements such as the inclusion of anti-ponding boards for roofs with pitches less than 20° are also described in this clause. Additionally, sarking cannot be used under tiles when mechanical venting from wet area rooms (bathrooms, laundries and the like) exhausts into the roof space – again this can cause moisture build up, rot and premature structural failure (Clause 3.8.5.2 of the NCC, Volume Two).

Sarking under metal roofs is not specified within the NCC. The decision to use it is the client's or builder's choice. The type to use is framed on the climatic considerations given earlier.

Forms of sarking

The two main descriptors of sarking have been stated previously; i.e. permeable or non-permeable. However, there are two other forms that have implications to the capacity to reduce water vapour and condensation in the roof space; these are:

- aircell membranes – these can be selected on the same basis as standard sarking; i.e. permeable or non-permeable. The main difference with aircell is its non-structural insulation and thermal break values.
- blanket membranes – are a reflective foil sarking faced with a woven insulation. This insulation 'blanket' can be of a range of thicknesses and is particularly designed for metal roof cladding. Blanket membranes are in most instances non-permeable.

Roof cladding

As stated previously, there are many variants of roof cladding available to the domestic home. In all cases, however, this cladding is governed by Part 3.5 of the NCC, Volume Two. There are a range of acceptable construction manuals listed at the beginning of this part covering tile, metal roofing and even asphalt shingles.

The roof cladding is generally installed by subcontractors such as roofing plumbers or roofing tilers. In ensuring compliance the task is one of inspection as the work is being carried out as well as when it is deemed to be finished. This includes inspection of the materials being used.

Roof tiling

Compliance is offered when all the relevant aspects of Clause 3.5.1.2 have been met. This includes evidence that the tiles meet AS 2049 in manufacture. Particular attention must be made to the manner and extent of mechanical fixings for the tiles based upon the wind classification and Figure 3.5.1.1 for the main roof sections, and Figure 3.5.1.2 for ridges, hips and valleys.

Metal roof sheeting

Metal sheet roofing requirements are stipulated in Clause 3.5.1.3. Key areas of compliance are material selections based upon corrosion prevention (Table 3.5.1.1a and 3.5.1.2). This same clause covers all issues of fixing and flashing, allowable spans and sheet laying sequences – paying particular attention to the overlapping of sheets being based upon prevailing weather directions.

This clause also stipulates minimum overlap of sheets in their length. Most contemporary installations will use full-length sheets where possible.

Flashings should be checked prior to installation to ensure capillary breaks are correctly folded and then checked again upon installation for correct overlaps and directions of flow. Where there are penetrations such as heater or venting flues, ensure that aprons have been correctly installed and screwed down.

The portal frame

The portal frame, shown in Figure 13.58 and discussed in Chapter 12, is neither a wall nor a roof, yet is the structural element behind both for many commercial, and occasionally domestic, structures. Commonly made of steel, large portal frames may also be made from engineered timber. The key to their stability, as outlined in the previous chapter, is in the rigidity of the joints. No matter the material, the portal frame must be engineered to withstand the various loads, live, dead and particularly wind. This includes the footings to which they are installed.

Portal frames are arrayed in 'bays', the horizontal distance between each frame. Any of the wall elements discussed previously may be used to infill these bays; i.e. between the columns of each frame. Alternatively, tilt-up panels (Figure 13.56) may be used as shear walls to brace the frames in place of steel or timber diagonal components.

When applying wall cladding only to the frames, the cladding will be supported on horizontal members referred to as girts. A purlin, effectively the same component, is used to support roof cladding. How all these components go together is critical to the long-term success of the structure as a whole. Small details, such as fillets (lateral restraints or compression stiffener) in the columns at the haunch, and struts or 'stays' between the purlins and the rafters, can be the difference between a safe and durable build, and total collapse under loads that it could well have catered for. Hence, the need for a skilled engineer experienced in this form of building design; and the need for you to carefully study the specifications and design produced by the engineer and build accordingly. Figures 13.57 and 13.58 give some examples of the details you must attend to.

Source: Shutterstock.com/Richard Thornton

FIGURE 13.56 Tilt-up shear wall panel

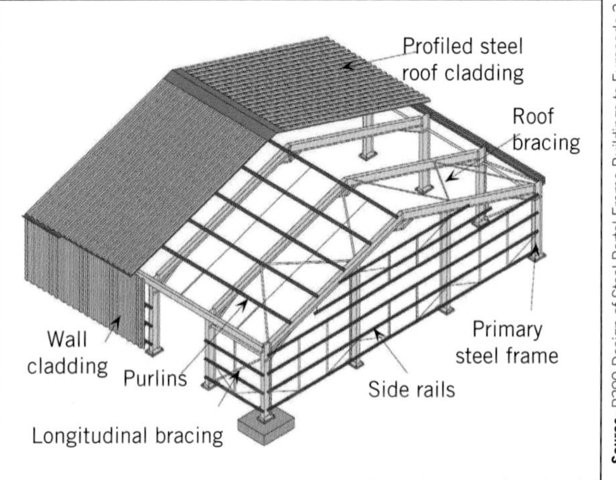

Source: P399 Design of Steel Portal Frame Buildings to Eurocode 3, SCI (the Steel Construction Institute), https://www.steel-sci.com/

FIGURE 13.58 Elements of a typical portal frame

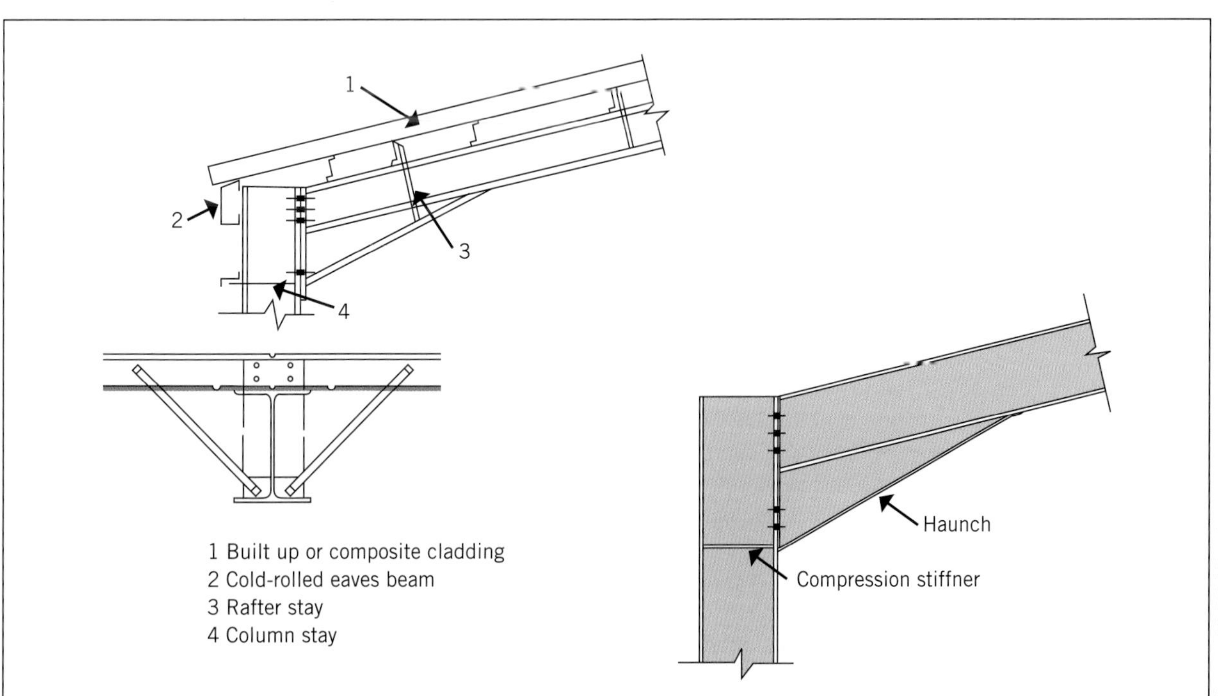

Sources: SCI P346 Best Practice for the Specification and Installation of Metal Cladding and Secondary Steelwork, SCI (the Steel Construction Institute), https://www.steel-sci.com/ P399 Design of Steel Portal Frame Buildings to Eurocode 3, SCI (the Steel Construction Institute), https://www.steel-sci.com/

FIGURE 13.57 Portal frame restraint details

LEARNING TASK 13.5

DETAILING CONSIDERATIONS IN ROOF CONSTRUCTION, READY FOR COVERINGS

1 Access AS 4440 (or a truss manufacturers manual), and using that information provide a checklist of items that should be progressively checked as part of the installation and certification of roof trusses to a simple rectangular-shaped building with a hip roof.

2 a Access NCC, Vol. 2, and explain when an anti-ponding device/board is required.
  b Explain or draw where ideally the anti-ponding device/board is located.
  c What clause in the NCC clarifies the requirements of anti-ponding boards?

Windows and wall cladding

The structural role of windows and cladding is often disregarded in the contemporary building. In brick veneer construction, for example, the cladding plays no role in carrying roof loads and in fact is supported against outward collapse by the timber or metal frame it shrouds. Likewise, this frame is designed to bridge openings for windows and doors which, with the exception of some major architectural bespoke designs, are inserted so as to be non-load-bearing.

While this is the case, there are still structural considerations that apply. Windows, though they may not hold up a roof, are today frequently a major part of the wall area. If the window fails then wind, rain, dust and the like can force its way into a building in a manner for which it was not designed to withstand: the structure as a whole then fails. The same holds for cladding: strip this away and the frame is exposed, internal linings will fail and again failure of the structure is the likely result.

As with any element of the building, codes and standards have been developed to stop these sorts of failures occurring; this section explores their application and what is required to ensure compliance. **Note**: The performance criteria in this section have been provided in reverse order to reflect the logical sequence of construction.

Window and door installation

Windows and doors are generally installed after the wall sarking has been installed, after the house is 'wrapped'. This tough but light-weight reflective insulation layer is then cut away at window and door openings allowing glazing elements and door frames to be inserted and fixed.

REFER TO AS 2047, AS 1288

The structural integrity of the door and glazed elements themselves is of importance and their selection is based upon a range of factors including wind category, bushfire (BAL) rating and location within the building, including likelihood of direct impact. These parameters are defined by Performance Requirements P2.1.1 and P2.2.2 of Volume Two of the NCC; and through Part 3.6 of the same volume, and the Australian standards AS 2047 and AS 1288.

These standards make it clear that it is up to the builder or designer to provide window and door manufacturers with the requisite design parameters by which they may manufacture these elements to specific contexts. Guidance on this matter may be found through the Australian Glass & Window Association's *A guide to window and door selection*, which may be found online at: https://www.agwa.com.au/documents/item/821

However, the focus of this section is upon the structural integrity of the fixing and flashing mechanisms employed. Unfortunately, neither the NCC nor the Australian standards address this issue in any specifics, Section 7 of AS 2047 stating effectively only:

- that windows or doors shall be fitted into that part of the building for which they were designed, and
- that they shall be fixed using recognised building practices.

AS 1288 is silent on the matter entirely.

Recognised building best practice

Due to this lack of advice in the codes and standards, the Australian Glass & Window Association (AGWA) have developed guides outlining best practice for both the fixing and flashing of windows and doors. These guides are framed around wind classifications (as per AS 4055), Ultimate Limit State wind pressures (ULS), the sizes of the elements being installed and the type of fixing being used.

These guides may be found at:

- fixing guide: https://www.agwa.com.au/documents/item/213
- flashing installation guide: https://www.agwa.com.au/documents/item/2124

Using the plan set offered earlier in this chapter (Figures 13.13 to 13.17), the use of these guides will be demonstrated by example.

EXAMPLE 13.12

FIXING

This example is based on the north wall lounge window W1. As per the schedule, this window is nominally 2100 mm in height x 2400 mm in width. The wind classification is N3 and the wall frame is timber. These guides also list ULS pressures; however, they are not required for domestic housing assessments.

In using this guide, you first determine how you intend to fix: by nail or by screw. You then identify the table and pages relating to the required wind category for this type of fixing. In this case the fixings will be by nail and the wind category is N3.

The guide requires that penetration into the timber frame be not less than 10 times the diameter of the fixing used (nail or screw). If a 65-mm gun nail of 2.87 mm Ø is used, it must penetrate the frame by approximately 30 mm. Given 20 mm for reveal and 10 mm for packing the nail is acceptable.

From this you then select the table that applies to the nail or screw size being used. As a 2.87 mm Ø nail is proposed, the following table (Figure 13.59) is appropriate.

ULS Wind Pressure: 1500 Pa, Nail Diameter: 2.8 mm

		Window Width										
		600	900	1200	1500	1800	2100	2400	2700	3000	3300	3600
Windows Height	600	4	4	4	4	4	4	4	4	6	6	6
	900	4	4	4	4	4	6	6	6	8	8	8
	1200	4	4	4	6	6	6	8	8	10	10	12
	1500	4	4	6	6	8	8	10	10	12	12	14
	1800	4	4	6	8	8	10	12	12	14	14	16
	2100	4	6	6	8	10	12	12	14	16	18	18
	2400	4	6	8	10	12	12	14	16	18	20	22
	2700	4	6	8	10	12	14	16	18	20	22	24

Source: Fixing - An Industry Guide to the Correct Fixing of Windows & Doors, Version 2.0, 2012 © Copyright Australian Window Association

FIGURE 13.59 Determining fixing requirements

From this table it shows that a minimum of 12 nails must be installed to fix this window in place and prevent its dislodgement through wind pressures. The arrangement of these nails is shown at the beginning of the document.

EXAMPLE 13.13

FLASHING INSTALLATION

Flashings to wall penetrations such as windows and doors are governed by Clause 3.5.3.6 of Volume Two of the NCC, the flashing material having to comply with AS/NZS 2904. While the diagrams offered in Figure 3.5.3.5 of the same volume are important, the guidance offered in the AGWA installation guide is more complete and covers more window types.

The information required to determine the flashing requirements in this case is:

- type of window – aluminium timber reveal
- type of cladding – timber weatherboard.

The general requirements for head and sill flashing are described in pages 9 and 10 of the AGWA guide. These should be read and adhered to. The guide is then broken up into window or door frame material types; i.e. timber, aluminium, or uPVC frames. These sections are then divided into typical installation scenarios; i.e. timber frame and timber clad, brick veneer, concrete block and the like.

In this case, page 14 is selected showing how the flashing is to be arranged for sill, head and jamb (side of windows).

When ensuring compliance, the flashing and fixings must be checked prior to cladding being installed.

Note: When installing head flashings, the flashing must go under the sarking; i.e. the sarking goes to the outside of the flashing. This is required, otherwise any draining moisture of condensation will drain behind the flashing and into the wall frame or onto the timber reveal of the window.

Cladding installation

The installation of wall cladding is governed by Part 3.5 of Volume Two of the NCC. More specifically, by clause 3.5.3. There are two acceptable construction manuals listed in this part: AS 1562.1 for metal claddings; and AS 5146.1 for Autoclaved aerated concrete (such as Hebel products).

Working through these standards and the NCC is similar to previous sections – seeking the key areas of compliance for any given cladding material. Demonstration of this will be by example, borrowing off the plan set found in Figures 13.13 to 13.17 of this chapter.

EXAMPLE 13.14

RESPONDING TO THE NCC

In this example, the cladding is to be timber weatherboards. No definition of these boards is given in the plans provided, so square-edged splayed Baltic pine boards will be assumed.

Information required:

- wind speed – N3
- type of timber cladding – Baltic pine, splayed boards
- governing clause: 3.5.3.2 Timber cladding.

In this example, the key factors are the wind classification and the shape and material of the weatherboard. The wind classification is important because the clause is only applicable up to N3, after which no design manuals have been listed and so fixings become dependent upon manufacturers' recommendations.

The ***timber species*** is important as this determines the amount of overlap required at each board based upon Clause 3.5.3.2 (a). In this case, the requirement for Baltic pine is 25 mm.

The shape or profile of the boards is important as this governs the arrangement of the boards on the wall and the fixing locations; in this case by Figure 3.5.4.1 (a) (ii) and (a) (iii), as shown in Figure 13.60.

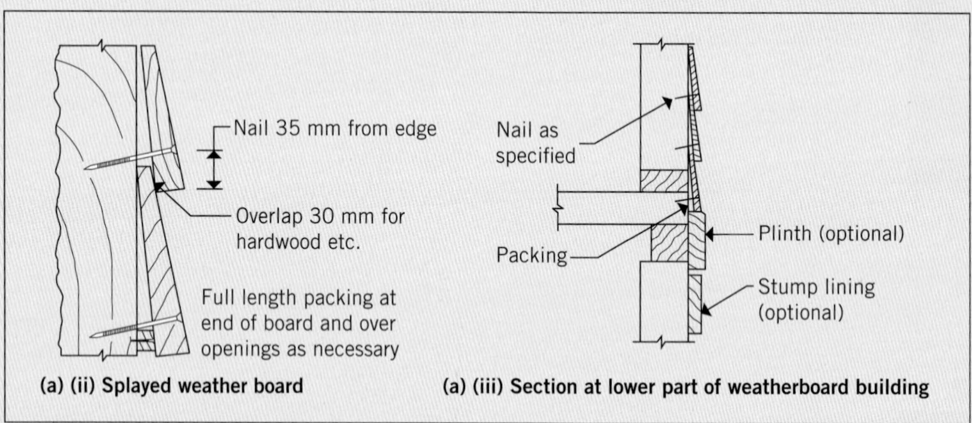

FIGURE 13.60 Weatherboard fixing requirements

The main checks taking place to ensure compliance are:

- that the nails do not penetrate the upper thin edge of the lower weatherboards – this constricts movement and causes the boards to split, meaning either inadequate effective cover, or water penetration through the face of the board
- that flashings as determined in the previous section have been correctly installed under or over the weatherboards and penetrations (such as windows and doors)
- cut ends of weather boards have been painted and sealed
- nails in ends of boards have been drilled or nailed such that no splitting has occurred
- the correct size and type of fixings are used; i.e. 50-mm × 3.15-mm galvanised bullet-head.

While there are no particular stipulations within the NCC or elsewhere regarding vertical flashing to internal and external wall corners, it is industry best practice to install either PVC or aluminium flashing over the sarking behind weatherboards in these locations.

Proprietary cladding systems: manufacturer's specifications

In the case of proprietary cladding systems, the manufacturer's specifications must be sourced and followed. These documents can be of immense detail and offer specific required solutions to particular framing and cladding contexts. These documents may also determine the requirements for flashings at penetrations, at corners and even behind individual joints.

Foam and autoclaved aerated concrete products will likewise be governed by their manufacturer's specifications, which often require specific batten types, specialised fixings and final coating procedures. In these cases, particular attention must be taken where provisions for expansion is stipulated.

Cladding as bracing: structural performance

Cladding, particularly structural plywood, is occasionally used as the main bracing against wind and other lateral loads. When cladding is used for bracing it must comply with either the NCC, Volume Two, Clause 3.5.3.4 and Table 3.5.3.4; AS 1684; and/or the manufacturer's specifications.

Cladding to AS 1684

AS 1684.2 shall be used in this discussion to allow specific clauses and diagrams to be discussed; the basic principles applying equally to AS 1684.3.

Volume Two of the NCC provides only very basic commentary on the use of cladding as bracing. And then only in regards to structural plywood. AS 1684.2 is therefore the main source of guidance in this matter. Table 8.18 (e) provides for the use of timber cladding with a minimum thickness of 12 mm and applied diagonally – at between 40° and 50° – to be regarded as a bracing unit. Dependent upon the nailing pattern used, each 2.1 m of wall is considered to have a raking resistance up to 3 kN/m (i.e. 6.3 kN).

Various plywoods, including decorative sheets, are also offered structural bracing performance via this table provisional on certified stress ratings of not less than F11. The specific nailing and tie-down procedures for these units are clearly identified in the table and their selection is the same as dealt with in the fifth section of this chapter. Likewise, the determination of required bracing is the same for cladding as bracing as it is for all other structural bracing. Compliance is therefore mapped in the same manner as in the fifth section of this chapter.

Proprietary cladding systems: manufacturer's specifications

Where a proprietary cladding system is used, such as fibre cement sheeting or the like, you must follow the manufacturer's specifications precisely. These specifications will reference the NCC and/or AS 1684 as required, but will be based upon their own testing in compliance with the verification methods stipulated for Performance Solutions.

A useful example manufacturer's installation guide may be found at: https://www3.jameshardie.com.au/uploads/file/20190116b65eb9-JH_Structural_Bracing_Application_Guide_October%202018%20V3.pdf

In the commercial sector, claddings, or facades as they are more commonly known, can vary markedly, and are seldom part of bracing unless forming an exterior shear wall element. These include site-poured tilt-up concrete panels, pre-tensioned panels manufactured off site, cross-laminated timber (CLT) panels, a vast array of glazed unitised systems, metal 'cassettes' (small panels individually fixed to the self-braced structure) as well as more stylised applications of timber, plastics and solid steel.

LEARNING TASK 13.6

WHAT MUST I CONSIDER WHEN CLADDING MY BUILDING?

1 Refer to the NCC, Vol. 2, and determine if autoclaved aerated concrete (AAC) is referenced. If referenced, provide the clause number for any acceptable construction manuals.
   - Not referenced
   - Yes referenced – Clause number is ....

2 Provide a detailed sketch and fixing details of a section through a timber-framed residential building external wall using autoclaved aerated concrete panels as the cladding material.

3 Refer to NCC 2016 and identify the requirements for window flashings to wall openings. Identify and specify the deemed-to-satisfy NCC clause number/s that deal with the specific requirements of window flashing for timber-framed windows in timber-framed walls clad with fibre cement sheet or weatherboard.
   a Does the most current NCC 2019 contain the same information?
   b If not, specify the technical requirements for detailing now required at wall openings.

4 Specify the NCC clause that details the requirements of how close to any external ground that cladding materials are able to extend.

SUMMARY

This chapter sought to develop your understanding of the application of structural principles to both residential and commercial buildings, in so doing covering all NCC building classifications up to two storeys in height and within the limitations of Type C. In many instances, it will have been noted that there are similarities across building classes, where the structural elements do not vary significantly. In other cases, there are structural forms common to commercial buildings that, although not completely foreign to the domestic residential sector, are seldom found in other than very alternative architectural designs. Examples of these included the extensive use of engineered timbers such as CLT, and prestressing systems for suspended concrete floors.

The most notable difference between standard timber or steel-frame housing and larger commercial works is the extended need for engineering advice. In the domestic sector, the likes of AS1684 and AS 2870 see you through most areas of your work. In the commercial sector, however, these no longer hold and a skilled and experienced engineer is invaluable.

REFERENCES AND FURTHER READING

Australian Building Codes Board 2019, *National Construction Code 2019 Volume 1, Building Code of Australia*, Australian Building Codes Board, Canberra, ACT.

Australian Building Codes Board 2019, *National Construction Code 2019 Volume 2, Building Code of Australia*, Australian Building Codes Board, Canberra, ACT.

National Association of Steel-Framed Housing 2014, *NASH Standard – Steel Framed Construction in Bushfire Areas, National Association of Steel-Framed Housing*, Hartwell, Vic.

Standards Australia 2001, AS 2601 Demolition of structures, Standards Australia, NSW.

Standards Australia 2010, AS 1684.2 Residential timber-framed construction – non-cyclonic areas, Standards Australia, NSW.

Standards Australia 2011, AS 2870 Residential slabs and footings, Standards Australia, NSW.

Standards Australia 2011, AS/NZS 1170.2 (R2016) Structural design actions – wind actions, Standards Australia, NSW.

Standards Australia 2011, AS 2870 Residential slabs and footings, Standards Australia, NSW.

Standards Australia 2012, AS 4055 Wind loads for housing, Standards Australia, NSW.

Standards Australia 2018, AS 3600 Concrete structures, Standards Australia, NSW.

Standards Australia 2018, AS 3959 Construction of buildings in bushfire-prone areas, Standards Australia, NSW.

PART 5

SUSTAINABILITY

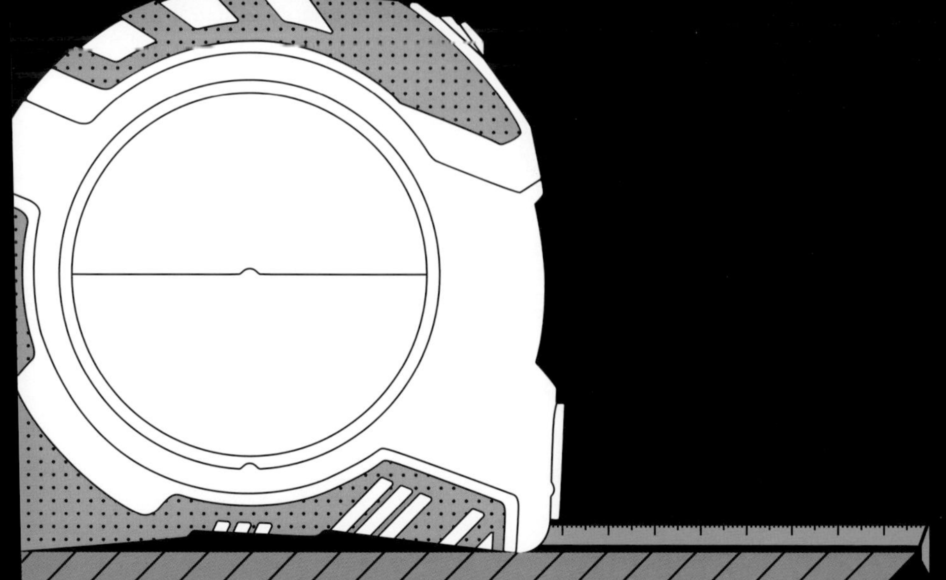

14 THERMAL EFFICIENCY AND SUSTAINABILITY

Chapter overview

The conservation of energy has become of critical importance to the sustainability of communities in this country and globally. Historically, it probably always was important, yet it is only now, as our population grows and becomes more demanding of energy, that we have been confronted with it as a factor in our survival as a species. The purpose of this chapter is therefore to provide you with the necessary tools and knowledge to engage informatively with designers and clients in the development and construction of more energy-efficient structures. In addition, the chapter provides guidance on material, design and construction choices based upon principles of sustainability: energy efficiency and sustainability being two different, not always parallel, construction concepts.

This chapter takes a 'best-practice' perspective of the engagement between yourself (the builder), the client and the designer – be it an individual designer or an architectural team. Best practice means one of mutual respect and recognition of skill and knowledge, deployed to the best interest of the client and the project as a whole. Hence, your input on issues of energy efficiency and sustainability is as crucial to project development as it is to project construction.

Elements

This chapter provides knowledge and skill development materials on how to:

1. apply legislative and planning requirements for thermal efficiency to the building process
2. review design solutions for effectiveness and compliance
3. manage the building process to ensure an effective outcome.

The Australian standards, codes and other instruments applicable to energy efficient and sustainable structures are:

- the National Construction Code (NCC) 2019
- the Nationwide House Energy Rating (NatHER) scheme
- Green Star rating system
- Window Energy Rating Scheme (WERS)
- Water Efficiency Labelling and Standards (WELS)
- Smart Approved WaterMark (WaterMark)

In addition, SAI Global have published a 'Guide to Standards – Energy Efficiency', which lists a large range of standards that have a bearing upon ensuring the construction and/or maintenance of energy-efficient structures. This document may be freely downloaded from the website listed at the end of this chapter.

Introduction

Energy supply and demand are critical economic concerns at the forefront of many nations' policy making. This is because supply via existing infrastructure is unable to keep up with ever-increasing demands. Despite a significant shift in recent years to renewable sources, and a global reduced reliance upon coal and oil, Australia's main energy resources remain non-renewable, with 37% by oil, 32% by coal, 25% by natural gas, and only 6% by renewables (wind/sun). This means one or more of following three things must occur:

- The ongoing maintenance of existing, and the creation of new, electrical generation plants. In Australia, this is generally hydro and associated dams, or coal and associated mining; but can include alternative sources such as wind farms, geothermal, tidal and solar plants.
- The development of mini grid, or domestic infeed systems, whereby power is generated through alternative sources such as roof-top solar, mini hydro plants, localised wind and geothermal systems.
- A reduction of power consumption by the community at large.

Both residential and commercial buildings have shown themselves to be massive drains upon energy over their life cycle: approximating 40% of the total energy use of nations such as Australia when transportation costs are included (https://www.usgbc.org/articles/green-building-101-why-energy-efficiency-important). Housing particularly comes into the spotlight given rapidly increasing home sizes – an estimated increase in floor area of 280% from 1986 to 2020. This means that any small reduction in energy usage per square metre through improved building shell efficiency is outpaced by increasing floor area; i.e. domestic housing energy consumption is rapidly rising. Coupled with the number of occupants per household reducing, the energy consumption per capita is also rising significantly. (See https://www.usgbc.org/articles/green-building-101-why-energy-efficiency-important; also, p. 7, executive summary of: https://www.energy.gov.au/sites/g/files/net3411/f/energy-update-report-2017.pdf.)

Changing climates, lifestyles and perceptions of acceptable living standards have also increased energy usage loads.

- Changing climate has meant that, except for the tropical north, all parts of Australia have undergone an increase in cooling loads and a decrease in heating: with cooling being the bigger energy drain. It is expected that this trend will continue to rise, giving grave concern due to its tendency to disrupt power supply during peak demand periods.
- Changed perceptions of living standards has led to whole of house heating strategies, with most being natural gas dependent; i.e. increased energy consumption per household, despite the increase efficiency of the systems themselves.
- Changing lifestyles have brought about a rapidly increasing demand for high-load entertainment items such as computers, large flat-screen TVs and audio equipment. In addition, working from home via this technology has increased, furthering the rise in apparent domestic home energy consumption.

Yet, depending upon the type of structure, the energy efficiency rating achieved and its location, the energy consumed in a building's construction – based upon the concept of **embodied energy** – will generally exceed its designed usage consumption over a 50-year period.

This means that getting it 'right' is not as simple as just putting in triple-glazed windows and lots of insulation batts; i.e. there can be a significant difference between a highly energy-efficient building and a sustainable one. This has led to a rise in the expectation by governments, councils and clients to construct not only more energy-efficient structures with regards to end usage, but also to a demand for 'cleverer' construction methods, and wiser material choices: sustainable material choices. From a government perspective, local or otherwise, this expectation is reflected in tightened standards, codes and legislation.

Apply legislative and planning requirements for thermal efficiency

The National Construction Code of Australia (NCC) is the primary document for implementing energy-efficient building practices in this nation. However, its authority is drawn from individual state or territory legislation. While this has worked reasonably well since the NCC's inception, from an **energy efficiency** and **sustainability** perspective it has been somewhat fragmented in application, and limited in success.

In 2015 the Council of Australian Governments (COAG) Energy Council developed the National Energy Productivity Plan (NEPP) providing a framework for achieving a 40% improvement in **energy productivity** by 2030. Energy productivity is defined as the economic return, as 'gross domestic product' or GDP in millions of dollars, divided by the total energy used by the economy to produce it. That is:

Energy productivity = economic output/energy used = GDP/petajoules where a petajoule is a measure of power that crudely equates to running 50,000 Australian homes with four occupants for a year (based on each home consuming about 5750 kWh annually).

Working with industry and the Sustainable Built Environment Council, COAG has refined a number of key points within the NEPP that affect the building and construction sector. These apply across both residential and commercial buildings and include:

- expanded and improved building energy rating schemes and disclosures with a national approach
- improvements in the energy performance and efficiency requirements of the NCC (both volumes)
- enhanced construction systems and methods
- enhanced skills.

The first instances of these changes have been implementation in the 2019 revision of the NCC.

It is to this code we now turn. The NCC has been dealt with in some depth in Chapter 1 Building codes and standards. That which follows revises some of this information and goes a fraction deeper into the relevant energy-efficiency parts. In addition, we shall look at how local climatic conditions and client requirements can challenge a design's capacity to perform, and how you must account for these within the parameters of both the NCC and the Australian standards.

State, territory and council requirements

Each state and territory of Australia has its own Acts and associated regulations that call up the National Construction Code (NCC) or Building Code of Australia (BCA) in a variety of ways. In the Victorian Building Regulations, for example, Division 2 is titled Building Code of Australia and is devoted to ensuring all elements of the BCA must be complied with in lengthy and explicit terms. In NSW, however, neither the Building Act nor the Regulations mention the NCC or BCA at all. Instead they are found in the Environmental Planning and Assessment Regulations; and only sporadically call up the BCA through clauses such as Clause 98, Division 8A of Part 6 and similarly obscure sections.

It is via such legislative paths that the NCC governs acceptable design and construction practices regarding energy efficiency and sustainability; doing so through:

- Section J of Volume One (p. 334, NCC, Vol. One, 2016)
- Parts 2.6 (p. 66) and 3.12 of Volume Two (p. 359)

Both Volume One and Volume Two of the NCC introduce a Nationwide House Energy Rating Scheme known as NatHERS, a computer software assessment tool developed by the CSIRO for use by certified practitioners. With minor variations, these sections apply to all states and territories except NSW. NSW energy efficiency is addressed instead through:

- NSW Appendix, Section J, Volume One (p. 453)
- NSW Addition, NSW 2 Part 2.6 of Volume Two (p. 431).

These sections and parts introduce the NSW-specific verification method known as BASIX. Both BASIX and NatHERS will be discussed in detail in later parts of this chapter.

Local councils are also heavily involved in the attempt to ensure their communities, and hence the buildings in which people live, work and play, are more energy efficient. Part of their role is to ensure, along with private building certifiers, that the NCC's requirements are met. However, they are also involved through local planning acts that may require suburban developers to ensure housing blocks are shaped and oriented in ways that ensure good solar access. Councils may also require buildings to have a certain amount of renewable energy supply (solar, wind, mini hydro, geothermal) and/or be equipped with voltage power optimisation units (VPOs). VPOs reduce high voltage, thereby saving energy and increasing the life span of electric motors and electrical equipment.

6 stars: the NCC and regional climate variation

The NCC address energy efficiency differently depending upon the classification of a building. In addition, it offers more than one path by which the requirements of the Code may be met. In most cases, state and territory variations are minor, but specific – so you must always check the application of any clause for relevance to your location. The paths available may be grouped into those applicable to residential buildings (and parts of buildings) and commercial structures.

Class 1 and Class 2 buildings, and Class 4 parts of buildings

In most states and territories there are four paths by which you can address the energy-efficiency requirements of the NCC for Class 1 buildings, the sole-occupancy units of Class 2 buildings and a Class 4 parts of other buildings. These are outlined in Volume Two of the NCC for Classes 1 and 2 buildings, and in Volume One for the Class 4 parts of other buildings. They include:

- the deemed-to-satisfy solutions as described in the NCC [3.12.0 (a)(ii)]
- a Performance Solution that meets the Performance Requirements of the NCC
- using housing energy rating software approved by the NCC; i.e. NatHERS [3.12.0 (a)(i)]
- using a reference building (V2.6.2.2).

Note: NatHERS only addresses the energy performance requirements *in part*. It must be used in conjunction with specific clauses of Part 3.12 (see flow chart NCC, Vol. Two, p. 363).

The first three of these paths award an energy rating based on stars. Star ratings run from 1–10, the higher the better. Currently all states and territories, except NSW, require a 6-star rating for residential buildings (with some minor variations based upon structural inclusions and locations). Compliance with the DTS provisions is deemed to equate to a 6-star rating, as is the development of an acceptable Performance Solution.

In NSW, Class 1, 2, the Class 4 parts of other buildings and even some Class 10 buildings are subject to the web-based sustainability indexing tool known as BASIX. This also applies to alterations and additions to these classes of buildings valued $50,000 or more and/ or include a pool or spa with a volume greater than 40,000 litres.

Class 3 and Class 5–9 buildings

The energy-efficiency requirements for these structures are held in Volume One of the NCC. For these classes of buildings there are three paths available:

- the deemed-to-satisfy solutions as described in the NCC
- a Performance Solution that meets the Performance Requirements of the NCC
- using a reference building.

In the 2019 edition it is proposed that compliance may be gained through two other paths depending upon class:

- For Class 5 buildings – JV1 NABERS Energy assessment for offices with a minimum 5.5 star rating. NABERS (National Australian Built Environment Rating System) is a system that ranks a building's energy efficiency, water usage, waste management and indoor environment quality, as well as its impact on the environment through greenhouse gas emissions, on a 6-star scale (6 being best). This is a partial compliance, requiring adherence to additional parts of the specification JVa.
- For Classes 3, 5–9 Buildings – JV2 **Green Star** compliance based upon set parameters for greenhouse gas emissions and thermal comfort. This is a partial compliance, requiring adherence to additional parts of the specification JVa.

Using the deemed-to-satisfy (DTS) and Performance Solutions have been discussed in some length in Chapter 1 of this book and you are directed to that chapter for detailed commentary. The use of a reference building, however, has not been discussed. As this method is open to all classes of buildings, and is specific to energy-efficiency performance, it needs some explanation before expanding upon NatHERS and BASIX.

Using a reference building

Within the NCC, a reference building is a hypothetical structure used to calculate the energy loads for the proposed building. The hypothetical building is provided with the various insulation levels, solar gains and energy losses via glazed elements, and the required temperature settings for internal spaces that match the proposed building and its climate zone. This is an approved verification method – under V2.6.2.2 of Volume Two for houses (p. 68), and JV3 of Volume One for other classes – designed to allow alternative building techniques that would otherwise fall outside of the scope of the NCC to be assessed. No star rating is provided under this assessment, it is pass or fail.

The reference building is modelled using building energy analysis software that is compliant with the NCC protocols. It is a complex procedure that costs much more to undertake than the other compliance paths available. However, particularly within the housing sector, this is a path that has gained a significant proportion of current housing approvals, to the concern of the CSIRO, federal, state and local governments, and the Australian Building Codes Board (ABCB) themselves.

The reason for the concern is that when used incorrectly, with the wrong software analysis tools and the wrong inputs, it is possible to 'pass' buildings with significantly less insulation values. Some homes passed using this system have been estimated to have a star rating less than 3, thereby costing the client less up front, but more in energy bills over the long term. It also means the purpose of home energy usage reduction is lost, requiring increased energy generation infrastructure by governments.

When using this approach for housing, it is therefore critical that you ensure those conducting the analysis for you are following the NCC protocols. For commercial structures using JV3, it is particularly complex as it is the energy used by services (heating, cooling) in response to the thermostat settings of buildings that is analysed. This differs from housing, where it is the energy gain and loss through the building's envelope that is assessed.

NatHERS

The Nationwide House Energy Rating Scheme (NatHERS) is a software-based energy-efficiency assessment tool for residential buildings developed by the CSIRO. The analysis is based upon the capacity of the building envelope (the outside structure including doors, windows, walls, flooring and ceiling/roof) to restrict energy flows; e.g. how much heat is gained or lost through a window and its frame, for example.

The software is used to model the house design in both structure and layout, and then estimate the amount of energy needed for heating and cooling.

In doing so, it is taking account of multiple (hundreds) of factors, including the construction types (brick veneer, timber or steel frame, and the like), the climate zone and the building's orientation. The outcome is a rating of 'stars'. This star rating is a nationally applicable system (excepting NSW) with zero (0) stars being the worst case, and 10 stars being a home that is completely self-sufficient, requiring no heating or cooling at any time of the year for that climate zone. Though there are frequent challenges by designers and architects to produce a 10-star home, 6 stars is the minimum requirement for most climate zones throughout Australia as per the NCC. Many homes owners now seek 7- and 8-star rated homes on the basis that this will save them energy costs in the long term.

NatHERS software may only be used by those trained and qualified to do so. In addition, the software – of which there are three different versions currently available on the market – only partially addresses the energy performance requirements of the NCC. It must be used in conjunction with, and the building designed in reference to, specific clauses of Part 3.12 (see flow chart NCC, Vol. Two, p. 363).

BASIX

BASIX is the NSW web-based software developed to ensure that not only energy efficiency and thermal comfort is designed into the home, but also water conservation measures. This tool is required to be used to assess all Class 1, Class 2 and the Class 4 parts of other buildings classes in NSW. In consultation with local councils, it will also be required for Class 10 buildings if converting them to a habitable space. If making alterations or additions costing more than $50,000 to existing buildings of these classes, or if these changes include a pool or spa with a volume greater than 40,000 litres, then BASIX again applies.

Unlike NatHERS software, BASIX may be completed online by anyone. No formal training is required. At the same time, however, it is a moderately complex tool with a large number of inputs required. Upon completion and payment of a fee a certificate is generated – assuming that the inputs have been approved by the software analysis. The fee varies by project scale and type. This certificate is then presented with the project documentation to council or a private certifier as part of any application for a development.

Although BASIX can be used as a complete certifying tool in itself, the Thermal Comfort Index offers two paths of assessment: DIY and Simulation. The first path, DIY, may be completed within the BASIX program; Simulation, on the other hand, must be undertaken by an accredited assessor using accredited software. This will be discussed further in the second section of this chapter.

Regional climate implications

Australia is a vast land mass with a highly varied climate: change location, and heating and cooling requirements may change dramatically. The energy efficiency DTS provisions within the NCC reflect this through changed requirements based on temperature, humidity and solar access variables. In so doing, eight zones have been developed around areas with similar climates, even though they may be separated by vast distances. Theses climate zones are as follows:

- Climate zone 1 – High humidity summer, warm winter
- Climate zone 2 – Warm humid summer, mild winter
- Climate zone 3 – Hot dry summer, warm winter
- Climate zone 4 – Hot dry summer, cool winter
- Climate zone 5 – Warm temperate
- Climate zone 6 – Mild temperate
- Climate zone 7 – Cool temperate
- Climate zone 8 – Alpine

The eight climate zones are illustrated within the NCC by way of a map created using Bureau of Meteorology data (see Figure 14.1). The zone boundaries align with local government areas and so may change over time.

Detailed maps are available online through the ABCB website for each state and territory. Being high resolution, you can zoom in on each area with a high degree of precision. In addition, you can obtain your regional climate zone from the local council.

In the hotter zones such as 1, 2 and 3, the NCC requirements tend towards limiting the cooling input loads required. In the cooler zones of 7 and 8 the focus is on reducing the heating loads needed to maintain comfort. In the zones regarded as temperate, heating and cooling will both be required over a year; keeping a balance between solar heat gain in winter, and reduced gain in summer is therefore the challenge to good design in these areas. The term 'design response' is one that has great value when evaluating a construction proposal in each of these areas.

The Australian government has produced a significant text that covers the key elements of such responses for each climate zone in great detail. This may be viewed online, or downloaded free as a PDF file from: http://www.yourhome.gov.au/passive-design/design-climate

You are advised to download and review the design responses for each climate zone. The design advice for each zone is offered in the following format:

- the main characteristics of the zone
  - temperature range
  - seasonal variation (based on four seasons: summer, autumn, winter, spring)
  - humidity

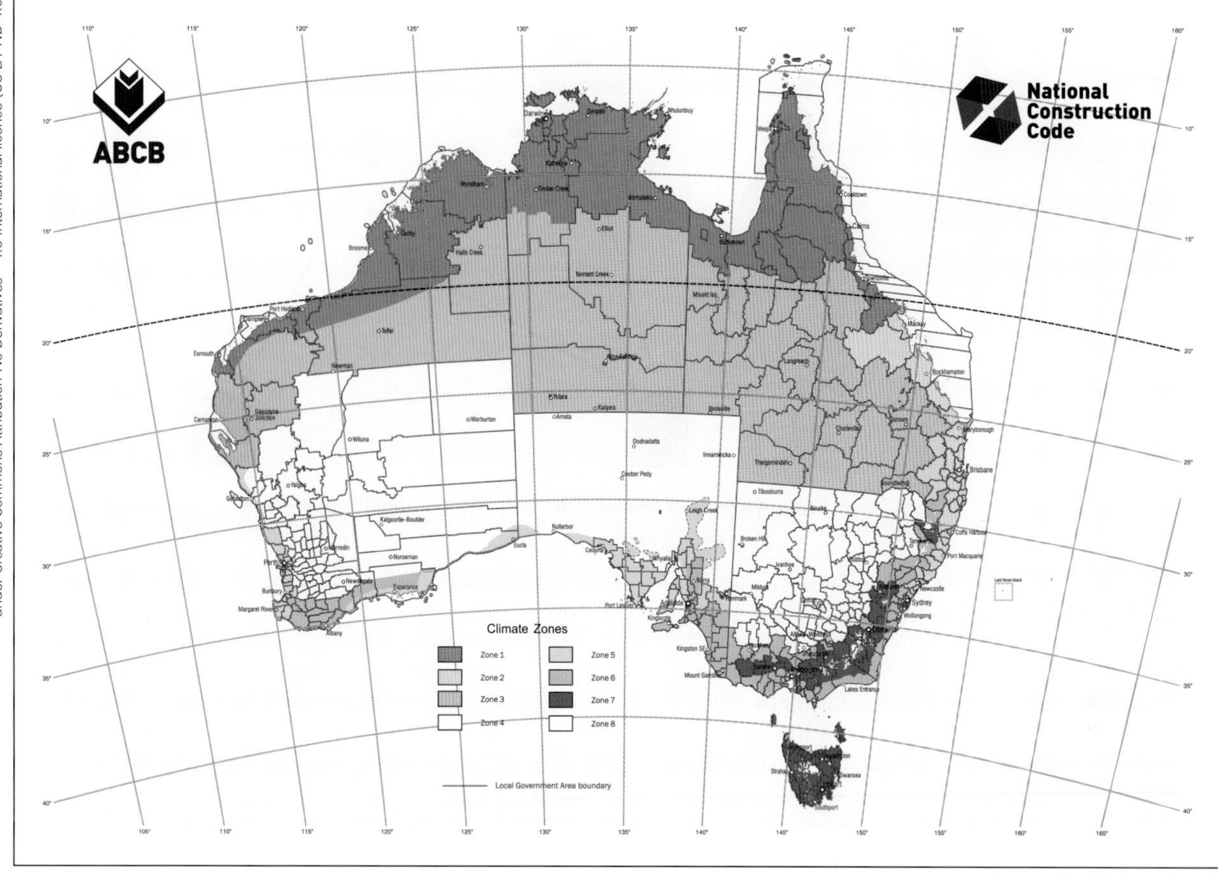

FIGURE 14.1 Climate zone map Australia-wide

- key design objectives
- key design responses and considerations including:
  - windows and shading
  - insulation
  - heating and cooling
  - approaches to design
  - construction systems.

In reviewing the material, keep in mind that which is not stated: that buildings in Zone 1 and much of Zones 2 and 3 are situated above the Tropic of Capricorn. This means that in summer and, depending how north the building is sited, some of spring and autumn, the sun is to the south of these buildings. The rest of the year the sun will traverse northly much as the rest of the country will experience it, only higher (see Figure 14.2).

Other complications arise when acknowledging the implications of ongoing climate change. This means that a zone's characteristics are likely to change over the average home's life span of 50 years. Further, climate zones alter the expected energy use of a home in a manner that the NatHERS energy rating system, indeed all energy rating systems, do not account for. A 6-star house in a northern location, like Cooktown or Cairns, for example, will use significantly more energy to remain comfortable than a home rated identically in Sydney; i.e. the star ratings are climate-zone specific.

Identifying and negotiating client needs and expectations

What a client needs, what they want and what they expect from what is delivered can vary greatly. This author's experience, supported by personal interaction with builders and architects from around Australia, suggests that the average client is still not well informed on energy-efficiency and/or sustainability when it comes to building designs; and yet it is the client that has the final say. Education is therefore part of your role when engaging with your client on potential design implications.

At the same time, you will find that it is necessary to be flexible in how a design might cater for the client's needs, while still reflecting requirements of the codes and standards. The common example is when a building site has magnificent views in totally the wrong direction from an energy-efficiency perspective. All the views may be to the south and west and the client and/or designer wish to capture these views. It would be ridiculous to tell the client to forget the view and focus upon traditional energy-efficiency orientation – north and east. Instead, you would explore the options that could capture the views, and yet reduce thermal losses through the

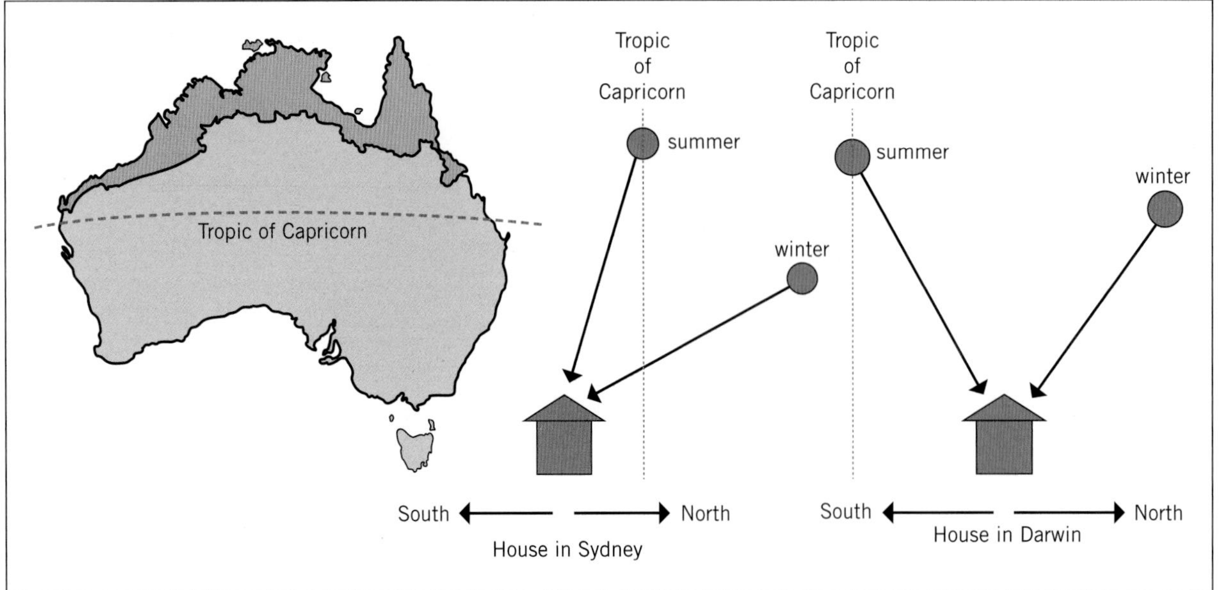

FIGURE 14.2 The Tropic of Capricorn and implications for solar access

windows; i.e. double or triple glazing, increased thermal efficiencies in other areas, northern light capture from clerestory windows, and the like. This may apply to office spaces, hospitals and store fronts as much to domestic residences.

Who the client is also needs to be considered. In some instances, the 'client' may be the designer or architect, in which case your input may be called upon with regards to material options and structural issues, as well as your interpretation of the codes and standards that must apply. On other occasions the client may be the end user, developer, or homeowner. Again, it is your expertise in codes, standards, materials, components and structure that will be called upon – as well as your experiential knowledge of energy-efficient building practices, particularly with regards to cost.

Common client requirements

While from an energy-efficiency and sustainability perspective, the informed client remains scarce, those seeking an independently designed and built home do tend to be more informed and have higher expectations than those buying one 'off the shelf'. Commonly, you will find that they will approach you with a sketch plan or design brief informed by:

- an examination of their current home or business premises, lifestyle and/or business practices and needs
- an understanding of their preferred lifestyle and hence how they would like the home to function
- a block or building site with the views they desire to capture, or a building they seek to renovate, adapt, alter, or extend
- a budget
- some perception of energy efficiency (sometimes quite highly informed), preferred materials and building methods
- a style, frequently a modern one, reflective of energy efficiency and therefore a 'green' outlook – be it a business, home or retail property.

Of all of these, the hardest one to satisfy is frequently that of budget. Chances are they have chosen you because word of mouth, your website, or past projects have led them to believe that you are capable of realising their vision of the desired premises. Unbeknown to them, your real strength is in that key word in the heading to this section: negotiating. Negotiating the needs of the client, in partnership with the designer or architect, with the requirements of the NCC, building practice, material availability and capabilities (with your insights on loads and spans), and Australian standards. Most importantly, your appreciation of costs and time.

Client expectations

In end use, the client expects that the structure you have created will run on the smell of a burnt battery. This is not uncommon, given the time and cost that they have put into their new building. The reality is unlikely to be quite that efficient. That is the other part of your negotiation in the planning and building process, particularly when it comes to variations of the plan part-way into the project.

You and the designer (if one is in play) will have developed a sound knowledge of the capabilities of the structure from an energy perspective as the design developed. You must be realistic when informing the client as to what the building is capable of, and the implications of any changes made part-way into the build. From an energy-assessment perspective, a change to a window, room size, use, shading, or construction material will frequently call up a new assessment. In BASIX this is a fairly simple process; through DTS, Performance, or even NatHERS, it can be somewhat more protracted.

In addition, you need to inform the client of what they need to do to ensure the building performs to expectation. It is somewhat strange that in the vast majority of cases, homeowners particularly, but also owners of business premises, are not provided with a 'users guide'. The money and keys change hands, but little information is passed to the client on how they should use the building to best effect.

Advising the client of end-use practice and expectations is also important in reducing call backs for things you know are going to happen. Natural timber floors, for example, when exposed to intense sunlight – which may be part of the energy efficiency plan – will move and probably open up to some degree. You know this, your client needs to know this. They need to know when to open windows, when to close them; the cost of running various lighting elements and other power drains. All this they need to learn if their home is to perform as they, and you, intended.

Expert consultation

Though the client will often come to you as the expert in all things construction, you need to be wise enough to realise your limitations. There are a variety of industry experts to whom you may turn for advice on energy efficiency, even in areas where you feel that you 'should' be the expert. This advice may come from a range of government, professional, industry, commercial and trade areas. What you must accept is that your expertise is in knowing who to ask, and what to ask, rather than in answering the question.

The main areas for which you will need to source some level of expert opinion, solution options, or advice – which in turn may be a redirection to another possible source – may be derived from the design responses to the specific climate zone in which the project is sited. As such they may include, but are not limited to:

- initial design analysis
- alternative materials
- footing designs and subfloor venting
- heating, cooling, and ducting options
- window materials and design
- shading options and design
- insulation
- flooring materials
- specific engineering solutions (for non-standard building solutions)
- bushfire responses that also respond to the energy-efficiency response
- lighting solutions
- landscape design and plant/tree selection when incorporated into the design response
- termite prevention when design response increases risk through ground contact
- condensation control.

To these must of course be added the need for energy auditing and rating. As has been already discussed, in all states excepting NSW, energy auditing must be undertaken by practitioners trained in the use of NatHERS approved software. In addition, that person cannot be the designer of the structure. That is, even if the designer is qualified as a NatHERS accredited assessor, it must be assessed by someone else.

In turning for advice, the internet is loaded with options and advice on every aspect of energy efficient materials, designs and engineering solutions. Frequently there is too much of it, thus it is not always easy to make it suit your specific design response. This is not to discount options identified in this manner; however, it is best if you run the ideas past those who may have already tested their feasibility. This is where first-hand expert knowledge and advice comes in.

However, who is an expert in any given field is not always easy to identify. Often the first advice you need is who to ask.

Industry groups

Members of industry groups such as the HIA or MBA have an advantage here in that these organisations hold a vast network of members, as well as links to other advisory networks, professionals, and representatives from other industries, as well as manufacturers and suppliers. As a member or not, you will need to replicate this network in your own area. That is, it is to manufacturers and suppliers whom you might turn, additional to specific professionals such as designers, architects and engineers.

In so doing, don't forget some of the best sources of experientially-based advice: other builders, subcontractors and trades people. From such sources you may not get specific expert advice, but you may well gain knowledge developed from valuable hands-on experience; particularly regarding timelines and labour costings, aside from what works, and what does not in your specific area.

Construction implications: Australian standards

As you know from your readings so far, Australian standards cover all aspects of a construction – be it a residential, commercial, or industrial structure. With regards to energy efficiency and sustainability there are no specific standards to address. However, there remain all the standards that apply to all the materials and components of any structure.

As the builder, your initial concerns will be with the obvious framing, footing and other structural standards that will always need to be addressed as per the NCC requirements. From an energy-efficiency perspective, however, you may find that certain aspects of these structural elements change. Framing may need to be designed differently to those approaches that you have normally used. The footings and slab may be similar, but point loads may differ, as might thickness of the slab as part of a thermal mass calculation by the designer, rather than a load-bearing capacity by the engineer. It is these subtleties that you, as the builder concerned with the ultimate compliance of the building, must check and advise on as the design proceeds.

Other key areas, most of which you will need expert advice on, may be:

- glazing and window frames
- proposed insulation types
- alternative claddings
- novel materials
- alternative building techniques
- bushfire survivability
- termite prevention.

Glazing and insulation will be dealt with in depth later in this chapter, the others we will touch on briefly here.

Alternative claddings

Many claddings that were once considered 'alternative' are now if not common, at least not uncommon, to contemporary Australian residential homes and commercial properties. Materials such as: rendered Styrofoam or fibre cement sheet, Alucobond and other lightweight cored aluminium sandwich panels (known as aluminium composite materials or ACMs), decorative ply sheeting, structural insulate panels (SIPs), plastic weatherboards and many more are now part of both our commercial and residential landscapes.

The problem for you, as the builder, is you must ensure that the material meets the applicable Australian standards. They may be perfect from an energy-efficiency perspective, have a low embodied energy or carbon footprint and even be low maintenance. But they must still satisfy the standards, both as a material, and in application. The two worst-case examples of getting this wrong are the Grenfell Tower in the United Kingdom, and the Lacrosse tower in Docklands, Melbourne. In both cases it was non-compliant 'use' of ACM cladding that allowed the fires to spread so quickly, rather than the materials themselves being non-compliant at the time of installation. These are the sort of issues you must keep in mind.

Most plastic and styrene or ethylene products will burn despite retardants; or if not burn – and thereby promote or maintain a fire – will melt. In either case, they release extremely large quantities of toxins, so care must be taken in both their selection and use. Given the increasing quantity and types of construction materials entering the country, you are strongly advised to check the 'credentials' of any product with the importer. If still not satisfied, low-cost sample analysis for substances such as asbestos are easily obtained from numerous NATA (National Association of Testing Authorities) accredited organisations. These may be found by calling or via the website of your state or territory WHS authority.

Identifying and accessing the appropriate standard has been covered in depth in Chapter 1; however, some of the key standards for claddings are listed below:

- AS 1562.1 – 2018 Design and installation of sheet roof and wall cladding – metal
- AS/NZS 1562.2 – 1999 Design and installation of sheet roof and wall cladding – corrugated fibre-reinforced cement
- AS 1562.3 – 2006 Design and installation of sheet roof and wall cladding – plastic
- AS 4040.2 – 1992 (R2016) Methods of testing sheet roof and wall cladding – resistance to wind pressures for non-cyclone regions
- AS 4256.2 – 2006 Plastic roof and wall cladding materials – unplasticized polyvinyl chloride (uPVC) building sheets
- AS 4256.3 – 2006 Plastic roof and wall cladding materials – glass fibre reinforced polyester (GRP)
- SA HB 39 – 2015 Installation code for metal roof and wall cladding.

Most of these cover installation, except AS 4040.2. This is just one of many standards that outline testing procedures. It is this type of standard that you will be seeking reference to when evaluating a material for compliance for use in Australia.

Novel materials

Aside from claddings, novel materials may be proposed to you for use as any construction component, including structural. Plastics, carbon fibre compounds, light-weight metal fittings, adhesives, fixings, engineered timber products, bamboo, low-carbon concretes, specialty glazing frames, glass and films, lighting and water conservation fittings will all make an appearance in high-end, energy-efficient buildings. Each must be assessed in the same manner as cladding, through a check of compliance to Australian standards testing procedures. Even blinds, curtains and external shading devices must comply with these standards.

Having identified that a component or material does comply you must also ensure that fixings, installation and, where applicable, spans and loads, are adequately addressed. This may be through the manufacturer, another standard or code (such as the SA HB 39 – 2015 for cladding – which is a guidance code or handbook, not a 'standard'), or by testing undertaken by yourself or commissioned by you.

As an example, a material that has yet to take off significantly in this country, but is becoming known in the energy-efficient and sustainability sector of construction, is cross-laminated timber or CLT. It's gaining traction with designers due to its light weight, capacity to span large distances and ability to be formed into wide sheets; unlike LVL (laminated veneered lumber), which can be long, but not overly wide relative to its thickness. Currently, however, there are no standards directly applicable to this material. Indirectly, the standards for LVL and associated approved adhesives can be used informatively, but not conclusively. It is to the manufacturers that you must therefore apply directly for evidence of suitability, capacities and fixing procedures – particularly in hold down.

Alternative building techniques

An alternative material may invoke an alternative construction method or technique. Or, common materials may be used in less common ways. These practices may fall outside the scope of the standards that you are used to using, or there may be no standard at all. For example, there are clear standards for the use of timber and steel in framing. AS 1684 has very standardised framing practices informing the engineering assumptions. Step outside of these practices and you have to tread with great care – from load bearing, hold down, and appropriate fixing perspectives. In some cases, you will still be able to use these standards, but you will have to read them with much greater care than perhaps you are used to.

On other occasions there will be no Australian standards or codes that govern the particular technique you will be employing. In these cases, the building is judged by the council based on the NCC Performance Solution approach; i.e. you will have to satisfy the council or relevant building authority (independent surveyor or group) that the structure is adequate. This means that it complies with all the performance requirements for a structure of that type and that location. This would need to include bushfire survivability and all other fire-resistance elements of the NCC.

Some of the more common 'alternative' building techniques are listed below, along with any standards, handbooks or guidance notes that may apply:

- **Rammed earth** – HB 195 – 2002 The Australian Earth Building Handbook
- **Mud brick** – HB 195 – 2002 The Australian Earth Building Handbook
- **Cob construction** – HB 195 – 2002 The Australian Earth Building Handbook
- **Strawbale construction** – no standards for this type of structure apply directly, but there is a wealth of practical material in the form of downloadable documents from such organisations as Ausbale
- **Insulated concrete forms** (ICFs) – again, no standards apply but there are multiple manufacturers in the country who can supply technical support.

There are numerous other building techniques out there, many that include off-site manufacture including structurally insulated panels (SIPs), pre- and post-tensioned concrete tilt-up panels and CLT or LVL components. Your task is to ensure that they comply with the standards, or that they can be shown to satisfy the NCC performance requirements.

Bushfire survivability

This is covered extensively in AS 3959, and given authority through the NCC. Having identified the applicable bushfire attack level (BAL) – via council overlay or by an experienced assessor – you must ensure that every element of your, possibly quite alternative, structure complies. This particularly applies to any exposed timber usage, gaps in decking, glazing elements and their frames.

The design of the structure must also reflect this code. For example, raised floors for airflow may need to be enclosed to prevent ember attack, windows may need steel shutters or screens; it will all depend upon the BAL rating imposed. Roof venting would also need to be carefully considered on the basis of ember attack.

Checking compliance with AS 3959 is not a particularly difficult task. Once a BAL rating has been identified you are directed to a particular section of the standard which outlines all the fire-resistance levels required for each element of the structure, design limitations, requirements and the like. But you, in partnership with the owner and design team (if such are involved), must be very thorough: the lives of the occupants, and possibly your business, depends upon it.

Termites

A three-part Australian standard applies to this element of construction design:

- AS 3660.1 Termite management – New building work
- AS 3660.2 Termite management – In and around existing buildings and structures
- AS 3660.3 Assessment criteria for termite management systems.

Alternative construction techniques may be challenged by some of the requirements of these standards, but as they are called up by the NCC, they must be strictly complied with. In regions where heat loss is the issue, then you may find the design calls for insulation of the concrete slab edge, for example.

This has high potential for undetected termite penetration of the main structure. You must find ways for preventing such access, which may be complicated by a resistance on your client's behalf to use chemical treatments; Figure 14.3 offers a simple example of how this may be achieved by use of an extended termite-proof flashing.

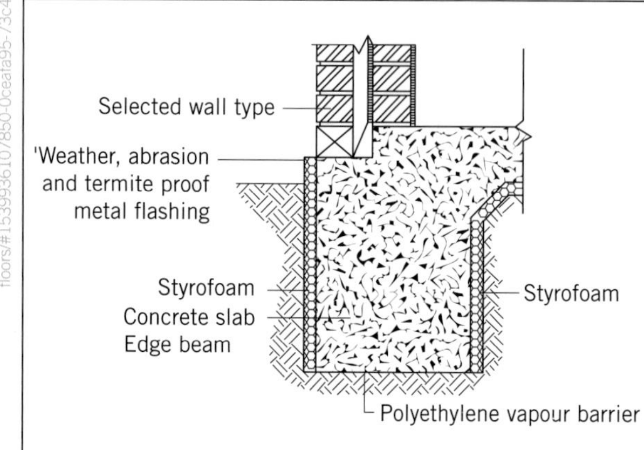

FIGURE 14.3 Insulation of floor slabs

Note: As with any standard, make sure you are accessing the most current version. In this case, only the 2014 edition is valid according to the current NCC. Amendments were issued in 2017 and others may have been applied after publication of this book – always check.

LEARNING TASK 14.1

GIVING ADVICE BASED ON REGULATORY INFORMATION ON SUSTAINABILITY PRINCIPLES

You have been sent a preliminary set of plans for a proposed single/double-storey brick veneer building.

- Provide a simple checklist of suggested options for a more sustainable single or double-storey building in your state or territory. This checklist will allow you to compare against what the current building plans have in place and ensure you can advise the client of alternatives that they might like to consider before their plans go to final approvals stage at council.

Reviewing the design: effectiveness and compliance

Assuming you have been an active part of the design process, you will have been actively reviewing the design throughout the design stage. Either way, once the design has been settled upon, you must carry out a thorough check of all the construction elements to ensure two things:

- effectiveness of the design with regards to the client's brief
- compliance with the NCC including state variations and associated Australian standards, and any specific council requirements, which may include heritage or environmental overlays.

This section provides guidance on each of these issues, paying particular attention to how the design might address thermal comfort, glazing and insulation. How the final design may be assessed for energy compliance with regards to star rating or BASIX systems is also discussed. The section closes with suggestions regarding your final consultations with the design team and clients in readiness for moving to the construction phase of the project.

Thermal comfort: radiation, convection, conduction and evaporation

Human thermal comfort depends on a range of factors, and may be accounted for through building design in numerous ways. Most contemporary 'off the shelf' homes or commercial buildings do so via mechanical heating and cooling, both of which tend to be high users of energy – about 40% of that style of building's energy usage annually. Knowing how the human body interacts with its surroundings to maintain what an individual considers to be comfortable helps us design more efficiently with regards to energy usage.

There are two aspects to human thermal comfort:

- psychological – how our mind perceives the current conditions with regards to our level of comfort
- physiological – how our senses react to the physical changes in the climate around us.

These two aspects may encourage our bodies to lose or produce heat as we strive to maintain a core temperature between 36.5° and 37.5 °C. Outside of this range we start to feel unwell. Prolonged periods at which the core temperature is 2 °C or more outside the normal range is likely to cause collapse or death. It stands to reason that we choose, or control, our environment such that our bodies don't have to work too hard to maintain this important core temperature.

What influences our sense of both physical and psychological comfort are:

- temperature
- humidity
- air movement
- radiant heat
- cold materials, objects, or surfaces that conduct heat away from us.

We need to address both physical and psychological aspects if we are to achieve not only a building capable of being energy efficient, but one that will be used by its occupants in a manner that will maximise its advantages.

Psychological or perceived thermal comfort

Unfortunately, due to the high variance in human perceptions, mostly based upon lived experience, we are not able to design with any certainty for the psychological or perceived sense of thermal comfort. For some, windows showing a hot or cold vista (desert or ice and snow), no matter how thermally efficient, can make some people 'feel' hot or cold. Others may react to simply the colour or materials of the interior surfaces, irrespective of the physical temperature of those surfaces.

However, there are some elements of the psychological or perceived level of thermal comfort that we can influence. These are:

- Radiation: If we feel warmth from a surface, material or other source such as a fire, then mentally we 'feel' warmer if we were previously feeling cold. We will be attracted to this source when cold, or repelled by it if feeling hot. That such sources are energy inefficient in the main is beside the point as far as the mind is concerned. They make us feel better (or worse) than just a room at an appropriate 22 °C.
- Conduction: The reversal of radiation, but the effects are similar – we will move towards conducting (cold) surfaces if feeling hot, and are repelled if feeling cold. Again, this is irrespective of the actual physical temperature of the room.
- Air movement: It takes very little movement for air to conduct heat away from the body, thereby making us feel colder than the room temperature actually is. What is commonly referred to as 'wind chill', or the 'chill factor', for outside temperatures is based upon this air movement. An easily survivable 0 °C, for example, will become a much more dangerous equivalent temperature of 13 °C with winds around 50 km/h. On the other hand, air movement can be used to advantage to feel cooler when we are too hot.
- Relative humidity: Relative humidity (RH) is the amount of moisture in the air as a percentage of the amount the air can actually hold at that specific temperature. Note that warm air can hold more moisture than cold hair. So air at 35 °C with an RH of 40% holds more water moisture than air with the same RH (40%) but at only 5 °C.

High and low humidity, like air movement, affects us psychologically as well as physically. Cold air that is high in moisture content (high humidity) will 'feel' colder than dry air of the same temperature. This is because water conducts better than air and 'wet' skin 'feels' colder than dry with the slightest air movement. Conversely, hot air with high humidity will make us feel hotter; this is because the air already has a significant amount of moisture in it, preventing it from taking up (evaporating) the sweat from our skin – the normal means by which we cool ourselves.

In an effort to achieve psychologically acceptable thermal comfort, building occupants will tend to raise or lower temperature settings on mechanical heating or cooling devices beyond what is required for their actual physical comfort – which for most people is 22 °C. Given that every 1 °C rise or reduction in room temperature leads to an approximate 10% rise in energy consumption, failure to address the psychological factor in thermal comfort has huge implications for energy efficiency over the life cycle of the building.

Physiological or actual thermal comfort

Our bodies physically lose or gain heat in a variety of ways, only some of which reflect how we perceive our thermal comfort level to be. We can, however, work with more confidence towards these physical aspects when attempting to account for them through good building design.

Heat loss

There three ways whereby heat may be lost:

- Evaporation is our most effective cooling mechanism. Our body's natural reaction to excessive heat is to push moisture to the surface which is then evaporated, taking the warm moisture with it and thereby reducing our temperature. The evaporation process is increased by air movement. Hence the breezes that serve to make us 'feel' cooler, also make us physically cooler. This cycle can be effectively endless, provided we can replenish our water supply.
- Radiation is a notable part of our psychological or perceived thermal comfort, but is also a manner by which we, albeit slowly, can physically lose heat. If the temperature of our surrounding environment is less than our own body, then we will radiate heat as the natural laws of physics ignores our biological attempts to stay warm and tries to equalise temperatures.
- Conduction is another factor in our perception of comfort that also acts directly on the body's physical comfort. Conduction occurs when we come in contact with objects, surfaces, or fluids that are colder than ourselves. Once more physics takes no account of biology's whim to stay warm, and attempts to equalise the temperature of the two bodies; i.e. transfer heat from our body to the other material by direct conduction. If the other material is extremely cold, we can actually burn our flesh as we transfer heat rapidly to that point: our biology attempting to keep the spot warm in response to the sudden heat loss; physics attempting to pull heat rapidly through that spot to equalise the heat differential.

Heat gain

Again, there are three ways by which we might gain heat:

- radiation
- conduction
- metabolism.

Two of these, radiation and conduction, are as before – only in reverse. That is, the other body or environment is hotter than our body temperature, thereby radiating heat towards us; or we conduct heat – draw heat into our body – from the other object, fluid, or surface with which we are in contact.

The third manner by which we gain heat is our metabolism – in simple terms, the conversion of foods to energy. Technically, we do not gain heat in this manner, but rather we produce it chemically. We are, in effect, our own heat source. Additionally, if we start to feel cold, we can shiver which causes the body to work harder and faster to generate more heat; or we can exercise which also warms the body, but burns our 'fuel' faster. Recent studies have proven that we can also control – elevate or lower – our body temperature though strict meditative practices; i.e. the trained controlled use of the mind only to cause the body to generate more or less heat on demand without external movement, such as shivering or exercise. Alternatively, we can put on more clothing and/or seek shelter – your projected building being but one option.

Thermal comfort: influence on building design

The earlier part of this chapter outlined the need for a building to respond to the climate in which it is sited. The requirements for human thermal comfort inform our design from the occupant's perspective. From the above information we know that our design must respond to both aspects influencing human thermal comfort: the physical and the psychological.

Designing for the physiological

The physical human needs can be dealt with in a fairly straightforward manner: your target being a temperature of around 22 °C in the living spaces, and a few degrees less in the sleeping areas. The rest is to some degree informed by the climate in which the structure is situated. Breezes for hot climates, slowly circulating air for cold. You can use radiant and conduction principles for any climate, noting in which direction you want the heat to flow – to the person, or away from them. In hot climates you could use cooling panels on the ceiling, from which cold air would 'fall' upon the occupants; in cold climates you could use heated flooring from which warmth would be conducted by the feet, or otherwise rise upwards around their bodies.

Glass and insulation types will also need to be carefully considered based upon the climate and these physiological influences. Evaporation from people can, in poorly ventilated structures and colder climates, lead to condensation. This can lead to mould and, in turn, ill health. These factors will be discussed more in the following sections.

Designing for the psychological

When designing a home for a specific family or client, they are best served by a survey of their needs. What sort of things influence their particular sense of thermal comfort: colours, views, textures, radiant heat, conducted heat, humidity levels and so on. Someone who has lived extensively in the tropics, or has just moved from a northern state to the south east, will probably want greater internal house temperatures and possibly higher humidity levels than those born and bred in colder climes. The reverse applies to those who have moved north. But this cannot be assumed, and so the questions should be asked.

In any building, take note of the need by some people for radiant warmth, not just warm inside temperatures. Note also the positive and negative influence of air movement and humidity depending upon climate and the 'sense' of thermal comfort you are attempting to convey (cooler or warmer).

To further your understanding in this area, download and review the document *Passive design: Design for climate*; available at http://www.yourhome.gov.au/passive-design/design-climate. The design principles in this document are climate relevant, and sound practice from an energy-efficiency perspective, but may not match the needs of your client based on what you have read above. This is where you must advocate for design, or temper design for client needs.

Design responding to both physical and psychological needs will be discussed more fully in the following sections, particularly those areas dealing with passive heating and cooling.

Orientation, glazing and thermal mass: evaluating effectiveness and compliance

Orientation is one of the most obvious considerations for any construction seeking energy efficiency. However, as has been discussed in the previous sections, thermal comfort will require more than just pointing the house towards or away from the sun. Likewise, the views, the whole point of building in some locations in the first place, may be completely at odds with the solar orientation. But still, if we are to make any efforts to reduce energy consumption, we must make best use of the free energy afforded by the climate in which our structures reside.

This latter point, the use of 'free' energy, is the basis of what is known as ***passive design***. Various aspects of this concept are discussed in the following pages; covering elements such as glazing, thermal mass, and the key design principles underpinning passive heat and cooling.

Building orientation

This is the arrangement of the building such that it takes best advantage of the sun's daily and seasonal variations in angle and intensity. At the same time, it takes into account the prevailing weather patterns, particularly the direction of wind. Careful orientation will reduce heating and cooling energy inputs from sources such as gas, electricity, or wood.

Here, however, is our first hurdle. The climate is changing: over the life span of the building, the period needed for heating is likely to retract, while the period needed for cooling will probably increase. With this in mind, the design should allow for adaptation over time. The reality, of course, is that most clients will be seeking solutions confronting the 'here and now', not future uncertainties. However, if the allowance for future adaptation comes at little or no additional cost, then they may well thank you in the years ahead. In the main this will be simply additional shading devices.

Basic principles

The orientation will depend upon a few basic factors:

- latitude: particularly the location of the building in relation to the Tropic of Capricorn (assuming your structure in located in the southern hemisphere)
- whether it is heating, cooling or a combination that you are targeting
- True North not Magnetic North
- prevailing winds.

Latitude is marked upon a globe by a series of parallel lines ringing the planet. The Tropic of Capricorn is a line of latitude marking the point on our globe at which the sun is directly overhead at noon on the southern hemisphere's summer solstice; i.e. 23.5 ° south of the equator on 22 December (occasionally the 21st or 23rd). This is as far south as the sun will come, after this date the sun will retreat to the north again. The Tropic of Cancer is where the sun does the same in the northern hemisphere; i.e. latitude 23.5 ° north of the equator (see Figure 14.4).

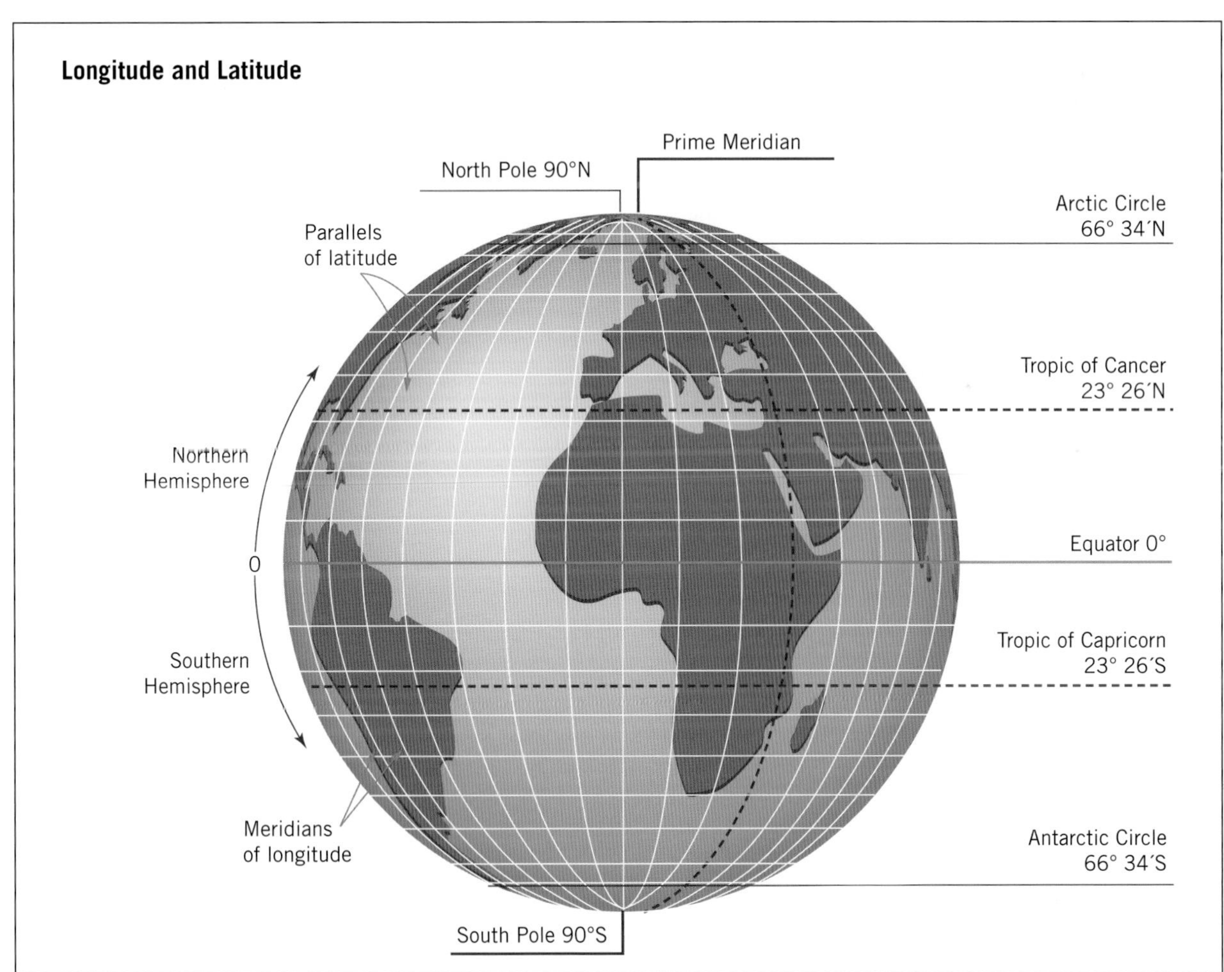

FIGURE 14.4 Latitude and longtitude

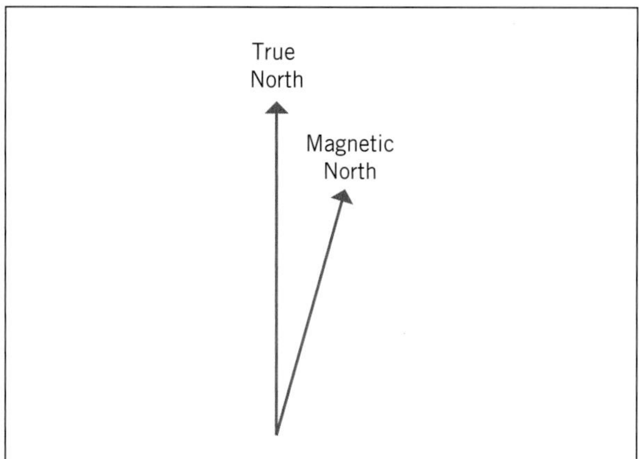

FIGURE 14.5 Magnetic North vs True North for most of Australia

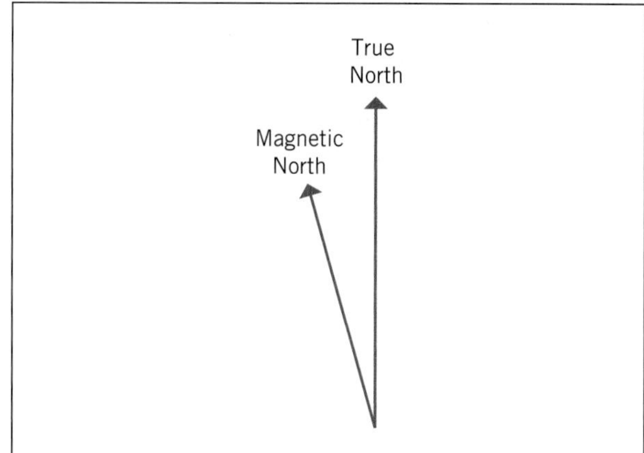

FIGURE 14.6 Magnetic North vs True North for part of Western Australia

As discussed earlier in the chapter, and described in Figure 14.2, once you move north of the Tropic of Capricorn, you will experience a portion of the year with the sun traversing to the south of your building. Anywhere south of this line, and the sun will always traverse to your north.

In climates where ***cooling*** is the main target– typically north of the Tropic of Capricorn – you will need to prevent sunlight from direct access to the interior of the building. Being north of the tropic means that the sun can hit all sides of the building at some time in the year – the southern walls receiving the sunlight at the hottest time. Shading is therefore required around the whole of the building, be it by extended eaves, verandahs, other buildings, or trees.

In areas where winter ***heating*** is essential; i.e. areas significantly south of the Tropic of Capricorn such as Victoria, Tasmania, or in alpine areas, you will need high levels of solar access for those walls facing to the north. Due to the angle of the sun in winter you can ensure maximum solar access when its needed, and limit it in summer, through careful design of eave or verandah widths. We will explore this further in the sections on calculating the sun's angle and shading below.

The concept of True North was discussed in Chapter 5, Plans and specifications, for exactly the reasons we discuss it here, correct orientation. Magnetic North, which is identified by way of a compass, will generally be significantly at variance to the True North in any part of Australia. In most cases, True North will be to the west of where a compass will point – except for part of Western Australia. A line running roughly from just west of Esperance via Southern Cross to Exmouth marks where there is no deviation; anything west of that and True North is east of Magnetic North.

You can check your project site's magnetic deviation by going to Geoscience Australia's website and putting in the appropriate coordinates for latitude and longitude – you can search for these through the same site by name or by pointing at an interactive map (see Figures 14.5 and 14.6).

The prevailing winds for a given site may also be found via the internet; in this case, the Bureau of Meteorology's Wind Rose interactive page: http://www.bom.gov.au/climate/averages/wind/selection_map.shtml

A wind rose as shown in Figure 14.7 offers the direction and speed of wind averaged over a number of decades. This wind rose is of Hobart, Tasmania. The very small circle in the centre denotes calm, listed on the left as being only 7% of the year. The majority of the wind (over 40%) is shown to come from the north-west, with rapidly decreasing amounts from all other directions; from this it may be seen that it hardly ever blows from the east.

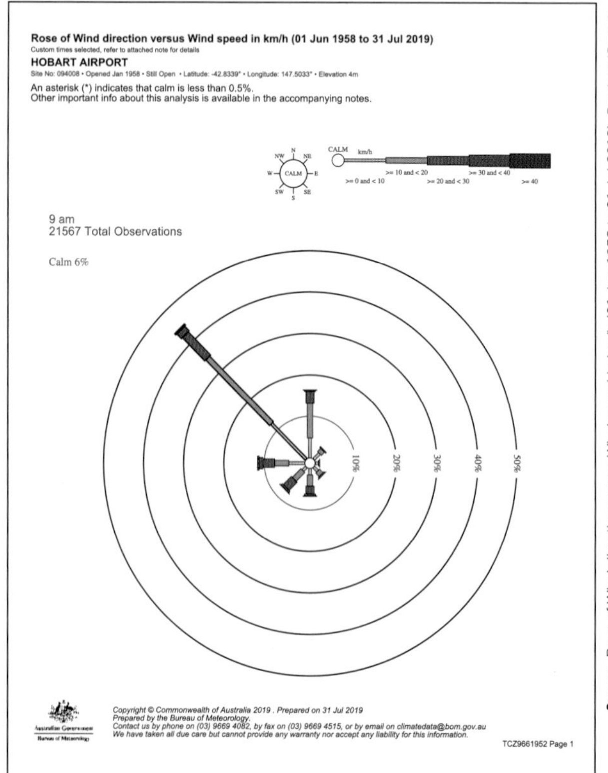

FIGURE 14.7 9 am wind rose for Hobart Airport

Source: Rose of Wind direction versus Wind speed in km/h (01 Jun 1958 to 31 Jul 2019). Product ID code: IDCJCM0021. Reproduced by permission of Bureau of Meteorology, © 2018 Commonwealth of Australia

Based on this and the other information above, you would seek to orient your structure to the north, but with significant shielding on the north-west side.

Tracking the sun and calculating its angle at midday

Knowing the angle at which sunlight will be striking your building at any particular time is important for good energy-efficient design. It is the basis of what is referred to as passive solar design. The diagrams shown in Figure 14.8 provide you with the basics of the sun's path in various parts of Australia. The main diagram is how the sun will traverse the skies for most of Australia lying south of the Tropic of Capricorn. The exact sunrise and sunset locations and its altitude at zenith vary the further south a site is from that tropic. The smaller diagram depicts the sun's transit at locations north of the Tropic of Capricorn.

Knowing these angles and directions allows us to determine the exact moment when the sun will enter a window and strike the floor for a given day or time of the year. In this manner we can control the solar heat gain for a building not only by window size and direction, but also with precise dimensioning of eaves, verandahs and other shading devices.

The calculations behind the web-based program are not overly complex, and it is worth your knowing at least how to determine the angle of the sun at midday for any given day of the year. For this you will need to know the latitude of the site – this may be obtained either from a map, or from any number of internet sites, including Geoscience Australia http://www.ga.gov.au/geodesy/astro/smpos.jsp. However, if you are on site then most phones have an app that can provide this for your current location

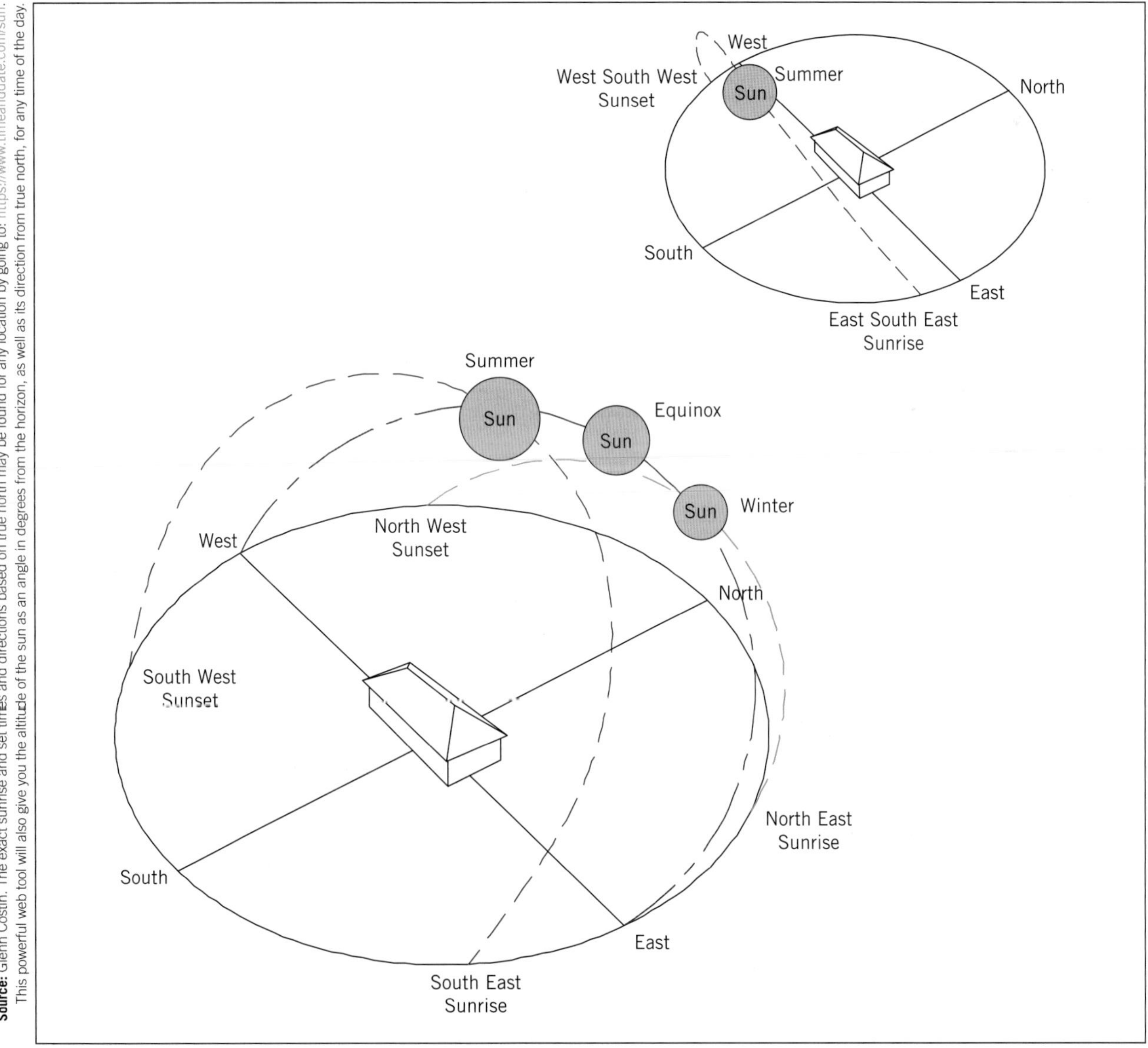

Source: Glenn Costin. The exact sunrise and set times and directions based on true north may be found for any location by going to: https://www.timeanddate.com/sun. This powerful web tool will also give you the altitude of the sun as an angle in degrees from the horizon, as well as its direction from true north, for any time of the day.

FIGURE 14.8 Typical sun paths for regions south of the Tropic of Capricorn

Note: For sites north of the Tropic of Capricorn, the summer path begins with sunrise slightly south of east (ESE), passes over the southern side of the building, before setting slightly to the south of west (WSW) as shown in the insert.

(make sure 'Location Services' are on in 'Settings', then open the compass app and your GPS coordinates are at the bottom of the screen).

The calculations are then as follows:

To find the sun's altitude at the equinoxes for your location (approx. 21 March and 23 September):

equinox altitude = 90° – Latitude

The altitude at each solstice is then found by:

summer solstice altitude = equinox altitude + 23.5°

winter solstice altitude = equinox altitude – 23.5°

Taking Adelaide as an example:

latitude = 34.92°

equinox altitude = 90° – 34.92° = 55.08°

summer solstice altitude = 55.08° + 23.5° = 78.58°

winter solstice altitude = 55.08° – 23.5° = 31.58°

The value in knowing these angles is clearly shown in Figure 14.9, which depicts the sun entering the window or being blocked by the eaves of the building.

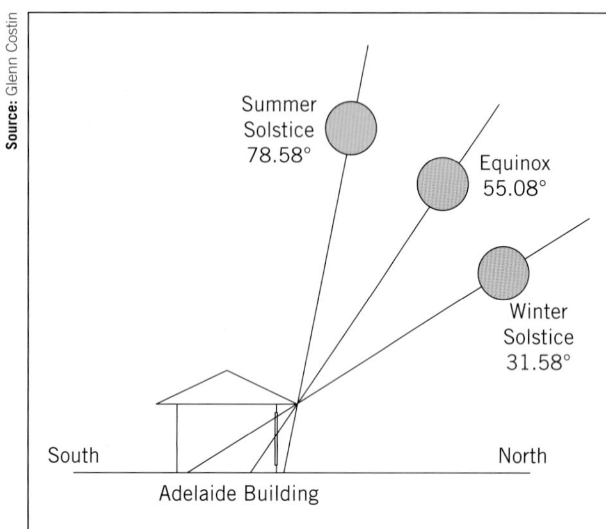

FIGURE 14.9 Seasonal solar angles for Adelaide, SA

To calculate the approximate angle for any given day at noon, you need to know the daily variation. This is found by:

Difference in angle between summer and winter solstices ÷ the number of days between solstices (1/2 a year or 365 ÷ 2)

Using Adelaide as an example again:

daily variation in altitude = (78.58 – 31.58) ÷ (365 ÷ 2)

= 47 ÷ 182.5

= 0.3°

You can now count backwards or forwards from the equinox or either of the solstices to the date you want to identify the sun's angle for at noon and multiple that number of days by 0.26°.

Again, using Adelaide:

date for sun angle to be determined – 13 April

equinox is 21 March, hence days variance = 23 (11 days of March, 13 days of April)

sun's altitude on 13 April = equinox altitude – (24 × 0.3)

= 55.08 – 7.2

= 47.88°

(note that it is 'minus' as the sun in getting lower after the autumn equinox)

Note: These calculations can only give approximate angles as there is a minor cyclic change in the daily angular variation (approximately 0.1 of a degree). Hence the actual solar altitude for the 13 April is closer to 46°.

Influence on orientation and shading

Determining the width of the eaves is based upon the height they are above the window *sill*, and the angle of the sun at the time relevant to the wall's orientation. It is then a matter of determining when warmth from the sun is wanted, vs when it is not. While climate and location – as discussed above – should be the main drivers, individual clients will vary in preference to some degree.

Following on from the Adelaide example, the temperature in this region becomes cooler around mid-March. Hence the autumn equinox is perhaps the time to allow direct solar access at midday – aware that you already gain access at earlier and later times of the day as the sun rises and sets. The diagram (Figure 14.10) below shows the width of the eave, including gutter, required such that the midday sun (1:22pm daylight saving time) first hits the floor on 21 March (equinox, approximately 55°).

This diagram also shows a plan view, with the times at which the sun will be striking the windows from 45° either side of N; i.e. from the NE and NW. It just so happens (a pure coincidence) that in Adelaide – and any other location at 34.92° latitude – these are also the times at which the sun will have an altitude of 45°.

This is not to suggest that a 1570-mm eave width is the perfect solution for all and every house in the Adelaide area, or at that latitude. For one, it is based upon a specific sill to under-eave height – in this case 2535 mm. At least three other factors must come into play: context, individual preference and available orientation.

- **Context** includes factors such as elevation, proximity to large bodies of water (reflection, breezes), surrounding landscape and other factors affecting general solar access such as trees and buildings.

Source: Glenn Costin

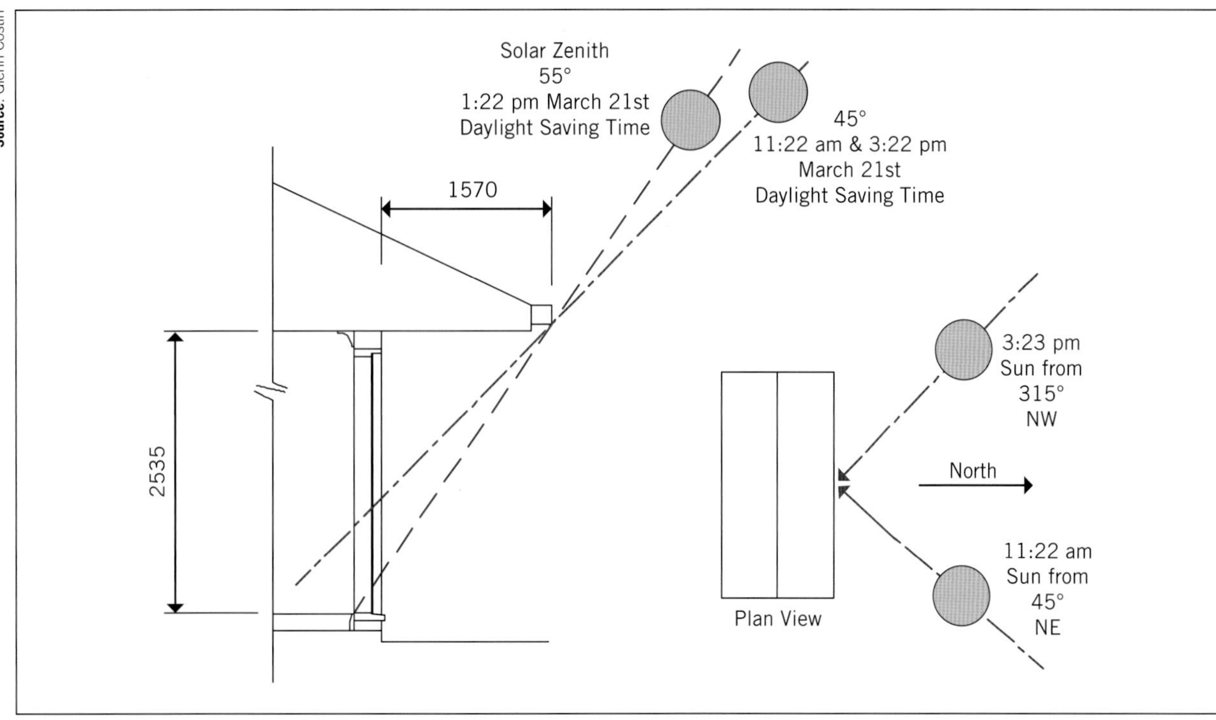

FIGURE 14.10 Plotting exact solar access by day and time

- **Individual preference** may reflect the psychological element of thermal comfort discussed earlier, or the intended end use of the building. Some clients, or building purposes, may require more direct sunlight than others. Some may prefer a hotter interior temperature, some cooler, some brighter, some darker. Your task, from an energy-efficiency perspective, is to juggle these requirements with other design implementations that can balance out, or at least reduce, any additional energy inputs that such requirements may induce.
- The **available orientation** refers to the site itself. Many (most ...) domestic sites are not well suited to the ideal orientation of due north +/- 15° as measured perpendicular off the long face of the building. Some buildings, by the nature of their use, also cannot match this preference. Other rural sites may be compromised by their location within the landscape, or a preference for a specific view – the whole purpose of the client picking that spot in the first place. Which brings us to the windows, glass and the frames that hold them – the focus of our next section.

Glazing systems

Windows offer character and elegance to a building, connect us with the landscape around us, as well as functioning as sources of light and fresh air. However, as much as 40% of a building's heat can be lost through poor glazing elements, and almost 90% of heat gained. A window's thermal performance is therefore critical to the overall performance of the building.

Making decisions on glazing and its arrangement in a building (fenestration) require careful consideration of all the aspects dealt with so far, as well as some that will be expanded upon later. These include: thermal mass, insulation and the properties of the glazing systems themselves, including individual size and orientation. In addition, the NCC has requirements governing glazing elements that capture many of these issues, including shading. We'll begin, however, by looking at two key thermal qualities of all glazed elements: conduction and solar heat gain.

Conduction: U-value (Uw)

U-value (Uw) is the measure of the rate at which the total window assembly – glass, frame, seals, etc. – transfers heat through it. This is not about solar gain or light transference, but simply heat conduction from one side to the other. It is effectively a measure of the window's insulation value. The *lower* the U-value, the better the insulation properties.

The specific U-value (tabled as ***Uw***) for the windows produced by over 450 registered manufacturers can be found through the Windows Energy Rating Scheme (WERS) website: https://www.awawers.net/index.php/en/res

Understanding the effect this value can have on energy efficiency is shown with the formula:

U-value (Uw) of window × temperature difference (°C) × window area (m^2) = energy in watts (W)

Assuming 75 m^2 of basic aluminium single glazed windows and doors of 5-mm clear glass with a Uw of

6.3, we can quickly calculate the energy loss for the building when temperature difference is 20 °C – i.e. a wintry 2 °C outside, and trying to maintain 22 °C inside:

$6.3 \times 20 \times 75 = 9450$ W; i.e. 9.45 kW

This is the power consumed by a standard slimline panel heater. Power prices around the country vary, but the average at the time of writing is around 35c/kWh. These glazing elements are costing the owners $3.30 every hour that night, and that is without factoring in wind chill if the windows are exposed.

Solar heat gain coefficient (SHGC and SHGCw)

This identifies the heat transmitted directly through a window by sunlight as a fraction of that which hits its surface. It includes the heat absorbed by the glass and released inwards, as well as that which passes straight through. **SHGC** may be thought of as a percentage of the possible amount of heat transferred through the *glass only* and is expressed as a number between 0 and 1; the lower the number, the less solar heat transmitted inwards.

For example: 5-mm clear float glass has an SHGC of 0.83. This means that 83% of the possible heat supplied by the sun gets through, 17% is reflected.

SHGCw is much the same, but applies to the *whole window assembly*. So different window types, with the same glass, may have different SHGCw factors, and can be manufacturer specific.

For example:

- 5-mm clear float glass in standard aluminium frame – SHGCw = 0.63
- 5-mm clear float glass in cedar frame – SHGCw = 0.58

In addition, the angle at which the sun strikes a window has a huge influence upon the amount of heat transmitted. The SHGCw supplied by manufacturers is based upon an '**angle of incidence**' (the angle at which the sun strikes the glass) of 0°; i.e. at 90°, or perpendicular to the face (see Figure 14.11).

The angle of incidence changes two factors in the calculation of direct solar heat transference. The first, as shown in Figure 14.11, is the effective area for solar catchment – this reduces as the sun's angle of incidence increases. The second is the SHGC of the glass is reduced – this is because the glass the sun must pass through, even allowing for refraction, has effectively got thicker.

Once the angle of incidence gets past 55° the solar heat gain drops off markedly. This applies to both the vertical and the horizontal planes.

Luckily, we do not need to do these calculations, they are covered by the software behind rating systems such as BASIX and NatHERS. But the principle is important to understand so that you can at least estimate what values a software system should provide (computer errors being all too common), as is how you use this information to inform design, your clients and the end users.

You also need to know from where to source SHGC and SHGCw factors, which in the main is from the window and glass manufacturers themselves. For SHGCw factors for literally thousands of different window frame and glass types listed by manufacturer, go to the Windows Energy Rating Scheme (WERS) web site: https://www.awawers.net/index.php/en/res

For glass SHGC values only, you need to go to a glass manufacturer's site, such as the Australian Glass Group or a smaller entity such as Highland Glass. The web addresses for their glass data tables are:

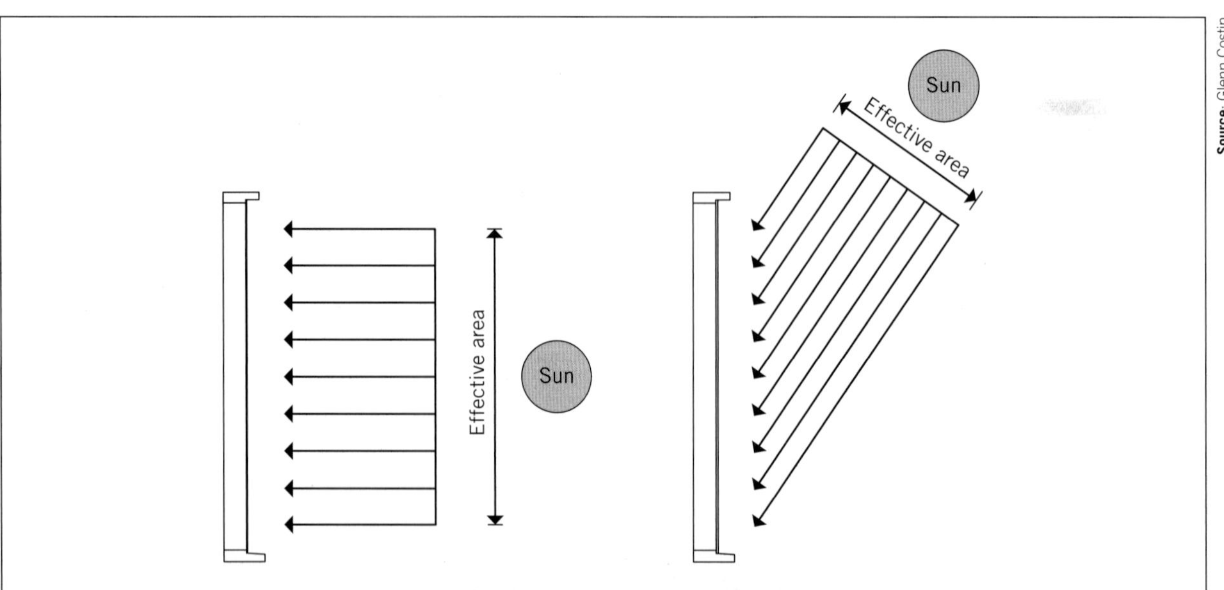

FIGURE 14.11 Angle of incidence and effective area for solar heat gain. Left is 0° incidence; right is 55° incidence.

- http://www.australianglassgroup.com.au/residential/products/insulglass-clear/

Glass types and arrangements

A quick review of the wesbites above informs you that there is more than one type of glass or glazing element. Each glass type offers various SHGC and U-values. Each glass type has its uses in different contexts. In Darwin or other northern parts of Australia, for example, solar heat gain is not what you want as a general rule, so a glass with a very low SHGC is preferred. Alternatively, in the more southern and alpine areas, a low SHGC is exactly what you don't want.

There are various ways by which a SHGC and U-values can be varied:

- toning (colouring) the glass
- laminations of glass and toned film in between
- low-E glass: thin metallic coatings
- insulated glass units (IGUs) – commonly known as double or triple glazing
- switchable glass.

Note: Glass thickness makes only limited difference to these values, though it can improve noise and safety factors.

Toning or colouring the glass changes its visible light transmittance (TVw or VLT), which in turn can serve to reduce the heat gain significantly – dropping the SHGC to as low as 0.41 for some dark tints. It does so, however, at the expense of visibility. That is, it is harder to see in or out, and there is less light in the building to see by.

Laminating glass is frequently done as a safety feature against impact. By using toned and/or low conductivity filaments in between instead of clear, reductions in both SHGC and U-values can be obtained.

Low-E glass is produced by applying a very thin layer of metal particles that reflect heat off the glass and back into the room, or outwards, depending upon how it is arranged in the frame. Production of this glass generally involves advanced pyrolytic techniques that bake the particles permanently to one side of the glass face; though it can be done by vacuum spraying. Low-E glass can achieve both high and low SHGC values:

- Low SHGC is achieved by clipping out the short wavelength ultraviolet (UV) and slightly longer infrared either side of the visible light spectrum.
- High SHGC is achieved by letting in the infrared additional to the visible light.

Sunlight is composed of UV, visible light and infrared. When this light hits an object and bounces back, it does so as long-wavelength light. Low-E glass (both high and low SHGC) reflect this long-wave infrared light and warmth back into the room.

Both low and high SHGC generally achieve very good (low) U-values.

Insulated glass units (IGUs), otherwise known as double or triple glazing units, can be created by any combination of these glass types depending upon desired function and costs (Figures 14.12, 14.13). These will always further reduce the SHGC value, but this may be marginal if required. It will significantly reduce conductivity (U-value). As can be seen from the diagrams, the glass panes are separated by low conductivity spacers, and the space formed between them is then filled with inert gas – usually argon.

Switchable glass is another form of laminated glass whereby the membrane between laminations may be

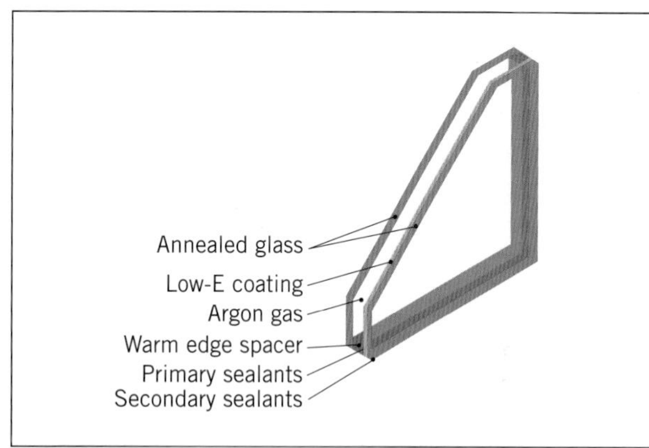

FIGURE 14.12 Double-glazed window

Source: Shutterstock.com/Golden Sikorka

FIGURE 14.13 Triple-glazed window

caused to change opaqueness and thereby reduce solar heat gain on a variable scale. The film changes its clarity by way of a low (5W/m^2) electric current that aligns microscopic droplets of liquid crystal such that light may be transmitted. This material does not generally fall into the energy-efficiency sector as it uses power to stay clear, becoming opaque (dark) when power is removed. It is expensive to create and if used as an external window, should be coupled with low-E glass to reduce direct heat on to the film which would otherwise deteriorate with time. It is also usually used in this configuration to prevent moisture getting to the edge of the glass and also degrading the switchable film.

Framing glazing units

As touched on briefly in discussing SHGCw values, what frames a pane of glass also has a significant influence upon the overall effectiveness of the glazing unit: either in allowing heat in, or letting it escape. You can have the most efficient glass in the world, but if it is framed or fitted poorly, or fitted into highly conductive materials, then significant amounts of energy will continue to be lost. Common framing materials include:

- aluminium and steel
- timber
- plastic
- composites.

Aluminium is an extremely good conductor, as is ***steel***. These two materials have been used extensively over recent decades in window manufacture as inexpensive alternatives to timber. Fitted with single panes of standard clear glass, these units are extremely poor value from an energy-efficiency perspective; both steel and aluminium can, however, be easily recycled. Aluminium is also a very light material, reducing transport costs to a degree relative to steel.

Timber is a natural insulator but, depending upon the species, tends to move significantly with changes in temperature and humidity. It also requires extensive maintenance over time to prevent earlier than expected end of life.

Plastics (uPVC) can be moulded to any shape required and has similar thermal properties to timber, but the capacity to produce better seals. However, timber, despite its maintenance requirements, remains the more sustainable of the materials.

Composite frames may be combinations of uPVC coated around timber, timber to aluminium, aluminium around either timber or plastic. These combinations are especially effective when creating frames which are 'thermally broken' or isolated from outside temperatures as discussed below.

Thermally broken glazing units

Figure 14.14 shows a composite aluminium and plastic thermally broken glazing unit. Note the low conductivity green polyamide pieces used to create the break between the interior and exterior elements of the window frame. Timber may be used to create a thermally broken frame as well, when aluminium or steel is used for the outside, and timber is applied as a facing on the inside.

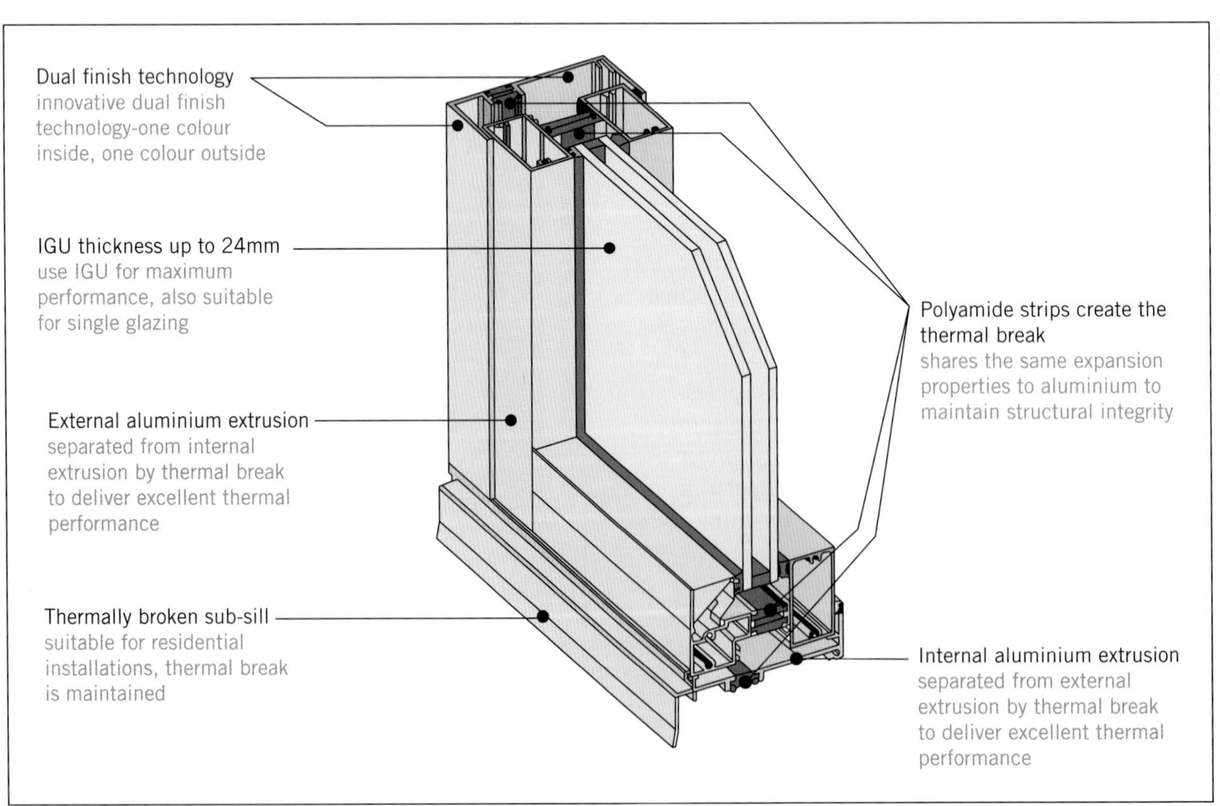

Dual finish technology
innovative dual finish technology-one colour inside, one colour outside
IGU thickness up to 24mm
use IGU for maximum performance, also suitable for single glazing
External aluminium extrusion
separated from internal extrusion by thermal break to deliver excellent thermal performance
Thermally broken sub-sill
suitable for residential installations, thermal break is maintained
Polyamide strips create the thermal break
shares the same expansion properties to aluminium to maintain structural integrity
Internal aluminium extrusion
separated from external extrusion by thermal break to deliver excellent thermal performance
Source: AWS Australia

FIGURE 14.14 Aluminium and plastic composite thermally broken window frame

Frames and ventilation

As discussed, windows and doors provide light, warmth, and connect us with the outside environment and landscape. But they also are our main source of ventilation. How this is achieved can vary dramatically depending upon the location of the window (or door), the type or style of window and the whether the ventilation needs to be fixed or variable.

Common frame types are listed below, along with the amount of openable space they provide as a percentage of their overall area. It should be noted that you cannot go on the percentage alone in determining the value of a particular style to its effectiveness as a means of ventilation. A double-hung window for example, though only offering 40% of its area as openable, gives more ventilation control than a sliding or casement sash. And though an awning sash is listed as only offering 30% as ventilation, both it and the casement sash, from an NCC perspective, are considered as 100% as the sash is openable, and the sash constitutes the whole of the window area.

- casement — 45%
- sliding — 45%
- awning — 30%
- double hung — 40%
- two panel sliding door — 45%
- louvre — 80%.

Door and window seals of a variety of forms (rubber, foam, plastic) provide much of the air, and hence thermal, leakage on both windows and doors. These must be inspected and maintained regularly to ensure passage of air is controlled.

Some windows frames may also come with fixed ventilation slots designed to allow the room to 'breathe' and so reduce condensation issues. There are complexities with such designs and they are not always effective. Where condensation is a possible issue, you should discuss the options with the window manufacturer.

Not all glazing elements are openable. Nor are they all in walls, while some are the walls. There are fixed windows as well as glass bricks, skylights (openable and fixed), as well as glass floors and floor 'windows'. In addition, there are doors that have windows, and windows that are doors. Fenestration is a very broad field.

Windows and thermal mass

Thermal mass refers to materials having the ability to absorb and retain energy; and to do so slowly. This latter point, known as thermal lag, is key to the role of thermal mass in energy-efficient design. A material that takes on energy slowly will tend to release that energy slowly. In addition, thermal mass can be used to take up unwanted energy. The benefits to our building's energy performance therefore include:

- storage and release of energy into the building
- maintenance of a desired temperature with progressively lower inputs
- reduced fluctuations in internal building temperatures
- capture and release of excess heat to the external environment or the ground.

High-density materials such as brick, concrete, rammed earth, stone, tiles and even water all have high thermal mass. Timber, plastic and fabric, on the other hand, are generally said to have low thermal mass.

Thermal mass is of particular benefit in climates where there is a high ***diurnal*** temperature range – the difference between day and night-time temperatures. In hot climates, materials with low thermal mass are generally considered best practice. However, as stated, thermal mass can be used to capture and release unwanted energy to the ground or atmosphere. Hence, concrete slabs can still be of use in high-temperature climates; as may shaded high-mass internal walls or features, provided that the heat they gain during the day can be released to the cooler air in the evenings.

With regards to windows, in cool to cold climates, they should be arranged such that materials with high thermal mass can capture the solar thermal energy these elements allow into a room. Concrete floors, tiled surfaces, water-filled structures and brick or stone walls can all be used to capture this energy simply by being exposed to direct sunlight. In the evenings, this energy will be slowly released back into the room. If using auxiliary energy (electric, gas, wood) for heating, the amount of additional energy need will be lessened.

In what is known as the 'thermal flywheel', thermal mass will also slowly take and give heat in such a way as to reduce peaks and troughs in internal temperatures. This is the influence of the thermal lag described earlier. With wood fires particularly, the surfaces will give back in such a way that heat fluctuations are notably reduced. Materials with low thermal mass tend to get hot then cold very quickly. Stop heating the room, open a door or window to the outside, and internal temperatures will drop rapidly.

In hot climates, therefore, windows and thermal masses must be isolated from each other, at least at times of high solar energy gain. That is, you must keep the sun off a concrete floor, brick, or stone wall, or any other high-mass element. For specific design responses to each of the NCC's climate zones you are advised once more to review the PDF documents available at: http://www.yourhome.gov.au/passive-design

Planning and NCC requirements

Developing an approach to window elements from the perspectives of energy efficiency and client or end-use preferences and requirements is critical, but is only part of the design process. The other is compliance. This has three distinct levels:

- compliance with the NCC energy-efficiency requirements reflected in part in the rating systems of either BASIX or NatHERS

- compliance with the NCC minimum natural light and ventilation requirements
- compliance with local council planning and development requirements.

The first of these will be covered shortly in the section 'Energy rating compliance'. The other two we shall look at briefly now.

NCC light and ventilation

This has been dealt with in some detail in various parts of Chapter 1 Building codes and standards. From this you are reminded that each volume of the NCC has performance requirements concerning light and ventilation. These may be found in Parts F4 (p. 267) of Volume One and P2.4 (p. 57) of Volume Two. This requirement may be summed up for both volumes in this statement from Volume One:

> Sufficient openings must be provided and distributed in a building, appropriate to the function or use of that part of the building so that natural light, when available, provides an average daylight factor of not less than 2%.
>
> **Source**: NCC 2019 Building Code of Australia - Volume One, © Commonwealth of Australia and the States and Territories 2019, published by the Australian Building Codes Board. Licensed under Creative Commons Attribution-NoDerivatives—4.0 International licence https://creativecommons.org/licenses/by/4.0/

and:

> A space in a building used by occupants must be provided with means of ventilation with *outdoor air* which will maintain adequate air quality.
>
> **Source**: NCC 2019 Building Code of Australia - Volume One, © Commonwealth of Australia and the States and Territories 2019, published by the Australian Building Codes Board. Licensed under Creative Commons Attribution-NoDerivatives—4.0 International licence https://creativecommons.org/licenses/by/4.0/

The deemed-to-satisfy (DTS) provisions, your simplest approach to compliance, may be found in each volume as follows:

- **Volume One**: F4.0 – F4.12 (NCC 2019, Vol. One, pp. 270–3); excluding F4.10, which is blank)
- **Volume Two**: 3.8.4 – 3.8.5 (NCC 2019, Vol. Two, pp. 283–9).

Again, these DTS provisions may be summarised for simplicity as follows (and so excluding allowances for mechanical intervention or accessing light or ventilation from other spaces):

3.8.4.2 Natural Light

- ***Natural light*** must be provided by ... ... windows, excluding roof lights that ... ... have an aggregate light transmitting area measured exclusive of framing members, glazing bars or other obstructions of ***not less than 10% of the floor area of the room*** ...

Source: NCC 2019 Building Code of Australia - Volume Two, © Commonwealth of Australia and the States and Territories of Australia 2019, published by the Australian Building Codes Board. licensed under a Creative Commons Attribution-NoDerivatives—4.0 International licence, https://creativecommons.org/licenses/by/4.0/

3.8.5.2 Ventilation requirements

- ***Ventilation*** must be provided to a *habitable room, sanitary compartment*, bathroom, shower room, laundry and any other room occupied by a person for any purpose by ... ... Openings, *windows*, doors or other devices which can be opened ... ... with a ventilating area not less than 5% of the *floor area* of the room *required* to be ventilated ...

Source: NCC 2019 Building Code of Australia - Volume Two, © Commonwealth of Australia and the States and Territories of Australia 2019, published by the Australian Building Codes Board. licensed under a Creative Commons Attribution-NoDerivatives—4.0 International licence, https://creativecommons.org/licenses/by/4.0/

You are advised to scrutinise the various additional clauses pertaining to light and ventilation in both volumes. The section on shared light calculations is of particular value when dealing with structures in alpine areas, or tropical extremes, where energy efficiency may suggest reduced glazing sizes.

You are also reminded that the NCC is a performance-based document. You can therefore come up with alternative solutions (Performance Solutions) provided that they are deemed to meet the Performance Requirements by the relevant building authority.

NCC windloads, human impact and safe movement

AS 4055 Windloads for housing defines 10 wind classes based upon designed gust wind speeds. They range from N1–N6 for non-cyclonic regions, and C1–C4 for cyclonic regions. Determination of a wind speed was covered in Chapter 1; for this section, it suffices to know that all windows have to be designed to meet the requirements of AS 4055 for the wind class applicable to the site on which the project is to be constructed. This standard in raised in NCC, Volume Two, Part 3.6. This section also covers issues of human impact with a glazing element – as does BP1.3 of Volume One, be it a door or window.

The strictures you may have to contend with will include

- size of glazing element – which will dictate thickness
- thickness of glass – based upon size and location (e.g. likelihood of human impact)
- type of glass – annealed, laminated or toughened – based upon location (e.g. overhead), size, and likelihood of human impact
- visibility markings on glass – based upon likelihood of human impact.

In addition, D2.24 of Volume One (p. 152) and Volume Two, Clause 3.9.2.6 and 7 (p. 320), require openable windows, where a fall of 2 m or greater is possible, be protected to prevent people falling out. While this does not have any implications on the NCC requirements for ventilation outlined in the previous section, it does from an energy-efficiency perspective when cross-breezes and the like affect part of the design.

Glass thickness requirements also have implications for your design, particularly with regards to specifying double- or triple-glazing units. For example, you may be required to include thicker glass for low-lying internal panes, which can alter critical SHGCw ratings.

There are numerous clauses and subclauses around each of these issues; you are advised to review these and become familiar with their implications.

NCC bushfire requirements

Bushfire attach levels (BALs) have been covered to some degree in Chapter 1. They apply to energy-efficient windows as much as any other. In choosing a window design, you must have access to the BAL rating of the project site, and from this make informed decisions on both frame and glass materials. You are advised to review the AS 3959 Construction of buildings in bushfire-prone areas, with particular reference to frame and sill materials.

There are six BAL ratings that may apply, and the material selections for each are outlined in the associated sections on construction requirements. Your main considerations will be increasing fire-resistance levels (FRLs) for the window system as a whole, the possible need for toughened glass, increased glass thickness, bushfire-resistant timber frames, and the possible inclusion of steel or bronze mesh or shutters. There are no restrictions on sizes.

Council planning and development requirements

Council involvement with the oversight of the NCC's provisions has already been outlined at the start of this chapter, as well as some of the ways by which they might encourage energy efficiency through various planning and development acts. However, some councils may also have provisions that may need to be negotiated to allow some of the more energy-efficient window designs or locations. The sort of things that can 'get in the way' of a good design can be:

- heritage overlays
- covenants
- overlooking neighbours' private space.

Heritage overlays may require that windows be of a particular type, style and/or shape and size. This can restrict your design options to some degree, requiring some level of negotiation between all stakeholders to ensure an outcome that reflects both community and energy-efficiency values – which, fortunately today, are becoming more aligned.

Covenants are another form of overlay, one that has been imposed and approved at the planning stage of a new development, such as a new housing estate or subdivision. Sometimes they are many decades old, but still have legal force. Covenants may require such things as:

- houses be constructed of specific material(s) – e.g. brick with terracotta tiles
- be of not less than a certain dollar value
- have windows of a certain type
- be of a certain range of colours
- only one storey high.

Depending upon the state or territory in which a project is situated, there will be an Act, such as the Victorian *Planning and Environment (Restrictive Covenants) Act 2000*, that governs what a council can and can't do regarding covenants. In the main, councils are prohibited from granting or amending a permit should it breach a covenant. The only paths for change involve lengthy and complex legal procedures that may need to be heard in the Supreme Court. In general, therefore, your design should respect the covenant and energy-efficiency proposals worked around these requirements.

Overlooking the existing private spaces of a neighbouring domestic dwelling is not permitted by most state building regulations and/or local council planning provisions. Below is an extract from the Victorian Building Regulations 2018:

Part 5, Division 2 - REG 84 Overlooking

(1) A habitable room window or raised open space of a building on an allotment must not provide a direct line of sight into a habitable room window or on to a secluded private open space of an existing dwelling on an adjoining allotment.

Source: Victorian Building Regulations 2018, http://www.ocpc.vic.gov.au/Domino/Web_Notes/LDMS/PubStatbook.nsf/93eb987ebadd283dca256e92000e4069/267BD1D0E0CDC73CCA25825D001EEFB9/$FILE/18-038sra%20authorised.pdf; http://www.legislation.vic.gov.au/Domino/Web_Notes/LDMS/PubStatbook.nsf/b05145073fa2a882ca256da4001bc4e7/267BD1D0E0CDC73CCA25825D001EEFB9/$FILE/18-038sra%20authorised.pdf

There are many provisions to this clause, covering line of sight, distances, etc, however, the implications are clear. Your window designs, energy efficient or not, must not impinge upon the amenity of an existing neighbour's dwelling.

Insulation: costs and effectiveness

The diagrams in Figure 14.15 demonstrate the heat gains and losses of a typical uninsulated structure in summer and winter. Suitably chosen and installed insulation can have a significant influence in reducing these gains and losses, thereby saving energy in cooling or heating.

The important word in the previous sentence is *suitable*. Choosing the wrong insulation, or incorrectly installing insulation, can actually make matters worse – usually through condensation, and through this, mould behind ceiling and wall linings. In addition, insulation can be expensive, so you need to be sure that what is being considered is cost effective, as well as effective as an insulator.

No matter the type or country of origin, insulation must comply with AS/NZS 4859.1 (currently under revision incorporating a draft companion standard AS/NZS 4859.2), so you must ensure this is the case.

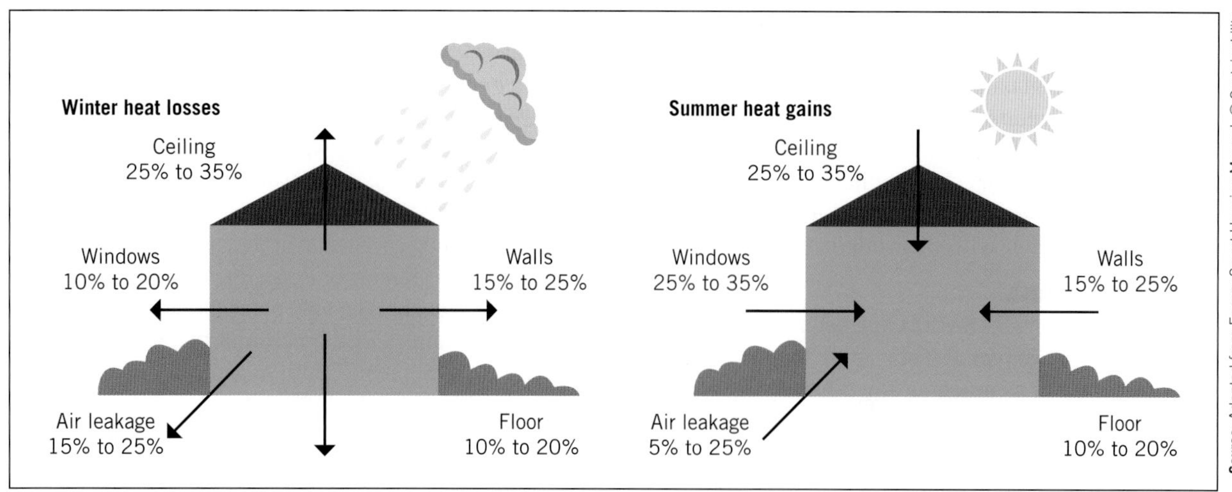

FIGURE 14.15 Thermal gains and losses in the standard house – winter and summer cycles

Insulation that does not comply can be dangerous, either in its constituents (e.g. asbestos), or in flammability.

Insulation and construction systems

As the diagrams show, insulation alone cannot solve all the heat loss or gain in a building, so its introduction to a design should be in tandem with all other passive design controls, particularly air leakage. To comply with NCC performance requirements on insulation values, you must look at elements of a structure, such as walls, floors, ceilings, as 'systems'. Thus, it is a wall construction 'system' that is measured, as against just the insulation that you choose to put in. Each part of a system – e.g. plasterboard, timber framing, reflective foil, cladding and even air spaces – has a measured insulative value: this is known as its R-value. Combined, they make up the ***Total R-value***, or insulation rating, for the building element being considered.

R-values

The R-value of a material is a measure of its resistance to thermal conductivity relative to its thickness; i.e. how much energy can pass through a given thickness of the material. The greater the R-value, the better its capacity to resist the passage of energy; i.e. the better the thermal performance. Today, each building material has an R-value that can be found specifically from manufacturers; or generically, either online through groups like Low Impact Development Consulting (LID, http://rvalue.com.au/) and, for some materials and systems, within the NCC itself.

Some insulation materials, particularly reflective sheets such as aircell and single-skin reflective foils (e.g. sarking), may seem to offer varying R-values dependent upon its location within a system. The reality is the material's R-value remains constant; however, as air spaces or voids also have an insulative effect, the creation of a space between the insulation foil and the exterior can therefore change the value of the system as a whole.

A material's R-value can also change with the direction the energy (heat) is applied to it. Thus, a material may have an Up, Out, or 'Winter' R-value, and a Down, In, or 'Summer' R-value. This value tends to change little for bulk insulation, but can be significant with reflective sheet types.

Types of insulation

As alluded to above, most insulation can be categorised as either 'bulk' or 'reflective', though they can be supplied as composites.

Bulk insulation is designed to resist the transfer of energy by convection or conduction. This form of insulation traps air in small voids within the material that must each be heated or cooled before transference to the next, and so on through the material; the slower this occurs, the better the insulation. The nature of bulk insulation means that there is little to no difference in its R-value irrespective of the direction from which the heat comes. Common bulk insulation materials include:

- glass fibre 'wool' – spun glass, may be new or recycled glass
- pure wool – sheep wool fibres treated for insects; may be recycled clothing or new
- cellulose fibre – timber fibres treated for flammability and insects; may be new or recycled
- polyester – may be recycled clothing or new
- polystyrene – expanded polystyrene sheets (EPS – open cell); extruded polystyrene (XPS – closed cell)
- other petroleum-based closed-cell foams – phenolic, polyisocyanurate foams (higher fire resistance and higher R-values)
- mineral fibres – rock 'wool'.

Most bulk insulation comes in the form of easily handled thick sheets known as 'batts'. Some, such as pure wool and mineral fibres, can be pumped into spaces as loose fill. Pure wool and polyester also come in combination with reflective insulation in the form of rolls or blankets. Polystyrene is being used both as a composite material, and as a non-structural cladding over which cementitious coatings are applied.

Reflective insulation, or **reflective foil laminate**, is designed to resist radiant heat: i.e. it reflects the radiated energy back in the direction from which it came. To do this, reflective insulation generally requires an air gap between it and any cladding or other material. This prevents the heat from being trapped, warming the foil, and then being transferred into or out of a structure through conduction. Such being the case, reflective insulation is directionally specific in its application. Hence, it has a very different 'up' or 'down' value depending upon which way it is laid, its location in a system, or from which direction the heat is applied to it.

This form of insulation is usually created by laying a metallic foil (generally aluminium or an alloy of this material) on to fibre-reinforced paper, plastic, or a combination of these. It is then supplied either in rolls, as commonly seen on house walls and roofs prior to cladding, or in 'concertina' sheets for use under floors.

Composites have been developed from many of the different insulation types mentioned above. Pure wool, glass fibre and polyester fibres have each been combined with reflective sheets to form 'blankets' that may be rolled out over ceiling and roof frames. Polystyrene is available in combination with reflective foil to produce thin sheets that may be laid under floors or behind walls. And thin sheets of bubbled plastic laid between two reflective sheets form what is commonly referred to as 'air-cell' insulation. This comes in rolls used under roof cladding and behind walls (replacing the traditional single-layer sarking) and under floors.

Insulation is also being 'built' into structural materials such as chipboard flooring sheets, and metal or timber-faced panels known as SIPs – structural insulated panels. To date, SIPs have tended to focus on polystyrene cores of varying thicknesses.

Installation

Many of the types listed above are injurious to the health of humans. Installation may therefore require protective equipment (PPE) such as gloves, masks, eyewear, full-cover clothing and, in some cases, disposable overalls. This is particularly the case with fibreglass and crushed rock products. You should check carefully all manufacturer's instructions before opening or handling the material, particularly if it is new to you.

Most insulation will require clearances around electrical fittings, while foil types must not be installed over or around electrical wiring. Downlights and fans generate heat and may either burn out or cause the insulation or other materials around them to catch fire. Modern LEDs are often marketed as being energy efficient because they do not waste power generating heat. This is only partially true; behind the lights there is a driver, and some form of heat sink to dissipate the excess energy lost in converting AC to DC power. Both can still get exceptionally hot if not allowed to vent their heat safely; again, you cannot cover them with insulation.

Having stated that gaps are important around certain fittings, it must also be understood that there should not be any gaps elsewhere. CSIRO testing has shown that even 5% of unfilled area created by poor fitting between members or batts can result in as much as 50% reduction in the efficiency of the system as a whole.

Whatever the material, and whatever the strictures on its location, correct installation will maximise the thermal value of the building elements and reduce the chance of failure over time.

Reflective insulation should always be applied with the 'shiny' face downwards, or vertical and facing inwards, even in hot desert or tropical climates. This is counter-intuitive, and even counter to some manufacturer's specifications. It is now recognised, and proven, that the reflective foil will 'reflect' energy equally well whether the heat comes at it from the front (reflective) or the back (dull) face. If, however, the reflective surface is faced upwards, dust will quickly settle upon the foil. The dust then absorbs the reflected heat, which in turn is transferred back through the foil by way of conduction, something aluminium will do very easily.

Reflective foil needs at least a 25 mm (preferably more) air space between it and any conductive surface. Bulk insulation can take the place of this gap as it is in effect multiple pockets of air.

Care must be taken when installing reflective insulation as it is highly conductive of electricity and so it should not be laid over ceiling joist or wiring in any location.

Composite insulation, when reflective foils are involved, should likewise be arranged with the reflective material facing down, not up. That stated, some manufacturers still suggest that for the tropics, foil outfacing is preferred for reasons of condensation control: though the effectiveness of this practice is not well documented or described (see 'Insulation and condensation' below). In most cases the manufacturer will have a printed stamp indicating the manner of its installation as a guide. As with foil insulations generally, these must not be laid over electrical cables or where contact with electricity is at all possible.

Bulk insulation can take the form of loose fill or batts. Loose fill is generally only applicable to ceilings and floor spaces where it will sit flat and not compact under its own weight (thereby reducing its R-value.) The value of loose fill insulation is in ensuring that there are no gaps between framing members or individual batts.

Batts can be supplied as wall/vertical types or ceiling/floor/horizontal types. The wall types tend to be denser, retaining their form when placed upright within stud framing. Depending upon the material, wall types may give a higher R-value to thickness ratio than horizontal types, but they are generally more expensive. In either case, to obtain their stated R-value they must have sufficient space to expand to their nominated thickness. For example, typically, an R1.5 horizontal batt will expand to 90 mm to obtain that rating; an R2.0 polyester vertical batt will need an identical space to achieve its rating – in either case, if this space is not available, then the R-value will be reduced. When installing batts, you must make sure that there are no gaps between each batt, or framing members.

Polystyrene and similar hard foam sheets must be tight fitting between framing members, or as per manufacturer's instructions when used as cladding. Particular attention must be paid to any instructions concerning venting of walls, floors or other building elements to prevent condensation build up; an issue discussed more fully below.

Insulation and condensation

Condensation has become one of the biggest issues in construction in recent years. This is because, in our efforts to regulate the climate within the home with air conditioners and heaters, we have progressively sealed the home and eliminated virtually all incidental ventilation. In so doing, water vapour in the air, either from outside or generated by our own bodies and activities inside, becomes trapped. Where this is trapped, and for how long, is the issue.

The dew point

The dew point is the point at which the water vapour in the air reaches saturation and becomes water: in buildings this is generally seen as condensation on walls, ceilings and windows. However, it can also occur within building materials, or behind them. This means that water can be collecting unseen within the building fabric and components. This is known technically as 'interstitial condensation' and is the main cause of concern for architects, designers, builders and clients alike. This is because it can lead to moulds, mildew, fungus and rust that are destructive to the building, and damaging to occupant health.

The dew point varies, dependent upon air and surface temperatures, and the humidity. Such being the case, you cannot limit condensation by simply controlling room temperature. And if the humidity is coming from outside, then this is totally outside your control or observation: unless it is built out, or means of dealing with it are built in.

The risks of insulation

Insulation is one part of a total design strategy that attempts to exert an influence over the water, air, water vapour and heat within a building. The issue is that in controlling heat gain or loss particularly, the other factors become more problematic. Insulation limits the heat lost to, or passing through, the building fabric. In so doing, it limits evaporation from the interior and drying of the building components and voids.

Figure 14.16 shows a typical suspended and insulated subfloor design. The problem identified is that after insulating to a total value of R3.25 with the aid of a membrane (high or low vapour resistance is actually not critical in this case) such as foil sarking – as required by the NCC for alpine climates (zone 8), the top surface of that membrane will be cool enough to increase humidity and evoke condensation. This water is then trapped and ponds under the timber joists. At the same time, the timber joist may be taking up moisture to the point of saturation. In such cases, mould and/or fungus becomes possible, and the components may fail.

A suggested solution is to remove the membrane, increase the bulk insulation R-value and while ventilation of the subfloor is maintained, bring the wall insulation to ground level. This, of course, means increased provisions for termite prevention, and possibly the need to seal the ground against rising moisture. Testing would be needed to confirm the viability of the approach and that it equally satisfied or exceeded the NCC Performance Requirements.

In facing such problems, the Australian Building Codes Board (ABCB) – the agency responsible for the publication of the NCC – acknowledges that the issues surrounding condensation are complex and the knowledge to date is very much in flux. As such, the NCC does not directly address condensation through specific deemed-to-satisfy solutions. Indeed, the only references are in the Objectives (FO1 [Vol. One] and O2.2 [Vol. Two]) which seek to:

> ... safeguard the occupants from illness or injury and protect the building from damage caused by – ... (iii) the accumulation of internal moisture in a building.
>
> **Source**: NCC 2019 Building Code of Australia - Volume Two, © Commonwealth of Australia and the States and Territories of Australia 2019, published by the Australian Building Codes Board. licensed under a Creative Commons Attribution-NoDerivatives—4.0 International licence, https://creativecommons.org/licenses/by/4.0/

After this, condensation appears but once by name in each volume, and then only as commentary.

However, the ABCB, in conjunction with the industry and the Australian Institute of Architects, has issued a non-mandatory handbook entitled *Condensation in Buildings*. This is an extensive document that outlines the knowledge to date and the

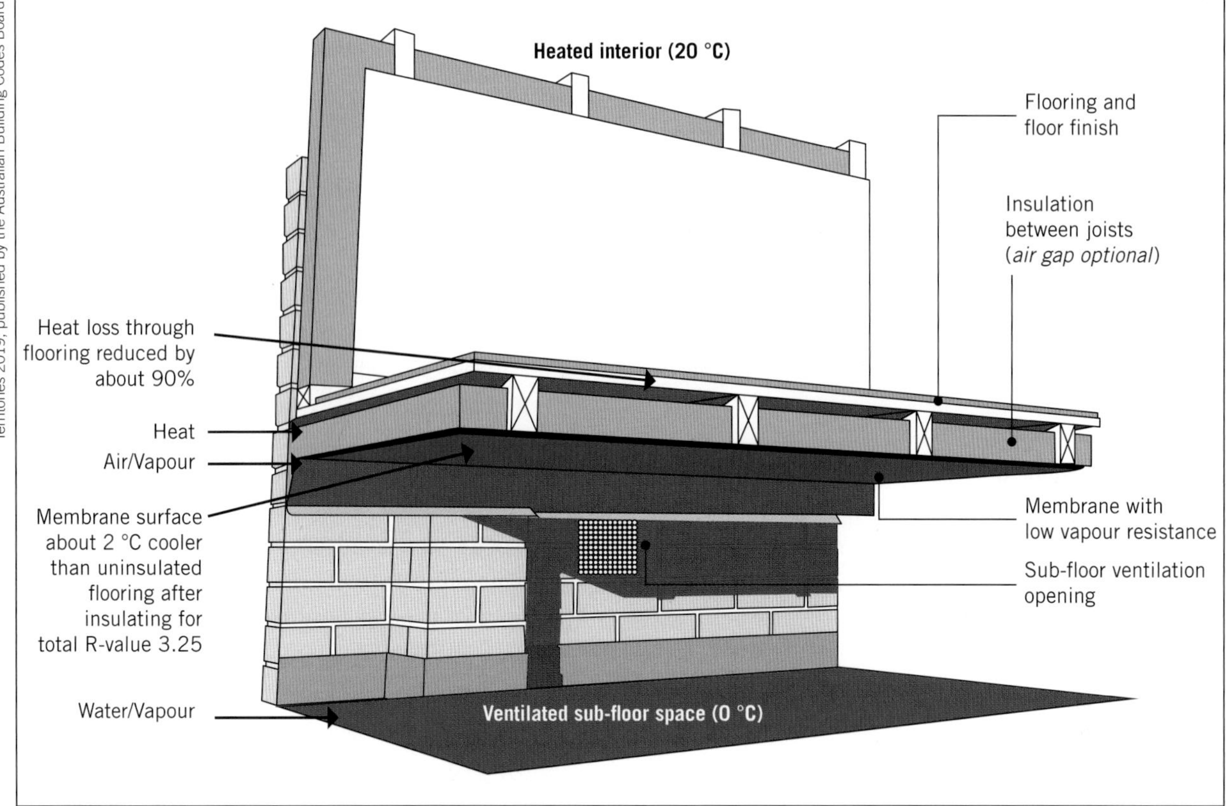

Source: Adapted from Handbook - Condensation in buildings © Commonwealth of Australia and States and Territories 2019, published by the Australian Building Codes Board.

FIGURE 14.16 Typical suspended floor insulation design

suggested best-practice solutions based upon climate zones. This is freely available from the ABCB website: http://www.abcb.gov.au/Resources/Publications/Education-Training/Condensation-in-Buildings

Safe buffering capacity: some materials, classed as hygroscopic, will take up water vapour as the relative humidity rises, and release it when it lowers. Each of these materials has a limit known as its 'safe buffering capacity' when applied to building solutions. This is the amount of water that the total quantity of a particular material in the structure can hold. Using a typical 185 m^2 brick veneer home as an example, this means that the safe buffering capacities for all the:

- brick cladding is approximately 1100 litres
- timber framing is 150 litres
- plaster lining is 15 litres.

This is a very useful factor in combating condensation, as the building structure itself is able to safely adsorb (note the 'd') a high amount of the water vapour present before reaching a point at which it will dangerously absorb (note the 'b') more. Unfortunately, in changing to more energy-efficient materials, we are making choices that frequently reduce this total capacity. Polystyrene claddings, steel frames and even engineered timbers all have significantly lower or non-existent safe buffering capacities; i.e. increased condensation issues.

Energy rating compliance

As identified in the first part of this chapter, the path to compliance depends upon the classification of the building under consideration, and if residential whether the project is in NSW or not. We will begin with residential compliance first, before moving to the more commercially-focused structures.

Building Classes 1, 2 and 4

To reiterate from the first part of this chapter, excepting NSW there are four paths to compliance for Class 1 buildings, the sole-occupancy units of Class 2 and the Class 4 parts of other buildings:

- the deemed-to-satisfy provisions – NCC, Vol. One (Part J0); Vol. Two (3.12.0)
- a Performance Solution that meets the Performance Requirements of the NCC
- in part, by using housing energy rating software approved by the NCC; i.e. NatHERS
- using a Reference Building, NCC, Vol. One (JV3); Vol. Two (V2.6.2.2).

With the exception of the reference building, these will each award a dwelling with a star rating, which must be 6 stars or higher to gain approval. The reference building approach is simply pass or fail.

In NSW, the path for these classes of buildings is via BASIX.

Deemed-to-satisfy

Determining compliance by this approach involves some 20 individual and complex calculations. These provisions are prescriptive, and so must be adhered to exactly for the structure to gain the performance required. In addition, this approach requires the use of Excel spreadsheets developed by the ABCB, and accessed via their website, for the evaluation of window thermal efficiency and lighting requirements.

Part of a submission using this approach may look something like the following (see extract below and Figure 14.17):

Climate Zone: 6 (Indigo)

Star Rating Required: 6

Assessed Star Rating: 6+

Basis for assessment:

"Deemed to Satisfy" As per NCC Vol 2, 2016 Clause 3.12.0 Energy Efficiency

Plus

Roof fed Water Tanks supplying both WC's and Shower/ Baths.

Breakdown of Assessment

Application of Part 3.12 (Victorian Variation)

(a) *Performance Requirement* P2.6.1 for the thermal performance of a building is satisfied by

(i) complying with—

(A) 3.12.0.1 for reducing the heating or cooling loads; and

(B) 3.12.1.1 for building *fabric*, thermal insulation; and

(C) 3.12.1.2(c) and 3.12.1.4(b) for thermal breaks; and

(D) 3.12.1.2(e) for compensating for a loss of ceiling insulation; and

(E) 3.12.1.5(c) and 3.12.1.5(d), for floor edge insulation; and

(F) Part 3.12.3, for building sealing; and

in the case of a new Class 1 building, having either a rainwater tank connected to all sanitary flushing systems, or a solar water heater system, installed in accordance with the Plumbing Regulations 2008; or

(ii) complying with—

(A) 3.12.1, for building *fabric*, and

(B) 3.12.2, for the external glazing and shading; and

(C) 3.12.3, for building sealing; and

(D) 3.12.4, for air movement; and

in the case of a new Class 1 building, having either a rainwater tank connected to all sanitary flushing systems, or a solar water heater system, installed in accordance with the Plumbing Regulations 2008.

NCC VOLUME TWO GLAZING CALCULATOR (first issued with NCC 2014) HELP

ENTER BUILDING NAME AND DESCRIPTION BELOW - identifying the particular part(s) covered by this assessment — Climate zone

CONSTANTS: C_U C_{SHGC}

Storey — Floor Construction — Area

Direct contact

Suspended

Air Movement

Wall insulation option chosen for 3.12.1.4

ALLOWANCES: C_U (only) C_{SHGC} x Area

Number of rows for table below 11 (as currently displayed)

GLAZING ELEMENTS, ORIENTATION SECTOR, SIZE and PERFORMANCE CHARACTERISTICS								SHADING		CALCULATION DATA			CALCULATED OUTCOMES			
Glazing element		Orientation	Size			Performance		P&H or device		Exposure		Size	Conductance		Solar heat gain	
ID	Description (optional)	Facing sector	Height (m)	Width (m)	Area (m²)	Total System U-Value (AFRC)	Total System SHGC (AFRC)	P (m)	H (m)	P/H	Es	Area used (m²)	U x area / winter access	Element share of % of allowance used	SHGC x Es x area	Element share of % of allowance used
1																
2																
3																
4																
5																
6																
7																
8																
9																
10																
11																

IMPORTANT NOTICE AND DISCLAIMER IN RESPECT OF THE GLAZING CALCULATOR

The Glazing Calculator has been developed by the ABCB to assist in developing a better understanding of glazing energy efficiency parameters.
While the ABCB believes that the Glazing Calculator, if used correctly, will produce accurate results, it is provided "as is" and without any representation or warranty of any kind, including that it is fit for any purpose or of merchantable quality, or functions as intended or at all.
Your use of the Glazing Calculator is entirely at your own risk and the ABCB accepts no liability of any kind.

Source: NCC Volume Two Glazing Calculator (first issued with NCC 2014)

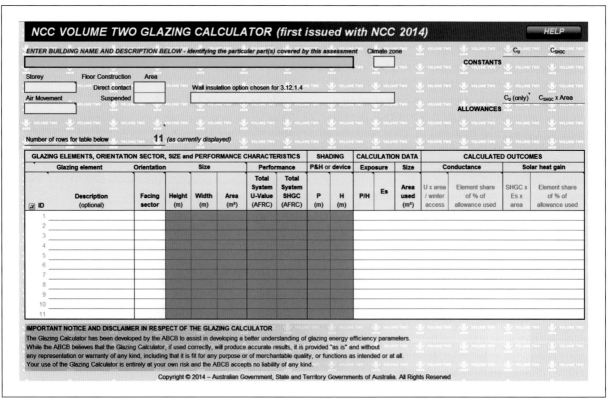

FIGURE 14.17 NCC glazing calculator

This Audit will deem-to-satisfy via compliance with clause (ii).

(A) Fabric (Part 3.12.1)

Roofs (3.12.1.2)

(From Figure 3.12.1.1 (d) Total R-Value for typical roof and ceiling construction)

Main: **Traditionally Pitched roof with flat ceiling, ventilated roof space (Metal Cladding solar absorptance value < 0.4: Light cream)**

Total R-Value required for Zone 6 (upwards):	R4.60
Insulation applied to ceiling:	R4.50
Roof materials value (figure 3.12.1.1(d) – Ventilated)	R0.21
Reflective insulation allowance:	R0.40
Total R-Value offered =	**R5.11**

- - - - - - - - - - - - - - - -

(B) External glazing (Part 3.12.2.1) *(See Table 1 Glazing Calculator)*

Proposed: **Timber Framed 5/12Ar/5 Glazing (Toughened as required).**

Conductance (Cu) [U-Value 3.2]

Allowed = 6.4 (Floor area 139; constant 6.418)

Calculated = **6.34 (99%)**

See Table 1 – Glazing Calculator

Deemed to satisfy.

Aggregate Solar Heat Gain (Cshgc) [SHGCw 0.48]

Allowed = 21.3 (Floor area 139; constant 0.153)

Calculated = **8.1 (38%)**

See Table 1 – Glazing Calculator

Deemed to satisfy

Source: Glenn Costin

The values offered in the above audit are from manufacturers and, in some cases, within the NCC itself. Solar absorptance values (roof cladding), for example, are dependent upon the roof colour, and can be obtained either from the NCC or the manufacturer. U-value and SHGCw value have been obtained from the manufacturer for the type of glazing system being provided. The two constants are from the NCC, Vol. Two (Table 3.12.2.1a and b, p. 389); likewise, the ***Required R-values*** for roof and ceiling are from Table 3.12.1.1a–g (NCC, Vol. Two, p. 369). The DTS provision audit would continue for each other building element, obtaining the relevant values in like manner.

An accredited assessor is *not* required to do these calculations, and some council and private certifiers still look favourably upon the process as it provides clear information on what has been considered and is easily checked. The alternatives, certificates provided by computer simulation software, are believed to allow room for results to be 'massaged' and are difficult to analyse.

To aid those wishing to go down this path, the ABCB has published two non-mandatory handbooks entitled *NCC Volume One Energy Efficiency Provisions 2016*, and similarly for NCC Volume Two, which are accessible by download from their website (https://www.abcb.gov.au/Resources/Publications/Education-Training/NCC-Volume-Two-Energy-Efficiency-Provisions). They also publish YouTube clips that talk you through the use of both the lighting and window spreadsheets.

If you choose to take this path, you are recommended to review Chapter 1 of this book, as well as the ABCB Handbook and YouTube clips. Alternatively, you can employ an experienced professional (accredited or not) to carry out the task, though the responsibility remains with yourself.

Performance Solutions

Performance Solutions respond to the Performance Requirements of each NCC volume. These are found at:

- NCC, Volume One: Section J, JP1 – JP3
- NCC, Volume Two: Section 2, Part 2.6.

As with the DTS provisions, the ABCB's non-mandatory handbook, NCC, Volume Two, Energy Efficiency Provisions and its partner text for the NCC, Volume One, are your first step here. As you are likely to be using non-standard glazing practices, you can use the guidance notes in these documents to assist in doing your own calculations on the energy efficiency of the glazing elements.

Again, you may do these calculations yourself, or employ an experienced professional, remembering that the responsibility still remains yours. The end result, however, will look significantly different to that provided above for the DTS approach. The main differences are in the clauses quoted, and the verification 'proofs' provided. This will probably include manufacturer's documentation and/or the certificates from testing facilities, engineer's reports and the like. You are directed to Chapter 1 of this textbook for further insights into this approach, including a brief example.

NatHERS accredited assessors

The vast majority of builders, designers and architects today will use a NatHERS assessor accredited with the Australian Building Sustainability Association (ABSA) for determining compliance. Such assessors will then use an approved rating software to determine the dwelling's star rating. In so doing, they will provide guidance on building materials, window and other products and energy-efficiency specifications.

This is particularly the case in reference to Class 2 buildings and Class 4 parts of buildings (see NCC, Vol One, J0.2). Unlike Class 1 dwellings, there are no direct DTS provisions to follow for the individual units.

Using NatHERS software, it is acceptable to obtain a minimum of 5 stars for each unit, with an average of 6 across the building as a whole. As long as the other NCC provisions for building sealing and energy savings are met, the structure will pass.

There are three software programs open to ABSA assessors to use when conducting their assessment:

- AccuRate v2.3.3.13 SP3
- BERS Pro v4.3.0.2a (3.13)
- FirstRate5 v5.2.6 (3.13).

Note the versions offered. These are accurate at the time of writing, but must be checked on the certificate issued and on the NatHERS website to ensure currency. Out of date software is not compliant. These develop a simulated model of the residence and derive from the expected indoor temperatures based on data provided by you, the designer/architect as well as the specifications and drawings. Thus, they demonstrate compliance with the thermal performance provisions of the NCC, effectively using the DTS path.

In making a determination, the following elements are considered:

- location, context, climate zone (NatHERS uses 69 subzones as against the NCC's eight zones)
- orientation
- construction type
- materials, windows, installed products
- size, location and function of rooms
- size and location of openings.

Other information that must be provided includes the design and layout of the building, and common patterns used. However, the only electrical appliances considered are those that heat or cool the building, and airflow devices such as ceiling fans.

From this data the amount of heating and cooling energy inputs required for occupants to remain comfortable may be modelled. This modelling is for a typical year based upon the climatic history of the identified zone. The software does not use an average temperature for the year, or even a month, but makes calculations based upon every hour of every day.

Within the software modelling, a dwelling's internal temperatures can be made to shift outside a specified comfortable range, and the energy needs for heating or cooling can be derived. Such calculations can be based on the assumption that occupants will open or close windows, blinds or awnings before deploying heating or cooling systems. From this, the total annual heating and cooling requirements may be estimated, and a star rating out of 10 can be awarded.

At the conclusion of the software simulation, a certificate is issued that provides the basis for the calculations (climate zone, dwelling type, limitations and the like) and a clear star rating. An approved ABSA certificate looks like that offered in Figure 14.18.

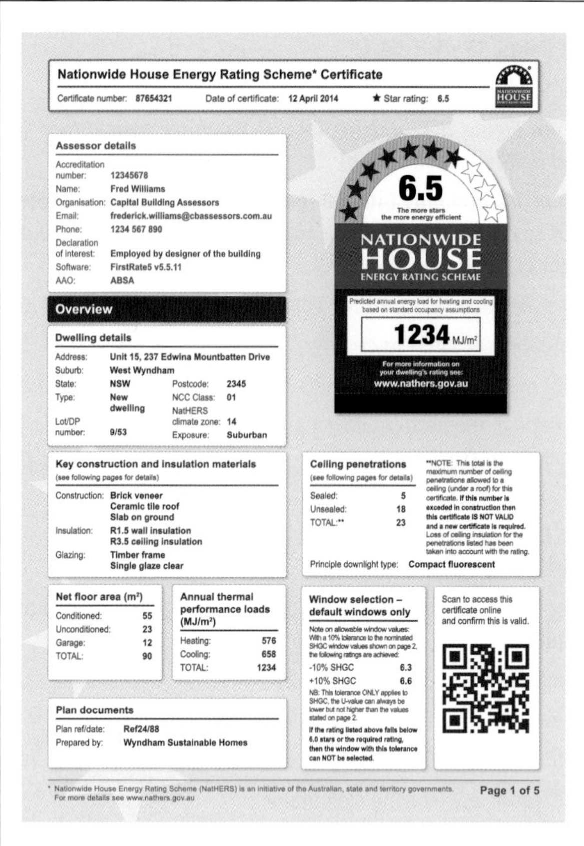

Nationwide House Energy Rating Scheme* Certificate

Certificate number: **87654321** Date of certificate: **12 April 2014** ★ Star rating: **6.5**

Assessor details

Accreditation number: **12345678**
Name: **Fred Williams**
Organisation: **Capital Building Assessors**
Email: **frederick.williams@cbassessors.com.au**
Phone: **1234 567 890**
Declaration of interest: **Employed by designer of the building**
Software: **FirstRate5 v5.5.11**
AAO: **ABSA**

6.5
The more stars the more energy efficient
NATIONWIDE HOUSE ENERGY RATING SCHEME
Predicted annual energy load for heating and cooling based on standard occupancy assumptions
1234 MJ/m²
For more information on your dwelling's rating see: www.nathers.gov.au

Overview

Dwelling details

Address: **Unit 15, 237 Edwina Mountbatten Drive**
Suburb: **West Wyndham**
State: **NSW** Postcode: **2345**
Type: **New dwelling** NCC Class: **01**
NatHERS climate zone: **14**
Lot/DP number: **9/53** Exposure: **Suburban**

Key construction and insulation materials
(see following pages for details)

Construction: **Brick veneer, Ceramic tile roof, Slab on ground**
Insulation: **R1.5 wall insulation, R3.5 ceiling insulation**
Glazing: **Timber frame, Single glaze clear**

Ceiling penetrations
(see following pages for details)

Sealed:	5
Unsealed:	18
TOTAL:**	23

NOTE: This total is the maximum number of ceiling penetrations allowed to a ceiling (under a roof) for this certificate. **If this number is exceded in construction then this certificate IS NOT VALID and a new certificate is required. Loss of ceiling insulation for the penetrations listed has been taken into account with the rating.

Principle downlight type: **Compact fluorescent**

Net floor area (m²)

Conditioned:	55
Unconditioned:	23
Garage:	12
TOTAL:	90

Annual thermal performance loads (MJ/m²)

Heating:	576
Cooling:	658
TOTAL:	1234

Window selection – default windows only

Note on allowable window values: With a 10% tolerance to the nominated SHGC window values shown on page 2, the following ratings are achieved:

-10% SHGC	6.3
+10% SHGC	6.6

NB: This tolerance ONLY applies to SHGC, the U-value can always be lower but not higher than the values stated on page 2.

If the rating listed above falls below 6.0 stars or the required rating, then the window with this tolerance can NOT be selected.

Scan to access this certificate online and confirm this is valid.

Plan documents

Plan ref/date: **Ref24/88**
Prepared by: **Wyndham Sustainable Homes**

* Nationwide House Energy Rating Scheme (NatHERS) is an initiative of the Australian, state and territory governments. For more details see www.nathers.gov.au

Page 1 of 5

FIGURE 14.18 Example NatHERs Certificate

Note that there are five pages to a typical accredited NatHERS certificate. The other pages provide evidence of window types, directions etc., as well as wall, ceiling and flooring structures, insulation materials and the like.

BASIX

BASIX is required to assess all Class 1, Class 2 and the Class 4 parts of other buildings' classes in NSW. BASIX may be completed online by anyone: no formal training or accreditation is required. Like NatHERS, a certificate is generated that forms part of the project documentation to council or a private certifier. The requirements for completing a BASIX assessment are available on the associated website: www.basix.nsw.gov.au

BASIX can be used as a complete certifying tool in itself; however, the Thermal Comfort Index offers two assessment paths entitled:

- DIY
- Simulation.

The first path must be completed within the BASIX program itself, but the Simulation approach must be undertaken by an accredited assessor using accredited software. That software is NatHERS, and accreditation is through the ABSA, albeit via BASIX. Despite this requirement, due to its capacity to improve performance and reduce costs overall, the Simulation path is preferred by over 75% of NSW applicants. A quick investigation of the purpose of the Thermal Comfort Index, and the two available paths, demonstrates why.

The Thermal Comfort Index is designed to assess energy loads placed on a dwelling by its fabric. No consideration is made in this index for heating and cooling appliance loads or fuel types; these are dealt with in the Energy Index. The Thermal Comfort Index seeks to:

- ensure thermal comfort of occupants relative to climate and season
- reduce the need for ancillary heating and cooling loads through appropriate design and materials
- thereby reduce the demand on energy infrastructure by reducing peak loads during extreme climatic conditions.

That is, the index is about the capacity of a house to reduce energy consumption through good material choices and design principles.

DIY: This path was developed as a simple approach to assessing house performance. In being simple, it works off a range of assumptions based upon sample existing dwelling designs. Although these samples cover a range of specifications and conditions, they remain limited in scope. One particular condition is that the dwelling cannot be greater than 300 m^2 in total floor area. Given that the average free-standing home in Australia is around 240 m^2, a lot of homes will exceed this limitation. In addition, to use the DIY method, the proposed dwelling must meet *all* the criteria of the sample chosen, not just most.

However, overriding all of the above issues is the fact that a home assessed using NatHERS software can achieve a 7-star rating, yet fail using the DIY method in BASIX. This is because of the lack of inputs in the DIY method compared with NatHERS programs. Hence, to better account for improved design, higher insulation levels and specialised construction features and materials – the expressed purpose of the Thermal Comfort Index – the Simulation approach is preferred.

Simulation: This method provides for broad selection of materials, shading solutions and insulation. In so doing it can reduce the need for expensive glazing elements and other costs. However, to use the method you must employ an accredited assessor trained in the use of approved software. The heating and cooling loads so obtained must then be entered into the BASIX program. These loads must be lower than those specified for the dwelling by BASIX based on the applicable climate zone.

Who is an accredited assessor and what programs they may use are defined in the BASIX thermal comfort protocol. In effect, this means an ABSA-accredited NatHERS assessor; and the programs are the same as listed above as accredited NatHERS software.

BASIX® Certificate

Building Sustainability Index www.basix.nsw.gov.au

Single Dwelling

Certificate number: 816995S_03

This certificate confirms that the proposed development will meet the NSW government's requirements for sustainability, if it is built in accordance with the commitments set out below. Terms used in this certificate, or in the commitments, have the meaning given by the document entitled "BASIX Definitions" dated 06/10/2017 published by the Department. This document is available at www.basix.nsw.gov.au

This certificate is a revision of certificate number 816995S lodged with the consent authority or certifier on 15 May 2017 with application DA2017/0132.

It is the responsibility of the applicant to verify with the consent authority that the original, or any revised certificate, complies with the requirements of Schedule 1 Clause 2A, 4A or 6A of the Environmental Planning and Assessment Regulation 2000

Secretary
Date of issue: Tuesday, 22 October 2019
To be valid, this certificate must be lodged within 3 months of the date of issue.

NSW GOVERNMENT | Planning, Industry & Environment

Project summary	
Project name	9 Dawn Street, Peakhurst NSW 2210_03
Street address	9 Dawn Street Peakhurst 2210
Local Government Area	Hurstville City Council
Plan type and plan number	deposited 31418
Lot no.	5
Section no.	-
Project type	separate dwelling house
No. of bedrooms	2

Project score		
Water	✔ 40	Target 40
Thermal Comfort	✔ Pass	Target Pass
Energy	✔ 43	Target 40

Certificate Prepared by
Name / Company Name: YJIREH
ABN (if applicable): 77168219734

BASIX Planning, Industry & Environment www.basix.nsw.gov.au Version: 2.3 / Certificate No.: 816995S_03 Tuesday, 22 October 2019 page 1/6

Source: NSW Government Department of Planning

FIGURE 14.19 Sample BASIX Certificate

Whichever path is taken, at the conclusion a certificate is generated (Figure 14.19) which must be appended to the planning proposal submitted to council or the independent certifier. Much like a NatHERS certificate, this is followed by six pages covering the commitments being made regarding energy, water, thermal comfort and the like.

Using a reference building

A reference building is a computer model used to calculate the energy loads for the proposed building, much like the BASIX DIY process mentioned above. The model is provided with the insulation levels, solar gains and energy losses via glazed elements, and the required temperature settings for internal spaces designed to match the proposed building and its climate zone. The concept is to allow for alternative building techniques otherwise outside the scope of the NCC. It is an approved verification method – under V2.6.2.2 of Volume Two for houses, and JV3 of Volume One for other classes.

This approach does not have to use any particular calculation method or software, provided it complies with the requirements of subclause V2.6.2.2(d). Many agencies have used NaTHERS software in the past; this is now expressly excluded by V2.6.2.2(a), as this only satisfies the temperature settings required by the clause in a very limited number of circumstances. Using other software tends to make the process more expensive again, and still it relies heavily on the correct inputting of data, including temperature settings, zoning and overshadowing.

Unlike the NatHERS accreditation process, which ensures the software and the data input procedures are current and align with best practice, no such back-up is available for this path. Such being the case, many structures, homes particularly, are being passed by accessors which should not be. Some are rated post-construction as low as 3 stars.

If choosing to use this approach for housing, it is critical that those conducting the analysis follow the NCC protocols as per the ABCB Advisory Note 2017-3. For commercial structures using JV3, this is particularly complex as the analysis is based upon the energy used by services (heating, cooling) in response to the thermostat settings of the building. In addition, due to the above concerns, the 2019 edition of Volume One of the NCC has much more extensive compliance requirements and specifications. Housing, on the other hand, is assessed by measuring the energy gain and loss through a building's fabric.

Class 3 and Class 5 – 9 buildings

The requirements for these classes of buildings are covered by sections Section J of NCC Volume One. The requirements for these classes are much broader than for residential dwellings, with the section holding nine parts running from J0–J8, of which two are blank. These include not only the building fabric, glazing, artificial lighting and sealing seen in housing Volume Two, but also:

- air-conditioning and ventilation systems
- heated water supply and swimming pool and spa pool plant
- facilities for energy monitoring.

Despite this breadth, as of 2019, only one Performance Requirement needs to be satisfied regarding energy efficiency: JP1. However, there are four Verification Methods (JV1–JV4) along with three specifications: JVa which holds additional requirements; JVb the parameters for modelling; and JVc that holds some modelling profiles.

The important shift in the Performance Requirement (JP1) is to defined maximum regulated energy consumption levels specific to building classes, measured in kJ/m^2/hr. Compliance with JP1 is possible by:

- **JV1 – NABERS** (National Australian Built Environment Rating System) **Energy for offices** assessment. This is for Class 5 buildings only. A minimum 5.5 star rating must be obtained while also matching specific data requirement of this Verification Method. This is a partial compliance, requiring also adherence to the specification **JVa**.
- **JV2 – Green Star**. This is for Classes 3 and 5–9 buildings only. Buildings must comply and be registered for a Green Star Design and As-Built rating. Compliance is based upon set parameters for greenhouse gas emissions and thermal comfort. This is a partial compliance, requiring adherence to additional parts of the specification **JVa**.
- **JV3 – Reference Building**. This is for Classes 3 and 5–9 buildings only. Compliance is based upon set parameters for greenhouse gas emissions and thermal comfort levels modelled on a reference building. This is a partial compliance, requiring adherence to additional parts of the specification **JVa**.

Note: all three of the above approaches must also comply with ANSI/ASHRAE Standard 140, Specification JVb, and the required modelling profiles of Specification JVc.

- **deemed-to-satisfy (DTS)** provisions J0 through to J8
- **Performance Solutions** in accordance with A0.7

Using the deemed-to-satisfy (DTS) and Performance Solutions have been discussed in some length in Chapter 1 of this book and you are directed to that chapter for a breakdown of these approaches. As with the residential sector, it is permitted to undertake these last two approaches by yourself. Indeed, the same may be said of JV3. However, with the complexity and scope of most commercial structures, this is not advised. Professional assistance by those experienced in undertaking this form of audit is the preferred path.

On major works, this analysis will be built into the building information modelling (BIM) systems so that any changes or variations, either during the design phase, or the construction phase, are correlated, and flow-on effects accounted for.

The final design

The final design should match the client's design brief as well as the Performance Requirements of the NCC, with specific reference to the regulations of your state or territory, and the local council planning provisions. Prior to handover to whichever certifying body (private or council) you have engaged, you should check that this is indeed the case with the designer/architect and the client. Alternatively, you may choose to include the certifying agent in this review. This requires a complete check of all drawings, specifications and compliance certificates. As the design to compliance timeline on major structures, and even some domestic dwellings, can be quite long, you should also ensure that the certificates obtained are still current (e.g. the changes in 2017 to BASIX, and the 2019 revision of the NCC).

The best approach for confirming the final design is by means of some form of checklist. Over the years there have been computer-based programs developed (e.g. DesignCheck) which have then been migrated into building information modelling (BIM) systems. These are particularly useful for the automated checking of compliance with codes and standards. They can also be used to automatically or manually inform all stakeholders of design changes and, in particular, any issues with compliance deriving from such changes that may need some level of design revision.

The advantage of the BIM approach is clear: it can check and recheck all aspects of the design with all aspects of the codes and standards – repeatedly, and with major or minor revisions to the design as a test for implications. It is significantly faster than any manual process and goes beyond a standard checklist.

However, on smaller projects, BIM has yet to gain traction in the construction community due in part to the steep learning curve and setup costs. In such cases, a simpler checklist approach should be considered. Today, that still should be a digital document for ease of dissemination, and indeed initial compilation. An example checklist and its possible contents is offered in the next section. The actual content will change to some degree with project type, state or territory and/or regional council. In creating a checklist, it must be remembered that energy efficiency, although only one element of the design, has an influence upon the whole of a design – hence your checklist will be for the whole, rather than the part.

As such, this form of checklist can only 'check that you have checked' each aspect. That is, have you obtained a BASIX or NatHERS certificate, and is it current? Has any changes been made to the design since obtaining such certificates? And so on. For residential works your checklist will not go back through all the codes and standards by individual clause. However, for commercial projects, such may well be required, hence the preference for the BIM approach.

Checklists

From an energy-efficiency perspective your first check is that you have the relevant documentation with which to carry out any meaningful assessment. That is, do you have:

- site plan
- floor plans and roof plans
- elevations, glazing and door schedules
- sections
- lighting plan and schedule
- specifications
  - floor coverings
  - construction materials including cladding types and colours (solar absorptance values)
  - insulation types, locations and R-values
  - mechanical ventilation and air movement (e.g. ceiling fans)
- water tanks' sizes, locations and what they service
- solar and other site-generated renewable energy inputs
- NCC building classification, elevation of building and climate zone.

If creating a generic checklist, the next section should cover what form of assessment(s) you have carried out. Type of energy efficiency assessment undertaken:

- DTS
- Performance Solution
- NatHERS
- BASIX
- JV1
- JV2
- JV3
- multiple paths (multi-class buildings).

Acknowledgement of additional clauses (e.g. JPb).

Glass assessment method:

- individual
- ABCB window calculator.

Next comes the types of certificates you should hold if the checks had been undertaken and the structure has passed.

- NatHERS performance star rating certificate: showing 6 stars or greater
- BASIX certificate: project scores in excess of targets
- thermal comfort certificate: for BASIX if not using the DIY procedure
- NABERS certificate

- Green Star certificate
- JV3 Certifier's certificate
- window assessment calculation
- engineers'/manufacturers' certificates on novel or new materials.

Like any contract documentation, you now need to ensure that all lead stakeholders have sighted and approved the final design. Aside from the client and designer/architect, you may need to include engineers and manufacturers of new materials – particularly if structural, but also from a fire safety perspective. You will need their signatures with dates demonstrating their approval of the design as a whole.

The Master Builders' Association produce an online `Six Star Workshop' covering a number of aspects of energy-efficient design that is free to access on their website: http://www.masterbuilderslowenergyhome.com.au/data/sus001/sus010/content.htm. While it addresses many of the aspects covered in this chapter, it does not do so from a regulatory perspective. As such, it is useful in confirming your approach only so far as energy-efficiency conceptual design goes rather than your necessary pre-submission checks. Likewise, there are council-supplied checklists, but these will only provide a single box to be ticked regarding ensuring some form of energy-efficiency report or certificate has been obtained and included.

Such being the case, it is best if you create your own modelled upon the information given above. How deep you go with regards to covering the regulations and NCC clauses previously checked during the design process is a risk-management decision which should be confirmed by all key stakeholders.

LEARNING TASK 14.2

GIVING USABLE SUSTAINABILITY ADVICE TO YOUR CLIENT

Your client has asked for your input (as the builder) on their project on how they might make their current house design (a simple project home) be more sustainable and thermally efficient, without needing to spend a large amount of money.

1. Provide a list of five to six options that could lead to improved building or land/site improvements that might/would have some support from the local council and other professionals to allow your suggestions to be included in the final building design submitted to council for approvals.
2. What options are offered as part of a BASIX (or similar) assessment that would promote sustainability in residential buildings in your state or territory?

Construction management: ensuring the outcome

This section goes back to the beginnings of the design phase, to the point at which you, as the builder, first entered into collaboration with your client and the design team. The purpose of the section is to outline best practice with regards to stakeholder communications and quality assurance. In addition, this section explores the cost factors involved when dealing with energy-efficiency concepts alternative to mainstream construction practice. The section then closes with a discussion on the life-cycle costs of materials, components and the structure as whole, relative to more traditional approaches.

Stakeholder communications from design to handover

Although there are many stakeholders in a construction project with whom effective communications must be maintained, the focus of this section is upon just the three leads:

- client
- architect/designer
- yourself.

In maintaining this loop, we will also speak of engineers, building certifiers and other professionals as necessary parts of what is effectively a coherent design team, if not a static one. This last point recognises that as the project unfolds, the importance of some members may diminish, others become more important, and some may simply come and go. Even though the core three remain, the importance of their roles, and hence their significance at any one point in the project timeline, will vary.

Identifying and prioritising stakeholders

If there are only the three of you this may sound like the easy bit; and it is, to a degree. Developing a list of stakeholders begins with a quick brainstorming session. Beginning with the three we know of (of which you are one), the client may be more than one person – e.g. a couple, an extended family group, a company, etc. To this may be added their financial backing – banks and the like, or other family members. You may be dealing with a single designer, or an architectural group with multiple drafting people. As to 'yourself', you might be in a partnership, or the representative of a large construction firm.

Once the list is made, you then start asking the questions of expectations and input – what is their stake and level of importance, and how does success or failure of the project affect them? Interests, rights, ownership, knowledge, influence and contribution are all issue of interest to the project. Personal and/or organisational gain including reputation are expectations from the project. Before mapping communication paths based upon the above, you need also to determine main lines of influence: up to lending bodies and other organisations to maintain commitment, inwards to tradespeople or other workers, outwards to suppliers and subcontractors, or sideways to clients and designers' subcontractors; remembering that this is seldom one-way traffic. As a typical sole operator residential builder, you may be many of these at the one time. Add a partner and you may share workloads or, more generally, split them.

Communication planning

The purpose behind this mapping is to understand the context in which you will be operating. Small or large project, the principles remain: you need to understand the communication paths and what may interrupt them; what 'strings' may be pulling on your client that may cause a reluctance in taking one path over another; what is driving your designer to take a particular approach which may make your work more difficult.

Most importantly, as the various aspects of the design unfold, or as variations arise during construction, the influence of these changes to energy efficiency, structural integrity, or any number of construction elements needs to be evaluated. Again, the right people need to know, so that they can inform the team as a whole, including yourself, of any issues that may arise. With regards to energy efficiency, the previous sections of the chapter have made you aware that even subtle changes can require a new energy audit certificate to be developed. As has been mentioned previously, this is the value of the BIM system for large projects.

Mapping these influences helps you determine correct communication paths; paths that will ensure those who need to know, do know, when they need to. Which of course includes you. They also help you to choose the right communication tools. In the world of mass communication, you have multiple choices:

- noticeboards (physical or virtual)
- reports
- emails and intranets
- web portals and other information repositories
- phone conversations and face-to-face (either direct or virtual)
- presentations
- team briefings and meetings
- focus and consultation groups and workshops (can be virtual).

The basis of sound communication planning in construction is very simple: what types of information does a certain person need, in what format, from which source, how often and how will feedback be obtained; and, finally, how is the communication plan monitored. In its simplest form, a typical communications plan may look like that shown in Table 14.1.

To this would be added the person who should manage this communications relationship. Even in a simple husband and wife building firm, it can be that the client may simply communicate more easily with one or the other of the partners. The more stress involved, the more important that this is identified early and appropriate communication paths set in place.

Communication plan monitoring

This means the monitoring of the communications plan, not the communication traffic itself. The purpose of monitoring is to evaluate the effectiveness or otherwise of the communication plan that is in place. If it's working, leave it alone; if it's not, then you need to determine why and fix it. Finding the why is often best served by simply asking the individuals themselves. Finding a solution can also come from these sources, but sometimes it needs to be discussed through a workshop or other type of group meeting. This is generally considered the best approach because if one area of communication is overtly failing, other relationships may be functioning at less than optimal levels.

Handover

This is where, surprisingly frequently, communications break down. From a business perspective, communications should be maintained for a number of clear reasons; that is to:

- maintain good customer relations that can lead to future work
- ensure that the client understands the workings of their new building

TABLE 14.1 Typical communications management document

Stakeholder	Message type	Format	Sender	Frequency	Requested Response time
Client	Draft energy assessment	Word doc Email	Energy assessor	As per project plan	5 days
	Design report	Word doc Email	Architect/builder	Fortnightly	5 days

- ensure that the new building is functioning as per the designed intent
- learn from the building's strengths and weaknesses for future projects.

All the information so learnt can now be fed into your company's knowledge bank. Communications should continue for the same reasons between yourself and the architect or designer. Through these sorts of relationships, improvements to buildings continue to be fed into our built environment and, as a community, our burden on available energy resources lessen.

Quality assurance and control

Quality assurance and quality control are two different but interwoven processes that must be got right if a project is to be fully successful. Fully, in that the product is to the highest standard achievable with the resources to hand – skills, knowledge, materials and finances – and that it is done in a timely manner with appropriate remuneration to all those involved: that includes profits and reputations. As such, both aspects serve as part of any viable approach to risk management in having clear implications for risks to:

- quality
- work health and safety (WHS)
- costs
- deadlines
- legislative infringement.

Quality assurance and quality controls are put in place to reduce exposure to these risks. Quality assurance (QA) is the selection and/or setting of standards by which the various elements and stages of the project and its conduct will be evaluated. Quality control (QC) is ensuring that these standards and the desired conduct are being adhered to. In simple terms, QA is the selection of the measuring instrument, QC is putting it to use.

Quality assurance

Depending upon the size of the company involved, quality assurance may be outsourced to the architectural firm who designed the project, or a completely external agency. In most cases, however, it is preferred that in-house quality assurance practices are developed even if external oversight is engaged. Best practice suggests the establishment of a third-party certified QA approach which meets or exceeds the International and Australian Quality Assurance Standard AS/NZS ISO 9001. It should also meet related guidelines for construction AS/NZS ISO 3905.2. The focus should be on integrated and systematic management techniques that allow all stakeholders to be informed of actions, changes and challenges in a timely manner as the project proceeds. It also should inform these stakeholders of the requirements – qualifications and expected standards of operations and outcomes – associated with their participation.

During the design phase, quality assurance focuses upon:

- identifying and developing standards and procedures to ensure the design meets the requirements of the client, the various stakeholders and the regulations
- the WHS considerations associated with the construction processes and materials implied or dictated by the design – to be fed back into the design to reduce risks where practicable
- the qualifications and/or experience base required by designers, engineers and architects
- standards for drawing preparation, revision tracking and approval.

As the design develops, the following will also be identified and incorporated into the QA plan:

- standards for proposed materials, construction elements and components
- credentials of suppliers
- qualifications and experience required (legislatively or otherwise desirable) by workers, subcontractors and their staff that will be required to complete the works
- stages for inspection and testing (legislatively required or strategic for QC purposes)
- internal auditing and managerial review intervals
- WHS strategies across all project phases.

Quality assurance is a constant cycle of internal review as the project proceeds through the rest of the phases leading to handover. It is conducted through a combination of formal internal audits and managerial reviews at specific intervals, with the capacity for rapid response change if need arises. Where possible, reviews should be conducted by those external to, but experienced in, the area being reviewed.

Specific on-site activities and processes and stages need to be identified as points for quality control inspections; examples being:

- points at which further inspection will not be possible due to enclosure (e.g. steel reinforcement to concrete footings)
- stages at which further works or activity must depend (e.g. scaffolding)
- work that sets the standard for replication (e.g. the first few wardrobes of multiple)
- product delivery (e.g. windows or doors).

These are then controlled through the provision of documented inspection and/or tests, the verification of which is included in the details of the QA plan. This will include details of the stage at which the inspections or tests should be conducted, who may carry tests out (qualifications) and the acceptance criteria.

With regards to energy-efficient and sustainable construction, much of the standard, code and regulation identification has been developed and discussed in the previous sections. The project specifics will then dictate most of the other elements of the QA plan based on what is described above. With the basis of a quality assurance plan developed – remembering, that it is an ongoing cycle of continuous improvement as the project proceeds – you are in a position to identify appropriate quality control practices.

Quality control

Quality control (QC) is the tool by which quality assurance measures are conducted. It checks that those things identified in the quality assurance process are actually in place, such as:

- people carrying out the work have the specified qualifications
- that companies supplying materials and products have the certifications or approvals identified
- products and materials match the specified standards or regulations
- drawings and specifications are signed properly
- drawing variations tracked and appropriately disseminated
- WHS is in place and constantly monitored as per the QA plan
- WHS considerations or issues are fed back into the QA plan
- tests and inspections are conducted as per the QA plan schedules and requirements.

With regards to energy efficiency particularly, QC measures will ensure that the plan's specifications are ready for approvals and certificates such as BASIX, NatHERS, NABERS, or GreenStar are obtained. At the same time, QC ensures that the commitments of such certifications are being adhered to as the construction proceeds.

BIM and quality management

Quality assurance and control may be significantly supported by a variety of software programs. In their simplest state these may take the form of computer spreadsheets and/or stand-alone proprietary programs, again often based on spreadsheets. As complexity increases, online tools may be used to aid in information dissemination, feedback and version control. Currently, the peak of computer aided quality management is though building information management (BIM) systems. While BIM will not eliminate face-to-face meetings, workshops and the like, it can aid in determining when such meetings need to be held, who should attend and what should be discussed. BIM can also ensure all stakeholders are engaging with the current plan sets and are in a position to feed in issues arising from new phases in the design, or variations during construction.

This section is entitled '… quality management' because BIM can form the repository and monitoring system of both quality assurance plans, processes and standards, as it will for the ongoing quality control inputs of tests, checks and audits conducted. It can then cross-reference these to ensure a match, and when they don't, issue alerts and show the implications of a failed assessment.

When dealing with energy efficiency particularly, due to the strictures of certification, even small variations to the design can lead to a failure to meet commitments and/or the need for revaluation. BIM makes this process of checking much quicker, and the finding of possible alternative offsets, such as materials, further design variations, upgraded window elements or the like, that much easier.

Costing the alternatives

The general belief is that the construction of an energy-efficient and sustainable home is going to be significantly more expensive than standard homes. It was also once believed that such homes did not look 'normal' and so have reduced on-sell prices. The reality is that the first remains marginally true, and the second may be disregarded – indeed, given the trend toward mandatory disclosure of as-built energy efficiency ratings, the reverse is true.

Keeping costs rational is one of the hardest parts of the design process, and flows into the construction phase as well. When presenting the cost of materials and design implications to clients, it is important to factor in the costs of:

- purchase transport
- installation
- maintenance
- energy savings
- alternatives.

It is also generally held that operational energy requirements exceed that which is needed to construct a standard building, particularly commercial buildings, in under 10 years. However, as buildings have become efficient, operational energy consumption has decreased. Meaning that the environmental impact of building materials used in construction once again comes into focus from a whole-of-life energy expenditure and environmental sustainability perspective. That is, materials have an impact upon the environment additional to energy consumption, such as:

- degradation of habitat for native species
- oxygen generation depletion
- transport emissions
- preferential growing; e.g. monoculture plantations of bamboo or pine forests supplanting diverse native forests
- greenhouse gas emissions from transportation, production, installation
- future demolition and disposal requirements
- water entrapment.

This being the case, when considering sustainable materials for energy efficiency we might begin by looking for those that fulfil the following criteria:

- from renewable sources where possible

- materials that are recyclable or reusable
- manufacture requires low energy inputs
- low water expenditure in manufacture and/or installation
- low water entrapment
- low maintenance
- high thermal qualities.

Procurement then becomes a question of availability, transportation cost and the skill base required for installation. This latter point is a necessary inclusion in any costings made. In some instances, a particular material may require both expertise and equipment from overseas to be imported. This can have massive implications to the viability of a project.

Example costing

Costing examples that are meaningful are difficult to produce. Many examples will work on a square-metre basis of total house built, suggesting a typical brick veneer at one price/m^2, and then offer the alternative at another price/m^2. In practice these figures seldom play out in regards to energy-efficient design comparisons. Partly because of the cost of the design process, in part that only certain materials are being changed, or because the build process itself changes too radically for the prices to have any validity.

When pricing, therefore, you need to consider all the implications offered previously, and match them to alternative materials used within the same element of the structure based upon the required R-value, or an improvement of that value. For example:

- The R-value for a ceiling and roof combination in climate Zone 6 is 4.6.
- The example offered in the section 'Energy rating compliance', earlier in this chapter, suggests that the use of the following materials and structure will provide an appropriate solution:

Traditionally Pitched roof with flat ceiling, ventilated roof space (Metal Cladding solar absorptance value < 0.4: Light cream)

Total R-value required for Zone 6 (upwards):	R4.60
Insulation applied to ceiling:	R4.50
Roof materials value (figure 3.12.1.1(d) –Ventilated)	R0.21
Reflective insulation allowance:	R0.40
Total R-value offered =	**R5.11**

Every component of this roofing design is open to alternatives.

- **Roof sheeting**: Tiles, slate, or even shingles may be suggested by the designer to the client. Pricing for each has implications that go beyond the material itself. Shingles and slate would add enormous expense, particularly when maintenance is factored into shingles, irrespective of the installation costs. Tiles mean additional loads that go all the way to footings and so change costs to structural components. The gain in R-value from tiles is a mere 0.02 (NCC, Vol. Two, Figure 3.12.1.1(d) – Ventilated).
- **Insulation**: The statement is for insulation with an R-value of 4.50. Choices are open as to what form this may take. Polyester means for quicker and easier installation as no particular personal protective equipment (PPE) is required. But it is typically more expensive than glass fibre or 'rock' wool batts. There are then polystyrene sheets; however, the installation time goes up significantly and hence the cost of this option. Sprayed-in insulation is another possibility to consider, particularly as it can ensure no gaps. However, there are installation implications around fans and ceiling lights where they exist.
- **Reflective insulation**: this is typically simple single-skin sarking, which is on offer here with an R-value of a mere 0.40. This is where a significant change and cost savings may be made. If this was changed to an 'aircell' type, the R-value may change such that lower rated – and hence cheaper – insulation batts may be provided. For example, the total roof system (less bulk insulation):

Roof materials value (figure 3.12.1.1(d) – Ventilated)	R0.21
Reflective insulation allowance:	R0.40
Total R-value offered =	**R0.61**

  vs

Using Kingspan Air-Cell Insulbreak 65	
Total R-value offered =	**R1.30**

  That is, bulk insulation can be dropped from R4.50 to R3.5 and still exceed the required R-4.60. The cost implications would then need to be calculated based upon availability, and regional pricing and transport. Labour would remain identical.
- **Ceiling lining**: There are different ways by which this may be lined, only one of which is using the plasterboard described. There are a range of insulated plasterboard sheets available, and there are sheets that do not use plasterboard at all. Each would need to be rated as per the previous example to allow a full comparison of total R-values followed by regionally applicable costings and availability.

The cost of costing

As any builder will tell you, there is no such thing as a free quote. Someone must pay. If you do 20 quotes a year, and only do five of those jobs, then five people are covering the cost of four quotes each. It is simply subsumed into your overheads. Costing for complex energy-efficient alternatives takes longer than normal, often significantly so if you have not done so before. More research is required and not all pricing systems such as Cordell or Quotefast will have the options you require already built in.

However, if you are brought into a project as part of the design team, then the workload diminishes to some degree. As part of a larger company working on commercial projects, these costs are factored into the initial planning and design phase. Working digitally, through either of the previously mentioned systems, BIM or your own spreadsheet approach, you will begin to build a knowledge base applicable to these types of projects which will further reduce time to you, and costs to your clients.

Life-cycle costing

Each of the decisions and choices made in the previous sections with regard to construction material choices and approaches has an implication on the life-cycle costs of the final building. **Life-cycle costing (LCC)** is the total of all costs of a structure from its design and construction, through ownership, usage and maintenance, to its ultimate disposal. Your task it to ensure that you and your client, and the designer, are best positioned to make informed choices at the design stage, and that any variations during construction are equally well informed.

Life-cycle costing often takes a triple bottom line (TBL) approach; that is, the influences upon the following aspects are considered:

- economic
- social
- environmental.

These are all considered when determining a costing. By doing so the cost benefits can be more accurately evaluated over the life expectancy of the building.

For example, consider the Council House 2 in Melbourne, Victoria. Significant emphasis was placed on it being an ecologically sustainable building, with the acknowledgement that initial capital costs would be proportionately higher, but with planned future capital expenditure being lower. This building included shower towers for cooling, water tanks, low-flow devices, a vaulted concrete ceiling and even its own micro-turbine power generation system.

The result is a building in which the tenants not only pay less for operation and maintenance, they tend to stay longer and are less open to volatile utility prices. In addition, with improved worker satisfaction based upon their high-quality environment, productivity increases were noted to the tune of almost 11% or approximately $2 million. Payback on the project was achieved within seven years.

From a TBL outlook, the building designers set aims for an overall economic outlook beyond the construction of the building itself. This was clearly successful. As were the environmental impacts – significantly lower greenhouse gas emissions and carbon footprint, and social influences – and improved occupant worker conditions. For more in-depth information about these buildings, see the 'Further reading' section at the end of this chapter.

Client negotiations

Initial costs are always going to be the limiting factor in the construction of a new building. Client's pockets are, ultimately, only so deep. Residential clients will be particularly limited in most cases. Such being the case your task is, in conjunction with the designer, to develop accurate life-cycle costings based upon the information available. Such information will include:

- ongoing ancillary energy inputs
- maintenance
- likely water usage
- occupant satisfaction.

The data for much of this will come from the development of energy audits such as BASIX and NatHERS, or for commercial buildings, NABERS and GreenStar. Maintenance costs will need to be derived from manufacturers and subcontractors experienced in particular materials in particular climates. Occupant satisfaction is a conversation in to which, logically, the client may be able offer some insights, alongside those drawn from occupants of similar structures. In domestic scenarios this may also be said of expected water usage, as well as light. Remember this is about actual expected usage as against the allowances provided in the rating systems. For example, the owner already knows that they intend to use the garage as a workshop in which significant power tool or welding work is likely to take place. These will influence your costings and proposals back to the client.

LEARNING TASK 14.3

MANAGING THE OBLIGATIONS OF SUSTAINABLE BUILDING PRINCIPLES

1 Describe in your own words the difference between quality control and quality assurance.
2 Complete the following in your own words:
The main aim of communicating to the client the way 'alternative sustainable building techniques' need to be used, managed and maintained over their anticipated design life is to ...

SUMMARY

This chapter set out to inform you on the various regulations and best practices surrounding energy-efficient and sustainable design. In so doing you will have become aware that sustainability and energy efficiency are two different, and sometimes divergent agendas. A chapter of this kind, mapped as it must be to the unit of competency, cannot cover much of this divergence. However, it is important that you, as the builder, be aware of the choices being made in a project and aid a client in making informed choices where you can. Being aware that materials that may aid a structure in being energy efficient need not be inherently sustainable is something that not many of your clients or even designers may understand.

Concrete is but one simple example of this issue. To make cement as we know it today, we must emit CO_2. It is a simple chemical reaction for which currently we have no option. Cement => CO_2. The amount produced is variably reported as between 5–8% of the CO_2 produced globally by humans, the second-largest source. It also consumes fresh water, water that is locked away for the life of the concrete item produced. It does not 'dry out' and return to the atmosphere as many believe. The amount is staggering, being approximately ¾ of the whole of Port Phillip Bay, as fresh drinking water is locked into the global production of concrete each year.

Many of our other strategies, such as LED light bulbs, solar arrays and battery back-up systems, use rare earth elements (REE). These are mined at significant human cost by people in economically underdeveloped nations on extremely poor wages and sometimes under military force. These REEs are not recovered from LEDs through recycling. Lithium, though not technically an REE, has significant environmental issues in being mined from very delicate ecosystems such as the Atacama Desert in northern Chile. The mining process alone using millions of litres of fresh water, to the detriment of local villages.

In addition, you will have learnt that some of the best options for energy efficiency, such as improved insulation, can actually have a detrimental effect upon the building and the occupants through increased condensation.

In the main, however, you should now have a clear understanding of the various rating systems and the paths by which you can meet their requirements. This applies for both residential and commercial buildings. You will also be aware that you are building for tomorrow in more than one way: the future sustainability of our community and the planet upon which we live; and the future contextual changes to the climate in which the building will reside and be used. It is not an easy path to follow, but its most assuredly a necessary one.

REFERENCES AND FURTHER READING

ABC News, **http://www.abc.net.au/news/2017-10-10/construction-loophole-leaving-buyers-with-higher-energy-bills/9033916**

City of Melbourne , Council House 2, **https://www.melbourne.vic.gov.au/building-and-development/sustainable-building/council-house-2/Pages/council-house-2.aspx**

Commonwealth of Australia, 'ESD design guide for office and public buildings', viewed 2011, **http://www.environarc.com.au/wp-content/uploads/2016/12/esd-design-guide.pdf**

Highton, Jemima 2012, 'Life-cycle costing and the procurement of new buildings: The future direction of the construction industry', *Public Infrastructure Bulletin*, Vol. 1, Iss. 8, Article 5.

USEFUL WEBLINKS

https://www.timeanddate.com/sun

ABCB: **https://www.abcb.gov.au/Resources/Tools-Calculators/Glazing-Calculator-NCC-2014-Volume-Two**

ABCB, *Condensation in Buildings*: **http://www.abcb.gov.au/Resources/Publications/Education-Training/Condensation-in-Buildings**

Australian Glass Group: **http://www.australianglassgroup.com.au/wp-content/uploads/2016/03/Performance-Data-Single-Glazed-AGG-Glass-June-2016.pdf?x86596**

Cradle to Cradle: **https://www.c2ccertified.org/drive-change/built-environment**

Design for climate: **http://www.yourhome.gov.au/passive-design/design-climate**

Ecospecifier: **http://www.ecospecifier.com.au/knowledge-green/setting-priorities/https://www.energy.gov.au/sites/g/files/net3411/f/energy-update-report-2017.pdf**

GECA: **http://www.geca.eco/our-industries/green-building/**

Geoscience Australia: **http://www.ga.gov.au/geodesy/astro/smpos.jsp**

Global Green Tag: **http://www.globalgreentag.com/about-greentag/**

Healthy Building Network: **https://homefree.healthybuilding.net/products**

Living Building Challenge (LBC product certification – Declare): **https://living-future.org/declare/**

SAI Global: **https://infostore.saiglobal.com/store/getpage.aspx?path=/publishing/shop/productguides/energy_efficiency.htm**

Windows Energy Rating Scheme (WERS) website: **https://www.wers.net/**

MINIMISING WASTE

15

Chapter overview

Waste minimisation should be a given on all building and construction projects in Australia as part of any sound sustainable construction practice. Unfortunately, the reality is somewhat different. Argument about what actually is waste, how to limit it and what to do with it – be it in the form of building materials, topsoil and even equipment – still rages, particularly on domestic sites. The purpose of this chapter is to outline the principles behind sound waste-minimisation practices as one of the key elements of a sustainable civilisation. The chapter ranges from legislation to industry best practice on minimising waste on the building and construction site.

Elements

This chapter provides knowledge and skill development materials on how to:

1. develop a waste-management strategy
2. manage materials procurement to minimise waste
3. manage the building process to reduce waste.

The Australian standards that relate to this chapter are:

- AS/NZS 3831 – 1998 Waste management – Glossary of terms

Introduction

According to the 2018 *Australian National Waste Report* (Pickin et al. 2018), the waste generated by Australians continues to increase, rising 6% in the year 2016–17; this, despite imposed levies, strategies, policies and fines. On a 'per capita' basis, due to the growth in population, the percentages look deceptively better, with a slight reduction in waste generated per person. On the other hand, over the past 11 years, recycling has increased by as much as 34% in some material categories, but with an average of less than 1% per year over that period. Yet this too is deceptive, given that shipping recyclables offshore has increased, with little control (or interest) over what is being shipped, and what happens to the material upon receipt. Also of note is that energy recovery from waste has fallen markedly over that last two years. Unfortunately, even the best of these statistics look better on paper than in real life, as reports continue to show that significant amounts of supposedly recycled materials end up in landfill.

In addition, much recyclable material is being dangerously stockpiled as China will no longer take refuse from other countries, and Australia is not yet geared for the recycling of many products, particularly plastics and metals.

While the construction and demolition sector has improved, it remains a significant contributor to landfill through poor waste management practices and resistance to change. In some states there is strong evidence of illegal dumping on as-yet empty housing lots, roadsides, state forests and even the fringes of national parks and beaches. The cost of this to the community and environment – and hence the community again – is high, and so continued improvement is paramount.

With improved surveillance, more targeted levies and higher penalties, construction and demolition waste-management strategies are now high on the agenda of all creditable organisations and small businesses. Indeed, most councils will demand a clear waste-management strategy (some councils refer to it as a 'plan') as part of any planning application prior to approval. This is particularly the case for major projects, or where large numbers of homes are to be constructed as part of a suburban development – despite those homes being constructed by a variety of separate small, medium and large building firms.

It is the purpose of this chapter to guide you in the development of such strategies by looking at what makes up a strategy that aligns with the regulations, how you might reduce waste through better procurement (purchasing) practices, and how best to align your plan with the building process.

Waste management strategies

Waste management involves three key principles: avoidance, reduction and recovery. Most people will know or recognise these principles as the three 'R's of waste minimisation:

- reduce
- reuse
- recycle.

Reduce the generation of waste in the first instance, through better design and material choices, and more accurate use of the materials chosen.

Reuse existing materials and fittings where possible and appropriate. Reuse can include the entire building through renovation and good design.

Recycle is the last-ditch defence before landfill. Recycling still consumes energy, but significantly less than when using raw materials; varying from around 20% reduction for glass, to almost 95% for aluminium (Milne and Reardon 2010, p. 139).

With up to 40% of landfill waste being generated by building and construction projects, the government has put in place a number of regulations in an attempt to make reductions. In addition, education around the necessity for waste minimisation, such as the unit you are studying now, has been increased. This education program focuses upon not just the negatives, but also highlights the benefits of more efficient building practices such as cost reduction, improved standards and reduced timeframes. We will begin by looking at the regulatory framework first.

The legislation

The key legislation that concerns us with regards to waste management or 'mitigation' is of both federal (national) and state and territory levels. The federal Act and regulations seek to protect and monitor issues of national importance – including activities of a nuclear or radioactive nature. The state and territory regulations deal in the issues of a more regional interest.

The *Environmental Protection and Biodiversity Conservation Act 1999*

The *Environmental Protection and Biodiversity Conservation Act 1999 (EPBC Act)* provides Australia's legal framework for the protection of the environment and ecological heritage. While states and territories also have their own legislation, as shown below, the EPBC Act overrides all others with regards to issues of national importance. It is therefore possible to obtain state-sanctioned development consent, but be liable under federal law framed by this Act. This was exampled when Coal and Allied Pty Ltd had to commit to expending over $2 million as part of an enforceable undertaking for activities in the Hunter Valley, NSW, in contravention of Part 9 of the federal EPBC Act – despite having development consent from the NSW state government.

The EPBC Act is enforced through the Environment Protection and Biodiversity Conservation Regulations 2000 (with current amendments to 2016). Depending upon the 'action' being considered (type of project, for example), the main sections of interest in this regulation for waste management plans are those concerning:

- environmental impact statements – when such statements are required and what they must contain
- waste disposal – disposal in this regulation includes air, water and soil pollution; disposal in a manner that endangers wildlife or vegetation, aquatic or marine species.

Your 'action' (project) will fall directly under these regulations is if it imposes itself in any way upon an area of national interest. Knowing if it does or does not affect a national interest is at times difficult and so you must seek advice from multiple stakeholders, including the community, the council, and at times state and federal ministers. In all cases, this will require you to produce an environmental impact statement. This statement will then be challenged in terms of content (declared scope of activities – the actions) and acceptability or appropriateness of any protective actions you may have formulated. This initial draft is effectively a discussion document sent out for comment.

State Acts and regulations

It is easy to get lost in state environmental protection regulatory frameworks. The Victorian Environmental Protection Agency (EPA), for example, administers no fewer than four Acts and seven separate sets of regulations. No one expects you to become an expert in the content of all of these, but you do need to know of their existence and to whom you may turn for advice. It is important that this advice is sound, as Acts and regulations are law, and so in breaching them you will be penalised. Further, regulations are frequently changed to some degree, so you need to be sure that the advice you are given is current.

Listed below by state and territory are the various government departments through which you can access those Acts and regulations relevant to your context (see Table 15.1). It is from these agencies that you would seek initial advice on state or territory regulations and, to some extent, federal.

Fact sheets and guidance material

Each of the authorities or departments listed in Table 15.1 publish or hold a range of fact sheets and other guidance material that can be of use to you in developing your management strategies. Remember, waste minimisation can include to some degree what we might ordinarily call pollution control. Topsoil runoff from building sites into waterways, for example, is both a waste of a valuable resource (a failure to reuse appropriately), and a pollutant to our streams that endangers fish, frogs, and other organisms through turbidity (clouding the water) and possibly the carrying in of toxins from the site. You will usually find these sorts of documents under the tag 'publications'.

Regional councils

City or shire councils will always be your first point of access for advice on any regulatory matter. In the main they are well informed of the current regulatory framework(s) and can offer some sound advice, even if it is only who else it is you should ask.

National Waste Policy

In 2009 a National Waste Policy was introduced by the federal government and updated in 2018 in response to the United Nations Sustainable Development Goals (SDGs); particularly SDG12. The aims of this policy are reflective of the three 'R's mentioned previously:

TABLE 15.1 State environment departments

State or Territory	Office or Department	Website
ACT	Environment, Planning and Sustainable Development Directorate – Environment	http://www.environment.act.gov.au/
New South Wales	Office of Environment and Heritage	http://www.environment.nsw.gov.au/
Northern Territory	NTEPA: Northern Territory Environment Protection Authority	https://ntepa.nt.gov.au/
Queensland	Department of Environment and Science	https://www.des.qld.gov.au/
South Australia	EPA South Australia	http://www.epa.sa.gov.au/
Tasmania	EPA Tasmania: Environment Protection Authority	http://epa.tas.gov.au/
Victoria	Environment Protection Authority (EPA)	http://www.epa.vic.gov.au/
Western Australia	Western Australian Environmental Protection Authority (EPA)	http://www.epa.wa.gov.au/

- avoid the generation, and reduce the amount, of waste for disposal
- manage waste as a resource rather than 'waste'
- ensure that waste treatment, disposal, recovery and reuse is undertaken in a safe, scientific and environmentally sound manner
- reduce greenhouse gas emissions, and improve energy conservation and production, water efficiency and land productivity.

The policy provides direction in five key areas and outlines 14 nationally coordinated priority strategies. You can find this policy at: http://www.environment.gov.au/protection/waste-resource-recovery/publications/national-waste-policy-2018. It is a useful resource in that the policy itself, particularly the 14 strategies, provides some level of framework from which you might model some elements of your own waste-management strategy.

The importance of this policy is that it recognises that regulations in themselves are inadequate. They set the minimum standards only – which will never suffice if waste generation increases through sheer weight of numbers. New initiatives will need to come from within the industry itself, material developers, designers, or clients. However, clients only tend to drive change when they perceive a cost benefit.

Cost–benefit analysis and the client

When first considering a waste management plan, it is easy to become idealistic and attempt to purchase the 'greenest' materials with the least waste, reuse everything possible and recycle the odd bit left over. The problem is cost. It doesn't always play out that this is going to be cost effective and ultimately it is the client who makes the decision because they are the one who is paying.

A cost–benefit analysis or CBA is a means of demonstrating both to yourself, and the client, the overall value or otherwise of a particular action. The terms 'cost' and 'benefit' are important as the analysis looks not just at dollar values, but also at what non-financial gains or losses are to be had over the long and short term. For example, stockpiling topsoil may incur a cost in time, labour and machinery hire, compared to a simple cut and dispose technique. However, if topsoil can be reused on site there is the front-end cost saving of not having to pay for transport and disposal; and the benefit of not having to purchase new topsoil and having the material already on site for future distribution.

Sometimes the analysis can show that the costs are higher, but the benefit might outweigh those costs. Purchasing higher quality components that have less packaging and last longer can be more expensive. On the other hand, replacing cheaper components after a short duration may ultimately cost more, and possibly cost more to run or maintain over the life of the building.

The purpose of the CBA is to bring all these factors to light, allowing you to advise your client on the possibilities and potentials of the options before you, along with any associated costs and risks. The client then is in a position to make an informed decision which may be fed back into the design.

Designing for waste minimisation

Depending upon the scale of the project, there may be architects, designers and engineers from varying sectors involved in the development. Depending upon their processes, you may find you are involved from the outset, or very early into, the design development. In such cases you have the ability to have immediate input into material choices, procurement practices and other waste-reduction mechanisms. This is best practice as designing for waste management becomes part of the design 'conversation', rather than an afterthought tacked on at the end. By 'conversation' we mean that good design is an ongoing discussion between all the stakeholders, including yourself as the builder. The result is better design, inclusive of waste minimisation and management strategies.

Alternatively, you may be handed the design after it has been approved by the client. This is harder, but your role still requires that you create effective, ongoing communication channels between yourself, the client and the designer, be they an architect or otherwise. On some occasions the designer is the client, which makes the path easier.

In both cases, your task is to ensure that the design is developed such that the building process, and the material and component choices, allow for fluid waste management practices – and where possible, reduce waste generation in the first place.

Building information modelling: BIM

Large commercial construction projects in Australia and overseas have been using building information modelling (BIM – also known as building information management) systems for some time. More recently some larger domestic home building groups have also begun incorporating these 3D information and modelling systems as a means of improved quantification and procurement of materials. The offshoot of this is reduced waste generation, and more control over the building process.

As the builder you will be invited into this process. Your role is to use your experience and knowledge base to guide the construction process such that foreseeable waste that cannot be eliminated is fluidly and appropriately handled. For more information on BIM see Chapter 5 Plans and specifications.

Designing out waste: the key principles

As stated, throughout the design process there is the opportunity to minimise waste generation and control the throughput of unavoidable waste materials. The Waste & Resources Action Program (WRAP 2016) of the United Kingdom have developed and tested a five-principle approach to designing that can aid in waste minimisation and waste recovery as outlined below.

- **Design for reuse and recovery**: this may include materials from an existing building on site as part of the demolition process, or materials from other previously demolished structures. It can also include new materials containing significant recycled content. Designers are encouraged to analyse a site with the builder to see what materials the site offers in this regard.
- **Design for off-site construction**: prefabrication is well documented as means of increased productivity and safety. It is also an excellent means of waste minimisation and management.
- **Design for materials optimisation**: this implies standardisation of materials and components. Room dimensions designed around the common material sizes, for example, can reduce the amount of plasterboard lining wasted, timber offcuts, steel framing and the like.
- **Design for waste-efficient procurement**: expanded upon in later in this chapter, this suggests that materials are sourced such that they reflect the previous principles; i.e. that construction efficiency is maximised and waste through offcuts minimised. It also includes materials that have less packaging waste, or are packaged with reusable or recyclable materials. Again, your role as the builder, and your access to the subcontractor team, is to provide advice on how to limit waste in the supply chain.
- **Design for deconstruction and flexibility**: This concept is seldom considered and hence deconstruction of buildings is often problematic. Likewise, renovations and maintenance are not always as easy as perhaps they could be. Your advice might be sought on the ease of disassembly with minimal waste generation, life span of components along with the associated maintenance schedules and, again, the waste generated by these programs.

Each of the above principles appear as factors in a construction concept that is still in its infancy in this country but will soon become as important as BIM; indeed, the two processes intertwine – this concept is known as 'lean' construction.

Lean construction

Lean construction is an approach that focuses on innovation, waste elimination and, through these, customer service and value. It factors in safe work practices, error eradication and continuous improvement through customer and contractor feedback and procurement partnerships. The origins of lean construction stem from 'lean production', a term coined by American researchers into the Japanese automotive industry during the late 1980s. It is led in Australia by the Lean Construction Institute of Australia and is an international movement supported by all levels of industry, engineers and designers (including architects).

Lean construction, and production, is a book in itself; however, some of the basic principles are reflected in the previous sections. The key point of 'lean' is the development and management of projects through respectful relationships in which knowledge and goals are shared. In so doing it uncovers wasted resources including materials, time, transport and even human potential. Lean looks more at process and flow rather than the simple measurement of results, though these are included. Its particular focus is therefore the minimisation of waste at the project level – but it views the project as a whole-of-life exercise, not just initial front-end design or delivery of the 'finished' product.

On large projects working through BIM platforms, this means a continuous flow derived from decentralised decision making. This in turn means that as technology, materials and processes external to the project change, these concepts can be fed into the project as it proceeds. It also means you, as the builder, are more empowered to influence the project as you are informed by, and in turn inform, the project team of which you are a part.

On small projects, you can simulate this without needing to resort to BIM, though BIM would still be a beneficial inclusion if you can factor in the training required for its deployment – and one day your competitors will. Your main focus would be in establishing procurement partnerships and ensuring the basic communications between subcontractors, designers and clients are well established, respectful and provide a multidirectional information flow. It is also through this interaction that you can ensure that your waste-minimisation strategies do not inadvertently breach the codes and standards to which your project must adhere.

Waste minimisation and building to Australian standards

Waste minimisation is not about cutting corners, arguably it is the reverse. Any action you pursue, or changes you invoke with respect to a structure's design, has implications regarding its compliance with the National Construction Code (NCC) and through its volumes, the Australian standards.

National Construction Code (NCC)

This code is discussed in depth in Chapter 1 of this book. Although the NCC frames our approach to structure rather than building management and waste mitigation, it is to this text you will refer when determining your material selections – and thereby waste reduction through better procurement strategies. It will also be your guide with regards to how you put the building together when making material or construction method choices that are less common, but are also aimed at reducing waste.

Many waste-minimisation strategies will have little to no impact on the standard construction practices as laid out in the deemed-to-satisfy provisions. In many cases the same construction practices are being followed – such as brick veneer or other timber and steel framing approaches – just with the design geared for less waste in offcuts. But it should be remembered that the NCC is a performance standard, not prescriptive. Such being the case, performance-based solutions are allowed, meaning that your building may be constructed out of any materials or in any manner provided that the result equals or exceeds the Performance Requirements.

This is where your knowledge of the NCC becomes critical to the paths taken by designers and clients as they attempt to make waste-minimisation choices, and as you develop your overall waste-management strategy.

For more information on Performance Standards and the NCC generally, see Chapter 1.

Developing and documenting the strategy

In developing and documenting your strategy you do not have to start with a blank sheet. Most councils, irrespective of in which state or territory they exist, will have waste-management templates for you to work from. In some cases, it is a requirement that you use these documents, while in others it is simply a suggested format. Most will also provide examples of how to complete them.

Parramatta Shire in NSW is just one of many good examples of both templates and waste-minimisation and management guides which may be accessed from the shire website. See Appendix 5, the waste-management template for this shire, at the back of this book. What follows is an outline of what the typically required contents of a waste-management strategy are for commercial projects, based on common NSW shire development control plans, or DCPs.

The waste-management strategy document

In applying for development approval for a commercial project in most parts of Australia you will be required to submit a waste-management strategy based upon four phases or stages of the project: design, ground works, construct, end usage. These will outline:

- material choices and any other waste-minimisation strategies determined in the design phase
- the expected volumes and types of waste to be generated from the site during the remaining three phases of the project:
  - demolition and/or excavation and ground works
  - construction
  - ongoing usage of the facility
- details of waste on-site storage
- waste separation including types of expected waste
- details of any specialised waste services – e.g. asbestos, toxic chemicals, or fluids
- ongoing waste-management strategies for the foreseeable future of the building
- nominated private waste contractor providing collection services.

The associated site plans and relevant construction drawings

Submitted along with the waste management plan, you will be required to provide a set of plans. These plans will usually be limited to a site plan, but may include any relevant construction drawings that outline the following:

- the location of any indoor waste and recycling receptacles within the building
- a waste storage room or rooms on the site if required by the shire. Generally, the specifications for these would include that they be:
  - sized sufficiently to accommodate all waste generated on the premises, including allowances for the separation of different waste types
  - located on the ground floor or basement with adequate room for access
  - constructed with a floor graded and drained to an approved drainage outlet connected to the sewer
  - of cement-rendered walls that are of a smooth, even surface
  - provided with at least cold water
  - adequately ventilated (either natural or mechanical) in accordance with the NCC
- location and design of any designated waste storage area sized sufficiently to accommodate all waste generated on the premises: including allowances for the separation of different waste types
- location of **grease traps** (where applicable)
- identification of collection points: includes travel paths to collection points, or vehicular access paths when there is on-property collection
- when on-site collection by council-operated services is requested, further details may be required outlining traffic plans, road surfaces, widths and the like.

On site

The documented strategy you provide to council is but one part of the equation. You must also have a document plan for yourself. It will be similar in many aspects but handled very differently. Council wants to know that you comply with their requirements. You need mechanisms that ensure that you carry out your plan and, in some projects, adapt the plan to suit new or emerging situations. The basic mechanisms are outlined as follows.

Copy of submitted waste-management strategy

You will need to retain a copy of the submitted waste-management plan both for your records and as a reference for future actions. You will need to compare this with the actual activities taking place to ensure ongoing compliance.

Site diary

You will need to insert into the diary the planned disposal routine for unavoidable waste. This may include multiple service providers depending upon the materials being removed, whether on-site separation has been adopted – some shires in some states prefer off-site separation as a means of ensuring compliance – and the capabilities of the service providers available.

The site diary will also be your means of documenting waste-minimisation actions that are taking place on site and any particular time or stage of the construction. To know more about site diaries and their usage, see Chapter 6.

Project schedule

Aside from the site diary, you will have a project schedule or plan that outlines all the activities and those elements that are critical to the timeline (see Chapter 6 On-site supervision). Your waste-management strategy must be overlaid upon this schedule.

Communication links: BIM or other

Because plans change and variations are made by clients, designers, or circumstance, your waste-management strategy may need adaption. Any changes need to be communicated to the various stakeholders involved, including council and service providers. It is easy to overlook the need to reschedule the removal of full skips, for example, and so prevent access to a particular part of the project that is now required due to a variation; or have a service provider turn up when access to the skips is not possible. This is where BIM can demonstrate its full worth as changes can be fluidly incorporated in the ongoing planning process.

LEARNING TASK 15.1

PLANNING AND PREPARING TO MANAGE MY SITE WASTE

AS 3831 – Waste management glossary of terms

1. Provide a list of all materials (nominate at least 15) and the construction work being performed leading to that waste, which you might expect to generate on a residential single-storey brick veneer house (slab on ground) build leading up to 'lockup' stage.
2. Using this list, create a table to anticipate/determine the amount of waste that might be individually generated as a result of the use of those materials for this job.
3. Having completed your 'site-waste analysis', research and document in the same table where that waste can be disposed of in your local area.
4. What guidance about building site waste management is being given by the regulatory authorities at this time when you search the internet with the key words 'construction building waste'?

Waste management and procurement

This section covers the basics of procurement, but with a focus upon waste minimisation. The strategies outlined below are framed around three key points that you should keep in mind:

1. That you are buying, and paying for, the waste material you throw out.
2. That all products and materials may be said to have an 'embodied' energy. This is the energy it takes to create and transport the item to you. This embodied energy increases with the work needed to install it, especially if this work requires consumable equipment such as grinders, sanders, saws and the like.
3. That products have a 'life cycle': this includes the point at which it is ultimately removed and disposed of either through reuse, recycling, decomposition (burning or crushing), or landfill.

With these three points in mind, material selection, and its procurement, holds a significant place in the minimisation of waste over not just the construction of the project, but also over the structure's life cycle including maintenance and deconstruction.

Waste minimisation through material selection

With regards to material selection, waste minimisation is more than simply limiting what is disposed of in landfill. Rather, it is inclusive of a broader world view and acknowledges a wide range of environmental factors such as: resource sustainability, climate change, and even the release of toxins into the environment, both internal and external to the structure.

With regards to material selection, you should start with the end of the product's life cycle in mind. Can it be reused, reconditioned, or recycled? Durability is not always critical, domestic suburban kitchens, for example, commonly have a life span of less than five years – this is because the home changes hands regularly and frequently the new buyer will replace the kitchen almost as a matter of course. In such instances, reusability or the capacity to recondition is more important.

On other occasions, durability will be a critical choice factor, likewise the maintenance needs to obtain that durability. Australian hardwood decking timbers, for example, need frequent coatings of timber oils to maintain and extend their life. Be aware also that mechanical fixing increases the end-of-life options due to the ease of dismantling and repair.

In making your choices, consider also the following factors:

- Material dimensions relative to the required dimension of the component when fitted; i.e. seek to reduce offcuts.
- Does the material have a high potential for reuse?
- Can the material be recycled easily?
- Is there a collection and recycling facility for unavoidable offcuts (e.g. plasterboard linings)?
- What toxins or gases emit from the material – when initially supplied, and over time? (Some treated timbers and ply material can have very high emissions initially, requiring gloved and masked handling procedures which means more waste.)
- Are there known carcinogens in the material, or in the processes needed to manufacture the material?
- When discarded, what is the timeline for disintegration and are there any toxin waste issues; e.g. is it biodegradable?
- What percentage of freely available and renewable raw materials are included?
- What verification comes with the material for any associated environmental claims?
- Does the material comply with all relevant Australian standards?

Remember, in asking these questions of your suppliers, you are making them more conscious of the types of materials your organisation requires. In so doing, it means they are more likely to seek out those types of materials from their sources – this 'cleans up' the supply line, increases sustainable choices and ultimately reduces waste derived from your projects. It is also conducive to the development of procurement partnerships as part of a multi-party project partnering contract.

Lean construction procurement strategies

One doesn't have to be fully embedded into a lean construction framework to appreciate or adopt some of its key strategies; particularly with regards to procurement. Some of the key phrases you will already have heard, such as: 'just in time', 'right first time' and 'zero defects'. Each of these are target parameters designed to reduce waste.

- **Just in time**, for example, means that materials do not sit around on site with the risk of being damaged and having to be replaced. It also reduces repetitive handling.
- **Right first time** also leads to less waste, for although the cost of faulty materials is borne by the supplier, there are always associated costs and sometimes wasted ancillary materials as well.
- **Zero defects** likewise limits waste through not having to redo work or replace materials.

Lean construction principles focus intensely upon respectful relationships and partnering for the long term. Project outcomes are always in focus; hence when suppliers are brought into such frameworks the wastage tends to drop as mutual aims and objectives are developed. On large public (government) projects, public private partnerships (PPPs) are sometimes a requirement of the tender process and so these concepts will have even greater relevance (see Chapter 8).

As stated previously, BIM and lean construction are frequently linked. This allows for greater flexibility in making variations a more fluid part of the construction process, particularly on large, long-term projects. Partnering with suppliers again makes sense as they will already know the types of materials you seek, how you want them delivered, and – through BIM – can schedule delivery appropriately, all of which reduces waste.

Recycled materials

Using recycled materials, either directly associated with a project site, or introduced from other deconstructions, is becoming increasingly common. Once the realm of the owner-builder, this option is finding favour particularly in the commercial sector, or in high-end domestic homes, as a means of introducing aesthetically pleasing design elements. Having stated that, you need to be aware that there may be limitations within your contract regarding what are effectively second-hand materials. Most contracts have a clause that includes something like:

> … all materials used will be new unless this contract expressly provides otherwise.

So, if the intent is to use second-hand timbers or components you must make sure that these are written into the contract before beginning the project. Likewise, if the intent is to reuse any existing materials from an on-site demolition process.

There are several factors you should take into consideration when considering the use of recycled or reused materials, particularly timbers:

- Strength ratings – large section beams and the like, when used as a structural or loadbearing component, must be carefully checked for fractures and penetrations, as well as the correct identification of species to ensure span capacities are correct.

- Toxins – many older structures may have had dangerous toxins applied as preservatives, insect elimination, or to improve aesthetics. Creosote, dieldrin and lead are all common toxic materials found on reused timbers.
- Recycled concrete is commonly used as a road base to gravel driveways: you need confirmation from the supplier that it is free of small pieces of steel reinforcement which will degrade or puncture tyres.

Recycled materials may also come in the form of new materials. Some 'timber-like' decking, fencing and cladding, for example, hold a high percentage of recycled plastics and on occasions timber. You should check with suppliers as to the binders (glues and resins) and what accreditation is available demonstrating adherence to standards.

Unboxing

Little needs be said in this area, the unboxing concept is simple: seek materials and items that are of a high standard yet come with minimal or no packaging materials. The issues surrounding packaging align with the three key points outlined at the beginning of this section; i.e. that you are paying for what you are discarding, for its disposal you are adding to its embodied energy just by discarding it, and you are paying for something that is probably at the end of its 'life cycle'.

Benefits of reducing packaging waste

Most builders do not realise the actual costs and issues surrounding packaging waste on site. These derive from:

- **Labour** – unpacking, collecting, segregation and sorting, storing and transporting packaging waste all have an associated labour cost.
- **Waste disposal** – skip hire, transport, collection and disposal fees are extremely high, even of recyclables.
- **Material purchase price** – unless the packaging is critical to the safe transportation of the material or item, then you are paying for something which is not required.

In themselves these can add up to a significant amount, particularly when involving the installation of multiple individually packaged items.

Other benefits

There are a range of other incidental benefits that may derive from the reduction of packaging waste arriving or being stored on your site. These include:

- **Improved procurement relationships** – as the main contractor, you are responsible for all waste material, including that of subcontractors. When suppliers and subcontractors are aware of your stance (waste minimisation) they are more likely to seek similar paths. Over time, this reduces unnecessary cost to all parties and improves relationships through changes of mindset and through this operational culture.
- **Tighter specifications** – procurement specifications can include packaging requirements as part of the contract. This means suppliers and subcontractors are informed upfront of your needs – clients also see this and tend to respond positively, even if only due to cost savings.
- **Marketing** – demonstrated environmental performance and intent is part of your reputation, it is marketable. Even cleaner building sites, with less visible packaging waste in need of disposal 'sells' your image to existent and prospective stakeholders and clients.
- **Legislation** – if you handle significant quantities of packaging waste you make it harder to comply with legislation and local council regulations. Packaging is also frequently very light. This means that in high winds it is easily picked up and distributed outside of the construction site – leaving you open to fines and other penalties.

As a business you must make decisions about packaging waste from a profit and loss or commercial perspective. However, the above points suggest that there is value in framing these decisions in light of the environmental considerations – as these too have commercial and legal implications. These include:

- reduced energy consumption
- lowered risk of pollution incidents emanating from your site
- lower greenhouse gas emissions
- less wastage of virgin materials.

While there are many more purely environmental reasons for reducing packaging waste, the above alone suggest that, for a business of any size, choosing materials or items that have less packaging can be both a commercial, and an environmentally responsible, consideration.

LEARNING TASK 15.2 HOW TO BETTER MANAGE AND REDUCE WASTE ON MY SITE

You are in the process of ordering timber lengths for your latest project, where the floor to ceiling wall heights will need to be 2.8 m high for a two-storey building. It is anticipated that 90-mm timber will work, according to AS 1684.2 requirements, and the stud spacing will be at 450 ctrs max. on both levels.

You will be building the frames on site and will need to inform the subcontracted carpenters at the site your reasoning for the timber lengths you have ordered for use at site.

Source: Shutterstock.com/Lev Kropotov

1 What is the most economical timber lengths to order? Provide a **complete explanation as to the logic** of why you have chosen this length and sectional size.
2 What width of plasterboard and cornice will you order to ensure that minimal waste is created from wall-sheeting activities? Again, provide a **complete explanation as to the logic** of why you have chosen the nominated sheet widths and cornice size/s.
3 Using the picture, list the items on this worksite that could be better managed so as to improve the current site waste-management strategy.

Waste management and the building process

Despite significant advancements in construction techniques, technology and materials, the building process continues to generate a significant quantity of waste matter. This waste arises from a range of issues yet to be overcome: orderable material sizes and shapes, building designs, demolition, excavation, cleaning of equipment and the like. This section of the chapter focuses upon the ways in which we can better manage the unavoidable waste stream through improved planning and further limitation strategies.

Deconstruction

This is the first shift in your thinking. Deconstruction and demolition both invoke a mental perception of action: demolition suggesting a fast knock-down with no particular sequence as long as it clears the zone ready for new works; **deconstruction**, on the other hand, suggests the reversal of construction – a sequenced dismantling of a building into its component parts.

Demolition is always going to create the greatest waste stream despite much of this waste being reusable, or at least recyclable. However, there are cost and time implications to be considered that, from a commercial perspective, sometimes have to be balanced with the benefits of its alternative. This is particularly so given that the advantages of deconstruction are not always obvious, or even realisable, despite them including:

- work health and safety (WHS)
- environmentally responsible
- company reputation – clean, safe, coordinated site, even visually
- revenue stream from salvaged materials
- reduced waste disposal and transportation expenditure
- marketing potential to environmentally conscious clients
- potentially lower building materials cost to community at large
- the reduced consumption of new resources
- reduced landfill.

In recognising that these potential benefits cannot always be realised, it should also be recognised that it does not have to be an all or nothing decision between demolition and deconstruction: you can choose to do selective deconstruction. In making your decision, you should consider first and foremost the WHS aspects; only after this, would you factor in the:

- key environmental benefits
- space on site for deconstruction process and sorting, including any large plant and equipment access
- regional opportunities and capacity to absorb the materials and resources derived from the deconstruction process
- site security
- company image
- timelines
- waste disposal costs.

This last point is going to be key in any suggested level of demolition. By way of example, four house types were used in a demolition vs deconstruction costing study based on 2010 pricing for used materials, waste disposal and labour rates. With waste disposal fees having climbed considerably in most states and territories in Australia since then, the lesson remains relevant. Figure 15.1 shows that in all cases the cost of demolishing a house of any type is *always* higher than deconstruction. The figure for the asbestos-clad home does not include the costs associated with the removal and disposal of this material as it can never be reused or recycled.

Timelines

The second-last point in the previous list was timelines. Deconstructing a house is invariably a longer process with more skilful labour involved. Likewise, there will be more equipment and the handling, sorting and storage done with greater care if any return from the sale of the components is to be realised. This becomes another factor in your consideration of how much, if any, level of deconstruction will take place: clearly the type of structure and the material make-up being key drivers in this part of the discussion, as are, of course, the project timelines themselves.

Where a structure has been designed with deconstruction in mind, things become simpler and the decisions more clear-cut. The steel industry particularly has a focus upon reclamation of componentry, recommending bolted components where possible. But as has been mentioned previously, deconstruction is not very much at the forefront of architect and designer briefs as a key to waste reduction more generally.

Excavation

In accessing any council waste-management strategy or development control plan, there will be a section requiring or recommending the retention and reuse of topsoil on site. It will always be very specific about topsoil because of the importance of this material. As just about anyone living in Perth, WA, will know, sand is not a great garden material; likewise, those in the eastern states will attest that most clays are pretty much useless too. In each case it is the capacity to hold moisture and provide nutrients that is the issue. Topsoil has this; sand, clay and stone do not.

Unfortunately, it takes anything from 500 to many thousands of years for nature to lay as little as 25 mm of topsoil cover over any size area; most people want their gardens established a bit ahead of this timeframe. Further, Australia has its own specific issues with regards to topsoil generation and loss. The estimated global average for topsoil generation is approximately 114 mm every thousand years; Australia varies from as little as 10 mm/1000 years in NSW, to our best rate of 75 mm in Arnhem Land, NT. Add to this the quality of our soils, which is generally very poor. This land is old, in that unlike most other continents, the soils have been exposed to harsh weathering without the protection offered by the last Ice Age: most of the nutrients have been leached by wind, rain and sun.

The soil formation rates tell us something else of importance that we have only recently discovered: in Australia, viable soils are being formed at a rate less than they are being eroded; i.e. the land is losing topsoil quicker than it can generate it. Figure 15.2 graphically shows this degradation between just the years 2006 and 2015 – a mere nine years. Needless to say, we try not to waste this precious material by using it indiscriminately as land fill, or in waste disposal sites.

Can we make topsoil? Yes, we can, even from sand, and grow things in it remarkably well. But this takes time and the right ingredients: on a commercial scale it is remarkably expensive to produce.

The message is therefore fairly simple: when excavation is being planned, plan for the stockpiling, relocation and/or on-selling of the topsoil. For some councils this will not only be a recommendation, it

How much does demolition cost per square metre?					
Type of demolition project	**Sydney**	**Melbourne**	**Brisbane**	**Adelaide**	**Perth**
Framed residential building	$53/m^2	$40/m^2	$40/m^2	$39/m^2	$35/m^2
Brick residential building	$64/m^2	$54/m^2	$46/m^2	$47/m^2	$40/m^2
One-storey retail building	$99/m^2	$78/m^2	$60/m^2	$65/m^2	$62/m^2
Two-storey retail building	$114/m^2	$94/m^2	$82/m^2	$100/m^2	$85/m^2
Three-storey retail building	$102/m^2	$101/m^2	$72/m^2	$77/m^2	$95/m^2
Industrial brick warehouse	$91/m^2	$68/m^2	$50/m^2	$57/m^2	$54/m^2
Bulk storage shed	$61/m^2	$48/m^2	$37/m^2	$44/m^2	$43/m^2
Commercial offices (3-7 levels)	$124/m^2	$100/m^2	$101/m^2	$118/m^2	$96/m^2

Source: ServiceSeeking.com.au

FIGURE 15.1 Costs of deconstruction across the four building types

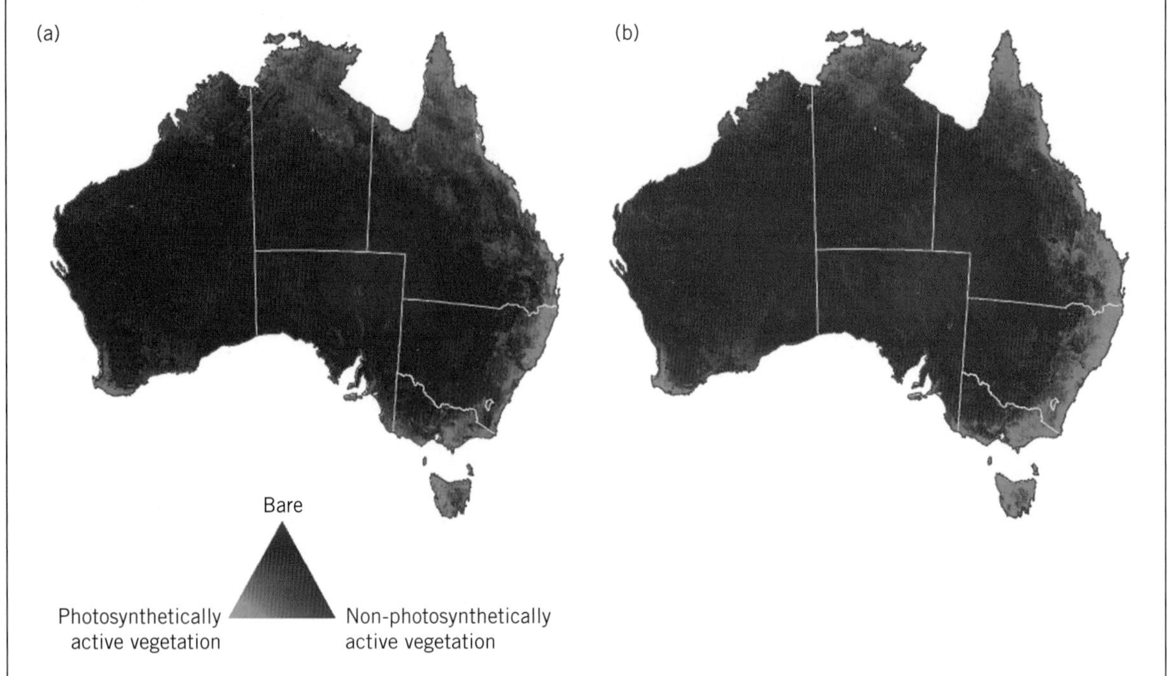

FIGURE 15.2 Images of Australia derived from remotely sensed data, showing the proportion of bare soil for April (a) 2006 and (b) 2015

will be a requirement. No doubt in the near future it will be a requirement in all regions of all states and territories.

Of course, topsoil is not the only material that will be excavated. Clays, rocks, tree roots, and on occasions old footings and other building materials are also dug up. Decisions must be made on each of these materials with regards to their disposal. Clean fill will be welcomed on many other sites if it is not of use on the site from which it has been dug. Clean fill includes clays, residual topsoil, and in some cases sand, and smaller stones and rocks. Rock is also often wanted on other sites as stabilising fill, garden surrounds and the like. Unfortunately, most mixed building materials excavated from a site will need to go to land fill.

Toxic excavated materials

These will generally be identified prior to excavation, in most cases before taking on the project as they form part of the contract. On occasions, however, toxic materials are discovered during excavation. In such cases, excavation should cease until the materials can be identified and appropriate cartage and disposal options developed. Examples can include asbestos sheeting, old farm garbage pits or small community 'middens', or buried petrol and chemical drums. In any of these cases, specialised dispersal or 'litter' abatement practices will need to be instigated to ensure the materials cannot escape the site boundaries: either by air, subsoil leaching, or through surface flows.

Litter abatement

When we think of litter, we tend to think of plastic containers, drink bottles, scrap paper and the odd chip wrapper. With regards to construction sites, litter is a far broader theme, including all of the above, but also:

- silt – eroded topsoil, clays, sands and other soils
- chemicals – paints, cement-based products, adhesives and coatings
- dusts – timber, asbestos, cement and metal dusts
- vegetation – discarded leaves, limbs, wood chips and the like
- packaging materials – plastic wraps, cardboard, strapping (metal and plastic) and composite materials such as plastic-lined paper cement bags.

Construction sites generate vast quantities of all of these litter types annually. And, unfortunately, water authorities around the nation have identified that the most common and damaging pollutants of our creeks, streams, rivers and bays is soil, sand and cement sediment and general litter from construction sites. For this reason, councils have a range of requirements for litter abatement, and for construction site litter generally; failure to meet these requirements can lead to severe fines and court prosecutions. Fortunately, councils also provide booklets and brochures advising on how to comply, along with a range of incidental and specific advice on litter reduction more generally.

When planning your project, you should therefore take this broader perspective on litter, and adopt strategies accordingly. Some of the more common approaches include:

- Perimeter fences – mesh fences surrounding the site are effective in preventing blown litter leaving the site. Solid hording does likewise, but also helps contain dusts and sediment.
- Sediment fencing – can be installed at the base of mesh fences, and around near to stormwater drains, gullies, and areas where water might run more directly into waterways.
- Containment cages and bins – these can be either mesh containment bays, or hired skips with lids that can hold unavoidable waste packaging and the like, or as sorting containers for reusable and/or recyclable materials.
- Cover or wet down dry materials – such as sand and soil that might be wind-blown beyond the site boundaries.
- Chemical storage – paints, cements and other chemicals should be securely stored to prevent accidental or mischievous spilling of these materials that might find their way into surface streams or our groundwater.
- Wet instead of dry cutting – of concrete, brick, tile or other such products to reduce dust.
- Hot knife or sharp blade (non-toothed) and vacuum cutting of styrofoam materials – to limit the dispersal of small beads of this material around the surrounding area.
- Washdown pits and rumble strips – for trucks and other vehicles leaving the site, this prevents mud and dusts being driven off the site and leaving a very ugly trail of evidence back to your project.
- Store materials within site boundaries – sands, gravels, commix and the like must always be stored on site, not on footpaths and road reserves.
- Skips – these should be stored on site also, never on footpaths, nature strips or the like. Likewise, any other form of waste storage pens from which litter may be seen, blown or distributed by animals.

In general, keeping the site tidy is going to be your best option for blown litter, while the above will serve you in containing all else. If you are uncertain, contact your local council who can send an enforcement officer out prior to any errors being made.

Toxic waste containment

Toxic waste removal is a very specialised area which will tend to use many of the filter and entrapment methods mentioned above. However, it will frequently require techniques specific to the material in question. Asbestos, for example, can require the complete containment of a structure inside an inflated plastic bubble. Oils, fuels, chemicals and the like found buried on a site will need identification prior to any further excavation being attempted.

Generally, these materials will have been identified within the contract and specialist teams engaged for removal and transport, as would be appropriate locations for disposal. In cases where the toxic material is a case of unplanned discovery, the same process will need to be followed through as if the materials have been planned for initially: i.e. materials identified; specialised personnel engaged for removal and transport; and correct disposal facilities contracted to receive the material.

Planning for waste

Throughout this chapter you have been developing the knowledge on how to minimise waste produced by or otherwise emanating from a construction project. In so doing, it has been noted that there will always be some level of unavoidable waste, either from the demolition and excavation process, or the building process more generally. This unavoidable waste must be quantified and material types identified, especially toxic materials including chemicals, asbestos and the like. Locations for the material's appropriate disposal then need to be found and contracted for receipt of approximate volumes at set points in the project schedule.

This is then fed into the documented waste-minimisation strategy submitted to council, held on site, and fed into the BIM systems, if such is in place. The content and format of this document has been outlined in the first section of this chapter, which is summarised here:

- council templates are available for use if they suit the nature of your project
- the waste management strategy is based upon four project stages
  - design
  - ground works
  - construct
  - end usage.

These four phases are then broken down to provide information on the following:

- waste minimisation strategies determined in the design phase
- expected volumes and types of waste to be generated
- demolition and/or excavation and ground works
- construction waste
- expected waste generated by ongoing usage of the facility
- details of waste on-site storage
- waste separation including types of expected waste
- details of any specialised waste services – e.g. asbestos, toxic chemicals or fluids
- nominated private waste contractor providing collection services.

Your waste-management strategy includes a plan set detailing:

- waste and recycling receptacles and/or rooms within the building with construction details aligning with the relevant sections of the NCC

- location of designated waste storage areas within the site external to the building
- location of grease traps
- identification of collection points and travel paths.

The information on this document is then used to inform site activities for the duration of the project unless it forms part of a BIM suite: in such cases the document is 'live' and may be updated with new information as project variations are accounted for. When not part of a BIM process, copies should be retained on site, and married into the project schedule and site diary.

LEARNING TASK 15.3 REGULATORY REASONS TO MANAGE THE WASTE PRODUCED FROM YOUR BUILDING SITE

1 Use the information from your local council or from the waste-management plan template at the website below:
   - www.shellharbourwaste.com.au (search 'waste management plan' and scroll down to the template)

   Document a formal waste-management plan ready for submission to your local council for a simple residential house build.

2 Access the NSW EPA website:
   - https://www.epa.nsw.gov.au

   or the regulatory authority in your state or territory, and respond to the following questions:
   a What is the legislation that governs construction waste management at your building site?
   b What fines might you expect to pay if the waste at your site is not properly managed and disposed of?
   c How might you ensure that you have appropriate evidence that the waste at your site has been legally disposed of?

SUMMARY

The purpose of this chapter has been to outline the principles behind sound waste-minimisation practices and the development of council-acceptable waste-management strategy documentation. As has been discussed, although waste minimisation is based upon the well-known principles of avoidance, reduction and recovery, better known as the three 'R's of reduce, reuse and recycle, it is much more than that. It is a systematic construction approach backed by a strong regulatory framework, high levels of surveillance, levies, and penalties for non-compliance.

On the positive side, construction firms are adopting these strategies as a means for commercial success: recognising the marketing potential as well as the moral imperative driven by heightened environmental awareness. In addition, as has been shown, careful planning, in both the design phase of the new structure and in the demolition of the old, can bring about significant improvements to the financial bottom line. Which brings us to the last point that may be made here: sound business practice in today's world actually has three bottom lines: people, the planet and profit – in that order. It's known as TBL or triple bottom line. This chapter attempts to reflect that premise as it wove through the strictures of the competency unit – it is now for you to expand upon these beginnings.

REFERENCES AND FURTHER READING

http://www.environment.gov.au/mediarelease/company-commit-spending-2-million-after-clearing-woodland

Milne, G. and Reardon, C. 2010, *Your Home: Australia's guide to environmentally sustainable homes*, 4th edn, Commonwealth of Australia, Canberra, ACT

National Waste Policy: **http://www.environment.gov.au/protection/national-waste-policy**

Pickin, J., Randell, P., Trinh, J. and Grant, B. (2018), *National Waste Report 2018*, Docklands, Victoria, available at: **https://www.environment.gov.au/system/files/resources/7381c1de-31d0-429b-912c-91a6dbc83af7/files/national-waste-report-2018.pdf, (accessed 17 October 2019).**

Waste & Resources Action Programme (WRAP) (2016) *Designing out Waste: A design team guide for civil engineering: Less waste, sharper design*, Banbury, UK, available at: **http://www.wrap.org.uk/sites/files/wrap/Designing%20out%20Waste%20-%20a%20design%20team%20guide%20for%20civil%20engineering%20-%20Part%201%20(interactive)1.pdf**, accessed 17 October 2019

APPENDIX 5

Waste Management template from Paramatta Shire (NSW), **https://www.cityofparramatta.nsw.gov.au/sites/council/files/2017-02/Waste%20Management%20Plan%20Application%20Template.doc**

APPENDIX 1

Home building contract for work over $20,000

Legislation obliges a contractor to give a copy of the *Consumer building guide* to consumers before entering into a home building contract.

This contract includes a copy of the Guide or it can be downloaded from our website.

www.fairtrading.nsw.gov.au

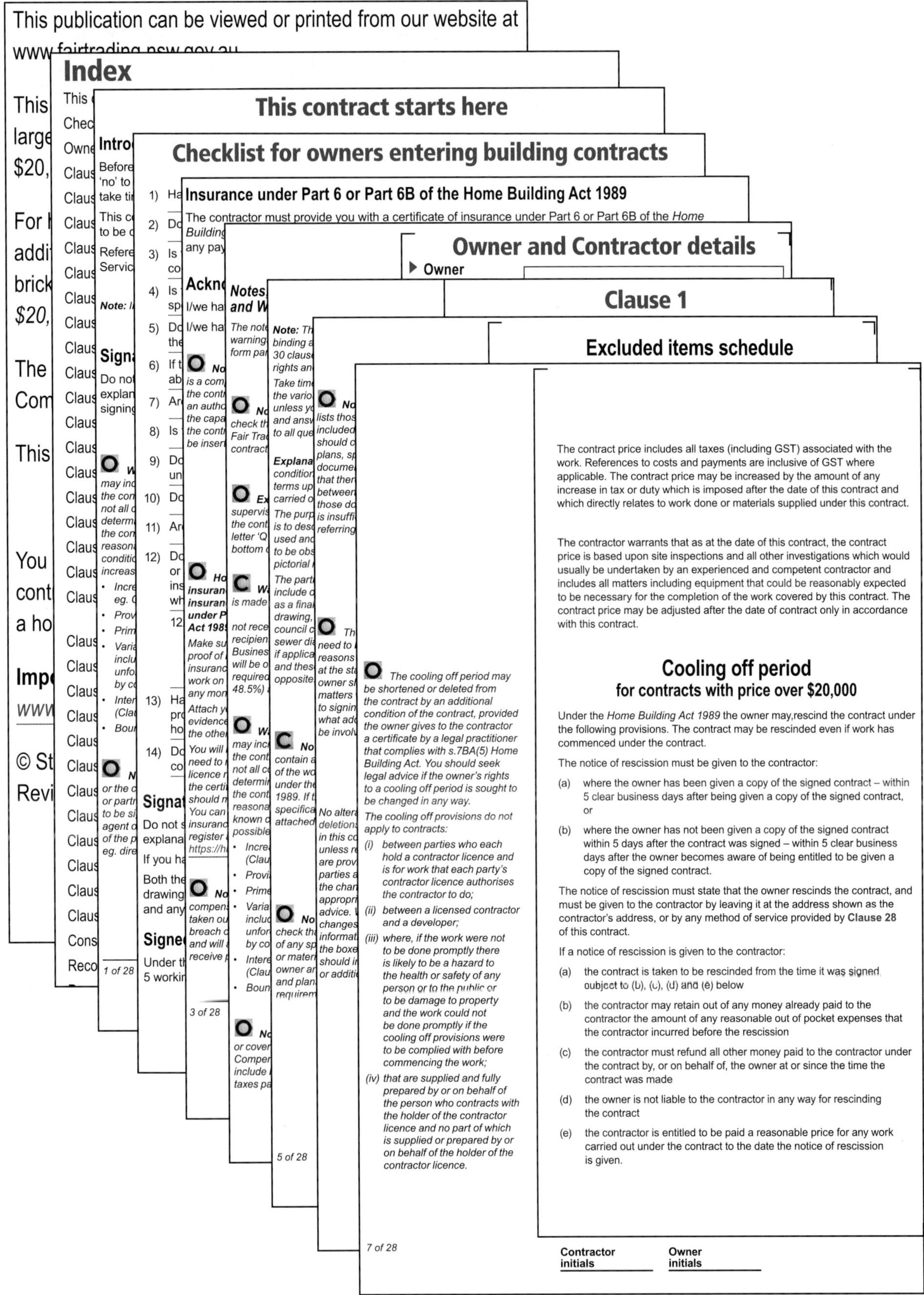

This publication can be viewed or printed from our website at www.fairtrading.nsw.gov.au

Index

This contract starts here

Checklist for owners entering building contracts

Insurance under Part 6 or Part 6B of the Home Building Act 1989

Owner and Contractor details

Clause 1

Excluded items schedule

The cooling off period may be shortened or deleted from the contract by an additional condition of the contract, provided the owner gives to the contractor a certificate by a legal practitioner that complies with s.7BA(5) Home Building Act. You should seek legal advice if the owner's rights to a cooling off period is sought to be changed in any way.

The cooling off provisions do not apply to contracts:

(i) between parties who each hold a contractor licence and is for work that each party's contractor licence authorises the contractor to do;

(ii) between a licensed contractor and a developer;

(iii) where, if the work were not to be done promptly there is likely to be a hazard to the health or safety of any person or to the public or to be damage to property and the work could not be done promptly if the cooling off provisions were to be complied with before commencing the work;

(iv) that are supplied and fully prepared by or on behalf of the person who contracts with the holder of the contractor licence and no part of which is supplied or prepared by or on behalf of the holder of the contractor licence.

The contract price includes all taxes (including GST) associated with the work. References to costs and payments are inclusive of GST where applicable. The contract price may be increased by the amount of any increase in tax or duty which is imposed after the date of this contract and which directly relates to work done or materials supplied under this contract.

The contractor warrants that as at the date of this contract, the contract price is based upon site inspections and all other investigations which would usually be undertaken by an experienced and competent contractor and includes all matters including equipment that could be reasonably expected to be necessary for the completion of the work covered by this contract. The contract price may be adjusted after the date of contract only in accordance with this contract.

Cooling off period
for contracts with price over $20,000

Under the *Home Building Act 1989* the owner may,rescind the contract under the following provisions. The contract may be rescinded even if work has commenced under the contract.

The notice of rescission must be given to the contractor:

(a) where the owner has been given a copy of the signed contract – within 5 clear business days after being given a copy of the signed contract, or

(b) where the owner has not been given a copy of the signed contract within 5 days after the contract was signed – within 5 clear business days after the owner becomes aware of being entitled to be given a copy of the signed contract.

The notice of rescission must state that the owner rescinds the contract, and must be given to the contractor by leaving it at the address shown as the contractor's address, or by any method of service provided by **Clause 28** of this contract.

If a notice of rescission is given to the contractor:

(a) the contract is taken to be rescinded from the time it was signed, subject to (b), (c), (d) and (e) below

(b) the contractor may retain out of any money already paid to the contractor the amount of any reasonable out of pocket expenses that the contractor incurred before the rescission

(c) the contractor must refund all other money paid to the contractor under the contract by, or on behalf of, the owner at or since the time the contract was made

(d) the owner is not liable to the contractor in any way for rescinding the contract

(e) the contractor is entitled to be paid a reasonable price for any work carried out under the contract to the date the notice of rescission is given.

7 of 28

Contractor initials ____ **Owner initials** ____

Download a full copy of the home building contract: https://www.fairtrading.nsw.gov.au/__data/assets/pdf_file/0009/389871/Home_building_contract_over_20000.pdf

APPENDIX 2

PROJECT SPECIFICATIONS

This document outlines the specifications for the project as outlined:

PROJECT:

Construction of a new extension and renovation and repair to existing house.

OWNERS

Amanda J. Owen & Clive S. Owen

ADDRESS OF BUILDING SITE

Street No.	*Street:*	*Town or suburb:*
1	Hidden Lane	Nine Mile Creek, Victoria

TITLE DETAILS

BUILDER

Fine Line Designs
Bayview Street,
Nine Mile Creek Vic 3658
Phone: 03 58258992 ABN: 62 3111 318 267

Building Practitioner	**Building Practitioner No.**
Uta Hartman	DB-U642 341 (Vic) 238501 – M (NSW)

These Specifications, in conjunction with the Construction Drawings, form part of the Building Contract dated the:

________________________ day of ________________________ 20_ _

Note – It is incumbent upon the Client, Principal, or Owners, to ensure that these specifications are complete, and accurately depict the specific project requirements.

SIGNATURES

Owners ________________________

Builder ________________________

Witness ________________________

Witness ________________________

Date ________________________

Fine Line Designs
Project Specification for
1 Hidden Lane, Nine Mile Creek, Victoria

PROJECT SPECIFICATIONS

1. PRELIMINARY

1.1 Specifications

These are the Specifications to be used in the construction and refurbishment of the building as shown and dimensioned in the accompanying drawings. The Specifications specify the materials and finishes to be used in the construction of the building.

Amendments to the specification must be initialled by both parties.

These Specifications shall be taken as being generally applicable to the drawings and other documents forming part of the building contract.

1.2 Regulations

All works shall comply with the Building Regulations, as legislated for and adopted by the State Government, inclusive of any Code or other document that is adopted by, or specified in the Building Regulations, and any other regulation provided, or administered by a State or Local authority, having jurisdiction over the building works.

Where in addition to the works referred to in the drawings and specifications any authority having jurisdiction over works requires additional work to enable the issue of a building approval/permit, or directs that additional work be performed, that work shall be at the Owners' expense, in accordance with the conditions of contract.

1.3 Notices & Permits

The Builder shall give and receive all notices, except fencing notices.

The Builder shall obtain and pay for the Building Construction Certificate, pay all fees legally required in connection with the works, and comply with the relevant regulations of municipal and other authorities having jurisdiction over the works, unless otherwise stated.

1.4 Plant & Labour

In accordance with the contract the Builder shall supply tools, scaffolding, plant and do or have done, works in all trades necessary to carry out the work indicated on the drawings and in these Specifications.

1.5 Materials

Unless otherwise specified, agreed to by the Owners and approved of by the Builder and Local Authority, materials used in the works shall be new, of good quality and in conformity with the drawings and these Specifications. Defective materials shall, as soon as practical, be removed from the site.

1.6 Allotment Identification

The Builder shall display on site, a conspicuous notice indicating the lot, street or identification number of the property and the name of the Builder.

1.7 Water & Electric Power Supply

Refer to Clause 2 regarding 'Services'.

The cost of power and water supplied to Site and used for construction purposes is at the Owners' expense.

1.8 Dimensions

Figured dimensions as shown on drawings shall be given preference to scaled dimensions, which should only be used where figured dimensions are not indicated.

Owner's Initials...................................... / .... Builder's Initials...

Fine Line Designs
Project Specification for
1 Hidden Lane, Nine Mile Creek, Victoria

1.9 Allotment Verification

The Owners shall be responsible for the accurate and clear delineation of all the allotment boundaries. In addition, the Owners shall supply to the Builder a current copy of Title of the allotment. If requested by the Builder the Owners shall further provide a survey plan showing the correct boundaries of the allotment, and its location that can be established from a fixed reference point.

Unless otherwise shown on the drawings the Owners shall be responsible for establishing the point from which the Builder will set out the building, and such set out shall be carried out by the Builder using the details as shown on the site plan.

1.10 Fencing

No fencing is required.

1.11 Instructions

All instructions given to the Builder are to be confirmed in writing. Any variations arising out of such instruction will be priced by the Builder and approved by the Owners before commencement of the work involved.

1.12 Insurances

The Builder carries a Contractors All Risk insurance policy with a minimum $5 million public liability content.

A Home Owners Warranty Insurance policy in the name of the Owners will be provided.

1.13 Final Completion

Where such work is in the scope of the contract the Builder shall remove all builder's equipment, excess materials and debris from the site, check satisfactory operation of installed equipment (doors, windows, locks); remove paint spots, clean windows, sweep floors, clean all plumbing fixtures and fittings, clean cupboards, ensure gutters and down pipes are clear to operate, and leave the building and the site in a clean and tidy condition.

The Builder shall obtain and give to the Owners all necessary certificates of final approval from the various authorities.

2. SERVICES

2.1 Water Supply

Not required / Available at existing premises

2.2 Electricity Supply

Not required / Electricity is connected to existing building

2.3 Telephone

Not required / Telephone is connected to existing building

2.4 Sewer/Septic

Not required / Connected to existing building

3. DEMOLITION

3.1 Partial Demolition of house to be carried out by Builder to make way for new works and refurbishment, as per plans supplied, exceptions:
A) Carport: Carport to be demolished by owners.

3.2 Demolition material from partial demolition will become the property of the Builder and is to be removed from site by builder as soon as practical and possible after demolition.

Owner's Initials...................................... / .... Builder's Initials..

Fine Line Designs
Project Specification for
1 Hidden Lane, Nine Mile Creek, Victoria

4. RUBBISH AND EXCESS BUILDING MATERIALS: REMOVAL FROM SITE

4.1 All rubbish, demolition material and any excess building materials will be removed from the construction site, by the Builder, as soon as practical and possible.

5. PRELIMINARY SITE WORKS AND FOUNDATIONS

5.1 Site preparation

Required/By builder

5.2 Excavation

General site excavation, grading and levelling to be carried out by Builder.

Stump hole boring and foundation works to be carried out by Builder in accordance with working plans supplied.

6. SUBFLOOR

6.1 Stumps to new works

Builder to supply and install 100 mm × 100 mm steel stumps to new extension works, placement and footings as per Footing Details plan and Stump Footings schedule as per working plans supplied.

6.2 Termite protection

Physical Barrier – Ant caps to be provided to all subfloor stumps.

7. EXTERNAL WALL FINISHES

7.1 External Cladding

Provide Baltic pine, square-edged, pre-primed weatherboards to external walls of new works.

7.2 West Elevation Gable

60/60/60 Fire rated wall of either:

Brick veneer from old red bricks

Hebel 200 mm block

100 mm Hebel Power Panel on 25 mm top hat battens to 90 mm timber frame.

8. CARPENTRY

(Note: Timber sizes and framing to be in accordance with AS 1684 – Residential timber-framed construction or Timber Framing Manual, or AS 1720 Timber structures - design methods or Timber Structures Code)

8.1 Timber Grade

Stress Grade of Timber	Placement
Cypress Pine F5	Extension floor frame and roof frame or Bed 1
Duragal as per plans	Robe/ Ensuite/Loft (Games Room) floors
MGP 10 LOSP (Termite Resistant)	Extension wall frames
LVL Hyspan	Rafters and Ridge beams to Games Room

8.2 Floor Framing

Provide F5 Cypress Pine 100 mm × 75 mm Bearers and 90 mm × 45 mm Joists.

Duragal subfloor framing as per plans provided.

Owner's Initials....................................... / .... Builder's Initials...

Fine Line Designs
Project Specification for
1 Hidden Lane, Nine Mile Creek, Victoria

8.3 Wall Framing to new works and relocated internal walls

Frame all loadbearing and non-loadbearing walls MGP 10 LOSP 90 × 45 studs at 450 mm centres, MGP 10 LOSP 90 × 35 Top and Bottom plates, 1 row of noggings.

Externally – provide sheet bracing as per plans.

Internally – provide bracing as per plans or otherwise required.

8.4 Roof Framing to new works

Frame all rafters in 150 mm × 45 mm LVL Hyspan.

8.5 Internal Timber Flooring

Provide Murray Pine 140 mm × 19 mm tongue and groove floorboards to:

- Extended section of Bedroom 1 and new Robe.
- Particle Board Flooring

Provide 19 mm particleboard flooring to Robe, Ensuite new Loft area.

8.7 Architraves and Skirtings

Skirtings – Provide throughout new works and demolished areas – Heritage Profiles' 'Richmond' profile, 190 mm in MDF.

Architraves – Provide throughout new works and demolished areas – Classic Architraves and Skirtings' 'Early Colonial – CB', 70 mm in MDF.

8.13 Fascia/Barges

Provide Treated Pine Fascia and Barges to all new work, to match existing.

Exclusion: Metal Barge to West Elevation gable for fire-rated wall.

9. WINDOWS AND GLAZED EXTERNAL DOORS

9.1 Window Schedule

Window/Door No.	Size mm (height × width)	Window/Door Type	Material	Glazing
W1	1200 × 900	Double Hung	Cedar	Double glazed/obscure
W2-4	1600 × 900	Double Hung	Cedar	Double glazed/clear
W5	1200 × 900	Double Hung	Cedar	Double glazed/clear
W6-9	1150 × 780	Velux M06 GHL	Hwd	Double glazed/clear
W10	1200 × 900	Double Hung (fixed)	Hwd	Toughened/clear
D1	2040 × 820	Solid External Colonial	Hwd	NA
D2-3	2040 × 720	Solid internal Colonial	Hwd	NA
D4-5	2040 × 820	Solid internal Colonial	Solid, pre-primed	NA
RD1	2200 × 3300	Automated B&D Lift Panel	Metal	NA
RD2	2200 × 2500	Manual B&D Roller	Metal	NA

9.2 Glazing

Provide glazing to AS 1288 throughout.

9.3 Timber Window Frames and Doors

Finish, painted or other, to timber window frames and doors by Owners.

Owner's Initials...................................... / .... Builder's Initials...

Fine Line Designs
Project Specification for
1 Hidden Lane, Nine Mile Creek, Victoria

9.4 Flywire Screens and Doors

Window Screens: Aluminium framed

Flymesh Material: Aluminium (not Nylon)

9.5 Roof Windows

Supply and install a Velux manually operated roof windows, includes a roller blind and fly screen (comes in 'Light Beige' colour only) and telescopic rod for window and blind operation.

10. ROOFING AND METAL

10.1 Roof Cladding

Supply and fix Colorbond galvanised custom orb corrugated roofing sheet to entire house, complete with ridge capping and flashings.

10.2 Sarking

Provide aluminium foil Sisalation to entire roof area.

10.3 Gutters

Supply and install Colorbond Ogee gutters to new sections only.

*Colour to be advised by Owners – refer 'External Colour Selection Schedule'.

10.4 Verandah and Pergola posts

Provide 90 mm × 90 mm Duragal galvanised posts as required.

11. ELECTRICAL

Refer 'Provisional Sum' allowance – Clause 21.1.

11.1 Supply and install

Includes 6 only double power points; 5 only light points, 1 only 2-way points; Tastic to Ensuite.

*N.B. Locations of power points, etc to be advised by Owners.

12. PLUMBING

12.1 Supply and install

Includes hot and cold plumbing to new extension; alter sewer line; connection to existing stormwater; 1 only dry wastes (Ensuite); connection of fixtures and fittings to services.

12.2 Sewer Line

Use existing line.

12.3 Downpipes

Supply and install zincalume downpipes as needed.

12.4 Stormwater

Connect new stormwater run to existing.

13. INSULATION

13.1 Walls

To external stud walls of new works provide vented double-sided foil sisalation fixed to external face of studs.

To external walls of new works provide R2.0 polyester batts, internal walls R1.5.

Owner's Initials...................................... / .... Builder's Initials..

Fine Line Designs
Project Specification for
1 Hidden Lane, Nine Mile Creek, Victoria

13.2 Ceilings

To ceilings of new works, except for Loft, provide R3.5 Polyester thermal insulation batts.

14. INTERNAL WALL LININGS

14.1 Generally

Provide and fix 10 mm plasterboard to stud walls throughout new works and to stud walls throughout demolished areas, in accordance with manufacturer's instructions.

14.2 Garage

13 mm Firecheck to partition wall between garage/workshop and Robe/Ensuite, also; 13 mm Firecheck to ceiling of garage/workshop.

14.3 Wet Area Walls

Provide Villaboard wall lining to ensuite as per manufacturer's specification.

15. CEILING LININGS

15.1 Generally

To ceilings throughout new works and demolished areas provide 10 mm plasterboard on steel battens in accordance with manufacturer's instructions.

15.2 Cornice

Provide 90 mm cornice to all wall /ceiling joints throughout new works (except for Loft/games room).

16. JOINERY

16.1 Ensuite

Supply and install vanity cupboards – *refer 'Prime Cost' allowance – Clause 21.1.*

*Layout and Colours to be advised by Owners – refer 'Internal Colour Selection Schedule'.

16.2 Robes to Bed 1

Robes – provide chrome hanging rail with a 19 mm chipboard shelf above and a bank of 8 (eight) draws constructed from hardwood.

17. DOORS & HARDWARE

17.1 Internal Door Schedule

Hinged Doors – Provide 4 no. Hume 'Oakfield', pre-primed doors.

Hardware to Doors – Provide 5 no. Gainsborough 'Governor' passage sets

*Colour of internal door furniture to be advised by Owners – refer 'Internal Colour Selection Schedule'.

17.2 External Door Schedule

External Entrance Doors – Provide 1 no. Malep solid timber external

Hardware to Entrance – Gainsborough Trilock

*Colour of door external furniture to be advised by Owners – refer 'Internal Colour Selection Schedule'.

18. FLOOR FINISHES

18.1 Internal Timber Floors

All internal floor finishes are to be by the Owners.

Owner's Initials...................................... / .... Builder's Initials..

18.2 Timber Decking Finishes

All external decking finishes are to be by Owners.

18.3 Floor Tiling

Refer 'Prime Cost' allowance – Clause 21.2

Supply and fix floor tiles to Laundry and Bathroom.

*Types and Colours of tiles and Colour of Grout to be advised by Owners – refer 'Internal Colour Selection Schedule'.

19. ALL FINISHES

19.1 Painting

External and Internal painting to be by Owners.

19.2 Wall Tiling

Refer 'Prime Cost allowance – Clause 21.3

Supply and fix wall tiles to the following tiling heights:

Ensuite: – Shower alcove to 2000 mm
– floor/wall junction (skirting tile) approx. 150 mm
– splashback above bathroom vanity and basin to approx. 600 mm high.

*Types and Colours of tiles and Colour of Grout to be advised by Owners – refer 'Internal Colour Selection Schedule'.

20. PRIME COST ALLOWANCES

Prime Cost allowances are based on a reasonable cost for supply of a fitting or fixture where the exact requirements or product has not yet been specified by the Owner or is unknown at the time of contracting. The following allowances have been included in the Contract Price and include GST.

20.1 Joinery (*Refer Clause 16*)

Allowance of $1200.00

To supply and install: and bathroom vanity with one bank of drawers.

Allowance of $3500.00

To supply only Hwd staircase to games room

Allowance of $3030.00 for 3 only Velux windows (note, plan shows 4 but only three required. Location of third window pending heritage approval for window on South Elevation)

Allowance of $6000.00 for 6 only Double Hung Windows.

20.2 Floor Tiling (*Refer Clause 18.3*)

Floor Tiles: Allowance of $35.00 per m^2 for purchase of tiles.

20.3 Wall Tiling (*Refer Clause 19.2*)

Wall Tiles: Allowance of $35.00 per m^2 for purchase of tiles.

21. PROVISIONAL SUM ALLOWANCES

Provisional Sum allowances are based on a reasonable cost for supply of work (including materials) required by the Working Drawings and Project Specification, where the extent of the work has not been ascertained. The following allowances have been included in the contract price and include GST.

Owner's Initials...................................... / .... Builder's Initials...

21.1 Electrical – (*refer Clauses 11*)

Allowance of $2000.00 general electrical.

Allowance of $1500.00 for relocation of solar panel array.

22. FITTINGS

Supply and installation of the following fittings are by the Builder.

22.1 Bathroom

a 1 no. Stylus 'Minuet' vanity basin – white
b 1 no. Stylus 'Minuet' china linked pan – white
c 1 no. Marbletrend polymarble, corner shower base – white
d 1 no. shower screen – clear, toughened glass with powder coated, aluminium frame.

*Colour of powder coat (from base range) to be advised by Owners – refer 'Internal Colour Selection Schedule'.

e Tapware – Phoenix 'Festival' – white with gold or silver trim
  - 1 no. shower set
  - 1 no. vanity set

f Ensuite Accessories – Heirloom 'Pastels' – white with gold or silver trim
  - 1 no. double towel rail – 610 mm long
  - 1 no. toilet roll holder
  - 1 no. soap dish (to shower)
  - 1000 mm × 1000 mm, 3 mm polished edge mirror

Owner's Initials...................................... / .... Builder's Initials..

APPENDIX 3

BOB'S CARPENTRY BUSINESS PLAN (EXCERPT)

Business history

The business is an existing operation which began four years ago. The following results summarise the business growth over the last four years (see **Table A3.1**).

TABLE A3.1 Business growth

Year	Annual Sales Turnover $	Annual Net Profit $
Year 1	208 600	5 900
Year 2	357 500	6 250
Year 3	395 000	5 800
Year 4	440 300	5 500

Legal requirements

The legal requirements to operate the business are:

- Builders licence
- ABN registration (for GST and PAYG tax) with the Australian Tax Office
- Building approval from the local council for each construction job.

Business objectives

The key objectives of the business for the next three years are summarised as follows (**Table A3.2**).

TABLE A3.2 Business objectives

	Year 1	Year 2	Year 3
Marketing			
Sales turnover	$455 000	$466 375	$478 034
Purchasing			
Average gross profit margin	57%	57%	57%
Financial			
Net profit	$6 000	$6 000	$6 000
Net cash flow	$15 000	$15 000	$15 000

Marketing strategies and controls

Bob's Carpentry focuses on the provision of a quality, tailored service for its clients. Its main marketing strategy is through word of mouth and the strategic placement of signage outside the property while a project is being completed. Bob has recently established a Facebook page and regularly updates it with photographs of the projects he has completed.

Bob's team present as professional, friendly and courteous, which is enhanced by the provision of team uniform in the form of a badged, collared t-shirt and coordinating shorts.

Marketing controls include the review of recorded sales against customer feedback received via Facebook and via questionnaires that he asks customers to fill in at the completion of a project. This review and evaluation take place quarterly.

Bob's Carpentry seeks to ensure that the marketing strategies are adaptable to change to allow for new opportunities and changes in the marketplace.

Professional advisers

Accountant (CPA)	Jeff Jones
Bookkeeper	Jenny Jones
Solicitor	Cindy Chang
Bank	NBA

Financial forecasts

Forecast Profit Statements			
INCOME			
Sales	455 000	466 375	478 034
Less Job materials purchases	204 750	209 869	215 115
GROSS PROFIT (at 55%)	**250 250**	**256 506**	**262 919**
Less OPERATING EXPENSES			
Advertising	3 000	3 075	3 152
Adviser's fees	2 000	2 050	2 101
Bank charges	500	513	525
Depreciation	4 250	4 250	4 250
Hire fees	1 500	1 538	1 576
Home office expenses	1 000	1 025	1 051
Insurances	4 800	4 920	5 043
Interest paid	1 000	500	500
Legal fees	800	820	841
Loose tools replaced	2 000	1 500	1 500
Motor vehicle running costs	5 000	5 125	5 253
Repairs & Maintenance – equipment	1 000	1 000	1 000
Stationery	500	500	500
Subcontract payments (gross)	125 000	128 125	131 328
Telephone and internet	1 800	1 845	1 891
Wages (gross)	70 000	71 750	73 544
Wages on-costs	17 500	17 938	18 386
Other expenses	500	500	500
Total operating expenses	242 150	246 973	252 941
NET PROFIT BEFORE TAX	**8 100**	**9 534**	**9 978**

Assumptions:
job materials are purchased according to job requirements
no allowance is made for GST
depreciation is 10% of the original cost of long-term assets
interest paid rate is 8% p.a.
motor vehicle running costs include petrol, repairs and maintenance, and insurance and registration
wages on-costs include superannuation, workers compensation insurance and staff amenities for employees
wages on costs are calculated at 25% of wages
unless determinable, each cost is increased by 2.5% per annum

Forecast Cash Flow Statements

Cash position – start of year	(12 512)	(4 162)	16 947
CASH RECEIPTS			
Cash sales	265 000	275 000	285 000
Cash from debtors	185 000	200 000	185 000
Capital contributions			
Borrowings			
Total Cash Receipts	450 000	475 000	470 000
Less CASH PAYMENTS			
Job materials purchases	204 750	209 869	215 115
Operating expenses	236 900	241 723	247 691
Capital expenditure			
Taxation		2 300	2 600
Total cash payments	441 650	453 891	465 406
NET CASH FLOW	8 350	21 109	4 594
Cash position – end of year	(4 162)	16 947	21 541

Assumptions:
existing overdraft at the start of year 1 is $12 512
exisiting debtors (Accounts Receivable) at the start of year 1 is $49 300 to be collected as follows:

July	25 000
August	15 800
September	8 500
Total	**49 300**

Sales are collected by progress payments – see accounts receivable policy
Purchases and operating expenses are paid for in cash
Operating expenses paid exclude depreciation
Operating expenses paid include interest on overdraft

Net profit margin	2%	2%	2%	2%

Financial records

A business bank account is operated at the NBA Bank. All receipts and payments are processed through the account. A manual system of financial records will continue to be maintained. These records include:

- Source documents: sales quotations, sales invoices, sales receipts, purchase orders, deposit slips, cheque butts and bank statements
- Cash journal books: cash receipts journal, cash payments journal and petty cash book
- Secondary books: subcontractors payment book, time and wages book.

Internal controls will be adopted to ensure the accuracy of financial records. These include:

- Using standardised pre-numbered source documents in consecutive number sequence
- Checking purchase orders against suppliers' invoices
- Paying suppliers invoices by EFT bank transfer or BPay
- Preparing monthly bank reconciliations to check cashbook totals with bank statement balances.

APPENDIX 4

MORE COMPLEX FORCE SYSTEMS: FURTHER SOLUTION EXAMPLES

The following examples of force systems extend the learning developed in Chapter 12. They have been provided to aid in extending your understanding of the complexity of the forces that might act upon the elements of a structure, such as a simple beam.

Example 1: Two offset point loads

In this example we have two point loads acting on the one beam to consider.

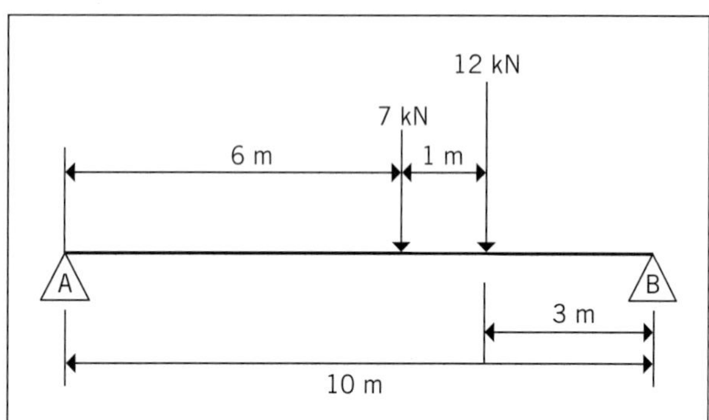

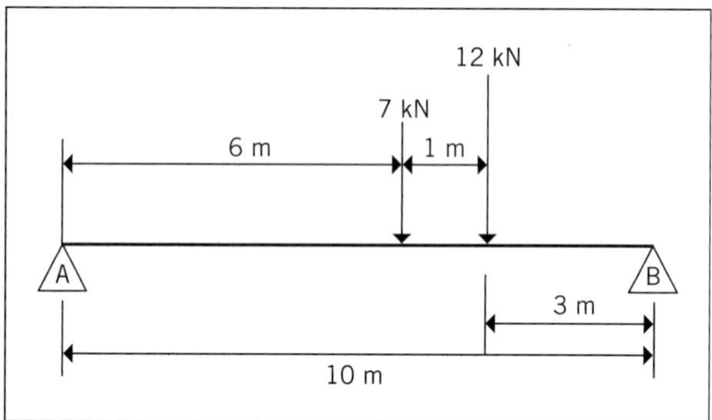

Solving for moments around A

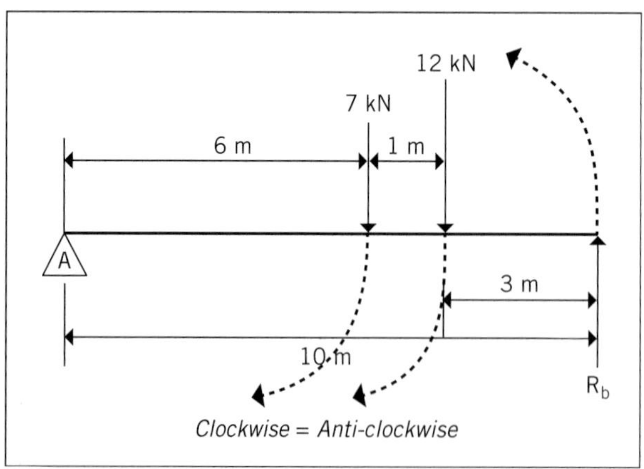

Clockwise = Anti-clockwise

(7 kN × 6 m) + (12 kN x 7 m) = RB × 10 m

42 kN.m + 84 kN.m = RB × 10 m

RB = 126 kN.m ÷ 10 m

RB = **12.6 kN**

Solving for moments around B

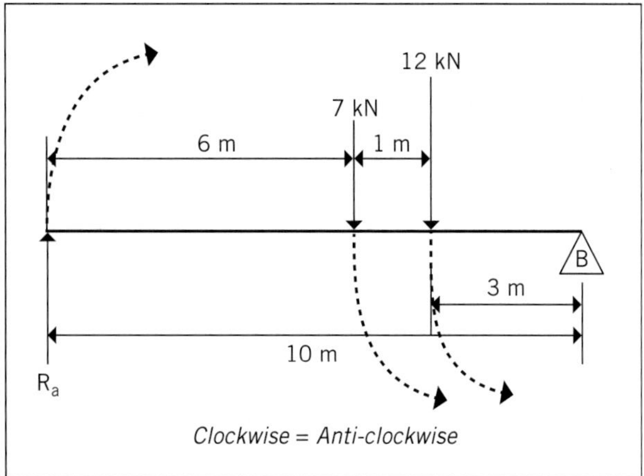

Clockwise = Anti-clockwise

RA × 10 m = (7 kN × 4 m) + (12 kN × 3 m)

10 RA = 64 kN.m

RA = 64 kN.m ÷ 10 m

RA = **6.4 kN**

Check:

Total loads = total reactions

12.6 kN + 6.4 kN = 12 kN + 7 kN

19 kN = 19 kN

Example 2: Solving of uniformly distributed loads (UDLs)

Although the load of a uniformly distributed load (UDL) is just that, uniformly distributed across the whole of a beam or platform, they are considered to act at their centreline. In this manner, they can then be treated as you would a point load.

Example 2a

A 5-metre beam has a self-weight of 2 kN/m.

This gives a UDL of:

UDL = 5 × 2

= 10 kN

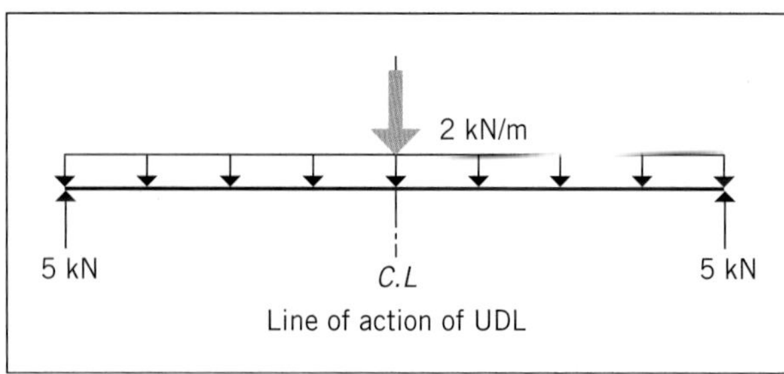

This is then considered to act on the centre line of the beam, giving an even balance of reactions at both supports as shown above. As usual, for equilibrium these reactions must total the same as the load itself; i.e.

5 kN + 5 kN = 10 kN

Example 2b

In this example the UDL acts over only a part of the beam.

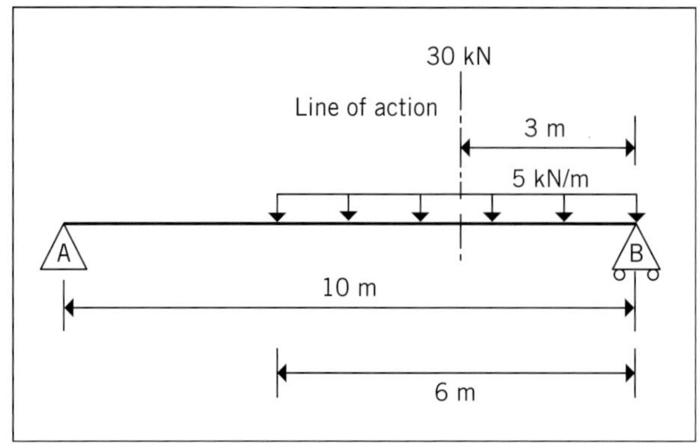

The UDL in this case acts over a 6 m area of the beam. However, the 'line of action' is the centreline of the width: i.e. 3 m from support 'B'. The UDL is offered as 5 kN/m.

Therefore:

6 × 5 = 30 kN

acting at 3 m from support 'B' as shown above

We can now solve for moments as in previous examples:

Solving for moments around A

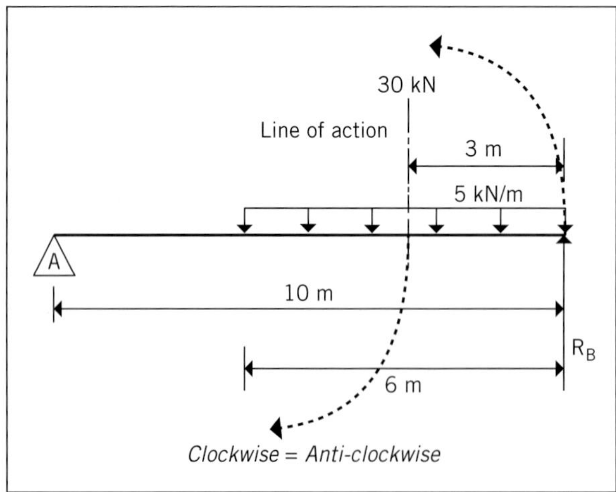

Clockwise = Anti-clockwise

30 kN × 7 m = RB × 10 m

210 kN.m = RB × 10 m

RB = 210 kN.m ÷ 10 m

RB = **21 kN**

Solving for moments around B

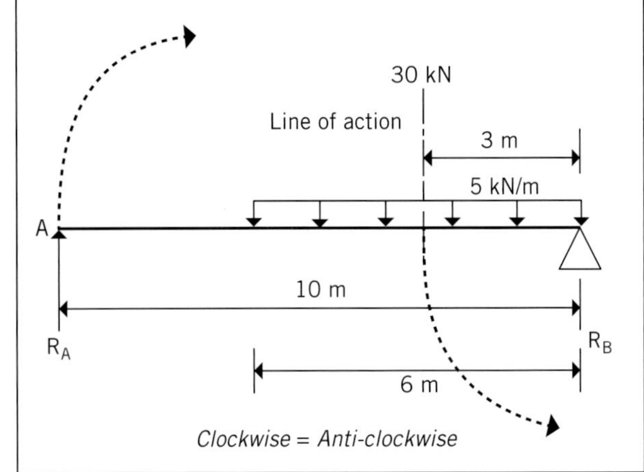

Clockwise = Anti-clockwise

RA x 10m = 30 kN × 3 m

10RA = 90 kN.m

RA = 90 kN.m ÷ 10 m

RA = **9 kN**

Check:

Total loads = total reactions

30 kN = 21 kN + 9 kN

30 kN = 30 kN

Example 3

This example puts all that you have done so far together. Although it looks more complex, the task is essentially the same: remove one support and determine which loads and reactions go clockwise, and which go anti-clockwise.

In this case we have a 12-metre beam with a self-weight of 1 kN/m, another UDL of 2 kN/m sited at one end of the beam, plus 5 kN and 10 kN point loads.

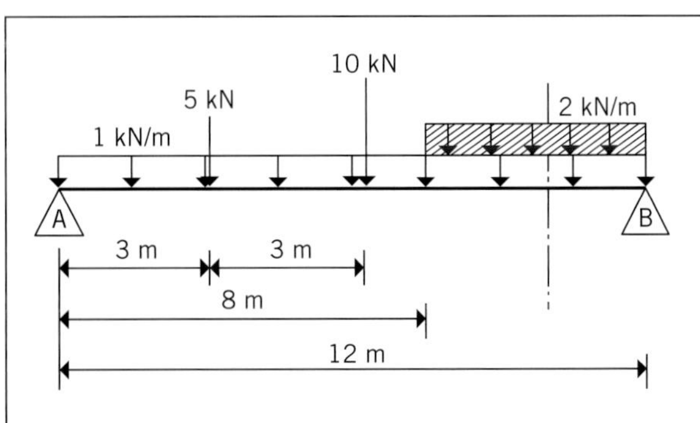

Solving for moments around A

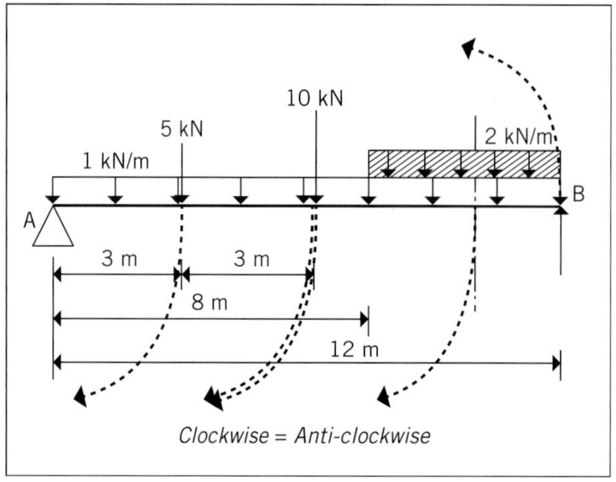

Clockwise = Anti-clockwise

(5 kN × 3 m) + (10 kN × 6 m) + (1 kN/m × 12 m × 6 m)
+ (2 kN/m × 4 m × 10 m) = RB × 12 m
15 kN.m + 60 kN.m + 72 kN.m + 80 kN.m = RB × 12 m
227 kN.m = RB × 12 m
227 kN.m ÷ 12 m = RB
18.92 kN = RB

Solving for moments around B

Clockwise = Anti-clockwise

RA × 12 m = (1 kN/m × 12 m × 6 m) + (5 kN × 9 m)
+ (10 kN × 6 m) + (2 kN/m × 4 m × 2 m)
RA × 12 m = 72 kN.m + 45 kN.m + 60 kN.m + 16 kN.m
RA × 12 m = 193 kN.m
RA = 193 kN.m ÷ 12 m
RA = **16.08 kN**

Check:

Total loads = total reactions

5 kN + 10 kN + (1 kN/m × 12 m)
+ (2 kN/m × 4 m) = 18.92 kN + 16.08 kN
35 kN = 35 kN

Example 4

This last example shows a beam with a cantilever: i.e. a section of the beam travels past the support, as might a deck joist.

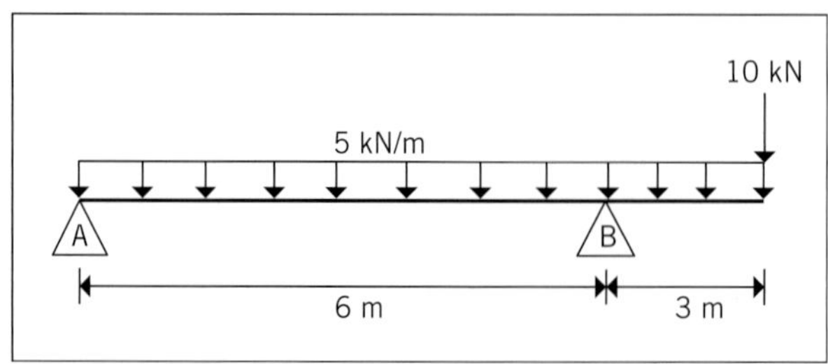

Once again, the same principles apply, remove a support and determine the required clockwise and anti-clockwise actions and reactions.

Solving for moments around A

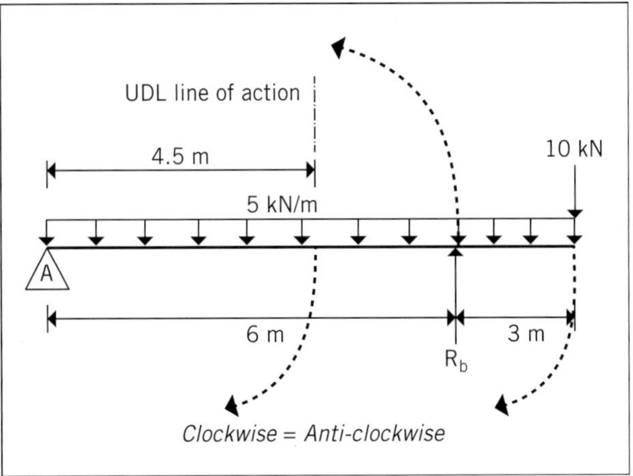

Clockwise = Anti-clockwise

(5 kN/m × 9 m × 4.5 m) + (10 kN × 9 m) = RB × 6 m

202.5 kN.m + 90 kN.m = RB × 6 m

292.5 kN.m = RB × 6 m

RB = 292.5 kN.m ÷ 6 m

RB = **48.75 kN**

Solving for moments around B

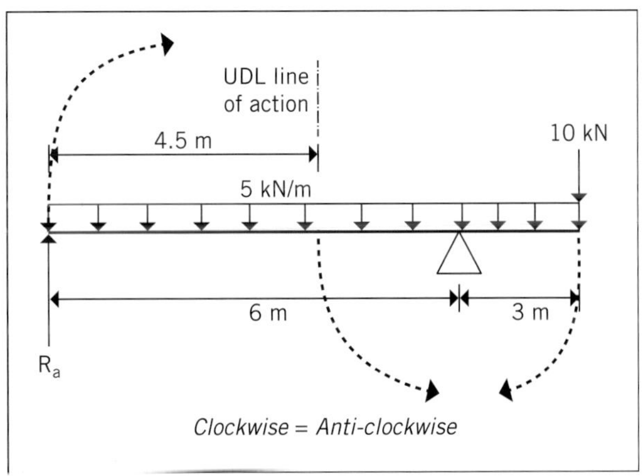

Clockwise = Anti-clockwise

(RA × 6 m) + (10 kN × 3 m) = (5 kN/m × 9 m × 1.5 m)

(RA × 6 m) + 30 kN.m = 67.5 kN.m

RA × 6 m = 67.5 kN.m – 30 kN.m

RA × 6 m = 37.5 kN.m

RA = 37.5 kN.m ÷ 6 m

RA = **6.25 kN**

Alternative

It is possible to view a problem like this as two separate UDLs acting either side of support B. It is an unnecessary complication as the following calculation shows that either way the result is the same.

$$(RA \times 6\text{ m}) + (10\text{ kN} \times 3\text{ m}) + (5\text{ kN/m} \times 3\text{ m} \times 1.5\text{ m}) = (5\text{ kN/m} \times 6\text{ m} \times 3\text{ m})$$

$$(RA \times 6\text{ m}) + 30\text{ kN.m} + 22.5\text{ kN.m} = 90\text{ kN.m}$$

$$RA \times 6\text{ m} = 90\text{ kN.m} - 30\text{ kN.m} - 22.5\text{ kN.m}$$

$$RA \times 6\text{ m} = 37.5\text{ kN.m}$$

$$RA = 37.5\text{ kN.m} \div 6\text{ m}$$

$$RA = \mathbf{6.25\text{ kN}}$$

Check:

Total loads = total reactions

$$(5\text{ kN/m} \times 9\text{ m}) + 10\text{ kN} = 48.75\text{ kN} + 6.25\text{ kN}$$

$$\mathbf{55\text{ kN} = 55\text{ kN}}$$

APPENDIX 5

WASTE MANAGEMENT PLAN

Demolition, construction, and use of premises

The applicable sections of this table must be completed and submitted with your Development Application.

Completing this table will assist you in identifying the type of waste that will be generated and will advise Council of how you intend to reuse, recycle or dispose of the waste.

Please refer to the City of Parramatta Waste Management Guidelines for new applications for the specific requirements for your type of application.

If you choose to provide an alternative waste management plan to the attached template please ensure all of the required information is addressed. Failure to provide all the required information may lead to further information being requested and a hold up in the final decision of your application.

The information provided will be assessed against the objectives of City of Parramatta Council Development Control Plan (DCP) 2011.

If space is insufficient in the table please provide attachments.

Outline of Proposal

Site address:

____

Applicant's name and address:

____

____

____

____

Phone: ____________________ **Mobile:** ____________________

Email: ____

Building and other structures currently on site:

____

____

____

Brief description of proposal: ______________________________

____

____

The details provided on these forms, plans and attached documents are the intentions of managing waste relating to this project.

Signature of applicant: ______________________________ **Date:** ____________

Demolition & construction

Council is seeking to reduce the quantity of waste and encourage the recycling of waste generated by demolition and construction works. Applicants should seek to demonstrate project management which seeks to:

1. Reuse excavated material on-site and disposal of any excess to an approved site
2. Green waste mulched and reused on-site as appropriate, or recycled off-site
3. Bricks, tiles and concrete reused on-site as appropriate, or recycled off-site
4. Plasterboard waste returned to supplier for recycling
5. Framing timber reused on site or recycled off-site
6. Windows, doors and joinery recycled off-site
7. All asbestos, hazardous and/or intractable wastes are to be disposed of in accordance with WorkCover Authority and EPA requirements
8. Plumbing, fittings and metal elements recycled off site
9. Ordering the right quantities of materials and prefabrication of materials where possible
10. Re-using formwork
11. Careful source separation of off-cuts to facilitate reuse, resale or recycling

How to estimate quantities of waste

- There are many simple techniques to estimate volumes of construction and demolition waste. The information below can be used as a guide by builders, developers & homeowners when completing a waste management plan:

To estimate Your Waste:

1. Quantify materials for the project
2. Use margins normally allowed in ordering
3. Copy these amounts of waste into your waste management plan

When estimating waste the following percentages are building 'rule of thumb' and relate to renovations and small home building:

Material	Waste as a Percent of the Total Material Ordered
Timber	5–7%
Plasterboard	5–20%
Concrete	3–5%
Bricks	5–10%
Tiles	2–5%

Converting Volume into Tonnes : A Guide for Conversion

Timber = 0.5 tonnes per m^3
Concrete = 2.4 tonne per m^3
Bricks = 1.0 tonne per m^3
Tiles = 0.75 tonne per m^3
Steel = 2.4 tonne per m^3

To improve/provide more reliable figures:

- Compare your projected waste quantities with actual waste produced
- Conduct waste audits of current projects
- Note waste generated and disposal methods
- Look at past waste disposal receipts
- Record this information to help estimate future waste management plans
- On a waste management plan amounts of waste may be stated in – m^2 or m^3 or tonnes (t).

IMPORTANT

- **The following tables should be completed by applicants proposing any demolition or construction work including the change of use, fit-out as well as alterations and additions of existing premises.**
- **The location of temporary waste storage areas and soil stockpiles during demolition and construction are to be shown on the submitted plans.**
- **Vehicle access to and from the site must be shown on the submitted plans.**
- **Stage three – Design of facilities should be completed by all applicants including change of use, fit-out as well as alterations and additions.**

Demolition Stage One – To be completed for proposals involving demolition

Materials On-Site		Destination		
		Reuse & Recycling		Disposal
Type of material	Estimated Volume (m^3) or Area (m^2) or weight (tonnes)	On-Site Specify how materials will be reused or recycled on-site	Off-Site Specify the contractor and recycling outlet	Specify the contractor and landfill site
*Example only * Bricks	*2m^3	* Clean and reuse for footings	*Broken bricks sent by XYZ demolishers to ABC Recycling company (including address and contact number)	* Nil to landfill *or sent by XYZ demolishers to ABC Recycling company (including address and contact number)
Excavation material				
Green waste				
Bricks				
Tiles				
Concrete				
Timber				
Plasterboard				
Metals				
Asbestos				
Other waste				

How will waste be separated and/or stored onsite for reuse and recycling? How will site operations be managed to ensure minimal waste creation and maximum reuse and recycling?
e.g. Staff training, selected deconstruction vs straight demolition, waste management requirements stipulated in contracts with sub-contractors, on-going checks by site supervisors, separate area set aside for sorted wastes, clear signage for waste areas, etc.
Note: Details of the site area to be used for on-site separation, treatment and storage (including weather protection) should be provided on plan drawings accompanying your application.

Construction Stage Two – To be completed for proposals involving construction

Materials On-Site		Destination		
		Reuse & Recycling		Disposal
Type of material	Estimated Volume (m^3) or Area (m^2) or weight (tonnes)	On-Site Specify how materials will be reused or recycled on-site	Off-Site Specify the contractor and recycling outlet	Specify the contractor and landfill site
*Example only * Bricks	 *$2m^3$	* Clean and reuse for footings	*Broken bricks sent by XYZ demolishers to ABC Recycling company (including address and contact number)	* Nil to landfill *or sent by XYZ demolishers to ABC Recycling company (including address and contact number)
Excavation material				
Green waste				
Bricks				
Tiles				
Concrete				
Timber				
Plasterboard				
Metals				
Other waste				

How will waste be separated and/or stored onsite for reuse and recycling? How will site operations be managed to ensure minimal waste creation and maximum reuse and recycling?
e.g. Staff training, recycled materials used in construction, waste management requirements stipulated in contracts with sub-contractors, on-going checks by site supervisors, separate area set aside for sorted wastes, clear signage of waste areas, etc.

___

___

___

___

Note. Details of site area to be used for on-site separation, treatment and storage (including weather protection) must be provided on plan drawings accompanying your application.

Design of facilities (Use of site) Stage three – To be completed for all proposals including change of use, fit-out as well as alterations and additions

- Applicants should refer to Council's document 'Waste Management Guidelines for new Development Applications' for specific requirements related to the type of development proposed. This is available on Council's website.
- In the case of change of use, fit-out as well as alterations and additions, if the proposal involves existing waste management practices then full details of current methods are to be provided.
- *All* proposals are to show the waste storage areas on plan drawings which should accompany your application.

Type of waste to be generated	Expected volume per week, number and size of bins	Proposed on-site storage and treatment facilities	Destination and contractor
Please specify. E.g. glass, paper, food waste, green waste, compost etc.	Volume (Litres – L)	For example: waste storage room, garbage chute, compaction equipment	For example: Recycling, landfill by council or private contractor (include name of contractor)
*Example only *Non-recyclable	*480L/week 2 x 240 L bins	*Waste storage room	*Landfill and recycling collected by XXX Collection company

Describe how you intend to ensure on-going management of waste on-site (e.g. lease conditions, caretaker, strata manager) as well as provide details of how the bin store area complies with council's bin storage area requirements relevant to the type of proposed development.

___

___

___

___

___

___

___

___

___

___

___

FINAL CHECK

Please read and tick the box to ensure all required information has been provided.

1 Have you checked the waste requirements for the proposed type of development in Council's document 'Waste Management Guidelines for new Development Applications' and provided all of the required information? ☐

2 Have you completed the relevant sections to your application of the above waste management plan template or provided an alternative waste management plan addressing the required information? ☐

3 Have you shown use of site waste storage areas, garbage chutes, bin pulls and compaction equipment on plans accompanying this application? ☐

4 Have you shown the location of temporary waste storage areas, soil stock piles and vehicle entry/exit points during construction and demolition on the plans accompanying this application? ☐

5 Have you shown the waste collection vehicle access to the collection point on-site (if applicable) on the plans accompanying this application? ☐

6 Have you shown the pathway taken to move the bins to and from the on-street collection point and the location of the on-street collection point on the plans accompanying this application? ☐

GLOSSARY

A

acceptance a clearly stated agreement by one party to the offer or promise made by another

Accounts Receivable the amount of money owed by debtors at the end of the financial year

activity ratios indicate the effectiveness with which financial resources are used

ag pipe a perforated pipe (usually covered with a geo-textile fabric) laid behind retaining walls and other areas to catch seeping stormwater

air brick ventilation built into brickwork to provide ventilation through the wall

air movement it takes very little movement for air to conduct heat away from the body, thereby making us feel colder than the actual room temperature. What is commonly referred to as 'wind chill', or the 'chill factor', for outside temperatures is based upon this air movement. On the other hand, air movement can be used to advantage to feel cooler when we are too hot.

aircell membranes these can be selected on the same basis as standard sarking; i.e. permeable or non-permeable. The main difference with aircell is its non-structural insulation and thermal break values.

allowance an estimated amount required to cover the cost of an element of a project that cannot otherwise be calculated accurately

alternative dispute resolutions (ADRs) processes designed to aid in the development of resolutions to disputes in a manner that is open and fair but are conducted outside of court proceedings

angel investors those who provide finance in exchange for equity or shares in the company

angle of incidence the angle at which the sun strikes the glass of a window

arbitration a dispute resolution process whereby parties present their evidence and arguments to a third party, who makes an independent and binding determination

arbitrator the independent third party who makes the final binding decision at the conclusion of the arbitration process

asset resources that the business owns or utilises for a future benefit; e.g. cash, motor vehicle, plant and equipment, debts owed by customers

audit the action of checking and assessing documented WHS procedures and requirements against observed actions and context

Australian Building Codes Board (ABCB) the standards writing body responsible for the development of the National Construction Code (NCC).

Australian Business Number (ABN) a unique identifier (number) issued by the government for each business irrespective of type for the purposes of tax and other legal liabilities

Australian Consumer Law (ACL) is a national law covering, and administered by, all states and territories. This law allows courts to limit or make void any term or condition in a contract seen by that court to be unfair. It takes into account the relative strength of each party in the bargaining process leading up to the creation of the contract.

award an agreed set of conditions and remuneration developed between an employee representative body (such as a union and/or government body) and a representative employer body (or government body) for stipulated and timely work-based outputs

axial loading allows a column to take up the compressive forces equally throughout its structure

B

back charge an industry term that, in itself, has no legal basis: one party, usually the client or principal, claims a reduction in payment due to money that they consider you owe them

backsight this is the first sighting of any survey. It is also the first sight taken when you change position of the instrument – known as changing station – and begin taking readings again. It is called a 'back' sight because you are looking back to a reference point of known elevation or RL, usually your datum or the last sighting you took before moving the instrument. Think of it as looking back to that which has gone before, and is known.

bagging a method of finishing brickwork involving the application of a thin mortar slurry using a hessian bag or sponge. Can be painted over or left to fade in an oxide finish. Usually completed by the bricklayer. Bagging varies greatly in texture and colour and is not uniform like render.

Balance Sheet a single report that provides a picture of the financial position of a business at a point in time

barge board the board covering the roof timbers on the gable or skillion end of a roof, fixed parallel to the roof slope

BASIX web-based software used in NSW, developed to ensure that not only energy efficiency and thermal comfort is designed into the home, but also water conservation measures

bearing stress occurs at the point where one material supports the load of another, such as a brick column supporting a lintel near its edge, or that part of the steel plate that the side of a bolt makes contact with

benchmark a level of performance used as a target by businesses to measure the success or otherwise of a company relative to past, desired, or competition achievement

bending stress is a combination of compressive, tensile and shear stress acting within the one piece of material

bill of quantities an itemisation of all knowable costs to be incurred during a construction project, such as plant, labour and material costs. The document may also contain provisional costs and preliminaries costs.

blackwater recycling systems the means of capturing and cleansing waste water from toilets, kitchens, dishwashers and the like that would otherwise have gone into a septic system or sewer

blanket membranes are a reflective foil sarking faced with a woven insulation. This insulation 'blanket' can be of a range of thicknesses and is particularly designed for metal roof cladding. Blanket membranes are in most instances non-permeable.

boundary line the boundary of the building lot, identified with length and direction relative to true north

bracing systems or components in structures designed to resist lateral loads or other forces that may cause walls to be pushed out of square or otherwise bring about damage or collapse of a building

brainstorming a method of idea generation whereby thoughts on an issue are put up at random, without order or discussion of value, until no more suggestions are forthcoming; only then is the potential of each statement, or combination of statements, considered for merit

breach where an obligation, condition, or duty expressed or implied within a contract has not been fulfilled

break-even point the point at which the income of your business equals the fixed expenses of your business

budget a financial plan that outlines an organisation's monetary and operational goals

building information modelling (BIM) also **building information management** a collection of computer applications such as CAD, spreadsheet, word processing, image viewing and data storage/manipulation programs accessible by multiple parties online ensuring all stakeholders have access to the most current documentation

building permit a document issued by local councils allowing the construction of a building to an agreed set of plans and specifications

bulk insulation designed to resist the transfer of energy by convection or conduction. This form of insulation traps air in small voids within the material that must each be heated or cooled before transference to the next.

bushfire attack level (BAL) the potential exposure of a building or building site to attack from embers, high temperature winds, radiant heat, or flame generated by bushfires. Level is derived from the context, inclusive of vegetation type, density and slope of ground. Expressed as kW/m^2, but bracketed into the levels Low, 12.5, 19, 29, 40, and FZ (Flame Zone). Used to identify appropriately resistive building materials, elements and designs.

Business Activity Statement (BAS) density a form submitted to the Australian Tax Office by businesses registered for GST and/or have employees disclosing relevant taxation information including GST, PAYG and other taxation-related monies

business capital the financial resources available to the company to pursue its aims and objectives

business plan the document that outlines the goals and objectives of a business, both in written statements and through the presentation of financial projections

C

camber a slight curve or bend that a designer may deliberately incorporate into a beam

capital costs this includes all items that your business will use to make a profit, such as tools, motor vehicle and any other significant purchases that will last for more than one year

case law *see* common law

cash at bank the amount of money, the balance, of the business bank account at the end of the financial year

cash flow the movement of money in and out of a business

cash flow report a document that traces the flow of money in and out of a business

cause the reason behind why one party has decided not to respond favourably to the request of the other

centroid is effectively a shape's centre of gravity

chainage refers to a distance, chainages to multiple distances, generally of the same nominated length for that particular survey

claim a request of some form made by one party to another

closed questions questions that require a yes or no answer only

common law law based upon the principle of precedent, or the outcome of previous legal challenges where such exist

compartmentation the division of large buildings in a manner that isolates one part from another with a wall with the required fire resistance level

complex ('action at a distance') forces are harder to discern than simple forces, but are a daily occurrence. Gravity and magnetic forces are the prime examples.

composite insulation has been developed from many of the different insulation types: pure wool, glass fibre and polyester fibres have each been combined with reflective sheets to form 'blankets' that may be rolled out over ceiling and roof frames

compressive stress is when forces try to squash the material to compress it. This is the second form of 'axial' stress as it too tends to change the length of a material.

computer aided drafting (CAD) computer applications that allow the development of two-dimensional plans and three-dimensional virtual models of buildings and other structures. Also known as **computer aided design**.

conciliation a dispute resolution process whereby objectors meet with an independent person who has a sound knowledge of the area of work in which the dispute has arisen and assists them in the development of a solution

conciliator a skilled professional, knowledgeable in the field in which a dispute has arisen, who enables parties to understand the issues that divide them, and aids them in the development of a solution based in part upon their expertise in the area

condition in a contract, outlines circumstances that may trigger an action or stipulate when an action can or should be taken, and within what time period

condition report reflects observations of a range of factors based upon the information in the documentation held, the type of project involved and the site and its neighbouring areas

conditions subsequent something that happens after contracted works have begun that may bring about the end of a contract

conduction occurs when we come in contact with objects, surfaces or fluids that are colder than ourselves. Once more physics takes no account of biology's whim to stay warm, and attempts to equalise the temperature of the two bodies; i.e. transfer heat from our body to the other material by direct conduction. If the other material is extremely cold, we can actually burn our flesh as we transfer heat rapidly to that point.

conforming tender a submission that meets all the document criteria provided by the principal

consent evidence that the full terms and conditions of a contract have been sighted and understood by the agreeing parties

consideration the promised return made by one party as payment or remuneration for the offer or promised activity made by another

continuously supported beams ones that are supported at three or more points along their length

contract an agreement between two or more parties whereby one party, the contractor, agrees to do or provide something for the other party for a stipulated consideration (usually money). May be written, verbal, or reciprocal.

control measures systems, procedures, process or equipment developed as a means of mitigating risks associated with an identified hazard or hazards

Cost of Sales in the construction industry this calculation represents the mark-up that you place on the materials included in your costing for a project.

cost plus contracts based upon the actual costs of undertaking the work plus an agreed percentage. There is no maximum price.

cost–benefit analysis a consideration of the overall costs versus the potential benefits of a particular waste-minimisation strategy

court hearing when a dispute is being determined in a court of law

covenants a form of overlay, one that has been imposed and approved at the planning stage of a new development, such as a new housing estate or subdivision

Current Assets represents the working capital of the business and consists of the short-term assets that are used within in a 12-month period to generate profit for the business

Current Liabilities these are the amounts owing by the business that must be paid in 12 months or less

D

damp-proof course (DPC) a continuous layer of an impervious material placed in a masonry wall or between a floor and wall to prevent the upward or downward migration of moisture.

datum the point from which all other measurements or 'levels' are taken; also to which all other heights are referenced on a building site

dead (permanent action) loads result from the mass of the component parts of a structure. As the phrase 'permanent action' suggests, these are forces considered to act continuously upon a structure with insignificant variation over time.

deck a raised platform, usually abutting a dwelling, constructed with an open subfloor and commonly floored with narrowly spaced timber or timber-like strips.

deconstruction the reversal of construction: a sequenced dismantling of a building into its component parts

deemed-to-satisfy (DTS) provisions commonly used solutions that have been tested and considered to comply with the performance requirements of a code or standard

deep beam footings these are an extension of the edge beam of a stiffened raft slab, with AS 2870 showing them as between 750 and 1500 mm from finished floor level to under the beam

deflection the amount a beam bends under a given load

demolition the planned destruction, or preferably deconstruction of a building or part thereof for the purposes of clearing a site

deposit monies taken by the builder in advance of beginning the contracted works

designated bushfire-prone area an area decreed by the relevant authority (local or state government) as likely to be subject to attack from bushfire. Bushfire attack is inclusive of embers, high temperature winds, radiant heat and/or flame.

detail references are used on plans, elevations and sections to refer the reader to other larger-scale construction details that describe particular construction elements more fully

dew point the point at which the water vapour in the air reaches saturation and becomes water

dimensions generally depicted in one of two ways: by arrow-headed lines running between two clearly demarked points; or by what are known as 'architectural ticks', small lines running at an angle through the beginning and end points of the dimension line

discharged the ending of a contract, be it early termination by default by one party, or by satisfactory completion or performance

discrimination treating or acting against an individual or group, based upon their gender, sexual alignment, religion, skin colour, age, marital status, nationality or race

dispute a difference of opinion between two or more parties over any number of issues concerning the construction or construction processes including but not limited to: timing, quality, variations, staff conduct, access, material selection and the like

dispute resolution procedure documented systems or procedures designed to develop acceptable resolutions to disputes

duress the use of force or the threat of force to ensure the signing and apparent consent to a contract

duty of care the legal responsibility of persons and companies not to cause harm, or allow harm to occur, to others when such harm could be reasonably foreseen

E

easements a section of a title of land over which parties other than the landowner hold certain rights, for certain purposes. As such, an easement can, and generally does, limit development over that particular area.

elastic while a material is operating within Hooke's Law it is said to be exhibiting elastic behaviour. This means that when the stress is released, it will return to its original length with no detrimental effect.

elevation shows the side views of a building or structure and therefore shows heights or vertical distances

embodied energy the total energy consumed in the process of producing a product. It includes all the associated processes and materials for the production of a building: mining, harvesting and processing of natural resources, as well as manufacturing, packaging and transportation.

end fixity the manner in which the ends of the beam are secured

endorsement the act of company management granting approval for a tender to be submitted

energy efficiency the use of energy such that is consumed at the lowest achievable rate

energy productivity defined as the economic return as 'gross domestic product' or GDP in millions of dollars, divided by the total energy used by the economy to produce it; i.e. energy productivity = economic output/ energy used = GDP/petajoules

***Environmental Protection and Biodiversity Conservation Act 1999* (EPBC Act)** provides Australia's legal framework for the protection of the environment and ecological heritage

equal employment opportunity the legally enforceable code of employment conduct that requires all peoples to be treated equally upon their merits for a position exclusive of their gender, sexual alignment, religion, skin colour, age, marital status, nationality or race

evaporation our body's natural reaction to excessive heat is to push moisture to the surface, which is then evaporated, taking the warm moisture with it and thereby reducing our temperature

executive that arm of government that enacts and thereby gives power to the laws

exit travel distances distance an occupant must journey to get from an identified location to an exit, emergency or otherwise

expenses the costs incurred in the course of running your business; that is, earning income, e.g. wages, advertising, insurance

express terms provisions in the contract that are clearly 'expressly' stated in writing

extension of time (EOT) the formal request and/ or agreement to push back the planned handover or practical completion of a project to a later date

F

fall a decrease in height between a reading and the one taken immediately before it

falsework supports the formwork and ensures it is capable of handling the loads

fenestration the arrangement, proportioning, and design of windows and doors in a building

financial budget *see* budget

financial ratios a range of measures by which financial performance may be evaluated

financial viability the capacity of a business to meet its monetary obligations and achieve its goals

financing activity money used to finance the activities of your business, such as loans received and loan payments

finished floor levels (FFLs) the height of a floor above a given datum

fire resistance the capacity of a building or its elements to resist the action and/or reduce the spread of fire, while maintaining stability – this includes the spread of fire between buildings

fire safety system (active/passive) a combination of passive and active fire safety measures designed to restrict the spread of fire, extinguish fire, alert people of fire and/or provide safe evacuation

fire-resistance level (FRL) the time in minutes that a material or building element (wall, ceiling, door, window, etc.) must withstand the flame, heat and/or smoke of a fire. Three criteria of resistance are stated in the following order: structural adequacy/integrity/insulation; e.g. 90/60/60. A '–' means no requirement; e.g. 60/ – / 60, no requirement for integrity.

fixed price contract also known as lump sum or fixed sum contracts, the most common type of contract in the construction industry for domestic work. Provides a set 'fixed' price for the agreed works. Price may only be varied through written agreement of both parties due to documented variations, or proven changes to provisional or prime cost sums.

fixed sum contract *see* fixed price contracts

footing slabs in this design the slab edge sits upon a strip footing, to which when constructed in Class A soils it needs not be connected. In other classes, it must be tied in by way of R10 reinforcement rod at a minimum of 600-mm centres.

footings that part of a structure that is in immediate contact with the earth or foundation material. Engineered to evenly distribute structural and imposed loads to the foundation while withstanding overturning and uplift.

foresight the last sighting taken before moving your instrument (changing station) or ending the survey. You are looking for(e)ward to the future.

formwork provides the shape, the 'form' of the finished concrete

fracture if stress increases until the material's ultimate strength is found, the material becomes significantly elongated (necking) and ultimately fractures

free-body diagram a graphical illustration of the various forces applied to a structural component

fringe benefits tax money paid to the government as a tax against forms of remuneration or 'benefits' enjoyed or received by employees in addition to their wage or salary

frustration occurs when a contract cannot be completed due to causes other than any fault of either party

G

General Requirements (GR) the mandatory set of rules governing the NCC, on how it must be used and the processes that must be followed

geology the study or science of the origin, history and structure of the Earth: rocks, minerals, stratification, oil, natural gas and fossils all inform this field of study

girt horizontal members to support the cladding when applying wall cladding only to the frames

glazing the glass elements or features of a structure, generally windows, doors and skylights

Goods and Services Tax (GST) monies paid to the government additional to that paid for things or services bought by individuals; currently 10%

grade at which a pipe or drain rises or falls refers to the angle of the pipe to the horizontal

grease trap a device in the shape if a box with baffle plates to slow the flow of liquid waste and prevent the passage if greasy substances into the drainage system

Gross Profit Gross Profit is calculated by subtracting the cost of sales from the total income you have earned, as follows: Total Income – Cost of Sales = Gross Profit

guaranteed maximum price (GMP) contract limits the price of works to an agreed maximum but may cost less

H

harassment behaviour that intimidates, offends or humiliates a person or group of persons. May take many forms including yelling/shouting, humiliation, unwanted sexual advances, physical threats, intimidation or assault, or discomfort through exposure to unwanted sexual humour

hazard something that may cause harm to you, other people, or animals

heritage overlay an area zoned by council to be of historical significance with regards to the types of buildings present, and those that may be built

high wind areas regions subject to wind speeds *greater* than those classified as N3 or C1, which equate to 50 m/s or 180 km/h

Hooke's Law when a material is stressed, the amount of strain it incurs is directly proportional to that stress up to a point

hydrostatic pressure through the build-up of water, significant lateral pressure can form behind a wall, either toppling it, fracturing it, or simply penetrating it

I

implied terms conditions that are not written in the contract but are implicit either by the expressed terms, or due to legislation brought to bear due to the contract's existence

in-house client someone or some section within a business that relies upon another person or section of the business to feed them information or products so that they may carry out their work effectively

in-house stakeholder someone or some section within a business that has an interest or will otherwise be affected or influenced by the actions of another in the pursuit or conduct of a project

income also referred to as revenue, represents the funds received or earned in the course of running a business; e.g. sales revenue, fee income, service income

informative (standard) a standard listed within the NCC (and/or within another standard) that provides advice or information but need not be followed

inspection and test plans (ITPs) a quality management document that describes how and when specific work being undertaken will be inspected

insulated glass units (IGUs) otherwise known as double or triple glazing units. The glass panes are separated by low conductivity spacers, and the space formed between them is then filled with inert gas, usually argon

intermediate sight all the sightings taken in between the backsight and the foresight

invert levels reduced level (RL) of the bottom inside level of a pipe, sometimes referred to as the 'floor' level of a pipe

investment activity involves the purchase and return on investments, such as property, fixed assets and equipment

isometric a drawing technique that offers a three-dimensional image. In this technique, vertical lines remain vertical; however horizontal lines are usually set at 30° to a horizontal base. Generally, all dimensions are full length.

J

job safety analysis (JSA) the systematic identification and documentation of hazards and associated risks associated with specific activities in the workplace

judiciary a legal system and persons that stand separate to all levels of government, who enforce the laws made by those governments

just in time means that materials do not sit around on site with the risk of being damaged and having to be replaced. It also reduces repetitive handling

K

key performance indicators (KPIs) the two main types of performance indicators used are: *financial indicators*, which use information from the financial records and are expressed in dollar terms; and *non-financial indicators*, which are commonly expressed in real terms and often make use of qualitative data

L

latitude marked upon a globe by a series of parallel lines ringing the planet

lean construction company practices that encourage innovation, waste elimination and continuous improvement with a particular focus upon improved procurement practices

legislature that arm of government that makes the laws of the nation, state or territory

level of fire resistance an interpretation of the various sections of the NCC, and the appropriate application of each clause as it applies in any given instance based upon class of building and the relevant Performance Requirements

leverage ratios indicate the extent to which debt funds are used in the business

liability amounts owed (debts) of a business to outsiders and representing a commitment to pay cash at some point in the future, e.g. amounts owed to suppliers, borrowings from the bank or other financial institution

licence permit gained from an appropriate authority to own, operate or sell something, or some practice or level of practice such as building works, business or the like

life-cycle costing (LCC) the total of all costs of a structure from its design and construction, through ownership, usage and maintenance, to its ultimate disposal

limited tender by which predetermined prospective contractors are approached directly

liquidated damages costs incurred by the client or principal when works are not completed on time or to the standards specified

liquidity ratios measure the ability of a business to meet its short-term financial obligations

litter abatement means of reducing the amount of any material exiting the construction site including dusts, chemicals, packaging materials, food wrappings, contaminated water, or the like

live (imposed action) loads are all those items and materials within a building that may move, or may be moved. AS/NZS 1170.0 defines an Imposed Action simply as a set of forces '... resulting from the intended use or occupancy of the structure...'.

loss leader tendering submitting a bid for a project knowing that little or no profit will be made, and potentially a loss, as a means of entering new market territory for the business

low-E glass produced by applying a very thin layer of metal particles that reflect heat off the glass and back into the room, or outwards, depending upon how it is arranged in the frame

lump sum contract *see* fixed price contracts

M

Magnetic North identified by way of a compass

magnitude (size/amount) of a force is measured in newtons (N); though in construction this is more generally kilonewtons (kN)

materials inventory the materials held by a company at the end of the financial year

mediation a dispute resolution process whereby objectors meet with an independent person who assists them to interpret each other's points of view, enabling the parties to come to an equitable solution themselves

mediator a skilled professional who enables two or more disputing parties to develop their own solution to that which divides them, but does not have input into that decision

misrepresentation where a person or persons, or an element within the contract, represents as factual but is not

moment the measure of the tendency to try to resist the turning effect of an applied force

N

NABERS (National Australian Built Environment Rating System) a system that ranks a building's energy efficiency, water usage, waste management and indoor environment quality, as well as its impact on the environment through greenhouse gas emissions, on a 6-star scale (6 being best)

National Construction Code (NCC) a three-volume suite that includes the Building Code of Australia (Volumes One and Two) and the Plumbing Code of Australia (Volume Three) with ancillary volumes as guides

National Employment Standards (NES) The minimum parameters under which employers may engage employees, inclusive of remuneration (wages and benefits), types of leave, public holidays, termination and redundancy clauses, superannuation and the like

National Waste Policy 2009 government initiative to avoid the generation and reduce the amount of waste, manage waste as a resource, and improve the safe and correct handling of waste while also bringing about a reduction in greenhouse gas emissions

Nationwide House Energy Rating Scheme (NatHERS) software-based energy-efficiency assessment tool for residential buildings developed by the CSIRO. The analysis is based upon the capacity of the building envelope (the outside structure including doors, windows, walls, flooring and ceiling/roof) to restrict energy flows

nature the type of request or 'claim' made by one party to another

negotiation a formal or informal discussion between two or more parties generally aimed at gaining an agreement acceptable to all

Net Assets when the Total Liabilities are deducted from the Total Assets, the difference is the Net Assets or net worth of the business as at 30 June

Net Profit uses the Gross Profit calculated and subtracts all other operating expenses, or overhead costs

nominal size in name only: the approximate but not exact dimensions of a material. Commonly used to describe the sectional size of rough-sawn timber and expressed as EX. E.g.: EX 100 x 50.

nomination an agreement between all parties to a contract that a particular dispute resolution process and/or agency or individual shall be the means by which disputes shall be resolved

non-axially loading of a column results in bending and compressive stress being applied to the column unequally

non-conforming tender a submission that does not meet all the document criteria provided by the principal

Non-current Assets assets such as your motor vehicle, tools and other fixed assets, which are purchased for use, not for sale

Non-current Liabilities: these are amounts owed by the business but not due for immediate payment. Non-current Liabilities usually consist of bank loans or other sources of long-term finance.

normative (standard) a standard listed within the NCC (and/or within another standard) that *must* be followed

novation *see* rescission – the substitution of an old contract with a new one

O

oblique drawing a three-dimensional drawing technique in which the front plane is drawn proportionally to scale and flat to the drawing plan while one side is drawn angling back at 45°. These side dimensions are usually foreshortened by 50% to give the impression of distance.

offer the promise by one party to another to do or not do a certain activity in return for a consideration (generally money)

Ombudsman an independent officer appointed by the government to investigate and resolve disputes between people and government agencies

open questions questions that allow or require the respondent to expand their answer beyond yes or no.

open tender effectively allowing anyone to make an offer, which is then judged on capability and experience as well as cost

operating activity this is all cash received, and cash paid out through the operation of the business

operating costs those costs that are recurring, such as rent, insurance, fuel, etc.

optical automatic ('dumpy') levels surveying instruments incorporating a telescope and a self-levelling system used to transfer, or determine differences in, heights

orientation the arrangement of the building such that it takes best advantage of the sun's daily and seasonal variations in angle and intensity. At the same time, it takes into account the prevailing weather patterns, particularly the direction of wind.

orthographic drawing offers the viewer a flat plane depiction of a subject. These drawings are useful for describing true lengths and proportions.

overdraft facility a banking arrangement allowing a business to withdraw an agreed amount more than its account balance

overlooking to look, or be able to, into a building or an otherwise private open space of a neighbouring building

Owners' Equity represents what the business owes to the owner of the business. The owner(s) of the business are the beneficiaries of any profits and bear any losses incurred.

P

pad or **blob footings** these are round or square mass pour concrete, generally without steel reinforcement. They are sized to carry the loads of stumps or piers as part of the subfloor system of bearer and joist construction, or for verandah posts, columns and other such point loads.

partnership a business made up of two or more people where profit or loss is distributed between themselves. Like sole traders, the assets of each individual are bound to the company.

passive solar design knowing the angle at which sunlight will be striking your building at any particular time is important to good energy-efficient design

Pay As You Go (PAYG) money withheld by employers from employees and paid directly to the government as the tax owing by employees from their wages. Also, instalments paid (usually quarterly) by sole traders and partnerships as tax presumed to be owed based on the previous year's earnings.

pedology the study of soils; their characteristics, origins and use

performance (code or standard) a code or standard that sets out the required performance of a solution but does not dictate what that solution might be. It allows for alternative 'Performance' Solutions.

performance appraisal a site report, or element of a site report, outlining the satisfactory or otherwise nature of work being undertaken by staff or subcontractors

performance report a documentary tool for monitoring and/or tracking the success or otherwise of a business in achieving its goals and objectives

Performance Requirements (PR) the mandatory specifications of the minimum performance level for all elements of buildings, including plumbing and drainage

Performance Solution an alternative solution to the common 'deemed-to-satisfy' approach that can be shown to meet or exceed the performance requirements of a code or standard

performance-based design brief (PBDB) a document developed by key project stakeholders that shapes research and development of a Performance Solution alternative to the deemed-to-satisfy options held within the NCC

pergola although referenced within the National Construction Code (NCC) as an unroofed structure, pergolas are generally columned structures with beams and rafters to support climbing plants, or a full roof. They have open sides providing a shaded open space.

personal protective equipment (PPE) protective clothing or equipment worn to reduce or protect against identified risks associated with work of a hazardous nature. The last option in the development of safe work procedures when no other reasonably practicable forms of mitigation exists.

perspective drawing an advanced form of pictorial drawing that provides a close approximation of how an observer sees an object. There are various forms of perspective drawing. In its simplest form, the viewer sees things vanish back to one point only. In its most realistic form, things vanish back to three points.

pictorial drawings attempt to overcome the flat plane issues of orthographic depiction by representing the missing third dimension: attempts to provide a sense of depth

pier or **pile** and **edge beam footings** these are governed by AS 2159 and are of multiple types and forms. AS 2159 makes no distinction between piers and piles. Piles, in this standard, cover all pier and pile types in that they may be bored and filled, drilled, screwed, jacked, vibrated, or driven.

planning permit a document issued by local councils allowing the development of a site for the purposes of a structure or building of agreed form and function or use

point loads those deemed to be acting upon a specific point

portal frame is neither a wall nor a roof, yet is the structural element behind both for many commercial, and occasionally domestic, structures. Commonly made of steel, large portal frames may also be made from engineered timber. The key to their stability is in the rigidity of the joints. No matter the material, the portal frame must be engineered to withstand the various loads, live, dead and particularly wind.

practical completion when a project or structure is fit to be occupied and used for its intended purpose despite some elements not actually being finished, or all defects resolved

precedent conditions something that should have happened, but did not, that precludes a contract from taking effect or work to begin, effectively terminating a contract before it begins, despite being signed by both parties

prequalified tender generally open only to those already qualified to be on a listing

prescriptive (code or standard) a code or standard that dictates what a solution must be for a given situation. Within itself it, does not allow for alternative solutions.

presentation drawings artistically 'rendered' sketch plans which help non-technical people understand what the project will look like when complete

prevailing winds the wind coming from the most common direction at a given location

prime cost (PC) an allowance given in a contract for items such as appliances, bathroom fixtures, tiles, etc.; the price of which may vary depending upon decisions or purchases made as the project nears completion

principal the individual, group or company that has called for the tender, and will receive the tender application. They are the initiator of the tender process.

private company an entity in its own right, capable of owning assets, and making profits and losses as a person may; but it is not a person, nor tied to a person

procurement the paths by which resources are purchased and transported to a project

professional indemnity a form of insurance that protects the individual from claims of negligence or breach of duty as a result of work undertaken or advice given.

Profit and Loss Statement (Income Statement) Also known as a revenue statement, a documented calculation of the income within a financial year matched against all expenditures such that an accurate depiction of total profit or loss may be determined.

profit report reports that provide an overview of the financial performance of the business

profitability ratios measure efficiency and indicate the ability of a business to generate profits from sales, assets employed and owners' investment

progress payment monies owed by the principal or client to the builder/contractor for works completed at contractually agreed stages of a project

projection a calculated assessment of future profits and losses

protection notices (also called *protection work notices*) notices to the owners of neighbouring properties that their land will need to be accessed for the purposes of work required to protect their assets from potential harm

provisional sum (PS) an allowance given in a contract for elements of a project that cannot be accurately estimated, such as excavation, due to a range of factors that may be unknown until work commences

public liability where a company or individual is deemed to be negligent or at fault resulting in injury or loss by a member of the public. Also, insurance that protects against such claims.

purlin effectively the same component as a girt, is used to support roof cladding

Q

qualitative the evaluation of an object, element, activity or performance through a judgement that is not easily translated into numbers, such as aesthetics, feelings, sense of security and the like

quality control a set of procedures designed to ensure the quality of a project, or elements of a project, meet the demands of the client or company

quality management manual a composite document holding and describing, and aiding in the use of, all elements of the quality control system

quality management systems (QMS) the overarching documentation of processes and procedures, inclusive of quality control, assurance and planning, developed to ensure that a project is completed safely, on time, and to the best possible quality that time and cost allows

quantitative the assessment of an object, element, activity or performance through the use of measurements, such as weights, time, length, temperature and the like

quote the price calculated to cover all labour, materials, overheads and related fees for a specific project, part project, or variation to a project. Differs from a tender in being smaller in scope and less detailed in the documentation.

R

R-value a measured insulative value, or insulation rating. It is a measure of a material's resistance to thermal conductivity relative to its thickness; i.e. how much energy can pass through a given thickness of the material. The greater the R-value, the better its capacity to resist the passage of energy.

radiation a manner by which we, albeit slowly, can physically lose heat. If the temperature of our surrounding environment is less than our own body, then we will radiate heat as the natural laws of physics ignores our biological attempts to stay warm and tries to equalise temperatures.

radius of gyration (r) finds which way a beam is likely to buckle when loaded on end

recycling the conversion of waste into other products and/or base material

reduced level (RL) a height derived from a datum, benchmark or other nominated starting point

reference building a hypothetical structure used to calculate the energy loads for the proposed building

reflective insulation or **reflective foil laminate** designed to resist radiant heat: i.e. it reflects the radiated energy back in the direction from which it came. To do this, reflective insulation generally requires an air gap between it and any cladding or other material.

reflective questions questions made in response to the statements made by another so as to elicit confirmation or clarification of that statement

refusal to decline a request or claim

registration being listed (registered) with the authority deemed by local, state or federal government to oversee the appropriate conduct of specific fields of work, operation or business. May also apply to equipment such as boats, motor vehicles, aircraft, and ownership or care of animals, domestic or otherwise

relative humidity (RH) is the amount of moisture in the air as a percentage of the amount the air can actually hold at that specific temperature

render the covering of a brick wall with one or more coats of cement mortar consisting of sand, cement and plasterers clay

request to require, or make a claim upon, one party to do, not do, or stop doing, something.

request for information (RFI) the call by the site supervisor to the client, architect or engineer, or from subcontractors to the supervisor, for clarification of specific aspects or details of plans or specifications

rescission the discharge of one contract for the purpose of replacing it with another; the repeal, cancellation, or revocation of a law, order, agreement or contract

residential occupancy the act of people or persons living or dwelling on a regular basis in a building designed and constructed for that purpose, such as a house, unit or cluster of units

retention money withheld by the principal or owner as a form of surety against the project not being completed or where works subsequently are shown to be substandard

ridged (fixed) joint system in timber, now usually achieved through flat steel 'fish' plates, which limits the amount of movement at the joint, transferring the load instead to the column and base

right first time leads to less waste, for although the cost of faulty materials is borne by the supplier, there are always associated costs and sometimes wasted ancillary materials as well

rise an increase in height between a reading and the one taken immediately before it

rise and fall amount the increase or decrease in the cost of labour, materials, transport or other procurement factors calculated by way of a contractually agreed formula

risk the likelihood or potential for a hazard to cause harm and the level of that harm

risk assessment the evaluation of the likelihood of a hazard causing injury, and the potential level of that injury in a worst-case scenario

risk control means of reducing or mitigating the level of risk from an identified hazard

S

safe buffering capacity the amount of water that the total quantity of a particular material in the structure can hold

safe work method statement (SWMS) similar to JSA, but also documents the agreed best practice or procedure for undertaking specific workplace activities

scale the relationship, usually expressed as a ratio, between the actual size of something and a drawing or model representation of that same thing

second moment of area (I) a statement of a shape's ability to resist bending when load is applied to one axis or the other

section drawings are used to show detail that otherwise could not be seen without this alternative perspective, produced by making a virtual 'cut', as if by a large blade, through a structure (or a portion of a structure) and then looking at the cut face exposed

section modulus (Z) offers a direct measure of a beam's strength

sectional properties are based solely on the shape of the section, not the material they are made of; how a material is shaped can make a significant difference to its performance under loads

Security of Payment Act (SOP Act) An act of parliament designed to ensure persons who carry out work, or supply goods or services, under a contract, are fully paid in a timely manner

set-back distance from the boundary of a site to the face of the building closest to that boundary

setdown the drop in height between two surfaces, generally floors

shear stress occurs when a force acting perpendicular to one surface is being counteracted by another equal and opposite force acting on the opposite side

shear wall is rigid in being a solid sheet that when pushed from the end must buckle, fold, or slide before the other wall can collapse

simple or **'contact' forces** occur when objects actually touch each other

simple beams those that are supported at each end only

site diary a daily record kept by the site supervisor of all activities that take place in relation to a given project

site instruction a brief document usually written by the site supervisor outlining the exact nature of the change, then dated and signed by the supervisor and those who gave the instruction. Also: pre-printed, pro-forma sheets, which include scaled grids; each box can represent any scale of your choosing

site report any one of a range of reports relating to site-based activities or context, such as site evaluations, progress reports, training outcomes and the like

sketch plans ('roughs') are generally done to scale. This offers true proportions for the client, council, or other key stakeholders to consider. These may be created by 'pencil and stick' (drawn by hand) or developed on a computer (computer aided design, otherwise known as CAD).

slip joint a joint designed to allow movement between two members, usually in the form of two layers of sheet metal with grease installed on top of a brick wall prior to installation of a concrete slab.

soffit the underside of a slab or eave

solar heat gain coefficient (SHGC and SHGCw) identifies the heat transmitted directly through a window by sunlight as a fraction of that which hits its surface. It includes the heat absorbed by the glass and released inwards, as well as that which passes straight through.

sole trader the simplest form of business structure where an individual is legally responsible for all debts, losses, assets, profits and decisions. The individual's assets are also bound to the business.

sole-occupancy unit (SOU) a room or an identifiable part of a building that is for the exclusive use of the owner, occupier or tenant; be that an individual, group or company

specifications document associated with a plan set that holds information on a range of structural and non-structural matters, such as: material choices, standards to which the work must comply, permits and application fees that may apply, the obligations of the client, such as ensuring access, or works that the client intends to undertake themselves

spoil the earth that has be removed through excavation of a site

stage checking the process of ensuring that no element of a new tender bid clashes with existing projects or other proposed or potential projects

standards documents that set out the authoritatively accepted performance requirements, qualities, or manner of being or doing with regards to products, processes, management or any other form of human endeavour

standards and tolerances guide state- or territory-produced documents that outline the minimum accepted quality of specific elements of construction work

state tribunal the main avenue by which a final determination is made on construction disputes in most states and territories. Functions in a similar manner to a court but is less expensive. Generally, the decision made may be legally challenged on points of law only.

statics (equilibrium) derives from Newton's first and third laws in application. Statics requires that for a building, or its component parts, to remain stationary – static – then the sum of all the forces acting upon it – wind, gravity, people, etc. – must equal zero. This must be so irrespective of the direction from which those forces are applied; be it up, down, sideways, or at an angle.

station the location of the surveying instrument. To 'change' station is to pick up the instrument and relocate it elsewhere. The new location becomes another station.

statutory warranty conditions implied through the existence of a contract by the Acts and Regulations of the state or territory within which the contracted works are undertaken

storey a space between one floor level and the floor level, roof, or ceiling, immediately above. See the 'Storey: NCC definition' heading in Chapter 1 for a more expansive definition.

strain is the response demonstrated by a material undergoing stress. Strain is effectively a measure of the effect that stress has upon a material or component. Strain is calculated by finding the ratio between how far the material has stretched or 'deformed' relative to its original length.

stress materials under load undergo stress. The level of force applied over a given area dictates the amount of stress a component is withstanding.

strip footings generally, a steel-reinforced strip of concrete that runs around the perimeter of a building to support the external walls and what is known as a dwarf wall

subcontract a contract developed between two parties, one of whom holds the main contract between themselves and the principal or client. I.e. the subcontract is developed as an extension of the main contract activity.

subcontractor an independent business engaged to complete works by way of a contract made subsequent by agreement to the main contract holder.

superannuation money paid by employers to nominated funds managers as a means of ensuring employees have some provision for their retirement

superintendent an agent of the principal responsible for ensuring the project's aims are being met through oversight of daily operations, making decisions on variations, assessing quality and directing contractors

sustainability the use of resources such that all peoples may achieve their potential and improve their quality of life without harming the Earth's life support system

switchable glass another form of laminated glass whereby the membrane between laminations may be caused to change opaqueness and thereby reduce solar heat gain on a variable scale. The film changes its clarity by way of a low ($5W/m^2$) electric current that aligns microscopic droplets of liquid crystal such that light may be transmitted.

T

tax report reports that cover all aspects of the taxation liabilities of a business including GST, income tax, payroll tax, PAYG and FBT. The most significant report that is prepared from this information is the Business Activity Statement (BAS) which is required to be submitted to the Australian Taxation Office (ATO) at regular intervals.

tender a formal offer in writing to undertake a project, supply goods, or carry out works in accordance with a prescribed set of documentation. It includes a definitive price or expected remuneration.

tenderer the party that submits the tender. These may be contractors, subcontractors, suppliers of materials or services such as labour (a non-exclusive listing). They are the respondents in the tender process.

tensile stress is when forces trying to stretch the material place it in tension. This is one of two forms of 'axial' stress – those that tend to change the length the 'axis', the line running down the centre, of a material.

termination the ending or closure of a contract

thermal comfort how our mind perceives the current conditions and how our senses react to the physical changes in the climate around us

thermal efficiency the reduction of the passage of energy through a material or building envelope

thermal mass refers to materials having the ability to absorb and retain energy; and to do so slowly

title blocks describe the project details and the project context for the specific sketches

torsional stress occurs when a material is twisted or rotated while one end remains relatively stationary

Total Income to calculate this you will need to consider which information to include. If your business operates solely on a cash basis, you will include all your cash sales as your total income earned throughout the year; however, if you offer credit facilities to your customers will need to calculate your total of cash sales plus the total of sales invoiced during the year; i.e., 1 July – 30 June, even if you have not yet received the cash from those customers.

toxic waste materials that are dangerous to humans or wildlife including insects, birds and fish

trade credit the delaying of payment to a seller or supplier until an agreed future date

transaction the recorded movement or exchange, or promised movement or exchange, of money, goods or assets between two parties.

triangulation squares and rectangles are locked from racking by something fixed diagonally from corner to corner. In so doing, a triangle is formed and the shape cannot change without twisting to the side or the brace or joints breaking.

Tropic of Capricorn a line of latitude marking the point on our globe at which the sun is directly overhead at noon on the southern hemisphere's summer solstice; i.e. 23.5° south of the equator on 22 December (occasionally the 21st or 23rd)

True North north as it is identified by the geographic North Pole. Differs from Magnetic North by the amount of magnetic deviation at any given location at any given time

truss to efficiently transfer loads outwards to its extremities where such loads and forces can be received by walls, posts, or beams, thus leaving a clear safe span underneath

trust where a person or persons holds and manages assets for the benefit of others

U

U-value (Uw) the measure of the rate at which the total window assembly, glass, frame, seals, etc., transfers heat through it

unboxing the practice of procuring materials and resources for a project with minimal packaging and/or wrapping

unconscionable conduct deceitful, underhanded, ruthless, or unethical action to bring about the signing and apparent consent to a contract

underpinning the construction of new footings or concrete piers under an existing footing to prevent its collapse or failure

undue influence the use of authority or presumed authority, position of power or wealth, to gain consent when no consent might have been given without such coercion

unfair dismissal the ending of an employee's work contract (verbal or otherwise) without due cause or on a basis not reflective of equal employment opportunity

uniformly distributed load (UDL) loads deemed to be evenly distributed over a given area or length

uniformly varying loads (UVLs) distribute a load across a surface or support member. The difference compared to UDLs is that UVLs have a load that is greater at one end than the other: the load tapers off as it is distributed along the beam.

unit mass the effect of gravity has not been included. This will be offered in kg. In such cases you must multiply the kg by 10 to convert to newtons (N).

unit weight the weight of a given material is generally provided as a force in N or kN; i.e. the effect of gravity is already factored in. Unit weight is usually expressed as a volume in m^3, or per lineal metre of a specific sectional size.

unsatisfactory work notice a written document used to inform subcontracting teams that their work is not to the standard expected by the supervisor as the representative of the builder, client, or principal

V

value of a moment found by multiplying the amount of force applied by the distance that this force is applied from the support

variance positive or negative differences between budgeted accounts and actual expenditure or income

variations a change to the originally agreed design and/or contracted works causing a change to the original, or then current, set of plans and/or specifications. May be requested by the architect, designer, engineer, client or builder.

vector magnitude, direction and sense of direction of a force

vent (vent pipe) a pipe provided to limit pressure fluctuations within a discharge pipe system by the induction or discharge of air and/or to facilitate the discharge of gases

verandah a generally narrow roofed and floored space running parallel to the external walls of a dwelling. Generally, verandahs are just wide enough to offer shade and shelter to windows and doors, and have sufficient space for a chair or table.

Verification Method a system of tests, calculations, inspections, or other such methods to determine if a Performance Solution satisfies the relevant Performance Requirements

visible light transmittance the measurable amount of light within the visible spectrum that a glazing material or element will allow to pass through

W

waffle pod slabs a system that borrows from the stiffened raft slab design, but uses polystyrene foam void forms (boxes) laid on the ground to create internal stiffening beams. By this means, excavation is reduced.

walk-through the inspection of a workplace by observation while passing (walking) through that space as normal and/or specialised work is being conducted

warranty 1: an agreement within the contract or implied through legislation that work undertaken will be free of defects for a given period of time; 2: the promise implied or expressed by the contract that works will be undertaken to a certain standard or in a certain manner.

waste minimisation a strategy or collection of strategies designed to reduce the amount of waste generated by a project

waste residual material or consumables from the construction process that cannot be subsequently used in future projects

weep holes vertical joints or perpends in brickwork left open above the flashing line to allow water from behind the wall to escape.

WHS management plan a document developed by or for management in consultation with workers to ensure systematic and timely evaluation of hazards, risks and mitigation processes, actions, and the like surrounding WHS

wind rose shows the direction and speed of wind averaged over a number of decades

working capital the amount of cash available to, or needed by, a business so that it may meet its short-term financial obligations

working drawings the detailed technical drawings used to describe to all stakeholders the exact size, form and nature of a proposed structure. Also known as construction drawings.

workplace agreement an agreed set of conditions and remuneration developed between an employee or collection of employees and an employer or employer body (or forum) for stipulated and timely work-based outputs

workplace health and safety (WHS) the phrase used to encapsulate work practices that reduce or limit harm that may come to persons or animals associated with activities in or around places of work. In all states and territories, except Victoria and Western Australia, WHS has replaced the phrase occupational health and safety.

Y

yield once a material, for example mild steel, has 'yielded', it will not return to its former length. It has stretched to the point of, if not no return, at best only partial return; i.e. Young's modulus no longer applies.

Young's modulus of elasticity (E) $E = \text{stress} \div \text{strain}$, a value or 'constant' (something that will always be the same for that particular material). This value is particular to a given material. It is a property of that material and so applies no matter how large or small a piece of that material you have.

Z

zero defects limits waste through not having to redo work or replace materials

INDEX

Note: Bold page numbers indicate defined terms; italic page numbers represent figures.

C

H

I

J

K

T

Y

Z